Solar Energy Sciences and Engineering Applications

Solar Energy Sciences and Engineering Applications

Napoleon Enteria
Enteria Grün Energietechnik, Davao, Philippines

Aliakbar Akbarzadeh
RMIT University, Melbourne, Australia

CRC Press
Taylor & Francis Group
Boca Raton   London   New York   Leiden

CRC Press is an imprint of the
Taylor & Francis Group, an **informa** business

A BALKEMA BOOK

Cover illustration:

Photovoltaic Installation in Demonstration Building, Yonsei University, Incheon, South Korea.
Photographer: Napoleon Enteria

CRC Press/Balkema is an imprint of the Taylor & Francis Group, an informa business

© 2014 Taylor & Francis Group, London, UK

Typeset by MPS Limited, Chennai, India
Printed and Bound by CPI Group (UK) Ltd, Croydon, CR0 4YY

Library of Congress Cataloging-in-Publication Data

Enteria, Napoleon.
    Solar energy sciences and engineering applications / Napoleon Enteria,
Enteria Grün Energietechnik, Davao, Philippines, Aliakbar Akbarzadeh, RMIT University, Melbourne, Australia.
        pages cm
    Includes bibliographical references and index.
    ISBN 978-1-138-00013-1 (hardback)
1. Solar energy.   I. Akbarzadeh, Aliakbar.   II. Title.
    TJ810.E58   2013
    621.47—dc23

                                                                          2013041799

Published by: CRC Press/Balkema
                P.O. Box 11320, 2301 EH Leiden, The Netherlands
                e-mail: Pub.NL@taylorandfrancis.com
                www.crcpress.com – www.taylorandfrancis.com

ISBN: 978-1-138-00013-1 (Hbk)
ISBN: 978-0-203-76205-9 (eBook PDF)

Table of contents

Preface

As the world's conventional energy supply nears its peak, and with the demand for that energy increasing year on year, it is expected that balancing supply and demand will become increasingly challenging. Consequently it is expected that non-conventional energy sources and renewable energy resources are likely to play a greater role in addressing the imbalance between supply and demand.

Many experts advocate increased harnessing of renewable energy as an important alternative energy source. Utilization of renewable energy resources is sometimes expensive and difficult to apply fully in particular sectors of society because of the location, intensity and nature of the applications. Therefore specific matching of the renewable energy source to the application is a very important aspect of maximizing the utilization of renewable energy.

Solar energy is available in differing intensities in different parts of the planet. Maximization of its potential as a primary alternative renewable energy source depends however on the specific usage made of it. Hence, this book was conceived to serve the purpose of identifying primary and secondary applications of solar energy in order to maximize their potential.

As solar energy applications can span almost the entire spectrum of human activity, including for example biological processes, chemical processes, mechanical processes and other aspects of our daily lives, preparation of a book that considers all these facets is very important in determining how existing sciences and technologies can further refine and expand solar energy utilization and applications.

Many experts in solar energy were invited to contribute to this book, with content ranging from basic to higher concepts of solar radiation, the thermodynamics of solar energy processes and applications, the application of solar energy to producing an alternative, renewable secondary energy source through hydrogen production, thermo-chemical processes to separate some greenhouse gases, how to apply solar energy in thermo-mechanical processes, maximization of solar energy use in energy-efficient housing and other buildings, and the role of solar energy in planning the outdoor environment.

International experts from the many different fields of science and technology to which solar energy has feasible application have collaborated in the preparation of this book. Consequently there is wide coverage of solar energy as an alternative energy source which can also offer low greenhouse gas emissions.

In this context, firstly the editors acknowledge with gratitude each of the global experts in solar energy who have fully supported and contributed chapters to this book and who are individually listed in the book chapters. Secondly, we are grateful to Janjaap Blom of the Taylor Francis-CRC Press for the support given from conceptualization through to the publication of this book. Thirdly, we thank our families for their support during the entire process of production of the book with our aim of supporting the prospect of sunnier and clearer skies in the future of our planet.

Napoleon Enteria
Aliakbar Akbarzadeh

About the editors

 **Napoleon Enteria** is the Managing Consultant of the Enteria Grün Energietechnik, Philippines. At the same time, he is a Visiting Researcher of the Faculty of Engineering, Tohoku University, Japan. He was a Research Staff of the Faculty of Engineering, Tohoku University, Japan, for the Industry-Academia-Government Collaboration. He was doing research in collaboration with different Japanese universities and companies with the prime support of Japanese government agencies in the area of solar energy, HVAC systems and building sciences. In addition, he provides technical and scientific advice to graduate and undergraduate students. He was a scientist with the Solar Energy Research Institute of Singapore, a component of the National University of Singapore, performing collaborative research with the Fraunhofer Institute of Solar Energy Systems in Germany, a German company and the Department of Mechanical Engineering of the National University of Singapore in the field of solar thermal energy, HVAC systems and membrane heat exchangers; the latter was supported by the Singaporean government agency during his stay in Singapore. Before going to Singapore, he was a Global Center of Excellence Researcher in the Wind Engineering Research Center of Tokyo Polytechnic University doing research in natural ventilation and air-conditioning systems in collaboration with Japanese universities, companies and the Global Center of Excellence Program of the Japan Ministry of Education, Culture, Sports, Science and Technology. In addition, he was a guiding instructor to two undergraduate students for theses research.

Napoleon has authored several scientific and engineering papers in books, review journals, research journals and conference proceedings. He has presented and submitted dozens of technical reports for collaborative projects with research institutes, universities and companies in different countries. He is regularly invited as reviewer for several international journals in the field of air handling systems, energy systems and building sciences. On occasion, he is invited to review research funding applications and gives technical and scientific comments on international scientific and engineering activities. He is a member of the American Society of Mechanical Engineers (ASME), the International Solar Energy Society (ISES) and an associate member of the International Institute of Refrigeration (IIR). He was awarded his Doctor of Philosophy (2009) in engineering, specializing in Building Thermal Engineering at the

Tohoku University, Japan, as Japanese Government Scholar; and his Master of Science (2003) and Bachelor of Science (2000) in the field of mechanical engineering from Mindanao State University at Iligan Institute of Technology, Philippines, as Philippine Government Scholar.

Aliakbar Akbarzadeh was born in Iran in 1944. He received his BSc degree in Mechanical Engineering from Tehran University in 1966. In 1972, he obtained his MSc and in 1975 his PhD, also in Mechanical Engineering and both from the University of Wyoming, USA. From 1975 to 1980 he was an Associate Professor and also Head of the Mechanical Engineering Department at Shiraz University in Shiraz, Iran. Later he worked at the University of Melbourne as a Research Fellow (1980–1986), primarily doing research on applications of solar energy as well as energy conservation opportunities in thermodynamic systems. Since June 1986, Aliakbar has been working as an academic at RMIT University in Melbourne, Australia. During this period, he also worked as a visiting Fellow for half-a-year at the Nuclear Engineering Department of the University of California at Berkeley, USA, where he did research on passive cooling of nuclear reactors through computer modelling as well as experimental simulations. At present, Aliakbar is a Professor in the School of Aerospace, Mechanical and Manu-facturing Engineering at RMIT University, and also the Leader of the Energy CARE (Conservation and Renewable Energy) Group in the same school. Aliakbar lectures in thermodynamics as well as remote Area power supply systems. He is the Principal Supervisor of ten full-time PhD postgraduate research students on energy conservation and renewable energy systems. He has also one post-doctoral research fellow working with him on geothermal energy utilization for power generation.

Aliakbar is a specialist in thermodynamics of renewable energy systems. His industry oriented research projects enrich his teachings and makes them relevant. He spends about half of his time in supervising industry supported research in energy conservation and renewable energy area, which also form a vehicle for postgraduate training of his PhD students. He has been the first supervisor of about 30 PhD candidates who have completed their degrees. Aliakbar has over 100 refereed publications and two books all in his area of specialization which is solar energy applications. One of his publications on solar energy won the ASME Best Paper of the year award in 1996. Aliakbar's industry-oriented research on energy systems has resulted in a number of Australia National Energy Awards for him, as well as a number of products, such as the Heat Pipe-based Heat Exchanger for waste heat recovery in bakeries, the Temperature Control of solar water heaters using thermo-syphons and an innovative system for simultaneous power generation and fresh water production using geothermal resources. Aliakbar has also been working on salinity gradient solar ponds as a source of industrial process heat and also for power generation. In the last 35 years he has developed several concepts related to salinity gradient maintenance, as well as efficient methods of heat extraction from solar ponds. At present, his research group is the world leader on applications of solar ponds.

Chapter 1

Physics of solar energy and its applications

Napoleon Enteria[1] *& Aliakbar Akbarzadeh*[2]
[1]*Enteria Grün Energietechnik, Davao, Philippines*
[2]*School of Aerospace, Manufacturing and Mechanical Engineering,*
RMIT University, Melbourne, Australia

1.1 INTRODUCTION

Solar energy has existed for millions of years and has been used by many non-living and living things for physical, chemical and biological changes and processes. For generation after generation, solar energy has been the main source of daily energy in many ways. The start of human modernity and civilization changed the utilization of solar energy. Previous human civilizations utilized solar energy for lighting, food drying and personal care. Modern humans utilize solar energy for everyday existence, work purposes and generally for living. As human demand for modern energy supply increases, attention to solar energy becomes more intense.

Because of the increasing population, demand for better comfort, urbanization and industrialization, the pressure on conventional energy sources is increasing. The rapid increase of energy demand rattles the chain of energy supply which has difficulty in meeting the rapid rise in energy demand. This rapid demand not only makes the world more volatile in terms of energy politics but also accentuates the environmental hazards associated with emission of greenhouse gases, particularly from chemical processes involved in utilizing the thermal energy obtained from conventional energy sources. Consequently there are active plans to utilize solar energy for different processes to minimize energy demand from conventional energy supply sources.

Conventional energy supplies are based on fossil fuel or carbon-based energy sources, based on the liberation of stored energy through combustion. Through combustion processes or chemical processes, thermal energy is generated. At the same time, radiant energy or solar energy is an available energy which can be used as replacement for the energy liberated/generated during combustion processes. Therefore, several methods, concepts and ideas are being pursued to maximize the utilization of solar energy as an alternative form of energy to minimize the usage of carbon-based energy sources. This chapter introduces the concept of solar energy and its applications as alternatives for conventional energy supply and demand.

1.2 SOLAR ENERGY AND ENERGY DEMAND

The Sun is the nearest and only star around which the Earth and other planets are constantly rotating. The energy generated by the Sun is utilized on Earth for the support of living organisms. As the Sun is radiating and transmitting energy in the form of radiant energy with a range of wavelengths and intensities, a nuclear fusion-fission-fusion reaction is happening inside the core of the Sun. The nuclear processes of the Sun create tremendous amounts of energy which are sometimes devastating due to the power of nuclear processes. The power generated by the Sun is almost endless and the human benefit from it is expected to last for countless generations to come.

The energy generated in the Sun is transmitted to its surface through convection and radiation. From the surface, the energy is transmitted to its surroundings through solar radiation. This is a consequence of space being a vacuum from which both conductive and convective modes of energy transmission are eliminated. The solar energy available in the upper atmosphere of the Earth is almost constant and depends on the motions and distance between the Earth and the Sun associated with the constant rotation and revolution of the Earth around the Sun and the nuclear activities of the Sun. As the Earth has a layer of atmosphere, the available solar energy in the upper atmosphere of the Earth is reduced in transmission before it reaches the surface of the Earth, depending on the weather and climatic conditions, surface locations and local activities such as heavy smokes from wildfires.

The net energy of the Sun, reaching the surface of the Earth in the form of radiant energy, has different intensities. The intensities depend on the radiant energy wave spectrum. The radiant energy wave spectrum is generally classified as short wave (0 to 300 nanometer), infrared (300 to 750 nanometer) and long wave (750 nanometer and above). With these different radiant energy wave spectrum energy intensities, utilization of solar energy is very interesting and for different applications depends on a variety of techniques and methods.

The radiant energy from the Sun has been naturally utilized by different terrestrial living organisms for millions of years. Early humans utilized solar energy in different applications including food production, cloth making and others. As modern civilization demands more energy, alternative sources can be investigated and attention to solar energy and its provider, the Sun, becomes an interesting topic for the modern world. There are many concepts, methods, ideas and practical solutions both for simple and for more advanced utilization of radiant energy from the Sun. As global demand is for clean energy sources which are potentially renewable and inexhaustible, solar energy becomes the center of attention.

Demand for conventional energy sources has increased tremendously since the start of the industrial revolution. Conventional energy sources are carbon-based, including coal, oil and gas, and burning these sources creates large amounts of carbon dioxide which is a so-called greenhouse gas. The greenhouses gases are responsible for the increase of the Earth's temperature through trapping the Sun's radiant energy as it reflects from the surface of the Earth. Hence, massive consumption of carbon-based conventional energy sources has major future effects – depletion of the energy stock resulting in increased prices, political issues related to energy shortage and increase of the Earth temperature as human demand for better comfort increases. Moreover there are other effects on biological and agricultural production.

As the population, urbanization, industrialization and demand for better human comfort increase, it is expected that the demand for energy consumption will increase. The rapid increase in demand for energy is associated with rapid industrialization of the developing world. China and India are expected to contribute a bigger share in the rapid increase of energy consumption. Latin America and South East Asia are also expected to contribute a bigger part of the increase of energy demand. The rapid demand for conventional energy sources has created a geopolitical energy tension due to the sizable amount of energy supply from major energy producing countries and regions instead. It is expected that the Middle East and Russia will play a major role as global energy producers.

The increase of greenhouse gas emissions is attributed to the increase of carbon-based energy consumption. With rapid demand for coal for power plants and oil and gas for transportation and other sectors, greenhouse gas emissions are increasing. The report of the Intergovernmental Panel on Climate Change (IPCC) shows that the increase of the global temperature is caused by the increase of greenhouse gas emissions (IPCC). As global warming and climate change have tremendous and complex effects, they will have greater significance for human survival. Therefore, minimization of greenhouse gas emissions without stifling the demand for progress and human comfort is crucial.

The reduction of greenhouse gas emissions is possible through reduction of the consumption of conventional energy resources. The reduction of the use of conventional energy resources is achievable by the means of energy efficiency, energy conservation and utilization of renewable energy sources. Existing processes, devices and operations can be energy efficient through application and development of new and novel technologies with the aid of basic sciences. The energy conservation measures are possible through the combined application of the passive (natural) and active (artificial) methods, of which in previous and present times active methods have dominated. Utilization of renewable energy sources, particularly solar energy which is available in most parts of the planet, can be advanced through investigation of the existing equipment, processes and sources of energy and energy requirements which can be alternatively sourced from solar energy. In this scenario, demand for and consumption of conventional energy sources or carbon-based energy sources will gradually decrease, resulting in the reduction of greenhouse gas emissions.

1.3   SOLAR ENERGY UTILIZATIONS

In the modern world, cities and other modern facilities are operated using electrical energy. This is because of easy and simple transmission and storage of electrical energy in different forms. In this situation it is expected that demand for electrical energy will increase in future generations.

Electricity production through solar energy is being achieved in two major processes – solar thermal and solar-photovoltaic. Solar thermal power plants employ concentrating solar collectors to produce high temperatures by focusing solar energy in order to produce steam for power plants. Solar-photovoltaic power plants employ semiconductor materials to convert radiant energy photons to electrons to produce electricity. The two technologies have both advantages and disadvantages depending on the point of view of the user. However, as solar energy is free and clean, the utilization of solar energy for electricity production is compelling and offers both energy security and environmental benefits.

There are many biological processes in which thermal energy is a requirement and which are called endothermic processes. In endothermic processes, the thermal energy is utilized to accelerate the biological processes and thus reduce the waiting time or increase the production cycle. In endothermic biological processes, application of solar energy directly or indirectly is possible using different solar energy collectors. Production of bio-fuels with the support of solar energy enhances the processes.

Hydrogen production from water and other sources has been demonstrated as an alternative clean energy source. However, production of hydrogen through thermo-chemical or molecular breaking or through electro-chemical processes are the main methods and require external energy input. Hydrogen production or any thermo-chemical processes can be done with the support of solar energy either as direct thermal energy or using electricity generated from solar energy. There are several large scale research facilities for the production of hydrogen through thermal energy. It has also been demonstrated that the production of hydrogen through electrochemical processes is feasible.

In agricultural industries, application of thermal energy for drying is most important to increase the shelf life of the products. Solar energy can also be used for the bio-chemical processes in agricultural production. There are many existing and under-demonstration technologies for agricultural applications. The most common and typical application is the solar dryer which has been demonstrated to be feasible and practical using simple design, local materials and unskilled workers. In addition, solar cookers and other food processing applications of solar energy have been demonstrated.

Machines minimize human effort with increase of production and operations in many different applications. There are many thermally operated mechanical processes including heating, ventilating and air-conditioning systems, heat engines, pumps and fans. Application of solar energy to support the thermal requirements of these machines and equipment is feasible. Thermally operated air-conditioning systems, pumps and fans have been shown and demonstrated to be feasible and practical. There are many concepts, designs and technologies readily available and being conceptualized for applications.

The occupants of buildings always demand thermal comfort conditions during summer time and winter time, both daytime and nighttime. The maintenance of indoor thermal comfort conditions in buildings consumes large amounts of energy. Also, the reduction of humidity in buildings consumes large amounts of energy in tropical climates. Indoor cooling and heating both in tropical and temperate climates consumes considerable amounts of energy. Furthermore, the lighting for buildings is another main consumer of energy particular for office buildings. Solar energy which is readily available can be utilized to support the day-to-day operation of buildings and to support the indoor comfort conditions for occupants. For instance, solar energy can be utilized to support the air flow rate requirement through natural ventilation. Solar energy can be used to support the thermal energy requirement of buildings through solar thermal collection. Solar energy can be used to support the electricity requirements of buildings through photovoltaic installations. Solar energy can be collected through thermal storage to support the nighttime thermal energy requirement of buildings.

The effect of a heat island is felt in big cities because of the increased utilization of air-conditioning systems and application of urban materials which absorb solar energy. Proper design through urban planning can minimize the effect of solar radiation including sun shading, alternative materials, air movement for natural ventilation and the general utilization of solar energy for building applications. There are conceptualizing technologies to capture the solar energy in urban areas through photovoltaic technologies and solar thermal technologies to reduce the solar energy contribution to the heat island effect. Application of technologies, which can effectively collect the solar energy in the different areas of cities or urban areas, minimizes its contribution to the heat island effect.

There are several day-to-day applications of solar energy which we are using that can be alternatively sourced. Day-to-day applications of light to illuminate us such as during reading or other indoor activities are common. Washing of clothes and drying can consume energy. The daily usage activities and personal usage of energy can be sourced alternatively using solar energy in the form of day lighting and thermal energy which can make us minimize the use of conventional and existing energy sources.

1.4 PERSPECTIVE

Solar energy is available in every part of the planet; however, utilization of the energy depends on our desire and needs and in most cases on economic factors. Depleting conventional energy sources and the issue of global energy politics have become very serious. Global warming and climate change present a serious situation which endangers humanity. Therefore, serious consideration of alternative approaches to the issue of energy resources is vital.

REFERENCE

IPCC Second Assessment Report: Climate Change 1995 (SAR). Working Group I: The Science of Climate Change. http://www.ipcc.ch/publications_and_data/publications_and_data_reports.shtml#1

Chapter 2

Exergy analysis of solar radiation processes

Ryszard Petela
Technology Scientific Ltd., Calgary, Alberta, Canada

2.1 INTRODUCTION

Radiation energy can be converted to work, heat, chemical energy or electricity. Direct conversion to work is so far not well developed, but potential examples are the idea of sailing in space via photon wind, a combination of gravity and buoyancy in a solar chimney power plant or utilization of radiation pressure. By conversion to heat the enthalpy of any operating fluid is usefully increased. Conversion to chemical energy is photosynthesis and conversion to electricity occurs in photovoltaic.

The present chapter contains problems of engineering thermodynamics of thermal radiation and thus is mainly based on the book by Petela (2010), until now the only one written on this subject. The chapter outlines the fundamentals of examining processes in which radiation takes place. Beside traditional methods of energy analysis of such processes, the full thermodynamic analysis, including the exergy analysis, is discussed and illustrated by examples of some typical utilization processes of solar radiation. The analysis is preceded by a basic description of exergy which is a property of any matter.

Everything which has mass is called *matter*. The matter appears in substantial or non-substantial forms. *Mass* is a property of matter which determines momentum and the gravitational interactions of bodies. *Substance* is matter for which the rest mass is not equal to zero. Thus, the substance is the macroscopic body composed of elemental particles (atoms, molecules). The matter for which the rest mass equals zero (e.g. radiation photons) appears in the form of different fields; e.g. fields of electromagnetic waves (radiation), gravity fields, surface tension fields, etc.

Substance can be the object of the conservation equation. Non-substantial matter (called sometimes *field matter*) can also be considered as the component in processes of energy conversion; however it does not fulfill the matter conservation equation. Processes considered here are composed of substance and field matter.

The chapter develops a methodology of examining thermodynamic processes under the assumption that the reader is familiar with the fundamentals of engineering thermodynamics and heat and mass transfer. The details of mechanisms of the considered processes and installations are discussed in other parts of the book together with relevant references.

2.2   EXERGY

2.2.1   Definition of exergy

Exergy is a thermodynamic concept and is one of several thermodynamic properties (functions of thermodynamic parameters) of states of matter which can appear as a substance or any field matter, e.g., as radiation. The functions are defined to make consideration easier, allow for interpretation of phenomena and, most important, most of them have practical applications in thermodynamic calculations. For example, for a substance the enthalpy is used to determine the energy of an exchanged substance with the considered system, whereas the internal energy expresses the energy of a substance remaining within the system during the time of the system's consideration. Entropy determines the thermodynamic probability of a given matter state. Exergy was introduced to express the practical energetic value of matter relative to the environment given by nature. This practical value is determined by the ability of matter to perform mechanical work. Work was selected as the measure not only due to the human inclination to laziness, but first of all, because work represents the energy exchange at the unlimited level. Work is a process in which the energy does not degrade.

However, the full utilization of energetic value of matter to perform work within the determined environment could not occur without the cooperation with this environment. For example, to utilize the energetic value of natural gas by its combustion, a certain amount of oxygen contained in the environmental air has to be taken. Another example: to fully utilize a compressed air at the environment temperature, the depressurizing of the air has to occur at constant temperature, and to keep this temperature steady, heat from the environment has to be taken. The full definition of exergy was given by Szargut which after modification, i.e. not only for substance but for any matter, including radiation, is as follows:

> Exergy of matter is the maximum work the matter could perform in reversible process in which the environment is used as the source of worthless heat and worthless substances, if at the end of the process all the forms of participating matter reach the state of thermodynamic equilibrium with the common components of the environment.

The environment is the natural reference state given by nature which consists of the worthless components available in arbitrary amounts. The environmental components of apparent energetic value, appearing however in limited amounts, are the exceptions and are recognized as natural resources, e.g. natural fuels.

The key component of exergy results from the values of the thermodynamic parameter of the considered matter, and such component is called the thermal exergy. The formula for thermal exergy can be derived as follows. The considered matter of a given temperature T and entropy S is delivered into the system from which this matter is exiting at the state of equilibrium with the human environment, i.e. at environmental temperature T_0 and appropriate entropy S_0. To obtain the maximum work W, which according to the exergy definition would be equal to exergy B of the matter $(W = B)$, the process in the system has to be ideal (reversible) and appropriately adjusted by heat Q exchanged with a heat source, which is the environment at temperature T_0. In the energy balance equation for the system, the energy E of the considered matter delivered to the system is equal to the energy E_0 of this matter leaving the system, to performed work W and to exchanged heat Q:

$$E = E_0 + B + Q \qquad (2.2.1)$$

The algebraic overall entropy growth for the reversible process is zero:

$$S_0 - S + \frac{Q}{T_0} = 0 \qquad (2.2.2)$$

When equation (2.2.2) is used to eliminate Q from equation (2.2.1) then the general formula for the exergy B of matter is:

$$B = E - E_0 - T_0(S - S_0) \qquad (2.2.3)$$

For example when the considered matter is a substance then energy E is interpreted as the enthalpy of the substance. In the case of radiation the appropriate radiation energy has to be taken for E as discussed in the next paragraphs.

The thermodynamic value of a matter determined by energy E is significantly different from the value determined by exergy B. This fact enables the interpretation of the value of a matter, as well as the process, from two viewpoints and the deeper understanding of the considered thermodynamic problem. However any engineering designing should be based on the results from energy calculation.

The exergy concept is also applied to the phenomena of energy exchange which are work and heat. Exergy of work, by definition, is directly equal to the work. Work, beside mechanical work, can also appear as work performed by an electrical current or magnetic field, etc.

Regarding heat, the thermodynamic concept of a heat source is applied. A heat source is defined as the body, at given temperature, which can absorb or release infinitely large amounts of heat without a change in the body temperature. Exergy B_Q of heat Q at temperature T is measured by the change of exergy of the heat source at temperature T absorbing heat Q. The maximum work which could be performed by heat Q is determined as the work performed in the ideal Carnot engine cooperating with two heat sources: one at temperature T and other at environment temperature T_0. Using the definition of Carnot efficiency:

$$B_Q = \left| Q\frac{T - T_0}{T} \right| \qquad (2.2.4)$$

For example, heat Q at temperature $T = T_0$ has practically zero value and from formula (2.2.4) the exergy B_Q of such heat is zero. This is a typical illustration of interpretation advantages of exergy over energy (heat). The value B_Q can be recognized either as a positive input to the system at $T > T_0$ or a positive output from the system if $T < T_0$.

2.2.2   Exergy annihilation law

In reality, there is no exergy conservation equation. Exergy can be conserved only in ideal processes (e.g. model processes), which are reversible because it is required they occur without friction and the heat and substance transfers are driven at the infinitely small differences of temperature and substance concentration. All real processes occur irreversibly, the energy is dissipated and thus the processes are accompanied by unrecoverable exergy loss.

Any process can be analyzed for determination of exergy loss. For example a thermodynamic medium of energy E_1 and entropy S_1, is entering any real thermal installation, performing work W, and then leaving the installation with the energy E_2 and entropy S_2. In general the installation can absorb valuable heat Q from the source at temperature T, which differs from the environmental temperature T_0. At the same time the installation extracts worthless heat Q_0 to the environment.

The work can be determined from the energy conservation equation:

$$W = Q + E_1 - E_2 - Q_0 \qquad (2.2.5)$$

The ideal comparable installation at the same heat Q and energies E_1 and E_2 can perform maximum work, W_{max}, if due to reversibility the real heat Q_0 adjusts itself to a value $Q_{0,i}$ for an ideal process:

$$W_{max} = Q + E_1 - E_2 - Q_{0,i} \qquad (2.2.6)$$

The exergy loss δB, caused by irreversibility in the real installation, is equal to the loss of work ($W_{max} - W$) which results from combining equations (2.2.5) and (2.2.6):

$$\delta B = W_{max} - W = Q_0 - Q_{0,i} \qquad (2.2.7)$$

According to the Second Law of thermodynamics, the algebraic overall entropy growth Π for the medium and heat sources participating in the considered real process is:

$$\Pi = -\frac{Q}{T} + S_2 - S_1 + \frac{Q_0}{T_0} \qquad (2.2.8)$$

and for the compared ideal process, ($\Pi = 0$):

$$0 = -\frac{Q}{T} + S_2 - S_1 + \frac{Q_{0,i}}{T_0} \qquad (2.2.9)$$

If ($Q_0 - Q_{0,i}$) is replaced from (2.2.7) with use of equation (2.2.8) and (2.2.9), then:

$$\delta B = \Pi T_0 \qquad (2.2.10)$$

and this result is known as the Gouy-Stodola law expressing the law of exergy annihilation due to irreversibility.

Example 2.2.2.1 Water is heated from an irradiated black surface. Cold water of entropy S_1 disappears (minus) and in its place there appears (plus) warm water of entropy S_2. Entropy S_r of heating radiation disappears (minus) because of absorption on the surface and the surface emits (plus) its own radiation of entropy S_e. Exergy loss can be calculated from formula (2.2.10) in which

$$\Pi = -S_1 + S_2 - S_r + S_e \qquad\qquad (2.2.11)$$

The overall entropy growth Π does not include entropies of any work as well as of effect of such fields like gravitational or surface tension of substance. These magnitudes, although they contribute to the disorder measured by entropy, have no thermodynamic parameters and act only indirectly by changing parameters of involved matters.

The overall entropy growth has to be positive even in the smallest step ($d\Pi > 0$) in the course of the process. For any theoretical model of reversible phenomenon $d\Pi = 0$; if however $d\Pi < 0$ then the whole phenomenon is impossible. For example, during design of a heat exchanger, care should be taken about the so-called pinch point for which, if locally $d\Pi < 0$, then the whole process of heat exchange is impossible. Thus, the entropy is very useful in verification of designed new processes from a reality viewpoint. The larger the overall entropy growth, the more irreversible is the considered process.

If the substance remains unchanged during a process (e.g. during physical process), then only the respective change in the substance entropy exiting and entering the system are taken into calculation of Π. If a substance disappears (chemical reaction) then its absolute entropy has to be taken with a negative algebraic sign. If a substance appears, the positive sign of entropy should be used.

The exergy loss expressed by formula (2.2.10) is called the *internal exergy loss*, because it occurs within the considered system. This loss is totally non-recoverable. Internal loss of exergy for a multi-component system is calculated by summing up the internal losses of exergy occurring in the particular system components.

Each exergy loss contributes to the increase in the consumption of the energy carrier which sustains the process or to the reduction of the useful effects of the process. One of the main engineering tasks is operating the processes in the way at which the exergy loss is at a minimum. However, most often, the reduction of the exergy loss is possible only by increasing the capital costs of the process. For example the reduction of exergy loss in a heat exchanger is reachable by costly increases of the surface area of heat exchange. Therefore, the economics of such reduction of exergy loss can be verified by economic calculations. The exergy analysis explains the possibilities of improvement of thermal processes; however only the economic analysis can finally motivate an improvement.

Usually, from a thermal process there is released one, or more, waste thermodynamic media (e.g. combustion products), of which the parameters are still different from the respective parameters of such a medium being in equilibrium with environment. The waste medium represents certain exergy unused in the process. Such exergy, if released to the environment, is destroyed due to irreversible equalization of parameters of the waste medium with the parameters of environmental components. The exergy loss of the system, caused by such a way, is called *external exergy loss* and its numerical value is equal to the exergy of the waste medium released by the system. External exergy loss is recoverable, at least in part, e.g., by utilization in other systems.

The irreversibility of radiation processes occurs due to such basic phenomena as emission and adsorption. The irreversibility mechanism of many radiation processes, e.g., diluting or attenuation of propagating radiation, can be explained based on the irreversibility of emission and absorption. Obviously, in the combined processes, in which substance and radiation take place, all the sources of irreversibility should be considered; those for substance together with those for radiation. The radiation entering or absorbed in the considered system has a negative sign of radiation entropy, whereas radiation leaving the system, or emitted, has positive entropy. The radiation entropy is recognized as absolute.

The problem of radiation irreversibility was considered for the first time by Petela (1961b) and later edited in the book (Petela 2010). Based on the overall entropy growth for considered processes it was proved that the emission alone (not accompanied by any adsorption) is possible ($\Pi > 0$), whereas the absorption alone is irreversible, and without accompanying emission of the considered surface is impossible ($\Pi \leq 0$). The simultaneous emission and absorption is always possible ($\Pi > 0$).

The exergy of radiation reaching any surface can be reflected (re-radiated) and the reflected radiation has its exergy at the temperature of the original radiation, which was not utilized by the absorbing surface. If the reflection process does not change the radiation temperature then this process is reversible, and not generating any exergy loss. However, the radiation emitted by the absorbing surface has its own exergy determined by the emissivity and temperature of the absorbing surface. This is the problem of the efficiency of the absorbing surface, or any other device utilizing the radiation somehow, in how much of the whole incident exergy b the surface, or the other device, can be grasped and utilized. The efficiency of the absorbing device or surface is an entirely different thing and does not depend on the theoretical potential represented by b. Acceptance of such interpretation is very important in correct reasoning on the theory of radiation exergy, because if not noticed by some researchers, it can mislead to strange conclusions.

2.2.3 Exergy of substance

2.2.3.1 Traditional exergy

A total exergy of substance is composed of some components as shown schematically in Figure 2.2.1. Usually, only these components are used, which vary during the analysis. Most often is used the thermal exergy, which is the sum of physical and chemical exergies. The physical exergy results from different temperature and pressure of the considered substance in comparison to its temperature and pressure in equilibrium with the environment (dead state). The chemical exergy results from the different chemical composition of the considered substance in comparison to the common substance components of the environment.

If the considered substance has significant velocity, then the kinetic exergy can be recognized as equal to the kinetic energy calculated for the velocity relative to the environment. Potential exergy is equal to the potential energy if it is calculated for the

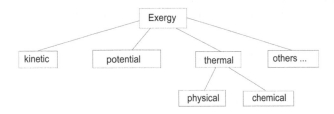

Figure 2.2.1 Exergy components.

reference level, which is the surface of the Earth. The "other" possible components, e.g., nuclear, or interfacial tension, are rarely used and are excluded from the present consideration. The sum $(B_{ph} + B_{ch} = B)$ of most important components for substance considerations; physical exergy B_{ph} and chemical exergy B_{ch} is called thermal exergy B and according to the general formula (2.2.3) is determined as:

$$B = H - H_0 - T_0(S - S_0) \qquad (2.2.12)$$

where B is the thermal exergy of the considered substance at enthalpy H and entropy S, and H_0 and S_0 are the enthalpy and entropy of this substance in eventual state of thermodynamic equilibrium with the environment parameters.

The thermal exergy B expressed by equation (2.2.12) is for a substance passing through the system boundary. The exergy B of substance can be positive or negative (e.g. for each medium flowing through the pipeline there could be a sufficiently low pressure at which thermal exergy B is smaller than zero). Especially for air, such negative exergy can happen for example when the air temperature is not much higher relative to environment, and the air pressure is lower than atmospheric. However always positive is the exergy B_s of any part of the system which remains within the system boundary under pressure p and occupying volume V:

$$B_s = B - V(p - p_0) \qquad (2.2.13)$$

where p_0 is the environment pressure. The concepts B and B_s suggest an analogy to enthalpy and internal energy.

In a particular case if a space of volume V is empty, then the pressure $p = 0$, so $B = 0$ (because there is no substance), then from formula (2.2.13) the exergy of the lack of substance is:

$$(B)_{p=0} = Vp_0 \qquad (2.2.14)$$

As shown later, a similar effect regarding lack of radiation (for $T \rightarrow 0$) in formula (2.2.31) shows the finite exergy value $b_{b,S} = a \cdot T_0^4/3$ in the vessel and the value from formula (2.2.42) for emissions.

In practice, the formula (2.2.3) can be rearranged for example to the frequently applied form for the exergy b, J/kg, of an ideal gas of temperature T and pressure p:

$$b = c_p(T - T_0) - T_0\left(c_p \ln \frac{T}{T_0} - R \ln \frac{p}{p_0}\right) \qquad (2.2.15)$$

where c_p is the specific heat of gas at constant pressure, R is the individual gas constant, T_0 is the absolute temperature of environment and p_0 is the partial pressure of the considered gas in equilibrium with environment.

2.2.3.2 Gravitational interpretation of exergy

Solar heating of a surface on Earth, combined with the gravity field, creates specific effects, e.g., a driving force for solar chimney power plant. The full role of the Earth's gravity field in such processes could be better analyzed by introducing the additional component of mechanical exergy b_m (called shortly ezergy) and by additional terms of gravity input G in the exergy balance, both proposed by Petela (2010).

The mechanical exergy concept b_m is derived from the difference between density ρ of considered substance and density ρ_0 of environment. Regardless of the temperature T and pressure p of the substance under consideration, the buoyancy of the substance is unstable, and thus the ability to work in the environment at respective parameters T_0 and p_0 is sensed if removed as either an anchor ($\rho < \rho_0$) or a supports ($\rho > \rho_0$). In the first case, the substance moves upwards, in other case, the substance sinks.

The altitude of the considered substance is measured from the actual level $x = 0$. In both the above cases the substance tends to achieve an equilibrium altitude ($x = H$), at which the density of local environment $\rho_{0,x}$ is equal to density of the considered substance; $\rho = \rho_{0,x}$. The motion of substance (remaining at constant T and p) to the equilibrium altitude would generate work which is called the buoyant exergy b_b, which does not depend on the kind of the substance. During repositioning of the substance, from actual altitude $x = 0$ to $x = H$, the gravity acceleration g_x is changing, e.g. decreasing with growing altitude x, thus:

$$b_b = \int_{x=0}^{x=H} g_x\left(\frac{\rho_{0,x}}{\rho} - 1\right)dx \qquad (2.2.16)$$

The solution of equation (2.2.16) is discussed by Petela (2010).

At level H the substance could be allowed to generate additional work, denoted by b_H, which would occur during a reversible process of equalization of parameters T and p with the respective local environment parameters $T_{0,H}$ and $p_{0,H}$. In case of a gas, during equalizing of the gas parameters T and p with the parameters $T_{0,H}$ and $p_{0,H}$ at the altitude H, based on Formula (2.2.15), the following work (exergy b_H) can be done:

$$b_H = c_p(T - T_{0,H}) - T_{0,H}\left(c_p \ln \frac{T}{T_{0,H}} - R \ln \frac{p}{p_{0,H}}\right) \qquad (2.2.17)$$

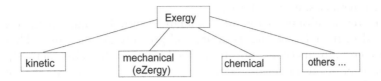

Figure 2.2.2 Exergy components including the mechanical exergy (ezergy).

On the other hand, the gas at the actual altitude ($x = 0$) has the traditional physical exergy b, expressed by formula (2.2.15), equal to the work which could be done by the gas during the equalizing of its parameters, T and p, with respective environment parameters T_0 and p_0.

Exergy definition postulates the exergy to be the maximum possible work. Therefore, the larger work of the two, $b_b + b_H$ or b, is the true exergy called the mechanical exergy.

$$b_m = \max[(b_b + b_H), b] \qquad (2.2.18)$$

Including the mechanical exergy (eZergy) into consideration, the scheme of the all components of the exergy of substance is now shown in Figure 2.2.2.

The eZergy is applied only for the substance (e.g. not for heat or radiation) and it replaces the two traditional exergy components: physical (B_{ph}) and potential (B_p). To better distinguish exergy of the substance from the eZergy of the substance, different symbols could be used: B for eXergy and Z for eZergy, (e.g. in paragraph 2.4.3: $B_m \equiv Z = f(B_{ph}, B_p)$.

2.2.3.3 Chemical exergy of substance

In a chemical process, in contrast to a physical process, substances change and only the chemical elements remain unchanged. Therefore, to calculate the chemical energy (or exergy) of substances the reference substances have to be assumed appropriately.

The most common methods for determination of the chemical energy of substances are *enthalpy formation* and *devaluation enthalpy*, which differ mainly by the definition of reference substances. In the enthalpy formation method, the reference substances are the chemical elements at standard temperature and pressure.

In the devaluation reaction method the number of reference substances is the same; however, they are not the chemical elements but the devaluated substances (compounds or chemical elements most commonly appearing in the environment). For example, the reference substance of C is gaseous CO_2, for H it is gaseous H_2O, and for O it is just O_2. Therefore, in any particular case, when a substance is composed only of C, O, H, N, and S, then the devaluation enthalpy of the substance is equal to its calorific value.

Contrary to the devaluation enthalpies, the values of the enthalpy of formation are a little illogical. For example, the enthalpy of formation for valuable pure carbon C is zero and the enthalpy of formation of not valuable CO_2 is significantly different from zero (-394 MJ/kmol). However, for comparison, the devaluation enthalpy of C is equal to its calorific value ~394 MJ/kmol, whereas the calorific value of CO_2 is zero (as it is the reference substance for C).

The reference substances for the devaluation enthalpy and chemical exergy are the same. Also, the reference temperature and pressure are the same. Thus, only the devaluation enthalpy method, contrary to the formation enthalpy method, allows for fair comparison of the values of chemical energy and chemical exergy. For example, the chemical exergy of C is ~413 MJ/kmol and devaluation enthalpy (calorific value) of C is only ~394 MJ/kmol. Therefore, only the devaluation enthalpy method should be used in thermodynamic analysis, because then the comparative energy and exergy analyses are simultaneously included.

More details on the devaluation method is discussed by Szargut et al. (1988). Only the significance of the concept of devaluation reaction and the resulting concept of devaluation enthalpy, used for calculation of chemical energy, is outlined here. The method allows calculation of the following quantities: enthalpy devaluation of a substance (appearing in the energy conservation equation), standard entropy of devaluation reaction (in the entropy considerations), and, consequently, chemical exergy of substance (appearing in the exergy balance equation).

Devaluation enthalpy is determined based on the stoichiometric devaluation reaction for a substance. The devaluation reaction has to be a combination only of the considered substance and the various reference substances. A good example of a devaluation reaction is reaction of photosynthesis:

$$6H_2O + 6CO_2 \rightarrow C_6H_{12}O_6 + 6O_2 \tag{2.2.19}$$

in which, besides the considered substance of sugar ($C_6H_{12}O_6$), only the reference substances appear; CO_2, H_2O and O_2. The devaluation enthalpy, d_n, is calculated from the energy conservation equation for the chemical process in which substrates are supplied, and products are extracted, all at standard temperature and pressure.

As shown, in comparison to the physical exergy b_{ph} of a substance, the calculation of the chemical exergy b_{ch} of a substance is more complex, depending on its composition, and is based on the devaluation reaction. The calculation procedure is discussed by Szargut et al. (1988), where the calculated standard values of the devaluation enthalpy (d_n) and chemical exergy (b_n) are tabulated for the standard environment temperature $T_0 = T_n$. If the environment temperature T_0 differs from the standard environment temperature T_n, then, when using the standard data on d_n and b_n, the formula for the chemical exergy of condensed substances (solid or liquid), should be corrected as shown, e.g., for the specific chemical exergy $b_{ch,SU}$ of sugar:

$$b_{ch,SU} = b_{n\,SU} + \frac{T_n - T_0}{T_n}(d_{n\,SU} - b_{n\,SU}) \tag{2.2.20}$$

where $b_{n,SU}$ and $d_{n,SU}$ are the standard tabulated values of the chemical exergy and devaluation enthalpy of sugar, respectively. If a substance has a temperature different from the surrounding environment, then also a physical component of energy or exergy has to be included as shown, e.g., again for the physical exergy $b_{ph,SU}$ of the sugar:

$$b_{ph,SU} = c_{SU}(T - T_0) - T_0 c_{SU} \ln\frac{T}{T_0} \tag{2.2.21}$$

where c_{SU} is the specific heat of sugar.

Note as well that based on the devaluation reaction, the so-called standard entropy σ_n of the devaluation reaction can be determined. For example, again for the photosynthesis reaction, based on Equation (2.2.19), the standard entropy of the devaluation reaction, $\sigma_{n,SU}$, is:

$$\sigma_{n,SU} = 6(s_{H_2O} + s_{CO_2})_n - 6(s_{O_2})_n - s_{n,SU} \qquad (2.2.22)$$

where s_{H_2O}, s_{CO_2}, and s_{O_2} are the absolute standard entropies of the respective gases. The stoichiometric factor 6 results from Equation (2.2.19).

2.2.4 Exergy of photon gas

Radiation has two meanings: it could be the process of a radiating body or it could be the product of this process. This process can be considered twofold – either as the propagating magnetic field or as the population of photons traveling in space. Then the photon populations can be imagined as the photons trapped in a system of limited space, or as the freely traveling flux of photons. The states of trapped or traveling photons are similar to the substance concepts of thermodynamic functions of state which are respectively the internal energy or enthalpy. The traveling photons in a form, e.g. of emitted radiation from a substance body (emission), or as a bundle of rays from many bodies of different temperatures (arbitrary radiation flux), are considered respectively in the next paragraph 2.2.5 and 2.2.6. In the present paragraph the large population of photons trapped in a space is considered.

The *temperature* concept in radiation problems can be applied only to a photon batch and the temperature T of thermal radiation can only be determined indirectly, i.e. by measuring the temperature of the substance with which the radiation is in equilibrium.

It can be derived that the internal energy U, in J, of the photon gas within a system of volume V, is:

$$U = aVT^4 \qquad (2.2.23)$$

where a is the universal constant ($a = 7.564 \cdot 10^{-16}$ J/(m^3 K^4)). The rest mass of photon gas is zero, therefore, energy U cannot be related to the mass of this gas but rather to its volume. Thus the photon gas energy density u, J/m^3, is:

$$u = \frac{U}{V} = aT^4 \qquad (2.2.24)$$

The entropy density s_S, J/(K m^3), of black radiation in the system is derived as:

$$s_S = \frac{4}{3}aT^3 \qquad (2.2.25)$$

Radiation transports a linear momentum and may exert radiation *pressure* on an object by irradiating it. Radiation pressure can be considered either as the effect of interaction between radiation and substance, or as the effect only within the internal structure of the radiation. In the first case, radiation pressure exerted on a substance of transmissivity $\tau = 0$ is zero, however this pressure grows with the growing transmissivity and is maximum for $\tau = 1$. In the second case, the photons can transfer momentum to other particles upon impact and such possibility of potential pressure exists regardless of the properties (e.g. transmissivity) of any target. As in the consideration of substance pressure, the equipartition theorem can be applied to the energy in a three-dimension system and accordingly the pressure p of black radiation is:

$$p = \frac{u}{3} \qquad (2.2.26)$$

and using (2.2.24) in (2.2.26):

$$p = \frac{a}{3} T^4 \qquad (2.2.27)$$

One of the possible processes of photon gas is the isentropic process during which the photon gas does not exchange heat with the surroundings. For example such a process can be imagined for a photon gas trapped in the space surrounded with an expandable wall of perfect reflectivity. The isentropic process occurs reversibly and the entropy in each elemental process stage remains constant. The entropy in J/K of the gas occupying volume V is determined based on formula (2.2.25) and the condition of constant entropy in the process is:

$$V \frac{4}{3} a T^3 = \text{const.} \qquad (2.2.28)$$

or eliminating temperature T by pressure p with use of formula (2.2.27):

$$p V^{4/3} = \text{const.} \qquad (2.2.29)$$

Exergy of black radiation was determined for the first time by Petela (1964), by consideration of the isentropic process in which the $V = 1\,\text{m}^3$ of radiation of temperature T changes the initial pressure p to the final pressure p_0 at temperature of the environment. The final state of the considered radiation is in equilibrium with the environment and the exergy of this radiation is zero. Therefore, according to the exergy definition, the initial density $b_{b,S}$, J/m³, of black radiation within the system is equal to the useful work performed during the process:

$$b_{b,S} = \int_V^{V_0} p\, dV - p_0(V_0 - V) \qquad (2.2.30)$$

Using (2.2.27) and (2.2.29) in (2.2.30):

$$b_{b,S} = \frac{a}{3}(3T^4 + T_0^4 - 4T_0 T^3) \qquad (2.2.31)$$

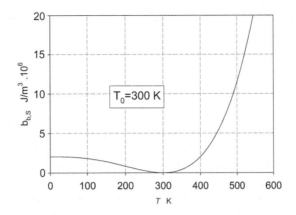

Figure 2.2.3 Exergy of photon gas as function of temperature.

For example, Figure 2.2.3 shows $b_{b,s}$ as a function of temperature T at $T_0 = 300\,\mathrm{K}$. The exergy of radiation is always non-negative and the zero value is achieved only at $T = T_0$.

2.2.5 Exergy of radiation emission

During radiation of substantial bodies (solid, liquid or some gaseous), a part of their energy (e.g. internal energy or enthalpy) transforms into the energy of electromagnetic waves of a length theoretically from 0 to ∞. These waves can travel in a vacuum because they do not require any medium for their propagation. Independently one can also imagine this radiation process as the energy non-continuously emitted in form of the smallest indivisible energy portions called photons. If the energy of a body is not simultaneously supplemented from an external source, then temperature of the radiating body decreases. The phenomenon of such radiation is called emission, which is the key problem in study of radiation, especially its simplified models.

Temperature of radiation is always equal to the temperature of the emitter. According to the Stefan-Boltzmann law the energy e_b of emission of a black surface at temperature T is:

$$e_b = \sigma T^4 \tag{2.2.32}$$

where $\sigma = 5.6693 \cdot 10^{-8}\,\mathrm{W/(m^2\,K^4)}$ is the Boltzmann constant for black radiation. Emissivity of the emitter, e.g., the emissivity of a solid surface, determines the surface ability measured by the rate at which the black radiation is produced. The perfect gray surface of emissivity $\varepsilon < 1$ emits black radiation of an amount determined by the emissivity ε. If the density of emission e_b expresses the amount of the emitted black radiation energy from $1\,\mathrm{m^2}$ of black surface at $\varepsilon = 1$, then density e, ($e = \varepsilon \cdot e_b$) expresses the amount of the emitted black radiation energy from $1\,\mathrm{m^2}$ of gray surface, at the rate ε:

$$e = \varepsilon \sigma T^4 \tag{2.2.33}$$

The black emission e_b has its exergy b_b, however as the rate of emission e of a gray surface is smaller ($e = \varepsilon \times e_b$) then the exergy b (of this gray surface emission e), is also reduced by ε:

$$b = \varepsilon b_b \tag{2.2.34}$$

However using the energy emissivity ε in formula (2.2.34) for exergy calculation is not precise and, as discussed in details by Petela (2010), the smaller is the precision the smaller is the emissivity ε. The definition of exergy can be applied to the emitted photon gas. Therefore, the exergy of the emitted photon gas is a function of an instant state of the gas and of the state of the gas in the instant of eventual equilibrium with the environment. Such equilibrium determines the reference state in calculation of the exergy of the photon gas.

The environment consists of many bodies of different temperatures and different radiative properties (e.g. emissivities or transmissivities). The dominating temperature of the environment's bodies can be assumed as the standard (averaged) environmental temperature T_0. As discussed previously the surface always emits black radiation, thus the environment surface at temperature T_0, regardless of the surface properties, emits black radiation at temperature T_0. The properties of the surface determine only the rate at which the emission occurs. Thus, the environment space permanently contains the black radiation at temperature T_0 and this radiation is in equilibrium with the environmental surfaces at T_0.

Therefore, the exergetic reference state for a photon gas (black radiation) is its state at temperature T_0, and such reference state depends only on the temperature T_0 and does not depend on diversified values of emissivities of the environmental bodies.

The black emission exergy b_b appearing in formula (2.2.33) is always a function only of temperature T of the considered surface and of the environmental temperature T_0, $b_b = f(T, T_0)$. No pressure has to be considered for establishing the exergy reference state for radiation because the pressure of radiation is determined only by the radiation temperature and any pressure of environmental substance does not affect the radiation, which is not a substance.

Detailed analyses confirming the independency of radiation exergy on emissivity of the environment, on the configuration of surrounding environment and on the presence of other surfaces of different temperatures, are gathered in the book by Petela (2010).

The exergy $b_{b,S}$ (J/m^3) of the photon gas enclosed within a system was discussed in Paragraph 2.2.4. However the exergy of the photon gas as being the propagating product of the emitting process, i.e., the exergy b_b (W/m^2) of emission density of a black surface is:

$$b_b = \frac{\sigma}{3}(3T^4 + T_0^4 - 4T_0 T^3) \tag{2.2.35}$$

The simplest way of obtaining formula (2.2.35) is multiplication of $b_{b,S}$ by factor $c_0/4$, making $b_b = c_0 \cdot b_{b,S}/4$, based on purely geometric considerations and keeping in mind that $\sigma = a \times c_0/4$, where c_0 is the light velocity. Another way to derive formula (2.2.35) is by application of the exergy definition formula (2.2.3) with interpretation of energy E and entropy S as respective emission e_b and its entropy s.

However for the confirmation of both ways and for the purpose of the well teaching demonstration of the analysis of radiation processes, an additional way can be

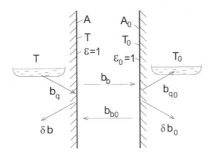

Figure 2.2.4 Radiating parallel surfaces.

also discussed as follows. Simple derivation of the emission exergy of a black surface, published for the first time by Petela (1961b) in Polish and then repeated in English, Petela (1964), is based on the balance of the emitting surface according to the model shown in Figure 2.2.4. The two surfaces A and A_0 which are black, flat, infinite, parallel, facing each other, enclose a space without substance (vacuum) and interchange heat by means of radiation. The model of such two-surfaces-only is often selected for consideration because the space is enclosed by the simplest possible geometry involving only two plane surfaces. Each surface is maintained at uniform and constant temperatures due to exchange of the compensating heat with the respective external heat sources. Surface A_0 at temperature T_0 represents the emitting environment whereas surface A, at arbitrary temperature T, emits the considered radiation. The simplicity of the model with the black surfaces is that there is no reflected radiation to be considered.

In order to derive the formula on the emission exergy density b_b of a black surface, the following exergy balance for surface A, is considered:

$$b_{b0} + b_q = b_b + \delta b \tag{2.2.36}$$

where the terms in equation (2.2.36) or in Figure 2.2.4, all in W/m², are:

b_b, b_{b0} – exergy of emission density of surfaces A and A_0, respectively,
b_q, b_{q0} – change in exergy of respective heat source,
δb, δb_0 – exergy loss due to irreversibility of simultaneous emission and absorption on the respective surface.

From the definition of exergy the radiation of a surface at environment temperature $b_{b0} = 0$. The change in exergy of heat source, based on formula (2.2.4):

$$b_q = q \frac{T - T_0}{T} \tag{2.2.37}$$

where q, W/m², is the heat delivered by the heat source of temperature T. This is the amount of heat which allows surface A to emit and maintain its constant temperature T. This is also the heat exchanged by radiation between surfaces A and A_0, which with use of formula (2.2.32) can be calculated from the energy balance of surface A:
$q = (e_b - e_0) = \sigma \cdot (T^4 - T_0^4)$.

The algebraic overall entropy growth Π due to simultaneous emission and absorption of heat taking place at surface A:

$$\Pi = -\frac{q}{T} + s - s_0 \tag{2.2.38}$$

where $-q/T$ is the decrease of entropy of heat source at temperature T, and the entropy s, J/(m^2 K), of emission density

$$s = \frac{4}{3}\sigma T^3 \tag{2.2.39}$$

is used for determination of s and $-s_0$, respectively for surface A and A$_0$.

The exergy loss δb is determined by the Gouya-Stodola law (2.2.10).

Making use of (2.2.37), (2.2.38) and (2.2.39) in equation (2.2.36), after some rearranging and using (2.2.34), the formula for the exergy of emission density b, W/m^2, of a perfectly gray surface of emissivity ε is obtained:

$$b = \varepsilon \frac{\sigma}{3}(3T^4 + T_0^4 - 4T_0 T^3) \tag{2.2.40}$$

The mathematical analysis of formula (2.2.40) (Petela (1964)), reveals first of all that exergy b is always nonnegative and it has the lowest value zero when $T = T_0$. The exergy b reaches also zero if the considered surface is white (i.e. perfectly reflecting, $\varepsilon = 0$).

Keeping in mind formula (2.2.33), it follows that the exergy of emission of black surface ($\varepsilon = 1$), for the environment temperature approaching absolute zero, becomes equal to the energy of emission:

$$\lim_{T_0 \to 0} (b_b) = \sigma T^4 = e_b \tag{2.2.41}$$

It could be noticed that the characteristic term in brackets of formula (2.2.31), appearing also in formula (2.2.40) was derived by Petela (1964) from the consideration without using the Stefan-Boltzmann law (2.2.32). The obtained equations (2.2.31) as well as (2.2.40) can be recognized as independent of equation (2.2.32). Therefore, the energy of emission e can be interpreted as the particular case of the exergy of this emission at the theoretical condition $T_0 = 0$, or other words, the Stefan-Boltzmann law is a particular case of the emission exergy law expressed by formula (2.2.31).

As the surface temperature T approaches absolute zero the exergy of emission expressed by formula (2.2.40) approaches the finite value:

$$\lim_{T \to 0} (b) = \varepsilon \frac{\sigma}{3} T_o^4 \tag{2.2.42}$$

Based on equation (2.2.40), Figure 2.2.5 illustrates the exergy b_b (solid thick line) of emission density of a black surface ($\varepsilon = 1$) at the constant value of the environment temperature $T_0 = 300$ K. For comparison, the energy e_b (solid thin line) of emission density according to equation (2.2.32) is presented. For a sufficiently small temperature T the exergy of black emission is larger than energy of such emission. The dashed line

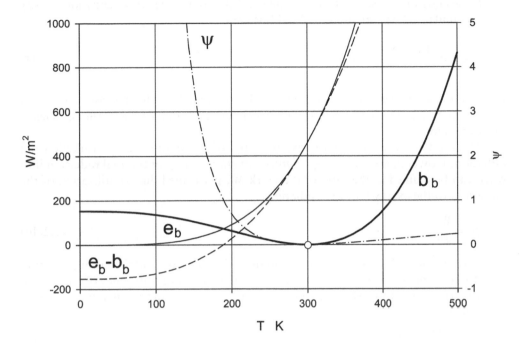

Figure 2.2.5 Emission e_b, exergy b_b, difference $(e_b - b_b)$, and the exergy-energy ratio ψ as function of surface temperature T, at $T_0 = 300$ K.

in Figure 2.2.5 shows the difference $(e_b - b_b)$ which with the growing temperature T from the negative values grows indefinitely.

The emission exergy of a surface of temperature lower than the environment temperature (so-called cold radiation), the effects of varying environmental temperature and the suggestions regarding calculation of radiation exergy of a surface of non-uniform temperature, are discussed in Petela (2010).

A real *energy conversion efficiency* η_E of thermal radiation into work can be defined as the ratio of work W, performed due to utilization of the radiation, to the energy e of this radiation:

$$\eta_E \equiv \frac{W}{e} \tag{2.2.43}$$

The maximum work W_{max} can be obtained from radiation energy in the ideal (reversible) conversion process. Such work in an ideal process is equal to the exergy of the radiation; $W_{max} \equiv b$, and its efficiency η_E changes to the maximum conversion efficiency $\eta_{E,max}$ called *exergy/energy radiation ratio* ψ, defined for the first time by Petela (1964):

$$\frac{b}{e} = \eta_{E,max} \equiv \psi \tag{2.2.44}$$

which can be larger $(\psi > 1)$, equal $(\psi = 1)$, or smaller than unity $(\psi < 1)$, (Fig. 2.2.5).

If the emission density e_b from formula (2.2.32) and exergy b_b of emission density from formula (2.2.35) are used in (2.2.44) then:

$$\psi = 1 + \frac{1}{3}\left(\frac{T_0}{T}\right)^4 - \frac{4}{3}\frac{T_0}{T} \qquad (2.2.45)$$

where T is the temperature of the considered radiation. This characteristic ratio ψ in thermodynamics of radiation has a significance similar to that of the Carnot efficiency for heat engines.

The ratio ψ represents the relative potential of maximum energy available from radiation. However, the real *exergy conversion efficiency* η_B of thermal radiation into work can be defined as the ratio of the work W, performed due to utilization of the radiation, to the exergy b of this radiation:

$$\eta_B = \frac{W}{b} \qquad (2.2.46)$$

Introducing (2.2.43) to (2.2.46) to eliminate the work W, and then using equation (2.2.44) to eliminate the exergy b, one finds that the real exergy efficiency of conversion of radiation exergy to work is equal to the ratio of the real and the maximum energy efficiencies:

$$\eta_B = \frac{\eta_E}{\eta_{E,max}} \leq 1 \qquad (2.2.47)$$

Using equation (2.2.44) in (2.2.47) to eliminate $\eta_{E,max}$, the ratio ψ becomes also the ratio of energy-to-exergy efficiency of the radiation conversion to work:

$$\psi = \frac{\eta_E}{\eta_B} \qquad (2.2.48)$$

Figure 2.2.5 presents the example of the ratio ψ (dotted line) for $T_0 = 300$ K. With the growing temperature T from zero to infinity the value ψ decreases from infinity to the minimum value zero for $T = T_0$ and then increases to unity. However in spite of ψ approaching unity for infinite temperature T, the difference $(e_b - b_b)$ does not approach the expected zero but approaches infinity.

Although the ψ is not defined as efficiency, it can be recognized like an efficiency of a maximum theoretical conversion of radiation energy to radiation exergy. For example, for any arbitrary radiation of the known energy and at certain presumable temperature T, the exergy of this radiation could be approximately determined as the product of the considered energy and the value ψ taken from (2.2.45) for this temperature T.

Example 2.2.5.1 The value $\psi = 0.2083$ for a black emission at temperature $T = 473$ K (200 C) is calculated from formula (2.2.45) at $T_0 = 300$ K. In paragraph 2.2.6, example 2.2.6.1, the ψ_{wv} value for water vapor at $T = 473$ K and $T_0 = 300$ K is calculated based on the radiation spectrum as $\psi_{wv} = 0.185$. The smaller value of ratio ψ_{wv} for water vapor, in comparison to black surface radiation ($\psi_{wv} < \psi$) results from a significant difference in spectra of the water vapor and black surface. However, in paragraph 2.4.1.1, example 2.4.1.2, for solar radiation, the difference between calculated $\psi_S = 0.9326$ for the considered solar spectrum and the value $\psi = 0.9333$, (example 2.4.1.1), for black surface at 6000 K is insignificantly smaller because the solar spectrum is not much different from the black surface spectrum.

2.2.6   Exergy of radiation flux

In paragraph 2.2.4 the energy u, J/m^3, of trapped radiation residing within a space is discussed. Radiation emission density e, W/m^2, of a surface at known temperature is discussed in paragraph 2.2.5. However generally, the radiation flux propagating in space can consist of many emissions from unknown surfaces and of unknown temperatures. Such radiation energy flux, discussed in the present paragraph, can be categorized as a radiosity j, W/m^2, of an arbitrary radiation flux of an arbitrary energy spectrum which can be determined, for example, from measurement.

Now we know that there are many different methods for derivation of the general formula for exergy of the arbitrary flux of radiation. Each method leads to the same result which could be also achieved simply e.g. by interpretation of equation (2.2.3) as shown by Petela (2010):

$$E \rightarrow j = \int\limits_{\beta} \int\limits_{\varphi} \int\limits_{\lambda} (i_{b,0,\lambda,\max} + i_{b,0,\lambda,\min})_T d^2C\,d\lambda \qquad (2.2.49)$$

$$E_0 \rightarrow j_0 = \int\limits_{\beta} \int\limits_{\varphi} \int\limits_{\lambda} (i_{b,0,\lambda,\max} + i_{b,0,\lambda,\min})_{T_0} d^2C\,d\lambda \qquad (2.2.50)$$

$$S \rightarrow s_j = \int\limits_{\beta} \int\limits_{\varphi} \int\limits_{\lambda} (L_{b,0,\lambda,\max} + L_{b,0,\lambda,\min})_T d^2C\,d\lambda \qquad (2.2.51)$$

$$S_0 \rightarrow s_{j,0} = \int\limits_{\beta} \int\limits_{\varphi} \int\limits_{\lambda} (L_{b,0,\lambda,\max} + L_{b,0,\lambda,\min})_{T_0} d^2C\,d\lambda \qquad (2.2.52)$$

where:

j, j_0 – radiosity density of considered radiation and environment, W/m^2,
s_j, $s_{j,0}$ – entropy of radiosity density of considered radiation and environment, W/(m^2 K),
T, T_0 – absolute temperature of radiating surface and the environment, K.

The monochromatic normal directional intensity $i_{b,0,\lambda}$, W/(m^2 sr), for linearly polarized black radiation propagating within a unit solid angle and dependent on wavelength λ, was established by Planck (1914):

$$i_{b,0,\lambda} = \frac{c_0^2 h}{\lambda^5} \frac{1}{\exp\left(\dfrac{c_0 h}{k\lambda T}\right) - 1} \qquad (2.2.53)$$

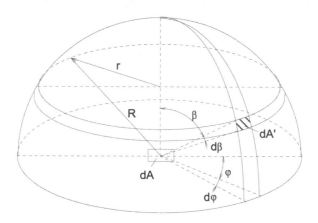

Figure 2.2.6 Geometry scheme for radiation flux (from Petela, 1962).

where: $c_0 = 2.9979 \cdot 10^8$ m/s is the speed of propagation of radiation in vacuum, $h = 6.625 \cdot 10^{-34}$ J s is the Planck constant, $k = 1.3805 \cdot 10^{-23}$ J/K is the Boltzmann constant.

The entropy $L_{b,0,\lambda}$, W/(m^3 K sr), of monochromatic directional normal radiation intensity and for linearly polarized black radiation propagating within a unit solid angle and dependent on wavelength λ according to Planck (1914) is:

$$L_{b,0,\lambda} = \frac{c_0 k}{\lambda^4}[(1 + Y)\ln(1 + Y) - Y \ln Y] \quad \text{where} \quad Y \equiv \frac{\lambda^5 i_{b,0,\lambda}}{c_0^2 h} \tag{2.2.54}$$

The total energy or entropy of radiation is respectively the same regardless whether the spectrum is expressed as function of the wavelength λ or frequency v. Therefore, e.g.:

$$\int_0^\infty i_{b,0,v}\, dv = \int_0^\infty i_{b,0,\lambda}\, d\lambda \tag{2.2.55}$$

To apply such recalculation formula (2.2.55), the formula (2.2.53) for $i_{b,0,\lambda}$ has to be used together with relation $\lambda \cdot v = c_0$. Based on such a possibility, the formulae on radiation exergy can also be presented as functions of the wavelength λ or frequency v.

It is assumed that the considered elemental flux propagates between any two control elementary surface areas dA and dA' separated by distance R, in a direction determined by the flat angles of β (called declination) and ϕ (called azimuth), as shown in Figure 2.2.6. The solid angle of propagation $d\omega = dA'/R^2 = \sin\beta \cdot d\beta \cdot d\varphi$ and the abbreviation:

$$d^2 C \equiv \cos\beta \sin\beta\, d\beta\, d\varphi \tag{2.2.56}$$

used, e.g., in the case of surface radiating to the forward hemisphere is:

$$\iint_{\beta \; \varphi} d^2 C \equiv \int_{\beta=0}^{\beta=\pi/2} \int_{\varphi=0}^{\varphi=2\pi} \cos \beta \sin \beta \, d\beta \, d\varphi = \pi \qquad (2.2.57)$$

Exergy of arbitrary polarized radiation. The exergy $b_{A'}$, W/m^2, of the arbitrary polarized radiation originating from unknown surface A$'$ and arriving in point P of the considered surface A per unit time and unit absorbing surface area, can be interpreted in equation (2.2.3) as $B = b_{A'}$. Developing the whole equation (2.2.3) by including also interpretation (2.2.49) – (2.2.52), after rearranging, it yields:

$$b_{A'} = \iiint_{\beta \; \varphi \; \nu} (i_{0,\nu,\min} + i_{0,\nu,\max}) \cos \beta \sin \beta \, d\beta \, d\varphi \, d\nu$$

$$- \iiint_{\beta \; \varphi \; \nu} (L_{0,\nu,\min} + L_{0,\nu,\max}) \cos \beta \sin \beta \, d\beta \, d\varphi \, d\nu + \frac{\sigma T_0^4}{3\pi} \iint_{\beta \; \varphi} \cos \beta \sin \beta \, d\beta \, d\varphi$$

$$(2.2.58)$$

In order to utilize formula (2.2.58) one has to know the solid angle ω within which the surface A$'$ is seen from point P on surface A, and to know (e.g. from measurements) $i_{0,\nu,\min}$ and $i_{0,\nu,\max}$ as a function of frequency ν and direction defined by β and ϕ.

The total exergy $B_{A' \to A}$ of the considered arbitrary radiation arriving to the all points of surface A is calculated as:

$$B_{A' \to A} = \int_A b_{A'} \, dA \qquad (2.2.59)$$

Exergy of arbitrary non-polarized radiation. The formula for such radiation is obtained after taking into account in formula (2.2.58), that for a non-polarized radiation $i_{0,\lambda,\max} = i_{0,\lambda,\min}$, thus $i_{0,\lambda,\max} + i_{0,\lambda,\min} = 2 \cdot i_{0,\lambda}$. Additionally also $L_{0,\lambda,\max} = L_{0,\lambda,\min}$ thus $L_{0,\lambda,\max} + L_{0,\lambda,\min} = 2 \cdot L_{0,\lambda}$.

$$b_{A'} = 2 \iiint_{\beta \; \varphi \; \nu} i_{0,\nu} \cos \beta \sin \beta \, d\beta \, d\varphi \, d\nu - 2 \iiint_{\beta \; \varphi \; \nu} L_{0,\nu} \cos \beta \sin \beta \, d\beta \, d\varphi \, d\nu$$

$$+ \frac{\sigma T_0^4}{3\pi} \iint_{\beta \; \varphi} \cos \beta \sin \beta \, d\beta \, d\varphi \qquad (2.2.60)$$

In order to utilize formula (2.2.60) one has to know the solid angle ω within which the surface A$'$ is seen from point P on surface A, and to know $i_{0,\nu}$ as function of frequency ν and direction defined by β and φ. Formulae (2.2.59) can be also useful.

Exergy of arbitrary, non-polarized and uniform radiation. The formula for such radiation results from (2.2.60) in which $i_{0,\nu}$ (and $L_{0,\nu}$) does not depend on angles β and φ:

$$b_{A'} = \left(2\int_\nu i_{0,\nu} d\nu - 2T_0 \int_\nu L_{0,\nu} d\nu + \frac{\sigma T_0^4}{3\pi}\right) \int_\beta \int_\varphi \cos\beta \sin\beta \, d\beta \, d\varphi \qquad (2.2.61)$$

and to utilize formula (2.2.61) the solid angle ω within which the surface A′ is seen from point P on surface A, as well as the radiation spectrum as function of frequency, $i_{0,\nu}(\nu)$, (e.g. determined by measurement), has to be known. Again, the formulae (2.2.59) can be applied.

Exergy of arbitrary, non-polarized and uniform radiation propagating within solid angle 2π. The formula for such radiation is derived by substituting equations (2.2.57) into (2.2.61):

$$b = 2\pi \int_\nu i_{0,\nu} d\nu - 2\pi T_0 \int_\nu L_{0,\nu} d\nu + \frac{\sigma}{3} T_0^4 \qquad (2.2.62)$$

To utilize formula (2.2.62) the function $i_{0,\nu}(\nu)$, has to be known. The total exergy of the considered radiation arriving to the all points of the surface A is calculated as follows:

$$B = bA \qquad (2.2.63)$$

Example 2.2.6.1   Figure 2.2.7 shows the measured monochromatic normal radiation intensity $i_{0,\lambda}$, (solid line) of radiation, as a function of wavelength λ, for the water vapor layer of equivalent thickness 1.04 m at temperature 200°C according to Jacob (1957). The product of the thickness and the partial pressure for the vapor is 10.4 m kPa. The monochromatic normal intensity $i_{b,0,\lambda}$ for black radiation, calculated from equation (2.2.53), is also shown for comparison (dashed line). For approximate calculation, instead of the surface area under the solid line, the area of seven rectangles (dotted line) is taken into account as the integral energy emitted by the vapor upon the hemispherical enclosure. The areas of these rectangles can be recognized as the absorption bands of width $\Delta\lambda$ spread symmetrically on both side of wavelength λ, of which values are given in Table 2.2.1.

Exergy of radiation arriving in 1 m² of the enclosing hemispherical wall can be calculated from formula (2.2.62), in which frequency, with interpretation explained by formula (2.2.55), is eliminated by wavelength and each integral can be replaced by the sum of appropriate products:

$$b = \frac{\sigma}{3} T_0^4 + 2\pi \sum i_{0\lambda} \Delta\lambda - 2\pi T_0 \sum L_{0\lambda} \Delta\lambda \qquad (2.2.64)$$

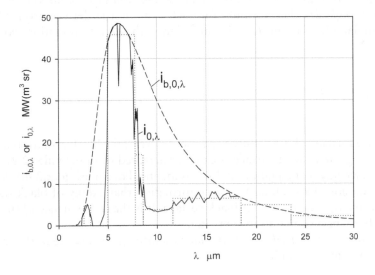

Figure 2.2.7 Radiation of water vapor layer of thickness 1.04 m at temperature 473.15 K and pressure 0.1 MPa (from Jacob, 1957).

Table 2.2.1 Radiation data for water vapor layer of thickness 1.04 m at temperature 473.15 K and pressure 0.1 MPa.

Successive rectangle number	λ μm	$\Delta\lambda$	$i_{0,\lambda} \times 10^{-6}$ $\dfrac{W}{m^3\,sr}$	$i_{0,\lambda} \times \Delta\lambda$	$L_{0,\lambda} \times 10^{-4}$ $\dfrac{W}{m^2\,K\,sr}$	$L_{0,\lambda} \times \Delta\lambda$
1	2.69	0.66	5.0	3.3	1.07	0.0071
2	6.15	2.8	45.7	128.0	11.72	0.3282
3	7.95	0.8	17.2	13.8	5.41	0.0433
4	9.8	2.9	3.7	10.7	1.56	0.0452
5	14.8	7.1	6.4	45.4	2.38	0.1690
6	21.0	5.3	5.1	27.0	1.83	0.0970
7	26.8	6.3	2.2	13.9	0.78	0.0491
Total				242.1	–	0.7389

For the assumed temperature $T_0 = 300$ K formula (2.2.64) yields:

$$b = \frac{5.6693 \times 10^{-8}}{3} 300^4 + 2\pi \times 0.2421 - 2\pi \times 300 \times 0.7389 \times 10^{-3}$$

$$= 0.153 + 1.521 - 1.393 = 0.281 \,\text{kW/m}^2$$

The ratio of the exergy of radiation of the vapor to its energy emission is $\psi_{wv} = b/e = 0.281/1.521 = 0.185$. More details of the considered example are discussed by Petela (1961a).

Exergy of arbitrary, non-polarized, black and uniform radiation, propagating within solid angle 2π. The formula for such radiation is derived from (2.2.62) in which formulae (2.2.53) and (2.2.54) are used with the interpretation explained by formula (2.2.55):

$$b_b = \frac{\sigma}{3}(3T^4 + T_0^4 - 4T_0T^3) \tag{2.2.65}$$

To utilize formula (2.2.65) only the temperature of the black radiation is required. The total exergy arriving at surface A can be calculated from formula (2.2.63).

It is noteworthy that equation (2.2.65) is identical to equation (2.2.40) derived for the black emission. This similarity is a confirmation that exergy of black radiation is equal to the exergy of emission of black surface. It is the consequence of the radiosity of a black body being equal to its emission.

Exergy of non-polarized, uniform, black radiation propagating within solid angle ω. The formula for such radiation is:

$$b_{b\omega} = \frac{b_b}{\pi} \int\limits_{\beta} \int\limits_{\varphi} \cos\beta \sin\beta \, d\beta \, d\varphi \tag{2.2.66}$$

where solid angle ω has to be determined by the appropriate ranges of variation of the flat angles β (declination) and ϕ (azimuth).

The magnitude ψ is defined as the ratio of exergy and energy of the same radiation. Both the exergy and the energy are functions of state, thus they do not depend on any geometrical configuration parameters. The angles β and φ do not have any geometrical meaning but are only the coordinates determining the solid angle in which the spectrum is considered.

For any arbitrary radiation the ratio ψ could be defined as the ratio of exergy determined by formula (2.2.58) to the first term of the right hand side of this formula which represents the radiation energy. Thus for a polarized radiation:

$$\psi = 1 - \frac{\dfrac{\sigma T_0^4}{3\pi} \displaystyle\int\limits_{\beta}\int\limits_{\varphi} \cos\beta \sin\beta \, d\beta \, d\varphi - \displaystyle\int\limits_{\beta}\int\limits_{\varphi}\int\limits_{\lambda} (L_{0,\lambda,\min} + L_{0,\lambda,\max})\cos\beta \sin\beta \, d\beta \, d\varphi}{\displaystyle\int\limits_{\beta}\int\limits_{\varphi}\int\limits_{\lambda} (i_{0,\lambda,\min} + i_{0,\lambda,\max})\cos\beta \sin\beta \, d\beta \, d\varphi} \tag{2.2.67}$$

The exemplary values of ψ are discussed in this book in examples 2.2.6.1 (for non-polarized, uniform water vapor radiation propagating within a solid angle 2π), and in example 2.4.1.2 (for non-polarized and uniform solar radiation).

2.3   THERMODYNAMIC ANALYSIS

2.3.1   Significance of thermodynamic analysis

The *thermodynamic analysis* is the method which can be applied to the examining of any energy conversion phenomenon. In the first step the analysis develops the conservation equations of mass and energy, based on the First Law of thermodynamics, which allow for the traditional *energy analysis* of the process. Next, the examining can apply the Second Law of thermodynamics, which allows for the *entropic evaluation* of the process irreversibility, and then it applies the crowning of the whole thermodynamic analysis with *exergy analysis*. Thermodynamic analysis based on developed balance equations for mass, energy, exergy and on the entropy growth equations, provides different (energy, entropy and exergy) views of the same phenomenon in terms of engineering quantity, probability and quality, respectively.

An energy balance, (based on the First Law of Thermodynamics), is developed to better understand any process, to facilitate design, operation and control, to point at the needs for process improvement, and to enable eventual optimization. The degree of perfection of energy utilization in the process, or its particular parts, allows for comparing the degree of perfection, and the related process parameters, to those in other respective processes. Comparison with the currently achievable values in the most efficient systems is especially important. Also the priorities for the required optimization attempts for the systems, or its components, can be established. Such establishing can be carried out either based on the excessive energy consumptions or on the particularly low degree of perfection.

Entropy analysis, (based on the Second Law of Thermodynamics), requires the complete data obtained from mass and energy considerations to allow for developing entropy relations to verify the correctness of a mathematical model of mass and energy results. The analysis allows for identification and location of the sources of irreversibility contributing to the overall unavoidable degradation of energy. Entropy can be used for process optimization by minimization of entropy generation. However the entropy has limited application for micro systems containing a denumerable number of independent particles. The smaller the number of particles, the less precisely the Second Law is fulfilled. For example for any microbiological system containing only a few components the Second Law may not be fulfilled.

Exergy is the concept derived from joint application of the First and Second Laws of Thermodynamics. Exergy balance is developed according to a similar methodology as for energy analysis, and with the same purposes. Whereas thermodynamic probability is expressed in units of entropy, exergy is expressed in energy units. Consequently exergy data are more practical and realistic in comparison to the respective energy values. Thus, the exergy analysis provides a more realistic view of process, which sometimes dramatically differs in comparison to the standard energy analyses. Exergy analysis can be compared to the energy analysis like the second different projection in a technical drawing disclosing additional details of the subject seen from a different side.

The knowledge about nature is continually studied with many methods and observations. The scale of approach may be microscopic (e.g. a microscopic observation or differential calculus) or macroscopic (phenomenological considerations or integral calculus). Usually the studies are organized by focusing attention on the particular system appropriately representing the aimed problem. Description and definition of the system is then a very important stage in the investigating approach. However, any analysis not based on the precisely defined system can lead to astonishing but incorrect results. The system has to be precisely determined by separating precisely the elements included from those excluded. This is usually effectively rendered by applying an imaginary system boundary which tangibly separates the system from its surroundings. The best practical way is to draw a scheme of contents of the system indisputably separated from surroundings by the drawn system boundary. Sometimes the investigated problem can be easily solved by introducing sub-systems, also precisely defined. Each balance equation allows for determination of an unknown variable or for establishing a relation between variables.

The radiation processes accompanying processing on substances can be non-negligible and often the systems in which radiation and substance play roles together, have to be considered.

As discussed, the variables obtained from mass and energy analyses are very important, thus they have to be carefully prepared. The variables can be measured, assumed or calculated. If the system is *over-determined*; i.e. the number of unknowns is smaller than the number of available independent equations, then all the variables can be corrected based on the probability reconciliation calculus e.g. like that discussed for radiation by Petela (2010).

2.3.2 Energy balance equations

The energy conservation equation is based on the substance balance equation. The principle of conservation of substance claims that the number of molecules in physical processes is constant, or a number of elements in chemical processes or a number of nucleons in the processes of split and synthesis of nuclei is constant. The substance conservation equation does not need to account for radiation or any other form of matter except for substance. Such an equation is developed for the system defined precisely by the system boundary. For the elementary process lasting a very short period:

$$dm_{in} = dm_S + dm_{out} \tag{2.3.1}$$

where m_{in} and m_{out}, kg, is the elementary amount of substance respectively delivered and extracted from the system, and m_S is the elementary increase of amount of substance within the considered system. The equation (2.3.1) can be appropriately modified for steady state ($dm_S = 0$), or for a certain instant with use of mass flow rates, or for a certain period of time. The equation can be separately applied for particular compounds (if there is no chemical reaction) or elements. The amount unit can be kg, kmol or the standard m^3 of the considered component.

A particular form of the substance conservation equation can be e.g. the equation summarizing fractions of components in a considered composite material: $\Sigma f_i = 1$, where f_i is the fraction of the i-th component of the material.

The energy conservation equation is the result of observations and cannot be proved or derived. From a long view of the history of mankind there are no phenomena recorded occurring in disagreement with the First Law of Thermodynamics.

Energy balance is the basic method for solving problems of thermodynamics. If sometimes one wants to start analysis of any problem but does not know how, the general advice is to try to make an energy balance of a system which would represent the problem subject. The energy balance can be applied to diversified problems which, however, require an appropriately well defined system for consideration. The system boundary should be the same for energy and substance balances because the substance balance is the basis for balance of energy. Sometimes only the specific definition of the system and particular tracing of the system boundary allows for the solution of the thermodynamic problem. In other cases the solution can be obtained by defining more different sub-systems.

Generally, the energy E_{in} delivered to system remains partly within the system as the increase ΔE_S of the system energy, and the rest is the energy E_{out} leaving the system. Thus, the general equation of energy balance is:

$$E_{in} = \Delta E_S + E_{out} \qquad (2.3.2)$$

Usually, for better illustration of the balance equation, the particular terms of the equation are shown in the bands diagram. The principle of such a diagram is shown by a simple example (Fig. 2.3.1), illustrating equation (2.3.2).

In principle, for energy considerations, the reference state for calculation of energy of the matters included in consideration can be defined arbitrarily; however, it is recommended to select this reference as for the exergy consideration, to make fair comparison of both, energy and exergy viewpoints.

Generally, application of an energy balance does not require analyzing of processes occurring within the system boundary. It is sufficient only to know (e.g. from measurements) the parameters determining components of the energy delivered and leaving the system as well as to know the parameters determining the initial and final state of the system. Obviously, if only the one unknown magnitude appears in the balance equation then the equation can be used to calculate this magnitude.

The energy balance can be differently tailored depending on the considered viewpoint and actual conditions. For example, there are some possibilities to categorize the case under consideration as: a) energy delivered is spent entirely for an increase of system energy at no energy leaving the system, b) energy leaving system comes entirely from the decrease of energy of system at no energy delivered to system, c) there is neither delivered nor leaving energy but only energy exchange (e.g. by work or heat) within the system, d) energy delivered is equal to energy leaving the system at no change of the system energy. Other possibilities are that some components of energy can be neglected either due to relatively small changes, or because they are unchanged at all. The balance equation can be written for the steady or transient systems, for the system considered on macro scale or micro scale for which differential equations are applied, etc.

For example, for the elemental process lasting an infinitely short time, the balance equation (2.3.2) can take the form:

$$dE_{in} = dE_S + dE_{out} \qquad (2.3.3)$$

In equation (2.3.3) only dE_S is the total differential and in order to demonstrate it clearly it is better to write equation (2.3.3) as follows:

$$\dot{E}_{in}(t)dt = dE_S + \dot{E}_{out}(t)dt \qquad (2.3.4)$$

where $\dot{E}_{in}$ and $\dot{E}_{out}$ are the respective fluxes (e.g. in W) of energy delivered and extracted from the system and t is the time.

Determination of dE_S requires not only accounting for the change of intensive parameters of the system state but also on the eventual change in the amount of matter in the system. If, e.g. the considered system contains only a homogeneous substance then:

$$dE_S = d(m_S e_S) = m_S de_S + e_S dm_S \qquad (2.3.5)$$

where m_S and e_S are, respectively, the amount of substance and its specific energy contained within the system.

Sometimes the subject of consideration can be recognized as moving in space (e.g. solar vehicle, radiometer vane). Then the simplest energy balance equation is obtained by assuming that the coordinates system determining velocity and location is moving together with the system boundary. However there are some consequences of such an assumption. Kinetic energy should be determined for the velocity relative to the moving system. The useful work done by the system does not appear in the energy balance because the forces acting on the system do not make replacements relative to the coordinates system. The useful work can be determined only for velocity and location relative to earth.

The components of an energy balance equation are the energy of the system and energy exchanged with the system.

Energy of system (E_S) depends on its state. An increase ΔE_S of an energy system, changing from its initial to final state, does not depend on the means of change between these states and is a difference of final $E_{S,fin}$ and initial $E_{S,inl}$ energy of the system:

$$\Delta E_S = E_{S,fin} - E_{S,inl} \qquad (2.3.6)$$

Generally, the system energy can consist of macroscopic components like $E_{macr,i}$ due to velocity (kinetic energy), surface tension (surface energy), gravity (potential energy), or any other energy of field nature (e.g. radiation). The remaining part of the system energy, containing microscopic components U_j, constitutes the internal energy:

$$E_S = \sum_i E_{macr_i} + \sum_j U_j \qquad (2.3.7)$$

where i and j are the successive numbers of the macro and micro components, respectively, of the system energy.

If the kind of substance, before and after process (e.g. physical process), is the same, then the reference state for calculation of the energy of the substance can be established with a certain degree of freedom. For example, the reference state can be assumed as the state of the substance entering the system. Thus, the substance energy entering the system is zero whereas the energy of this substance exiting the system is equal to the energy surplus relative to the reference state. This surplus has to account for the latent heat of eventual change of the substance phase.

In addition, the components of negligible or constant value must not be taken into account. For example, the energy of surface tension can be included only in consideration of a fluid mechanics process of liquid atomization or of a mechanical process of solid material comminution. Both processes have been analyzed by Petela (1984a,b).

Energy exchange (E_{in} and E_{out}) with system can occur on different ways.

Electrical energy can be delivered for heating the system, for driving an electric motor, or generating an electromagnetic effect within the system (e.g. a strong electric field affects combustion). In reverse processes the electric energy can be obtained, e.g. with use of an electric generator the energy is obtained from the system. The energy flux of electric energy (power) is measured by a wattmeter.

Mechanical work can be exchanged with a system by means of a piston rod of reciprocal motion or with a rotating shaft.

The energy balance of a system should comprise the mechanical work performed by all forces acting on system boundary. Therefore, if a *substance flux* passes through the boundary then the work performed by the force acting in the place of passing should be taken into account. Such work of transportation of substance through the boundary is expressed by enthalpy. For some kind of substance the enthalpy can be calculated with specific formulae, e.g., formulae for plasma are discussed by Petela and Piotrowicz (1977).

If the considered system is moving relative to the coordinate system determining the location and velocity, then the work done by the forces causing the system displacement, has to be accounted for. The energy balance should also include the work done by deformation of the system boundary if its shape changes during consideration.

Kinetic energy should be accounted if substance passes the system boundary with significant velocity relative to the boundary.

Potential energy of a substance exchanged with a system is included in the energy balance if the substance has significant elevation above the reference level. This energy component results from the presence of the gravity field.

Energy transferred by *heat* occurs by direct contact of system with body at a temperature different from the system temperature, or can occur without contact by radiation. The effect of contact during heat exchange appears in heat conduction as well as in heat convection. The model of *pure conduction* occurs when the particles of the contacted body do not change their location (solids). The energy is then transferred by free electrons and oscillations of atoms in the crystal lattice. Still pseudo pure conduction can be recognized between fluids of very laminar flow; conduction occurs in the direction perpendicular to the ordered motion of particles at component velocity only in the flow direction. In such cases, excluding the possibility of diffusion, there is no perpendicular substance flow and in spite of the medium flow this heat is transferred by conduction.

The essence of heat *convection* is the motion of substance (fluids), during which mixing of hot and cold fluids occur. However, the micro-mechanism of this mode of heat transfer, also depends on direct effective contacts (conduction) between the hot and cold fluids portions being replaced. If mixing is caused by non-uniform distribution of density (temperature profile) then convection is called *natural convection*.

In contrary, if the mixing is a result of acting of pump or ventilator, etc., then a *forced convection* occurs.

Energy can be also exchanged with the system due to a *diffusive substance flux*. Then, the enthalpy of the diffusing substance has to be taken into account. For example, consider a system boundary demarcated over the laminar zone of a mixture of gases of a non-uniform temperature distribution. If it is assumed to be a laminar (no convection) mode of transparent gases (no radiation), then the energy E_L, W, exchanged through the boundary due to the heat conduction and enthalpy of diffusing substance, is composed of two respective terms. The first term represents the heat conducted according to Fourier's law and the second term expresses enthalpy of diffusing gas according to Fick's law. Thus:

$$E_L = -A\left(k\frac{\partial T}{\partial y} + T\sum c_{p\,i}D_{L\,i}\frac{\partial c_i}{\partial y}\right) \tag{2.3.8}$$

where A, m^2, is the surface area, k, W/(m K), is the overall conductivity of gas mixture, T, K, is the temperature of the gas at the boundary, y, m, is the space coordinate perpendicular to the system boundary surface and perpendicular to the gas flow direction, $c_{p,i}$, J/(kg K), D_i, m^2/s, and c_i, kg/m^3, are respectively the specific heat at constant pressure, the laminar diffusion coefficient, and the concentration of gas component, where i is the successive number of the gas mixture component.

Real processes occur with friction on which a friction work has to be spent. The friction work increases the energy of system due to absorption of heat in amount equivalent to friction work. Friction causes dissipation of energy which can be only partly recovered. The *friction heat* does not appear as a member of the energy balance equation; however it affects the final system energy and the components of exiting energy.

Chemical energy is assumed to be the same for the substance considered as the component of the system and for the substance component separately exchanged with the system. The enthalpy and internal energy include generally physical and chemical components.

2.3.3 Exergy balance equations

The exergy balance equation is the basis of the exergetic part of thermodynamic analysis. Exergy analysis can be applied to diversified problems which, however, like the energy analysis, require an appropriately well-defined system for the analysis. The system boundary should be the same as for the matter balance.

The exergy conservation equation can be applied only to reversibly occurring processes. For real processes the exergy conservation equation is fulfilled only when the unavoidable exergy loss, due to irreversibility of the process, is taken into account. Thus, corresponding to energy equation (2.3.2), the following *traditional* exergy balance equation is applied:

$$B_{\text{in}} = \Delta B_S + B_{\text{out}} + \delta B \tag{2.3.9}$$

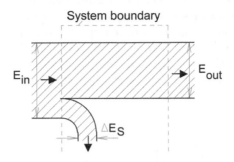

Figure 2.3.1 Bands diagram of energy balance.

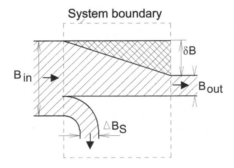

Figure 2.3.2 Bands diagram of the traditional exergy balance.

where B_{in} and B_{out} are the respective sum of exergy delivered and released from the system, ΔB_S is the change in the exergy of the system and δB is the exergy loss due to the process irreversibility, calculated from the Guoy-Stodola law, equation (2.2.10).

The bands diagram for exergy balance is shown in Figure 2.3.2. In comparison with the respective diagram for energy balance (Fig. 2.3.1), the exergy diagram shows the exergy δB disappearing within the system.

Like the energy balance, the exergy balance can be differently tailored depending on the considered problem and actual conditions. For example, some components of exergy can be neglected either due to relatively small changes, or because they are unchanged at all. The balance equation can be written for steady or transient systems, for system considered on the macro scale or micro scale using differential equations, etc. Obviously, for calculation of exergies, there is no freedom in defining the reference state, which is only the environment, as determined by exergy definition.

For any elemental process lasting an infinitely short time, the exergy balance equation can take the form:

$$\dot{B}_{in}dt = dB_S + \dot{B}_{out}dt + \delta B \tag{2.3.10}$$

where $\dot{B}_{in}$ and $\dot{B}_{out}$ are the respective fluxes of exergy delivered and extracted from the system and dB_S is the total differential exergy growth of the system.

The differential dB_S should be determined analogously to equation (2.3.5).

If the subject of consideration is moving in space, then the simplest exergy balance equation is obtained by assuming, as for the energy balance, that the coordinates system determining velocity and location is moving together with the system boundary.

An *increase* ΔB_S *of the exergy system*, changing from its initial to final state, does not depend on the method of change between these states and is equal to the difference of final and initial exergy components:

$$\Delta B_S = \left[\sum_i (B)_{fin,i} - \sum_j (B)_{inl,j} \right]_S \tag{2.3.11}$$

where the sum of initial or the sum of the final components is:

$$(B)_S = B_k + B_p + B_S + B_b + \cdots \tag{2.3.12}$$

and where i and j are the successive numbers of the final and initial (respectively) exergy components, B_k is the kinetic exergy, B_p is potential exergy, B_S is the thermal exergy of the system calculated with use of formula (2.2.13), and B_b is the exergy of photon gas (black radiation) calculated for example based on equation (2.2.31). Also the other eventual components in equation (2.3.12), as shown in Figure 2.2.1, can be added if necessary, e.g. exergy of surface tension which is equal to the energy of surface tension, etc.

Exergy fluxes (B_{in} and B_{out}) exchanged with system can occur on different ways described for the energy balance.

Electrical exergy is equal to electrical energy. *Exergy of mechanical work* is equal to work. *Exergy of substance flux* is calculated with use of formula (2.2.12), however *kinetic exergy* should be taken separately, (calculated as the kinetic energy for absolute velocity), and *potential exergy* (equal to potential energy relative to the Earth surface level). *Exergy of heat* exchanged with the system is determined by formula (2.2.4).

Exergy can be exchanged with the system also due to a *diffusive substance flux*. Then, the exergies of diffusing substances is taken into account as the exergy determined by formula (2.2.12) interpreted for the partial pressure of the substances.

Any internal exergy loss δB_F caused by friction is determined e.g., by assumption that the friction heat Q_F, equal to the *friction work*, is entirely absorbed by the substance at temperature T. For the heat absorption process, assuming the entropy growth $\Pi_F = Q_F/T$, the exergy loss can be calculated from formula (2.2.10) as follows:

$$\delta B_F = Q_F \frac{T_0}{T} \tag{2.3.13}$$

The exergy loss by friction is the smaller the higher is the temperature T of absorbing substance. The exergy loss δB_F can be smaller or larger than the friction heat Q_F dependently on the temperature ratio T_0/T. This observation is particularly important for refrigerating processes where often $T < T_0$.

The *chemical exergy*, discussed in paragraph 2.2.3.3, is assumed to be the same for the substance considered as the component of the system and for the substance component separately exchanged with the system.

An exergy balance is usually carried out on the assumption of constant parameters of environment during the consideration time. The effects of *varying environmental parameters* are usually small and the assumption of the mean environmental parameters is sufficient for the exergy analysis. The inclusion of the variations of the environment parameters would make the analysis more difficult because the considered exergy values should be taken for instantaneous environment parameters, i.e., the values used in the balance equation would need calculations by integration over assumed time period.

Moreover, usually the balance equation would need the introduction of an additional member without which the equation is not fulfilled. For example, such a need appears when a perfectly insulated container has been closed while being filled with a substance in equilibrium with the environment and if, meanwhile, the environment parameters after closing are changed. Then it can be deduced that the enclosed substance gains the positive exergy and the exergy balance equation is not fulfilled (Petela (2010)). In another example, ice stored during a frosty winter has exergy close to zero, whereas the exergy of the same ice existing in hot summer time is relatively large and again the exergy balance equation for the ice taking part in such a season change ris not fulfilled.

Therefore generally, the exergy balance equation for the process occurring at the varying environmental parameters should contain the compensation term ΔB_e which modifies equation (2.3.9) as follows:

$$\int\limits_{inl}^{fin} \dot{B}_{in}\, dt = \Delta B_S + \int\limits_{inl}^{fin} \dot{B}_{out}\, dt + \int\limits_{inl}^{fin} \dot{\Pi} T_0\, dt + \Delta B_e \tag{2.3.14}$$

where ΔB_e is the exergy gain due to variation of the environment from the initial state (*inl*) to the final (*fin*) state, $\dot{B}$ is the respective rates of exergy and $\dot{\Pi}$ is the overall entropy growth. As mentioned, the ΔB_e can be positive or negative or zero. The bands diagram for exergy balance at varying environment parameters is shown in Figure 2.3.3. Direct calculation of ΔB_e is not easy, thus the best method is to calculate this value as a completion of the balance equation.

The Gouy-Stodola law, expressed by equation (2.2.10), was derived for a constant environment temperature T_0. If T_0 is varying, then the law can be applied only for the infinitely short process as is expressed by the presence of the appropriate integral term in equation (2.3.14). The Gouy-Stodola law cannot be applied to the processes which occur at the varying environment temperature if such variation is caused by the considered process. An example calculation illustrating the effect of the varying environment temperature is given by Petela (2010).

The variation of environment parameters can instantaneously generate or destroy exergy at no role of internal processes of the examined system. Therefore, if the environment variations are significant, then wherever possible, the process should be organized to utilize the instantaneous positive value of ΔB_e. The predicting data on the change in

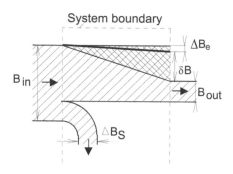

Figure 2.3.3 Bands diagram of exergy balance at varying environment with positive compensation term ($\Delta B_e > 0$), (from Petela, 2010).

the environment parameters could then be helpful. Fluctuation of environment parameters is potentially one of the natural low-value resources, like e.g., a waste heat at very low temperature. A particular problem is a variation of the effective *sky temperature*, discussed by Duffie and Beckman (1974), which determines the radiant heat lost from the Earth's surface. The fluctuation of environment parameters has relatively insignificant influence on the high-values natural resources, e.g. natural fuels.

It is also possible to consider the variation of environment parameters with altitude and this effect is taken into account in consideration of the concept of mechanical exergy, discussed in paragraph 2.2.3.2. However, application of mechanical exergy (eZergy) leads to the *exergy balance with gravity input* term. Petela (2009a) proposed to insert an appropriate term, called gravity input G, as an additional exergy input in the left hand side of the exergy balance equation. Thus equation (2.3.9) for the constant environment parameters becomes:

$$B_{in} + G = \Delta B_S + B_{out} + \delta B \tag{2.3.15}$$

The gravity input G can be positive, zero or negative. The bands diagram for the exergy balance with included positive gravity input is shown in Figure 2.3.4. The value of G is calculated from exergy balance equation.

The gravity input G can appear only when a substance is considered in the exergy balance and if eZergy is applied to the substance. The interpretation of the algebraic sign of G from exergy viewpoint could be proposed as follows.

In the case $G < 0$; due to the effect of a gravity field on the considered process, the process product expressed by the total exergy value of the right hand side of the exergy balance equation diminishes and has to be balanced by the negative gravity input G added to the left hand side of the equation. The considered process can be recognized as opposing the effect of the gravity field.

In the case of $G > 0$; the presence of a gravity field during the considered process generates a certain "surplus" of exergy disclosed by the right hand side of the exergy balance equation. This surplus has to be balanced by a positive gravity input G added to the left hand side of the equation. The gravity field favors the process by contributing with some exergy input.

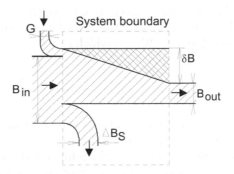

Figure 2.3.4 Bands diagram of exergy balance interpretation including gravity input in case $G > 0$, (from Petela, 2010).

In case of $G = 0$; there is no change in the traditional exergy, and it means that the work of substance during theoretical expansion at altitude H (considered in paragraph 2.2.3.2) to obtain the equilibrium of densities has no accountable importance.

The exemplary calculation and analysis including gravity input in the case of the waste combustion products in a chimney were carried out by Petela (2009b). The other example of calculation of gravity input, for adiabatic expansion of air in a turbine and for drawing air through a throttling valve followed by a fan are discussed by Petela (2009a). Further application of gravity input interpretation is also discussed in paragraph 2.4.3.

2.3.4   Process efficiency

2.3.4.1   Carnot efficiency

By observation of nature we know that the continuous generation of a useful effect (e.g. work or heat) or conversion of energy is possible only in a situation when at least two heat sources of different temperatures are available. Such situation can then be utilized in an installation in which a fluid (e.g. gas, liquid or photon gas) is circulating and cyclically exchanging heat and performing work.

In a search for the best effectiveness of the cycle process, which would occur reversibly without any losses, the ideal model was established by Carnot (1824). The real cycles could be then designed possibly close to the model by application of different "carnotization" attempts. The model cycle should consist of only ideal (reversible) processes. Thus the cycle processes of releasing and absorbing heat should occur reversibly (at an infinitely small temperature difference between the heat source and circulating fluid) and the flow of fluid should be frictionless. The other cycle processes, during which work is generated or consumed, should also occur reversibly (at constant entropy) which is possible if the fluid does not exchange heat (adiabatic) with the surroundings when with no friction (isentropic) it expands or is compressed.

Based on the energy conservation law the net work W performed in the Carnot cycle is equal to heat absorbed Q_I and released Q_{II} by fluid: $W = Q_I - Q_{II}$. The overall entropy growth for such a reversible cycle is zero and takes into account only entropies of exchanged heat at the respective heat source temperatures T_I and T_{II}: $0 = Q_{II}/T_{II} - Q_I/T_I$. The efficiency η_C of the considered Carnot cycle is the ratio of work W to the cycle input Q_I; $\eta_C = W/Q_I$, which is:

$$\eta_C = 1 - \frac{T_{II}}{T_I} \qquad (2.3.16)$$

The commonly called Carnot efficiency is in fact the efficiency of the Carnot cycle and is the most important efficiency in thermodynamics. All other defined efficiencies are less general, mostly arbitrary or specifically adjusted to the objects or situations. One of the most significant properties of the Carnot efficiency is that it is valid independently of the nature of the working fluid and can be applied to any material or field matter used as the working fluid.

The Carnot efficiency can be used as a reference value for calculation of exergy efficiency of a thermal engine. The energetic and exergetic efficiency of an engine are respectively $\eta_{E,eng} = W/Q_I$ and $\eta_{B,eng} = W/B_{QI}$. Based on formulae (2.3.16) and (2.2.4) the ratio of energetic and Carnot efficiencies is:

$$\eta_{B,eng} = \frac{\eta_{E,eng}}{\eta_C} \qquad (2.3.17)$$

The exergy efficiency $\eta_{B,eng}$ of engine demonstrates how much the real energy efficiency departs from the ideal efficiency represented by the Carnot efficiency. In the ideal case ($\eta_{E,eng} = \eta_C$) the exergy efficiency approach 100%.

2.3.4.2 Perfection degree of process

Practically, process efficiency can be defined in different ways. For example, energy or exergy can be used for expressing the numerator and denominator of the efficiency. However, the best method for reviewing the process seems to be the application of the degree of perfection recommended by Szargut et al. (1988) for measuring the thermodynamic perfection of process.

The energy and exergy degrees of perfection are defined analogously for convenient comparison. To determine the degree of perfection, all terms of the energy (or exergy) balance equation are categorized either as useful product, or process feeding, or loss. The perfection degree is then defined as the ratio of useful product to the process feeding. The loss is not disclosed in the perfection degree formula because it is a compensation of the perfection degree to 100%.

As was discussed in paragraph 2.2.2, the exergy losses can be internal and external. The energy balance can disclose only energy external loss, whereas the exergy balance can contain the terms of the external and internal exergy losses. Internal exergy loss is calculated from the Guoy-Stodola law. External loss is equal to the energetic or exergetic value of the unavoidably released waste heat or matter which however, can be somehow utilized beyond the considered system in an additional process of "waste recovery". In multi-processes systems, in contrary to external losses, only the internal exergy losses can be summed.

The concept of perfection degree can include also exergy change due to the varying of environment parameters and the specific terms (e.g. gravity input). Thus, in the modified version it can be proposed that the denominator of the degree of perfection represents the feeding terms, gravity input and exergy change due to the environment variation, whereas the numerator expresses the useful products. For example, for the steady process in which numerous fluxes of energy are exchanged the exergy degree η_B of perfection could be proposed as follows:

$$\eta_B = \frac{\sum_i B_{use,i} + \sum_k B_{Q,use,k} + W_{use}}{\sum_j B_{feed,j} + \sum_m B_{Q,feed,m} + W_{feed} + G - \Delta B_e} \tag{2.3.18}$$

where
i is the number of useful exergy fluxes B_{use} obtained from the process, including substance and radiation,
k is the number of useful exergy fluxes $B_{Q,use}$ of heat
j is the number of entering exergy fluxes B_{feed}, including substance and radiation,
m is the number of entering exergy fluxes $B_{Q,feed}$ of heat,
W_{use} is the total work produced,
W_{feed} is the total work consumed,
G is the gravity input, considered if eZergy is applied,
ΔB_e is the exergy gain in case of variation of environment parameters.

Formula (2.3.18) can be applied also for combined processes in which more than one intended product is obtained (e.g. the combined generation of heat and power). A particular example of application of the energy and exergy perfection degrees, with no work, G and ΔB_e, is discussed in paragraph 2.4.4 for photosynthesis.

The exergy balance should be developed possibly with most detailed distribution of the internal losses in order to obtain the most exact information on the possibility of perfection improvement of the considered system. For example the internal exergy loss could be divided into the components corresponding to friction, heat transfer at finite temperature difference, radiation emissions and absorptions, etc. If in any part of the considered system there occur several irreversible phenomena then, in principle, it is possible to calculate only the overall internal exergy loss caused by the phenomena. The splitting of the effects of these irreversible phenomena, occurring simultaneously at the same place and time is impossible because these phenomena interact mutually. The splitting of the exergy loss in such case can be based only on the assumed agreement. For example for combustion process the radiative heat exchange occurs between the flame and surrounding wall. In order to split the effects of irreversible chemical reactions of combustion from the irreversible radiation exchange, it can be assumed that first the combustion occurs and then the heat exchange takes place. However, with such an assumption the temperature differences in heat exchange are larger than the real.

Therefore it is better to split exergy losses according to time and location of occurrence, instead of according to the causes, unless the examined causes occur in different locations and different instants.

2.3.4.3 Specific efficiencies

Generally, the efficiency of a process can be arbitrarily defined to expose the most important aspect. For example, the exergy effect of the hot water heated in pipe by solar radiation can be related either to the exergy of heat q at the temperature of the Sun's surface T_{Sun}; $q \cdot (1 - T_0/T_{Sun})$, or to the exergy b_{Sun} of the Sun's radiation, or to the exergy of heat q absorbed at the water pipe temperature T_W; $q \cdot (1 - T_0/T_W)$. The exergy efficiency increases successively through the above three possibilities due to the decreasing values of the denominators in the efficiency formulas: $q \cdot (1 - T_0/T_{Sun}) > b_{Sun} > q \cdot (1 - T_0/T_W)$. An exergy efficiency which relates the process effect to the decrease of the Sun's exergy, $q \cdot (1 - T_0/T_{Sun})$, is unfair because the exposed surface of pipe obtains only the solar radiation exergy and the pipe is independent of irreversible emissions at the Sun's surface. Relating the process effect to the exergy of heat absorbed, $q \cdot (1 - T_0/T_W)$, favors the exposed surface by neglecting its imperfection during the absorption of heat q. Thus, from these three possibilities, the relating the heating water effect to the exergy b_{Sun} of the Sun's radiation is the only best estimation in this analysis.

2.3.4.4 Consumption indices

Sometimes instead of efficiency the specially defined indices are used for estimation of processes. For example there are some processes which occur spontaneously due to interaction with the environment. Drying, cooling, vaporizing or sublimation, are the examples of such processes in which the self-annihilation of exergy takes place. Often these processes, especially in industrial practice, are accelerated with use of the appropriate driving input. Exergy application for estimation of perfections of these processes reveals some problems.

For example applying the common exergy efficiency definition, effect and input ratio, leads to a negative or infinite value of the efficiency. Therefore instead of the efficiency some specially defined criteria have to be used for the evaluation and comparison of processes perfection. For example for drying processes the unit exergy consumption index is defined as the ratio of the exergy of the used in the drying medium to the mass of the liquid extracted in form of the vapor. In the case of the application of solar energy for drying, the index would express the exergy of absorbed radiation per mass of the vaporized moisture.

Another index can be used for process occurring in a water cooling tower. Szargut and Petela (1968) propose the evaluation of the process with the index defined as ratio of the sum of exergy lost in the tower and the heat extracted from water. The typical value of the index for cooling tower of steam power station is about 0.088 kJ of exergy per kJ of heat.

Petela (1990) proposes a specific approach to the exergy annihilation due to spontaneous processes. He considers the natural exergy annihilation rate which expresses the ability of the environment to spontaneously reduce the exergy of a substance or radiation. The natural wind velocity, the temperature and composition of environment air, particularly humidity, as well as the solar radiation, the local surrounding surfaces configuration and surfaces' emissivities taken into account together can determine the available exergy effect for annihilation of exergy in the spontaneous processes of drying, cooling, etc. A so called "windchill" factor is the example of the concept expressing certain ability of environment air.

2.4   SOLAR RADIATION PROCESSES

2.4.1   Conversion of solar radiation into heat

2.4.1.1   Calculation the exergy of solar radiation

Solar energy is the most important renewable source of energy on the Earth. Solar energy is a high temperature source, however it's harvesting occurs inefficiently due to extensive degradation of the energy. The degradation of solar energy is well demonstrated by exergy analysis. Therefore the engineering thermodynamics of thermal radiation addresses mainly exergy analyses of diversified problems of utilization of solar radiation.

Generally, the solar radiation passing through the atmosphere is absorbed, scattered and reflected not only by air molecules but also by e.g. water vapor, clouds, dust, pollutants, smoke from forest fires and volcanoes. These factors cause diffusion (called also dilution) of solar radiation. The portion of solar radiation which reaches the Earth's surface without being diffused is called direct beam solar radiation. Thus, the global solar radiation (global irradiance) consists of the diffuse and direct solar radiation. For example, during thick cloudy days the atmosphere reduces direct beam radiation to zero.

Only a part of scattered sunlight reaches the Earth because some sunlight is scattered back into space. Also some radiation from the Earth, together with sunlight scattered off the Earth's surface is re-scattered to the atmosphere. This effect can be significant e.g. when the Earth's surface is covered with snow.

Solar radiation energy is difficult to calculate because, as discussed, the radiation energy reaching the surface on Earth, is composed of direct and diluted radiation components, and depends on geographic location, time of day, season of year, local weather and even on local landscape. The relatively effective method of determining of solar radiation is by carrying out spectral measurement and application of the obtained results in the formulae derived in paragraph 2.2.6.

A certain basis for the evaluation of solar radiation reaching the Earth's surface can be the energy of solar radiation incident outside the Earth's atmosphere which is called extraterrestrial radiation, and its average value is about $1367\,W/m^2$. The exergy of extraterrestrial radiation is determined with the following examples.

Example 2.4.1.1   The exergy of the extraterrestrial solar radiation, when recognized as non-polarized, black, uniform and propagating within solid angle ω, may be approximately calculated by means of equation (2.2.66). The required exergy b_b of emission density can be calculated from (2.2.65) for the Sun's surface temperature $T = 6000\,K$ and for the environment temperature $T_0 = 300\,K$ as follows:

$$b_b = \frac{5.6693 \times 10^{-8}}{3}(3 \times 6000^4 + 300^4 - 4 \times 300 \times 6000^3) = 68.5\,MW/m^2 \quad \text{(a)}$$

Approximately, the radius of the sun is $R = 695,500\,km$ and the mean distance from the Sun to the Earth is $L = 149,500,000\,km$. The integral in formula (2.2.66) expresses the solid angle ω and is equal to the area of circle of radius R divided by the square distance L, thus:

$$\int\limits_{\beta}\int\limits_{\varphi} \cos\beta \sin\beta \, d\beta \, d\varphi = \frac{R^2\pi}{L^2} = 2.16 \times 10^{-5}\pi \text{ sr} \tag{2.4.1}$$

By substitution of (a) and (2.4.1) into (2.2.66):

$$b_{b,\omega} = \frac{68500}{\pi} 2.16 \times 10^{-5}\pi = 1.48 \text{ kW/m}^2 \tag{b}$$

The emission $e_{b,\omega}$ of black solar radiation arriving in the solid angle ω is:

$$e_{b,\omega} = \frac{e_b}{\pi} \int\limits_{\beta}\int\limits_{\varphi} \cos\beta \sin\beta \, d\beta \, d\varphi \tag{2.4.2}$$

where $e_b = 73.47 \text{ MW/m}^2$ is determined by formula (2.2.32). The ratio of exergy to energy of emission is $\psi = b_{b,\omega}/e_{b,\omega} = 68.5/73.47 \approx 0.93$.

Example 2.4.1.2 More exact computations of the exergy of solar radiation were carried out by Petela (1961a) based on the extraterrestrial radiation spectrum determined experimentally by Kondratiew (1954). Calculations are based on equation (2.2.61) for non-polarized and uniform radiation. Table 2.4.1 presents some exemplary Kondratiew data on intensity of radiation $i_{0,\lambda}$ (column 2) as a function of wavelength λ (column 1). The part of the spectrum is shown in Figure 2.4.1 together with three spectra (dashed lines), for comparison, for black radiation at absolute temperatures 6000, 5800 and 5600 K. The $i_{0,\lambda}$ values in Table 2.4.1 are assumed constant for the respective ranges of wavelengths $\Delta\lambda$ (column 5). Correspondent ranges of frequency $\Delta\nu$ calculated based on equation $d\nu = c_0 \cdot d\lambda/\lambda^2$, for $c_0 = 2.9979 \cdot 10^8$ m/s, are shown in column 6 whereas values $i_{0,\nu}$ in column 4 were determined from equation $i_{0,\lambda} = c_0 \cdot i_{0,\nu}/\lambda^2$. The ν values in column 3 were determined from equation $\lambda \cdot \nu = c_0$. The $L_{0,\nu}$ values of column 8 are calculated from equation (2.2.54). Columns 7 and 9 are calculated as respective products of columns 4 and 6; ($i_{0,\nu} \cdot \Delta\nu$), and 6 and 8; ($L_{0,\nu} \cdot \Delta\nu$).

Formula (2.2.61) is applied in the following numerical form:

$$b_{A'} = \left(2\sum i_{0,\nu}\Delta\nu - 2T_0 \sum L_{0,\nu}\Delta\nu + \frac{\sigma T_0^4}{3\pi}\right) \int\limits_{\beta}\int\limits_{\varphi} \cos\beta \sin\beta \, d\beta \, d\varphi \tag{2.4.3}$$

Assuming the environment temperature $T_0 = 300$ K, substituting formula (2.4.1) into (2.4.3) and using data from Table 2.4.1:

$$b_{A'} = \left(2 \times 10079300 - 2 \times 300 \times 2263.3 + \frac{5.6693 \times 10^{-8} \times 300^4}{3\pi}\right) \times \pi \times 2.16 \times 10^{-5}$$

$$= 1367.9 - 92.151 + 0.0033 = 1275.8 \text{ W/m}^2$$

Table 2.4.1 Spectrum of the extraterrestrial solar radiation, (from Petela, 1962).

$\lambda \cdot 10^{10}$	$i_{0,\lambda} \cdot 10^{-10}$	$\nu \cdot 10^{-11}$	$i_{0,\nu} \cdot 10^{12}$	$\Delta\lambda \cdot 10^{10}$	$\nu \cdot 10^{-12}$	$i_{0,\nu} \cdot \Delta\nu$	$L_{0,\nu} \cdot 10^{13}$	$L_{0,\nu} \cdot \Delta\nu$
m	$\dfrac{W}{m^3 sr}$	$\dfrac{1}{s}$	$\dfrac{J}{m^2 sr}$	m	$\dfrac{1}{s}$	$\dfrac{W}{m^2 sr}$	$\dfrac{J}{m^2 s\, sr}$	$\dfrac{W}{m^2 K\, sr}$
1	2	3	4	5	6	7	8	9
2200	10	13627	15	100	62.0	960	0.03	0.205
2300	26	13035	47	100	56.7	2650	0.10	0.540
2400	31	12492	59	100	52.1	3090	0.12	0.639
.	.	.	.	.	.	.	.	.
.	.	.	.	.	.	.	.	.
60000	1	500	1765	10000	0.714	1260	7.19	0.514
70000	1	428	1201	10000	0.535	640	5.48	0.293
Total						10079300		2263.306

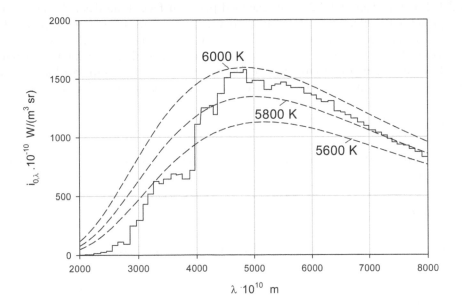

Figure 2.4.1 Spectrum of extraterrestrial solar radiation (from Petela, 1962).

The obtained result $1275.8 \, W/m^2$ is the exergy of the extraterrestrial solar radiation arriving in a $1 \, m^2$ surface which is perpendicular to the direction of the sun. The obtained ratio of radiosity (equal to emission) to exergy is $\psi_S = 1275.8/1367.9 = 0.9326$.

2.4.1.2 *Possibility of concentration of solar radiation exergy*

Solar energy, although rich, is poorly concentrated and thus it requires a relatively large surface to harvest the Sun's radiation. From this viewpoint solar radiation is especially valuable for those countries which have lot of unused areas (e.g. deserts).

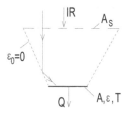

Figure 2.4.2 Scheme of concentrated radiation (from Petela, 2010).

The poor concentration of energy needs intensive theoretical studies in order to obtain acceptable efficiency of energy utilization. Effective method for such purpose could be exergy analysis.

The concentration possibility of radiation can be illustrated with use of a simple model shown in Figure 2.4.2. The imagined plane surface (long dashed line) of area A_S represents the black ($\varepsilon_S = 1$) solar irradiance IR at constant temperature T_S. The other plane surface (solid line) of area A is gray, at emissivity ε, and its temperature T is controlled by the cooling heat Q. The vacuum space between the two surfaces is enclosed by other cone-shape surface (short dash line) which is mirror-like ($\varepsilon_0 = 0$). The surface areas ratio $a_S = A_S/A$. The energy balance of the cooled surface A is:

$$a_S \varepsilon IR = \varepsilon \sigma T^4 + k(T - T_0) \tag{2.4.4}$$

where k is the heat transfer coefficient at which heat Q is extracted from surface A. The heat rate

$$q = k(T - T_0) \tag{2.4.5}$$

can be used to express the total heat Q absorbed by surface A:

$$Q = \frac{A_S}{a_S} q \tag{2.4.6}$$

The energy efficiency η_E of concentration of solar radiation can be measured as the ratio of absorbed heat Q and the solar irradiance IR:

$$\eta_E = \frac{Q}{IR} \tag{2.4.7}$$

For comparison, exergetic efficiency η_B can also be considered based on the following definition:

$$\eta_B = \frac{B_Q}{IR\psi} \tag{2.4.8}$$

where ψ is the exergy-energy ratio discussed in paragraph 2.2.5 and the exergy B_Q of heat absorbed by surface A is:

$$B_Q = Aq \left(1 - \frac{T_0}{T} \right) \qquad (2.4.9)$$

The reality of the discussed effect of concentration of solar radiation can be evaluated by the calculated value of the overall entropy growth Π, which consists of: the positive entropies of heat Q, the emission of surface A, and the negative entropy of absorbed solar radiation:

$$\Pi = \frac{Q}{T} + A\varepsilon \frac{4}{3}\sigma T^3 - A_S \varepsilon SR \qquad (2.4.10)$$

where SR is the entropy of irradiance IR. It has to be noted that using the energy emissivity ε in formula (2.4.10) for entropy calculation is not precise and, as discussed in details by Petela (2010), the smaller is the precision the smaller is the emissivity ε. The magnitude SR can be evaluated from the assumed ratio SR/IR to be equal the ratio s/e of the black emission entropy and emission energy, $SR/IR=s/e$. With use of formulae (2.2.32) and (2.2.39), the following relation can be derived:

$$SR = \frac{4}{3} \frac{IR}{T_S} \qquad (2.4.11)$$

The overall entropy growth determined from equation (2.4.10) should be positive ($\Pi > 0$). Otherwise, (when $\Pi \leq 0$), the concentration of solar radiation is impossible as being against the Second Law of Thermodynamics.

Example 2.4.1.3 The concentration of solar radiation can be considered, e.g., at $IR = 800$ W/m^2 arriving at the imagined surface of area $A_S = 1$ m^2 shown in Figure 2.4.2. Assuming $k = 3$ W/(m^{2}K) and the environment temperature $T_0 = 300$ K equation (2.4.4) allows for determining the temperature T of surface A as function of the surface ratio a_S. As is shown in Figure 2.4.3, with the increase in a_S, the temperature T grows (thin solid line) and the heat rate q is also increasing (long-dashed line), determined by formula (2.4.5).

However, according to formula (2.4.6), with growing a_S the total heat Q is varying (short-dash line) with a maximum of about 134 W at about $a_S \approx 2$. The maximum appears because with growing a_S its effect becomes stronger than the effect of growing heat rate q.

The energy efficiency η_E of concentration of solar radiation, based on definition (2.4.7) is varying as shown with the thick-dashed line in Figure 2.4.3. The efficiency η_E has the maximum of about 16.8% appearing also at about $a_S \approx 2$, correspondently to the maximum of Q.

Exergy B_Q of absorbed heat is determined by (2.4.9) and shown in Figure 2.4.3 with a dash-dot line. The exergy B_Q varies and has a maximum of about 45.8 W, which appears in the surface area ratio about $a_S \approx 6$. The maximum is a result of two factors varying with growing a_S: one is growing exergy of heat due to growing temperature T, other is due to diminishing of the absorbed heat Q.

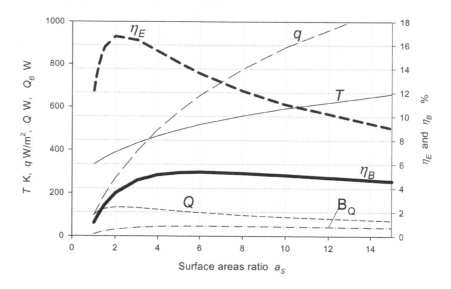

Figure 2.4.3 Exemplary effects of concentration of solar radiation (from Petela, 2010).

Assuming $\psi = 0.933$ like for the black radiation at temperature $T_S = 6000\,\mathrm{K}$, the efficiency η_B can be determined from formula (2.4.8) and shown in Figure 2.4.3 (thick solid line). The efficiency η_B has a maximum about 5.34% which also corresponds to value of $a_S \approx 6$.

The overall entropy growth determined from equation (2.4.10) for the data used in the example is always positive ($\Pi > 0$) and with growing a_S diminishes to zero ($\Pi = 0$) for $a_S = 91,843$ corresponding to temperature $T = 6000\,\mathrm{K}$. For further growing of a_S the overall entropy growth becomes negative ($\Pi < 0$) i.e. the further concentration of solar radiation is impossible.

Based on the calculations the process of "de-concentration" of solar radiation, which would correspond to reducing a_S below 1, is irreversible and can occur but heat absorbed by the surface A is negative which means that the surface would be heated.

The data used in the present example were also used for the computation of the results shown in Table 2.4.2 which illustrates the trends of the output data in response to changes in some input parameters. The values in column 3 of Table 2.4.2 are considered as the reference values for studying the influence of the varying input parameters in the output. Therefore each of the next columns (4 to 6) corresponds to the case in which the input is changed only by the values shown in a particular column, whereas the other input parameters remain at the reference level.

For example, column 4 corresponds to a change in the emissivity ε, which increases from 0.9 to 1. This 10% ε increase causes the increase of: temperature T from 517.8 to 519.9 K, q from 653.3 to 659.5 W/m^2, Q from 108.9 to 109.9 W, η_E from 13.61 to 13.74%, B_Q from 45.79 to 46.49 W and η_B from 5.34 to 5.42%.

Column 5 and 6 can be similarly interpreted. For example increasing the heat transfer coefficient k from 3 to 5 W/(m^2 K) causes increase in exergetic efficiency from 5.34 to 8.04%, which is the result of increased heat rate q from 653.3 to 1021 W/m^2.

Table 2.4.2 Responsive trends of output to change of some input parameters (for $IR = 800 W/m^2$, $T_S = 6000$ K, $\psi = 0.933, A_S = 1$ m^2), (from Petela, 2010)

Quantity	Units	Reference value	Mono-variant changes of input parameters and resulting outputs		
1	2	3	4	5	6
Input					
ε		0.9	1		
k	W/(m^2 K)	3		5	
T_0	K	300			270
Output					
T	K	517.8	519.9	504.2	514.8
q	W/m^2	653.3	659.6	1021	734.6
Q	W	108.9	109.9	170.2	122.4
B_Q	W	45.79	46.49	68.94	58.23
η_E	%	13.61	13.74	21.28	15.30
η_B	%	5.34	5.42	8.04	6.79

2.4.1.3 Global warming effect

Solar radiation is the energy source for life on the Earth and this radiation establishes the temperatures of the atmosphere and the Earth's surface. However, human activity seems to change the conditions of the energy exchange between the Sun and Earth and it seems that the observed tendency is a gradual increase of these temperatures. This effect of temperature growth is called the global warming effect. Some exergy insights are discussed in the present paragraph.

The atmosphere absorbs some of visible radiation directly from the Sun. From the Earth the atmosphere receives some infrared radiation and exchanges convective heat. The Earth, besides energy exchanged with the atmosphere, receives radiation from the Sun and reflects some radiation to the space. Assuming that the extraterrestrial solar radiation directed to the Earth renders 100%, the energy flow diagram for the thermal equilibrium for any Earth temperature T_0, can be shown schematically in Figure 2.4.4a. It is usually estimated that the annual average temperature T_0 of the Earth surface is about 14°C (287 K) and at this temperature the Earth's energy emission is ~192%. The atmosphere, at a certain assumed effective temperature T_A, emits energy to the Earth (~138%) and space (~83%).

The global warming effect is not a subject well exposed by the exergy concept because the exergy relates to environment temperature T_0 with no regards of how high this temperature is. For example, for any Earth temperature T_0 the exergy radiated from the Earth's surface will be always zero, as shown in Figure 2.4.4b in which 100% is assumed for the exergy of extraterrestrial solar radiation directed to the Earth.

Example 2.4.1.4 To illustrate the global warming effect a very rough consideration of exchanged radiative and convective heat fluxes are analyzed in a simplified model of a polluted air layer between the space and the Earth's surface. The air layer of transmissivity τ is assumed as a body of reflectivity $\rho = 0$ and of absorptivity equal to

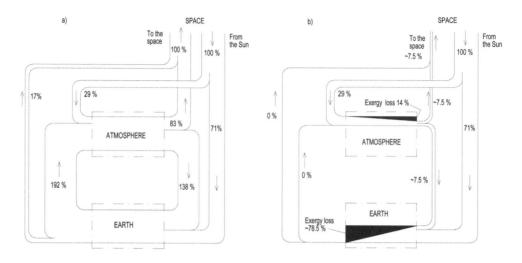

Figure 2.4.4 Simplified scheme of radiation energy (a) and exergy (b), (from Petela, 2010).

emissivity $\alpha = \varepsilon$, thus the body emissivity is $\varepsilon = 1 - \tau$. The air layer has temperature T_A, transmissivity $\tau_S = 0.71$ for the high temperature solar radiation and $\tau_0 = 0.17$ for a low temperature radiation. The yearly average solar irradiance $S = 100$ W/m² arriving in the air layer is assumed. The Earth's surface is black, does not transfer energy to the ground, and has temperature $T_0 = 287.16$ K (14°C).

The air layer absorbs: solar energy $(1 - \tau_S) \cdot S$, radiation from sky and radiation from earth surface, whereas it releases the energy by radiation and convection to the sky and the Earth surface:

$$(1 - \tau_S)S + (1 - \tau_0)\sigma(T_{sky}^4 + T_0^4) = 2(1 - \tau_0)\sigma T_A^4 + k[(T_A - T_0) + (T_A - T_{sky})] \tag{2.4.12}$$

and the Earth's surface absorbs: solar energy, radiation from the air layer and radiation from sky, whereas it releases the energy by radiation and convection:

$$\tau_S S + (1 - \tau_0)\sigma T_A^4 + \tau_0 \sigma T_{sky}^4 = \sigma T_o^4 + k(T_0 - T_A) \tag{2.4.13}$$

where T_{sky} is the sky temperature representing black space above the air layer, σ is the Boltzmann constant for black radiation, and k is the convective heat transfer coefficient assumed equal for all convections. For $k = 3$ W/(m² K), from the two equations (2.4.12) and (2.4.13) the two unknown temperature can be calculated: $T_A = 279.6$ K and $T_{sky} = 267.4$ K.

The global warming effect could be considered based on the influence of changing the factor τ_0 describing pollution of air, on the change of environment temperature T_0. From equation (2.4.12) the partial derivative

$$\frac{\partial T_0}{\partial \tau_0} = \frac{T_0^4 + T_{sky}^4 - 2T_A^4}{4T_0^3(1 - \tau_0) + \frac{k}{\sigma}} = 44.087 \, \text{K/\%} \tag{2.4.14}$$

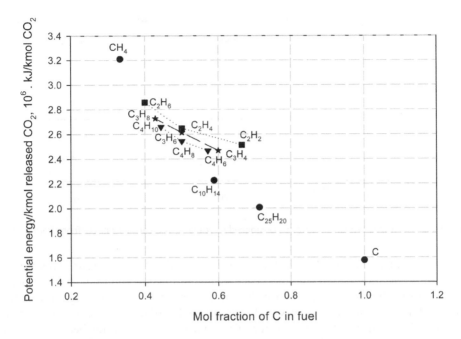

Figure 2.4.5 Effect of combustion of hydrocarbons on CO_2 generation.

For example, due increased pollution, the relatively significant increase from 17% do 17.5% of the transmissivity τ_0 of the air for the low temperature radiation, causes, based on formula (2.4.14), a relatively small increase in environment temperature T_0 from 287.16 K to: $44.087 \cdot (0.175 - 0.17) + 287.16 = 287.38$ K.

The main factor considered in global warming is the concentration of CO_2 in the atmosphere. The CO_2 is a product of combustion of substances containing carbon. It is noteworthy to emphasize that regardless which kind of compound (fuels or biomass, etc.), when carbon is combusted there is always CO_2 released into the atmosphere. Therefore combustion of any compounds containing carbon (e.g. hydrocarbons) is releasing CO_2 proportionally to the content of carbon in the compound. The effect of combustion of some exemplary hydrocarbons is illustrated in Figure 2.4.5. Data were obtained by the assumption that the calorific value of combusted hydrocarbon is utilized by 40% for power generation. Generally, the larger the molar content of C in combusted material, the smaller the energy per released amount of CO_2. The smallest value corresponds to pure carbon C. However, for example, pure hydrogen cannot be shown in the diagram because the corresponding potential energy amount is infinity. This should be emphasized that such infinity value corresponds also to utilization of solar energy in production of power, heat or photosynthesis.

2.4.1.4 Canopy effect

The global warming effect often is compared to the effect of a greenhouse in which the solar radiation is trapped at the Earth by using transparent canopy over the surface

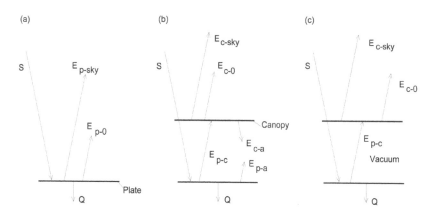

Figure 2.4.6 The three typical situations considered in study of the canopy effect (from Petela, 2010).

absorbing solar radiation. The real greenhouse is built of transparent walls through which solar radiation penetrates and heats the ground inside and then the confined air gets heat from this ground. Based on the rough analogy the term 'greenhouse effect' is referred also to the process in which the infrared radiation is exchanged between the atmosphere and the Earth's surface. These two processes differ because in a real greenhouse air is trapped whereas the environment air, when warmed from the ground, rises and mixes with cooler air aloft. The analogy could be eventually recognized in the fact that the glass roof of the greenhouse traps the infrared radiation to warm the greenhouse air and the greenhouse roof plays the role of the huge and thick layer of the atmosphere.

It is estimated that in the absence of the greenhouse effect the Earth's surface temperature would be decreased from about 14 to about $-19°C$. It is believed that the recent warming of the lower atmosphere is the result of an enhancing of the greenhouse effect by an increasing of amount of gaseous, liquid and solid ingredients of different radiative properties than air.

Using exergy, the effect of the canopy located above a considered surface to screen the surface from the direct solar radiation can be considered, based on the three situations presented schematically in Figure 2.4.6.

A black horizontal plate, of surface area A, located on the Earth's surface can be exposed to direct solar radiation as shown schematically in Figure 2.4.6a. In the thermodynamic equilibrium state the irradiance S is spent on heat Q extracted at constant plate temperature T_p and on the convective (E_{p-0}) and radiative ($E_{p\text{-sky}}$) heat fluxes from the plate to the surroundings. The plate temperature T_p is controlled by the appropriately arranged amount of heat Q. The energy balance equation for the plate is:

$$S = Q + E_{p\text{-sky}} + E_{p-0} \tag{2.4.15}$$

where

$$E_{p\text{-sky}} = A\sigma(T_p^4 - T_{\text{sky}}^4) \tag{2.4.16}$$

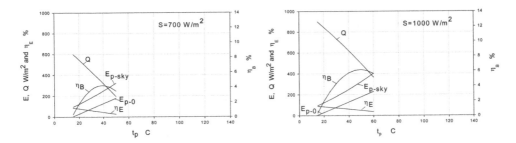

Figure 2.4.7 Plate exposed to solar radiation, $S = 700\,\text{W/m}^2$ (left) and $S = 1000\,\text{W/m}^2$ (right), (from Petela, 2010).

$$E_{p-0} = A k_{p-0}(T_p - T_0) \tag{2.4.17}$$

and where k_{p-0} is the convective heat transfer coefficient. The harvest of the solar energy can be determined by the energetic efficiency η_E:

$$\eta_E = \frac{Q}{S} \tag{2.4.18}$$

or by the exergetic efficiency η_B:

$$\eta_B = \frac{Q}{\psi_c S}\left(1 - \frac{T_0}{T_p}\right) \tag{2.4.19}$$

where ψ_c is the exergy/energy ratio discussed in paragraph 2.2.5. It can be shown that, for the direct radiation of the Sun at its surface temperature 6000 K, the theoretical value of the ratio is $\psi = 0.933$. According to Gueymard (2004), the irradiance of the direct solar radiation arriving at the Earth is 1366 W/m². As the irradiance values applied in the present canopy consideration are smaller then the exergy/energy ratio, they could be taken as for a smaller irradiance temperature, e.g., $\psi_c = 0.9$.

For example, assuming $A = 1\,\text{m}^2$, $k_{p-0} = 5\,\text{W/(m}^2\,\text{K)}$ and determined by the Swinbank (1963) formula: $T_{sky} = 0.0552 \cdot T_0^{1.5}$ in which $T_0 = 287.16\,\text{K}$ (14°C), Figure 2.4.7 shows the calculation results for the two different values of irradiance, $S = 700\,\text{W/m}^2$ and $S = 1000\,\text{W/m}^2$. With the increasing plate temperature T_p, $(t_p^\circ\text{C})$, the energetic efficiency η_E decreases whereas the exergetic efficiency η_B is maximum.

In the second situation, shown in Figure 2.4.6b, the plate of the surface area 1 m² is a fragment of a very large and flat surface of the same uniformly distributed values of temperature T_p and radiative properties. The plate is screened from the solar radiation with a very large, flat and horizontal canopy. Material of the canopy can transmit the whole solar radiation to the plate (canopy transmissivity $\tau_{SOL} = 1$), although the low temperature emission from the plate is entirely absorbed by the canopy (canopy transmissivity $\tau_{PLA} = 0$). The extreme values of these two transmissivities are assumed to show better the effect of the canopy on the exchanged radiative heat. Due to a very small thickness of the canopy the both its surfaces, that exposed to the Sun as well as that exposed to the plate, have the same canopy temperature T_c.

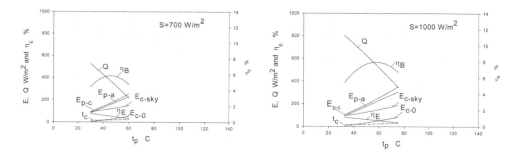

Figure 2.4.8 Plate under a canopy in the environment air, $S = 700\,\text{W}/\text{m}^2$ (left) and $S = 1000\,\text{W}/\text{m}^2$ (right), (from Petela, 2010).

In the thermodynamic equilibrium state of the situation shown in Figure 2.4.6b, the irradiance S is spent on heat Q extracted at constant plate temperature T_p and on the convective (E_{p-a}) and radiative (E_{p-c}) heat fluxes from the plate to the canopy. The plate temperature T_p is controlled by the appropriately arranged amount of heat Q. The canopy temperature T_c is constant for the given plate temperature T_p and distributed uniformly over the surfaces of the canopy. The energy balance equation for the plate is:

$$S = Q + E_{p-c} + E_{p-a} \tag{2.4.20}$$

where

$$E_{p-c} = \sigma(T_p^4 - T_c^4) \tag{2.4.21}$$

$$E_{p-a} = Ak_{p-a}(T_p - T_a) \tag{2.4.22}$$

and where k_{p-a} is the respective convective heat transfer coefficient and T_a is the temperature of air between the plate and the canopy. Simplifying, it is assumed $T_a = T_0$. The energy balance of the canopy is:

$$E_{p-c} = E_{c-\text{sky}} + E_{c-0} + E_{c-a} \tag{2.4.23}$$

where

$$E_{p-a} = E_{c-0} = k_{p-a}(T_p - T_0) \tag{2.4.24}$$

and where $k_{p-a} = k_{c-a}$ are the respective convective heat transfer coefficients. The harvest of the solar energy in the considered situation can be again determined by the energetic efficiency η_E and exergetic efficiency η_B determined respectively from formulae (2.4.18) and (2.4.19).

For example, assuming $k_{p-a} = k_{c-a} = 5\,\text{W}/(\text{m}^2\,\text{K})$, Figure 2.4.8 shows the calculation results for the two different values of irradiance, $S = 700\,\text{W}/\text{m}^2$ and $S = 1000\,\text{W}/\text{m}^2$. As in situation (a) also in situation (b), with the increasing plate

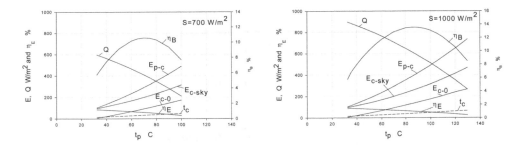

Figure 2.4.9 Plate under a canopy with a vacuum between the canopy and the plate, $S = 700\,\text{W/m}^2$ (left) and $S = 1000\,\text{W/m}^2$ (right), (from Petela, 2010).

temperature T_p, the energetic efficiency η_E decreases whereas the exergetic efficiency η_B is at maximum.

The third possible situation, shown in Figure 2.4.6c, is the same as in the previous situation (b), except that between the plate and canopy is a vacuum, thus in this space heat convection does not occur. The energy balance equations for the plate and the canopy are:

$$S = Q + E_{p-c} \tag{2.4.25}$$

$$E_{p-c} = E_{c-sky} + E_{c-0} \tag{2.4.26}$$

The energetic efficiency η_E and exergetic efficiency η_B are determined respectively also from formulae (2.4.18) and (2.4.19). Figure 2.4.9 shows the calculation results for the two different values of irradiance, $S = 700\,\text{W/m}^2$ and $S = 1000\,\text{W/m}^2$. As in situations (a) and (b), also in situation (c), with the increasing plate temperature T_p the energetic efficiency η_E decreases whereas the exergetic efficiency η_B is at maximum.

The comparative discussion of the three models (Fig. 2.4.6) can be summarized as follows. The irradiated black plate (a), the plate under the canopy (b) and the plate under the vacuum and canopy (c), were considered under simplifying assumptions of extreme values of surface properties to better emphasize the canopy idea. The comparison of Figures 2.4.7, 2.4.8 and 2.4.9 illustrates benefits of application of canopy for increasing effect of trapping solar radiation. The amount of exergy (practical value) of absorbed heat grows gradually through the three considered situations from (a) to (c).

2.4.1.5 Evaluation of solar radiation conversion into heat

Solar radiation can be converted to heat in many various applications. Generally, in each application the solar radiation is absorbed by certain designed surfaces at a temperature controlled by appropriate amount of heat extracted. Unfortunately the high quality of solar energy, e.g., measured by exergy, is significantly degraded during a conversion to heat.

A simple introduction to evaluation of the non-concentrated solar radiation potential for heating and determination of heat temperature can be considered with use of the model of 1 m² of absorbing surface shown in Figure 2.4.10. The considered surface

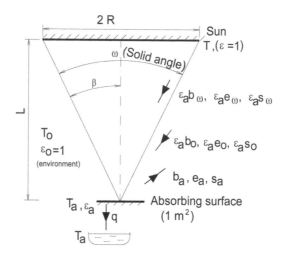

Figure 2.4.10 Scheme of the fluxes in energy balance of the surface on earth absorbing radiation from sun, (from Petela, 2003).

is on the Earth and is perpendicular to the Sun's direction. From the Sun, the black ($\varepsilon = 1$) radiation of exergy b_ω, energy e_ω and entropy s_ω, all within the solid angle ω, arrives in the absorbing surface. (For simplification the subscript "b" for black is omitted in this paragraph). These three fluxes are absorbed by the absorbing surface at temperature T_a and emissivity ε_a. The absorbing surface, in the solid angle 2π, emits its own radiation fluxes of exergy b_a, energy e_a entropy s_a and obtains, in the solid angle $2\pi - \omega$, the radiation fluxes of exergy b_0, energy e_0 and entropy s_0 from the environment at temperature T_0 (assumed to be equal to the sky temperature, $T_{sky} = T_0$) and at assumed emissivity $\varepsilon_0 = 1$.

For the conversion of solar radiation the energy efficiency η_E and the exergy efficiency η_B are:

$$\eta_E = \frac{q}{e_\omega} \tag{2.4.27}$$

$$\eta_B = \frac{b_q}{b_\omega} \tag{2.4.28}$$

In the present considerations the solar radiation is considered as non-polarized, uniform and black, at temperature $T = 6000\,\text{K}$, arriving in the Earth within the solid angle ω. Exergy b_ω of such radiation can be calculated from formulas (2.2.66) and (2.4.1) in which the double integral represents angle $\omega = \pi \cdot R^2 / L^2$ with use of the radius of the Sun R and the mean distance L from the Sun to the Earth:

$$b_\omega = \frac{b}{\pi} \frac{\pi R^2}{L^2} = b \frac{R^2}{L^2} \tag{a}$$

Analogously, the energy emission e_ω arriving from the Sun at the absorbing surface within the solid angle ω, is:

$$e_\omega = e \frac{R^2}{L^2} \tag{b}$$

where the Sun density emission e and the exergy b of the black radiation density emitted by the Sun are:

$$e = \sigma T^4 \tag{c}$$

$$b = \frac{\sigma}{3}(3T^4 + T_0^4 - 4T_0 T^3) \tag{d}$$

The heat q, absorbed by the surface at temperature T_a, is extracted in the amount determined from the following energy conservation equation for the absorbing surface:

$$q = \varepsilon_a e_\omega - e_a + e_0 \tag{e}$$

and the exergy b_q of this heat, absorbed by the heat source temperature T_a, is:

$$b_q = q \frac{T_a - T_0}{T_a} \tag{f}$$

Determination of e_a and e_0 in equation (e) is required. The absorbing surface of emissivity ε_a and temperature T_a, radiates its own emission e_a to the whole hemisphere:

$$e_a = \varepsilon_a \sigma T_a^4 \tag{g}$$

and the respective exergy b_a which is:

$$b_a = \varepsilon_a \frac{\sigma}{3}(3T_a^4 + T_0^4 - 4T_0 T_a^3) \tag{h}$$

The considered absorbing surface, beside emission from the Sun, obtains from the remaining part of environmental hemisphere the black ($\varepsilon_0 = 1$) radiation energy e_0, ($e_0 \equiv e_{0,(2\pi\omega)}$), at temperature T_0, which is absorbed in amount determined by emissivity ε_a of the adsorbing surface:

$$e_0 = \varepsilon_a \sigma T_0^4 \left(2 - \frac{R^2}{L^2}\right) \tag{i}$$

However, regarding exergy, according to the definition, the exergy of environment radiation is zero:

$$b_0 = 0 \tag{j}$$

Using equations (b), (c), (e), (g) and (i) in (2.4.27):

$$\eta_E = \varepsilon_a \left[2 - \frac{T_a^4 - T_0^4 \left(1 - \dfrac{R^2}{L^2} \right)}{T^4 \dfrac{R^2}{L^2}} \right] \tag{2.4.29}$$

Using formulae (a)–(i) in (2.4.28), the exergy conversion efficiency of solar radiation into heat can be determined as follows:

$$\eta_B = 3\varepsilon_a \frac{T_a - T_0}{T_a} \frac{T^4 - T_0^4 - (T_a^4 - 2T_0^4)\dfrac{L^2}{R^2}}{3T^4 + T_0^4 - 4T_0 T^3} \tag{2.4.30}$$

The larger the ratio L/R, the smaller are both efficiencies. The increasing emissivity ε_a of the absorbing surface will increase the conversion efficiencies.

The larger the T_a, the smaller is the energy efficiency η_E, however, the exergy conversion efficiency η_B is at maximum. The optimal temperature $T_{a,opt}$ can be calculated based on (2.4.30) from the condition:

$$\frac{\partial \eta_B}{\partial T_a} = 0 \tag{2.4.31}$$

For example, if the solar radiation is considered at $\varepsilon_a = 1$, $T_0 = 300\,\mathrm{K}$, $T = 6000\,\mathrm{K}$, $R = 6.955 \cdot 10^8$ m and $L = 1.495 \cdot 10^{11}$ m, then $T_{a,opt} \approx 363\,\mathrm{K}$ (90°C). If the environment temperature drops to $T_0 = 273\,\mathrm{K}$ then $T_{a,opt} \approx 383\,\mathrm{K}$ (110°C).

The T_a optimum, at the unchanged exergy b_ω of solar radiation, results from the fact that with increasing T_a, which increases the heat quality (b_q), the amount of this heat decreases. The emissivity value ε_a does not appear in equation (2.4.31) so this emissivity has no effect on the optimal temperature $T_{a,opt}$.

The universal traveling of the human population motivates considering the environment temperature in a wide range, theoretically for $0 < T_0 < T$. This aspect is shown in Figure 2.4.11. With a decreasing environment temperature T_0, the optimal temperature $T_{a,opt}$ of the absorbing surface continuously diminishes and the exergy conversion efficiency η_B grows approaching 80% for $T_0 \to 0$. At the same time the Carnot efficiency, $\eta_{C,a} = 1 - T_0/T_a$, also grows and reaches 100% for $T_0 \to 0$.

For further analysis of solar radiation conversion, the energy and exergy balances equations for the absorbing surface (Fig. 2.4.10) are:

$$e_\omega - (1 - \varepsilon_a)e_\omega + e_0 = e_a + q \tag{k}$$

$$b_\omega - (1 - \varepsilon_a)b_\omega + b_0 = b_a + b_q + \delta b \tag{l}$$

To calculate exergy loss in equation (l) some entropy formulae has to be used. Entropy s_ω of the solar radiation arriving at the absorbing surface:

$$s_\omega = s\frac{R^2}{L^2} \tag{m}$$

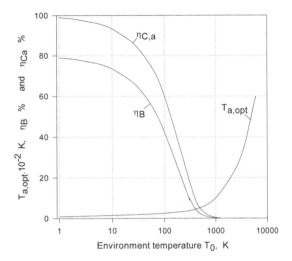

Figure 2.4.11 Effect of varying environment temperature T_0, at constant $T = 6000$ K and $\varepsilon_a = 0.8$, (from Petela, 2003).

where the entropy s of solar radiation considered for the whole hemisphere, is:

$$s = \frac{4}{3}\sigma T^3 \tag{n}$$

The entropy fluxes, s_a for emission of absorbing surface and s_0 for absorbed environment radiation, are respectively:

$$s_a = \varepsilon_a \frac{4}{3}\sigma T_a^3 \tag{o}$$

$$s_0 = \varepsilon_a \frac{4}{3}\sigma T_0^3 \left(2 - \frac{R^2}{L^2}\right) \tag{p}$$

The overall entropy growth Π for all processes involved consists of the entropy of generated heat $(+)$, disappearing entropy of solar radiation $(-)$, emitted entropy of absorbing surface $(+)$, and disappearing entropy of environment radiation $(-)$:

$$\Pi = \frac{q}{T_a} - \varepsilon_a s_\omega + s_a - s_0 \tag{r}$$

The overall process occurring at the absorbing surface is irreversible and the respective exergy loss δb is determined by using (r) in formula (2.2.10).

Some exemplary calculation data is shown in Table 2.4.3 for $T_0 = 300$ K, $T_a = 350$ K, $T = 6000$ K, $b_\omega = 1.484$ kW/(m^2 sr), $e_\omega = 1.59$ kW/(m^2 sr) and two different values $\varepsilon_a = 0.8$, and $\varepsilon_a = 1$.

Table 2.4.3 Comparison of the exergy and energy balances of a surface absorbing solar radiation, for $\varepsilon_a = 0.8$ and $\varepsilon_a = 1$, (from Petela, 2003).

Item	Terms	$\varepsilon_\alpha = 8.0$		$\varepsilon_\alpha = 1$	
		% energy	% energy	% energy	% energy
Input	Sun	100	100	100	100
	Environment	23.1	0	28.88	0
	Total	123.1	100	128.88	100
Output	Reflection	20	20	0	0
	Emission	42.8	1.7	53.5	2.12
	Heat, (efficiency)	60.3	9.23	75.38	11.54
	Loss	0	69.07	0	86.35
	Total	123.1	100	128.88	100

For the considered case in Table 2.4.3, the values of efficiencies can be interpreted as $\eta_E = 60.3\%$ and $\eta_B = 9.23\%$ for $\varepsilon_a = 0.8$, or, respectively, $\eta_E = 75.38\%$ and $\eta_B = 11.54\%$ for $\varepsilon_a = 1$. The energetic and exergetic efficiencies values differ significantly and, except for reflected radiation by absorbing surface, other items of the both balances are also very different.

The temperature T_a of absorbing surface can be considered practically only in the range $T_0 \leq T_a \leq T_{a,max}$. Temperature T_a smaller than T_0 requires additional energy to generate surroundings colder than the environment, whereas for $T_a > T_{a,max}$ the heat q becomes negative because the radiation of absorbing surface to the environment is larger than the heat received from solar radiation.

Some more computation results for analyzing effect of T_a, $(t_a C)$, are shown in Figure 2.4.12. An increase in T_a, $(t_a\ C)$, will decrease exergy loss δb and heat q, whereas η_B and b_q reveal maxima as results of growing of the Carnot efficiency: $\eta_{C,a} = 1 - T_0/T_a$.

2.4.2 Solar cylindrical-parabolic cooker

The most common devices for utilization of solar radiation are cookers of different types. The simple solar cylindrical-parabolic cooker (SCPC), shown schematically in Figure 2.4.13, is used to demonstrate the methodology of exergy analysis of the cooker and the distribution of the exergy losses. Also explained is the general problem of how the exergy loss at any radiating surface should be determined, if the surface absorbs many radiation fluxes of different temperatures. Additionally a possibility of introduction of an imagined surface to complete the cooker surfaces system is shown.

The cylindrical cooking pot filled with water is surrounded with the cylindrical-parabolic reflector. The considered system of exchanging energy consists of three long surfaces of length L. The outer surface 3 of the cooking pot has an area A_3. The inner

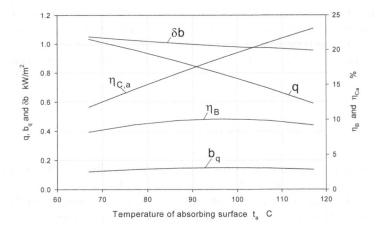

Figure 2.4.12   Effect of varying temperature T_a of absorbing surface, at constant $T = 6000\,\text{K}, T_0 = 300\,\text{K}$ and $\varepsilon_a = 0.8$ (Petela, 2003).

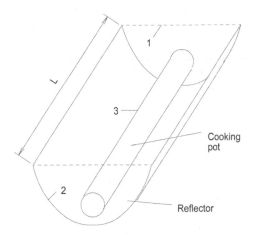

Figure 2.4.13   The scheme of the SCPC, (from Petela, 2005).

surface 2 of the reflector has an area A_2. The system is completed with the imagined plane surface 1 of area A_1.

The imagined surface 1, which represents the ambience and the irradiation supplied to the considered system, is defined by transmissivity $\tau_1 = 1$ (and thus reflectivity $\rho_1 = 0$), absorptivity $\alpha_1 = 0$ and emissivity $\varepsilon_1 = 0$.

The effective emission of the imagined surface 1, can be determined as the irradiation I calculated as follows:

$$I = 2.16 \cdot 10^{-5} A_1 \varepsilon_S \sigma T_S^4 \qquad (2.4.32)$$

where $2.16 \cdot 10^{-5}$ accounts on the solid angle within which the Sun has been seen from the Earth, ε_S is the emissivity of the Sun surface, (assumed $\varepsilon_S = 1$), σ is the Boltzmann constant for black radiation, and T_S is the absolute temperature of the Sun surface. Formally, it can be assumed that the energy emission of surface 1 is $E_1 = I$.

It is assumed that the surfaces 2 and 3 have uniform temperatures T_2 and T_3, respectively, uniform reflectivities, respectively, ρ_2 and ρ_3, different from zero, and the emissivities of the surfaces, $\varepsilon_2 = 1 - \rho_2$ and $\varepsilon_3 = 1 - \rho_3$. Thus, the emissions of surfaces 2 and 3 are:

$$E_2 = A_2 \varepsilon_2 \sigma T_2^4 \tag{2.4.33}$$

$$E_3 = A_3 \varepsilon_3 \sigma T_3^4 \tag{2.4.34}$$

The geometric configuration of the SCPC can be described by the value φ_{i-j} of the nine mutual view factors for the three surfaces 1, 2 and 3.

The considerations are carried out only for the 1 m section of the SCPC length which remains in thermal equilibrium. The known input data are:

- outer diameter D of cooking pot and its geometric location,
- dimensions of the parabolic reflector,
- surfaces areas A_1, A_2, A_3, and all view factors, ϕ_{i-j},
- heat transfer coefficients (including conductivity) k_2 and k_3, for surfaces 2 and 3,
- emissivities ε_2 and ε_3 of surfaces 2 and 3,
- reflectivities ρ_2 and ρ_3 (defined by respective emissivities ε_2 and ε_3),
- absolute temperature of the Sun's surface $T_S = 6000$ K,
- absolute water temperature T_w (average of the inlet and outlet temperatures),
- absolute environment temperature $T_0 = 293$ K.

The unknown output data are:

- emissions E_2, E_3 and irradiation I, $(I = E_1)$,
- convective heat $Q_{2,c}$ from reflector to environment,
- radiative heat $Q_{2,r}$ from outer side of reflector to environment,
- convective heat $Q_{3,c}$ from surface 3 to environment,
- radiosity of the three surfaces J_1, J_2 and J_3,
- absolute temperatures T_2 and $T_3$of surface 2 and 3,
- energy efficiency of the SCPC, expressed by the water enthalpy change $Q_{3,u}$,
- all respective quantities related to exergy.

The *energy analysis* is based on the following energy conservation equations for each involved surfaces:

$$J_1 = \varphi_{2-1}J_2 + \varphi_{3-1}J_3 + Q_{2,c} + Q_{2,r} + Q_{3,c} + Q_{3,u} \tag{2.4.35}$$

$$\varepsilon_2(\varphi_{1-2}J_1 + \varphi_{2-2}J_2 + \varphi_{3-2}J_3) = E_2 + Q_{2,c} + Q_{2,r} \tag{2.4.36}$$

$$\varepsilon_3(\varphi_{1-3}J_1 + \varphi_{2-3}J_2 + \varphi_{3-3}J_3) = E_3 + Q_{3,c} + Q_{3,u} \tag{2.4.37}$$

The magnitudes J_1, J_2 and J_3 are the radiosity values for surfaces 1, 2 and 3, respectively, and the values of φ_{i-j} are the respective view factors. The radiosity expresses the total radiation which leaves a surface and includes emission of the considered surface as well as all reflected radiations arriving from other surfaces of the system. The concept of radiosity is convenient for energy calculation, however it cannot be used for exergy considerations because it does not distinguish the temperatures of the components of the radiosity. The radiosity J_1 of the imagined surface 1 equals the irradiation I:

$$J_1 = I \tag{2.4.38}$$

and another independent equation on the radiosity is also included into calculations:

$$J_2 = E_2 + \rho_2(\varphi_{1-2}J_1 + \varphi_{2-2}J_2 + \varphi_{3-2}J_3) \tag{2.4.39}$$

It is assumed that the reflector is very thin so the uniform temperature T_2 prevails through the whole reflector thickness and on the inner and outer side of the reflector. Thus the heat $Q_{2,c}$ transferred from both sides of the reflector is:

$$Q_{2,c} = 2A_2k_2(T_2 - T_o) \tag{2.4.40}$$

and heat radiating from the outer side of reflector to the environment is:

$$Q_{2,r} = A_2\varepsilon_2\sigma(T_2^4 - T_o^4) \tag{2.4.41}$$

where k_2 is the convective heat transfer coefficient and T_0 is the environment temperature.

Heat $Q_{3,c}$ transferred by convection from surface 3 to the environment is:

$$Q_{3,c} = A_3k_3(T_3 - T_o) \tag{2.4.42}$$

and the useful heat $Q_{3,u}$ transferred through the wall of the cooking pot is:

$$Q_{3,u} = A_3k_3'(T_3 - T_w) \tag{2.4.43}$$

where k_3 is the convective heat transfer coefficient, T_w is the absolute temperature of water in the cooking pot and k_3' is the heat transfer coefficient, which takes into account the conductive heat transfer through the cooking pot wall and convective heat transfer from the inner cooking pot surface to the water.

The equations system (2.4.32)–(2.4.43) can be solved by successive iterations. The energy analysis of the SCPC can be carried out based on the evaluation of the terms in the following energy conservation equation for the whole SCPC:

$$\varphi_{2-1}J_2 + \varphi_{3-1}J_3 + Q_{2,c} + Q_{2,r} + Q_{3,c} + Q_{3,u} = I \tag{2.4.44}$$

The first two terms in equation (2.4.44) represent radiation energy escaping from the SCPC due to the radiosities of surfaces 2, $(\varphi_{2,1} \cdot J_2)$, and 3 $(\varphi_{3,1} \cdot J_3)$. Dividing the both sides of equation (2.4.44) by I, the percentage values β of the equation terms can be obtained, e.g., for heat $Q_{2,c}$ respectively is $\beta_{2,c} = Q_{2,c}/I$, however the term with $Q_{3,u}$ determines the energetic efficiency

$$\eta_E = \frac{Q_{3,u}}{I} \tag{2.4.45}$$

Therefore, equation (2.4.44) can be also written as:

$$\sum \beta + \eta_E = 1 \tag{2.4.46}$$

The *exergy analysis* enables the additional quality interpretation of the cooker. For the considered SCPC, the exergy of radiating fluxes, overall exergy efficiency of the SCPC process and the exergy losses during irreversible component phenomena occurring in the SCPC are calculated.

Sometimes, it is convenient to determine an exergy B of radiation emission at temperature T by multiplying its emission energy E by the characteristic exergy/energy ratio ψ, defined by formula (2.2.45), e.g., $B = E \cdot \psi$. Thus, the exergy efficiency η_B of the SCPC is the ratio of the exergy of the useful heat $Q_{3,u}$, at temperature T_3, and of the exergy of solar emission at temperature T_S:

$$\eta_B = \frac{Q_{3,u}\left(1 - \frac{T_0}{T_3}\right)}{I\psi_S} \tag{2.4.47}$$

where ψ_S is the exergy/energy ratio for the solar emission of temperature T_S, which for $T_S = 6000\,\text{K}$ and $T_0 = 293\,\text{K}$ is $\psi_S = 0.9348$.

Reflection and transmission of radiation are reversible so the exergy losses in the SCPC are considered only for the following component phenomena:

- Simultaneous emission and absorption of radiation at surfaces 2, (δB_2) and 3, (δB_3). There is no exergy loss at the imagined surface 1, $(\delta B_1 = 0)$, because neither absorption nor emission occurs but only transmission of radiation which is reversible. Other surfaces, 2 and 3, are solid and thus produce the irreversible effects of radiation.
- Irreversible transfer of convection heat Q_{2c} from the both sides of surface 2 to the environment, (δB_{Q2c}).
- Irreversible transfer of radiation heat Q_{2r} from the outer side of surface 2 to the environment, (δB_{Q2r}).
- Irreversible transfer of heat Q_{3u} from surface 3 to water, (δB_{Q3u}), due to temperature difference $T_3 - T_w$,
- Irreversible transfer of convection heat Q_{3c} from surface 3 to environment, (δB_{Q3c}).

The exergy δB_{1-0} escaping through surface 1, results from reflections from the SCPC surfaces to the environment. This loss is sensed only by the SCPC and is not irreversible because theoretically it can be used elsewhere. This loss consists of the radiation exergies B_{1-1}, B_{2-1}, B_{3-1} at the three different temperatures (T_S, T_2 and T_3)

$$\delta B_{1-0} = B_{1-1} + B_{2-1} + B_{3-1} \tag{2.4.48}$$

Analogously to the energy conservation equation, the exergy balance equation can be applied. When relating all the equation terms to the exergy input, which is the exergy $I \cdot \psi_S$ of solar radiation entering the SCPC system, the following conservation equation for the whole SCPC, can be written:

$$\xi_{B11} + \xi_{B21} + \xi_{B31} + \xi_{BQ2c} + \xi_{BQ2r} + \xi_{Q3u} + \xi_{Q3c} + \xi_{B2} + \xi_{B3} + \eta_B = 100 \quad (2.4.49)$$

where any percentage exergy loss ξ is calculated as the ratio of the loss to the exergy input, e.g., for the convection heat $Q_{2,c}$ one obtains $\zeta_{Q2c} = \delta B_{Q2c}/(I \cdot \psi_S)$.

In the considered system of non-black surfaces the radiation energy striking a surface is not totally absorbed and part of it is reflected back to other surfaces. The radiant energy can be thus reflected back and forth between surfaces many times. To simplify the effect of further such multi reflections, it is assumed that surface 3 is black, $(\varepsilon_3 = 1)$. Thus, as the imagined surface 1 was previously assumed to be black $(\varepsilon_1 = 1)$, the only non-black surface in the exergetic analysis of the SCPC system is surface 2, $(\varepsilon_2 < 1)$.

The nine exergy losses appearing in equation (2.4.49) can be categorized in three groups. The first group contains the exergy losses (external) related to heat transfer, $(\xi_{BQ2c}, \xi_{BQ2r}, \xi_{Q3u}, \xi_{Q3c})$. The second group $(\xi_{B11}, \xi_{B21}, \xi_{B31})$ determines the exergy fluxes (external losses) escaping from the SCPC and the third group contains the exergy losses (internal) due to irreversible emission and absorption on surface (ξ_{B2}, ξ_{B3}).

First group losses. The exergy loss δB_{Q2c}, due to the convection transfer of heat $Q_{2,c}$ from the each of the two sides of the reflector to the environment, is equal to the exergy of heat $Q_{2,c}$:

$$\delta B_{Q2c} = Q_{2,c}\left(1 - \frac{T_0}{T_2}\right) \quad (2.4.50)$$

The exergy loss δB_{Q2r}, due to the radiation transfer of heat $Q_{2,r}$ from the outer side of the reflector to the environment, is equal to the exergy of heat $Q_{2,r}$:

$$\delta B_{Q2r} = Q_{2,r}\left(1 - \frac{T_0}{T_2}\right) \quad (2.4.51)$$

The external exergy loss δB_{Q3u}, due to the transfer of the useful heat $Q_{3,u}$ from surface 3 through the cooking pot wall to water, is equal to the difference of the exergy of this heat at temperature T_3 and at temperature T_w:

$$\delta B_{Q3u} = Q_{3,u}\left(\frac{T_3 - T_0}{T_3} - \frac{T_w - T_0}{T_w}\right) \quad (2.4.52)$$

The exergy loss δB_{Q3c}, due to convective heat transfer from surface 3 to the environment, is determined similarly:

$$\delta B_{Q3c} = Q_{3,c}\left(\frac{T_3 - T_0}{T_3} - \frac{T_0 - T_0}{T_0}\right) \quad (2.4.53)$$

where, obviously, the second fraction in the brackets of equation (2.4.53) is zero.

Second group losses. As results from formula (2.4.48) the external exergy loss δB_{1-0} is equal to the escaping exergy of three unabsorbed emissions at temperatures $T_1 = T_S$, T_2 and T_3 reflected to the environment. By multiplying these emissions respectively by the exergy/energy ratio the three losses can be expressed as follows:

$$B_{1-1} = Q_{1-1}\psi_S \tag{2.4.54}$$

$$B_{2-1} = Q_{2-1}\psi_2 \tag{2.4.55}$$

$$B_{3-1} = Q_{3-1}\psi_3 \tag{2.4.56}$$

where ψ_S, ψ_2 and ψ_3 are calculated from formula (2.2.45) for T_S, T_2 and T_3, respectively, and Q_{1-1}, Q_{2-1} and Q_{3-1} are the sums of the unabsorbed portion of the respective emissions of surfaces 1, 2 and 3. Thus, heat Q_{1-1} represents energy portions from many reflections of emission E_1 of temperature T_S, at the concave surface 2, and arriving in surface 1:

$$Q_{1-1} = E_1\varphi_{1-2}\rho_2\varphi_{21} + E_1\varphi_{12}\rho_2\varphi_{2-2}\rho_2\varphi_{2-1} + E_1\varphi_{12}\rho_2\varphi_{2-2}\rho_2\varphi_{2-2}\rho_2\varphi_{2-1} + \cdots \tag{2.4.57}$$

The portions in equation (2.4.57) can be expressed as the sum of the terms of the infinite geometric progression with the common ratio $\phi_{2-2}\cdot\rho_2$, thus

$$Q_{1-1} = E_1\varphi_{1-2}\rho_2\varphi_{2-1}\frac{1}{1 - \varphi_{2-2}\rho_2} \tag{2.4.58}$$

Heat $Q_{2-1}(T_2)$ represents the portion $E_2\cdot\phi_{2-1}$ of emission E_2 of surface 2 which directly arrives at surface 1 and the portions in results of many reflections of emission E_2 at surface 2, arriving at surface 1:

$$Q_{2-1} = E_2\varphi_{2-1} + E_2\varphi_{2-2}\rho_2\varphi_{2-1} + E_2\varphi_{2-2}\rho_2\varphi_{2-2}\rho_2\varphi_{2-1} + \cdots \tag{2.4.59}$$

and

$$Q_{2-1} = E_2\varphi_{2-1}\frac{1}{1 - \varphi_{2-2}\rho_2} \tag{2.4.60}$$

Heat $Q_{3-1}(T_3)$ represents the portion $E_3\cdot\phi_{3-1}$ of emission E_3 of surface 3 which arrives at surface 1 as the direct radiation, and the portions as a result of many reflections of emission E_3 at surface 2:

$$Q_{3-1} = E_3\varphi_{3-1} + E_3\varphi_{3-1}\rho_2\varphi_{2-1} + E_3\varphi_{3-1}\rho_2\varphi_{2-2}\rho_2\varphi_{2-1}$$
$$+ E_3\varphi_{3-1}\rho_2\varphi_{2-2}\rho_2\varphi_{2-2}\rho_2\varphi_{2-1} + \cdots \tag{2.4.61}$$

and

$$Q_{3-1} = E_3\left(\varphi_{3-1} + \varphi_{3-2}\rho_2\varphi_{2-1}\frac{1}{1 - \varphi_{2-2}\rho_2}\right) \tag{2.4.62}$$

Third group losses can be calculated based either on the determination of the overall entropy growth used in the Guoy-Stodola equation (2.2.10) or determined from the exergy balance equation for the considered surface in the steady state. The latter method will be used.

To develop the analysis, an imagined heat source connected to each considered surface has to be assumed. For each surface, 2 or 3, there are the arriving emissions at three different temperatures to be taken into account. It is assumed that these emissions are absorbed by the surface and transferred as heat to the imagined heat source. Then immediately this heat is taken from the source to generate the emission of the surface at its temperature.

Thus the exergy balance equation for surface 2 can be interpreted as including: *the input exergy entering the surface 2 and represented by terms due to:*

- emission $\psi_S \cdot Q_{1-2}$ arriving from surface 1,
- emission $\psi_2 \cdot Q_{2-2}$ arriving from surface 2,
- emission $\psi_3 \cdot Q_{3-2}$ arriving from surface 3,
- heat $E_2 \cdot (1 - T_0/T_2)$ needed for emission of surface 2 and delivered from the heat source.

the output exergy leaving the surface 2 and represented by terms due to:

- emission $Q_{1-2} \cdot (1 - T_0/T_2)$ of surface 1 converted as the heat absorbed by the heat source,
- emission $Q_{2-2} \cdot (1 - T_0/T_2)$ of surface 2 converted as the heat absorbed by the heat source,
- emission $Q_{3-2} \cdot (1 - T_0/T_2)$ of surface 3 converted as the heat absorbed by the heat source,
- emission $E_2 \cdot \psi_2$ of surface 2.

The exergy balance equation can be written in form of the exergy loss δB_2 equal to the difference of the exergy input and output:

$$\delta B_2 = \psi_3 Q_{3-2} + \psi_S Q_{1-2} + \psi_2(Q_{2-2} - E_2) - (Q_{3-2} + Q_{1-2} + Q_{2-2} - E_2)\left(1 - \frac{T_0}{T_2}\right)$$

(2.4.63)

Analogically, the exergy loss δB_3 is:

$$\delta B_3 = \psi_2 Q_{2-3} + \psi_S Q_{1-3} + \psi_3(Q_{3-3} - E_3) - (Q_{1-3} + Q_{2-3} + Q_{3-3} - E_3)\left(1 - \frac{T_0}{T_3}\right)$$

(2.4.64)

Applying respectively again the formula for the sum of the terms of the infinite geometric progression, the values or required heat can be determined as follows.

Heat Q_{1-2}, at temperature T_S, is the sum of the portions of emission of surface 1 reaching surface 2, thus:

$$Q_{1-2} = E_1 \varphi_{1-2} \varepsilon_2 \frac{1}{1 - \varphi_{2-2} \rho_2} \tag{2.4.65}$$

Heat Q_{2-2} at temperature T_2, is the sum of the portions of emission of surface 2 reaching surface 2, thus:

$$Q_{2-2} = E_2 \varphi_{2-2} \varepsilon_2 \frac{1}{1 - \varphi_{2-2} \rho_2} \tag{2.4.66}$$

Heat Q_{3-2} at temperature T_3, is the sum of the portions of emission of surface 3 reaching surface 2, thus:

$$Q_{3-2} = E_3 \varphi_{3-2} \varepsilon_2 \frac{1}{1 - \varphi_{2-2} \rho_2} \tag{2.4.67}$$

Heat Q_{1-3}, at temperature T_S, which is the sum of the totally absorbed irradiation which reaches surface 3 at view factor φ_{1-3}, and the totally absorbed irradiation parts reflected from surface 2, can be determined as follows:

$$Q_{1-3} = I \varphi_{1-3} + I \varphi_{1-2} \rho_2 \varphi_{2-3} \frac{1}{1 - \varphi_{2-2} \rho_2} \tag{2.4.68}$$

Heat Q_{2-3} at temperature T_2, is the sum of the portions of emission of surface 2 reaching surface 3, thus:

$$Q_{2-3} = E_2 \varphi_{2-3} \frac{1}{1 - \varphi_{2-2} \rho_2} \tag{2.4.69}$$

Heat Q_{3-3} at temperature T_3, is the sum of the portions of emission of surface 3 reflected from surface 2 to surface 3:

$$Q_{3-3} = E_3 \varphi_{3-2} \rho_2 \varphi_{2-3} \frac{1}{1 - \varphi_{2-2} \rho_2} \tag{2.4.70}$$

As shown in the exergy part of the present paragraph even the assumption on the black surfaces 1 and 3 for exergetic consideration required far more equations in comparison to the respective energetic considerations developed for the system in which only one black surface 1 was assumed. Obviously, exergetic consideration of the system with only one black surface 1 would require developing of significantly more equations in comparison to the considerations presented in this paragraph.

Comparison of the energy and exergy balances for the considered SCPC, at the assumed $\varepsilon_3 = 1$, is presented in Table 2.4.4. In the both analyses the radiation escaping from the SCPC is estimated at a relatively high level, (energy: $68.34 + 4.43 = 72.77\%$ and exergy $57.069 + 0.026 + 0.132 = 57.226\%$). The energy analysis allows for splitting the escaping radiation loss according to the radiosity of surface 2 and 3, whereas

Table 2.4.4 The comparison of the energy and exergy balance terms for the considered SPC, $(\varepsilon_3 = 1)$, (from Petela, 2005).

Description	Energy expression	%	Exergy expression	%
Input:				
Irradiation (radiosity $J_1 = I$)	$(I = 1586.4\,\text{W})$	100	$(I \cdot \psi_S = 1483.1\,\text{W})$	100
Total		100		100
Output:				
Escaping radiosity from surface 2	$\beta\varphi_{21J2}$	68.34		
Escaping radiosity from surface 3	$\beta\varphi_{31J3}$	4.43		
Escaping fraction of emission $E_1, (E_1 = I)$			ξ_{B11}	57.069
Escaping fraction of emission E_2			ξ_{B12}	0.026
Escaping fraction of emission E_3			ξ_{B13}	0.132
Radiation irreversibility on surface 2			ξ_{B2}	20.433
Radiation irreversibility on surface 3			ξ_{B3}	20.528
Transfer of convective heat from surface 2	β_{Q2c}	13.41	ξ_{BQ2c}	0.619
Transfer of radiative heat from surface 2	β_{Q2r}	1.36	ξ_{BQ2r}	0.063
Transfer of convective heat from surface 3	β_{Q3c}	3.22	ξ_{BQ3c}	0.292
Irreversibility of transferred useful heat Q_{3u}			ξ_{BQ3u}	0.004
Useful heat Q_{3u} delivered to water	η	9.24	η_B	0.834
Total		100		100

the exergy analysis makes this split according to the temperature $(T_S,\ T_2$ and $T_3)$ of the escaping emissions.

According to the energy analysis the heat losses to the environment are relatively high; by convection 13.41% $(Q_{2,c})$, 3.22% $(Q_{3,c})$, and by radiation 1.36% $(Q_{3,r})$, whereas the exergy estimation of these losses is relatively very low; (0.619% and 0.292% and 0.063%), respectively.

As shown in Table 2.4.4, the energy analysis does not reveal any degradation losses at surfaces 2 and 3 and during transfer of the useful heat, in contrast to the exergy analysis which, respectively, estimates the first two losses relatively high 20.433%, 20.528%, and the third loss as a very small; 0.004%.

More details of the energy and exergy balances, as well as optimization possibilities, were analyzed by Petela (2010).

2.4.3 Solar chimney power plant

The considered solar chimney power plant (SCPP) is one of many possible examples of the power plant driven by solar radiation. The overall process is very complex and, up to date, only some selected aspects of the SCPP have been studied. The outline of the methodology of simplified exergy analysis applied to the SCPP and the possible different thermodynamic interpretations of processes occurring in the SCPP were developed including the following characteristic elements:

-   formulation of energy balance of total SCPP,
-   application of exergy balance for interpretation of component processes,

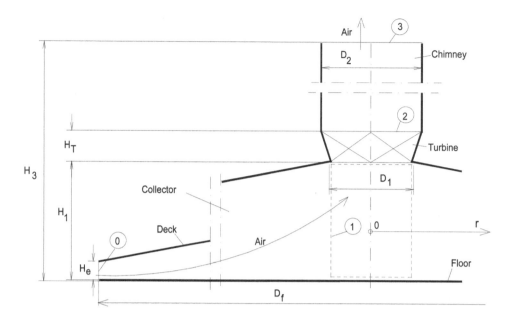

Figure 2.4.14 Scheme of the considered SCPP, (from Petela, 2009).

– application of eZergy balance for estimation of effect of gravity,
– involving exchange of radiation energy and exergy between chimney and deck,
– distinguishing the energy, exergy and eZergy losses to the environment and sky,
– proposing the convective-radiative effective temperature concept for the surfaces.

As shown in Figure 2.4.14, the considered SCPP consists of a circular greenhouse type collector and a tall chimney at its centre. Air flowing radially inwards under the collector deck is heated from the collector floor and deck, and through a turbine enters the chimney. A draft-driven environmental air (point 0) enters the collector through the gap of height H_e. The collector floor of diameter D_f is under the transparent deck which declines appropriately to ensure a constant radial cross-section area for the radially directed flow of the air. The assumption of constant cross-section area in the collector means that $\pi \cdot D_f \cdot H_e = \pi \cdot D_1 \cdot H_1 = \pi \cdot D_1^2/4$, and so, the assumed value H_e allows for calculation of the inlet turbine diameter $D_1 = (4 \cdot H_e \cdot D_f)^{0.5}$ and height $H_1 = D_1/4$. The collector floor preheats air from state 0 to state 1 (state 1 prevails in the zone denoted with a dashed line). Preheated air (state 1) expands in the turbine to state 2. The turbine inlet and outlet diameters are D_1 and D_2, respectively. The height of turbine is H_T; $(H_1 + H_T = H_2)$. Expanded air leaves the SCPP (at point 3) through the chimney at height H_3.

For the established geometrical parameters of the collector-turbine-chimney system, and for the constant thermodynamic input data, like solar radiation intensity and environment parameters, the system spontaneously self-models in response to the actual situation. This means that the buoyancy effect determines the flow rate of air through the system and all the air parameters; temperature and pressure along the air flow path.

The present study attempts to develop analysis of total SCPP process. The complexity of such a thermodynamic object enforced many simplifying assumptions and the main are:

(i) Floor has no heat loss to the soil; is perfectly insulated, and is perfectly black (emissivity $\varepsilon_f = 1$). Thus, there is no solar energy reflected from the floor.

(ii) Deck material is prepared in such a way that it is almost perfectly transparent for solar radiation (transmissivity $\tau_d = 0.95$) and the remaining part (5%) of solar radiation arriving at the deck is reflected. However, the deck material absorbs perfectly (absoptivity $\alpha = 1$) low temperature radiation from the floor. Thus, consideration of multi-reflected radiation fluxes is simplified. In addition, the deck is thin enough that heat conducted through the deck occurs at zero temperature gradient.

(iii) A certain "effective temperature" T_{eff} of floor or deck is applied, which expresses potential to the heat transfer by conduction and radiation:

$$A(k_{average}T_{eff} + \sigma T_{eff}^4) = \int_A k_{local}T \, dA + \sigma \int_A T^4 dA \qquad (2.4.71)$$

where $k_{average}$ and k_{local} are the average and local convective heat transfer coefficients, respectively, σ is the Boltzmann constant for black radiation, A is the surface area and T is the local surface temperature.

(iv) The chimney material is perfectly black. The chimney wall is thin thus there is no temperature gradient along the wall thickness and both sides of chimney (inner and outer has the same temperature (average) constant along the chimney height.

(v) Distribution of air temperature is represented by a certain effective temperature defined according to equation (2.4.71), however with excluded radiative heat transfer.

(vi) Air is considered as an ideal gas whose parameters fulfill the state equation; $p = \rho \cdot R \cdot T$, and the specific heat is assumed constant, (average, not varying with temperature).

(vii) Air is almost perfectly transparent for radiation, (transmissivity $\tau_a \approx 1$ and emissivity $\varepsilon_a \approx 0$). Air can exchange heat only by convection and conduction.

(viii) Air flow in the whole SCPP is frictionless. According to investigation by von Backström and Fluri (2006), the relative air pressure drop r_T during expansion in turbine: "for maximum fluid power, the optimum ratio" is 2/3, thus

$$\frac{p_1 - p_2}{p_1 - p_3} \equiv r_T = \frac{2}{3} \qquad (2.4.72)$$

The exemplary distribution of environment pressure and the pressure of air along its flow within the SCPP is shown in Figure 2.4.15. The air pressure of environment (solid thin line) drops from p_0, at the zero level, to p_3, at the level H_3 of chimney inlet. The air pressure inside the SCPP drops from p_0 to p_1 at

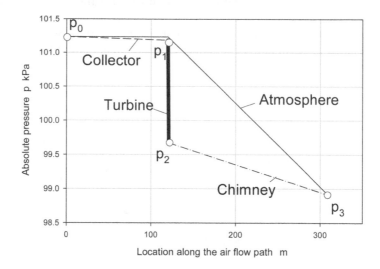

Figure 2.4.15 Distribution of the absolute pressure in the considered SCPP, (from Petela, 2009).

the collector outlet (dashed line) and it is assumed that the same pressure p_1 prevails also at the inlet to the turbine. Then, within the turbine, air pressure (thick solid line) drops from p_1 to p_2 during adiabatic (isentropic) expansion generating power. Air from the turbine flows upward and its pressure (dotted line) achieves value p_3 at the chimney exit.

(ix) Using average values of gravitational acceleration and air density along the height H_3:

$$p_3 = p_0 - \frac{g_0 + g_{H3}}{2} \frac{\rho_0 + \rho_{H3}}{2} H_3 \qquad (2.4.73)$$

where the following approximations, used by Petela (2008), (Table 1), were applied: $g_{H3} = g_0 - 3.086 \cdot 10^{-6} \cdot H_3$ and $\rho_{H3} = \rho_0 - 9.973 \cdot 10^{-5} \cdot H_3$. At the Earth's surface the atmospheric pressure $p_0 = 101.235$ kPa and gravity acceleration $g_0 = 9.81$ m/s^2.

(x) Momentum conservation equation for the air flow within collector is derived as:

$$p_0 - p_1 = \rho_{a1} w_1^2 \qquad (2.4.74)$$

where ρ_{a1} and w_1 are the density and flow velocity of air at point 1.

(xi) Deck and chimney radiate to the space of sky temperature T_{sky} determined by the Swinbank (1963) formula: $T_{sky} = 0.0552 \cdot T_0^{1.5}$ for a clear sky.

(xii) In order to obtain a fair comparison basis, the reference state for calculation of energy is the same as for exergy: environment temperature $T_0 = 288.14$ K, (15 C), and environment pressure $p_0 = 101.235$ kPa.

Energy analysis is based on the energy conservation equations. The energies E are used in six equations written successively for: floor surface, air in collector, collector (including floor, air and deck), turbine, chimney and chimney surface:

$$E_{S\text{-}f} = E_{f\text{-}a} + E_{f\text{-}d} \tag{2.4.75}$$

$$E_{f\text{-}a} + E_{d\text{-}a} = E_{a1} + E_{w1} + E_{p1} \tag{2.4.76}$$

$$E_{S\text{-}f} = E_{a1} + E_{w1} + E_{p1} + E_{d\text{-}sky} + E_{d\text{-}0} + E_{d\text{-}ch} \tag{2.4.77}$$

$$E_{a1} + E_{w1} + E_{p1} = E_{a2} + E_{w2} + E_{p2} + E_P \tag{2.4.78}$$

$$E_{a2} + E_{w2} + E_{p2} + E_{d\text{-}ch} = E_{a3} + E_{w3} + E_{p3} + E_{ch\text{-}0} + E_{ch\text{-}sky} + E_{ch\text{-}gr} \tag{2.4.79}$$

$$E_{a\text{-}ch} + E_{d\text{-}ch} = E_{ch\text{-}0} + E_{d\text{-}sky} + E_{ch\text{-}gr} \tag{2.4.80}$$

Energies E have the following subscripts:

S-f – solar radiation arriving at the floor,
f-a – convection heat from floor to air,
f-d – energy exchanged by radiation between floor and deck,
d-a – convection heat from deck to air,
d-sky – energy exchanged by radiation between deck and sky,
d-0 – convection heat from deck to atmosphere,
d-ch – energy exchanged by radiation between deck and chimney,
ch-0 – convection heat from chimney surface to atmosphere,
ch-sky – energy exchanged by radiation between chimney surface and sky,
ch-gr – energy exchanged by radiation between chimney surface and ground,
a-ch – heat transferred from chimney air to the chimney surface,
a1, a2, a3 – enthalpy of air at points 1, 2 and 3,
w1, w2, w3 – kinetic energy due to the air flow velocity w_1, w_2 and w_3,
p1, p2, p3 – potential energy of air at points 1, 2 and 3,
P – turbine power.

Kinetic energies are calculated as $E_w = m \cdot w^2/2$, where m is the air mass flow rate; $m = 0.25 \cdot \pi \cdot D_1^2 \cdot w_1 \cdot \rho_{a1}$. Enthalpy of air is $E_a = m \cdot c_p \cdot (T_a - T_0)$ where c_p is the specific heat of air at constant pressure.

The potential energy of the considered air, at its constant density ρ, depends on the altitudinal variation of atmospheric air density and gravity acceleration. The solution of the differential formula (2.2.16) on potential energy E_p, J/kg, equal to potential exergy B_p, is determined by Petela (2010):

$$E_p = m\left\{ -\frac{1}{a_4 \rho}\left[\frac{a_2}{6a_4}(\rho - a_3)^3 + \frac{a_1}{2}(\rho - a_3)^2 \right] \right\} \tag{a}$$

where the constant values are $a_1 = 9.7807$ m/s^2, $a_2 = -3.086 \times 10^{-6}$ 1/s^2, $a_3 = 1.217$ kg/m^3, and $a_4 = -9.973 \times 10^{-5}$ kg/m^4.

Total solar energy received by the floor is:

$$E_{S\text{-}f} = \tau_d \varepsilon_f S A_d \tag{b}$$

where S, W/m^2, is the solar radiosity at the Earth's surface, τ_d is the transmissivity of deck, and $\varepsilon_f = 1$ is the floor emissivity.

Energy exchanged by radiation between deck and chimney:

$$E_{d\text{-}ch} = \varepsilon_d \varphi_{d\text{-}ch} \frac{\pi}{4} [D_f^2 - (c_D D_2)^2] \sigma (T_{d,eff}^4 - T_{ch}^4) \tag{c}$$

where σ is the Boltzmann constant for black radiation, $T_{d,eff}$ is the effective temperature of the deck, and c_D is the factor to account on thickness of the chimney wall. The view factor $\varphi_{d\text{-}ch}$ can be calculated from reciprocity relation:

$$\varphi_{d\text{-}ch} \frac{\pi}{4} [D_f^2 - (c_D D_2)^2] = \varphi_{ch\text{-}d} \pi c_D D_2 (H_3 - H_2) \tag{d}$$

It can be derived that $\varphi_{ch\text{-}d} = 0.5 \cdot (90 - \beta)/90$ where the angle β is determined by $\tan\beta = 2 \cdot H_3/D_f$.

Energy exchanged by radiation between floor and deck:

$$E_{f\text{-}d} = A_d \sigma (T_{f,eff}^4 - T_{d,eff}^4) \tag{e}$$

where $T_{f,eff}$ is the effective temperature of the floor and surface area $A_d = \pi \cdot (D_f^2 - D_1^2)/4$.

The following formulae are applied for convection heat transfer from:

floor to air:

$$E_{f\text{-}a} = A_d k_{f\text{-}a} (T_{f,eff} - T_{a,eff}) \tag{f}$$

deck to air:

$$E_{d\text{-}a} = A_d k_{d\text{-}a} (T_{d,eff} - T_{a,eff}) \tag{g}$$

deck to environment:

$$E_{d\text{-}0} = A_d k_{d\text{-}0} (T_{d,eff} - T_0) \tag{h}$$

chimney to environment:

$$E_{ch\text{-}0} = A_{ch} k_{ch\text{-}0} (T_{ch} - T_0) \tag{i}$$

and chimney air to chimney wall:

$$E_{a\text{-}ch} = \pi D_2 (H_3 - H_2) k_{a\text{-}ch} \left(\frac{T_{a,2} + T_{a,3}}{2} - T_{ch} \right) \tag{j}$$

where k is the respective coefficient and the chimney surface $A_{ch} = \pi \cdot c_D \cdot D_2 \cdot (H_3 - H_2)$. It is assumed that the coefficient $k_{a\text{-}ch} = Nu \cdot \lambda/D_2$ where $\lambda = 0.0267\,W/(m\,K)$ is thermal conductivity of air and the Nusselt number $Nu = 0.023 \cdot Re^{0.8} \cdot Pr^{0.4}$ and where the Prandtl number for air is $Pr = 0.7$ and the Reynolds number $Re = w_2 \cdot D_2/\nu$, (kinematic viscosity coefficient for air $\nu = 1.6 \cdot 10^{-5}\,m^2/s$).

The coefficient $k_{f\text{-}a}$ is determined in a similar way. Although the air flow is driven by buoyancy effect, the forced convection mechanism of the air flow is assumed. Thus the calculations are based on the Reynolds number instead of the Grashof number. In calculations the average flow velocity of air is assumed. The effective diameter D_{eff} for the air flow can be assumed as the average ratio of the respective flow cross-section area A_1 multiplied by four, to the respective average perimeter lengths; $D_{eff} = (4/\pi) \cdot A_1/(D_f + D_2)$. It was assumed that $k_{a\text{-}d} = k_{f\text{-}a}$.

The following formulae are applied for energy exchange by radiation between:

floor and deck:

$$E_{f\text{-}d} = A_d \sigma(T^4_{f,eff} - T^4_{d,eff}) \tag{k}$$

deck and chimney:

$$E_{d\text{-}ch} = \varphi_{d\text{-}ch} A_d \sigma(T^4_{d,eff} - T^4_{ch}) \tag{l}$$

deck and sky:

$$E_{d\text{-}sky} = \varphi_{d\text{-}sky} A_d \sigma(T^4_{d,eff} - T^4_{sky}) \tag{m}$$

chimney and sky:

$$E_{ch\text{-}sky} = \varphi_{ch\text{-}sky} A_{ch} \sigma(T^4_{ch} - T^4_{sky}) \tag{n}$$

chimney and ground beyond the floor:

$$E_{ch\text{-}gr} = \varphi_{ch\text{-}gr} A_{ch} \sigma(T^4_{ch} - T^4_{gr}) \tag{o}$$

where the view factors fulfill the following relations:

$$\varphi_{d\text{-}sky} + \varphi_{d\text{-}ch} = 1 \tag{p}$$

$$\varphi_{ch\text{-}sky} + \varphi_{ch\text{-}d} + \varphi_{ch\text{-}gr} = 1 \tag{q}$$

The view factors $\varphi_{ch\text{-}d}$ and $\varphi_{d\text{-}ch}$ are determined based on equation (d), whereas the configuration of chimney relative to sky determines view factor $\varphi_{ch\text{-}sky} = 0.5$.

Calculation of temperature $T_{a,2}$ is based on the equation for the isentropic expansion in the turbine at assumed isentropic exponent κ for air and the internal efficiency of the turbine η_T. Conversion of the energy of air into electric power occurs at an overall efficiency η_o which would include additionally mechanical and electric efficiencies of the turbine-generator unit.

Additionally it was assumed that the air temperature distribution in the collector is linear and thus $T_{a,eff} = (T_0 + T_{a1})/2$. The diameter ratio $D_1/D_2 = 0.95$. The air temperature drop in the chimney can be tentatively estimated as proportional to the chimney surface and inversely proportional to the air mass rate; $T_{a,2} - T_{a,3} \approx 0.154 \cdot D_2 \cdot H_3/m$.

The presented mathematical model of SCPP is illustrated with exemplary computation results in which, for comparison to the 36-kW pilot SCPP plant in Manzares, near

Madrid, Spain, the floor diameter is $D_f = 240$ m and the chimney height $H_3 = 195$ m. Other data are as follows:

$S = 800$ W/m^2 $T_{gr} = T_0$ $c_D = 1.015$ $c_p = 1000$ J/(kg K)
$\kappa = 1.4$ $\eta_T = 0.7$ $R = 287.04$ J/(kg K) $H_T = 1$ m
$k_{ch-0} = 7$ W/(m^2 K) $k_{d-0} = 5$ W/(m^2 K) $H_e = 0.3$ m

The computation results are shown in the bands diagram (Figure 2.4.16) in which the values e are expressed in %, and the solar radiation energy arriving at the deck $E_S = 39.05$ MW is assumed as 100%. This amount, reduced by the 5% reflection, is distributed between five SCPP components; collector air, floor, deck, turbine and chimney. The floor (black body) fully absorbs the solar radiation (95.00%) transmitted through the deck and converts this radiation energy to the energy at the level of temperature $T_{f,eff}$. Part of this energy ($e_{f-d} = 77.19\%$) radiates to the deck and the rest $e_{f-a} = 17.81\%$ is transferred by convection to heated air in the collector. The power performed by the turbine is relatively small ($E_P = 0.23$ MW) mostly due to the small mass flow rate of air ($m = 276$ kg/s) and due to small pressure drop during the air expansion. The percentage power of the turbine $e_P = 0.64\%$ represents the energy efficiency η_E of the SCPP. The exhausted energy (enthalpy) of air from chimney is $e_{a3} = 20.75\%$ whereas the exhausted potential and kinetic energies are small; $e_{p3} = 0.52\%$ and $e_{w3} = 3.87 \times 10^{-4}\%$, respectively. The other SCPP energy losses are by radiation and convection heat transferred from deck and chimney to the sky and environment. Solar energy reflected from the deck is assumed $e_R = 5.00\%$.

Exergy analysis is based on the exergy balance equations. Exergy B in these equations has the subscripts respectively to E in equations (2.4.76)–(2.4.80) for energy analysis. The five separate exergy equations can be written for floor, deck, air in collector, turbine and chimney. The exergy equations are analogical to energy equations and differ by the additional members, ΔB, representing the respective irreversible exergy losses:

$$B_{S-f} = B_{f-a} + B_{f-d} + \Delta B_f \tag{2.4.81}$$

$$B_{f-d} = B_{d-a} + B_{d-sky} + B_{d-0} + B_{d-ch} + \Delta B_d \tag{2.4.82}$$

$$B_{f-a} + B_{d-a} = B_{a1} + B_{w1} + B_{p1} + \Delta B_a \tag{2.4.83}$$

$$B_{a1} + B_{w1} + B_{p1} = B_{a2} + B_{w2} + B_{p2} + B_P + \Delta B_T \tag{2.4.84}$$

$$B_{a2} + B_{w2} + B_{p2} + B_{d-ch} = B_{a3} + B_{w3} + B_{p3} + B_{ch-0} + B_{ch-sky} + B_{ch-gr} + \Delta B_{ch} \tag{2.4.85}$$

Exergy of solar radiation can be estimated for the radiation temperature slightly smaller than 6000 K, e.g., $B_S \approx \psi \cdot E_S$, where $\psi \approx 0.9$.

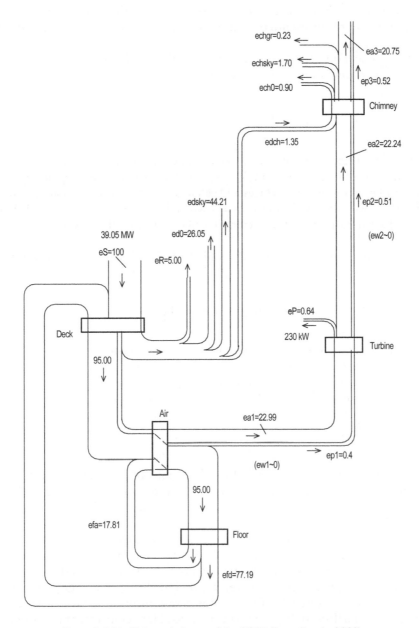

Figure 2.4.16   ENergy balance of the SCPP, (from Petela, 2009).

Generally, in determined geometrical configuration, the radiation exergy B of a surface at its temperature T, emissivity ε and surface area A, is determined based on formulas e.g., (2.2.33), (2.2.40) and (2.2.63) applied for the whole area A:

$$B = \varphi A \varepsilon \frac{\sigma}{3}(3T^4 + T_0^4 - T_0 T^3) \qquad (2.4.86)$$

where φ is the view factor accounting for geometrical configuration of the considered surface in relation to an eventual surface at which the considered radiation would arrive.

Based on equation (2.4.86) the exergy of radiation $B_{x\text{-}y}$ exchanged between any two different surfaces at different temperature T_x and T_y can be determined according Petela (2010):

$$B_{x\text{-}y} = B_x - B_y = \varphi_{x\text{-}y} A_x \varepsilon_{x\text{-}y} \frac{\sigma}{3}[3(T_x^4 - T_y^4) - 4T_0(T_x^3 - T_y^3)] \tag{2.4.87}$$

where A_x is the surface area of one of two considered surfaces, φ_{x-y} is the view factor for configuration of surfaces x and y, ε_{x-y} is the effective emissivity depending on emissivities ε_x and ε_y of respective surfaces and calculated as for radiation energy exchange. The effective emissivity simplifies to $\varepsilon_{x\text{-}y} = 1$ when the emissivities $\varepsilon_x = \varepsilon_y = 1$. Formula (2.4.87) is used appropriately for calculations of the five radiation exergies: $B_{f\text{-}d}$, $B_{d\text{-}sky}$, $B_{d\text{-}ch}$, $B_{ch\text{-}sky}$, and $B_{ch\text{-}gr}$.

The physical exergy of air (B_{a1}, B_{a2} and B_{a3} in W), is calculated for the air mass flow rate m with use of formula (2.2.15) for specific exergy, (J/kg):

$$B_a = m\left[c_p(T_a - T_0) - T_0\left(c_p \ln \frac{T_a}{T_0} - R \ln \frac{p}{p_0}\right)\right] \tag{s}$$

where c_p and R are the specific heat and individual gas constant. Obviously, exergy of air entering the collector is zero, ($B_{a0} = 0$), because air is taken from environment.

Exergy B of convective heat E transferred from a surface at temperature T to air (environmental or heated) is calculated based on formula (2.2.4):

$$B = E\left(1 - \frac{T_0}{T}\right) \tag{t}$$

Formula (t) is used appropriately for calculations of the four exergies $B_{f\text{-}a}$, $B_{d\text{-}a}$, B_{d-0}, and B_{ch-0}. Potential exergies of air are equal to potential energies, ($B_{p1} = E_{p1}$, $B_{p2} = E_{p2}$ and $B_{p3} = E_{p3}$). Kinetic exergies of air are equal to kinetic energies, ($B_{w1} = E_{w1}$, $B_{w2} = E_{w2}$ and $B_{w3} = E_{w3}$).

The computation results are shown in the bands diagram (Figure 2.4.17). The solar radiation exergy arriving at the deck $B_S = 32.41$ MW, assumed as 100%, is distributed between five SCPP components; collector air, floor, deck, turbine and chimney. In the diagram the exergy streams B, W, are represented by their percentage values b related to the solar radiation exergy B_S. Exergy considerations disclose large degradation of solar radiation. The floor fully absorbs the received high temperature radiation exergy and converts it to the exergy at the lower temperature $T_{f,eff}$. Part of this $T_{f,eff}$ exergy ($b_{f\text{-}d} = 17.24\%$) radiates to the deck and another part $b_{f\text{-}a} = 5.10\%$ is transferred by convection to heated air in the collector. The remaining large part ($\Delta b_f = 72.16\%$) is lost during irreversible processes of absorption and emission at the floor surface.

The power B_p performed by turbine is the same as in the energy balance; $B_P = E_P = 0.23$ MW. The percentage power of the turbine $b_P = 0.70\%$ represents the exergy efficiency η_B of the SCPP. Exergy efficiency is slightly higher than the energy efficiency because the same power is related to the radiation exergy which is smaller than the

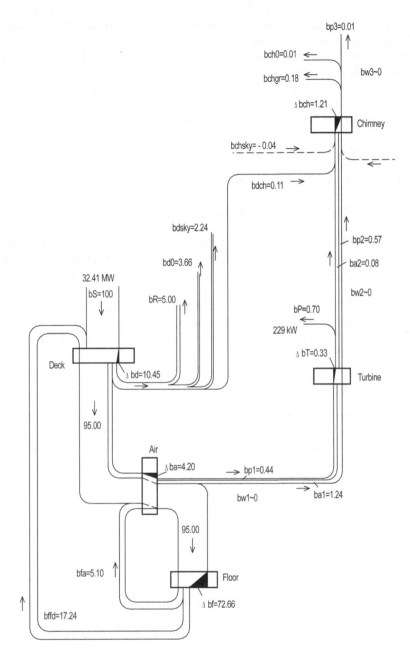

Figure 2.4.17 EXergy balance of the SCPP, (from Petela, 2009).

radiation energy. The exhausted exergy of air from the chimney is negative $b_{a3} = -0.61\%$ whereas the exhausted potential and kinetic exergies are small; $b_{p3} = 0.01\%$ and $b_{w3} = e_{w3}$, respectively. The SCPP is losing the exergy due to irreversibility and by radiation and convection heat transferred from deck and chimney to the sky and environment. Solar exergy reflected from the deck is $b_R = e_R = 5\%$. The possibility of negative value of physical exergy (b_a) of air was discussed in paragraph 2.2.3.1.

EZergy analysis uses mechanical exergy component for air. Exergy balance equations for floor and deck do not contain terms for substance. These two equations, (2.4.81) and (2.4.82), remain unchanged because they contain only terms of the radiation and convection heat for which the gravitational effect is not considered. However, the exergy balance equations, (2.4.83), (2.4.84) and (2.4.85), considered for heating air in collector, turbine and chimney, are modified by adding the gravity input.

$$G_a + B_{f\text{-}a} = Z_{1a} + B_{w1} + B_{a-d} + \Delta B_a \tag{2.4.88}$$

$$G_T + Z_{1a} + B_{w1} = Z_{2a} + B_{w2} + B_P + \Delta B_T \tag{2.4.89}$$

$$G_{ch} + Z_{2a} + B_{w2} + B_{f-ch} = Z_{3a} + B_{w3} + B_{cv} + B_{ch-0} + \Delta B_{ch} \tag{2.4.90}$$

where G_a, G_T and G_{ch} are the gravity inputs in eZergy balance for the collector air, turbine and chimney, respectively. Note that in eZergy balance equation the potential exergy does not appear as a separate term because this exergy is interpreted by eZergy of substance.

To more easily distinguish it from the traditional exergy (B or b), the eZergy is denoted by Z, W, or $z,\%$. Denotations of other exergy magnitudes remain unchanged because their values are unchanged. However, eZergy of air generally differs from exergy of air; $Z_a \geq B_a$:

$$Z_a = \max(B_p + B_H, B_a) \tag{2.4.91}$$

where B_p is the potential exergy, ($B_p = E_p$), B_a is the traditional physical exergy of air calculated from formula (s). Magnitude B_H is the physical exergy calculated also based on equation (s), however, for the environment parameters (temperature T_H and pressure p_H) prevailing at the altitude H:

$$B_H = m\left[c_p(T_a - T_H) - T_H\left(c_p \ln \frac{T_a}{T_H} - R \ln \frac{p}{p_H}\right)\right] \tag{2.4.92}$$

According to the interpolation formulae given by Petela (2009b):

$$H = 1.215485 \cdot 10^6 - 1.214 \cdot 10^6 \rho_a^{6.02353 \cdot 10^{-3}} \tag{2.4.93}$$

and the interpolated atmospheric parameters at altitude H:

$$T_H = 288.16 - 0.0093H + 3.2739 \cdot 10^{-7}H^2 - 2.9861 \cdot 10^{-12}H^3 \tag{2.4.94}$$

$$p_H = 101235e^{1.322 \cdot 10^{-4}H} \tag{2.4.95}$$

The computation results of exergy balances with use of eZergy are shown in the bands diagram (Figure 2.4.18). The solar radiation exergy $B_S = Z_S = 32.41\,\text{MW}$, assumed as 100%, arriving at the deck is distributed between five SCPP components in the case of using substance eZergy. The eZergy streams Z, W, are represented by

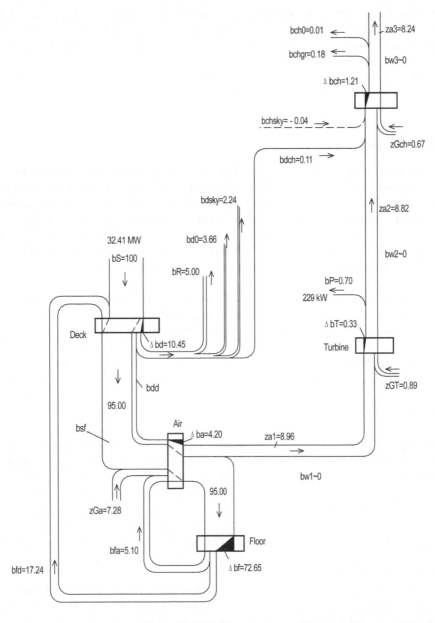

Figure 2.4.18 EZergy balance of the SCPP, (from Petela, 2009).

their percentage values z, related to the solar radiation exergy Z_S. The part of the diagram (Fig. 2.4.18) related to floor and deck is the same as in Figure 2.4.17, because substance does not appear in balances of the floor and deck. Also degradations of solar radiation and convective heat are the same like shown in Figure 2.4.17 and also the power performed by turbine is unchanged (0.23 MW). The percentage power of the turbine $z_P = b_P = 0.70\%$ represents the eZergy efficiency of SCPP. Specificity of the diagram in Figure 2.4.18 is showing the relatively large ezergies of air $z_{a1} = 8.96\%$, $z_{a2} = 8.82\%$ and $z_{a3} = 8.24\%$. As a result, the gravity inputs are: $z_{Ga} = 7.28\%$ for the air in collector, smaller for turbine $z_{GT} = 0.89\%$ and the smallest for chimney $z_{Gch} = 0.67\%$

More details of the energy and exergy balances, especially for different input parameters, were analyzed by Petela (2010).

2.4.4 Photosynthesis

The very simplified threefold study of photosynthesis process was developed by Petela (2008a) including a) an energy analysis (the energy conservation equation developed to estimate the energy effects of the process); b) entropy analysis (the changes of entropy were used to estimate the irreversibility of the component processes); and c) exergy analysis (developed for thermodynamic evaluation of involved matters). In the present paragraph only the outline of energy (a) and exergy (c) analyses is discussed based on engineering thermodynamics to propose the methodology of exergetic consideration of photosynthesis.

Photosynthesis is the process by which the energy of the photosynthetically active radiation (PAR), i.e. within the wavelength range 400–700 nm, is used to split gaseous carbon dioxide and liquid water and recombine them into gaseous oxygen and a sugar called glucose. The photo-chemical reaction of photosynthesis cannot occur without the presence of chlorophyll and is a complex two stage process. For the present analyses only the following endothermic overall reaction of the photosynthesis is considered:

$$6H_2O + 6CO_2 \rightarrow C_6H_{12}O_6 + 6O_2 \qquad (2.4.96)$$

A simplified scheme of the considered system shown in Figure 2.4.19 is defined by the system boundary and contains a leaf surface layer in which biomass is created at temperature T. Diffusion of gaseous substances and convective heat transfer occurs through the gaseous boundary layer at the leaf surface. The boundary layer is not considered for radiation fluxes because it is assumed that air in this layer is transparent to radiation. The leaf surface absorbs part of the incident solar radiation and emits its own leaf radiation of temperature T. The absorbed radiation is expended on the metabolism processes of the leaf and on maintaining the leaf temperature T above the environment temperature T_0.

Liquid water, at temperature T, from the leaf body enters the considered system. A relatively small amount of this water is used for the assimilation of the CO_2, which diffuses into the leaf from the external environment. The large excess of water is transpired in the form of vapor diffusing from the leaf to environment. Oxygen produced during the photosynthesis also diffuses into the environment. The water vapor and oxygen exiting the boundary layer, as well as CO_2 entering the boundary layer, have environment temperature T_0 at the respective environment mole concentrations $z_{H2O,0}$, $z_{O2,0}$, and $z_{CO2,0}$.

Only the chemical and physical components of the energy and exergy of the substances are considered. Also only the overall effects described by equation (2.4.96),

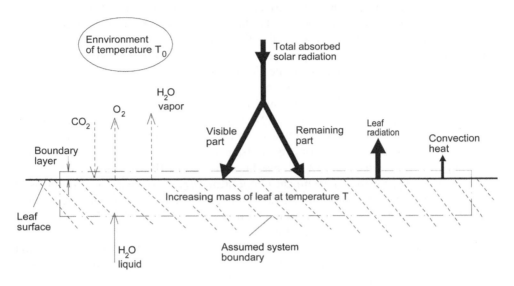

Figure 2.4.19 Simplified scheme of substances and radiation fluxes in photosynthesis, (from Petela, 2008a).

and the matter fluxes observed around the leaf (Figure 2.4.19), are analyzed based on the following *main assumptions*:

(i)   The considered area is a conventional horizontal unitary $(1\,m^2)$ surface of the leaf in a certain instant at the determined constant conditions during which the input is equal to output and change in the system.

(ii) To determine the actual energy arriving at the leaf surface, the solar radiation energy of the spectrum measured at the highest layer of atmosphere, is multiplied by a certain weakening factor γ; the larger the γ, the smaller the weakening, $(\gamma \leq 1)$. In the proposed simplified model the weakening factor is not studied, nor is it concretely defined. Only certain possible values of γ are used in calculations. The radiation arriving at the leaf could be accurately determined by measuring of the radiation spectrum directly at the leaf surface so the factor γ would be given. However the purpose of the present considerations is analysis of photosynthesis for a given γ.

(iii) Cloudy situations are not analyzed.

(iv) The solar radiation arrives, directly from the Sun in the zenith (solar radiation is perpendicular to the horizontal surface of the considered leaf), within the solid angle determined by the diameter of the Sun and its distance from the Earth. The reduced effect due to the non-perpendicular radiation could be expressed e.g. by an appropriate value of factor γ.

(v)   Sufficient chlorophyll necessary for the photosynthesis process is available. Any change in the chlorophyll concentration during photosynthesis, and the thermodynamic effect of such change, are neglected.

(vi)   The surroundings of the considered leaf consists only of the surfaces at temperature T_0 and of absorptivity $\alpha_0 = 1$, (the Sun's surface area is neglected as being seen within a relatively very small solid angle).

(vii)   Mixtures of substances in the system are ideal; the components do not mutually interact. Therefore, mixture properties are the respective sums of the component properties. For example, the biomass contained within the leaf structure is an ideal solution of solid $C_6H_{12}O_6$ and water.

(viii) The environment air contains only N_2, O_2, CO_2 and H_2O. The dry environment air contains 79.07% N_2, 20.9% O_2, and 0.03% CO_2. The sum of all mole fractions of the air components is: $z_{N_2,0} + z_{O_2,0} + z_{CO_2,0} + z_{H_2O,0} = 1$, where $z_{H2O,0}$ is determined by the relative humidity φ_0 of the air and the saturation pressure p_{s0} for the environment temperature T_0: $z_{H2O,0} = \varphi_0 \cdot p_{s0}$. In terms of radiation it is assumed that the diatomic gases have transmissivity 100% and the concentration of the triatomic components (CO_2 and H_2O) is relatively small and they also have transmissivity 100%.

(ix)   The considered leaf has uniform temperature T; there is no heat transfer within the considered surface layer. According to the evaluation by Jørgensen and Svirezhev (2004) there is a several degree difference ΔT between the leaf temperature and environment temperature.

(x)   The liquid water required for photosynthesis is available in sufficient amount.

(xi)   In the considered conditions the rate of sugar production is limited by the effectiveness of diffusion of gases; not by the reaction kinetics depending on temperature.

(xii)   The generated sugar has only chemical exergy b_{ch} resulting from chemical reaction (2.4.96). The component of the sugar exergy gained as a result of ordering (structure of the biomass in according to genetic plan) is neglected.

It could be expected that some simplifications may not qualitatively affect the final conclusions although the quantitative results could be affected remarkably.

The considered substances in the system (Fig. 2.4.19) are gaseous CO_2, O_2, H_2O (assumed to be ideal), liquid water and the leaf substance (biomass). The enthalpies of the gases are zero because at the system boundary they have environment temperature T_0. However, for the liquid H_2O the water vapor is the reference phase. Therefore, the enthalpy of liquid water is equal to the sum of the negative value of the latent heat of vaporization at temperature T_0 and of the temperature difference $\Delta T = T - T_0$ multiplied by the specific heat of water.

The generated biomass is assumed to be a mixture of liquid water and sugar. The enthalpy of this mixture is calculated as the sum of the component enthalpies. (For liquid and solid bodies the enthalpy is practically equal to the internal energy). The enthalpy of sugar is the sum of the physical enthalpy and devaluation enthalpy. The physical enthalpy of sugar is calculated for the temperatures range from T_0 to T at the constant specific heat $c_{SU} = 430.2$ kJ/(kmol K).

The values of devaluation enthalpies d_n for the standard parameters (pressure $p_n = 101.325$ kPa and temperature $T_n = 298.15$ K) are tabulated by Szargut et al. (1988). The varying of the devaluation enthalpy within the considered temperatures range is negligible.

Due to lack of data, the devaluation enthalpy $d_{n,C_6H_{12}O_6} \equiv d_{n,SU} = 2,529,590 \, kJ/$ (kmol of sugar) is assumed as for the α-D-galactose, predicting that the devaluation enthalpy of the real substance generated in the leaf differs insignificantly. Such an assumption can be supported by the fact that the devaluation enthalpies tabulated for the substances of the same chemical formula (α-D-galactose and L-sorbose), differ insignificantly.

The physical b_{ph} and chemical b_{ch} exergy, are taken into account to calculate the total exergy $b = b_{ch} + b_{ph}$ of any substance. The exergy of each gas (CO_2, H_2O and O_2) is zero because in the considered case their states are in full equilibrium with environment.

The exergy of liquid water b_w, kJ/kmol, is the sum of the physical part $b_{w,ph}$ and chemical part $b_{w,ch}$, where $b_{w,ch} = R \cdot T_0 \cdot \ln(1/\varphi_0)$. Using the Szargut and Petela (1965b) diagrams the interpolation formula for calculation of the physical exergy $b_{w,ph}$ of liquid water is $b_{w,ph} = a + b \cdot t + c \cdot t^2$, where $a = -23.22 + 2.718 \cdot t_0 + 0.0675 \cdot t_0^2$, $b = 2.689 - 0.5787 \cdot t_0 + 0.00767 \cdot t_0^2$, and $c = 0.117 - 1.05 \cdot 10^{-3} \cdot t_0 + 2.7 \cdot 10^{-4} \cdot t_0^2 - 7.5 \cdot 10^{-6} \cdot t_0^3$ and where $t_0 = T_0 - 273$.

The exergy of the generated biomass is the sum of the exergy of the components (liquid water and sugar). The specific chemical exergy of sugar is determined based on the standard tabulated value $b_{n,SU} = 2,942,570 \, kJ/kmol$, to which the correction on the difference of temperatures T_n and T_0 is added according to formula (2.2.20). The specific physical exergy $b_{ph,SU}$ is determined from formula (2.2.21).

The radiation arriving at the leaf surface from the Sun is recognized as non-polarized and uniformly propagating within the solid angle under which the Sun is seen from the Earth. The radiosity j_S of such solar radiation of the real spectrum as function of wavelength λ is:

$$j_S = 2 \left(\int_\beta \int_\varphi \cos \beta \sin \beta \, d\beta \, d\varphi \right) \int_\lambda i_{0,\lambda} d\lambda \qquad (2.4.97)$$

The double integral in the bracket of equation (2.4.97) was calculated in Example 2.4.1.1; formula (2.4.1), and if the single integral in equation (2.4.97) is presented in a numerical form, then:

$$j_S = 4.329 \cdot 10^{-5} \pi \sum_n (i_{0,\lambda} \Delta\lambda)_n \qquad (2.4.98)$$

where $i_{0,\lambda}$ is the measured monochromatic intensity of radiation depending on the wave length λ, and n is the successive number of the wavelength interval within the considered wavelength range. For the 0 to ∞ wavelength range $j_S = 1.3679 \, kW/m^2$, as shown in Example 2.4.1.2. For the PAR arriving only within the wavelengths range (400–700 nm) the radiosity of the PAR calculated from equation (2.4.98), is $j_V = 0.5446 \, kW/m^2$.

The energy emission of the leaf surface propagates in all directions of hemisphere and it is assumed that the radiation of the environment arrives at the leaf surface from all directions of the hemisphere. Therefore, the energy e_L exchanged between the leaf and the environment is: $e_L = \alpha_{L,a} \cdot \sigma \cdot (T^4 - T_0^4)$, where $\alpha_{L,a}$ is the average absorptivity of the leaf surface, σ is the Boltzmann constant for black radiation and T_0 is the environment temperature. To simplify the consideration the sky temperature is assumed to be equal to the environment temperature. At small temperatures (T or T_0) the energy of PAR is relatively small e.g. in comparison to the case of radiation at the Sun temperature. Therefore, the assumption that the average absorptivity $\alpha_{L,a}$ equals the leaf absorptivity α_L for the non-PAR wavelengths range: $\alpha_{L,a} \approx \alpha_L$, slightly affects the value of the calculated energy e_L.

Exergy b_S for a non-polarized, uniform and direct solar radiation arriving in the Earth's atmosphere, is calculated based on formula (2.4.3) in which frequency ν is replaced by wavelength λ and the double integral value is according to (2.4.1):

$$b_S = 4.329 \cdot 10^{-5} \pi \left(\sum_n (i_{0,\lambda} \Delta \lambda)_n - T_0 \sum_n (L_{0,\lambda} \Delta \lambda)_n + \frac{9.445 \cdot 10^{-12}}{\pi} T_0^4 \right) \quad (2.4.99)$$

Using $T_0 = 293\,\text{K}$ in equation (2.4.99), the exergy of total solar radiation $b_S = 1.2835\,\text{kW/m}^2$ and the exergy of PAR, $b_V = 0.5155\,\text{kW/m}^2$, are calculated.

Emission exergy b_L of leaf surface at temperature T can be determined by the formula (2.2.40) with introduction of absorptivity α_L (assumed equal to emissivity ε_L):

$$b_L = \alpha_L \frac{\sigma}{3} (3T^4 + T_0^4 - 4T_0 T^3) \quad (2.4.100)$$

whereas the exergy of environmental radiation arriving at the leaf surface is zero.

Mass conservation equations are the basis for further considerations. The mass fluxes, $\text{kmol/(m}^2\,\text{s)}$, of CO_2 and O_2 are determined by the stoichiometric factors of equation (2.4.96): $n_{CO_2} = 6n_{SU}$, $n_{O_2} = 6n_{SU}$, where n_{SU} is the amount of produced sugar (kmol) within period of 1 s and per 1 m^2 of irradiated leaf surface. The mass flux n_w of water entering the leaf contains: a) water n_{wL} within the generated biomass, b) water $6 \cdot n_{SU}$ entering the chemical reaction, and c) water n_{H2O} vaporized into environment. Thus, $n_w = n_{wL} + 6n_{SU} + n_{H2O}$ where $n_{wL} = n_{SU}(1 - z_{SU})/z_{SU}$ and where z_{SU} is the mole fraction of sugar in the biomass composed of sugar and water.

As discussed by Jørgensen and Svirezhev (2004), an important factor in the determination of the effectiveness of photosynthesis is the mole ratio $r = n_{H2O}/n_{HCO2}$ of the water vapor and carbon dioxide rates. Water vapor diffuses from the internal surface of the leaf, through the stomata and intercellular space, towards the external surface of the leaf, and then diffuses through the boundary layer to the atmosphere. The water rate, n_{H2O}, is proportional to the generalized coefficient D_{H2O} of diffusion and to the difference ($z_{H2O,L} - z_{H2O,0}$) where $z_{H2O,L}$ is the initial mole concentration of vapor at the inner surface and $z_{H2O,0}$ is the final mole concentration in the environment. Diffusion of carbon dioxide occurs in the opposite direction and is also proportional to the generalized CO_2 diffusion coefficient D_{CO2}, and to the respective difference of mole concentrations $z_{CO2,0}$ and $z_{CO2,L}$. Therefore the rates ratio is:

$$r = \frac{D_{H_2O}}{D_{CO_2}} \frac{z_{H_2O,L} - z_{H_2O,0}}{z_{CO_2,0} - z_{CO_2,L}} \frac{M_{H_2O}}{M_{CO_2}} \quad (2.4.101)$$

where M_{H2O} and M_{CO2} are the molecular masses of H_2O and CO_2. The diffusion coefficients ratio was estimated by Jørgensen and Svirezhev (2004) as $D_{H2O}/D_{CO2} \approx 1.32$ and according to Budyko (1977): $z_{CO_20} - z_{CO_2L} \approx 0.1 z_{CO_20}$.

It is also assumed that the concentration of water vapor within the leaf corresponds to the saturation pressure $p_{s,T}$ at temperature T: $z_{H2OL} = p_{s,T}/p_0$. Thus equation (2.4.101) can be written as:

$$r = 5.4 \frac{p_{sT} - \varphi_0 p_{s0}}{p_0 z_{CO_20}} \qquad (2.4.102)$$

The ratio r is determined by the diffusion processes which control the rate of reaction (2.4.96), accordingly to assumption (xi).

Energy conservation equation for the system at instant steady state shown schematically in Figure 2.4.19 includes the energy delivered of absorbed solar radiation and the enthalpies of carbon dioxide and liquid water. The energy increase of the system is determined by the rates of the sugar substance and liquid water in the produced biomass. The extracted energy consists of the enthalpies of oxygen and water vapor as well as convective heat and emission exchanged by the leaf surface:

$$\gamma[\alpha_V j_V + \alpha_L(j_S - j_V)] + n_{CO_2} h_{CO_2} + n_w h_w$$
$$= n_{SU} h_{SU} + n_{wL} h_w + n_{O_2} h_{O_2} + n_{H_2O} h_{H_2O} + q_k + e_L \qquad (2.4.103)$$

where α_V and α_L are the absorptivities of the leaf within and beyond the PAR wavelength range, respectively.

According to assumption (vii), the biomass is an ideal solution of sugar and water and the total enthalpy of biomass is the sum of the respective components; $n_{SU} \cdot h_{SU}$ and $n_{wL} \cdot h_w$.

The heat transferred by convection from the leaf surface to the environment is: $q_k = k \cdot (T - T_0)$, where k is the convective heat transfer coefficient. Equation (2.4.103) is used to calculate the unknown rate n_{SU}. The leaf temperature T is higher than the environment temperature T_0 by the difference ΔT, i.e. $T = T_0 + \Delta T$, according to assumption (ix).

Exergy balance equation according to the scheme (Fig. 2.4.19) is:

$$\gamma[\alpha_V b_V + \alpha_L(b_S - b_V)] + n_{CO_2} b_{CO_2} + n_w b_w$$
$$= n_{SU} b_{SU} + n_{wL} b_w + n_{O_2} b_{O_2} + n_{H_2O} b_{H_2O} + b_{qk} + b_{eL} + \delta b \qquad (2.4.104)$$

where δb is the total exergy loss due to the sum of every irreversibility occurring within the system and is determined by formula (2.2.10). Again as for enthalpy, the total exergy of biomass is the sum of the respective components; $n_{SU} \cdot b_{SU}$ and $n_{wL} \cdot b_w$.

Perfection degrees of photosynthesis are calculated based on the assumption that the produced sugar represents the useful product and the feed is determined by radiation, CO_2 and liquid water. Other components of the balance equations are categorized as waste. Based on the definition discussed in paragraph 2.3.4.2, the energy degree of perfection η_E, of the considered photosynthesis:

$$\eta_E = \frac{n_{SU}h_{SU}}{\gamma[\alpha_V j_V + \alpha_L(j_S - j_V)] + n_{CO_2}h_{CO_2} + n_w h_w} \qquad (2.4.105)$$

whereas the exergy degree of perfection η_B, of the photosynthesis based on formula (2.3.18):

$$\eta_B = \frac{n_{SU}b_{SU}}{\gamma[\alpha_V b_V + \alpha_L(b_S - b_V)] + n_{CO_2}b_{CO_2} + n_w b_w} \qquad (2.4.106)$$

Example 2.4.4.1 The following input values have been used in the exemplary computations for the system presented in Figure 2.4.19:

- Environment temperature $T_0 = 293$ K,
- Temperature difference $\Delta_T = 5$ K,
- Relative humidity of environment air $\varphi_0 = 0.4$,
- Environment pressure p_0 equal to the standard pressure $p_0 = p_n = 101.325$ kPa,
- Weakening radiation factor $\gamma = 0.7$,
- Leaf absorptivity within PAR wavelength range $\alpha_V = 0.88$,
- Leaf absorptivity beyond the PAR range $\alpha_L = 0.05$,
- Convective heat transfer coefficient $k = 0.003$ kW/(m^2 K),
- Mole fraction of sugar in biomass $z_{SU} = 0.08$.

The leaf temperature is $T = T_0 + \Delta T = 298$ K and from equation (2.4.104) the sugar production rate $n_{SU} = 3.21 \cdot 10^{-9}$ kmol/(m^2 s). The percentage terms of energy and exergy equations (2.4.103) and (2.4.104), respectively, are shown in Fig. 2.4.20. The 100% reference for the output terms is assumed as the input sum due to the absorbed radiation and the substances of CO_2 and liquid water.

Thus the perfection degree value $\eta_E = 35.4\%$ is larger than $\eta_B = 2.608\%$ (by about $35.4/2.608 \approx 14$ times) mainly because of the denominators in equations (2.4.105) and (2.4.106). The exergy of liquid water, in the denominator of equation (2.4.106), is positive whereas the energy of this water in the denominator of equation (2.4.105) is negative, (the water vapor is assumed as the reference phase for enthalpy calculation).

The energy terms (Fig. 2.4.20, energy balance) show that the input consists of the positive radiation energy ($1459.8 + 125.4 = 1585.2\%$) and the negative ($-1485.2\%$) liquid water enthalpy. The energy of the consumed carbon dioxide is zero because it enters the system at the reference temperature T_0. The output energy terms show no irreversible loss and the zero energy of both; the produced oxygen and released water vapor leave the system at the reference temperature. Heat transferred by convection and radiation are 65.3% and 6.4%, respectively. The energy of liquid water contained in the produced biomass is negative (-7.1%) also because the vapor phase was assumed as the reference substance for water.

The exergy input terms (Figure 2.4.20, exergy balance) are the absorbed radiation ($87.76 + 7.43 = 95.19\%$) and the water of positive value 4.81%. The exergies of the delivered CO_2, released O_2 and water vapor are zero because these gases, at the

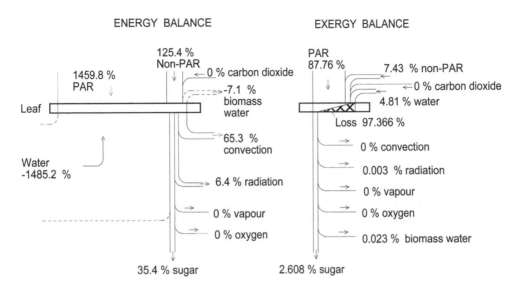

Figure 2.4.20 Band diagram of energy and exergy balances for the considered photosynthesis process shown in Figure 2.4.19 (from Petela, 2010).

external side of the system boundary layer, have the parameters equal to the environment parameters. The exergy of convective heat is zero because it is released to the environment. The exergy of the leaf radiation (0.003%) is small due to relatively low temperature of the leaf and is significantly smaller than the respective energy (6.4%). The exergy of liquid water contained in the produced biomass is 0.023%, (positive). Unlikely the enthalpy and energy analyses, exergy analysis shows the irreversibility loss, which in this case is relatively very large (97.366%).

More aspects of the energy and exergy balances and some inspired problems were analyzed by Petela (2010).

2.4.5 Photovoltaic

The present paragraph gives an outline of the simple energy and exergy analysis of the simultaneous generation of heat and power by photovoltaic (PV) technology. This double conversion of radiation energy can categorize the PV technology into the general systems of cogeneration of power and heat.

In the considerations only direct solar radiation is accounted for the case of a clear sky (at temperature T_{sky} assumed as environment temperature T_0, $T_{sky} = T_0$), and assuming the Sun is a black surface. As there is no motion of substance in the gravitational field, the eZergy consideration is not introduced. The 1 m^2 surface of a solar cell at temperature T_C is shown in Figure 2.4.21. Generally, the heat q_S transferred from the Sun's surface at temperature T_S to the outer surface of the solar cell on the Earth is distributed to the generated electrical energy E, reflected solar radiation q_r, useful heat q_C absorbed by the solar cell, and to convection and radiation heat

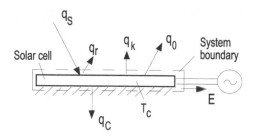

Figure 2.4.21 Energy streams of solar cell (from Petela, 2010).

q_k and q_0, respectively, both transferred to the environment. Therefore, the energy conservation equation for the considered system, defined by the system boundary is:

$$q_S = q_r + q_k + q_0 + q_C + E \tag{2.4.107}$$

where

$$q_S = 2.16 \times 10^{-5} \sigma T_S^4 \tag{i}$$

$$q_r = \rho_C q_S \tag{ii}$$

$$q_k = k(T_C - T_0) \tag{iii}$$

$$q_0 = \varepsilon_C \sigma(T_C^4 - T_0^4) \tag{iv}$$

and where 2.16×10^{-5} is the view (Sun-Earth) factor, σ is the Boltzmann radiation constant of a black surface, k is the convective heat transfer coefficient and T_0 is the environment temperature. The solar cell surface is assumed to be perfectly gray at emissivity ε_C, (reflectivity $\rho_C = 1 - \varepsilon_C$).

The useful heat q_C can be determined from equation (2.4.107) if the electrical energy E is known, e.g., from the measurement.

The solar cell can be evaluated by the energy electrical efficiency: $\eta_{E,el} = E/q_s$, by the energy heating efficiency: $\eta_{E,q} = q_c/q_s$, or by the energy cogeneration efficiency: $\eta_{E,cog} = (E + q_c)/q_s$

According to exergetic interpretation the exergy b_S incoming to the considered surface from the Sun is split into the exergy b_r of reflected solar radiation, the exergy of heat b_0 radiating to the environment, exergy of heat b_k transferred to the environment by convection, exergy of useful heat b_C transferred from the solar cell to its interior, electric energy E and the exergy loss δb due to the irreversibility of the considered system. Thus, the exergy equation for the system shown in Figure 2.4.21 is:

$$b_S = b_r + b_k + b_0 + b_C + E + \delta b \tag{2.4.112}$$

where:

$$b_S = 2.16 \times 10^{-5} \frac{\sigma}{3}(3T_S^4 + T_0^4 - 4T_0 T_S^3) \tag{v}$$

Table 2.4.5 Results of the energy and exergy calculations of the solar cell.

Term	Related symbol	Energy %	Exergy %
INPUT:			
solar radiation	e_S, b_S	100	100
Subtotal		100	100
OUTPUT:			
reflection	(ρ_C)	5	5
convection	q_k, b_k	6.21	0.62
radiation	q_r, b_r	12.43	0.67
useful heat	$q_c, b_c, (\eta_{E,q}, \eta_{B,q})$	65.88	6.61
electricity	$E, (\eta_{E,el}, \eta_{B,el})$	10.48	11.16
irreversibility loss	δb	–	75.94
Subtotal		100	100

$$b_r = \rho_C b_S \tag{vi}$$

$$b_k = q_k \left(1 - \frac{T_0}{T_C}\right) \tag{vii}$$

$$b_0 = \varepsilon_C \frac{\sigma}{3}(3T_C^4 + T_0^4 - 4T_0 T_C^3) \tag{viii}$$

$$b_C = q_C \left(1 - \frac{T_0}{T_C}\right) \tag{ix}$$

The exergy loss δb can be calculated by completion of equation (2.4.112). From the exergetic viewpoint the solar cell can be evaluated by: the exergy electric efficiency $\eta_{B,el} = E/b_s$, the exergy heating efficiency $\eta_{B,q} = b_c/b_s$, or the exergy cogeneration efficiency $\eta_{B,cog} = (E + b_c)/b_s$.

As the solar radiation energy is always larger than the solar radiation exergy ($q_S > b_S$), and the electrical exergy and electrical energy are equal, thus the energy electrical efficiency of the solar cell is always smaller than the exergy electrical efficiency, ($\eta_{B,el} > \eta_{E,el}$).

Example 2.4.5.1 A $1\,m^2$ surface of the photovoltaic cell at temperature $T_c = 318\,K$, emissivity $\varepsilon_C = 0.95$, (reflectivity $\rho_C = 0.05$), generates $E = 152\,W$ of electrical energy. The temperature of the Sun is assumed $T_S = 5800\,K$. The convective heat transfer coefficient $k = 3\,W/(m^2\,K)$.

The results calculated with the formula of the present paragraph are shown in Table 2.4.5. The values of the solar radiation energy and exergy: $e_S = 1386\,W/m^2$ and $b_S = 1294\,W/m^2$, were respectively assumed as the 100% in the energy and exergy balances.

The instant energy electric efficiency $\eta_{E,el} = 10.48\%$ is smaller from the exergy electric efficiency $\eta_{B,el} = 11.16\%$, however the energy cogeneration efficiency $\eta_{E,cog} = 10.48 + 65.88 = 76.36\%$ is significantly larger than the exergy cogeneration efficiency $\eta_{B,cog} = 11.16 + 6.61 = 17.77\%$. Table 2.4.5 illustrates also that the low temperature heat (convection, radiation and useful heat) has small exergy value.

Nomenclature

A	surface area, m^2
a	universal constant $a = 7.564 \cdot 10^{-16}$ J/(m^3 K^4)
B	exergy, J
b	exergy of emission, W/m^2
c_0	velocity of light in vacuum $c_0 = 2.9979 \cdot 10^8$ m/s
c_p	specific heat at constant pressure, J/(kg K)
D	diffusion coefficient, m^2/s or diameter, m
d_n	enthalpy of devaluation, J/kmol
E	energy, J or W/m^2
e	density of emission energy, W/m^2
G	gravity input, J
g	gravity acceleration, m/s^2
H	enthalpy, J or height, m
h	Planck's constant $h = 6.625 \times 10^{-34}$ J s or enthalpy, J/kg
I	irradiance, W/m^2
IR	irradiance (par. N.4.1.2), W/m^2
i_λ	monochromatic directional radiation intensity, W/(m^2 m sr)
J	radiosity, W
j	radiosity density, W/m^2
k	Boltzmann constant $k = 1.3805 \times 10^{-23}$ J/K
k	convective heat transfer coefficient, W/(m^2 K)
L	mean distance from the Sun to the Earth $L = 149,500,000$ km.
L_λ	entropy of i_λ, W/(m^2 K sr)
m	mass, kg
n	mass, kmol
PAR	photosynthetically active radiation
p	pressure, Pa
Q	heat, J
q	heat flux, W/m^2
R	individual gas constant, J/(kg K), or radius of the Sun $R = 695,500$ km
r	mole ratio of the water vapor and carbon dioxide rates
r_T	relative air pressure drop
S	entropy, J/K or irradiance (par. N.4.1.3 and N.4.1.4), W/m^2
SR	entropy of irradiance, W/(K m^2)
s	entropy, J/(K m^2) or J/K kmol)
s_j	entropy of radiosity density, W/(m^2 K)
t	time s, or temperature, C
T	absolute temperature, K
U	internal energy, J
u	specific internal energy, J/m^3
V	volume, m^3
W	work, J
x	distance, m
Z	eZergy, J
z	mole fraction or exergy, %

Greek

α	absorptivity of radiation,%
β	declination, deg
ε	emissivity,%
φ	azimuth, deg or view factor
γ	radiation weakening factor
λ	wavelength, m
η	efficiency,%
Π	overall entropy growth, J/K
ψ	exergy/energy radiation ratio
ν	oscillation frequency, 1/s
ρ	density, kg/m^3 or reflectivity of radiation,%
σ	entropy of devaluation reaction, J/(K kmol)
σ	Boltzmann constant for black radiation $\sigma = 5.6693 \times 10^{-8}\,W/(m^2\,K^4)$
τ	transmissivity of radiation,%
ω	solid angle, rd

Subscripts

B	exergetic
ch	chemical
E	energetic
k	convection
m	mechanical
n	standard (normal)
ph	physical
r	reflection
S	Sun
sky	sky
0	environment

REFERENCES

Budyko, M. I. (1977) Global Ecology. *Mysl'*, Moscow, 328 pp.

Carnot, S. (1824) *Reflections on the motive power of fire, and on machine fitted to develop that power*. Bachelier, Paris.

Duffie, J. A. & Beckman, W. A. (1991) *Solar Engineering of Thermal Processes*, 2nd edn. J. Wiley and Sons, New York.

Gueymard, C.A. (2004) The Sun's total and spectral irradiance for solar energy applications and solar radiation models. *Solar Energy*, 76, 423–453.

Jacob, M. (1957) *Heat Transfer*. Vol. II, John Wiley, New York.

Jørgensen, S.E. & Svirezhev, Y.M. (2004) *Towards a thermodynamic theory for ecological systems*. Elsevier Amsterdam.

Kondratiew, K. Ya. (1954) *Radiation energy of the Sun*. GIMIS, (in Russian).

Petela, R. (1961a) *Exergy of heat radiation*. PhD. Thesis, Faculty of Mechanical Energy Technology, Silesian Technical University, Gliwice (in Polish).

Petela, R. (1961b) Exergy of radiation of a perfect gray body. *Zesz. Nauk. Pol. Sl.* (30), Energetyka 5, 33–45, in Polish.

Petela R. (1962) Exergy of radiation radiosity. *Zesz. Nauk. Pol. Sl.* (56), Energetyka 9, 43–70, (in Polish).

Petela, R. (1964) Exergy of heat radiation. *ASME Journal of Heat Transfer,* 86, 187–192.

Petela, R. & Piotrowicz, A. (1977) Exergy of plasma. *Archiwum Termodynamiki i Spalania,* 3, 381–391.

Petela, R. (1984a) Exergetic analysis of atomization process of liquid. *Fuel,* 3, 419–422.

Petela, R. (1984b) Exergetic efficiency of comminution of solid substances. *Fuel,* 3, 414–418.

Petela, R. (1990) Exergy analysis of processes occurring spontaneously. *CSME Mechanical Engineering Forum, June 3–9, University of Toronto,* Vol. I, pp. 427–431.

Petela, R. (2005) Exergy analysis of the solar cylindrical-parabolic cooker. *Solar Energy* 79, 221–233.

Petela, R. (2008) Influence of gravity on the exergy of substance. *International Journal of Exergy,* 5, 1–17.

Petela, R. (2008a) An approach to the exergy analysis of photosynthesis. *Solar Energy,* 82, 311–328.

Petela, R. (2009) Thermodynamic study of a simplified model of the solar chimney power plant. *Solar Energy,* 83, 94–107.

Petela, R. (2009a) Gravity influence on the exergy balance. *International Journal of Exergy,* 6, 343–356.

Petela, R. (2009b) Thermodynamic analysis of chimney. *International Journal of Exergy,* 6, 868–880.

Petela R. (2010) *Engineering Thermodynamics of thermal radiation for solar power utilization,* McGraw Hill, New York.

Planck, M. (1914) *The theory of heat radiation.* Dover, New York, translation from German by Morton Mausius.

Swinbank, W. C. (1963) Long-wave radiation from clear skies. *Quarterly Journal of Royal Meteorological Society,* 89, 339.

Szargut, J. & Petela R. (1968) Exergy. *Energija,* Moscow, in Russian.

Szargut, J., Morris, D.R. & Steward, F.R. (1988) Exergy analysis of thermal, chemical, and metallurgical processes. *Hemisphere Publishing Corporation,* New York.

Von Backström, T.W. & Fluri, T.P. (2006) Maximum fluid power condition in solar chimney power plants – An analytical approach. *Solar Energy,* 80, 1417–1423.

Chapter 3

Exergy analysis of solar energy systems

Ibrahim Dincer & Tahir Abdul Hussain Ratlamwala
Faculty of Engineering and Applied Science, University of Ontario Institute of Technology, Oshawa, Ontario, Canada

3.1 INTRODUCTION

Extensive use of fossil fuels in past decades has led us to the era of global warming and depletion of the stratospheric ozone layer. Fossil fuels such as gasoline, diesel, natural gas etc. when used emit harmful greenhouse gases such as CO, CO_2, SO_2, NO_x etc. Since the realization of the harmful effect of using fossil fuels, researchers have started looking for alternative sources of energy which are renewable and environmentally friendly. One of the very promising alternative and renewable sources of energy is solar energy. Solar energy converts solar flux entering the Earth's surface into electricity or heat. The electricity generated by the solar energy system is in direct current format which can be used as it is or can be converted to alternative current based on end user requirements. Major benefits of using a solar energy system include (a) environmentally benign operation, (b) no moving parts, (c) no wearing of parts if the system is carefully protected from the environment, (d) energy output can vary from watts to megawatts based on the size of the system, (e) can be used to power phones or to power a community, and (f) module by module construction so that the size of the system can be altered based on the requirements.

Solar energy systems are characterized as passive solar energy systems or active solar energy systems. Solar energy systems are assigned to these categories based on the way they capture, convert and distribute solar energy. Passive solar techniques include (a) designing a building in a way that it uses solar energy for day lighting, (b) selecting materials with favorable thermal mass or light dispersing properties, and (c) designing spaces such that they naturally circulate air.

Active solar energy systems are divided in to three categories which are (a) photovoltaic systems which generate electricity, (b) thermal systems which generate heat, and (c) photovoltaic/thermal (PV/T) systems which generate both electricity and heat.

There are two types of photovoltaic panels available in the market, which are rigid photovoltaic panels and flexible photovoltaic panels. Rigid photovoltaic panels have higher power to area density and flexible photovoltaic panels have low power to area density. Solar thermal systems are sub-divided into following categories: (a) low temperature such as flat plate collectors, (b) medium temperature such as concentrated collectors or solar dishes, and (c) high temperature such as heliostat fields.

Low-temperature thermal sources are usually used for direct application such as providing hot water or heat to the building. Medium temperature thermal sources are used for either producing cooling with the help of absorption cooling systems or producing power using Organic Rankine Cycles (ORCs). High temperature thermal sources are used to produce huge amounts of power by running a steam cycle.

Integrated solar energy systems provide an attractive way of producing multiple outputs such as power, heat, hydrogen, cooling etc. in an environmentally benign manner. As it is expected that future economy will be dominated by hydrogen fuel, many researchers have studied integrated solar energy systems for hydrogen production. Thomas and Nelson (2010) stated that hydrogen fuel can be produced by using solar electric energy from photovoltaic (PV) modules for the electrolysis of water without emitting carbon dioxide or requiring fossil fuels. The results of analyses conducted by different researchers such as (Ratlamwala et al., 2011; Koroneos et al., 2004; Yilanci et al., 2009) related to solar hydrogen production show that an integrated solar production system is very promising technology as it produces hydrogen in an environmentally friendly and cost effective manner. Also studies have shown that using solar for hydrogen production enhances the efficiency of the overall system. Solar PV/T systems can also be used for multi-generation purposes as studied by Ratlamwala et al. (2011) where energy and exergy analyses show that integrated solar energy systems are suitable for hydrogen and cooling production. In this chapter, the aim is to discuss energy and exergy related aspects of solar energy systems, consider various solar energy based systems for analysis, assessment and comparison, and evaluate them for practical applications from the exergy point of view.

3.2 ENERGY AND EXERGY ASPECTS AND ANALYSES

The relationship between energy and economics was a prime concern in the 1970s. At that time, the linkage between energy and the environment did not receive much attention. As environmental concerns, such as acid rain, ozone depletion and global climate change, became major issues in the 1980s, the link between energy utilization and the environment became more recognized. Since then, there has been increasing attention for this connection, as it has become more clear that energy production, transformation, transport and use all impact the Earth's environment, and that environmental impacts are associated with the thermal, chemical and nuclear emissions which are a necessary consequence of the processes that provide benefits to humanity.

Many suggest that mitigating the environmental impact of energy resource utilization and achieving increased resource utilization efficiency are best addressed by considering exergy. The exergy of an energy form or a substance is a measure of its usefulness or quality or potential to cause change. The latter point suggests that exergy may be, or provide the basis for, an effective measure of the potential of a substance or energy form to impact the environment. In practice, the authors feel that a thorough understanding of exergy and the insights it can provide into the efficiency, environmental impact and sustainability of energy systems are required for the engineer or scientist working in the area of energy systems and the environment. The need to understand the linkages between exergy and energy, sustainable development and environmental impact has become increasingly significant.

The solar PV and PV/T systems are one of the most significant and rapidly developing renewable-energy technologies, and its potential future uses are notable. During the last decade, solar PV and PV/T applications have increased in many countries and are observed throughout the residential, commercial, institutional and industrial sectors. The clean, renewable and in some instances economical features of solar PV and PV/T systems have attracted attention from political and business decision makers and individuals. Advances in solar PV and PV/T technology have also driven the trend to increased usage.

Solar energy resources and technologies have three main benefits which are (a) they generally cause no or less environmental impact as compared to other energy sources, (b) they cannot be depleted; in contrast, fossil fuel and uranium resources are diminished by extraction and consumption, and (c) they favor system decentralization and local solutions that are somewhat independent of the national network, thus enhancing the flexibility of the system and providing economic benefits to small isolated populations. Also, the small scale of the equipment often reduces the time required from initial design to operation, providing greater adaptability in responding to unpredictable growth and/or changes in energy demand.

Exergy, as a potential tool, has several qualities that make it suitable as a common quantifier of the sustainability of a process (Dincer and Rosen, 2004; Sciubba, 2001). These qualities are (a) exergy is an extensive property whose value is uniquely determined by the parameters of both the system and the reference environment, (b) if a flow undergoes any combination of work, heat and chemical interactions with other systems, the change in its exergy expresses not only the quantity of the energy exchanges but also the quality, (c) the value of a product of a process, expressed in terms of 'resource use consumption,' may be obtained by adding to the exergy of the original inputs all the contributions due to the different streams that were used in the process.

The higher performance, lower cost and better reliability demonstrated by today's solar PV and PV/T systems are leading many potential users to consider the value of these systems for particular applications. Together, these applications will likely lead industry to build larger and more cost-effective production facilities, leading to lower solar PV and PV/T costs. Public demand for environmentally benign sources of electricity will almost certainly hasten adoption of solar PV and PV/T. The rate of adoption will be greatly affected by the economic viability of solar PV and PV/T with respect to competing options. Many analysts and researchers believe that it is no longer a question of if, but when and in what quantity, solar PV and PV/T systems will see widespread adoption.

In Table 3.2.1, the general thermodynamic quantities and general energy and exergy balance equations as well as energy and exergy efficiencies are listed as they will be employed for system analyses and performance assessment.

Table 3.2.1 General thermodynamic quantities and balance equations.

Specific enthalpy	$h = u + vP$
Specific entropy	$s_2 - s_1 = c \ln \dfrac{T_2}{T_1} - R \ln \dfrac{P_2}{P_1}$
Specific exergy*	$ex_i = [(h_i - h_0) - T_0(s_i - s_0)]$
Thermal exergy	$\dot{Ex}_{th} = \left(1 - \left(\dfrac{T_0}{T}\right)\right) \times Q''$
Energy balance	$\sum (\dot{m}_i h_i)_{in} = \sum (\dot{m}_i h_i)_{out} + \sum \dot{Q} + \sum \dot{W}$
Exergy balance	$\sum (\dot{m}_i ex_i)_{in} = \sum (\dot{m}_i ex_i)_{out} + \dot{Ex}_{de} + \sum \dot{Ex}_{th}$
Energy efficiency	$\eta_{en} = \dfrac{\sum \dot{En}_{out}}{\sum \dot{En}_{in}}$
Exergy efficiency	$\eta_{ex} = \dfrac{\sum \dot{Ex}_{out}}{\sum \dot{Ex}_{in}}$

*Based on changes in chemical formulation specific chemical exergy is added.

3.3 CASE STUDIES

In this section several studies are presented which highlight the use of solar energy systems. Detailed energy and exergy analyses are also presented in order to help model integrated solar energy systems. In the first case study a solar energy system is integrated with an Organic Rankine Cycle (ORC) for producing power, in the second study a solar PV/T system is modeled for power and heat production, in the third study a solar PV/T system is integrated with an electrolyzer and absorption system for hydrogen and cooling production.

3.3.1 Case study 1: Exergy analysis of an integrated solar, ORC system for power production

In this case study, a detailed energy and exergy model of an integrated solar ORC is presented for power production. Operating parameters such as ambient temperature, area of solar energy system and pressure at state 4 are varied to see their effect on energy and exergy efficiencies.

3.3.1.1 System description

An integrated solar thermal ORC system for power production studied in this case study is shown in Figure 3.3.1. Air at state 1 returning from the boiler is passed through the solar thermal collector. In the solar thermal collector, the solar flux hitting the collector is absorbed by the air passing through the collector. The air at high temperature leaving the solar collector at state 2 enters the boiler where it loses heat to the isobutane coming in at state 4. Isobutane coming in at state 4, after gaining heat from state 2 leaves the boiler at state 3 to enter the turbine. In the turbine, the pressure and temperature of the isobutane is dropped and isobutane leaves at state 6 to enter

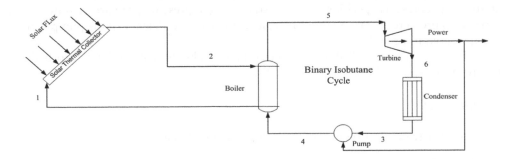

Figure 3.3.1 Schematic of solar thermal integrated with binary cycle.

the condenser. In the condenser, isobutane at state 6 loses heat to the environment to leave at state 3 as saturated liquid in order to enter the pump where its pressure is increased to that of state 4. The part of the power produced by the turbine is supplied to the pump and remaining power is available for later use.

3.3.1.2 Energy and exergy analyses

The rate of heat gained by the air passing through the collector is calculated as

$$\dot{Q}_{so} = \left(\frac{I \times A}{1000}\right) \tag{3.3.1}$$

where $\dot{Q}_{so}$ represents rate of heat gained by the air passing through the collector, I represents solar flux, and A represents solar collector area.

The exergy destruction rate in the solar thermal collector is calculated as

$$\dot{Ex}_1 + \dot{Ex}_{so} = \dot{Ex}_2 + \dot{Ex}_{de,so} \tag{3.3.2}$$

where

$$\dot{Ex}_1 = \dot{m}_1\big((h_1 - h_0) - T_0(s_1 - s_0)\big)$$

$$\dot{Ex}_{so} = \left(1 - \frac{T_0}{T_{sun}}\right)\left(\frac{I \times A}{1000}\right)$$

$$\dot{Ex}_2 = \dot{m}_2\big((h_2 - h_0) - T_0(s_2 - s_0)\big)$$

where $\dot{Ex}_1$ represents exergy rate at state 1, $\dot{Ex}_{so}$ represents exergy rate of solar flux, $\dot{Ex}_2$ represents exergy rate at state 2, and $\dot{Ex}_{de,so}$ represents exergy destruction rate in the solar collector.

The power consumed by the pump is defined as

$$\dot{W}_p = \dot{m}_3\left(\frac{v_3(P_4 - P_3)}{\eta_p}\right) \tag{3.3.3}$$

where $\dot{W}_p$ represents power consumed by the pump, $\dot{m}_3$ represents mass flow rate at state 3, v_3 represents specific volume at state 3, P_4 represents pressure at state 4, P_3 represents pressure at state 3 and η_p represents isentropic efficiency of the pump which is considered to be 80%.

The rate of heat supplied to the boiler is taken to be the same as heat absorbed by the air in the solar collector as shown below

$$\dot{Q}_{bo} = \dot{Q}_{so} \qquad (3.3.4)$$

The exergy destruction rate in the boiler is calculated as

$$\dot{E}x_4 + \dot{E}x_{bo} = \dot{E}x_5 + \dot{E}x_{de,bo} \qquad (3.3.5)$$

where

$$\dot{E}x_4 = \dot{m}_4\big((h_4 - h_0) - T_0(s_4 - s_0)\big)$$

$$\dot{E}x_{bo} = \left(1 - \frac{T_0}{T_{bo}}\right)\dot{Q}_{bo}$$

$$T_{bo} = \frac{T_4 + T_5}{2}$$

$$\dot{E}x_5 = \dot{m}_5\big((h_5 - h_0) - T_0(s_5 - s_0)\big)$$

where $\dot{E}x_4$ represents exergy rate at state 4, $\dot{E}x_{bo}$ represents exergy rate carried by heat entering the boiler, $\dot{E}x_5$ represents exergy rate at state 5, and $\dot{E}x_{de,bo}$ represents exergy destruction rate in the boiler.

The power produced by the turbine is found using

$$\dot{W}_t = \dot{m}_5(h_5 - h_6) \qquad (3.3.6)$$

where $\dot{W}_t$ represents power produced by the turbine, $\dot{m}_5$ represents mass flow rate at state 5, h_5 represents specific enthalpy at state 5, and h_6 represents specific enthalpy at state 6.

The exergy destruction rate in the turbine is calculated as

$$\dot{E}x_5 = \dot{E}x_6 + \dot{E}x_{de,t} + \dot{W}_t \qquad (3.3.7)$$

where

$$\dot{E}x_5 = \dot{m}_5\big((h_5 - h_0) - T_0(s_5 - s_0)\big)$$

$$\dot{E}x_6 = \dot{m}_6\big((h_6 - h_0) - T_0(s_6 - s_0)\big)$$

where $\dot{E}x_5$ represents exergy rate at stat 5, $\dot{E}x_6$ represents exergy rate at state 6, and $\dot{E}x_{de,t}$ represents exergy destruction rate in the turbine.

The rate of heat rejected by the condenser is defined as

$$\dot{Q}_{con} = \dot{m}_6(h_6 - h_3) \tag{3.3.8}$$

where $\dot{Q}_{con}$ represents heat transfer rate rejected by the condenser, $\dot{m}_6$ represents mass flow rate at state 6, h_6 represents specific enthalpy at state 6, and h_3 represents specific enthalpy at state 3.

The exergy destruction rate in the condenser is calculated as

$$\dot{Ex}_6 = \dot{Ex}_3 + \dot{Ex}_{con} + \dot{Ex}_{de,con} \tag{3.3.9}$$

where

$$\dot{Ex}_6 = \dot{m}_6\left((h_6 - h_0) - T_0(s_6 - s_0)\right)$$

$$\dot{Ex}_{con} = \left(1 - \frac{T_0}{T_{con}}\right)\dot{Q}_{con}$$

$$T_{con} = \frac{T_6 + T_3}{2}$$

$$\dot{Ex}_3 = \dot{m}_3\left((h_3 - h_0) - T_0(s_3 - s_0)\right)$$

where $\dot{Ex}_6$ represents exergy rate at stat 6, $\dot{Ex}_{con}$ represents exergy rate carried by heat exiting the condenser, $\dot{Ex}_3$ represents exergy rate at state 3, and $\dot{Ex}_{de,con}$ represents exergy destruction rate in the condenser.

The energy and exergy efficiencies are calculated as

$$\eta_{en} = \frac{\dot{W}_t - \dot{W}_p}{\left(\frac{I \times A}{1000}\right)} \tag{3.3.10}$$

$$\eta_{ex} = \frac{\dot{W}_t - \dot{W}_p}{\dot{Ex}_{so}} \tag{3.3.11}$$

where η_{en} represents energy efficiency of the system and η_{ex} represents exergy efficiency of the system.

3.3.1.3   Results and discussion

Effects of variation in parameters such as ambient temperature, solar thermal area, and pressure at state 4 on energy and exergy efficiencies are studied in this section. It is observed that rise in ambient temperature doesn't affect the energy efficiency but does result in a slight increase in exergy efficiency as seen in Figure 3.3.2. The energy efficiency is found to be constant at 16.37% while exergy efficiency increases from 17.19% to 17.3% with a rise in ambient temperature from 275 K to 310 K. This result is important as it shows that from the energy perspective changes in ambient temperature have no effect on the performance of the system, but we know that in real

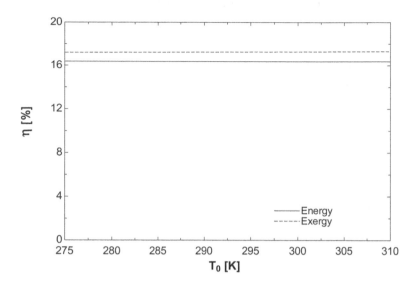

Figure 3.3.2   Effect of rise in ambient temperature on energy and exergy efficiencies of the system.

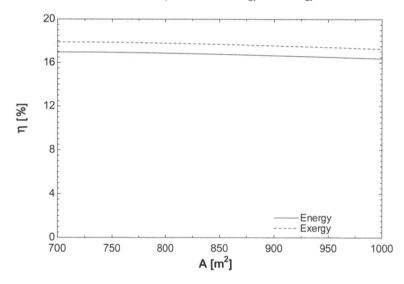

Figure 3.3.3   Effect of area of solar collector on energy and exergy efficiencies of the system.

life ambient temperature does play a role in performance determination of any system as it is shown by exergy efficiency.

The effect of an increase in solar thermal area on energy and exergy efficiency of the system is shown in Figure 3.3.3. It can be seen that an increase in solar thermal area results in degrading performance of the system. The energy and exergy efficiencies are found to be decreasing from 16.97% to 16.37% and 17.89% to 17.26%, respectively with increase in solar thermal area from 700 m^2 to 1000 m^2. Such results are obtained

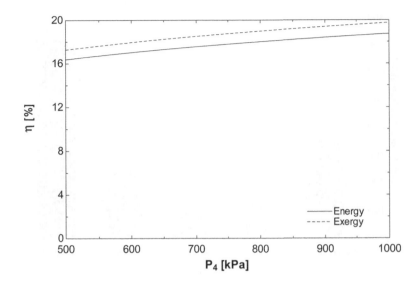

Figure 3.3.4 Effect of pressure at state 4 on energy and exergy efficiencies of the system.

because working fluid passing through the solar collector, as air in this case, has a certain limit up to which it can get heated and therefore an increase in area does not necessarily mean better performance of the system from the efficiency perspective.

The pressure of the working fluid entering the turbine at state 4 plays an important role in the performance of the system as shown by energy and exergy efficiencies in Figure 3.3.4. The energy and exergy efficiencies are found to be increasing from 16.37% to 18.745 and 17.26% to 19.76%, respectively with increase in pressure at state 4 from 500 kPa to 1000 kPa. Increase in pressure of the fluid entering the turbine means that the stream entering the turbine has higher energy content as compared to the low pressure stream and therefore for the same exit pressure the performance of the system is enhanced.

Finally, a bar chart is provided to illustrate which component of the system has the greatest amount of exergy destroyed at a constant ambient temperature and pressure of 298 K and 101 kPa as shown in Figure 3.3.5. It can be seen that the maximum amount of exergy is destroyed by solar thermal collector followed by condenser, boiler and turbine. This bar chart shows that in order to enhance the energy and exergy efficiencies one should first try to reduce the exergy destruction rate in the solar thermal collector as it is the source of maximum losses.

3.3.2   Case study 2: Exergy analysis of solar photovoltaic/thermal (PV/T) system for power and heat production

In this case study, a detailed energy and exergy model of solar PV/T system is presented. Operating parameters such as ambient temperature and solar flux are varied to see their effect on energy and exergy efficiencies.

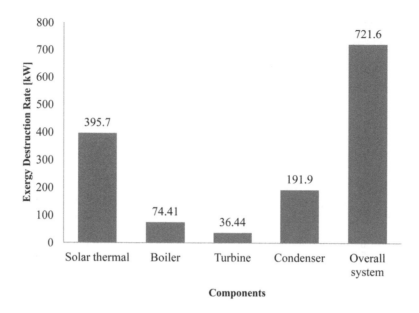

Figure 3.3.5 Exergy destruction rate in individual components and overall system.

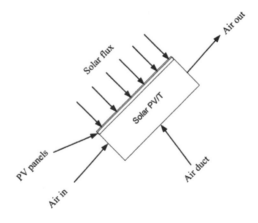

Figure 3.3.6 Schematic of solar PV/T system.

3.3.2.1 System description

A solar PV/T system studied is shown in Figure 3.3.6. Solar flux hits the surface of the system which is photovoltaic (PV) panels. As solar flux comes in contact with the PV panels, the molecules inside the panels start to vibrate and they produce power. Due to high solar flux falling on the PV panels, the back surface of the PV panels is heated up and that is the heat which is utilized to produce hot air. The air is allowed to pass through the duct of the PV/T system in order to act as a cooling medium for PV panels. The air passing through the duct gains heat from the PV panels and its temperature

Table 3.3.1 Solar PV/T parameters.

Parameter	Values
A	$100\,m^2$
B	$0.45\,m$
h_{p1}	0.88
L	$1.2\,m$
β_c	0.83
η_c	0.12
τ_g	0.95
U_b	$0.62\,W/m^2\,K$
U_t	$2.8\,W/m^2\,K$
α_c	0.90

increases. The heated air leaving the duct can be used for different heating and cooling purposes based on the temperature of the air leaving the duct and the operating system.

3.3.2.2 Energy and exergy analyses

The equations which are used to solve the mathematical model of the PV/T system are derived from Joshi et al. (2009a, and 2009b). Moreover, the assumptions and constants of solar PV/T system are listed in Table 3.3.1.

The equation to calculate power produced by the PV module is given as

$$\dot{W}_{so} = \eta_c \times I \times \beta_c \times \tau_g \times A \tag{3.3.12}$$

where $\dot{W}_{so}$ represents power produced by the solar panel, η_c represents cell efficiency, I represents solar flux, β_c represents the packing factor of the solar cell, τ_g represents transitivity of glass, and A represents the area of the solar PV/T system

The heat transfer rate from solar panels to air passing through the duct is given by

$$\dot{Q}_{so} = \frac{\dot{m}_a \times cp_a}{U_L} \times ((h_{p2G} \times z \times I) - U_L \times (T_{ai} - T_0)) \times \left[1 - \exp\left(\frac{-b \times U_L \times L}{\dot{m}_a \times cp_a}\right)\right] \tag{3.3.13}$$

where

$$z = \alpha_b \times \tau_g^2 \times (1 - \beta_c) + h_{p1G} \times \tau_g \times \beta_c \times (\alpha_c - \eta_c)$$

where $\dot{Q}_{so}$ represents heat transfer rate from PV panels to air, $\dot{m}_a$ represents mass flow rate of air, cp_a represents specific heat at constant pressure of air, U_L represents overall heat transfer coefficient from solar cell to ambient through the top and back surface of the insulation, h_{p2G} represents a penalty factor due to presence of an interface between glass and working fluid through an absorber plate for a glass-to-glass PV/T system, z represents a variable, T_{ai} represents temperature of air entering the duct, T_0 represents ambient temperature, b represents breadth of the single PV panel, L represents length of the single PV panel, α_b represents absorptance of a painted black surface, h_{p1G}

represents a penalty factor due to the presence of solar cell material, glass and EVA for a glass-to-glass PV/T system, and α_c represents absorptance of the solar cell.

The solar exergy rate is calculated by

$$\dot{Ex}_{so} = \left(1 - \frac{T_0}{T_{sun}}\right)\left(\frac{I \times A}{1000}\right) \tag{3.3.14}$$

where $\dot{Ex}_{so}$ represents solar exergy rate, and T_{sun} represents temperature of the Sun.

The thermal energy and exergy efficiencies of the solar thermal system are defined as

$$\eta_{th,en} = \frac{\dot{Q}_{so}}{\left(\frac{I \times A}{1000}\right)} \tag{3.3.15}$$

$$\eta_{th,ex} = \frac{\dot{Ex}_{PV/T}}{\dot{Ex}_{so}} \tag{3.3.16}$$

where

$$\dot{Ex}_{PV/T} = \left(1 - \frac{T_0}{T_c}\right) \times \dot{Q}_{so}$$

$$T_c = \frac{\tau_g \times \beta_c \times I \times (\alpha_c - \eta_c) + U_t \times T_0 + h_t \times T_{bs}}{U_t + h_t}$$

$$T_{bs} = \frac{z \times I + (U_t + U_{tb}) \times T_0 + h_{ba} \times T_{air}}{U_b + h_{ba} + U_{tb}}$$

$$T_{air} = \left[T_0 + \frac{h_{p2G} \times z \times I}{U_L}\right] \times \left[1 - \frac{1 - \exp\left(\frac{-b \times U_L \times L}{\dot{m}_a \times cp_a}\right)}{\frac{b \times U_L \times L}{\dot{m}_a \times cp_a}}\right] + T_{ai} \times \left[\frac{1 - \exp\left(\frac{-b \times U_L \times L}{\dot{m}_a \times cp_a}\right)}{\frac{b \times U_L \times L}{\dot{m}_a \times cp_a}}\right]$$

where $\dot{Ex}_{PV/T}$ represents exergy rate of PV/T, T_c represents cell temperature, U_t represents the overall heat transfer coefficient from solar cell to ambient through the glass cover, h_t represents the heat transfer coefficient from a black surface to air through glass, T_{bs} represents back surface temperature, U_{tb} represents the overall heat transfer coefficient from glass to black surface through solar cell, h_{ba} represents the heat transfer coefficient from black surface to air, T_{air} represents temperature of air flowing through the duct, and U_b represents the overall heat transfer coefficient from bottom to ambient.

The electrical energy and exergy efficiencies of solar PV system are defined as

$$\eta_{el,en} = \frac{\dot{W}_{so}}{\left(\frac{I \times A}{1000}\right)} \tag{3.3.17}$$

$$\eta_{el,ex} = \eta_c \times (1 - 0.0045 \times (T_c - 25)) \tag{3.3.18}$$

The overall energy and exergy efficiencies of the solar PV/T system are defined as

$$\eta_{ov,en} = \frac{\dot{W}_{so} + \dot{Q}_{so}}{\left(\frac{I \times A}{1000}\right)} \tag{3.3.19}$$

$$\eta_{ov,ex} = \frac{\dot{W}_{so} + \dot{E}x_{PV/T}}{\dot{E}x_{so}} \tag{3.3.20}$$

3.3.2.3   Results and discussion

This section highlights the importance of conducting a parametric study to see the effect of variation in operating conditions such as ambient temperature and solar flux on energy and exergy efficiencies of the system.

The ambient temperature plays an important role in the performance of any system. Figure 3.3.7 shows how the energy and exergy efficiency of electrical and thermal systems vary with a rise in ambient temperature. The electrical energy efficiency is seen to be a constant at 9.46% whereas electrical exergy efficiency is found to be decreasing from 9.04% to 7.74%, respectively with an increase in ambient temperature from 280 K to 310 K. The thermal energy and exergy efficiencies are found to be increasing from 5.89% to 61.65% and 1.28% to 11.6%, respectively with a rise in ambient temperature. The overall energy and exergy efficiencies are found to be increasing with a rise in ambient temperature as displayed in Figure 3.3.8. The overall energy and exergy efficiencies are found to be increasing from 15.36% to 71.11% and 11.22% to 21.59%, respectively with a rise in ambient temperature. These figures show that a rise in ambient temperature affects the performance of the system a lot. It is also observed that having a solar PV/T system is better than having an individual solar PV or thermal system, as a solar PV/T system has higher energy and exergy efficiencies as

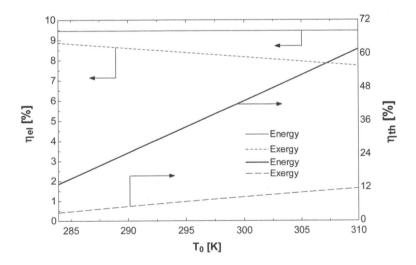

Figure 3.3.7 Effect of rise in ambient temperature on electrical and thermal energy and exergy efficiencies.

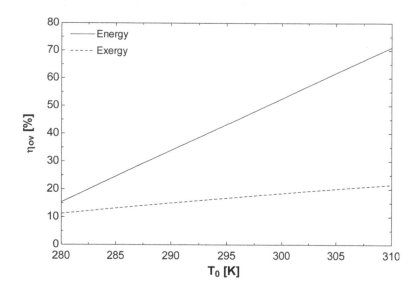

Figure 3.3.8 Effect of rise in ambient temperature on overall system energy and exergy efficiencies.

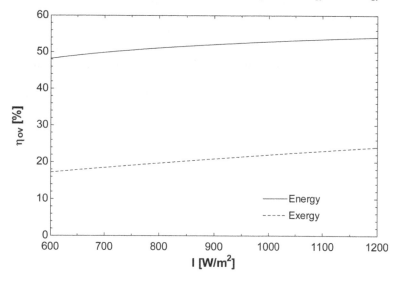

Figure 3.3.9 Effect of increase in solar flux on electrical and thermal energy and exergy efficiencies.

compared to individual systems. It is also observed that electrical energy efficiency is not affected by a rise in ambient temperature but electrical exergy efficiency is affected hence indicating that it is always better to conduct an exergy analysis.

The second parameter studied in this case study is solar flux. The effect of increase in solar flux on energy and exergy efficiencies of an electrical and thermal system is shown in Figure 3.3.9. The electrical energy efficiency remains constant at 9.46% whereas electrical exergy efficiency decreases from 8.53% to 5.13%, respectively with

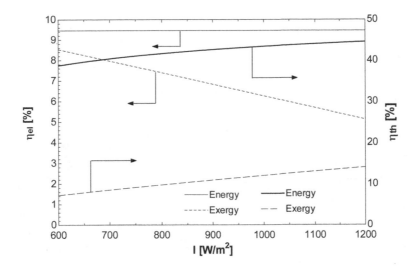

Figure 3.3.10 Effect of increase in solar flux on overall system energy and exergy efficiencies.

an increase in solar flux from 600 W/m^2 to 1200 W/m^2. The thermal energy and exergy efficiencies are found to be increasing from 38.73% to 44.57% and 7.23% to 14.06%, respectively. The effect of an increase in solar flux on overall energy and exergy efficiencies is also studied and is illustrated in Figure 3.3.10. The overall energy and exergy efficiencies increase from 48.19% to 54.03% and 17.21% to 24.03%, respectively with an increase in solar flux. The results show that energy and exergy efficiencies are dependent on solar flux received by the solar PV/T system.

3.3.3 Case study 3: Exergy assessment of an integrated solar PV/T and triple effect absorption cooling system for hydrogen and cooling production

In this case study, a detailed energy and exergy model of an integrated concentrated solar PV/T and triple effect cooling system is presented. Operating parameters such as solar flux, area, and air inlet temperature are varied to see their effect on overall energy and exergy efficiencies, the power produced by the solar PV panels, the rate of hydrogen produced, and energy and exergy COPs.

3.3.3.1 System description

In this case study, we have studied an integrated concentrated solar PV/T absorption cooling system for the cooling and hydrogen production as shown in Figure 3.3.11. In this integrated system, concentrated solar PV/T is used to produce power and heat. Power produced is supplied to the electrolyzer and the pump in the cooling cycle. The electrolyzer is utilized to break the water molecule bond. As the water molecule breaks, it splits into hydrogen and oxygen. The hydrogen molecules are taken out of the electrolyzer and are stored in a tank for later use as an energy source. The high temperature air coming out of solar PV/T is fed into the HTG of the absorption cooling

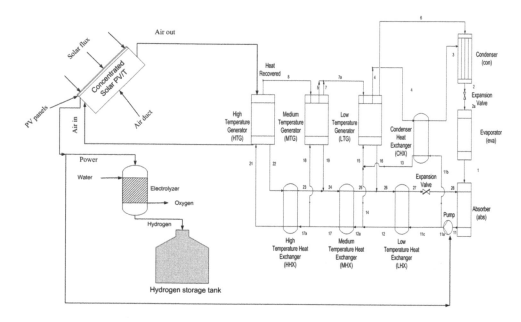

Figure 3.3.11 Schematic of integrated system.

system and is used as an energy source for the absorption cooling system. A detailed description of this system can be found in Ratlamwala et al. (2011).

3.3.3.2 Energy and exergy analyses

Detailed energy and exergy analyses of solar PV/T system are presented in case study 2 or can be found in Ratlamwala et al. (2011).

The electrolyzer is used to split water molecules into hydrogen and oxygen molecules, where hydrogen molecules are stored in the tank for later usage as an energy provider to the HTG or as a fuel to produce power using PEMFC. The amount of hydrogen produced depends on the efficiency of the electrolyzer, the heating value of the hydrogen, and power input to the electrolyzer.

$$\eta_{elec} = \frac{\dot{m}_{H_2} \times HHV}{\dot{W}_{solar} - \dot{W}_{pump}} \tag{3.3.21}$$

After getting the heat and power output from the PEM fuel cell, equations for HTG were written in order to run the TEACS to achieve the cooling load. The amount of energy provided to the HTG is shown below

$$\dot{Q}_{HTG} = \dot{Q}_{so} \tag{3.3.22}$$

The mass balance equations for HTG are given as follows

$$\dot{m}_{21}x_{21} = \dot{m}_{22}x_{22} + \dot{m}_8 x_8 \tag{3.3.23}$$

$$\dot{m}_{21} = \dot{m}_{22} + \dot{m}_8 \tag{3.3.24}$$

The energy balance equation for HHX is given below

$$\dot{m}_{17a}h_{17a} + \dot{Q}_{HHX} = \dot{m}_{21}h_{21} \tag{3.3.25}$$

$$\dot{m}_{22}h_{22} = \dot{Q}_{HHX} + \dot{m}_{23}h_{23} \tag{3.3.26}$$

The mass and energy balance equations for the condenser are given below

$$\dot{m}_2 = \dot{m}_6 + \dot{m}_3 \tag{3.3.27}$$

$$\dot{m}_2 h_2 + \dot{Q}_{con} = \dot{m}_6 h_6 + \dot{m}_3 h_3 \tag{3.3.28}$$

Below mentioned equation is for energy balance of evaporator

$$\dot{m}_{2a}h_{2a} + \dot{Q}_{eva} = \dot{m}_1 h_1 \tag{3.3.29}$$

The following energy balance equation is used to calculate the heat rejected by the absorber.

$$\dot{m}_{11}h_{11} + \dot{Q}_{abs} = \dot{m}_1 h_1 + \dot{m}_{28}h_{28} \tag{3.3.30}$$

The thermal exergy of evaporator and HTG are defined as

$$\dot{E}x_{th,eva} = \left(1 - \frac{T_0}{T_{eva}}\right) \times \dot{Q}_{eva} \tag{3.3.31}$$

$$\dot{E}x_{th,HTG} = \left(1 - \frac{T_0}{T_{HTG}}\right) \times \dot{Q}_{HTG} \tag{3.3.32}$$

The energy and exergy COPs are calculated as

$$COP_{en} = \frac{\dot{Q}_{eva}}{\dot{Q}_{HTG} + \dot{W}_P} \tag{3.3.33}$$

$$COP_{ex} = \frac{\dot{E}x_{th,eva}}{\dot{E}x_{th,HTG} + \dot{W}_P} \tag{3.3.34}$$

Overall energy and exergy efficiencies are defined as

$$\eta_{ov,en} = \left(\frac{\dot{m}_{H_2} \times HHV + \dot{Q}_{eva}}{I \times b \times L}\right) \times 100 \tag{3.3.35}$$

$$\eta_{ov,ex} = \left(\frac{\dot{E}x_{H_2} \times \dot{E}x_{th,eva}}{\dot{E}x_{so}}\right) \times 100 \tag{3.3.36}$$

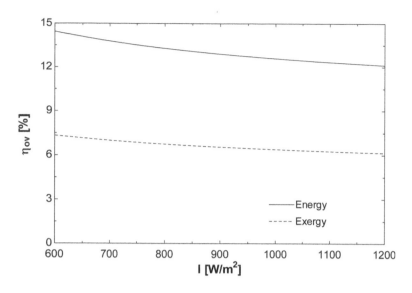

Figure 3.3.12 Effect of increase in solar flux on overall system energy and exergy efficiencies.

3.3.3.3   Results and discussion

In Figure 3.3.12, the study of the effect of solar radiation on the overall energy and exergy efficiencies has been carried out. It is noticed that the overall energy and exergy efficiencies decrease when the solar radiation is increased while keeping air inlet temperature, area of the PV/T and time for which solar radiation is available constant at 25°C, 10 m², and 12 hr, respectively. The energy and exergy efficiencies drop from 14.44% to 12.13% and 7.33% to 6.15% respectively, as the solar radiation increases. As the solar radiation increases the power production capacity of PV module increases and at the same time heat transfer rate also increases. The increase in power means that more water molecules are broken down to produce hydrogen. On the other hand, increase in the rate of heat results in a higher amount of heat given to the absorption system to provide the fixed amount of cooling. This increase in rate of energy fed into the cooling system results in the degraded performance of the cooling system as the cooling system rejects more heat through the condenser to achieve the desired cooling. These degrading performances of the cooling system result in lower energy and exergy COPs of the system. As the performance of the cooling system degrades the overall efficiency of the system decreases because more energy is being consumed to acquire the required outputs.

The increase in area of the PV module results in a higher power output from the PV module and higher hydrogen production. The solar radiation, operating hours, and air inlet temperature are kept constant at 608 W/m², 30.8°C, and 12 hr, respectively. The power output and hydrogen production increase from 0.28 kW to 0.86 kW and 5.24 kg to 15.7 kg, respectively as shown in Figure 3.3.13. The power output of the PV module is directly related to the solar radiation and the area on which it is concentrated. As the solar radiation increases, the molecules in the PV module vibrate at a higher pace

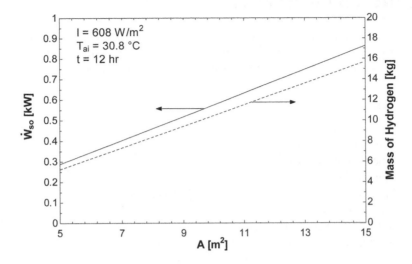

Figure 3.3.13 Effect of increase in area of solar cell on amount of power and hydrogen produced.

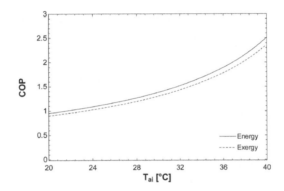

Figure 3.3.14 Effect of increase in air inlet temperature on energy and exergy COPs.

and as a result more and more bonds are broken into protons and electrons. As the power produced by the PV module increases, the hydrogen production rate increases because more power is being fed into the electrolyzer in order to break the bonds of water molecules at a higher rate to produce hydrogen. When the water molecules are broken into hydrogen and oxygen, a greater amount of hydrogen is available which can be taken out and stored in a cylinder for later use as an energy provider by burning or using PEMFC.

The effect of an increase in air inlet temperature of energy and exergy COPs of the TEACS is showing in Figure 3.3.14. The energy and exergy COP are found to be increasing from 0.95 to 2.51 and 0.89 to 2.36, respectively with increase in air inlet temperature from 20°C to 40°C. This increase is seen because, as the rate of heat input to the cooling system decreases to a certain practical limit for achieving certain cooling load, the performance of the system increases and a lesser amount of heat is being rejected through condenser.

3.4 CONCLUDING REMARKS

This chapter discusses energy and exergy related aspects of solar energy systems, considers various solar energy-based systems for analysis, assessment and comparison, and evaluates them for practical applications from the exergy point of view.

Nomenclature

A	Area of solar module, m^2
b	Breadth of PV module, m
COP	Coefficient of performance
c	Specific heat, kJ/kg·K
cp	Specific heat at constant pressure, kJ/kg·K
$\dot{E}_x$	Exergy rate, kW
ex_i	Specific exergy at state i, kJ/kg·K
ex_{ph_i}	Specific physical exergy at state i, kJ/kg·K
ex_{ch_i}	Specific chemical exergy at state i, kJ/kg·K
h	Specific enthalpy, kJ/kg
h_{ba}	Heat transfer coefficient from black surface to air, $W/m^2 \cdot K$
h_t	Heat transfer coefficient from black surface to air through glass, $W/m^2 \cdot K$
h_{p1G}	Penalty factor due to the presence of solar cell material, glass and EVA for glass to Glass PV/T system, $W/m^2 \cdot K$
h_{p2G}	Penalty factor due to presence of interface between glass and working fluid through absorber plate for glass to glass PV/T system, $W/m^2 \cdot K$
I	Solar flux, W/m^2
L	Length of the PV module, m
$\dot{m}$	Mass flow rate, kg/s
P	Pressure, kPa
$\dot{Q}$	Heat transfer rate, kW
s	Specific entropy, kJ/kg·K
T	Temperature, K
u	Specific internal energy, kJ/kg
U_b	Overall heat transfer coefficient from bottom to ambient, $W/m^2 \cdot K$
U_L	Overall heat transfer coefficient from solar cell to ambient through top and back surface of insulation, $W/m^2 \cdot K$
U_t	Overall heat transfer coefficient from solar cell to ambient through glass cover, $W/m^2 \cdot K$
U_{tb}	Overall heat transfer coefficient from glass to black surface through solar cell, $W/m^2 \cdot K$
v	Specific volume, m^3/kg
$\dot{W}$	Work rate, kW
z	Variable

Greek letters

α_c	Absorptance of solar cell
α_b	Absorptance of painted black surface

β_c	Packing factor of solar cell
η	Efficiency
τ_g	Transitivity of glass

Subscripts

a	Air
ai	Air inlet
abs	Absorber
bo	Boiler
bs	Back surface of photo-voltaic panels
c	Solar cell
con	Condenser
de	Destruction
el	Electrical
elec	Electrolyzer
en	Energy
ex	Exergy
eva	Evaporator
H_2	Hydrogen
HHX	High temperature heat exchanger
HTG	High temperature generator
i	State i
ov	Overall system
p	Pump
PV/T	Photo-voltaic/thermal
so	Solar
t	Turbine
th	Thermal
1...23	State numbers
0	Ambient or reference condition

Acronyms

CHX	Condenser heat exchanger
con	Condenser
eva	Evaporator
HHX	High temperature heat exchanger
HHV	Higher heating value
HTG	High temperature generator
LHX	Low temperature heat exchanger
LTG	Low temperature generator
PV	Photo-voltaic
PV/T	Photo-voltaic/thermal
MHX	Medium temperature heat exchanger
MTG	Medium temperature generator
TEACS	Triple effect absorption cooling system

REFERENCES

Dincer, I. and Rosen, M.A. (2007) *Exergy, energy, environment and sustainable development.* Oxford: Elsevier.

Joshi, A.S., Tiwari, A., Tiwari, G.N., Dincer, I. and Reddy B.V. (2009a) Performance evaluation of a hybrid photovoltaic thermal (PV/T) (glass-to-glass) system. *International Journal of Thermal Sciences*, 48, 154–164.

Joshi, A.S., Dincer, I. and Reddy, B.V. (2009b) Performance analysis of photovoltaic systems: A review. *Renewable and Sustainable Energy Reviews*, 13, 1884–1897.

Koroneos, C., Dompros, A., Roumbas, G. and Moussiopoulos, N. (2004) Life cycle assessment of hydrogen fuel production processes. *International Journal of Hydrogen Energy*, 29, 1443–1450.

Ratlamwala, T.A.H., Gadalla, M.A. and Dincer, I. (2011) Performance assessment of an integrated PV/T and triple effect cooling system for hydrogen and cooling production. *International Journal of Hydrogen Energy*, 36, 11282–11291.

Sciubba, E. (2001) Beyond thermoeconomics? The concept of extended exergy accounting and its application to the analysis and design of thermal systems. *Exergy-An International Journal*, 1, 68–84.

Thomas, L.G. and Nelson, A.K. (2010) Predicting efficiency of solar powered hydrogen generation using photovoltaicelectrolysis devices. *International Journal of Hydrogen Energy*, 35, 900–911.

Yilanci, A., Dincer, I. and Ozturk, H.K. (2009) A review on solar-hydrogen/ fuel cell hybrid energy systems for stationary applications. *Progress in Energy and Combustion Science*, 35, 231–244.

Chapter 4

Solar energy collection and storage

Brian Norton
Dublin Energy Laboratory, Dublin Institute of Technology, Dublin, Ireland

4.1 SOLAR THERMAL ENERGY COLLECTORS

4.1.1 Overview

Different types of solar thermal collectors provide energy in the form of heated

- water for direct use (Rabl, 1985)
- aqueous solutions, usually of glycols for freeze damage prevention (Norton and Edmonds, 1991; Norton et al., 1992), for hot water production
- air (Norton, 1992), usually for space heating
- specialized heat transfer fluid, mainly in solar thermal power generation (Duffie and Beckman, 1991)
- steam, also in solar thermal power generation systems (Kalogirou, 2003)
- inherent energy storage as in integral passive solar water heaters (Bainbridge, 1981; Smyth et al., 2006) and solar ponds (Leblanc et al., 2011)
- fluid with electricity production from a photovoltaic module (Norton et al., 2011) employed as the absorber
- refrigement or volatile fluid (Shreyer, 1981; Ong and Haider-E-Alathi, 2003)
- material undergoing solid to liquid phase change (Sion et al., 1979; Rabin et al., 1995).

The two principal solar thermal collector designs employed for space heating and hot water supply are the flat-plate solar collector and the vacuum tube solar collector the latter may be employed with line-axis concentrators . Flat-plate solar collectors have now been overtaken in total global numbers by the increasingly popular vacuum tube solar collector. The latter has a higher efficiency at higher temperatures and has become relatively inexpensive due to high production volumes of the all-glass type, particularly in China (Weiss and Mauthner, 2010). Integral collector-storage systems range from small-scale domestic water heaters to very large scale solar ponds. The former were the first mass-produced solar water heaters (Butti and Perlin, 1980) whereas the latter are site-specific designed civil engineering projects (Leblanc et al., 2011). Though most systems produce heated fluids for domestic, industrial (Kulkarni et al., 2008) or power generation applications, more esoteric users of solar heat include sterilization of medical equipment (Bansal et al., 1988) to the passive protection of grape vines from frost damage (Smyth and Skates, 2009).

4.1.2 Flat plate solar energy collectors

Flat plate collectors can absorb solar energy inident from a direct beam component, a diffuse component and ground reflected albedo of insolation. Flat plate collectors' inclusion of the latter two insolation components means that in many climates there are only a few instances where any viable long-term performance advantage is gained by tracking a flat plate collector to follow the sun's daily path across the sky. Most flat plate collectors have south-facing fixed mountings that usually provide a static inclination that is cogniscent of the maximum average diurnal solar energy collection period's duration, the annual durations of utilizable solar energy (Reddy, 1987), the prevalence of diffuse conditions and any diurnal or annual bias of the predominant times of hot water withdrawal.

A flat-plate solar collector usually consists principally of (Lenel and Mudd, 1984)

- tubes through which a heat transfer fluid is conveyed connected with good thermal contact with an
- an absorber plate that, often solar selectively coated, absorbs incoming solar radiation and
- an aperture cover plate that inhibits outgoing long-wave thermal radiation losses and traps an insulating air layer so inhibiting convective heat losses
- thermal insulation to maximize heat loss and a casing to provide weather protection together with mechanical integrity to the sides and back

A typical example of a flat plate solar energy collector is illustrated in Figure 4.1.1. The tubes are usually integrated fully in the absorber plate by a variety of

Figure 4.1.1 A flat plate solar energy collectors in close coupled thermosyphon systems.

manufacturing techniques (Duffie and Beckman, 1991) to ensure good thermal contact between the tubes and the plate.

4.1.3   Evacuated tube collectors

Evacuated-tube collectors are fabricated from either concentric glass tubes or a metal tube end-sealed to and within a glass tube. An enclosed evacuated annular space and a selective absorber surface provide a very low overall heat loss particularly when operated at higher temperatures. The evacuated space between the glazing and absorber eliminates convective loss and long-wave thermal radiative heat loss is inhibited by the deposition of the spectrally selective absorber coating on the absorber surface (Morrison, 2001). Evacuated-tube solar collectors generally have low thermal mass. The ability to heat-up rapidly (often from higher maintained overnight temperatures than a flat plate collector) gives low utilizable insolation thresholds providing good low insolation performance. Heat removal in evacuated tube collectors can be indirect often using a volatile fluid in the absorber via a closed heat pipe (Tabassum et al., 1988). More frequently water, as a heat transfer fluid, moving in a thermosyphon through the collector is employed (Morrison, 2001, Budihardjo and Morrison, 2007). Evacuated tube collectors can have copper absorbers or use a selectively coated glass inner glass tube as the absorber. The latter all-glass type is now commonplace in China with over 15 million square metres of collector area installed by 2010 (Norton, 2011). A typical example of an evacuated tube collector is show in Figure 4.1.2.

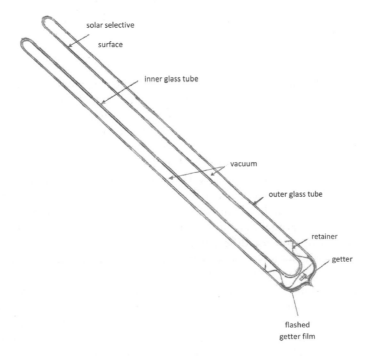

Figure 4.1.2 An all-glass "wet type" evacuated tube solar energy collector.

4.1.4   Collector components

4.1.4.1   Absorber plate

The plate and tubes of both a flat-plate and metal-in-glass evacuated tube solar energy collector absorber are usually made of a metal with high thermal conductivity such as copper or aluminum. Good heat transport is thus provided through the plate to the heat transfer fluid. An ideal absorber plate has a high solar absorbance surface to selectively absorb as much as possible of the incident insolation, together with low emittance to long-wave thermal radiation so that long-wave radiative heat losses are low (Norton, 1992). Such solar selective absorbers often consist of two layers with appropriately different optical properties; often a thin upper layer that exhibits high absorptance to solar radiation whilst being relatively transparent to thermal radiation is deposited on an underlying surface whose high reflectance provides low emittance to thermal long-wave radiation. Alternatively a heat selective mirror that has high solar transmittance with high infrared reflectance can be placed on top of a non-selective high absorbtance material. An example of this latter of selective surface combination is "black chrome" that consists of microscopic chromium particles deposited on a metal substrate; the chromium particles reflect long-wave thermal radiation, but shorter wavelength insolation passes between the chromium particles.

In very low cost low temperature solar energy collectors for applications, such as swimming pool heating (Ruiz and Martinez, 1992), all these guidelines for absorber materials and fabrication are inappropriate as the minimum initial cost usually dominates design choices. Black paint which has a high absorptivity but not being selective is equally high emittance is used on absorbers or is the dark pigment in unglazed often ground-mounted plastic coils through which water is solar heated.

4.1.4.2   Aperture cover

Most glasses are almost completely transparent to the shortwave radiation the associated with insolation, but nearly opaque to long-wave thermal radiation. When employed as an aperture cover, glass inhibits successfully radiative heat loss from an absorber plate to ambient or to the lower temperature radiative heat sink that the sky often constitutes. Glass is thus the aperture cover material employed most commonly for solar energy collectors (Norton, 1992). It is desirable that a large part of the incoming direct, diffuse and ground reflected solar radiation is transmitted through the cover and used efficiently to heat the transfer fluid. This means that the transmittance of an aperture cover must be high which requires both low reflectance and absorptance. The reflectance of a material depends on its refractive index and the angle of insolation incidence. It can vary for different wavelengths of insolation transmittance and decreases with increasing angle of insolation incidence (Duffie and Beckman, 1991). An aperture cover glazing with a smaller refractive index exhibits a lower reflectance and higher transmittance. The overall solar transmittance of an aperture cover glazing is the normalized sum of each quantized wavelength of the solar radiation spectrum transmitted by the aperture cover. Since iron absorbs light in the visible part of the spectrum, the transmittance of glass for the solar spectrum decreases with increasing iron concentrations. Thus low-iron glass typically with an iron content below 0.06 percent is preferred for flat plate collector aperature cover glazing.

Glass reflectance can be reduced either by coating the glass with a thin film with a low refractive index material, or by etching the surface to create a porous lower refractive index medium, reducing the reflectance of the glass. For improved insulation of the collector, multiple glasses that trap multiple insulating air layers can be used, though the optical losses associated with each additional glass sheet reduce the insolation transmitted.

Many clear plastics, without specific treatments, become yellow and brittle after long-term exposure to ultra-violet part of the insolation spectrum. Many plastics can also be damaged by the high temperatures that can be reached in solar energy collectors, particularly if fluid stagnation occurs in the solar energy collector. Another disadvantage of most plastics is that, compared to glass, transmittance to thermal radiation is high at longer thermal wavelengths. Altering the optical properties of plastics to reject heat gain under the influence of heating or an electric field has been studied as a means of avoiding overheating damage when plastics are used as absorbers in solar energy collectors (Resch and Wallner, 2008).

Both glass and plastic aperture covers do, despite their different solar optical properties, will enable absorber heat losses to be ameliorated as they both trap a largely stagnant insulating air layer between the glazing and absorber (Rommel and Wagner, 1992).

4.1.4.3 Choice of heat transfer fluid

The choice of the heat transfer fluid is determined by maximum operating temperature, initial and operating costs, toxicity flammability and environmental impact. For high temperature applications water and aqueous solutions are often inappropriate. Hydrocarbon and synthetic-based heat transfer oils may be used up to their maximum temperatures of around 450°C. They are however flammable. The special safety systems together with the environmental issues in their post-use disposal measure the operational costs of the use of synthetic heat transfer rate. Steam has been studied for many central receiver applications with maximum temperature applications 550°C. Water used to generate steam must be deionized in order to prevent scale buildup on the inner surfaces of the receiver. Liquid sodium and nitrate salt mixtures can also be used as both a heat transfer fluid and storage medium, with a maximum operating temperatures of 600°C and 560°C.

4.1.4.4 Combined photovoltaic – thermal collectors

A photovoltaic panel may convert typically 10%–15% of the incident insolation to electricity. The rest of the solar energy absorbed heats the panel reducing photovoltaic electrical energy conversion efficiency. Combined photovoltaic-thermal collectors have thus been developed with either water (Ji et al., 2006) or air as the heat transfer fluid. The many such systems that have been proposed all require displacement of both suitable electrical and heat loads if economic viability is to be achieved. Hot water produced by such combined systems can be used for space heating or domestic hot water. When warm air is produced it can be used for pre-heating of ventilation air. The heat can also be used in absorption chillers to cool a building. A typical PV/T collector is shown in Figure 4.1.3.

Figure 4.1.3 A PV/T solar energy collector.

A solar heat pump uses low-temperature, low-quality energy from the outside air, transferring heat from the cold reservoir to a warmer reservoir (Morrison, 1984). The working fluid is alternately evaporated and condensed. A heat pump involves isentropic compression in the compressor, constant-pressure heat rejection in the condenser, throttling in an expansion device and constant-pressure heat absorption in an evaporator (Charters et al., 1980). Solar heat pumps for small-scale heating and cooling applications are referred to as Combi[+] systems (Troi et al., 2008).

4.2 INTEGRAL COLLECTOR STORAGE SYSTEMS

4.2.1 Integral passive solar water heaters

Integral storage solar collectors are a tank, glazed lying on top of insulation (Bainbridge, 1981; Arthur and Norton, 1988; Smyth et al., 1998; 1999; 2001a; 2001b; 2003; 2004; 2005; 2006). The water inside the collectors is heated by absorbed incident insolation. The heated water may be pumped directly to a demand or a hot storage tank for later use. At night or during periods of low insolation, the water in the collectors may be drained back to tank, thereby conserving the heat collected (Dickinson et al., 1976). A typical example is illustrated in Figure 4.2.1. Many techniques are available to determine the optimal design of such systems for particular climates and hot water loads (Bar-Cohen, 1976; Chauhan and Kadambi, 1976).

4.2.2 Salt gradient solar ponds

A salt-gradient solar pond employs a salt concentration gradient to suppress natural convection. The physical processes are illustrated in Figure 4.2.2. Heated water

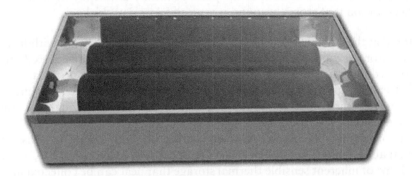

Figure 4.2.1 An integral collector storage solar water heater.

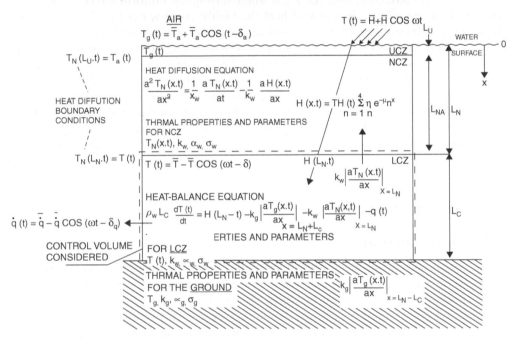

Figure 4.2.2 Heat transfer in a non-convecting solar panel (Norton, 1992).

dissolves holds more salt than does cooler water. Salty, heated water being heavier remains at the base of a solar pond. Insolation penetrating through the top layers of a pond is absorbed by the layer with heat loss inhibited by the intervening non-convecting layer (Leblanc et al., 2011).

Salt-gradient solar ponds consist of three zones:

- a surface convecting zone of low-salinity water, typically 0.2 m–0.4 m thick;
- a non-convecting or salinity-gradient zone beneath the surface zone, which thermally insulates a lower convecting heat-storage stratum in which dissolved salt concentration increases with depth, typically 1.0 m–1.5 m thick; and

- a storage zone at the bottom of the pond of ideally uniformly high dissolved salt concentration that stores heat and is typically 1 m–3 m thick.

Heat stored at the bottom of the pond, is removed by hot brine being withdrawn from the lowest storage zone of the pond by a pump passed through a heat exchanger and then returned back to the lower of the storage zone. For power production applications that employ an organic Rankine cycle, the engine's condenser cooling water is withdrawn from the top of the pond, from where it is passed through the condenser before being returned back to the surface layers of the pond.

Non-convecting solar ponds are energy collectors with "built-in" seasonal heat-storage capabilities that can provide heat at temperatures in excess of 90°C with such a large volume of inherent sensible thermal storage that heat can be collected in summer and stored for use during.

The lower convecting zone (LCZ), is where the highest uniform salt concentration occurs. In it the solar radiation will heat the highly-saline water but, because of its high relative density due to its salt content, this heated water will not rise into the lower salinity layers. Thus the heat is stored yet inhibited from being transferred by convection. In order to establish a conventional non-convecting solar pond for power production, it must have a large surface area (i.e. extend over several square kilometres), and so vast excavations and site preparations are inevitably necessary: these operations can usually account for more than 40% of the total capital cost of the non-convecting solar pond. Construction of economically viable solar ponds requires the ready availability of inexpensive flat land; accessibility to water; and an inexpensive source of salt or brines (Norton, 1992).

4.3 CONCENTRATORS

4.3.1 Introduction

Solar energy is concentrated to produce higher output temperatures and/or reduce collector cost by replacing an expensive absorber area with less expensive reflector. Concentration ratio is the factor by which an intervening concentrating mirror, lens or luminescent system increases the insolation flux on a solar energy absorbing surface. A geometric concentrator ratio is the collector aperture area (A_a) is divided by the absorber surface area (A_R) (Rabl, 1985):

$$C = \frac{A_a}{A_r} \tag{4.3.1}$$

4.3.2 Concentration systems

Thermal losses are dependent largely on the heat loss characteristics of, and proportional to, absorber area. Insolation can be concentrated by a two-dimensional system to a theoretical upper limit given by:

$$C_{\max,2D} = \frac{n}{\sin(\varepsilon/2)} \tag{4.3.2}$$

where ε is the angular size of the suns disc, n is the refractive index of the last material insolation traversed for the refractive index of air $C_{max,2D}$ is approximately 216. For a three-dimensional concentrator, the equivalent upper limit is:

$$C_{max,3D} = \frac{n^2}{\sin^2(\varepsilon/2)}$$

(4.3.3)

$C_{max,3D}$ in typically 46000 for concentrators operating in the refractive index of air. To maintain high concentration high concentration ratio two-dimensional concentrators need accurate solar tracking systems as concentration decreases sharply at off-normal insolation incidence. As they have much larger maximum concentration ratios three-dimensional systems need less-precise tracking accuracy to yield acceptable optical performance.

Both the earth's diurnal rotation about its axis and annual rotation about the sun mean that a tracking solar energy collector must move continually in two axis to maintain the direct insolation component precisely at normal incidence to the aperture plane (Rabl, 1985). Two-axis solar tracking is essential to achieve concentration ratios from 100–1000. Such concentrators can elevate fluids to temperatures at which it is usually possible to generate electricity using steam turbines or Stirling cycles or to provide high grade thermal energy for industrial processes. Single axis trackers are usually oriented either horizontal east-west, or inclined north-south achieving normal incidence once a day or twice a year respectively.

Stationary concentrating systems with a concentration ratio in the range from 1 to 3 can be integrated into buildings as they obviate the need for moving parts, complex mounting and associated mechanical systems. Most systems are based on two-dimensional non-imaging compound parabolic concentrators (CPCs). Building integration of high temperature solar thermal applications is attractive as a means of lowering installation cost. The initial investment cost will be reduced if lower cost reflector materials replaced more extensive use of evacuated tube collectors.

The two dimensional CPC is termed an ideal concentrator, with a concentration ratio of $1/\sin\theta_{max}$, since all the light incident at angles less than the angle of acceptance will arrive at the absorber. As can be seen in Figure 4.3.1, the CPC is deep in comparison with the width of the absorber. The arrangement is both impractical and costly in the practical fabrication of a concentrator (Tripanagnostopoulos et al., 2002; 2004a; 2004b). Increasing the concentration ratio reduces the angle of acceptance resulting in a considerably deeper trough.

Truncating the height of a CPC does not diminish the aperture significantly, if a third of the length of a low concentration ratio CPC trough were truncated it would only reduce the aperture area by about 3%. Thus in most practical solar energy collectors that employ CPCs, they are truncated significantly.

As the maximum concentration ratio is $n/\sin\theta_{max}$ it is possible to increase the concentration ratio by replacing the air in the trough with a dielectric medium. When a dielectric medium with a refractive index greater than $\sqrt{2}$ is used, total internal reflection will ensue at each reflection. A dielectric-medium concentrator without reflectors can thus be constructed, that will have no reflection losses. This will be of lower cost if eliminating the cost of the reflector is not offset by the cost of the dielectric material.

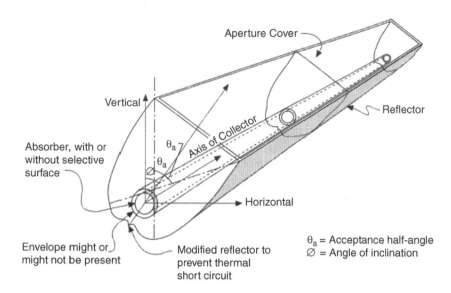

Figure 4.3.1 A compound parabolic concentrator.

When using a dielectric medium with a refractive index of >1, as the concentrator, light will be refracted to a smaller angle of incidence rendering it possible to accept insolation across a larger acceptance angle. A concentrator filled with a high refractive index low-iron dielectric would increase the concentration ratio by 52%.

4.4 SOLAR WATER HEATING

4.4.1 Overview

The diverse different forms of solar energy water heaters that are now available (Norton, 2011) include:

- thermosyphon solar water heaters (Norton and Probert, 1986) operate by natural convection, can be direct or indirect and can use a variety of flat-plate or evacuated tube collectors.
- pumped-circulation solar water heaters (Reddy, 1987; Duffie and Beckman, 1991) usually indirect and have larger collector area than thermosyphon systems, they also can use either flat plate or evacuated tube collectors. The pump is usually activated by a set collector outlet-inlet temperature difference though photovoltaic modules have also been used to power pumps (Parker, 1976; Al-Ibrahim et al., 1998). Most pumped systems are indirect (Frazer et al., 1995; Parent and Van der Meer, 1990; Yazdanshenas et al., 2008; Too et al., 2009).
- combisystems (Weiss, 2003) (a contraction of "combined systems") these provide both space heating and hot water. These, invariably indirect, systems require care-ful design of heat exchanger arrangement and central strategies of the optimal

solar savings fraction is to be provided. (Yazdanshenas and Furbo, 2007; Letz et al., 2009; Yazdanshanas et al., 2008). The annual variations of space and water heating loads can be quite different.

- integral collector storage (Smyth et al., 2006; 1998; 1999; 2001a, 2001b; 2003; 2004) in which the heat water store is also the solar energy collector.
- Combi$^+$; these are solar assisted heat pumps (Morrison, 1984; Troi et al., 2008).
- large scale interseasonal energy storage systems (Schmidt et al., 2004).
- swimming pool heating (Ruiz and Martinez, 2010); employing flat plate collectors often unglazed using a low-cost plastic pipe absorber.
- photovoltaic solar water heaters (Fanney and Daugherty, 1997; Fanney et al., 1997) where a electrical heating element immersed the water to be treated is powered by a photovoltaic array.

In addition to evacuated tube and flat plate collectors, solar water heating can also be accompanied using photovoltaic/thermal collectors. Such latter collectors use the 85–95% of the incident on a photovoltaic array not connected to electricity (Ji et al., 2006; Chow et al., 2008) as heat.

4.4.2 Applicability of particular collector types to specific outlet temperatures and diffuse fractions

Solar energy availability can be diurnally or annually out-of-phase with a heat load. When the heating needs are at their peak, the supply of solar energy can be at its lowest. However, domestic and industrial hot water needs often do not vary over the year and solar collectors can be used for producing hot water during the summer.

Solar energy water heaters can be categorised as either active or passive. An active system requires a pump to drive the heated fluid through the system, whereas a passive system requires no external power. Distributed systems (Prapas et al., 1993) comprise a solar collector, hot water store and connecting pipework; they may be either active or passive. In the former, a pump actuated by temperature sensors via a control circuit temperature difference between collector inlet and outlet is actuated by the required to convey the fluid from the collector to the store (Prud'homme and Gillet, 2009; Wuestling et al., 1985). The pump may be powered by a photovoltaic module giving automony from mains power (Al-Ibrahim et al., 1988). In a passive thermosyphon solar water heater; fluid flow is due to buoyancy forces occurring in a closed circuit comprising a collector, hot water store and the connecting pipework produced by the difference in densities of the water in the collector and that of the cooler water in the store.

Concentrating solar energy collectors require high direct components of insolation to be effective. Extensive use of solar power generation is likely to be most viable where the annual direct fraction is above 60%. This delineates their geographical applicability in Figure 4.4.1. Similarly solar water system productivity is a function of insolation and ambient temperature. For identical load profiles and temperatures solar savings fractions and continuous autonomous provision can be determined (Yohanis et al., 2006a; 2006b). A geographical distribution of the latter is shown in Figure 4.4.2.

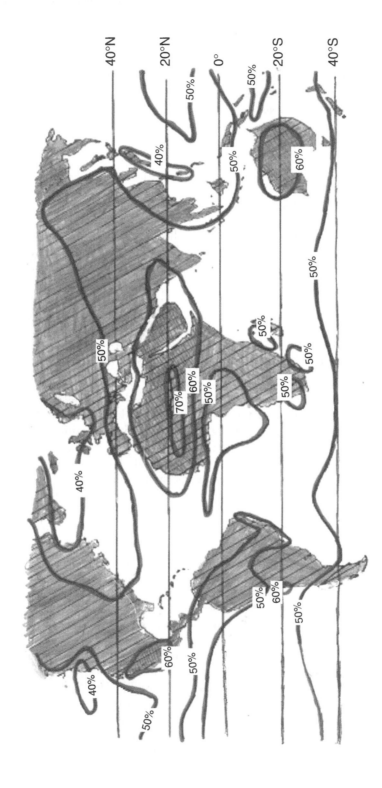

Figure 4.4.1 Annual direct fractions of insolation, areas suitable to solar thermal electrical power generation have >60% direct fractions.

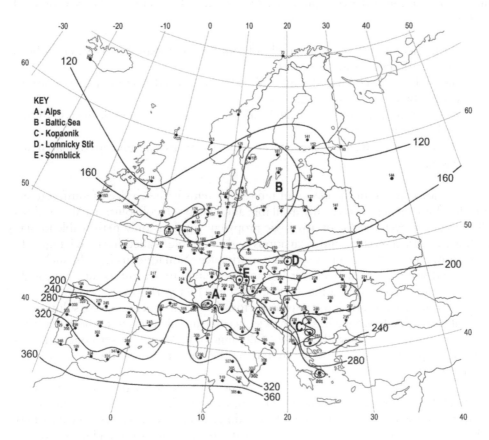

Figure 4.4.2   Distribution of continuous days of hot water production at 37°C is across Europe (Yohanis et al., 2006a; 2006b).

4.4.3   Freeze protection methods

Freeze protection of a water collector in winter can be gained by either draining the collector circuit using multiple glazings (Bishop, 1983) evacuation to eliminate convective heat loss. (Mason and Davidson, 1993; Hongchuan and Guangming, 2001), internal convection suppression (Smyth, 2001a) or heat transfer fluid choice. The heat transfer fluid used in a solar collector during winter sign should not allow freeze expansion damage. The most common heat carrier in solar collectors is an aqueous solution of water and glycol together with appropriate additives that inhibit corrosion. Regulatory compliance has meant that ethylene glycol has been superseded by less toxic propylene glycol. Glycol-filled indirect system can produce warm water in winter unlike a direct system which would still remain drained down (the latter being usually a biennial operation). The advantage of a longer operating period is counteracted in summer by the lower thermal efficiency of an indirect system with a, more viscous, propylene glycol solution when compared with a direct system in which water is the heat transfer fluid. For *thermosyphon* solar water heaters, the sources of an indirect system's relative inefficiency are, in order of significance (Norton and Edmonds, 1991; Norton et al., 1992):

(1) the higher viscosity of aqueous glycol solutions (compared with water) reducing the natural-circulation flow rate,
(2) the additional (compared with a direct system) fluid frictional flow resistance caused by the heat exchanger,
(3) heat transfer resistance associated with the, frequently double-walled, in the heat exchanger used, and
(4) the lower specific heat capacity of aqueous propylene glycol compared with water.

A major concern in the design of indirect systems is the choice of antifreeze heat transfer fluid. Being both readily available and cheap, glycol solutions are used usually. However, even the use of low-toxic propylene glycol is only permissible in many building codes, standards and regulations with a double-walled heat exchanger. The latter, however, further reduces system performance. Nonaqueous liquids, such as silicones and some hydrocarbons, are low to moderately toxic; but as they have higher viscosities (compared with aqueous solutions) together with lower specific heats and thermal conductivities, their use impairs performance. An aqueous-glycol solution may be used with a single-walled heat exchanger in some jurisdictions if it is both effectively nontoxic and stable chemically over the full range of the water heater's operating conditions, propylene glycol solutions containing the appropriate inhibitors satisfy these requirements.

Many solar energy water heaters act as a preheater in series with an auxiliary heater (DGS, 2005). In a preheat system, the mass of water withdrawn from the solar system forms the hot water consumption in the building, so energy is always gained from the solar energy system even when it raises the water temperature only slightly above main supply temperature. An auxiliary heater raises this water to a temperature above 55°C sufficient to eradicate any Legionnella bacteria. The auxiliary outlet water temperature is thus above usual bathing temperatures and is normally mixed with cold at the faucet, shower or bath to achieve the temperature desired. If water is drawn off at temperatures above the auxiliary heater set point, then it can be argued that the draw-off volume should be reduced because a greater proportion of cold water will be used to give the desired temperature at the point of use. In practice this will depend on whether a two-tank system is used and on heat losses from the auxiliary heater tank.

The freezing point of commercially available propylene-glycol solutions varies with concentration. The ambient temperature distributions for each of the months when frost is recorded also changes with location. Ensuant critical concentrations for system survival are thus location dependent.It is usually worthwhile to continue operating a solar energy water heater in winter if produces a *net* heat output. In London, UK, for example there is no month in which a solar energy water heater if designed appropriately containing propylene glycol at the critical concentration necessary for survival, would not produce heat output.

It could be preferable to use a direct solar energy hot water system, drain the collector in winter, and forego the energy that is available during the winter months in order to benefit from greater hot water production efficiency during the summer.

To determine if this approach is valid the direct solar energy water heating system performance thus needs to be compared with that of an indirect system. The latter should contain the optimum aqueous propylene-glycol solution concentration that yields the maximum total output during all the months where it can survive the minimum ambient temperature.

4.4.4   Sensible and latent heat storage

Energy storage is employed in solar thermal energy systems to enable excess energy produced during high insolation to be available when insolation is low or non-existent (at night). Energy storage may be needed; for where some of the solar thermal energy produced during the day is stored for use later during the night or to provide energy over sequences of cloudy days. There is a broad range of heat storage media for solar thermal energy systems. However, practical design considerations limit thermal energy storage options to sensible and latent heat energy storage in liquids/solids and phase change materials respectively. Thermochemical energy storage has considerable promise but as yet to realise practical deployment.

In sensible heat thermal energy storage, cold fluid in an insulated store is heated to a higher temperature by the hot fluid from the solar energy collectors. Colder fluid is withdrawn from the bottom of the store and is heated in the collector. The hot fluid from the collector returns to top of the hot water store. The less dense hot storage fluid will form a stratified layer delineated by a thermocline on top of the cold fluid (Hollands and Lightstone, 1989). Water heated by the solar energy collector can enter the store at a temperature lower than that of the local stored heated water. This can ensure of, for example, high insolation has been succeeded by cloudier skies. Fluid inlet arrangements have been devised to reduce destratification by water solution of colder water to the warmer layers of a store (Lavan and Thompson, 1977; Davidson and Adams, 1994). When the hot fluid is withdrawn from the store, the latter is usually replenished by cold mains water. The introduction of such cold water is best accomplished via a low velocity flow that does not disrupt the stratification of the hot water store (Eames and Norton, 1993; Furbo and Fan, 2008). This can cause an unacceptable rate of hot water delivery in both industrial and domestic (e.g. showers) applications that will require intermediate stores to accomplish end-use effectiveness. Large-scale sensible heat storage systems (Lund, 1986) have been built to supply district heating (Dalenback, 2010). The majority of large-scale interseasonal storage systems serve housing via distinct heating networks. Systems with output of 7.0 MW are currently most prevalent in Denmark as can be seen from Table 4.4.1 which summarises such systems with an output >4 MW$_{th}$ in operation in 2010 (Dalenback, 2010).

The cost of thermal energy storage systems is dominated by the initial cost of the storage medium. The use of water or steam (assuming a low cost pressurized storage tank) as a storage medium reduces storage fluid costs. In addition, the use of water or steam as a storage fluid in a solar thermal electric system using a steam-driven power generation unit obviates the need for an oil/water steam generator.

In a latent heat energy storage system, the

- latent heat storage materials must be of low cost and available readily
- latent heat storage material, if a mixture, must not separate into component materials after repeated phase change cycles.

Table 4.4.1 Large-scale interseasonal solar heating systems with output $>4\,MW_{th}$ (adapted from Dalenback, 2010).

Collector Area (m^2)	Nominal Power Output MW_{th}	In operation since 1996	Location	Country
18,300	12.8	1996	Marstal	Denmark
10,700	7.5	2009	Broager	Denmark
10,073	7.0	2009	Gram	Denmark
10,000	7.0	2000	Kungälv	Sweden
8,012	5.6	2007	Braedstrup	Denmark
8,012	5.6	2008	Strandby	Denmark
7,300	8.1	2003	Grailsheim	Germany
7,284	5.1	2009	Torring	Denmark
6,000	4.2	2008	Soenderberg	Denmark
5,670	4.0	1997	Neckarsulm	Germany

- latent heat storage material must not corrode or react with heat-transfer surfaces or solar energy collector materials.
- supercooling behaviour of the latent heat storage material on solidification should usually be limited and consistent over numerous freeze/thaw cycles.
- toxicity and flammability must satisfy regulating requirements

In a thermochemical energy storage system thermal energy separates chemical bonds reversibly. The displacement of the chemical bond energy requires absorbs heat energy resulting in thermal energy storage. The chemical products a useful thermochemical heat storage reaction are unreactive at ambient temperatures. As temperatures increase the chemical bonds are reestablished forming the original chemical with the release of stored heat. Highly endothermic chemical reactions can achieve very dense energy storage per unit material mass. The energy remains stored until it is recovered by an exothermic reaction.

4.4.5   Analytical representation of thermosyphon solar energy water heater

In a thermosyphon solar water heater a hot water store is located above and connected by dawncomer and upriser pipes to the solar collector. The height difference between the collector and store inhibits nocturnal reverse circulation (Norton and Probert, 1983).

A finite difference transient heat transfer and momentum analysis in which the thermal capacitances of both the collector plate and the cover are implicit can be applied to the liquid (i.e., water) circulating through the collector, upriser, hot water store, and downcomer of a thermosyphon loop (Hobson and Norton, 1988). All fluid properties and heat-transfer coefficients can be assumed to be temperature-dependent based on either the ambient or mean component temperatures. These coefficients can be updated at each timestep in a numerical calculation. An appropriate store model includes a simulation of buoyancy-induced mixing between stratified layers that occurs when warmer fluid is introduced below a cooler layer. Friction factors can be calculated using correlations appropriate to both the nonisothermal thermally-destabilized low Reynolds number flow and the isothermal developing laminar flow present in the lengths of straight pipeline. Empirically-determined laminar heat loss coefficients for the pipe bends are employed usually. Time variations of insolation, ambient temperatures and hot water withdrawal are the inputs to most simulation, models with the transmission of the glass collector cover being a function of the sun-hour angle.

A two-dimensional finite difference approach takes account of the thermal capacitance of the collector. In the derivation of the energy equations for the collector model, a glass cover, opaque to long wave radiation, and parallel fin-and-tube collector plate are treated usually as two large, parallel, grey bodies for the analysis of radiative heat exchanges. In addition, the glass cover is assumed to be at a uniform temperature at each moment in time and is therefore represented by a single node. It is also assumed usuallythat the temperature gradient through each thin fin is constant so that two-dimensional planar conduction is assumed to prevail. Conduction within the collector fluid in the direction of the main flow is taken to be negligible in most analysis as are the thermal capacities of the thermal insulations applied to the hot-water store, collector, and connecting pipes. An energy balance on an incremental volume of the fin gives, for a two-dimensional plate temperature distribution,

$$\rho_f C_f \delta_f \frac{\partial T_f}{\partial t} = k_f \delta_f \left(\frac{\partial^2 T_f}{\partial x^2} + \frac{\partial^2 T_f}{\partial y^2} \right) + h_{f \cdot g}(T_g - T_f) + U_{f,a}(T_a - T_f)$$

$$+ \left(\frac{\sigma}{\varepsilon_g^{-1} + \varepsilon_f^{-1} - 1} \right) (T_g^4 - T_f^4) + (T_\alpha)_e I. \tag{4.4.4}$$

The boundary conditions are: (i) from symmetry of adjacent nodes

$$\frac{\partial T_f}{\partial x} \bigg|_{(\frac{\omega}{2},o,t)} = \frac{\partial T_f}{\partial x} \bigg|_{(w,L,t)} = 0 \tag{4.4.5}$$

and, (ii) as there is no heat flux through ends of the plate

$$\frac{\partial T_f}{\partial y} \bigg|_{(x,o,t)} = \frac{\partial T_f}{\partial y} \bigg|_{(x,L,t)} = 0 \tag{4.4.6}$$

The boundary condition relating the temperatures of the fluid in the risers and the plate is obtained from a heat balance or an incremental volume of the fluid. The temperature of the pipe wall is assumed to be that of the fin at $x = 0$. Thus:

$$\rho_w C_w \left(\frac{\pi D_r^2}{4} \right) \frac{\partial T_w}{\partial t} + \frac{C_w \dot{m}_c}{N} \frac{\partial T_w}{\partial y} = h_{rw} \pi D_r \times (T_{f(o,y,t)} - T_w). \tag{4.4.7}$$

The cover temperature is obtained via

$$\rho_g \delta_g C_g \frac{\partial T_g}{\partial t} = h_{gf}\left(\bar{T}_f - T_g\right) + \left(\frac{\sigma}{\varepsilon_g^{-1} + \varepsilon_f^{-1} - 1}\right) \times \left(\bar{T}_f^4 - T_g^4\right)$$
$$+ h_{wind}(T_a - T_g) + \sigma \varepsilon_g (T_{sky}^4 - T_g^4). \tag{4.4.8}$$

To simulate accurately the transient response time of the collector, it is essential to take into account the thermal capacity of the water in the collector's header pipes. The transient response of the fluid in this component has been shown to account for the experimentally-observed the time lapse between a change in the temperature of the fluid leaving the end of the riser pipes and when a corresponding perturbation appears at the outlet of the header pipe. Assuming a uniform flow distribution between the collector risers, a heat balance on the nth section of the header pipe gives

$$\frac{W \rho_w}{\dot{m}}\left(\frac{\pi D_h^2}{4}\right) \frac{\partial T_n}{\partial t} = n(T_{n-1} - T_n) + (T_0 - T_{n-1}). \tag{4.4.9}$$

for the connecting pipes, a heat balance for an element of fluid within the pipes gives

$$\rho_w C_w \left(\frac{\pi D_h^2}{4}\right) \frac{\partial T_w}{\partial t} + C_w \dot{m}_c \frac{\partial T_w}{\partial y} = U_{s,a} \pi D_\rho (T_a - T_w). \tag{4.4.10}$$

for the hot water store, an energy balance on an incremental section of fluid which is remote from the end sections of the tank gives

$$\rho_w C_w A_s \frac{\partial T_w}{\partial t} + C_w \dot{m}_s \frac{\partial T_w}{\partial y} = k_w A_s \frac{\partial^2 T_w}{\partial y^2} + U_{s,a} P_s (T_a - T_w) \tag{4.4.11}$$

where $\dot{m}_s = \dot{m}_c - \dot{m}_L$.

The boundary conditions for the hot-water store are determined by considering incremental sections of fluid in contact with either the top or the base of the tank as $\delta y - 0$, i.e., (i) for the top of the tank,

$$U_{s,a,T}(T_a - T_w) + k_w \frac{\partial T_w}{\partial y} = 0 \tag{4.4.12}$$

(ii) for the base of the tank

$$U_{s,a,B}(T_a - T_w) - k_w \frac{\partial T_w}{\partial y} = 0 \tag{4.4.13}$$

In a simple mixing model, when warm fluid is introduced below cooler water, it is assumed that complete mixing ensues and the two adjacent nodes attained a single temperature. This process is repeated down the tank until a stable themocline is restored.

A transient momentum balance on the four components comprising the thermosyphonic loop, assuming one-dimensional incompressible flow, gives

$$\left(\frac{L_r}{A_p} + \frac{L_2}{A_p} + \frac{L_s}{A_s} + \frac{L_d}{A_p}\right)\frac{\partial \dot{m}_c}{\partial t} = g\phi\rho_w \sin(\theta)dy - \Sigma P_m$$

$$-\frac{\dot{m}_c}{2}\left[\frac{f_r L_r}{\rho_w D_r A_r^2 N^2} + \frac{f_u L_u}{\rho_w D_p A_p^2} + \frac{f_u L_u}{\rho_w D_p A_p^2}\right] + \frac{f_d L_d}{\rho_w D_p A_p^2}. \qquad (4.4.14)$$

Equations expressed in finite difference forms are solved simultaneously using their appropriate boundary conditions. When simultaneous equations are solved using a Gauss-Seidel iterative method in order to find $T_{(x,y,t+\Delta t)}$ for each node, the solution is stable unconditionally and the size of the time step is only limited by the accuracy required.

4.4.6   Solar water heater design

From a transient heat balance on a generic directly heated thermosyphon solar-energy water heater, the following dimensionless parameters Y, Z and X, designated the Heywood, Yellot and Brooks numbers respectively, have been identified (Hobson and Norton, 1988b):

Heywood number: $Y = \dfrac{F_{AV}A_C(\tau\alpha)_e H}{M_S C_W(T_a - T_m)}$ \qquad (4.4.15)

Yellot number: $Z = \dfrac{[F_{AV}A_C U_L + (UA)_S]t}{(M_S C_W)}$ \qquad (4.4.16)

Brooks number: $X = \dfrac{f Q_{tot}}{M_L C_W(T_a - T_m)}$ \qquad (4.4.17)

$$\frac{X}{1 - \exp(-Z)} = \frac{Y}{Z} + 1 \qquad (4.4.18)$$

The mean daily circulation number is Np and

Specific load: $W = \dfrac{M_L}{M_S}$ \qquad (4.4.19)

The Bailey number, K, which represents the system parameters effecting flow within the system is defined as:

Bailey number: $K = \dfrac{\rho\beta g \Delta T_{ref}[h_3 - h_2/2]}{v\dot{m}_{ref}[L_r/N(D_r^4) + L_\rho/D_\rho^4)]}$ \qquad (4.4.20)

where ΔT_{ref} and m_{ref} are given the values $10°C$ and $10^{-1}\,\text{kg}^{-1}$ respectively.

The relationships between the Yellot, Z, Bailey, K, Heywood, Y, Brooks, K, numbers and the specific load, W, may be summarised as a nomogram shown in Figure 4.4.3. The Heywood, Y, and Yellot, Z, numbers and the specific load, W, are functions

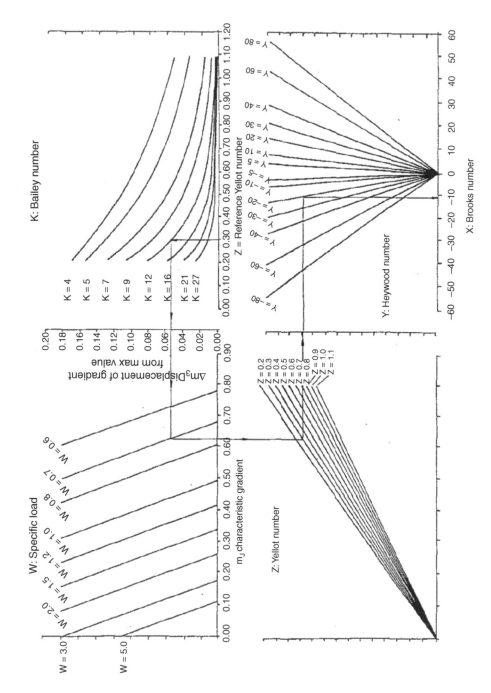

Figure 4.4.3 Nomogram representing the design formulae given by Equations 4.3.8, 4.3.9 and 4.3.5.

Table 4.4.2 Details of the configuration, operating conditions and thermal properties of the thermosyphon solar-energy water-heater used in the sample calculation.

System parameters	$A_c = 2.0$	(m^2)
	$F_{av} = 0.9$	
	$(\tau\alpha)_e = 0.72$	
	$U_L = 3.5$	$(W\,m^{-2}\,K^{-1})$
	$(UA)_s = 3$	$(W\,K^{-1})$
	$N = 8$	
	$M_2 = 297$	(kg)
	$h_3 = 1.8$	(m)
	$h_2 = 1.7$	(m)
	$L_r = 1$	(m)
	$D_r = 0.015$	(m)
	$L_p = 8.72$	(m)
	$D_p = 0.025$	
	$H_{td} = 19.2$	$(MJ\,m^{-2})$
Weather conditions	$T_a = 16$	$(°C)$
	$T_m = 15$	$(°C)$
	$t = 59,220$	(s)
Hot-water demand	$M_L = 208$	(kg)
	$T_L = 46$	$(°C)$
Fluid properties	$\rho_w = 998$	$(kg\,m^{-3})$
	$\mu_w = 10^{-3}$	$(Ns\,m^{-2})$
	$v_w = 1.00 \times 10^{-6}$	$(m^2 s)$
	$C_w = 4190$	$(J\,kg^{-1}\,K^{-1})$
	$\beta_w = 2.1 \times 10^{-4}$	(K^{-1})

of the applied conditions, whereas the Bailey number, K, is a function essentially of the system design. However, all these dimensionless groups include information available readily to a designer who, using the nomgram, can thus determine the Brooks number X, and thus the solar fractions.

A worked example of using the correlations to predict the daily solar fraction The component specifications of a thermosyphon solar- energy water heater and climatic conditions (for a typical day in June) used in the following example are given in Table 4.4.2. Evaluating the parameters K, W, I and Z gives $K = 12$, $W = 0.7$, $Y = 20$, and $Z = 0.3$ respectively. Also, since the thermal performance is being determined for the reference month of June, $Z_J = Z = 0.3$. A three-stage algorithm is used to determine the dimensionless Brooks number, X, from which the solar fraction can be calculated. Stage I is to determine the deviation, Δm_J (due to circulation number effects) of the characteristic gradient from the maximum value, $m_{J,max}$.

$m^* = 0.055$, giving $\Delta m_J = 0.051$. This stage corresponds to the first quadrant of the nomogram shown in Figure 4.3.4. Stage two is to determine the maximum gradient, $m_{J,max}$ for the system and subtract the gradient displacement, Δm, to give the actual gradient, m_J of the characteristic curve. $m_{J,max}, = 0.673$, the actual gradient, m_J, is then given by

$$m_J = m_{J,max} - (\Delta m_J) = 0.622 \qquad (4.4.21)$$

or using the nomogram, the value of m_J can be read off from the second quadrant (moving anticlockwise). Stage three is using the characteristic curve, X from which the solar $= 10.71$. This calculation corresponds to the third and fourth quadrants of the nomogram. The total thermal load, Q_{tot}, when M_L kg of water are heated from the mains cold water supply temperature, T_m to the required temperature of T_L is

$$Q_{tot} = M_L C_w (T_L - T_m) = 27.02 \times 10^6 \, J \tag{4.4.22}$$

From the definition of the Brooks number, X the daily solar fraction can be determined:

$$f = \frac{X M_L C_w (T_a - T_m)}{Q_{tot}} = 0.345 \tag{4.4.23}$$

4.5 SOLAR ENERGY COLLECTION AND STORAGE FOR DRYING CROPS

High crop losses can ensue from inadequate drying, fungial attacks, and rodent and insect encroachment in traditional "open-sun" drying. Solar-energy tropical-crop dryers that enclose the crop and enable air to circulate around it compete economically with traditional open-sun drying because they

(i)   require a smaller area of land in order to dry similar amounts of crop that would have been dried traditionally in the open,
(ii)   will yield a relatively high quality of dry crop, because insects and rodents are unlikely to infest it during the drying process,
(iii) have a shorter drying period,
(iv)   they afford protection from sudden rain,
(v)   incur relatively low capital and running costs, and
(vi)   give improved crop quality achieved after drying.

In an integral type natural-circulation solar-energy dryer, the crop to be dried is placed in a drying chamber with transparent walls; heat is supplied to the crop by direct absorption of solar radiation and by convection from the heated internal surfaces of the chamber. The heat abstracts the moisture from the product, while also lowering the relative humidity of the resident air mass thus increasing its moisture carrying capability. The direct absorption of solar radiation makes these dryer particularly appropriate for greenish fruits as during dehydration, the decomposition of residual chlorophyll enhances the proper colour "ripening". For certain varieties of grapes and dates this direct exposure to sunlight is considered essential for the development of the required colour in the dried products. Similarly exposure to sunlight of arabica coffee develops full flavour of the bean. Conversely, for some fruits, exposure to sun reduces considerably the vitamin content or blemishes pigments; such produces are thus best dried in enclosed opaque-wlled cabinets or silos to which solar heated air is provided.

Integral type natural-circulation solar-energy dryers are both simple and cheaper to construct than those of the distributed type for the same loading capacity. However the potential drawbacks of the former are (i) a liability to localised over-heating and (ii) relatively slow overall drying rates. To overcome these limitations, a "solar chimney" is employed to increase the buoyant force on the air stream and thus provide an increased rate of moist-air removal. Two generic dryer types can thus be identified, namely: the cabinet dryer and the ventilated greenhouse dryer.

A natural-circulation solar-energy cabinet dryer is simply a single or double-glazed insulated container referred to as a cabinet at small scales or a silo at large scales. Solar radiation is transmitted through the cover and is absorbed on the blackened interior surfaces as well as on the product itself, thus raising the internal temperature. Vents at both the base and lower parts of the cabinet or silos enable air ventilation, with warm air leaving via upper apertures under the action of buoyant forces, drawing in replenishing fresh air at the base. Shallow layers of the product are placed on perforated trays inside the enclosure. Solar cabinet dryers, constructed from cheap locally-available materials, are usually relatively small units used to preserve "household" quantities of fruits, vegetables, fish and meat.

The major drawback of small-scale cabinet dryers is the poor air circulation often that reduces the drying rate, and incurs very high internal temperatures that can overheat crops. Drying air temperatures above 70°C are excessive, particularly for perishables fruits and vegetables. Relatively large air inlet ducts with appropriately-designed solar chimneys are essential for effective circulation within the cabinet to minimise temperature elevations.

Natural-circulation solar cabinet dryers are probably the most widely used type of solar dryer. A cabinet dryer equipped with a solar chimney shows higher efficiency than that of a natural-circulation distributed type which the incoming air was heated as it passed through a solar-energy collector and for more efficient than open sun drying.

Natural circulation solar-energy greenhouse dryers are larger than most cabinet dryers and are characterised by extensive glazing on their sides. Usually the glazing is on the front side (i.e. sun facing side) of the dryer while the rear side is insulated. Insulant panels may be drawn over the glazing at night to reduce heat losses and heat storage may also be provided. Designed properly, a solar greenhouse dryer allows a greater degree of control over the drying process than the solar cabinet dryer and are more appropriate for large-scale drying.

Typical later designs of natural-circulation solar greenhouse dryers include the widely-reported polyethylene-tent fish dryer built (Doe et al., 1977) that consists of a ridged tent-like bamboo framework clad with clear polyethylene sheet both on the side orientated towards the sun and on the ends. The rear side was clad with black polyethylene sheet, which was also spread on the floor. The cladding at one end was arranged to allow access into the drying chamber for loading and unloading. The clear plastic cladding at the bottom edge of the front side was rolled around a bamboo pole which could be adjusted to control airflow into the chamber, while the vents at the top of the ends served as the exit for the moist exhaust air. The maximum temperature for the drying of fish is 50°C: above which the fish will cook. Dryer temperatures of about 45°C are however desirable as flies and larvae infestations within the fish are boiled. The high 5°C temperature merger is difficult to maintain with thermometry connected to vent aperature controls.

4.6 SOLAR ENERGY COLLECTOR AND STORAGE FOR THERMAL POWER GENERATION

Parabolic troughs, the most deployed widely solar energy concentrator, consist usually of long curved mirror-surfaced troughs, which concentrate direct insolation on a tube at the focal axis of a parabolic mirror (Fernandez-Garcia et al., 2010). Parabolic trough concentrators are made usually of back-silvered glass for high reflectance and durability. A stainless steel tube receiver is usually coated with a highly solar selective absorbing ceramic and metal blend that is durable at high temperatures material. The absorber is surrounded usually covered by a geometrically concentric and evacuated borosilicate glass tube envelope.

In a central receiver system a large array of heliostats mirrors individually tracks the sun to reflect insolation onto a fixed receiver mounted on a tower, that absorbs the heat. The heated fluid (usually molten salt) convectively removes the receiver's heat energy and is then transported from the receiver to drive a turbine generator or stored.

Parabolic dish concentrators reflect solar energy onto a receiver mounted at the focal point. Parabolic dishes typically use multiple curved reflective panel segments made of glass or laminated films. These concentrators are mounted on a two-axis solar tracking system. The concentrated sunlight at the receiver may be utilized directly by a cycle heat engine mounted on the receiver, or simply heats a fluid that is transported for storage.

Using multiple smaller engines means: (i) smaller engines can be replaced readily (ii) a plant can deliver close to rated power while engines are being repaired, and (iii) the system can be easily expanded by adding modules to accommodate growth.

The thermal efficiency of an engine is proportional to the difference between the maximum collector and heat rejection temperatures. Most real engines operate with efficiencies of just over half of the ideal Carnot efficiency. Whilst engine efficiency is higher at increased operating temperature, the efficiency of a solar collector decreases as its operating temperature increases. The optimal operating temperature thus depends on the particular efficiency trends of the specific engine and collector employed. Typically organic Rankine cycle or Sterling engines are used with evacuated tubes in parabolic troughs or two axis tracking parabolic dishes respectively.

Rankine and Brayton cycles both have constant-pressure heat-addition processes readily applicable to solar heating. Stirling engines use a reciprocating piston design can be solar heated directly.

A working fluid can pass through the absorber directly, or there can be an intermediate heat-transfer fluid flowing in a closed loop between the absorber and a heat exchanger to heat another fluid via a heat exchanger. Incorporating an intermediate heat-transfer fluid requires another pump, heat exchanger as well as two fluids. Utilizing an intermediate heat-transfer fluid a lower vapor pressure reduces the receiver's mass and obviates the need for high-pressure piping.

4.7 OVERALL SYSTEM OPTIMIZATION

Solar water heating systems can be designed to meet hot water heating (DGS, 2005), industrial process heating (Gordon and Rabl, 1982); Kalogirou, 2003; Kulkarni et al.,

2008), domestic space and water heating (Letz et al., 2009; Yazdanshanas and Furbo, 2008; Yazdanshanas et al., 2008). They can be augmented by the inclusion of heat pumps (Troi et al., 2008; Huang and Lee, 2005; Morrison, 1984).

A wide variety of methodologies are available for the sizing of system components and determining the optimal operating parameters to satisfy a known set of characteristics of the energy load (Norton et al., 2001). These methodologies include: utilizability (Collares-Pereira et al., 1984) empirical correlations (Gopffarth et al., 1968), simplified analysis, semi-analytical simulation, stochastic simulation, simplified representative-day simulations (Garg et al., 1984) and detailed hour-by-hour simulations (Morrison and Braun, 1985; Hobson and Norton, 1988a). Each of these will be considered individually.

There is a minimum threshold insolation at which the solar heat gained by a collector corresponds to its heat losses at a particular ambient temperature. Only above this minimum insolation threshold does the collector supply a useful heat yield. Utilizability is a statistical attribute of the location-specific variation of insolation over a given duration. For example hourly utilizability is the fraction of hourly incident insolation that can be converted to heat by a collector with ideal heat removal and no optical losses. As all solar collectors have heat losses, utilizability always has a value of less than one. Utilizability can be related to other statistical properties of diurnal and annual patterns of insolation (Reddy, 1987) to produce mathematical terms to which specific collector parameters can be attached. Various generalised expressions have thus been be derived, for example, for the yearly total energy delivered by flat plate collectors whose tilt angles equalled the latitude of their notional location (Rabl, 1985). This methodology can be very useful in initial design, but limitations include possible inaccuracy of underlying insolation data correlations particularly when extended to new locations collector inclinations and orientations. The technique has been applied to interseasonal storage (Braun et al., 1981), where due to the very large thermal store mass required, collector inlet temperatures are invariant.

Correlation-based system design techniques are predicated on the high probability that for a given solar energy process heat system, in a given period. More insolation will lead to solar energy satisfying a larger share of the heat load. A dimensionless or normalised solar energy input has usually been plotted against a similarly parameterised output for a given system configuration from which correlations were obtained.

Simplified analyses consider solely the key driving parameters of system performance assuming all other variables remain constant (Braun et al., 1981). For solar industrial heat loads that over the operating period have largely constant flow rates and temperatures, simplified analysis have been developed that can employed for feasibility and initial design of industrial hot water system with heat storage (Collares-Pereira et al., 1984). Simplified analyses maintain a physical basis for the relationships between parameters that is lost in empirical correlations whose equations are polynominal curve fits.

Semi-analytical simulation use detailed numerical models. However rather than undertaking hour-by-hour calculations using insolation, ambient temperature and load data, in this approach sinusoidal and linear functions are used to describe the insolation and load respectively with ambient temperature either varying sinusoidally or remaining constant. This approach has largely been superseded by hour-by-hour analysis as the computing resources required to successfully undertake the latter have become widely available.

To represent insolation, ambient temperature and load characteristics in stochastic simulations, Markov chain models can be produced from several years hour-by-hour data for a specific location. Though the transition probability matrices would give the long-term system behaviour, the technique has yet to be established in a software environment that would render it a practical proposition for use in system design.

A variety of detailed hour-by-hour simulation models are available to determine the outputs of systems with differing layouts, component specifications and control regimes (Klein et al., 1996). Either an overarching optimisation algorithm or multiple simulations are required to ascertain an economically optimal combination of system components. To determine economically optimum designs for solar industrial process heat systems artificial intelligence methods have be employed.

REFERENCES

Al-Ibrahim A.M., Beckman W.A., Klein S.A. and Mitchell J.W. (1998) Design procedure for selecting an optimum photovoltaic pumping system in solar domestic hot water system. *Solar Energy*, 64, 227–239

Arthur A.C. and Norton B. (1988) Factors affecting the performance of integral passive solar energy water heaters. *In: Proceedings of the 6th International Solar Energy Forum, Berlin, Germany*, pp. 189–194.

Bainbridge D.A. (1981) *The integral passive solar water heater book*. The Passive Solar Institute, Davis, California, USA.

Bansal N.K., Sawhney R.L., Misra A. and Boettcher A. (1988) Solar sterilization of water. *Solar Energy*, 40, 35–39.

Bar-Cohen A. (1976) Thermal optimisation of compact solar water heaters. *Solar Energy*, 20, 193–196

Bishop R.C. (1983) Superinsulated batch heaters for freezing climates. *In: Proceedings of the 8th National Passive Solar Conference, Sante Fe, New Mexico, USA*, pp. 807–810.

Braun, J.E., Klein, S.A. and Beckman, W.A. (1981) Seasonal storage of energy in solar heating, *Solar Energy*, 26, 403–411.

Budihardjo I. and Morrison G.L. (2007) Natural circulation flow through water-in-glass evacuated tube solar collectors. *Solar Energy*, 12, 1460–1472.

Butti K. and Perlin J. (1980) *A Golden Thread*. Van Nostrand Reinhold Co., New York, USA.

Chauhan R.S. and Kadambi V. (1976) Performance of a collector-cum-storage type of solar water heater. *Solar Energy*, 18, 327–335.

Charters W.W.S., de Forest L., Dixon C.W.S. and Taylor L.E. (1980) Design and performance of some solar booster heat pumps. *In: Annual Conference of the Australia and New Zealand Solar Energy Society, Melbourne, Australia*.

Chow T.T, W. He and J. Ji (2008) Hybrid photovoltaic-thermosyphon water heating system for residential application, *Solar Energy*, 80, 298–306.

Collares-Pereira, M., Gordon, J.M., Rabl A. and Zarmi Y. (1984) Design and optimisation of solar industrial hot water systems with storage. *Solar Energy*, 32, 121–133.

Dalenbäck J.D. (2010) Take off for solar district heating in Europe. *Polska Energetyka Sloneczna*, 1/2010, 9–13

Davidson J.H. and Adams D.A. (1994) Fabric stratification manifolds for solar water heater. *ASME Solar Energy Engineering*, 116, 130–136.

Deutsche Gesellschaft für Sonnenenergie (2005) *Planning and installing solar thermal systems: a guide for installers*, Earthscan, London, UK.

Dickinson W.C., Clark A.F., Day J.A. and Wouters L.F. (1976) The shallow solar pond energy conversion system. *Solar Energy*, 18, 3–10

Doe, P.E., Ahmed, M., Muslemuddin, M. and Sachithanthan, K. (1977) A polythene tent drier for improved sun drying of fish, *Food Tech*, 437–441

Duffie J.A. and Beckman W.A. (1991) *Solar Engineering of Thermal Processes*. 2nd Edition, Wiley Interscience, New York.

Eames P.C. and Norton B. (1998) The effect of tank geometry on thermally stratified sensible heat storage subject to low Reynolds number flows. *International Journal of Heat and Mass Transfer*, 41, 2131–2142

Fanney A.H. and Daugherty B.P. (1997) A PV solar water heating system. *ASME Journal of Solar Energy Engineering*, 119, 126–133

Fanney A.H., Daugherty B.P. and Kramp K.P. (1997) Field performance of PV solar water heating system. *ASME Journal of Solar Energy Engineering*, 119, 265 – 272

Fernandez-Garcia, A., Zarza, E., Valenzuela, L. and Perez, M. (2011) Parabolic-trough solar collectors and their applications. *Renewable and Sustainable Energy Reviews*, 7, 1695–1721.

Fraser, K.F., Hollands, K.G.T. and Brunger, A.P. (1995) An empirical model for natural convection heat exchangers in SDHW systems. *Solar Energy*, 55, 75–84.

Furbo S. and Fan J. (2008) Experimental investigations on small low flow sdhw systems with different solar pumps. *In: Proceedings of EuroSun 2008 October Lisbon, Portugal*.

Garg H.P., Datta G. and Bhargava A.K. (1984) Studies on an all plastic solar hot water bag. *International Journal of Energy Research*, 8, 3, 291–296

Gopffarth W.H., Davidson R.R., Harris W.B. and Baird M.J. (1968) Performance correlation of horizontal plastic solar water heaters. *Solar Energy*, 12, 183–196

Gordon, J.M. and Rabl, A. (1982) Design, analysis and optimization of solar industrial process heat plants without storage *Solar Energy*, 28, 519–530.

Hobson P.A. and Norton, B. (1988b) A design nomogram for direct thermosyphon solar-energy water heaters. *Solar Energy*, 43, 85–93

Hobson P.A. and Norton, B. (1988a) Verified accurate performance simulation model of direct thermosyphon solar-energy water heaters *ASME Journal of Solar Engineering*, 110, 282–292.

Hollands, K.G.T. and Lightstone, M.F. (1989) A review of low-flow, stratified-tank solar water heating systems *Solar Energy*, 43, 97–105.

Hongchuan, G. and Guangming, X. (2001) A study on the evacuated-tubes integral-collector-storage solar water heater with two different coated absorbers. *In: Proceedings of the International Conference on Energy Conversion and Application, Wuhan, China*. pp. 1227–1230

Huang B.J. and Lee C.P. (2005) Long-term performance of solar-assisted heat pump water heater. *Journal of Renewable Energy*, 29, 633–639.

Ji, J., Hun, J., Chow, T-T., Yi, H., Lu, J., He, W. and Sun, W. (2006) Effect of fluid flow and packing factor on energy performance of a wall-mounted hybrid photovoltaic/water-heating collector system. *Energy and Buildings*, 38, 1380–1387.

Kalogirou, S. (2003) The potential for solar industrial process heat applications. *Applied Energy*, 76, 337–361.

Klein, S.A. et al. (1996) TRNSYS: a transient systems simulation program, Version 14.2, Solar Energy Laboratory, University of Wisconsin, Madison.

Kulkarni, G.N., Kedere, S.B. and Bandyopadhyay, S. (2008) Design of solar thermal systems utilising pressurised hot water storage for industrial applications. *Solar Energy*, 82, 686–699.

Lavan Z. and Thompson J. (1977) Experimental study of thermally stratified hot water storage tanks. *Solar Energy*, 19, 519–524

Lenel U.R. and Mudd P.R. (1984) A review of materials for solar heating systems for domestic hot water. *Solar Energy*, 32, 109–120

Leblanc, J., Akbarzadeh, A., Andrews, J., Lu, H. and Golding, P. (2011) Heat extraction methods from salinity-gradient solar panel and the introduction of a novel systems of heat extraction for improved efficiencies. *Solar Energy*, 85, 3103–3142

Letz T., Bales C. and Perers B. (2009) A new concept for combisystems characterisation: The FBC method. *Solar Energy*, 83, 1540–1549.

Lund P.D. (1986) Computational simulation of district solar heating systems with seasonal thermal energy storage. *Solar Energy*, 36, 397–408.

Mason A.A. and Davidson J.H. (1993) Stratification and mixing in an evacuated-tube integral-storage solar collector. *In: Proceedings of the ISES Solar World Conference, Budapest, Hungary.* pp. 103–108

Morrison, G.L. (1984) Simulation of packaged solar heat pump water heaters, Solar Energy, 53, 249–257.

Morrison G.L. (2001) *Solar Water Heating, In Solar Energy; the stage of the art* ISES Position Papers, (J. Gordon, Ed) James and James, London

Morrison G.L. and Braun, J.E. (1985) System modeling and operation characteristics of thermosyphon solar system. *Solar Energy*, 30, 341–350.

Norton, B. and Edmonds, J.E.J. (1991) Aqueous propylene glycol concentrations for the freeze protection of thermosyphon solar energy water heaters *Solar Energy*, 47, 375–382.

Norton B., Edmonds, J.E.J. and Kovolos, E. (1992) Dynamic simulation of indirect thermosyphon solar energy water heaters *Renewable Energy*, 2, 283–297.

Norton B., Eames, P.C. and Lo, S.N.G. (2001) Alternative Approaches to Thermosyphon Solar-Energy Water Heater Performance Analysis and Characterisation *Renewable and Sustainable Energy Reviews*, 5, 79–96.

Norton B. and Probert S.D. (1986) Thermosyphon solar energy water heaters. *Advances in Solar Energy*, 3, 125–170.

Norton B. (1992) *Solar Energy Thermal Technology*, Springer-Verlag, Heidelburg, Germany.

Norton B. (2001) *Heating Water by the Sun*, 3rd Edition, Solar Energy Society, Oxford, UK.

Norton B. and Probert, S.D. (1983) Achieving Thermal Rectification in Natural-Circulation Solar-Energy Water Heaters *Applied Energy*, 14, 211–225.

Norton B., Eames P.C., Mallick T.K., Huang M.J., McCormack S.J. Mondol J.D. and Yohanis, Y.G. (2011) Enhancing the performance of building integrated photovoltaics *Solar Energy*, 85, 1629–1664.

Norton, B. (2011) Solar Water Heating: A review of systems research and design innovation, *Green. The International Journal of Sustainable Energy Conversion and Storage*, 2, 189–208.

Ong K.S. and Haider-E-Alathi, M. (2003) Performance of a 134a-filled thermosyphon. *Applied Thermal Engineering*, 23, 2373–2381

Parent, M.G. and Van Der Meer, Th.H. (1990) Natural convection heat exchangers in solar water heating systems: Theory and experiment. *Solar Energy*, 45, 43–52.

Parker G.J. (1976) A forced circulation system for solar water heating. *Solar Energy*, 18, 475–479.

Prapas D.E., Tsiamouis, S.G. Giannaras, V.D. and Sotiropoulos, B.A. (1993) Storage tanks interconnection and operation modes in large DHW solar systems. *Solar Energy*, 51, 83–91.

Prud'homme T. and Gillet D. (2001) Advanced control strategy of solar domestic hot water system with a segmental auxiliary heater. *Energy and Buildings*, 35, 463–475.

Rabl, A. (1985) *Active Solar Collectors and their Application* Oxford University Press.

Rabin Y., Bar-Niv I., Korin E. and Mikic B. (1995) Integrated solar collector storage system based on a salt-hydrate phase-change material. *Solar Energy* 55, 435–444

Reddy, T.A. (1987) *The Design and Sizing of Active Solar Thermal Systems*. Clarendon Press, Oxford.

Resch, K. and Wallner, G.M. (2008) Thermotropic layers for flat plate collectors – A review of various concepts for overheating protection with polymeric materials. *Solar Energy Materials and Solar Cells*, 93, 119–128.

Rommel M. and Wagner A. (1992) Application of transparent insulation materials in improved flat-plate collectors and integrated collector storage. *Solar Energy*, 49, 371–380

Ruiz E. and Martinez P.J. (2010) Analysis of an open air swimming pool solar heating system using an experimentally validated TRNSYS model. *Solar Energy*, 84, 116–123

Schmidt T., Mangold, D. and Müller-Steinhagen, H. (2004) Central solar heating plants with seasonal storage in Germany. *Solar Energy*, 76,165–174

Shreyer, J.M. (1981) Residential application of refrigerant charged solar collectors. *Solar Energy*, 26, 307–312.

Sion R., Sangameswar Rao, K., Rao, D. and Rao, K. (1979) Performance of a flat plate solar collector with fluid undergoing phase change. *Solar Energy*, 23, 69–73.

Smyth M., Eames, P.C. and Norton, B. (1998) A novel high performance integrated collector/storage solar water heater design. *Renewable Energy*, 16, 2217–2222.

Smyth M, Eames P.C. and Norton B. (1999) A comparative performance rating for an integrated solar collector/storage vessel with inner sleeves to increase heat retention. *Solar Energy*, 66, 291–303.

Smyth M., Eames P.C. and Norton B. (2001a) Evaluation of a freeze resistant integrated collector/storage solar water heater (ICSSWH) for Northern Europe. *Applied Energy*, 68, 265–274.

Smyth M., Eames P.C. and Norton B. (2001b) Annual performance of heat retaining integrated collector/storage solar water heaters in a northern maritime climate. *Solar Energy*, 70, 391–401.

Smyth M., Eames P.C. and Norton B. (2003) Heat retaining integrated collector/storage solar water heaters. *Solar Energy*, 75, 27–34

Smyth M., Eames, P.C. and Norton, B. (2004) Techno-economic appraisal of an integrated collector/storage solar water heater *Renewable Energy*, 29, 1503–1514

Smyth M., McGarrigle, P., Eames, P.C. and Norton, B. (2005) Experimental comparison of alternative convection suppression arrangements for concentrating integral collector storage solar water heaters *Solar Energy* 78, 223–233.

Smyth M., Eames, P.C. and Norton, B. (2006) Integrated collector storage solar water heaters *Renewable and Sustainable Energy Reviews*, 10, 503–538

Smyth, M. and Skates, H. (2009) A passive solar water heating system for vineyard front protection *Solar Energy*, 83, 400–408.

Tabassum S.A., Norton, B. and Probert, S.D. (1988) Heat removal from a solar-energy collector with a heat-pipe absorber *Solar and Wind Technology*, 5, 141–145.

Too Y.C.S, Morrison G.L and Behnia M (2009) Performance of solar water heater with narrow mantle heat exchangers *Solar Energy*, 83, 350–362.

Tripanagnostopoulos Y., Souliotis, M. and Nousia, T.H. (2002) CPC type integrated collector storage systems. *Solar Energy*, 72, 327–350.

Tripanagnostopoulos Y. and Souliotis, M. (2004a) ICS Solar systems with horizontal (E-W) and vertical (N-S) cylindrical water storage tank. *Renewable Energy*, 29, 73–96.

Tripanagnostopoulos, Y. and Souliotis M. (2004b) Integrated collector storage solar systems with asymmetric CPC reflectors. *Renewable Energy*, 29, 223–248.

Troi, A., Vangiouklakis, Y., Korma, E., Jähnig, D., Wiemken, E., Franchini, G., Mugnier, D., Egilegor, B., Melograno, P. and Sparber, W. (2008) Solar Combi+; identification of most promising markets and promotion of standardized system configurations for small scale solar heating and cooling applications. *In: Proceedings of EuroSun 2008, October, Lisbon, Portugal.*

Weiss W. (2003) Solar heating for houses – A design handbook for solar combisystems, James and James, London.

Weiss, W. and Mauthner, F. (2010) *Solar Heat Worldwide* Report of the International Energy Agency, Solar Heating and Cooling Programme.

Wuestling M.D., Klein S.A. and Duffie J.A. (1985) Promising control alternatives for solar water heating systems. *ASME Journal of Solar Energy Engineering*, 107, 215221.

Yazdanshenas E. and Furbo, S. (2007) Theoretical comparison between solar water/space - heating combisystems and stratification design options. *ASME Journal of Solar Energy Engineering*, 129, 438–448.

Yazdanshenas E., Furbo, S. and Bales, C. (2008) Theoretical comparison between solar combisystems based on bikini tanks and tank-in-tank solar combisystems. *In: Proceedings of EuroSun 2008: 1st International Congress on Heating, Cooling and Buildings, Lisbon, Portugal, October.*

Yohanis Y.G., Popel, O., Frid, S. and Norton, B. (2006a) Geographic variation of solar water heater performance in Europe. *In: Proceedings of the Institution of Mechanical Engineers, Part A, Journal of Power and Energy* 220, 395–407.

Yohanis Y.G., Popel, O., Frid, S.E. and Norton, B. (2006b) The annual number of days that solar heated water satisfies a specified demand temperature. *Solar Energy* 80, 1021–1030.

Chapter 5

Basics of the photovoltaic thermal module

Krishnan Sumathy
Department of Mechanical Engineering, North Dakota State University, Fargo, USA

5.1 INTRODUCTION

Energy plays a crucial role in social and economic development. Remarkable increases in oil prices have stimulated research in the renewable energy field as it contributes to the diversity in energy supply. Compared to other types of energy, renewable energy reduces the dependence on fossil fuel resources and carbon emissions to the atmosphere. Nobuyuki (2006) gives the latest data which show that renewable energy technologies provide 13.3% of the energy need around the world (Baños et al., 2010) and will be greatly dependent on the future, considering their sustainability and wide public acceptance. The total incoming solar radiation is about 3.8 million EJ per year, which can meet the entire energy demand if harnessed properly.

Solar energy can be either utilized for producing electricity directly or to produce heat. The combination of photovoltaic and thermal systems (PV/T) is particularly attractive because of its efficiency in directly converting solar energy into electricity and heat, concurrently. A photovoltaic/thermal (PV/T) module mainly comprises a photovoltaic panel and a solar thermal collector, forming the core components of a photovoltaic/thermal (PV/T) system.

Solar thermal systems and PV systems are essentially different because solar systems generate heat energy while PV systems generate electricity. The differences between PV system, PV thermal system, and solar collector are demonstrated in Figure 5.1.1. Solar energy can be captured in various ways: in a PV system, only a small fraction of solar radiation absorbed by the panel is converted into electricity. That is to say, a subsection of absorbed solar radiation is used by PV cells to generate electricity; letting most of the radiation turn into waste heat. This results in the decrease of the

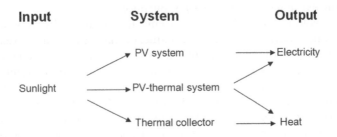

Figure 5.1.1 Comparison between PV, PV/T and solar thermal systems.

PV module's efficiency due to the rise of temperature of PV cells. Natural or forced circulation of air or other fluids can be used to cool PV cells which heats up the photovoltaic thermal systems (PV/T) and can be used as a substitute for the conventional PV modules. A photovoltaic thermal system (PV/T) integrates heat extraction devices to the PV cell to reduce the temperature of the PV module and to increase electricity production as well as useful heat simultaneously, thereby increasing the overall efficiency of PV/T system.

For more than thirty years, many researchers have used and discussed the idea of PV/T both experimentally and numerically. The initial focus was on glazed collectors utilizing air and liquid as the heat transport media. Later, unglazed heat pump-based collectors were also introduced. PV/T systems are mainly classified based on the type of fluid being utilized for heat removal. Compared to PV/T air systems, PV/T water systems relatively have a higher efficiency, because water has a higher thermal capacity and conductivity (Prakash, 1994). But considering the leakage and corrosion that can be caused by water, more robust construction must be included in a PV/T water system to make it water-tight and corrosion-free. Therefore, the easiest way of heat extraction from PV modules is by circulating natural or forcing air through an air channel either on the top or rear-side of the PV surface.

Other than giving rise to diverse ways of heat extraction, the types of PV modules also influence the operational state of the PV/T system. Crystalline-silicon (c-Si), polycrystalline silicon (pc-Si), and recently exploited thin films of amorphous-silicon (a-Si) cells can be used in PV module construction. The c-Si cells are exorbitantly expensive due to the comprehensive and energy-intensive procedures needed to produce them, even though they are highly efficient. In the past two decades, attempts have been made to lower the price of c-Si cells through various manufacturing methods to further improve module efficiency.

The PV/T systems have a wide range of applications. One of the feasible ways to utilize photovoltaic system is to incorporate the PV modules into the building envelope, as they can be incorporated into the façade and roof of buildings easily. Such building integrated PV (BIPV) technology is one of the most widespread applications of photovoltaic systems in urban buildings. Hence these systems have received a great interest from engineers and architects. The PV/T systems have a grander prospect by producing green, safe and tactically significant alternatives to power generation. The superiorities of grid-connected PV electricity to customers are reflected in both economic and environmental concerns. Customers can partially fulfill their electricity needs while using utility-generated power at night and on gloomy days by using a grid-connected PV system where utility power is attainable. The domestic or commercial buildings can also use the extracted heat for heating purpose.

Most photovoltaic thermal (PV/T) systems are set up in districts or areas where grid and telephone networks are rarely available, or may be difficult to reach. The PV/T systems have a relatively long service lifetime with nearly no maintenance. Therefore, autonomous or stand-alone PV/T systems are more welcome among remote rural areas. Furthermore, researchers are trying to improve solar output and power production by using solar trackers and concentrator photovoltaic thermal systems, which are most feasible when space is scarce. One of the main advantages of a concentrator integrated PV/T system is to reduce the number of solar cells, as well as increasing power output. A concentrator collects solar radiation that is received on a relatively large surface area and converts it to a concentrated radiation onto a small area with solar cells. The concentrators can be made from simple plastic lenses or conventional mirrors, or polished light water surfaces. Fins are added to the back of the PV panel to further improve the performance of air PV/T systems.

Numerous research has been attempted to advance the performance of photovoltaic thermal systems. Generally speaking, the conversion rate of PV modules available in the commercial market is very low. They can only convert about 6–18% of solar radiation incident on the PV panel to electricity. The rest of the solar energy is turned into waste heat affecting the cell competence and some is reflected back to the atmosphere.

A PV/T system has the following advantages:

(i)   Dual function: the integrated system can generate both electricity and heat;

(ii)   Flexible and efficient: with limited roof surface area, the integrated system has a higher efficiency than two independent systems, especially using the BIPV technology;

(iii) Wide application: the heat generated can not only be used for heating but also for cooling, for instance it can be used for desiccant evaporative cooling;

(iv)   Economical and practical: the PV/T system is simple to be retrofitted or integrated into different types of buildings with insignificant remolding. A PV/T system can be used as roofing material and has a shorter payback period.

Compared to parallel connecting of photovoltaic panels to solar thermal collectors, the integrated PV/T modules are not only able to produce more energy per unit surface area, but also have less initial and production cost. The collector area can be reduced up to 40% by using PV/T system producing the same amount of energy according to an ECN report (Energy research Centre of the Netherlands) (IEA, 2007). Furthermore, PV/T modules cater to the same aesthetic need as PV.

This chapter will focus on different technologies and module aspects of heat transformation of the PV/T systems. Modifications are suggested with the main objective being to improve heat transfer in the PV/T systems. A detailed description on hybrid PV/T solar systems is included in a recently published Roadmap (Zondag et al., 2005; Affolter et al., 2006), where several aspects regarding technology, present status and future perspectives of these solar energy conversion systems are presented (Arif Hasan and Sumathy, 2010).

5.2   PV/T DEVICES

PV/T devices can vary in design to suit various applications, ranging from PV/T domestic hot water systems to ventilated PV facades and actively cooled PV concentrators. The markets for both solar thermal and PV are growing rapidly and have reached a very substantial size. For PV/T, a similar growth can be expected, since the technical feasibility is proven and as such it can be integrated with other domestic applications. PV/T has broad range of applications, that is, it is not only suitable for domestic hot water heating (glazed PV/T collectors), but also for commercial buildings (ventilated PV to preheat ventilation air during winter and to provide the driving force for natural

Table 5.2.1 Recommendation of the collector type based on the type of demand.

Demand		Recommendation
Water	High temperature	Use glazed liquid collector. Also, an unglazed collector can be used if PV/T has to be integrated to a heat-pump.
	Low temperature	To meet only summer demand, use unglazed liquid collector. On the other hand, to meet both summer and winter demands, use glazed liquid collector; an unglazed collector can be chosen if PV/T has to be integrated to a heat-pump.
Air	High temperature	Use glazed air collector or unglazed collector. Ventilated PV can be used as a heat source if PV/T has to be integrated to a heat-pump.
	Low temperature	To meet only summer demand or for the place receiving high irradiation in winter, use unglazed air collector or ventilated PV. On the other hand, to meet both summer and winter demands, use glazed air collector; an unglazed collector can be a choice if PV/T has to be integrated to a heat-pump.

ventilation during summer). Hence, the market for PV/T might even be larger than the market for thermal collectors.

Depending on the application, the required thermal demand can be covered by choosing appropriate PV/T system. There exist various forms of PV/T system which depend on the type of PV module as well as its design, type of heat removal fluid (water/glycol or air) and on the concentration of the incoming radiation.

The existing PV/T designs can be classified as:

(i)   Liquid PV/T collector
(ii)   Air PV/T collector
(iii) Ventilated PV with heat recovery
(iv)   PV/T concentrator.

Irrespective of the type of collector, the absorber of each PV/T collector is provided with a glass cover to reduce the thermal losses. If such a cover is present, the collector is referred to as "glazed", otherwise as "unglazed".

Glazed collectors have less thermal losses, and hence they could higher collector fluid temperatures. For medium to high temperature applications, this significantly improves the annual thermal yield. However, glazed collectors result in high stagnation temperatures that may be critical for certain types of PV encapsulant (risk of yellowing and delamination) resulting in hot spots. In addition, bypass diodes may get overheated due to the additional insulation. Reflection losses at the glazing further reduce electrical performance. Increased temperature levels lower the electrical yield.

In summary, whether the collector should be glazed or not, it is important to find a good balance (illustrated in Table 5.2.1) between the increased thermal yield on one hand, and the reduction in electrical yield and the issues related to possible degradation on the other hand.

5.2.1 Liquid PV/T collector

In order to improve the energy performance of the photovoltaic system, much effort has been spent on research and development of the hybrid PV/T technology. One of the design modifications is to increase the PV module performance by circulating water to extract the heat using water as the coolant. These liquid PV/T collectors are similar to conventional flat-plate liquid collectors; an absorber with a serpentine tube or a series of parallel risers is applied, onto which PV has been laminated or glued as an adhesive epoxy joint.

Two common configurations used in PV/T systems are: "Parallel plate configuration", and "Tube-in-plate configuration". Prakash (1994), Huang et al. (2001), Tiwari and Sodha (2006) and Tiwari et al. (2006) have worked on the parallel plate design, while Zondag et al. (2002), Chow (2003), Chow et al. (2006), Kalogirou (2001), Huang et al. (2001) and Tiwari and Sodha (2006) have carried out an in-depth study on tube-in-plate design. Within the first works on PV/T water system, Bergene and Lovvik (1995) initially conducted a theoretical study on PV/T water system composed of flat-plate solar collector with solar cells. Their proposed system is particularly suitable to preheat the domestic hot water.

More recently, Zondag et al. (2003) grouped the design concepts of water-type PV/T collectors into four main types: sheet and-tube collectors, channel collectors, free-flow collectors, and two-absorber collectors. These collector types are designed for pump (forced) circulation (Figure 5.2.1). Based on numerical analysis it has been suggested that a channel should be provided to effect liquid flow below the transparent PV module to effect higher collector efficiency (Chow et al., 2006). Nevertheless, from the viewpoint of good overall performance and structural simplicity, single-glazing sheet and tube hybrid PV/T collector is regarded as the most promising design.

Dubey and Tiwari (2008) designed an integrated photovoltaic (glass-to-glass) thermal (PV/T) solar water heater system and tested it in outdoor conditions of India. Similarly, Erdil et al. (2008) constructed and tested a hybrid PV/T system for energy collection at geographical conditions of Cyprus, where they used water as the cooling fluid. It was reported that the payback period for their proposed modification was less than 2 years which made their hybrid system economically attractive. Also, Daghigh et al. (2011) used water as the working fluid and had presented the advances in liquid based photovoltaic/thermal (PV/T) collectors. The liquid-based photovoltaic thermal collector systems are practically more desirable and effective than air-based systems. Temperature fluctuation in liquid based PV/T is much less than the air-based PV/T collector. The future direction of water-cooled and refrigerant hybrid photovoltaic thermal systems was also presented. Their study revealed that the direct expansion solar-assisted heat pump system could achieve a better cooling effect than the PV/T collector.

Similar to the above modeling work, Chow et al. (2006) had developed a numerical model of a photovoltaic-thermosyphon collector system using water as a working fluid and verified the model's accuracy by comparing with measured data. The energy performance of the collector system was examined, through reduced-temperature analysis and the study was further extended to analyze the performance of the system in the "hot summer and cold winter" climate zone of China. The numerical results were found to be very encouraging, and according to them the equipment is capable of extending the

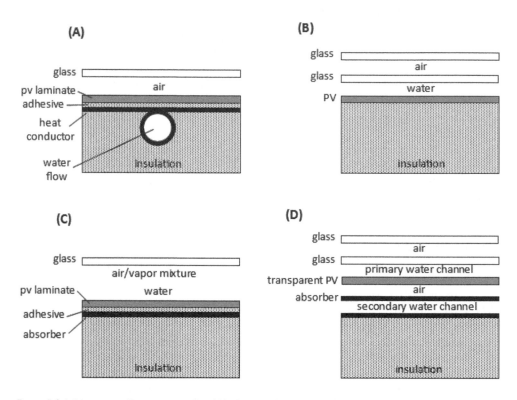

Figure 5.2.1 Various collector concepts: (A) sheet-and-tube PVT, (B) channel PVT, (C) free flow PVT, (D) two-absorber PVT (insulated type) (Zondag et al., 2003).

PV application potential in the domestic sector. Apart from the said study, Chow et al. (2009) also carried out analytical simulation to investigate the annual performance of building-integrated photovoltaic/water-heating system for the Hong Kong climate and found that annual thermal and cell conversion efficiencies were about 37.5% and 9.39%, respectively. Based on the results, they confirmed that PV/T systems could be applicable even to hot-humid regions.

Though the liquid collectors have proven to be technically feasible, economic feasibility is yet questionable. Compared to the air heating PV/T system, not many developments are seen in the literature on liquid-heating systems due to their inherent limitations such as: additional cost of the thermal unit pipes for the water circulation, and the inherent freezing problem of working fluid when used in low temperature regions, etc.

5.2.2 Air PV/T collector

The PV/T air collectors are similar to a conventional air collector with a PV laminate functioning as the top cover of the air channel. PV/T air collectors are cheaper than the PV/T liquid collectors because of the flexibility that conventional PV modules can be easily converted to a PV/T system, with very few modifications. PV/T air collectors can either be glazed or unglazed. In general, air collectors are mostly applied if the end-users have a demand for hot air, space heat, dry agricultural products, or to condition the indoor air (air cooling). At present, air heating systems are mainly designed to directly use the air for space heating. However, the opportunity for this application depends directly on the market share of air heating systems, which is low in most countries. A niche market is given by preheating of ventilation air for large volume buildings (stores, sport halls, schools and other commercial buildings) where temperatures in the range of 15 to 25°C are desirable. With the very same air systems, hot water preparation is often possible through an air/water heat exchanger, which is generally done during the summer season in order to increase the overall performance of the system.

The application of air as a heat transport medium compared to liquid, has significant advantages along with few inevitable disadvantages. Using air as a medium of heat transport avoids the damage caused by leakages, freezing or boiling. The disadvantages that brought up are the low heat transfer efficiency and high volume transfer demand due to the lower heat conductivity and density. It also loses more heat energy if leakage happens.

Tonui et al. (2007) carried out an experimental study on air-cooled PV/T solar collectors in which a few low cost performance improvements were introduced. Both water and air have been used for PV cooling through a thermal unit attached to the back of the PV module. Compared to water, air is preferred due to minimal use of material and low operating cost despite its poor thermo-physical properties. The study has investigated the performance of two low cost heat extraction improvement modifications in the channel such as the use of a thin flat metal sheet suspended at the middle or finned back wall of an air channel in the PV/T air configuration. A theoretical model was developed and validated against experimental data, where a good agreement between the predicted results and measured data were achieved. The validated model was then used to study the impact of various design parameters such as the channel depth, channel length and mass flow rate on electrical and thermal efficiency, PV cooling and pressure drop. The study had confirmed that the suggested modifications positively improve the performance of the PV/T air system.

As the heat transfer in the air cooled PV/T system is much more critical than in the liquid cooled PV/T system, it is important to model the heat transfer properly. Sopian et al. (1996) presented a performance analysis of single-pass and double-pass PV/T air systems. The performance of single-pass and double-pass combined photovoltaicthermal collectors are analyzed with steady-state models, with air as the working fluid. The performances of the two types of combined photovoltaicthermal collectors were compared. The results show that the new design, the double-pass photovoltaicthermal collector, has superior performance over the conventional design.

For a flow through a tube or duct, entrance-effect plays an important role in the heat transfer. Eicker (2003) presented an overview of entrance-effect heat transfer relations for air-collectors, showing a variation of about 10% on the average Nusselt number when integrated over the entrance length and reported that for a sufficiently wide channel, the hydraulic diameter should be twice the channel height. Hegazy (2000) analyzed four types of air PV/T model design including single glazed collectors with air flow over (Model I) or below (Model II) the absorber; with air flow on both sides of the absorber in a single pass (Model III); and a double pass model (Model IV) as shown is Figure 5.2.2. The effects of air specific flow rate and the selectivity

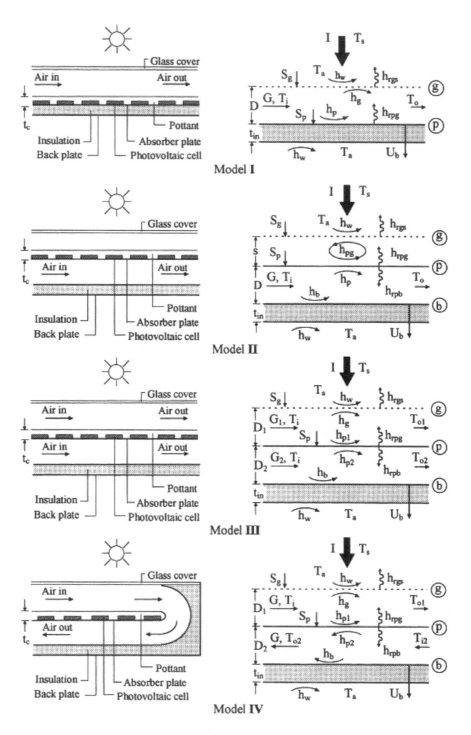

Figure 5.2.2 Schematics of the various PV/T models along with heat transfer coefficients (Hegazy, 2000).

of the absorber plate and PV cells on the performances were examined. The results showed that the Model I collector has the lowest performance under similar operational conditions. Other collectors achieved comparable thermal and electrical yields, among which the Model III collector requires the least fan power, followed by Models II and IV.

The impact of air flow induced by buoyancy and heat transfer through a vertical channel heated from one side by the PV module on the PV/T performance was investigated numerically and experimentally by Moshfegh and Sanberg (1998) and Sanberg and Moshfegh (2002). The study reports that the induced velocity increases the heat flux non-uniformly inside the duct and its impact depends on the exit size and design. More analysis and modeling on passively cooled PV/T air systems continue to appear (Tiwari et al., 2006; Tiwari and Sodha, 2006; Naphon, 2005; Garg et al., 1994; Tripanagnostopoulos et al., 2006) and a substantial amount of research has been specifically carried out (Brinkworth and Sandberg, 2006; Benemann et al., 2001; Hodge and Gibbons, 2004; Pottler et al., 1999; Tiggelbeck et al., 1993; Tonui et al., 2007; Tripanagnostopoulos, 2007) to improve heat transfer to the air of both buoyancy-driven and forced air flow systems. Their studies were focused generally on channel geometry, creation of more turbulence in the flow channel and increasing the convective heat transfer surface area in the channel. Most of these studies used simulation models for their experimental work where the PV module was simulated by a heated foil.

Similar to the liquid collectors, various types of solar air systems exist and an overview has been given by Hastings and Morck (2000). The main concepts on air-cooled PV/T systems were presented in the works of Kern and Russel (1978), Hendrie (1979), Florschuetz (1979), Raghuraman (1981) and Cox and Raghuraman (1985). The exclusive theoretical aspects of PV/T systems with air as the heat extraction fluid are detailed by Bhargava et al. (1991), Prakash (1994) and Sopian et al. (1996).

5.2.3 Ventilated PV with heat recovery

In general, for building integrated photovoltaic panels, to ensure the modules are not overheated, the ambient air is circulated by thermosiphoning beneath or at the rear end of the PV panel, which is commonly referred to as "ventilated PV."

If this waste heat can be harvested and be used for secondary purposes, it functions as a PV/T collector, providing additional benefits:

(i)   A PV-facade may limit the thermal losses in a building by infiltration. Also the PV facade has the advantage of shielding the building from solar irradiance, thereby reducing the cooling load. Hence, such facades are especially useful for retrofitting poorly insulated existing offices.

(ii)   If there is no demand for the generated heat, then air collectors and PV-facades can use their buoyancy induced pressure difference to assist the ventilation.

(iii) Facade integration of PV has additional cost incentives of substituting expensive facade cladding materials.

However, the ventilated PV-façade may contribute to the building's cooling load in hot summer, which is not desirable. To overcome this issue, a desiccant cooling cycle can be employed which can be energized with an additional collector. It is a novel open driven system introduced by Li et al. (2006), in which the required room space is affected by evaporative cooling and the PV-driven air heating system provides the necessary regeneration of air. With such systems, it is possible to achieve a solar fraction of 75% with an average COP of 0.52. Such hybrid systems have proven to be most effective to offset the capital costs involved with BIPV. Detailed studies on such systems have been carried out by several researchers. For instance, Ricaud and Capthel (1994) have recommended improved air heat extraction methods and Yang et al. (1994) have exclusively worked on roof integrated air-cooled systems. Infield et al. (2006) presented a methodology to evaluate the thermal impact on building performance of an integrated ventilated PV facade. This was based on an extension of the parameters to take account of the energy transfer to the facade ventilation air. Four terms describing ventilation gains and transmission losses in terms of irradiance and temperature components were defined to characterize the performance of the facade in total. Steady state analysis has been applied in order to express these four parameters in terms of the detailed heat transfer process within the facade. This approach has been applied to the ventilated facade of the public library at Mataró, Spain and was used for validating their developed model.

Several researchers (Posnansky et al., 1994; Ossenbrink et al., 1994; Moshfegh et al., 1995) have worked extensively on the building integrated PV/T systems. Later, Brinkworth et al. (1997), Brinkworth (2000), Brinkworth et al. (2000) and Krauter et al. (1999) presented design and performance studies regarding air type building integrated hybrid PV/T systems. In addition, Eicker et al. (2000) have presented on the performance of a BIPV PV/T system which was operated during winter for space heating applications and during summer for active cooling.

Yet another comprehensive examination of PV and PV/T in built environments has been presented by Bazilian et al. (2001). The study highlighted the fact that PV/T systems are well suited to low temperature applications. Furthermore, they pointed out that the integration of PV systems into the built environment could achieve "a cohesive design, construction and energy solution". However, it should be noted that there exists a need for further research in the said field, before combined PV/T systems become a successful commercial reality.

The building integrated photovoltaic is going to be a sector which would serve as a wider PV module application. The works of Hegazy (2000), Lee et al. (2001) and Chow et al. (2003) as well as Ito and Miura (2003) have given interesting modeling results on air-cooled PV modules. Recent work on building integrated air-cooled photovoltaic includes the study on the multi-operational ventilated PVs with solar air collectors (Cartmell et al., 2004), the ventilated building PV facades (Infield et al., 2004; Guiavarch and Peuportier, 2006; Charron and Athienitis, 2006) and the design procedure for air cooling ducts to minimize the loss in PV module efficiency. On the other hand, according to Elazari (1998), smaller size PV and PV/T systems, using an aperture surface area of about 3–5 m^2 and a water storage tank of 150–200 l, could be installed for small (one family) domestic houses, while large sized systems of about 30–50 m^2 and 1500–2000 l water storage are more suitable for multi-flat residential buildings, hotels, hospitals and various food processing industries. Further, Charalambous et al. (2007) suggested that the building-integrated PV/T collectors are most suited for climatic regions with low ambient temperatures so that the heat from PV surface can be put into effective use for space heating. Battisti and Corrado (2005) investigated the EPBT (energy payback period) for a conventional multi-crystalline building integrated with a PV system (retrofitted on a tilted roof) in Rome receiving an annual solar insolation of about $1530\,kWh/m^2/year$. The study reported that EPBT was reduced from 3.3 years (standalone system) to 2.8 years by integrating the PV modules to the building. Despite these improvements, commercial application of PV/T air collectors is still marginal, but it is expected to be wider in the near future with many building facades and inclined roofs expected to be covered with photovoltaics.

PV facades are already well established and are closely identical to PV/T facades. Hence, by replacing expensive facade cladding materials by PV facades, it is expected that the costs on a module level will be low compared to all other applications. However, on a system level the situation may be different; since PV facades are often unglazed, the temperature levels that can be reached are limited, and the costs of the additional infrastructure required may outweigh the benefits of the use of this heat, so it is essential to come up with alternative low-cost system designs. Yet another issue is that these systems are not yet standardized. However, due to the current strong link between this type PV/T systems and the existing building projects, efforts are made to formulate codes based on an architectural point of view and PV manufacturing constraints (Butera et al., 2005).

5.2.4 PV/T concentrator

The combination of solar radiation concentration devices with PV modules appear to now be a viable method to reduce system cost, replacing the expensive cells with a cheaper solar radiation concentrating system. By concentrating, a (large) part of the expensive PV area is replaced by a less expensive mirror area, which is a way to reduce the payback time. This argument serves as the main driving force behind PV concentrators. Concentrating photovoltaics present higher efficiency than the typical ones, but this can be achieved only when the PV module temperature is maintained as low as possible (Othman et al., 2005). The concentrating solar systems use reflective and refractive optical devices and are characterized by their concentration ratio (CR). Concentrating systems with CR >2.5 must use a system to track the sun, while for systems with CR <2.5, stationary concentrating devices can be used (Winston, 1974). The distribution of the solar radiation on the absorber surface (PV module) and the increase in its temperature are two problems that affect the electrical output. The uniform distribution of the concentrated solar radiation on the PV surface and the suitable cooling mode together can contribute to an effective system operation and the achievement of high electrical output. PV/T absorbers can be combined with low, medium or high concentration devices, but so far, only low CR PV/T systems have been mainly developed so far.

Reflectors of low concentration, either of flat type (Sharan et al., 1985; Al Baali, 1986; Garg et al., 1991) or of Compound Parabolic Concentrator (CPC) type (Othman et al., 2005; Garg and Adhikari, 1998; Garg and Adhikari, 1999; Garg and Adhikari, 2000) have been suggested. Tripanagnostopoulos et al., (2002) suggested a diffuse reflector to increase both electrical and thermal output of PV/T systems. Garg et al. (1991) presented a simulation study of the single-pass PV/T air heater with plane reflector. They further extended their work on a hybrid PV/T collector with integrated CPC troughs (Garg and Adhikari, 1998; Garg and Adhikari, 2000). Both the studies confirmed that the total efficiency of a PV/T collector with a reflector was marginally higher compared to the systems without concentrators. Due to the increase in solar radiation, the average plate as well as solar cell temperatures had shown a sharp rise, as expected. Hence, the system performance in terms of electrical efficiency was low, due to the fact that the cell performance is dependent on its temperature. To overcome such overheating issues, Othman et al., (2005) designed a new double-pass photovoltaic-thermal air collector with fins to enhance the heat extraction. It was observed that the cell temperature was reduced by a few degrees which had a positive influence on the cell efficiency.

A simple low concentrating water-cooled type PV/T collector of the building integrated type investigated by Brogren and Karlsson (2001). It incorporates PV/T string modules with low cost aluminum foil reflectors with a CR of 4.3 times. With reference to medium concentration devices, PV/T systems based on linear parabolic reflectors (Chemisana et al., 2011) or linear Fresnel reflectors (Rosell et al., 2005) have been investigated. Although concentrators of low or medium CR are interesting devices to be combined with photovoltaics, 3D Fresnel lens or reflector type concentrators have been recently developed, aiming at the market of concentrating photovoltaics. The concept of combined linear Fresnel lenses with PV/T absorbers has also been attempted (Tripanagnostopoulos et al., 2007). Chemisana et al. (2011) carried out a study on a photovoltaic-thermal module for Fresnel linear concentrator. An advanced solar unit was designed to match the needs of building integration and concentrating photovoltaic/thermal generation. The unit contained three basic components: a domed linear Fresnel lens as primary concentrator, a compound parabolic reflector as secondary concentrator and a photovoltaic-thermal module. Models for the electrical and thermal behavior of the system were developed and validated experimentally and were found that the predicted results showed a good agreement with experimental measurements.

Even though the PV efficiencies of concentrated PV/T systems have proven to be high, the market share for such systems is very minimal, which is mainly due to the fact that these systems are rather bulky, disqualifying them for many PV applications. Also, since concentrating devices require tracking either one axis or two axes, it makes building integration impossible. Furthermore, not all climates are suitable for high ratio concentration, because it depends on the amount of direct irradiation received. In the aesthetic point of view, the concentrating systems provide different reflections and optical effects, which are unusual to the built environment and also they might prevent such systems from being placed visibly in the facade construction. One of the feasible options may be to install the concentrator on a horizontal roof (e.g. PV/T systems integrated with booster reflector in parallel rows). One more additional point worthy to note is, though the small cell area allows the use of more efficient and expensive PV material, the combination of glazing and reflectors increases the stagnation temperature which may in turn lead to degradation of materials. For electrical performance, the uniformity of the irradiance may be compromised, increasing mismatch losses. However, this drawback might be overcome to certain extent by using diffuse reflectors.

5.3 PV/T MODULE CONCEPTS

In a PV cell, part of the solar spectrum does not contribute to the electricity production. Photons with energy lower than the band gap do not have enough energy to create

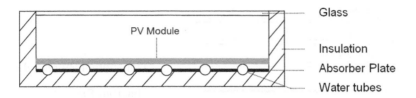

Figure 5.3.1 Cross section of a basic PV/T collector.

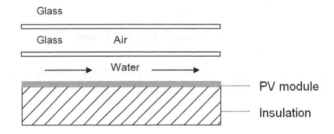

Figure 5.3.2 PV/T concept with liquid flowing on top of the PV module.

Figure 5.3.3 Channel PV/T concept with liquid flow beneath the PV cells.

photon-hole pairs and could in principle fully contribute to the generation of heat. This generation can take place either in the cell or outside the cell if the cell material does not absorb light at these wavelengths. The position of the heat generation in the device determines the possible PV/T device geometries. In most concepts the entire heat is generated in a simple device. In a two-absorber PV/T collector, however, part of the heat is generated outside the PV cells (van Helden et al., 2004).

5.3.1 Different types of PV/T modules

The design concepts can be categorized into four types based on most other articles about the PV/T systems. The most simple and basic design is a PV module attached on the back of a metallic heat absorber plate as shown in Figure 5.3.1. In this module concept, the distance between the heat generation device and heat collector determines the performance of the PV/T modules. The second type of PV/T modules has water that flows over the photovoltaic panel as shown in Figure 5.3.2. The third type of modules, which is demonstrated in Figure 5.3.3, also uses a liquid, but to improve its performance, the water flows through multiple channels underneath the PV panel to remove generated heat. The fourth type of design makes the PV cells transparent and

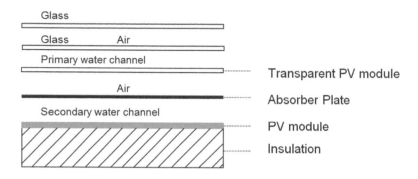

Figure 5.3.4 Two-absorber PV/T model.

applies two-absorber geometry. As shown is Figure 5.3.4, this design can lower the temperature of the PV cells on average, but it is relatively complicated to manufacture.

5.4 TECHNIQUES TO INPROVE PV/T PERFORMANCE

There are numerous methods to enhance the performance of PV/T air collectors such as the use of fins attached to the PV rear surface, corrugated sheet or wire mesh in the air channel or providing air circulation on both front and rear surfaces of the PV module.

Elements of several geometries can be placed between PV module and opposite channel wall, as well as on the back wall, by which air heat extraction can be effected more efficiently (Tripanagnostopoulos, 2007). Roughening the opposite channel wall with ribs or/and using a wall surface of high emissivity, which is a considerably lower cost air heating improvement, has also been adapted (Figure 5.4.1a). In addition, a corrugated sheet inside the air channel along the air flow can be attached on the PV rear surface as well as on the opposite channel wall surface (Figure 5.4.1b). An alternative modification is to insert lightweight pipes along the air flow in the air channel, with slight elasticity to ensure satisfactory thermal contact with the PV rear surface and channel wall (Figure 5.4.1c). These pipes can effectively extract heat from the PV panel by all three modes of heat transfer (conduction, convection, and radiation) thereby restricting the opposite channel wall surface temperature being overheated.

Tiwari et al. (2011) presented four types of photovoltaic modules and their applications. They are crystalline PV modules, thin film PV modules, single and multi-junction PV modules. Based on their cost analysis results, the BIPVT systems are reported to be more favorable than the conventional BIPV systems. In extension to the said work, they have also evaluated the overall performance of four types hybrid PV/T model, which are unglazed hybrid PV/T with tedlar, unglazed hybrid PV/T without tedlar, glazed hybrid PV/T with tedlar, and glazed hybrid PV/T without tedlar (Tiwari and Sodha, 2007). Experiments were conducted to validate the thermal model for unglazed PV/T air heating systems for summer conditions and the study concluded that the glazed hybrid PV/T without tedlar had the best performance, especially when in operation

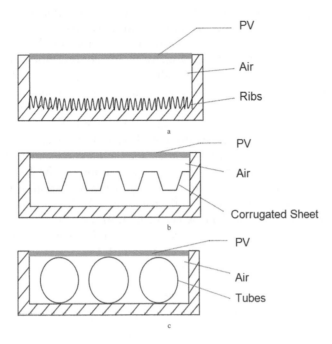

Figure 5.3.1 Modified PV/T dual systems provided with two TMS (Arif Hasan and Sumathy, 2010).

during the summer. Also, Tripanagnostopoulos et al. (2001a) carried out a comparison between different types of air PV/T systems. PV module provided with a float glass on either sides of a tedlar integrated to the rear end of the PV module and compared its performance to a pc-Si PV module using transparent tedlar on the front where only float glass is integrated to the rear end of the PV module. The experimental work revealed that the latter system has a higher electrical efficiency by reducing the temperature of the PV significantly. They also presented hybrid PV/T systems with dual heat extraction modes (Tripanagnostopoulos et al., 2001b). Three different design modes of PV/T systems were tested, with (i) a heat exchanger element was provided on the rear surface of the PV module, (ii) heat exchanger element provided in the middle of an air channel, and (iii) heat exchanger element provided on an opposite air channel surface. Results show that for both air and water circulation, PV with a heat exchanger on its rear surface produces the best thermal performance of the system.

Joshi et al. (2009) evaluated a hybrid photovoltaic thermal (glass-to-glass) system. They compared the performance of two types of photovoltaic module, which were PV module with glass-to-glass and glass-to-tedlar respectively. In the glass-to-glass case, the insulated base has both a black surface and solar cells to absorb the solar radiation, and then the heat from both the black surface and solar cell is transferred to the flowing air underneath the insulated base. In the glass-to-tedlar case, both the solar cell and the ethylene vinyl acetate absorb the solar radiation, and then transfer the heat to the flowing air underneath the base of the tedlar for thermal heating. The results obtained from both PV modules compared for a composite climate show that the hybrid air collector with glass-to-glass PV module has an approximately 2% higher overall thermal efficiency than the PV module with glass-to-tedlar.

Dubey et al. (2009) presented an analytical expression for electrical efficiency of PV/T hybrid air collector. They tested four different configurations of photovoltaic modules which are: glass-to-glass PV module with duct, glass-to-glass PV module without, glass-to-tedlar PV module with duct, and glass-to-tedlar PV module without duct. For electrical efficiency, the results show that the differences between PV modules with glass-to-glass and PV modules with glass-to-tedlar with and without duct are 1.24% and 0.086% respectively; the difference between the electrical efficiency of PV modules with glass-to-glass with and without duct is 0.66%. Similar to Dubey et al.'s work, Dupeyrat et al. (2011) had also worked to improve the PV module optical properties specifically to suit for hybrid PV/T collector application. It was shown that "the design of a PV module for a PV-T collector allows the use of alternative materials for the encapsulation process and a new encapsulation setup has been developed. This is a combination of a front layer with a low refractive index instead of glass cover and a low UV absorbing layer instead of the conventional EVA material." The results showed that the said configuration of the PV/T encapsulation module increased the generated current density at least 2 mA/cm^2. Compared to equivalent mc-Si cells laminated as a conventional standard glass/EVA/mc-Si/EVA/Tedlar module, the PV/T encapsulation module also has higher solar absorption coefficient. They designed, built and tested a prototype of PV/T collector based on experiments, and the results show 79% thermal efficiency and 8.7% electrical efficiency, which is 87% total efficiency.

Low cost performance improvements of PV/T solar collectors for natural air flow operation were introduced by Tonui and Tripanagnostopoulos (2007) and Tonui and Tripanagnostopoulos (2008). Two low-cost modifications of heat extraction were investigated in PV/T air system channel to cool down the PV as well as increase the thermal yield. They suggested using either finned back wall or thin flat metal sheet suspended at the middle of an air channel in the PV/T air configuration. The system consists of the PV module with a simple air channel attached on the back. It is very similar to using the PV module as an absorber plate with conventional air collectors. For the improved systems, the channels are modified by attaching a rectangular pro-file fins on the other side of the wall to the PV rear surface or by suspending a thin flat aluminum metal sheet in the middle of the air channel (Arif Hasan and Sumathy, 2010). Compared to attaching fins to the PV rear surface, attaching fins at the back wall was relatively easier, because attaching fins at the back of the PV module requires special designated features during the PV modules production. They pointed out that since the fins and metal sheets are easily obtained from the existing material which is not expensive, and the fabrication and modification are not complicated, being incorporated in the middle or on the opposite wall of the air channel, the total cost of the design models for the PV/T sir collectors is low. They also modified the duct geometry by changing the hydraulic diameter of the duct. The test results show that, when the hydraulic diameter of the duct is decreasing, the heat transfer surface area in the channel and convection heat transfer coefficient increased, thus more heat can be transferred from the PV panel to the air stream. Both the PV cooling capability and thermal efficiency of the system is increased. Tripanagnostopoulos et al. (2000) have also presented low cost improvements in integrated air cooled hybrid PV thermal systems specifically for buildings. They noted that the heat exchanging surface area of the air channel is increased by attaching fins of about 15 cm on to the opposite air channel surface. To form fin plate elements, they applied the 1.5 cm aluminum and 4 cm of π profile respectively.

5.5 CONCLUSION

The feasibility of the PV/T system will be dependent upon its technical and economic competitiveness with respect to other alternatives. The technical feasibility can be evaluated by comparing the electrical module efficiency and thermodynamic efficiency of such systems with those of the conventional ones, while the economic feasibility (energy metric analysis) can be tested by balancing the capital cost of the solar system against the savings in conventional fuel costs. As the economic feasibility is heavily dependent on the financial parameters based on some assumptions (e.g. the inflation rate of conventional fuel costs), it is certain that the viability of such solar systems will be more pronounced when the environmental costs of conventional electricity production are factored in.

As referred to in earlier sections, several studies (both theoretical and experimental) shows that most of the systems could only achieve a maximum thermal efficiency of about 60% for air-cooled and a slightly higher for water-cooled PV/T systems. The reduction in thermal efficiency might be due to reflection losses (since PV surfaces are not spectrally selective), and also due to the fact that the heat resistance between the absorbing surface and the heat transfer medium is increased because of the additional layers of material (e.g. tedlar). Hence, it is necessary to keep all layers between the PV panel and the absorber as thin as possible. It should be pointed out that several researchers (Tiwari and Sodha, 2007; Dubey et al., 2009) have confirmed to use glass instead of tedlar, as tedlar becomes a barrier for extracting thermal energy, in turn reducing both the electrical and overall efficiency of the system.

Poor thermal contact was also reported to be a problem by Sudhakar and Sharon (1994) who found a temperature difference of about 15°C between PV laminate and water output temperature. Hence, the objective of future research should aim to optimize the air channel geometry of the PV/T system and to simulate the PV/T collector characteristics, and further investigate the influence of various heat transfer promoters on the cell temperature of the PV module for different operating conditions. It is also essential to establish an analytical expression for the electrical efficiency of the PV module with and without air flow as a function of climatic and design parameters, which can be derived based on a detailed energy balance of each component of the chosen configuration. For the case of PV/T liquid collectors, though the sheet-and-tube design performs efficiently, the channel plate constructions may provide interesting ways of further increasing the heat transport, provided that the channels are made sufficiently thin. For an unglazed PV/T water collector, a heat pump can be integrated to the PV/T system, which may be a promising development for the future.

To make solar energy devices more attractive for potential applications, it is essential to develop a thermal model of integrated photovoltaic and thermal solar systems, which could be used to analyze the overall system performance under various climatic as well as design conditions. The possibility of generating electricity and heat energy from PV/T solar collector with either forced or natural flow (using water or air) has been demonstrated by various researchers. PV/T systems contribute immensely towards energy savings and mitigation of energy supply of buildings and consequently lower CO_2 emission among other social benefits. The choice of technique depends on the location and its application which dictates the usage of appropriate design considerations. Hybrid PV/T systems are especially suitable in regions with a cold climate since PV/T systems integrated to building integrated applications lower the temperature of the PV's with air and can supply hot air for space heating. However, based on the overview of research conducted to date, it is apparent that there is still a large amount of work that needs to be undertaken in terms of design aspects before PV/T systems can be successfully implemented and integrated into domestic and commercial applications. With an optimal design, PV/T systems can supply buildings with 100% renewable electricity and heat in a more cost-effective manner than separate PV and solar thermal systems and thus contribute to the long-term international targets on implementation of renewable energy in the built environment.

REFERENCES

Affolter, P., Eisenmann, W., Fechner, H., Rommel, M., Schaap, A. and Sorensen, H. (2006) *PVT Roadmap*. Edition of ECN; Netherlands, http://www.pvforum.org/.

Al Baali, A.A. (1986) Improving the power of a solar panel by cooling and light concentrating. *Solar and Wind Technology*, 3, 241–245.

Arif Hasan, M. and Sumathy, K. (2010) Photovoltaic thermal module concepts and their performance analysis: A review. *Renewable and Sustainable Energy Reviews*, 14, 1845–1859.

Baños, R., Manzano-Agugliaro, F., Montoya, F.G., Gil, C., Alcayde, A. and Gómez, G. (2010) Optimization methods applied to renewable and sustainable energy: A review. *Renewable and Sustainable Energy Reviews*, 15, 1753–1766.

Battisti, R. and Corrado, A. (2005) Evaluation of technical improvements of photovoltaic systems through life cycle assessment methodology. *Energy*, 30, 952–967.

Bazilian, M., Leeders, F., van der Ree, B.G.C. and Prasad, D. (2001) Photovoltaic cogeneration in the built environment. *Solar Energy*, 71:57–69.

Bhargava, A.K., Garg, H.P. and Agarwal, R.K. (1991) Study of a hybrid solar system-solar air heater combined with solar cell. *Energy Conversion and Management*, 31, 471–479.

Benemann, J., Chehab, O. and Schaar-Gabriel, E. (2001) Building-integrated PV modules. *Solar Energy Materials and Solar Cells*, 67, 345–354.

Bergene, T. and Lovvik, O.M. (1995) Model calculations on a flat-plate solar heat collector with integrated solar cells. *Solar Energy*, 55, 453–462.

Brinkworth, B.J. & Sandberg, M. (2006) Design procedure for cooling ducts to minimize efficiency loss due to temperature rise in PV arrays. *Solar Energy*, 80, 89–103.

Brinkworth, B.J., Cross, B.M., Marshall, R.H. and Yang, H.X. (1997) Thermal regulation of photovoltaic cladding. *Solar Energy*, 61, 169–178.

Brinkworth, B.J. (2000) Estimation of flow and heat transfer for the design of PV cooling ducts. *Solar Energy*, 69, 320–413.

Brinkworth, B.J., Marshall, R.H. and Ibarahim, Z.A. (2000) Validated model of naturally ventilated PV cladding. *Solar Energy*, 69, 67–81.

Brogren, M., Nostell, P. and Karlsson, B. (2000) Optical efficiency of a PV thermal hybrid CPC module for high latitudes. *Solar Energy*, 69, 173–185.

Brogren, M. and Karlsson, B. (2001) Low-concentrating water-cooled PV-thermal hybrid systems for high latitudes. In: *Proceedings of the 17th EU-PVSEC*, 22–26 October, Munich, Germany.

Butera, F., Aste, N., Adhikari, R.S. and Bracco, R. (2005) Ecomensa project at CRF: Performance of solar facade. In: *Proceedings 600 Congresso Nazionale ATI*, 13–15 September, Rome, Italy.

Cartmell, B.P., Shankland, N.J., Fiala, D. and Hanby, V. (2004) A multioperational ventilated photovoltaic and solar air collector: Application, simulation and initial monitoring feedback. *Solar Energy*, 76, 45–53.

Charalambous, P.G., Maidment, G.G., Kalogirou, S.A. and Yiakoumetti, K. (2007) Photovoltaic thermal (PV/T) collectors: A review. *Applied Thermal Engineering*, 27, 275–286.

Charron, R. and Athienitis, A.K. (2006) Optimization of the performance of double-facades with integrated photovoltaic panels and motorized blinds. *Solar Energy*, 80, 482–491.

Chemisana, M., Ibáñez, J.I. and Rosell. (2011) Characterization of a photovoltaic-thermal module for Fresnel linear concentrator. *Energy Conversion and Management*, 52, 3234–3240.

Chow, T.T (2003) Performance analysis of photovoltaic-thermal collector by explicit dynamic model. *Solar Energy*, 75, 143–152.

Chow, T.T., He, W. and Ji, J. (2006) Hybrid photovoltaic-thermosyphon water heating system for residential application. *Solar Energy*, 80, 298–306.

Chow, T.T., Hand, J.W. and Strachan, P.A. (2003) Building-integrated photovoltaic and thermal applications in a subtropical hotel building. *Applied Thermal Engineering*, 23, 2035–2049.

Chow, T.T., Chan, A.L.S., Fong, K.F., Lin, Z., He, W. and Ji, J. (2009) Energy and exergy analysis of photovoltaic-thermal collector with and without glass cover. *Applied Energy*, 86, 310–316.

Cox, C.H. and Raghuraman, P. (1985) Design considerations for flat-plate photovoltaic/thermal collectors. *Solar Energy*, 35, 227–241.

Daghigh, R., Ruslan, M.H. and Sopian, K. (2011) Advances in liquid based photovoltaic/thermal (PV/T) collectors. *Renewable and Sustainable Energy Reviews*, 15, 4156–4170.

Dubey, S. and Tiwari, G.N. (2008) Thermal modeling of a combined system of photovoltaic thermal (PV/T) solar water heater. *Solar Energy*, 82, 602–612.

Dubey, S., Sandhu, G.S. and Tiwari, G.N. (2009) Analytical expression for electrical efficiency of PV/T hybrid air collector. *Applied Energy*, 86, 697–705.

Dupeyrat, P., Ménézo, C., Wirth, H. and Rommel, M. (2011) Improvement of PV module optical properties for PV-thermal hybrid collector application. *Solar Energy Materials and Solar Cells*, 95, 2028–2036.

Eicker, U. (2003) *Solar Technologies of Buildings*. New York, John Wiley & Sons, Inc.

Eicker, U., Fux, V., Infield, D. and Mei, L. (2000) Heating and cooling of combined PV-solar air collector's facades. In: *Proceedings of International Conference of 16th European PV Solar Energy*. 1–5 May, Glasgow, UK. pp. 1836–1839.

Elazari, A. (1998) Multi Solar System-Solar multimodule for electrical and hot water supply for residentially building. In: *Proceedings of 2nd World Conference on Photovoltaic Solar Energy Conversion*, 6–10 July, Vienna, Austria. pp. 2430–2433.

Erdil, E., Ilkan, M. and Egelioglu, F. (2008) An experimental study on energy generation with a photovoltaic (PV)-solar thermal hybrid system. *Energy*, 33, 1241–1245.

Florschuetz, L.W. (1979) Extension of the Hottel-Whillier model to the analysis of combined photovoltaic/thermal flat plate collectors. *Solar Energy*, 22, 361–366.

Garg, H.P., Agarwal, R.K. and Joshi, J.C. (1994) Experimental study on a hybrid photovoltaic thermal solar water heater and its performance prediction. *Energy Conversion and Management*, 35, 621–633.

Garg, H.P., Agarwal, R.K. and Bhargava, A.K. (1991) The effect of plane booster reflectors on the performance of a solar air heater with solar cells suitable for a solar dryer. *Energy Conversion and Management*, 35, 543–554.

Garg, H.P. and Adhikari, R.S. (1998) Transient simulation of conversional hybrid photovoltaic/thermal (PV/T) air heating collectors. *International Journal of Energy Research*, 22, 547–562.

Garg, H.P. and Adhikari, R.S. (1999) Performance analysis of a hybrid photovoltaic/thermal (PV/T) collector with integrated CPC troughs. *International Journal of Energy Research*, 23, 1295–1304.

Garg, H.P. and Adhikari, R.S. (2000) *Hybrid Photovoltaic/Thermal Utilization Systems*. Final report submitted to All India Council of Technical Education. New Delhi, India.

Guiavarch, A. and Peuportier, B. (2006) Photovoltaic collector's efficiency according to their integration in buildings. *Solar Energy*, 80, 65–77.

Hastings, S.R. and Morck, O. (2000) *Solar Air Systems, A Design Handbook*. James & James, London.

Hegazy, A.A. (2000) Comparative study of the performance of four photovoltaic/thermal solar air collectors. *Energy Conversion and Management*, 41, 861–881.

Hendrie, S.D. (1979) Evaluation of combined photovoltaic/thermal collectors. In: *Proceedings of ISES International Congress and Silver Jubilee*, May 28-June 1, Atlanta, Georgia. pp. 1865–1869.

Hodge, E. and Gibbons, C. (2004) Convective cooling of photovoltaics. In: *Proceedings of ISES 2004 EuroSun*, June 20–24, 2004, Freiburg, Germany.

Huang, B.J., Lin, T.H., Hung, W.C and Sun, F.S. (2001) Performance evaluation of solar photovoltaic/thermal systems. *Solar Energy*, 70, 443–448.

IEA (2007) *Solar Heating and Cooling Program*. International Energy Agency.

Infield, D., Eicker, U., Fux, V., Li, M. and Schumacher, J. (2006) A simplified approach to thermal performance calculation for building integrated mechanically ventilated PV facades. *Building and Environment*, 41, 893–901.

Infield, D., Mei, L. and Eicker, U. (2004) Thermal performance estimation for ventilated PV facades. *Solar Energy*, 76, 93–98.

Ito, S. and Miura, N. (2003) Usage of a DC fan together with photovoltaic modules in a solar air heating system. *In: Proceedings of (CD-ROM) ISES World Congress*, 14–19 June, Goteborg, Sweden.

Joshi, A.S., Tiwari, A., Tiwari, G.N., Dincer, I. and Reddy, B.V. (2009) Performance evaluation of a hybrid photovoltaic thermal (PV/T) (glass-to-glass) system. *International Journal of Thermal Sciences*, 48, 154–164.

Kalogirou, S.A. (2001) Use of TRYNSYS for modeling and simulation of a hybrid PV-thermal solar system for Cyprus. *Renewable Energy*, 23, 247–260.

Karlsson, B., Brogren, M., Larsson, S., Svensson, L., Hellstrom, B. and Sarif, Y. (2001) A large bifacial photovoltaic-thermal low-concentrating module. In: *Proceedings of 17th PV Solar Energy Conference*, 22–26 October, Munich, Germany. pp. 808–811.

Kern, E.C. and Russell, M.C. (1978) Combined photovoltaic and thermal hybrid collector systems. *In: Thirteenth IEEE Photovoltaic Specialists Conference*, June 5–8, 1978, Washington DC, USA. pp. 1153–1157.

Krauter, S., Araujo, R.G., Schroer, S., Hanitsh, R., Salhi, M.J., Triebel, C. and Lemoine, R. (1999) Combined photovoltaic and solar thermal systems for facade integration and building insulation. *Solar Energy*, 67, 239–248.

Lee, W.M., Infield, D.G. and Gottschalg, R. (2001) Thermal modeling of building integrated PV systems. In: *Proceedings of 17th PV Solar Energy Conference*, 22–26 October, Munich. pp. 2754–2757.

Li, M., Infield, D., Eicker, U., Loveday, D. and Fux, V. (2006) Cooling potential of ventilated PV façade and solar air heaters combined with a desiccant cooling machine. *Renewable Energy*, 31, 1265–1278.

Moshfegh, B. and Sandberg, M. (1998) Flow and heat transfer in the air gap behind photovoltaic panels. *Renewable and Sustainable Energy Reviews*, 2, 287–301.

Moshfegh, B., Sandberg, M., Bloem, J.J. and Ossenbrink, H. (1995) Analysis of fluid flow and heat transfer within the photovoltaic facade on the ELSA building. In: *Proceedings of JRC ISPRA 13th European PV Solar Energy Conference*, 23–27 October 1995, Nice, France.

Naphon, P. (2005) On the performance and entropy generation of the double-pass solar air heater with longitudinal fins. *Renewable Energy*, 30, 1345–1357.

Nobuyuki, H. (2006) *Renewable Energy*: RD&D Priorities: Insights from IEA Technology Programs.

Ossenbrink, H.A., Rigolini, L., Chehab, O. and van der Venne, O. (1994) Building integration of an amorphous silicon photovoltaic facade. In: *Proceedings of IEEE First World Conference on Photovoltaic Energy Conversion*, 5–9 December, Hawaii, USA. pp. 770–773.

Othman, M.Y.H., Yatim, B., Sopian, K. and Bakar, M.N.A. (2005) Performance analysis of a double-pass photovoltaic/thermal (PV/T) solar collector with CPC and fins. *Renewable Energy*, 30, 2005–2017.

Posnansky, M., Gnos, S. and Coonen, S. (1994) The importance of hybrid PV Building integration. In: *Proceedings of IEEE First World Conference on Photovoltaic Energy Conversion*, 5–9 December 1994, Hawaii, USA. pp. 998–1003.

Pottler, K., Sippel, C.M., Beck, A. and Fricke, J. (1999) Optimized finned absorber geometries for solar air heating collectors. *Solar Energy*, 67, 35–52.

Prakash, J. (1994) Transient analysis of a photovoltaic thermal solar collector for cogeneration of electricity and hot air water. *Energy Conversion and Management*, 35, 967–972.

Raghuraman, P. (1981) Analytical prediction of liquid and air photovoltaic/thermal flat-plate collectors performance. *ASME Journal of Solar Energy Engineering*, 103, 291–298.

Ricaud, A. and Capthel, R.P. (1994) A 66% efficient hybrid solar module and the Ecothel co-generation solar system. In: *Proceedings of IEEE First World Conference on Photovoltaic Energy Conversion*, December 5–9, 1994, Waikoloa, Hawaii. pp. 1012–1015.

Rosell, J.I., Vallverdu, X., Lechon, M.A. and Ibanez, M. (2005) Design and simulation of a low concentrating photovoltaic/thermal system. *Energy Conversion and Management*, 46, 3034–3036.

Sandberg, M. and Moshfegh, B. (2002) Buoyancy-induced air flow in photovoltaic facades: effect of geometry of the air gap and location of solar cell modules. *Building and Environment*, 37, 211–218.

Sharan, S.N., Mathur, S.S. and Kandpal, T.C. (1985) Economic evaluation of concentrator-photovoltaic systems. *Solar and Wind Technology*, 2, 195–200.

Sopian, K., Yigit, K.S., Liu, H.T., Kakac, S. and Veziroglu, T.N. (1996) Performance analysis of photovoltaic thermal air heaters. *Energy Conversion and Management*, 37, 1657–1670.

Sudhakar, S.V. & Sharon, M. (1994) Fabrication and performance evaluation of a photovoltaic/thermal hybrid system. *Journal of the Solar Energy Society of India*, 4, 1–7.

Tiggelbeck, S.T., Mitra, N.K. and Fiebig, M. (1993) Experimental investigations of heat transfer enhancement and flow losses in a channel with double rows of longitudinal vortex generators. *International Journal of Heat and Mass Transfer*, 26, 2327–2337.

Tiwari, A. and Sodha, M.S. (2006) Performance evaluation of hybrid PV/thermal water/air heating system: a parametric study. *Renewable Energy*, 31, 2460–2474.

Tiwari, A., Sodha, M.S., Chandra, A. and Joshi, J.C. (2006) Performance evaluation of photovoltaic thermal solar air collector for composite climate of India. *Solar Energy Materials and Solar Cells*, 90, 175–189.

Tiwari, A. and Sodha, M.S. (2006) Performance evaluation of solar PV/T system: an experimental validation. *Solar Energy*, 80, 751–759.

Tiwari, G.N., Mishra, R.K. and Solanki, S.C. (2011) Photovoltaic modules and their applications: A review on thermal modeling. *Applied Energy*, 88, 2287–2304.

Tiwari, A. and Sodha, M.S. (2007) Parametric study of various configurations of hybrid PV/thermal air collector: Experimental validation of theoretical model. *Solar Energy Materials & Solar Cells*, 91, 17–28.

Tonui, J.K. and Tripanagnostopoulos, Y. (2007) Air-cooled PV/T solar collectors with low cost performance improvements. *Solar Energy*, 81, 498–511.

Tonui, J.K. and Tripanagnostopoulos, Y. (2007) Improved PV/T solar collectors with heat extraction by forced or natural air circulation. *Renewable Energy*, 32, 623–637.

Tonui, J.K. and Tripanagnostopoulos, Y. (2008) Performance improvement of PV/T solar collectors with natural air flow operation. *Solar Energy*, 82, 1–12.

Tripanagnostopoulos, Y., Souliotis, M., Battisti, R. and Corrado, A. (2006) Performance, cost and Life-cycle assessment study of hybrid PVT/AIR solar systems. *Progress in Photovoltaics: Research and Applications*, 14, 65–76.

Tripanagnostopoulos, Y., Nousia, T.H., Souliotis, M. and Yianoulis, P. (2002) Hybrid photovoltaic/thermal solar systems. *Solar Energy*, 72, 217–34.

Tripanagnostopoulos, Y., Siabekou, C.H. and Tonui, J.K. (2007) The Fresnel lens concept for solar control of buildings. *Solar Energy*, 81, 661–675.

Tripanagnostopoulos, Y. (2007) Aspects and improvements of hybrid photovoltaic/thermal solar energy systems. *Solar Energy*, 81, 1117–1131.

Tripanagnostopoulos, Y., Nousia, T.H. and Souliotis, M. (2001a) Test results of air cooled modified PV modules. In: *Proceeding 17th PV Solar Energy Conference*, 22–26 October, Munich, Germany. pp. 2519–2522.

Tripanagnostopoulos, Y., Tzavellas, D., Zoulia, I. and Chortatou, M. (2001b) Hybrid PV/T systems with dual heat extraction operation. In: *Proceeding 17th PV Solar Energy Conference*, 22–26 October, Munich, Germany. pp. 2515–2518.

Tripanagnostopoulos, Y., Nousia, T.H. and Souliotis, M. (2000) Low cost improvements to building integrated air cooled hybrid PV-Thermal systems. In: *Proceedings of 16th European PV Solar Energy Conference*, 1–5 May, Glasgow, UK. pp. 1874–1899.

van Helden, W.G.J., van Zolingen, R.J.C. and Zondag, H.A. (2004) PV Thermal Systems: PV Panels Supplying Renewable Electricity and Heat. *Progress in Photovoltaics: Research and Application*, 12, 415–426.

Winston, R. (1974) Principles of solar concentrators of a novel design. *Solar Energy*, 16, 89–95.

Yang, H.X., Marshall, G.H. and Brinkworth, B.J. (1994) An experimental study of the thermal regulation of a PV-clad building roof. In: *Proceedings of 12th European Photovoltaic Solar Energy Conference*, April 11–15, Amsterdam, The Netherlands. pp. 1115–1118.

Zondag, H., Bakker, M., van Helden, W.G.J., Affolter, P., Eisenmann, W. and Fechner, H. (2005) PVT roadmap: A European guide for the development and market introduction of PVT technology. In: *Proceedings of (CD) 20th European Photovoltaic Solar Energy Conference*, June 6–10, 2004. Barcelona, Spain.

Zondag, H.A., de Vries, D.W., van Helden, W.G.J., van Zolengen, R.J.C. and Steenhoven, A.A. (2002) The thermal and electrical yield of a PV thermal collector. *Solar Energy*, 72, 113–128.

Zondag, H.A., de Vries, D.W., van Helden, W.G.J., van Zolengen, R.J.C. and Steenhoven, A.A. (2003) The yield of different combined PV-thermal collector designs. *Solar Energy*, 74, 253–269.

Chapter 6

Thermal modelling of parabolic trough collectors

Soteris Kalogirou

Department of Mechanical Engineering and Materials Science and Engineering,
Cyprus University of Technology, Limassol, Cyprus

6.1 INTRODUCTION

As shown in Figure 6.1.1, in a parabolic trough collector (PTC) a sheet of reflective material is formed in parabolic shape. At the focal point of the parabola a metal black pipe, covered with a glass tube to reduce thermal losses, is placed. When the axis formed by the centre of the parabola and the receiver faces the sun, the parallel rays incident on the reflector are reflected and focused onto the receiver tube. In this way the concentrated radiation reaching the receiver tube heats the receiver pipe and thus the fluid circulating through it, transforming the solar radiation into useful heat.

As long collector modules are usually produced, a single axis tracking of the sun is used (Kalogirou 2004, 2009). The collector can be orientated in a north-south direction, tracking the sun from east to west, or orientated in an east-west direction, tracking the sun from north to south. The advantages of the latter tracking mode is that very little collector adjustment is required during the day and the full aperture always faces the sun at noon time but the collector performance during the early and late hours of the day is greatly reduced due to large incidence angles (cosine loss). North-south orientated troughs have their highest cosine loss at noon and the lowest in the mornings and evenings when the sun is due east or due west. Over the period of one year, a horizontal north-south trough field usually collects a little more energy than a horizontal east-west one. However the north-south field collects a lot more energy in summer and much less in winter. On the other hand the east-west field collects more energy in winter than a north-south field and less in summer, providing a more constant

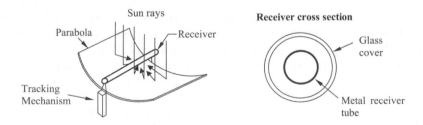

Figure 6.1.1 Schematic diagram of a parabolic trough collector and receiver detail. The collector comprise a sheet of reflective material in parabolic shape together with a metal black pipe, covered with a glass tube to reduce thermal losses, placed along its focal line.

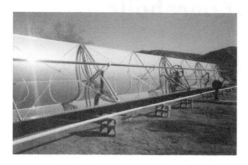

Figure 6.1.2 Photographs of parabolic trough collectors (left picture Industrial Solar Technology collector, right picture Eurotrough collector).

annual output. Therefore, the choice of orientation should depend on the application and whether more energy is needed during summer or winter (Kalogirou 2004, 2009).

Parabolic trough collector technology is the most advanced of the solar thermal technologies because of considerable experience gained so far and the development of a small commercial industry to produce and market these systems. Parabolic trough collectors are built in modules that are supported from the ground by simple pedestals at specific intervals or at either end if the collector is short. Photographs of two parabolic trough collectors are shown in Figure 6.1.2.

Parabolic trough collectors are the most mature solar technology to generate heat at temperatures up to 400°C for solar thermal electricity generation or process heat applications. This is due to the application of this type of system in the Southern California power plants, which have a total installed capacity of 354 MWe (Kearney and Price 1992) and are known as Solar Electric Generating Systems (SEGS). SEGS I is 14 MWe, SEGS II–VII are 30 MWe each and SEGS VIII and IX are 80 MWe each. Recently, new systems have been installed in Spain and the USA, and new plants are under development in many Middle East countries.

New developments in the field of parabolic trough collectors are focussed on cost reduction and improvements of the technology. In one such development, the collector is washed automatically during the night, thus reducing drastically the maintenance cost, as this is the most used maintenance process (Kalogirou 2009).

A linear receiver is used in a parabolic trough which is a metallic tube placed along the focal line of the parabola surrounded by a glass cover envelope (see Figure 6.1.1 detail). The size of the tube, and therefore the concentration ratio, is determined by the size of the reflected sun image and the manufacturing tolerances of the trough. The surface of the metal receiver is usually plated with selective coating that has a high absorptance for solar irradiation but a low emittance for thermal radiation.

The purpose of the glass cover tube placed around the receiver tube is to reduce the convective heat loss from the receiver, so as to decrease the heat loss coefficient. A disadvantage resulting from the use of the glass cover tube is that the reflected light from the concentrator must first pass through the glass to reach the receiver, and in doing so a transmittance loss is added of about 0.9, when the glass is clean. The glass envelope usually has an anti-reflective coating to improve transmissivity. Particularly for high temperature applications, the space between the glass cover tube and the receiver is evacuated to further reduce convective heat loss from the receiver tube and thereby increase the performance of the collector. The total receiver tube length of PTCs is usually from 25 m to 150 m.

In this chapter a detailed thermal model of the receiver of the collector is presented. Many researchers have published studies of energy models of parabolic trough collectors. The most important ones are the following.

Edenburn (1976) predicted the efficiencies for focusing collectors which consist of a cylindrical parabolic reflector and a collector tube surrounded by a transparent envelope and which heat a fluid flowing through the collector tube. These efficiencies have been predicted using analytical heat transfer methods. The analysis considers visible radiation transfer, IR radiation exchange, conductive and convective losses and energy transferred to a fluid flowing through the collector tube. The collector may have a tilted north-south axis, an east-west axis or it may fully track the sun and geometric parameters associated with tracking the sun are considered. Both evacuated and non-evacuated cases are considered and the predicted results are in excellent agreement with collector performances measured using Sandia Laboratories' collector test facility.

Clark (1982) analyzed the effects of design and manufacturing parameters that influence the thermal and economic performance of parabolic trough receivers. This is achieved by an identification of the principal design factors that influence the technical performance of a parabolic trough concentrator and which relate directly to design and manufacturing decisions. These factors include spectral-directional reflectivity of the mirror system, the mirror-receiver tube intercept factor, the incident angle modifier and absorptivity-transmissivity product of the receiver tube and cover tube, the end loss factor and a factor describing the effect of tracking errors and receiver tube misalignment. Each of these factors has been quantified in terms of design and manufacturing tolerances and associated performance degradation. Other design considerations that relate to thermal loss from the receiver tube are low emissivity coatings, evacuation and anti-reflection coating. The analysis of energy costs using the parabolic trough concentrator determines both the break-even, current metered cost of energy and the annual cash flow over periods of investment ranging from 5 to 15 yr. The economic factors include investment tax credit, energy equipment tax credit, income tax bracket, cost of auxiliary system, foundations and controls, cost of collector, costs of maintenance and taxes, costs of fuel, cost of capital, general inflation rate and fuel escalation rate.

Karimi et al. (1986) applied a piecewise two-dimensional model of the receiver, in which the receiver of the collector is divided into longitudinal and isothermal nodal sections as shown in Figure 6.1.3, performed by considering the circumferential variation of solar flux and applying the principle of energy balance to the glazing and receiver nodes.

Heidemann et al. (1992) studied the temperature field in the absorber tube of a direct steam generating parabolic trough collector. Steady-state and transient operating conditions are considered. They formulated a two dimensional heat transfer model for calculating the absorber wall temperature of a DSG collector under both conditions. A universal program was developed for solving the two-dimensional transient temperature field using a modular nodal point library. The temperature field is extremely asymmetric due to the variation of the heat transfer coefficient at the inner surface and the solar irradiation at the outer surface of the absorber tube. High temperature peaks

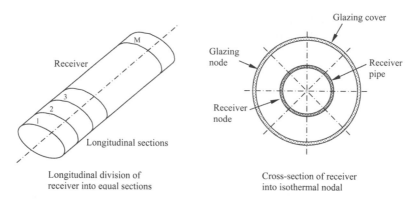

Figure 6.1.3 Piecewise two-dimensional model of the receiver assembly with the receiver divided into longitudinal and isothermal nodal sections (Karini et al., 1986).

are found, especially in stratified flow at higher void fractions. The numerical solution showed that a sudden drop of irradiation induces a very high temperature gradient inside the absorber tube in a short period of time.

Thomas and Thomas (1994) studied the design data required for the computation of thermal loss in the receiver of a parabolic trough concentrator for specific absorber tube diameters, various ambient temperatures, wind velocity and absorber temperatures from 50 to 350°C in steps of 10°C. Curve-fitting equations based on a numerical heat transfer model for the heat losses for the above parameters are given to enable the designer to generate the required data for any absorber temperature, absorber diameter, ambient temperature, wind velocity and emissivity of the solar selective coating of the absorber.

Dudley et al. (1994) developed an analytical model of SEGS LS-2 parabolic solar collector. The thermal loss model for the heat collection element was a one dimensional steady state model based on thermal resistance analysis. This model was validated with experimental data collected by Sandia National Laboratories (SNL) for different receiver annulus conditions: vacuum intact, lost vacuum (air in annulus), and broken annulus cover (bare tube). The results showed a reasonable agreement between the theoretical and experimental heat losses.

Odeh et al. (1998) studied the thermal performance of a parabolic trough solar collector used as direct steam generator for different solar radiation levels and geometric configurations. This heat transfer model showed better agreement with the in-focus test results than the polynomial curve fit equation obtained by Dudley et al. (1994). The thermal losses calculated for water were based on the receiver wall temperature and the results showed that thermal losses calculated for steam as heat transfer fluid were lower than those obtained for synthetic oil.

Forristall (2003) built and analysed both a 1-D and a 2-D heat transfer models of a PTC receiver implemented in EES software. For this purpose a detailed heat transfer solar receiver model was used. A one-dimensional energy balance for several segments was used for short and long receivers respectively. This model was used to determine the thermal performance of parabolic trough collectors under different operating conditions.

Garcia-Valladares and Velasquez (2009) developed a detailed numerical model for a single pass and double pass solar receiver and validated it. The governing equations inside the receiver tube, together with the energy equation in the tube walls and cover wall and the thermal analysis in the solar concentrator were solved iteratively in a segregated manner. The single-pass solar device numerical model has been carefully validated with experimental data obtained by Sandia National Laboratories (SNL). The effects of recycling at the ends on the heat transfer are studied numerically and show that the double-pass arrangement can enhance the thermal efficiency compared with the single-pass.

Cheng et al. (2010) in their contribution examined the solar energy flux distribution on the outer wall of the inner absorber tube of a parabolic solar collector receiver by adopting the Monte Carlo Ray-Trace Method (MCRT Method). They found that the non-uniformity of the solar energy flux distribution is very large. Three-dimensional numerical simulation of coupled heat transfer characteristics in the receiver tube is calculated and analyzed by combining the MCRT Method and FLUENT software, in which the heat transfer fluid was the Syltherm 800 liquid oil and the physical model was the LS2 parabolic solar collector from the testing experiment of Dudley et al. (1994). Temperature-dependent properties of the oil and thermal radiation between the inner absorber tube and the outer glass cover tube are also taken into account. Compared with test results from three typical testing conditions, the average difference is within 2%.

Gong et al. (2010) presented an optimised model and tested China's first high temperature parabolic trough receiver. The model is written in Matlab and computes the receiver's major heat loss through the glass envelope, and then systematically analyzes the major influence factors of heat loss in both 1-D and 3-D. Comparison shows the original 1-D model agrees with the 'ends of the receiver covered test' while remarkably deviating from the 'ends exposed' test. For the purpose of identifying the influence of the receiver end on total heat loss, an additional 3-D model was built using a CFD software to further investigate the different heat transfer processes of receiver's end components. The 3-D end model is verified by heating power and IR temperature distribution images in the test. Combining the optimized 1-D model with the new 3-D end model, the comparison with test data shows a good agreement.

He et al. (2011) used a coupled simulation method based on Monte Carlo Ray Trace (MCRT) and Finite Volume Method (FVM) to solve the complex coupled heat transfer problem of radiation, heat conduction and convection in a parabolic trough solar collector system. A coupled grid checking method is established to guarantee the consistency between the two methods and the validations to the coupled simulation model were performed. The heat flux distribution curve could be divided into 4 parts: shadow effect area, heat flux increasing area, heat flux reducing area and direct radiation area. The heat flux distribution on the outer surface of absorber tube was heterogeneous in the circumferential direction but uniform in the axial direction. Finally, the concentrating characteristics of the parabolic trough collectors (PTCs) were analyzed by the coupled method, the effects of different geometric concentration ratios (GCs) and different rim angles were examined. The results show that both variables affect the heat flux distribution.

Padilla et al. (2011) in their paper present a detailed one-dimensional numerical heat transfer analysis of a PTC. The receiver and envelope were divided into several segments and mass and energy balances were applied in each segment. The partial differential equations developed were discretized and the nonlinear algebraic equations were solved simultaneously. Finally, to validate the numerical results, the model was compared with experimental data obtained from Sandia National Laboratory (SNL) and other one-dimensional heat transfer models.

The model presented in this chapter takes into consideration all modes of heat transfer: forced convection into the receiver pipe and from the glass cover to ambient air (usual case when there is wind); natural convection in the annulus between the receiver and the glass cover; conduction through the metal receiver pipe and glass cover walls; and radiation from the metal receiver pipe to glass cover and from glass cover to the sky.

6.2 THE ENERGY MODEL

Although for low-temperature applications a bare tube receiver can be used, as for this kind of applications low technology collectors the flat-plate can be used, in this chapter only a glazed receiver is considered, which is the usual case for PTCs. For the annulus between the receiver and the glass cover two conditions are considered: the vacuum, usually used in high temperature applications, and the air case, which is used for lower temperature applications, and for cases when the vacuum is lost from the former design.

The model is written in Engineering Equation Solver (EES) software (Klein 2002). This is done for two reasons; the EES includes routines to estimate the properties of various substances by specifying any two properties, such as temperature and pressure, and EES can be called from TRNSYS which allows the development of a model which can use the capabilities of both programs. The model is validated with known performance of existing collectors, and subsequently is used to perform an analysis of the collector installed at Archimedes Solar Energy Laboratory at Cyprus University of Technology.

The collector performance model uses an energy balance between the fluid flowing through the receiver, usually a heat transfer fluid (HTF), and the atmosphere. It includes all equations necessary to predict the various expressions of the energy balance, which depend on the ambient conditions and the collector receiver optical properties and condition.

A cross-section of the collector receiver and the subscript definitions are shown in Figure 6.2.1a whereas Figure 6.2.1b shows the energy balance of the receiver and Figure 6.2.1c the steady-state thermal resistance model. The model assumes that all temperatures, heat fluxes, and thermodynamic properties are uniform around the circumference of the receiver. This assumption is not very accurate as it is well known that the radiation profile is not uniform, and the bottom part receives much higher solar flux than the top part because of the radiation reflected by the parabolic mirror. For small solar collectors however, this simplification does not introduce severe inaccuracies. Additionally, all flux directions shown in Figure 6.2.1b are positive. It should be noted that in the resistance model the incoming solar energy and optical

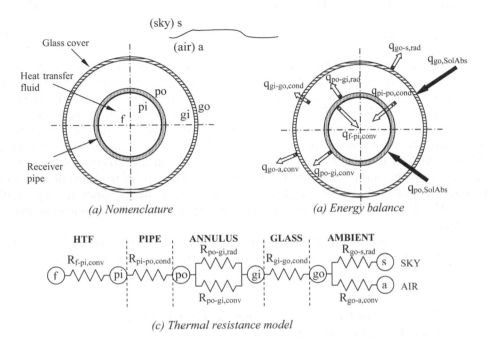

Figure 6.2.1 Collector receiver model a) nomenclature, b) energy balance and c) thermal resistance network for the cross-section of the receiver.

losses have been omitted for clarity. The optical losses are due to imperfections in the collector mirrors, tracking errors, shading and cleanliness of the mirror and receiver glazing. The incoming solar energy, which effectively is equal to the solar energy input minus any optical losses, is absorbed by the glass envelope ($q_{go,SolAbs}$) and receiver pipe ($q_{po,SolAbs}$). Most of the energy that is absorbed by the receiver is conducted through the receiver pipe material ($q_{pi-po,cond}$) and eventually transferred to the HTF by convection ($q_{f-pi,conv}$). The remaining energy is transmitted back to the glass envelope by convection ($q_{po-gi,conv}$) and radiation ($q_{po-gi,rad}$). The energy reaching the glass cover from radiation and convection then passes through the glass envelope wall by conduction ($q_{gi-go,cond}$) and along with the energy absorbed by the glass envelope wall ($q_{go,SolAbs}$) is lost to the environment by convection to ambient air ($q_{go-a,conv}$) and radiation towards the sky ($q_{go-s,rad}$).

The energy balance equations are determined by considering that the energy is conserved at each surface of the receiver cross-section, shown in Figure 6.2.1. Therefore:

$$q_{f-pi,conv} = q_{pi-po,cond} \tag{6.2.1}$$

$$q_{po,SolAbs} = q_{po-gi,conv} + q_{po-gi,rad} + q_{pi-po,cond} \tag{6.2.2}$$

$$q_{po-gi,conv} + q_{po-gi,rad} = q_{gi-go,cond} \tag{6.2.3}$$

$$q_{gi\text{-}go,cond} + q_{go,SolAbs} = q_{go\text{-}a,conv} + q_{go\text{-}s,rad} \tag{6.2.4}$$

$$q_{HeatLoss} = q_{go\text{-}a,conv} + q_{go\text{-}s,rad} \tag{6.2.5}$$

It should be noted that the solar absorption at the outside pipe, $q_{po,SolAbs}$ and outside glass, $q_{go,SolAbs}$ surfaces are treated as heat flux expressions, which simplifies the solar absorption expressions as it considers the heat conduction through the receiver pipe and glass envelope wall to be linear. Actually, the solar absorption in the glass envelope wall (semitransparent material) and receiver pipe (opaque metal material) are volumetric phenomena. However, it is well known from heat transfer textbooks (Cengel 2006) that most of the absorption in a metallic surface (receiver pipe) occurs very close to the surface (within a few μm) and although solar absorption occurs throughout the thickness of the glass envelope wall, its absorptance is very small ($\alpha = 0.02$). Thus, the error in treating solar absorption as a surface phenomenon is very small.

The various heat transfer interactions are analysed in different sections below, starting from the heat transfer fluid inside towards the ambient air and sky outside the receiver assembly.

6.2.1 Convection heat transfer between the HTF and the receiver pipe

Newton's law of cooling states that the convection heat transfer from the inside surface of the receiver pipe to the HTF is given by $hA(T_s - T_\infty)$. Therefore in the case of the PTC model and using the nomenclature adopted in Figure 6.2.1:

$$q_{f\text{-}pi,conv} = h_f \pi D_{pi} \left(T_{pi} - T_f \right) \tag{6.2.6}$$

The convection heat transfer coefficient at the inside pipe diameter, h_f is given by:

$$h_f = Nu_{D_{pi}} \frac{k_f}{D_{pi}} \tag{6.2.7}$$

where: $h_f = $ HTF convection heat transfer coefficient at T_f (W/m^2-°C); $D_{pi} = $ inside diameter of the receiver pipe (m); $T_{pi} = $ inside surface temperature of receiver pipe (°C); $T_f = $ mean (bulk) temperature of the HTF (°C); $Nu_{D_{pi}} = $ Nusselt number based on D_{pi}; and $k_f = $ thermal conductivity of the HTF at T_f (W/m-°C).

In Equation 6.2.6, both T_f and T_{pi} are independent of angular and longitudinal directions of the receiver. The same applies for all temperatures and properties in the energy model.

The Nusselt number depends on the type of flow through the receiver pipe. Although the flow in the receiver pipe is well within the turbulent flow region at typical operating conditions, the model includes conditional statements to determine the type of flow. When the Reynolds number is lower than 2300, laminar flow exists in the receiver pipe and the Nusselt number is constant. For pipe flow, the constant value, assuming constant heat flux, as in the case of a PTC, is equal to 4.36 (Cengel 2006). Turbulent and transitional cases occur at Reynolds number >2300. Therefore, the following Nusselt number correlation developed by Gnielinski (1976) is used for the convective heat transfer from the receiver pipe to the HTF:

$$
\mathrm{Nu}_{D_{pi}} = \frac{f_{pi}/8 \left(\mathrm{Re}_{D_{pi}} - 1000 \right) \mathrm{Pr}_f}{1 + 12.7 \sqrt{f_{pi}/8} \left(\mathrm{Pr}_f^{2/3} - 1 \right)} \left(\frac{\mathrm{Pr}_f}{\mathrm{Pr}_{pi}} \right)^{0.11}
$$

$$
\text{For} \quad 0.5 < \mathrm{Pr}_f < 2000 \quad \text{and} \quad 2300 < \mathrm{Re}D_{pi} < 5 \times 10^6 \tag{6.2.8}
$$

with

$$
f_{pi} = \left[1.82 \log \left(\mathrm{Re}_{D_{pi}} \right) - 1.64 \right]^{-2} \tag{6.2.9}
$$

where: f_{pi} = friction factor for the inside surface of the receiver pipe, D_{pi}; Pr_f = Prandtl number evaluated at the HTF temperature, T_f; and Pr_{pi} = Prandtl number evaluated at the receiver pipe inside surface temperature, T_{pi}.

Except for Pr_{pi}, all fluid properties are evaluated at the mean HTF temperature, T_f. The correlation assumes that the receiver pipe has a smooth inside surface and that the heat flux and temperature are uniform.

The above equations are valid for both turbulent pipe flow and the transitional flow which occur for Reynolds numbers between 2300 and 4000 (Cengel, 2006). Furthermore, the above correlations are adjusted for fluid property variations between the receiver pipe wall temperature and the bulk fluid temperature. The program will display a warning message if the correlation is used out of the range of validity, shown in Equation 6.2.8.

6.2.2 Conduction heat transfer through the receiver pipe wall

Conduction heat transfer through the receiver pipe wall is determined by Fourier's law of conduction through a hollow cylinder (Cengel, 2006) given by:

$$
q_{pi\text{-}po,cond} = \frac{2\pi k_{pipe} \left(T_{pi} - T_{po} \right)}{\ln \left(\dfrac{D_{po}}{D_{pi}} \right)} \tag{6.2.10}
$$

where: k_{pipe} = receiver pipe thermal conductivity at the average receiver pipe temperature $(T_{pi} + T_{po})/2$ (W/m-°C); T_{pi} = receiver pipe inside surface temperature (°C); T_{po} = receiver pipe outside surface temperature (°C); D_{pi} = receiver pipe inside diameter (m); and D_{po} = receiver pipe outside diameter (m).

In this equation the thermal conductivity is considered as constant, and evaluated at the average temperature between the inside and outside receiver pipe surfaces.

The thermal conductivity depends on the receiver pipe material type. The receiver performance model includes one copper and three types of stainless steels (304L, 316L, and 321H), which can be chosen by the user at the beginning. If copper is chosen, the thermal conductivity is constant and equal to 385 W/m-°C. If stainless steel 304L or 316L is chosen, the thermal conductivity is calculated from:

$$
k_{pipe} = (0.013)T_{pi\text{-}po} + 15.2 \tag{6.2.11}
$$

and if stainless steel 321H is chosen the thermal conductivity is calculated from:

$$k_{pipe} = (0.0153)T_{pi\text{-}po} + 14.775 \tag{6.2.12}$$

Both equations were determined by linearly fitting data from Davis (2000).

6.2.3 Heat transfer from the receiver pipe to the glass envelope

As is mentioned before, between the receiver pipe and the glass envelope heat transfer occurs by convection and radiation. Convection heat transfer depends on the annulus pressure (KJC, 1993). At low pressures ($\sim$ <0.013 Pa), heat transfer is by molecular conduction, whereas at higher pressures the heat transfer is by free convection. Radiation heat transfer also occurs because there is a difference in temperature between the outsider receiver pipe surface and the inside glass envelope surface. The radiation heat transfer calculation is simplified by assuming gray surfaces, for which ($\rho = \alpha$) and that the glass envelope wall is opaque to infrared radiation. All these are examined separately in the following sections.

6.2.3.1 Convection heat transfer

As mentioned above, two heat transfer mechanisms are considered in the determination of the convection heat transfer between the receiver pipe and glass envelope wall ($q_{po\text{-}gi,conv}$). These are the free-molecular and natural convection (KJC, 1993). The cases of vacuum and pressure in the annulus are examined separately.

a) *Vacuum in annulus* When the annulus is under vacuum (pressure $\sim$ <0.013 Pa), the convection heat transfer between the receiver pipe and glass envelope occurs by free-molecular convection (Ratzel et al., 1979) and is given by:

$$q_{po\text{-}gi,conv} = \pi D_{po} h_{po\text{-}gi}(T_{po} - T_{gi}) \tag{6.2.13}$$

where

$$h_{po\text{-}gi} = \frac{k_{std}}{\frac{D_{po}}{2\ln\left(\frac{D_{gi}}{D_{po}}\right)} + b\lambda\left(\frac{D_{po}}{D_{gi}} + 1\right)} \quad \text{For: } Ra_{Dgi} < (D_{gi}/(D_{gi} - D_{po}))^4 \tag{6.2.14}$$

and

$$b = \frac{(2-a)(9\gamma - 5)}{2a(\gamma + 1)} \tag{6.2.15}$$

$$\lambda = \frac{2.331 \times 10^{-20}(T_{po\text{-}gi} + 273)}{(P_a \delta^2)} \tag{6.2.16}$$

where: D_{po} = outside receiver pipe diameter (m); D_{gi} = inside glass envelope diameter (m); $h_{po\text{-}gi}$ = convection heat transfer coefficient for the annulus gas at $T_{po\text{-}gi}$ (W/m²-°C); T_{po} = outside receiver pipe surface temperature (°C); T_{gi} = inside glass envelope surface temperature (°C); k_{std} = thermal conductivity of the annulus gas at standard temperature and pressure (W/m-°C); b = interaction coefficient; λ = mean-free-path between collisions of a molecule (cm); a = accommodation coefficient; γ = ratio of specific heats for the annulus gas (air); $T_{po\text{-}gi}$ = average temperature $(T_{po} + T_{gi})/2$ (°C); P_a = annulus gas pressure (mmHg); and δ = molecular diameter of annulus gas (cm).

This correlation slightly overestimates the heat transfer for very small pressures ($\sim$ <0.013 Pa). The molecular diameter of air, δ, is obtained from Marshal (1976) and is equal to 3.55×10^{-8} cm, the thermal conductivity of air is 0.02551 W/m-°C, the interaction coefficient is 1.571, the mean-free-path between collisions of a molecule is 88.67 cm, and the ratio of specific heats for the annulus air is 1.39. These are for an average fluid temperature of 300°C and pressure equal to 0.013 Pa. Using these values, the convection heat transfer coefficients ($h_{po\text{-}gi}$) obtained from Equation 6.2.14 is equal to 0.0001115 W/m²-°C.

b) *Pressure in annulus* If the receiver is filled or partially filled with ambient air ($\sim$ pressure > 0.013 Pa) or if the receiver annulus vacuum is lost, the convection heat transfer between the receiver pipe and glass envelope occurs by natural convection. For this purpose the Raithby and Holland's correlation for natural convection in an annular space (enclosure) between horizontal concentric cylinders is used, given by (Cengel, 2006):

$$q_{po\text{-}gi,conv} = \frac{2\pi k_{eff}}{\ln(D_{gi}/D_{po})}(T_{gi} - T_{po})$$

$$\text{For: } 0.7 \leq Pr_{po\text{-}gi} \geq 6000 \quad \text{and} \quad 10^2 \leq F_{cyl}Ra_{po\text{-}gi} \geq 10^7 \quad (6.2.17)$$

$$\frac{k_{eff}}{k_{ag}} = 0.386 \left(\frac{Pr_{po\text{-}gi}}{0.861 + Pr_{po\text{-}gi}}\right)^{1/4} (F_{cyl}Ra_{Dpo})^{1/4} \tag{6.2.18}$$

$$F_{cyl} = \frac{[\ln(D_{gi}/D_{po})]^4}{L_c^3(D_{gi}^{-3/5} - D_{po}^{-3/5})^5} \tag{6.2.19}$$

In these equations the critical length is given by: $L_c = \frac{(D_{gi} - D_{po})}{2}$

where: k_{ag} = thermal conductivity of annulus gas at $T_{po\text{-}gi}$ (W/m-°C); T_{po} = outside receiver pipe surface temperature (°C); T_{gi} = inside glass envelope surface temperature (°C); D_{po} = outside receiver pipe diameter (m); D_{gi} = inside glass envelope diameter (m); $Pr_{po\text{-}gi}$ = Prandtl number for gas properties evaluated at $T_{po\text{-}gi}$; Ra_{Dpo} = Rayleigh number evaluated at D_{po}; and $T_{po\text{-}gi}$ = average temperature, $(T_{po} + T_{gi})/2$ (°C).

This correlation assumes long, horizontal, concentric cylinders at uniform temperatures, which is perfectly applied for a PTC. All physical properties are evaluated at the average temperature $(T_{po} + T_{gi})/2$.

6.2.3.2 Radiation heat transfer

Several assumptions were made in deriving an equation for the radiation heat transfer, as follows:

- The surfaces are gray,
- Diffuse reflections and irradiation,

- Non-participating gas in the annulus,
- Long concentric isothermal cylinders, and
- The glass envelope is opaque to infrared radiation.

These assumptions are not all completely accurate as the glass envelope wall is not completely opaque for the entire thermal radiation spectrum and the glass envelope wall and the selective coatings are not gray (Touloukian and DeWitt, 1972). However, any errors associated with the assumptions are relatively small.

The radiation heat transfer between the receiver pipe and glass envelope ($q_{po\text{-}gi,rad}$) is estimated with the following equation, applied for infinitely long concentric cylinders (Cengel, 2006):

$$q_{po\text{-}gi,rad} = \frac{\sigma\pi D_{po}(T_{po}^4 - T_{gi}^4)}{\left(\dfrac{1}{\varepsilon_{po}} + \left(\dfrac{(1 - \varepsilon_{gi})D_{po}}{\varepsilon_{gi}D_{gi}}\right)\right)} \tag{6.2.20}$$

where: $\sigma =$ Stefan-Boltzmann constant $(=5.67 \times 10^{-8}\ W/m^2\text{-}K^4)$; $D_{po} =$ outside receiver pipe diameter (m); $D_{gi} =$ inside glass envelope diameter (m); $T_{po} =$ outside receiver pipe surface temperature (K); $T_{gi} =$ inside glass envelope surface temperature (K); $\varepsilon_{po} =$ receiver pipe selective coating emissivity; and $\varepsilon_{gi} =$ glass envelope emissivity.

6.2.4   Conduction heat transfer through the glass envelope

The anti-reflective treatment on the inside and outside surfaces of the glass envelope is assumed not to introduce any thermal resistance or to have any effect on the glass emissivity. This is reasonably accurate since the treatment is usually a chemical etching which does not add any additional elements to the glass surface (Forristall, 2003). The conduction heat transfer through the glass envelope uses the same equation as the conduction through the receiver pipe wall described in Section 6.2.2. As in the receiver case, the temperature distribution is assumed to be linear. Furthermore, the thermal conductivity of the glass (k_{glass}) is assumed constant – as explained in Section 6.2.1 – with a value of 1.04, which corresponds to Pyrex glass (Touloukian and DeWitt, 1972). In equation form this is given by:

$$q_{gi\text{-}go,cond} = \frac{2\pi k_{glass}(T_{gi} - T_{go})}{\ln\left(\dfrac{D_{go}}{D_{gi}}\right)} \tag{6.2.21}$$

6.2.5   Heat transfer from the glass envelope to the atmosphere

The heat transfer from the glass envelope to the atmosphere occurs by convection and radiation. Depending on whether there is wind the convection will either be forced or natural. Radiation heat loss occurs due to the temperature difference between the glass envelope and sky. All these are examined separately below.

6.2.5.1   Convection heat transfer

The convection heat transfer is determined by knowing the Nusselt number, which depends on whether the convection heat transfer is natural (no wind) or forced (wind case). When there is wind, the convection heat transfer from the glass envelope to the atmosphere gives a much bigger heat loss. This is estimated from Newton's law of cooling:

$$q_{\text{go-a,conv}} = h_{\text{go-a}} \pi D_{\text{go}} (T_{\text{go}} - T_{\text{a}}) \tag{6.2.22}$$

and

$$h_{\text{go-a}} = \frac{k_{\text{air}}}{D_{\text{go}}} Nu_{D_{\text{go}}} \tag{6.2.23}$$

where: T_{go} = glass envelope outside surface temperature (°C); T_{a} = ambient air temperature (°C); $h_{\text{go-a}}$ = convection heat transfer coefficient for air at $(T_{\text{go}} - T_{\text{a}})/2$ (W/m²-°C); k_{air} = thermal conductivity of air at $(T_{\text{go}} - T_{\text{a}})/2$ (W/m-°C); D_{go} = glass envelope outside diameter (m); and $Nu_{D_{\text{go}}}$ = average Nusselt number based on the glass envelope outside diameter D_{go}.

a) *No wind*   When there is no wind, the convection heat transfer from the glass envelope to the environment occurs by natural convection and the correlation developed by Churchill and Chu is used to estimate the Nusselt number (Cengel, 2006):

$$\overline{Nu}_{D_{\text{go}}} = \left[0.60 + \frac{0387 R_{D_{\text{go-a}}}^{1/6}}{\left\{ 1 + (0.559/Pr_{\text{go-a}})^{9/16} \right\}^{8/27}} \right]^{2} \quad 10^5 < Ra_{D_{\text{go}}} < 10^{12} \tag{6.2.24}$$

$$Ra_{D_{\text{go}}} = \frac{g\beta(T_{\text{go}} - T_{\text{a}}) D_{\text{go}}^3}{\nu_{\text{go-a}}^2} Pr_{\text{go-a}} \tag{6.2.25}$$

$$\beta = \frac{1}{T_{\text{go-a}}} \tag{6.2.26}$$

$$Pr_{\text{go-a}} = \frac{\nu_{\text{go-a}}}{\alpha_{\text{go-a}}} \tag{6.2.27}$$

where: $Ra_{D_{\text{go}}}$ = Rayleigh number for air based on the glass envelope outside diameter, D_{go}; g = gravitational constant (=9.81 m/s²); $\alpha_{\text{go-a}}$ = thermal diffusivity for air at $T_{\text{go-a}}$ (m²/s); β = volumetric thermal expansion coefficient (ideal gas) (1/K); $Pr_{\text{go-a}}$ = Prandtl number for air at $T_{\text{go-a}}$; $\nu_{\text{go-a}}$ = kinematic viscosity for air at $T_{\text{go-a}}$ (m²/s); and $T_{\text{go-a}}$ = film temperature $(T_{\text{go}} + T_{\text{a}})/2$ (K).

This correlation assumes a long isothermal horizontal cylinder. Also, all the fluid properties are determined at the mean film temperature, $(T_{\text{go}} + T_{\text{a}})/2$.

b) *Wind* When there is wind, the convection heat transfer from the glass envelope to the environment occurs by forced convection. The Nusselt number in this case is estimated with Zhukauskas' correlation for external forced convection flow normal to an isothermal cylinder (Incropera et al., 2007):

$$\overline{Nu}_{D_{\text{go}}} = CRe_{D_{\text{go}}}^m Pr_{\text{a}}^n \left(\frac{Pr_{\text{a}}}{Pr_{\text{go}}} \right)^{1/4} \quad 0.7 < Pr_{\text{a}} < 500 \quad \text{and} \quad 1 < Re_{D_{\text{go}}} < 10^6$$

$$\tag{6.2.28}$$

Table 6.2.1 Constants for Equation 6.2.27.

Re_D	C	m
1–40	0.75	0.4
40–1,000	0.51	0.5
1,000–200,000	0.26	0.6
200,000–1,000,000	0.076	0.7

The constants C and m are given in Table 6.2.1, obtained from Incropera et al. (2007) whereas the constant n is equal to 0.37 for Pr <= 10 and is equal to 0.36 for Pr > 10.

All fluid properties are evaluated at atmospheric temperature, T_a, except Pr_{go}, which is evaluated at the glass envelope wall outside surface temperature.

6.2.5.2 *Radiation heat transfer*

In this model, only the useful solar irradiation is considered in the solar absorption expressions. Therefore, the radiation transfer between the glass envelope wall and sky is caused by the temperature difference between the glass cover and the sky. This is done by assuming that the cover is a small convex gray object in a large blackbody cavity, i.e., the sky. The net radiation transfer between the glass envelope and sky is given by (Cengel, 2006):

$$q_{go-s,rad} = \sigma \varepsilon_{go} \pi D_{go}(T_{go}^4 - T_s^4) \tag{6.2.29}$$

where: D_{go} = outside glass envelope diameter (m); ε_{go} = emissivity of the glass envelope outside surface; T_{go} = glass envelope outside surface temperature (K); and T_s = effective sky temperature (K).

It should be noted that the sky, especially during non-clear conditions, does not act as a blackbody; however, it is common practice to model it as such and to use an effective sky temperature to compensate for the difference (Kalogirou, 2009). Despite the fact that several relations have been proposed to relate the effective sky temperature for clear skies to measured meteorological data, to simplify the model, an approximate relation is used for the effective sky temperature as $T_a - 8°C$.

6.2.6 Solar irradiation absorption

In this model the optical efficiency terms are estimated and combined to form an effective optical efficiency, which is subsequently used to determine the optical loss and solar absorption expressions. The optical properties used in the collector performance model were obtained from a combination of sources, i.e., the parameters used to estimate effective optical efficiencies are generated from the National Renewable Energy Laboratory (NREL) report (Price et al., 2002), which was based on field tests conducted by Dudley et al. (1994), and software performance modelling. These are combined in the intercept factor and the actual values are as follows:

e_{sh} = Receiver shadowing (bellows, shielding, supports), 0.974
e_{tr} = Tracking error, 0.994

e_{ge} = Geometry error (mirror alignment), 0.98

ρ_{cl} = Clean mirror reflectance, 0.935

e_{dm} = Dirt on mirrors (reflectivity/ρ_{cl}) [reflectivity is an input parameter, usual value: 0.88–0.93]

e_{da} = Dirt on receiver, $(1 + e_{dm})/2$

e_{un} = Unaccounted, 0.96.

It should be noted that these parameters are valid only for normal solar incidence irradiation. To account for incident angle losses, the incident angle modifier is used, which accounts for end shading of the trough, reflection and refraction loses, and selective coating incident angle effects. The terms, e_{sh}, e_{tr}, e_{ge}, and e_{un}, shown above are estimates. The clean mirror reflectance ρ_{cl} is a known value, and the two dirt effects e_{dm} and e_{da} are obtained from recommendations by Duffie and Beckman (1991).

The above list of parameters account for collector geometric effects (shadowing, tracking, alignment), mirror and glass envelope transmittance effects (mirror reflectance and dirt), and a parameter for unexplained differences between field test data and modelled data. All these values can be altered by the user if in the future better and more accurate values become available.

Generally, the incident angle modifier is used to account for cases when the solar irradiation is not normal to the collector aperture (Kalogirou 2004, 2009). This is a function of the solar incidence angle (θ) to the normal of the collector aperture. The equation determined from a collector testing carried out at Sandia National Laboratory (SNL) is given by (Dudley et al., 1994):

$$K_\theta = \cos(\theta) + 0.000884\theta - 0.00005369\theta^2 \tag{6.2.30}$$

Other optical properties required include the selective coating absorptance and emittance, and the glass envelope transmittance, absorptance and emittance. The glass envelope absorptance and emissittance are constant (independent of temperature) and independent of selective coating type. The values used in the model are $\alpha = 0.02$ and $\varepsilon = 0.86$ and can be changed by the user if it is required. The glass envelope transmittance and the selective coating absorptance and emittance depend on the type of selective coating. Both the envelope transmittance and the coating absorptance are constants; whereas the coating emittance is a function of temperature. The properties of the Luz cermet selective coating type used in the model are as follow (Forristall, 2003):

- Envelope transmittance = 0.935
- Coating absorptance = 0.92
- Coating emittance = 0.06 at 100°C and 0.15 at 400°C.

The emittance equation used for the selective coating considered, which coincide with the emittance values given above (Forristall, 2003):

Coating Emittance, $\varepsilon_{po} = 0.000327(T + 273.15) - 0.065971 \tag{6.2.31}$

It should be noted that the temperature in Equation 6.2.31 is in degrees Celsius and that the emittance values between the two reference points, of 100°C and 400°C, are nearly linear.

6.2.6.1   Solar irradiation absorption in the glass envelope

As stated in Section 6.2.1, to simplify the model and although physically this is not true, the solar absorption into the glass envelope wall is treated as a heat flux. In fact, the solar absorption in the glass envelope wall is a heat generation phenomenon and as such is a function of the glass wall thickness. However, this assumption introduces an insignificant error since the glass envelope wall is relatively thin and the solar absorptance coefficient for glass is very small, 0.02 (Touloukian and DeWitt, 1972). Additionally, the optical efficiency is used to calculate the solar absorption in the glass envelope given by:

$$q_{go,SolAbs} = q_{sol}\eta_{env}\alpha_{env} \tag{6.2.32}$$

with

$$\eta_{env} = \gamma\rho_{cl}K_{\theta} \tag{6.2.33}$$

where: $q_{sol} =$ solar irradiation per receiver length (W/m); $\eta_{env} =$ effective optical efficiency of the glass envelope; $\alpha_{env} =$ absorptance of the glass envelope (Pyrex glass); $K_{\theta} =$ incident angle modifier, as defined by Equation 6.2.30; and $\gamma =$ intercept factor $[\gamma = e_{sh}e_{tr}e_{ge}e_{dm}e_{da}e_{ms}]$.

All parameters in Equation 6.2.33, except the incidence angle modifier (K_{θ}), are taken from the list presented before. Furthermore, the solar irradiation term (q_{sol}) in Equation 6.2.32 is determined by multiplying the direct normal solar irradiation (DNI) by the projected normal reflective surface area of the collector, i.e., aperture area, and dividing by the receiver length. In both equations, all terms are assumed to be independent of temperature.

6.2.6.2   Solar irradiation absorption in the receiver pipe

As stated before, the solar energy absorbed by the receiver pipe occurs essentially at the surface; therefore, it is treated as a heat flux (see Section 6.2.1). Therefore, the equation for the solar absorption in the receiver pipe is given by:

$$q_{po,SolAbs} = q_{sol}\eta_{abs}\alpha_{abs} \tag{6.2.34}$$

with:

$$\eta_{abs} = \eta_{env}\tau_{env} \tag{6.2.35}$$

where: $\eta_{abs} =$ effective optical efficiency at receiver pipe; $\alpha_{abs} =$ absorptance of receiver pipe; and $\tau_{env} =$ transmittance of the glass envelope.

In Equation 6.2.34, the effective optical efficiency of the glass envelope, η_{env} is obtained by Equation 6.2.33 and as before, all terms are assumed to be independent of temperature.

6.3 CODE TESTING

The code developed is tested using performance measurements for known collectors from test carried out at SNL and presented in Dudley et al. (1994). The information required to input to EES code is the following:

1   Direct normal irradiation (DNI) [W/m^2]
2   Wind speed [m/s]
3   Ambient temperature [°C]
4   Solar incidence angle [°]
5   Coating absorptance [−]
6   Coating emittance at 100°C [−]
7   Coating emittance at 400°C [−]
8   Mirror reflectivity [−]
9   Glass envelope transmittance [−]
10   Annulus pressure or vacuum [−]
11   Annulus absolute pressure [kPa]
12   HTF flow rate [m^3/s]
13   Type of heat transfer fluid
14   Receiver inside diameter [m]
15   Receiver outside diameter [m]
16   Glass envelope inside diameter [m]
17   Glass envelope outside diameter [m]
18   Collector aperture area [m^2]
19   Shadowing [−]
20   Tracking error [−]
21   Dirt factor on glass envelope [−]
22   Dirt factor on mirror [−]
23   Collector inlet temperature [°C].

A comparison of the performance of the code developed and the tests conducted at SNL is shown in the following figures. Figures 6.3.1 and 6.3.2 show a comparison of the actual efficiency and heat loss of the collector with the values determined from the EES code developed, when vacuum exists in the receiver annulus.

Similar results for air in the receiver annulus are presented in Figures 6.3.3 and 6.3.4. In all cases the agreement between the experimental results and those obtained by the EES code is acceptable. The agreement is better for the air case whereas in both cases the difference increases with increasing operating temperature. It should be noted that the relatively high percentage differences shown in Figure 6.3.2 are due to the magnitude of the actual numbers considered and generally the model underestimates the heat loss at high receiver temperatures and overestimates it at low receiver temperatures. In general, the reason for the deviation presented in the case of heat loss (Figures 6.3.2 and 6.3.4) is the possible dependence of the optical properties on the temperature.

Finally, the code developed is used with the characteristics of the collector erected at the premises of the Cyprus University of Technology and in particular at the Archimedes Solar Energy Laboratory. The collector, shown in Figure 6.3.5, is supplied

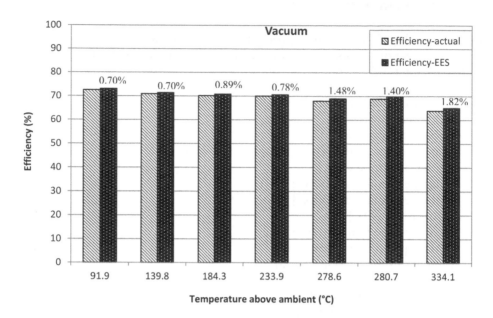

Figure 6.3.1 Comparison of measured against predicted thermal efficiency with vacuum in receiver annulus.

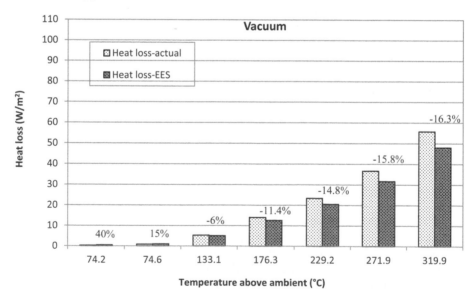

Figure 6.3.2 Comparison of measured against predicted heat loss with vacuum in receiver annulus.

from the Australian company NEP-SOLAR and has the characteristics presented in Table 6.3.1.

The collector is installed at the roof of the laboratory and has a length of 12.2 m (one-half of the standard module). It consists of galvanised steel mounts, lightweight,

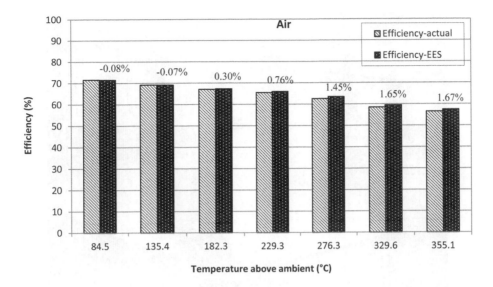

Figure 6.3.3 Comparison of measured against predicted thermal efficiency with air in receiver annulus.

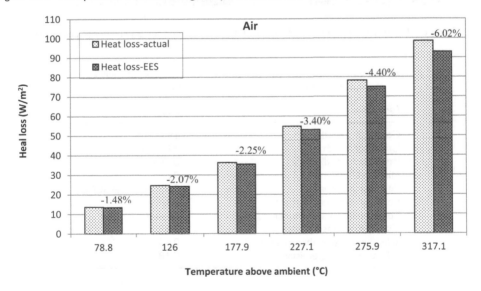

Figure 6.3.4 Comparison of measured against predicted heat loss with air in receiver annulus.

stiff and precise parabolic reflector panels manufactured from reinforced polymeric material, a structurally efficient galvanised steel torque tube, a tubular receiver and an accurate solar tracking system.

As the collector is able to operate up to about 200°C the selective coating properties are assumed to be constant to all possible temperature range. The results of the program are shown graphically in Figure 6.3.6 and give a thermal efficiency of 58.16% at 200°C, which is very satisfactory. These results were obtained at a solar radiation of

Figure 6.3.5 Photograph of the collector installed at Archimedes Solar Energy Laboratory.

Table 6.3.2 Characteristics of the NEP solar collector installed at Archimedes Solar Energy Laboratory.

Parameter	Value
Collector Length	1993 mm
Collector width	1208 mm
Parabola focal distance	647 mm
Mirror reflectivity	93.5%
Receiver material	Stainless steel 304 L
Receiver external diameter	28 mm
Receiver internal diameter	25 mm
Glass tube transmittance	0.89
Selective coating absorptance	0.93
Selective coating emittance	0.18

900 W/m^2, wind speed of 0.45 m/s, flow rate of 8.8 kg/s, ambient temperature of 25°C and ambient air at atmospheric pressure in receiver annulus. It should be noted that the manufacturer gives for these conditions an efficiency equal to 58%, which is very close to the obtained result.

It should be noted that if vacuum exists in the receiver annulus the efficiency is increased to 66.76% and the heat gain is equal to 735.8 W/m.

A major objective of the work carried out at Archimedes Solar Energy Laboratory is to develop improved selective coatings for parabolic trough collector receivers.

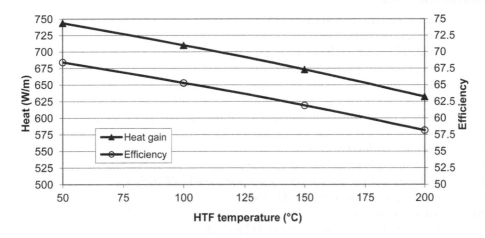

Figure 6.3.6 Performance of the collector installed at Archimedes Solar Energy Laboratory.

By using conservative properties of these coatings as receiver absorptance $= 0.93$ and emittance $= 0.18$ (both at 100 and 400°C) the performance obtained is; heat gain $= 341.0$ W/m, heat loss $= 417.2$ W/m and collector efficiency $= 31.36\%$. By using improved properties of receiver absorptance $= 0.95$ and emittance $= 0.1$ (both at 100 and 400°C) the performance obtained is; heat gain $= 427.1$ W/m, heat loss $= 347.0$ W/m and collector efficiency $= 39.29$ (7.93% improvement). These were obtained by using in the program the weather properties mentioned above, receiver temperature of 400°C and air in the receiver annulus. This is a significant improvement and the author believes that such values can easily be obtained with coatings based on diamond like carbon (DLC).

6.4 CONCLUSIONS

In this chapter a detailed thermal model, which can be used for the analysis of a parabolic trough collector receiver is presented. The model is written in the Engineering Equation Solver (EES) software and takes into consideration all modes of heat transfer; convection into the receiver pipe, in the annulus between the receiver and the glass cover, and from glass cover to ambient air; conduction through the metal receiver pipe and glass cover walls; and radiation from the metal receiver pipe to the glass cover and from glass cover to the sky. The validation of the model is done using the known performance of existing collectors tested at Sandia National Laboratories, and its performance is very satisfactory. Finally, the model is used to perform an analysis of the collector installed at the Archimedes Solar Energy Laboratory of the Cyprus University of Technology with good prediction of the efficiency compared to the manufacturer data. Therefore, the model can be used in the future for system optimisation studies and performance prediction under variable weather conditions. It is also planned to modify the model so as to be called from TRNSYS and thus used for the annual performance prediction of the collector.

Nomenclatures

a	accommodation coefficient (−)
b	interaction coefficient (−)
D_{pi}	inside diameter of the receiver pipe (m)
D_{po}	outside receiver pipe diameter (m)
D_{gi}	inside glass envelope diameter (m)
D_{go}	outside glass envelope diameter (m)
f_{pi}	friction factor for the inside surface of the receiver pipe, D_{pi} (−)
h	convection heat transfer coefficient (W/m^2-°C)
k	thermal conductivity (W/m-°C)
K_θ	incident angle modifier (−)
h_f	HTF convection heat transfer coefficient at T_f (W/m^2-°C)
g	gravitational constant (=9.81 m/s^2)
Nu	Nusselt number (−)
T_f	mean (bulk) temperature of the HTF (°C)
T_{pi}	inside surface temperature of receiver pipe (°C)
P_a	annulus gas pressure (mmHg)
q_{sol}	solar irradiation per receiver length (W/m)
Pr	Prandtl number (−)
Ra	Rayleigh number (−)
T_a	ambient air temperature (°C)
T_{pi}	receiver pipe inside surface temperature (°C)
T_{po}	receiver pipe outside surface temperature (°C)
T_{gi}	glass envelope inside surface temperature (°C)
T_{go}	glass envelope outside surface temperature (°C)
$T_{po\text{-}gi}$	average temperature $(T_{po} + T_{gi})/2$ (°C)
T_s	effective sky temperature (K)

Greek symbols

α_{abs}	absorptance of receiver pipe (−)
α_{air}	thermal diffusivity for air at $T_{go\text{-}a}$ (m^2/s)
α_{env}	absorptance of the glass envelope (−)
β	volumetric thermal expansion coefficient (ideal gas) (1/K)
γ	ratio of specific heats for the annulus gas (−)
δ	molecular diameter of annulus gas (cm)
ε_{po}	receiver pipe selective coating emissivity (−)
ε_{gi}	glass envelope emissivity (−)
ε_{go}	emissivity of the glass envelope outside surface (−)
η_{abs}	effective optical efficiency at receiver pipe (−)
η_{env}	effective optical efficiency of the glass envelope (−)
θ	solar incidence angle (°)
ν_{air}	kinematic viscosity for air at $T_{go\text{-}a}$ (m^2/s)
λ	mean-free-path between collisions of a molecule (cm)
σ	Stefan-Boltzmann constant (=5.67 × 10^{-8} W/m^2-K^4)
τ_{env}	transmittance of the glass envelope (−)

REFERENCES

Cengel, Y. A. (2006) *Heat transfer and mass transfer: A practical approach*. McGraw Hill book company.

Cheng, Z., He, Y., Xiao, J., Tao, Y. and Xu, R. (2010) Three-dimensional numerical study of heat transfer characteristics in the receiver tube of parabolic trough solar collector. *International Communications of Heat and Mass Transfer*, 37, 782–787.

Clark, J. (1982) An analysis of the technical and economic performance of a parabolic trough concentrator for solar industrial process heat application. *International Journal of Heat Mass Transfer* , 25, 1427–1438.

Davis, J. R. (2000) Editor, *Alloy Digest, Sourcebook, Stainless Steels*. Materials Park, OH: ASM.

Dudley, V. E., Kolb, G. J., Sloan, M. and Kearney, D. (1994) *Test Results: SEGS LS-2 Solar Collector*, SAND94-1884, Albuquerque, NM.

Duffie, J.A. and Beckman, W. A. (1991) *Solar Engineering of Thermal Processes*. John Wiley & Sons, New York.

Edenburn, M. W. (1976) Performance analysis of a cylindrical parabolic focusing collector and comparison with experimental results. *Solar Energy*, 18, 437–444.

Forristall, R. (2003) *Heat transfer analysis and modelling of a parabolic trough solar receiver implemented in Engineering Equation Solver*. NREL/TP-550-34169.

García-Valladares, O. and Velazquez, N. (2009) Numerical simulation of parabolic trough solar collector: improvement using counter flow concentric circular heat exchangers. *International Journal of Heat and Mass Transfer*, 52, 597–609.

Gnielnski, V. (1976) New equations for heat and mass transfer in turbulent pipe and channel flow. *International Chemical Engineering*, 562, 359–363.

Gong, G., Huang, X., Wang, J. and Hao, M. (2010) An optimized model and test of the China's first high temperature parabolic trough solar receiver. *Solar Energy*, 84, 2230–2245.

He, Y., Xiao, J., Cheng, Z. and Tao, Y. (2011) A MCRT and FVM coupled simulation method for energy conversion process in parabolic trough solar collector. *Renewable Energy*, 36, 976–985.

Heidemann, W., Spindler, K. and Hahne, E. (1992) Steady-state and transient temperature field in the absorber tube of a direct steam generating solar collector. *International Journal of Heat Mass Transfer*, 35, 649–657.

Incropera, F., DeWitt, D., Bergman, T. L. and Lavine, A. S. (2007) *Fundamentals of Heat and Mass Transfer*. Sixth Edition, New York, John Wiley and Sons.

Kalogirou, S. A. (2009) *Solar energy engineering: Processes and systems*. Academic Press, Elsevier Science.

Kalogirou, S. A. (2004) Solar thermal collectors and applications. *Progress in Energy and Combustion Science* , 30, 231–295.

Karimi, A., Guven, H. M. and Thomas, A. (1986) Thermal analysis of direct steam generation in parabolic trough collectors. *In: Proceedings of the ASME Solar Energy Conference*, 13–16 May, Anaheim, California, USA. pp. 458–464.

Kearney, D. W. and Price, H. W. (1992) Solar thermal plants – LUZ concept (current status of the SEGS plants). *In: Proceedings of the 2nd Renewable Energy Congress*, 13–18 September, Reading, UK. pp. 582–588.

Klein, S. A. (2002) *Engineering Equation Solver for Microsoft Windows*, Professional Version, Madison WI, F-Chart Software.

KJC Operating Company (1993) *Final Report on HCE Heat Transfer Analysis Code*, SANDIA Contract No. AB-0227.

Marshal, N. (1976) *Gas Encyclopedia*, New York, Elsevier.

Odeh, S., Morrison, G. and Behnia, M. (1998) Modeling of parabolic trough direct steam generation solar collectors. *Solar Energy*, 62, 395–406.

Padilla, R. V., Demirkaya, G., Goswami, D. Y., Stefanakos, E. and Rahman, M. M. (2011) Heat transfer analysis of parabolic trough solar receiver. *Applied Energy*, 88, 5097–5110.

Price, H., Lupfert, E., Kearney, D., Zarza, E., Cohen, G. and Gee, R. (2002) Advances in parabolic trough solar power technology. *Journal of Solar Energy Engineering*, 124, 109–125.

Ratzel, A., Hickox, C. and Gartling, D. (1979) Techniques for reducing thermal conduction and natural convection heat losses in annular receiver geometries. *Journal of Heat Transfer*, 101, 108–113.

Thomas, A. and Thomas, S. (1994) Design data for the computation of thermal loss in the receiver of a parabolic trough concentrator. *Energy Conversion and Management*, 35, 555–568.

Touloukian, Y. S. and DeWitt, D. P. (1972) Editors, *Radiative Properties, Nonmetalic Solids*, Thermophysical Properties of Matter, Vol. 8, New York, Plenum Publishing.

Chapter 7

Salinity gradient solar ponds

Abhijit Date & Aliakbar Akbarzadeh

School of Aerospace, Manufacturing and Mechanical Engineering,
RMIT University, Melbourne, Australia

7.1 INTRODUCTION

The most common form of solar pond is a salt-water solar pond. Salt water ponds exist naturally in a variety of locations, the first ponds being discovered in Eastern Europe at the beginning of the 20th century at a natural salt lake in Transylvania then part of Romania. Most of the salt water ponds operated today, however, are artificial, simulating natural solar ponds but taking advantage of engineering technologies to advance their operation and application for practical purposes.

In the case of fresh water ponds all the solar radiation that falls on the surface is absorbed by the top 3 meters of fresh water and this thermal energy is rapidly lost to the atmosphere through natural convection heat transfer. So the temperature of a fresh water pond never rises and is almost constant throughout the fresh water pond depth.

A solar pond utilizes a large body of salinity gradient water to absorb the radiation from the sun and store it in form of heat at the bottom. Figure 7.1.1 shows the schematic of a salinity gradient solar pond, which consists of three regions. The cold upper layer or Upper Convective Zone (UCZ) is a homogeneous thin layer of low salinity brine or fresh water. The middle gradient layer or Non-Convective Zone (NCZ) has a salinity gradient with salinity increasing from top of NCZ to the bottom of NCZ, this helps suppress the natural convection heat loss. The bottom layer or Lower Convective Zone (LCZ) has salinity close to saturation (high concentration brine) that absorbs and stores solar radiation that reaches the LCZ in form of thermal energy. Out of the 100% of solar radiation that is incident on the surface of the solar pond, around 5% is reflected back to the atmosphere; around 45% is absorbed by the water in the UCZ and eventually is lost to the atmosphere by convection; around 20% is absorbed by the water in the NCZ and eventually is conducted to the top UCZ and then lost to atmosphere; around 25% is absorbed by the water in LCZ and the remaining 5% is lost to the ground. Heat loss upwards in the pond from the storage zone is prevented since natural convection currents in the gradient zone are suppressed. This suppression and hence insulating effect occurs because of the density gradient present (Weinberger, 1964; Tabor, 1980).

Experiments show the formation of separate salinity/density gradient layers in the NCZ (seen in Figure 7.1.2). When a particular layer of solution is heated its density is slightly reduced, but remains higher than that of the layer above. Hence there can be no movement upwards by the 'buoyancy' effect that drives natural convection in

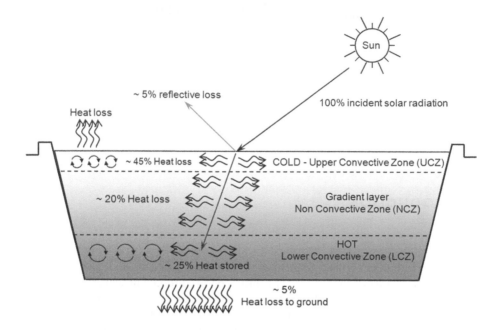

Figure 7.1.1 Schematic diagram of salinity gradient solar pond.

Figure 7.1.2 Experimental setup shows formation of separate salinity/density gradient layers in the NCZ.

a constant density water body. The only mode of heat transfer from the lower layers to the upper layers is by pure conduction and hence NCZ is sometimes also called as insulation layer. During peak summer the temperatures of the LCZ can reach close to boiling temperatures if no heat is removed.

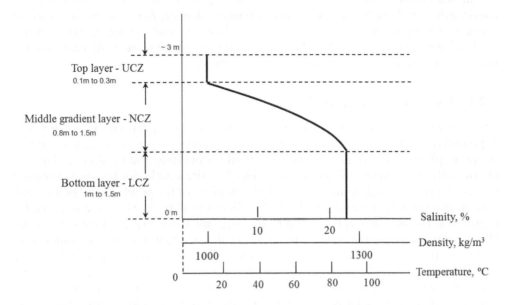

Figure 7.1.3 Typical salinity and density profiles in a solar pond.

Figure 7.1.3 shows the typical solar pond profiles of salinity, density and temperature. The UCZ is typically 0.2 m–0.3 m thick and is maintained at temperature close to local daily ambient. The UCZ requires continuous washing with fresh or low salinity water to remove the diffused salt from the saline layers below and to compensate for the evaporation water loss. In order to maintain the salinity gradient, salt crystals or saturated brine must be added to the LCZ to compensate for the salt lost by diffusion (Akbarzadeh et al., 2005).

7.2 SOLAR POND – DESIGN PHILOSOPHY

7.2.1 Sustainable use of resources

Solar ponds are very sensitive to different applications and hence knowing the application before design starts is very crucial. Knowing the end application of the solar pond would help estimate the energy requirements and hence an optimised solar pond could be designed. For example, the characteristics of a solar pond to supply hot water at say 35°C to an aquaculture facility will be very different from that of a solar pond to be used to generate electricity where sustained performance at higher temperatures of 80°C or above is essential. Obviously the solar pond must be located as close to its application as possible (Akbarzadeh et al., 2005).

Solar ponds would be more economical if constructed using local labour, materials and other resources. It is essential to have a local supply of salt or brine and low salinity water. Flat land is better with high solar radiation for easy construction and optimum operation of a solar pond. Since construction of the solar pond basically involves earth moving and plumbing, it makes good sense to use local contractors.

In order to match the thermal output of a solar pond to the energy and temperature requirements end application, it is very important that the temperature of the lower convective zone is always maintained 3–5°C above the end use temperature. When the LCZ temperature is much higher than the delivery temperature the heat loss will increase and some economic penalties would increase.

7.2.2 Best site characteristics

Site selection is very important for easy construction and operation of a solar pond. A potential site for solar pond should have easy access to salt or brine, low salinity water, ample flat land, consistent soil to be used for building pond walls and the most importantly the land should not be cultivable. The site should not be windy for most part of the year, as high wind can disturb the stability of the pond. The sites for small solar ponds should not be surrounded by buildings or tall trees. The site should receive plenty of solar radiation as this will directly affect the performance of the solar pond. However, it is still possible to build ponds that will operate well in high latitudes, with increased area compensating for less available radiation per unit area of surface. The performance of the solar pond would also depend on the local evaporation rates and depth of the natural underground water table. High evaporation and a shallow water table would make the solar pond performance drop as the heat loss to the atmosphere and ground would increase. Heat loss to the ground water can be reduced by insulating the floor of the solar pond, but this would add to the construction cost (Tabor, 1980; Hull, 1989; Akbarzadeh et al., 2005).

An ideal site for solar pond would have free draining soil, free salt available nearby to reduce costs, easy access to water, flat land to minimize earthmoving requirements, easily compactable soil for structural stability, low prevailing wind speeds to minimize wave-induced mixing and the depth of the top mixed zone, an environmentally acceptable disposal method or recycling ability for closed-salt inventory balancing, dry soil for good thermal insulation, high incident solar radiation for good thermal performance, low evaporation to minimize the need for make-up water, soil with good cohesion for forming stable walls for above-ground ponds, a low amount of wind-borne debris to easily maintain cleanliness, a stationary or deep groundwater table to minimize heat loss within the ground, most importantly proximity to end use application.

7.2.3 Performance and sizing

The thermal performance of a solar pond mainly depends on the absorption of solar radiation in the layers of the ponds. Sun light attenuation as it passes through the top layers of a solar pond puts an upper limit to the amount of solar radiation that can reach the lower convective zone. Further, the amount of sunlight that can reach the lower convective zone would decrease with an increase in turbidity, so it is very important to maintain high water clarity in a solar pond. The more radiation that penetrates, the higher the energy efficiency and operating temperature of the pond will be. In a well-designed and set up solar pond, upward heat losses from the LCZ are small. Therefore most of the solar radiation that gets through to the LCZ is stored there, apart from the small amount lost by conduction to the ground. A well maintained solar pond with a total depth of 3 metres, with 1 m deep LCZ (storage zone) would receive around

20–25% of the radiation incident upon the pond's surface. After accounting for losses to the ground, in practice around 15–20% of the incoming radiation is available for extraction to an application, with the heat delivered 40 to 50 degrees above the local daily average temperature (Weinberger, 1964).

Thermal energy output capacity of a solar pond further depends on the surface area and depth of the storage zone. For a given application with a known heat load the size of a solar pond that can meet the load can be approximated through the following steps, for example the heat load is 3000 GJ/year, (for this approximation it is assumed that the solar pond is clear and well maintained):

1   Find the annual solar energy incident per square metre on a horizontal surface for the proposed solar pond site, for example 6 GJ/m²/year.
2   Now estimate the solar energy that will reach the storage zone and be available for extraction, divide the annual solar energy value found in the previous step by 4 and 8, for example $6/4 = 1.5$ GJ/m²/year; $6/8 = 0.75$ GJ/m²/year.
3   Now estimate the solar pond surface area, divide the heat load by the fraction of solar radiation that is available for extraction as estimated from the previous step, for example solar pond surface area = 3000/1.5 = 2000 m²; area = 3000/0.75 = 4000 m².

The size of a real pond to supply the heat load will depend upon the brine clarity (transparency) and the rate of heat extraction. For continuous heat extraction the pond will be larger, while rapid peak load heat extraction the pond will be smaller. Thus the important design parameters will include: information on ground thermal conductivity, requiring site specific figures for soil thermal conductivity and permeability; pond configuration requirements, requiring knowledge of pond site characteristics; local weather features, including long-term weather and solar data; and pond depth characteristics.

The temperature and quantity of the heat storage in a solar pond depends upon and can be estimated by knowing the salinity gradient of the pond, transparency and the depth of the pond. The depth determines such factors as maximum pond temperature, heat losses to the surrounding soil and atmosphere, and temperature decay time for the storage zone. There are lot of assumptions to be made, and the experience of practical solar pond operations provide a very useful guide for design sizing purposes.

A solar pond with a deep storage zone (typically of the order of 2 to 5 m) will store a large quantity of heat for a long time. Heat losses will be lower and the collection and storage efficiencies of the pond will be high. In contrast, a shallower storage zone (typically of the order of 1 to 2 m) can readily attain higher temperatures (since there is not so much thermal mass in storage), but will then have higher heat losses to the air and ground and a shorter storage capacity. Hence the design of a solar pond would mainly depend on the application.

7.2.4 Liner, salt and water

Liners are a very important component of a solar pond as they prevent saline water from leaking into the soil underneath the pond and endangering the purity of the aquifer. Hot brine leaking out from a pond will carry away with it salt and heat, which in most places is not environmentally acceptable. In order to minimise the heat losses to the ground it is desirable that the underground water table is 5 metres or more below the natural ground surface. If the water table is shallower, then insulating the bottom of the pond may be considered using insulation materials such as sheets of polystyrene.

The liner material should be able to withstand the anticipated maximum pond temperature, be resistant to ultraviolet radiation, and should not react with salt. Above all it should be mechanically strong. Failure of liners has been one of the main problems encountered with working solar ponds. More environmentally friendly liners for solar ponds can be made from compacted clays. Not all clays are suitable to be used as natural liners for solar ponds. Hot NaCl brine can cause some clay to flocculate, making them more porous. Where unlined ponds can be operated effectively, the cost of solar ponds is lowered significantly since lining is one of the main cost components. However, in many locations pond lining is necessary, for both environmental as well as performance reasons (Almanza and Castaneda, 1993).

Different polymeric liner materials can be used for lining the floor of a solar pond. Very often low-density polyethylene (LDPE) and high-density polyethylene (HDPE) are used along with natural clay for lining the floor of a solar pond. For small solar ponds commercially available 10 m wide standard liners can be used. But for large solar ponds with a surface area of a few hundred hectares it is advisable to make the polymer liners on site so that they will cost less and also can be made to desired widths, depending on the available liner-making technology. Liners made from LDPE and HDPE should be protected against ultraviolet radiation and hence should be covered with a thin layer of soil or sand. The best liner laying practice is to make sandwich layers of clay and polymer liners; this will help achieve good leakproofing (Almanza and Castaneda 1993; Akbarzadeh et al., 2005).

Sodium Chloride (NaCl), often called common salt, is the most commonly used in salinity-gradient solar ponds for construction of a salinity gradient. Magnesium Chloride ($MgCl_2$), also known as bittern, is the second most common salt used in the construction of solar ponds. Bittern is a by-product of a NaCl salt production factory.

The density of sodium chloride solution can be increased up to $1300 \, kg/m^3$. However, this density can be increased to more than $1500 \, kg/m^3$ if the salt used is mainly magnesium chloride. It can be seen from Figure 7.2.1 that the solubility of the sodium chloride is fairly constant with temperature, while that of magnesium chloride increases (IUPAC, 2007). It would be possible to construct and maintain stable gradients with both salts. When setting up a solar pond as an integral part of a commercial salt production facility, it makes good economic and environmental sense to set up the solar pond using bittern, and save the more valuable sodium chloride for salt making.

Due to the large salinity difference between the lower convective zone and the upper convective zone, there is upward salt diffusion. To maintain the salinity gradient the top layer of the pond is continuously washed with fresh or low salinity water, while salt is added to the storage zone. It is important to recycle the salt extracted from the pond by surface washing to have minimum economic and environmental impact. An evaporation pond that is at least equal if not twice the total surface area should be constructed next to the solar pond in order to recycle the washed salt. Alternatively the flushed salt solution can enter a sequence of evaporation ponds in a salt production facility.

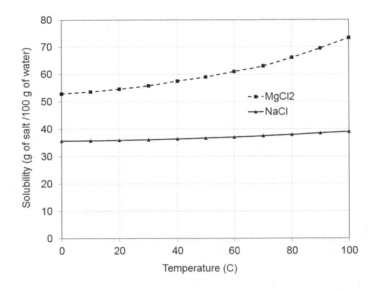

Figure 7.2.1 Solubility of NaCl and MgCl₂ in water at different temperatures.

One of the most important criteria in determining the viability of a solar pond is the availability of fresh or low-salinity water (less than 50,000 ppm salt concentration or a density of less than $1050 \, kg/m^3$). The amount of low-salinity water required to establish a pond is about the volume of the water in the pond measured from the surface to the middle of the gradient layer. The amount of water needed to maintain the gradient depends on evaporative losses, and the flow rate of the overflow system removing surface washing water containing the salt that has diffused upwards. As a rule of thumb, a surface washing water flow rate of two to three times the yearly average rate of evaporation is required. The rate of adding water to the surface zone must exceed the rate of removal through the overflow by the rate of evaporation (Akbarzadeh et al., 2005).

For large ponds (greater than about $1000 \, m^2$), it is best to construct the pond by establishing the walls using the soil excavated from the inner periphery of the pond. The bottom of the pond will thus be below the surrounding ground level. This arrangement provides the head required for gravity feeding of the surface water from the solar pond to adjacent evaporation ponds, as well as siphoning off brine samples from different depths in the pond for the required analysis.

7.2.5 Transient performance prediction

To predict the thermal performance of large solar ponds, the thermal process in these solar ponds can be treated as one-dimensional unsteady conduction loss with heat generation from incoming solar radiation and heat addition from any other source of thermal energy; here the finite difference method is used. As shown in Figure 7.2.2 the NCZ of the solar pond is divided into a number of small divisions (divisions can be non-equal) and the location of the node for determination of temperature is assumed

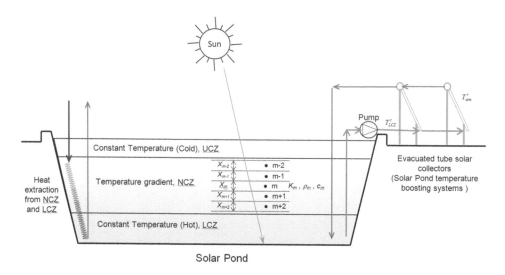

Figure 7.2.2 Schematic illustrating the vertical divisions of solar pond and external heat extraction and addition.

to be at the middle of each division. By applying an energy balance to the divisions in the different layers of the solar pond when extracting heat as shown in Figure 7.2.2, a finite difference model can be developed for the temperatures of three adjacent nodes (Wang and Akbarzadeh, 1982).

To estimate the performance of the initial temperature of the solar pond, density, conductivity, specific heat capacity, thickness, depth, incoming solar radiation and the time increment for each node are required as input boundary conditions. The solar radiation that penetrates the solar pond surface is calculated using the formula discussed by Bryant and Colbeck (Bryant and Colbeck, 1977)

$$h = H[0.36 - 0.08 \ln(x)]. \qquad (7.2.1)$$

Here H is the solar radiation reaching the top surface of the solar pond after deducting the reflective losses; h is the solar radiation that penetrates the solar pond surface and reached the depth of x.

7.3 SOLAR POND – CONSTRUCTION AND OPERATION

7.3.1 Set-up and maintenance

Initial steps in setting up a solar pond are very similar to that of constructing an artificial fresh water pond, similar to the rain water collection pond used for irrigation. The land is excavated and the excavated soil is used to build the side walls of the pond with a slope of about 1:2 (Hull et al., 1989). The newly exposed pond floor is compacted with heavy rollers and small sharp stones and dried soil clusters are removed before laying the liners. The liners are then covered by a thin layer of locally available clay.

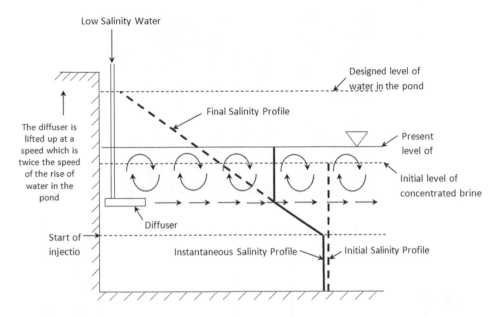

Figure 7.3.1   The process of setting up the salinity gradient is shown schematically. Fresh or low salinity water is injected horizontally into the body of water in the solar pond. Note that the mixing occurs only in the region above the level of injection.

There are a number of different methods for setting up a salinity gradient in a solar pond. The method described here is simple and used all over the world. The solar pond is filled with fresh water up to 1.3 m from bottom; this is approximately equal to the thickness of the storage zone and half of the gradient layer. Now solid salt crystals are added to the bottom of the pond and allowed to dissolve to create concentrated brine with around 25% salinity; this requires about 325 kg of salt per m². The first location of fresh water injection determines the thickness of the storage zone. If the fresh water is injected from 1 m above the bottom then the storage zone gets a thickness of 1 m and so on. The fresh water is injected using a diffuser as shown in Figure 7.3.1 and Figure 7.3.2 while the diffuser is gradually moved upwards. The mixing of rising fresh water with the high concentration brine creates the gradient layer. The diffuser should rise 2 cm for every 1 cm rise in the level of solar pond water (Tabor, 1980; Alagao et al., 1993).

The salt steadily diffuses upwards due to salinity differences between the storage zone and top layer. The rate of salt diffusion depends upon the temperature and the salinity gradient present in the solar pond. As a guideline the rate of diffusion in a sodium chloride pond can be up to 20 kg/m²/year. In order to maintain the difference in salinity between the bottom and top of the pond, salt must be continuously removed from the surface zone by surface flushing, and an equal amount added to the storage zone as shown in Figure 7.3.3.

A 'stability margin' number has been defined to establish an operational safety limit for the salinity gradient within a pond. Basically the density gradient must be kept above a certain minimum level to prevent convection currents starting and the

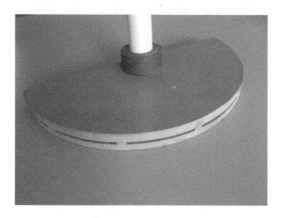

Figure 7.3.2 Diffuser used to setup gradient.

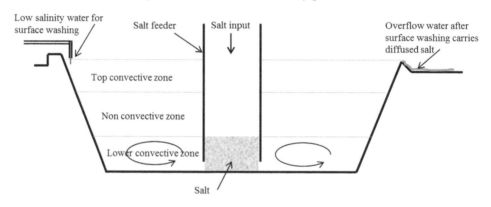

Figure 7.3.3 Passive salt replenishment and surface washing.

salinity profile of the pond becoming unstable (Tabor, 1980). To maintain the stability of the pond one should monitor and limit the stability margin number within the predetermined safety limit. The use of a scanning injection diffuser system allows local instabilities in the gradient layer to be eliminated.

7.3.2 Turbidity control

Turbidity is a measure of the clarity of water or in other words the degree of transparency; its unit of measurement is nephelometric turbidity units (NTU). Desired values of turbidity are less than 0.5 NTUs.

Dust, live or dead organic material and any other suspended material reduces the clarity of the water and hence reduces the thermal efficiency of a solar pond. Heavy dust particles would sink to the bottom of the pond and will not affect the pond clarity, while the dead organic material like leaves, pollen etc. if not removed on time can reduce the thermal performance of the pond and also help the growth of algae in the pond. All dead organic material acts as food material for the growth of algae. So

Figure 7.3.4 In-pond (left) and external (right) heat exchangers at RMIT University, Melbourne, Australia.

the best practice will be to prevent and outbreak of algae pond in the first place. Earlier studies have shown that adding copper sulphate to the solar pond water or making the solar pond water more acidic will help reduce the algae growth (Gasulla et al., 2011). Acidification of the pond provides a simple and reliable maintenance method for preventing or inhibiting algal blooms and maintaining high transparency. The pH level of the pond should be monitored and hydrochloric acid added to keep the level below 5.5. Another approach is to add brine shrimp to the pond to feed off and hence control the algae level (Wang and Seyed-Yagoobi, 1995; Malik et al., 2011).

7.3.3 Heat extraction

The main purpose of a solar pond is to supply heat and this heat is stored in the storage zone of the pond, and this is in addition to the heat that is stored in the gradient layer. There are different ways in which heat can be extracted from the storage zone of the solar pond. In the case of a large solar pond external heat exchangers are used with hot brine being extracted from the storage zone; after it passes through the external heat exchanger this brine is pumped back. In the case of small solar ponds in-pond heat exchangers are better suited, made from plastic tubes or copper nickel tubes (Jaefarzadeh, 2006). Figure 7.3.4 shows the in-pond and external heat exchangers in the 50 m² solar pond at RMIT University, Melbourne, Australia. The most common method of heat extraction from solar ponds is pumping hot brine from the storage zone through a diffuser located just below its interface with the gradient zone to a heat exchanger located near the pond. After delivering heat to the heat exchanger, the cooler brine is returned to the bottom of the pond. This brine removal and return can be accomplished without causing instability in the gradient layer. The pond must also always be operated with the salt concentration in the storage zone below the saturation level. Otherwise there can be some crystallisation of salt in the pipes, pumps and other equipment used in the heat extraction system, especially when it is not operating. The resulting blockages can be difficult to remove. Alternatively heat can be extracted using in-pond heat exchangers. For instance, use of plastic pipes connected to weights on the pond bottom by ropes to overcome buoyancy forces has proven to be a simple and reliable method of heat extraction from the 3000 m² demonstration solar pond in Pyramid Hill, Australia.

Recent investigations have shown that heat extraction from the gradient zone in addition to the storage zone helps improve the overall efficiency of the solar ponds. In-pond heat exchangers for heat extraction from the gradient zone are only effective in case of small ponds. To extract heat from the gradient zone of large solar ponds, a selective withdrawal method should be used. Here hot saline water from different depths is extracted and is passed through a respective external heat exchanger where the heat transfer fluid is preheated using the gradient zone (Andrews and Akbarzadeh, 2005; Leblanc et al., 2011; Yaakob et al., 2011).

7.3.4 Performance monitoring

For optimum operation and steady performance of a solar pond it is very important to have a reliable monitoring system and procedures. Physical parameters such as temperature, salinity profiles in the solar pond should be routinely monitored. It is also very important to monitor the water clarity and for this measurement of turbidity of water is essential, at the same time monitor the pH of water. Control of algae growth is very critical to the efficient performance of a solar pond as discussed in the earlier section. The temperature, density, turbidity and pH of the solar pond should be monitored once every two weeks. Failure to do this will lead to a decrease in the thermal efficiency of the solar pond. Temperature and salinity profiles help to examine the stability of the gradient zone of the solar pond.

The simplest method of monitoring these parameters is by using a sample extraction device as shown in Figure 7.3.5 (top) (Malik et al., 2011). This sample extraction device can be made from a 3.5 m long 25 mm diameter plastic tube. The sampling tube is used to withdraw brine samples from different levels in the pond (guideline: 5 cm intervals). Thermocouples are connected to the inlet of the tube such that the temperature at the point in the pond from which the sample has been withdrawn can be measured. An automated sampling system as shown with a schematic in Figure 7.3.5 (bottom) would be more practical for large solar ponds.

It is also useful to monitor the total global solar radiation incident upon a horizontal surface near the pond surface to keep records of solar energy input for calculation of the solar pond thermal efficiency. Inspection of temperature and salinity profiles is a direct and simple way to locate the depth of the gradient zone with the surface zone and storage zone, as well as the presence of any convective layers.

7.3.5 EEE (Energy, Environmental and Economic) benefit evaluation

It is simple to calculate the thermal utilisation coefficient of the solar pond: it is defined as the ratio of the useful heat extracted from the pond to the actual solar radiation received by the pond. The rate of heat extraction from the solar pond and the temperature of the heat delivered should be monitored and recorded. So to accurately calculate the thermal utilisation coefficient of a solar pond in operation, the total thermal energy provided by the pond over a period of several months should be determined. Similarly

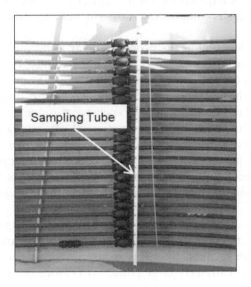

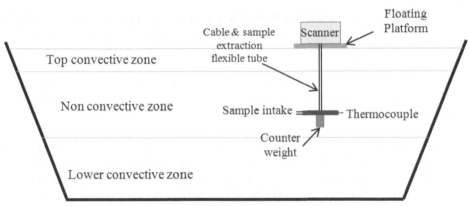

Figure 7.3.5 Manual sampling device (top) and automated sampling system (bottom).

calculate the average collective global horizontal solar radiation incident upon the top surface of the pond. Now the ratio of these two quantities will provide the value of the average thermal utilisation coefficient or thermal efficiency of the pond over the period of measurement.

Solar ponds can be environment friendly and economical when constructed at a suitable location and using local resources. For these situations solar ponds are lot cheaper than any other large-scale solar thermal collectors. The main advantage of the solar pond compared with other solar thermal collectors is that they have an integrated thermal energy storage system. So the solar ponds can supply a substantial amount of thermal energy on a continuous basis.

The costs of a solar pond vary widely according to location and application, so it is essential to perform an economic analysis of the specific design, site and application in mind. Using correct local quotes for the average costs of the construction and operation of a solar pond is very important; this will help to make a sound economic judgement.

Irrespective of the location of the proposed solar pond the economics of a solar pond will depend on the costs associated with land purchase or lease, operating and maintenance costs, and capital equipment, the lifetime of the pond and the associated depreciation rate, and the real discount rate or return on investment sought.

In addition to the land the other major expensive components of a solar pond are its liners (installed), solid salt or brine, excavation of earthen dam walls, monitoring system, heat exchangers. There are good economic benefits of scale, for example, in excavation and setting up nearby evaporation ponds for salt recycling, so that larger solar ponds are more economically favourable where the land is available. However, the maintenance of pond stability and heat extraction may become more difficult as the area increases.

Although solar ponds from hundreds of square meters to thousands of square kilometers are feasible, the sizes likely to find greatest application are in the range of 1 to 10 hectares. Larger facilities than this would probably use unit sizes of 10–20 hectares rather than one extremely large pond, for reasons of operational safety and reliability.

Solar ponds can be economically viable for industrial process heating, including manufacturing processes requiring low-temperature heat and aquaculture and drying applications, at sites where land and water (brackish or sea water) are available, and solar radiation is high. In the right situation solar ponds for heating can readily be economical in areas where natural gas is not available and the only alternative fuels are LPG or oil, and may even compete against natural gas where the price of the latter is high. Of course, a key benefit of a solar pond is a zero greenhouse emissions source of heat.

Solution mining uses low temperature heat for extracting minerals from the mines. The by-product of such mining operations is plenty of high salinity or brackish water. One of the major requirements for constructing a solar pond is a large and continuous supply of saline water and in return the solar pond can continuously supply a large amount of low temperature heat. So the economics would be more favourable when a solar pond is integrated with a solution mining site.

The use of solar ponds as the source of heat for thermal desalination processes such as multiple effect evaporation or multistage flash processes is also potentially one of the most economic zero-emission options for providing fresh water from saline ground water or sea water. With the world running desperately short of fresh water, solar ponds for desalination is a potentially important area of application.

Life cycle analysis, covering embodied energy and emissions in the materials used and in construction, indicate that solar ponds have one of the lowest greenhouse gas emissions per unit of thermal energy produced over their lifetime of any of the renewable energy options. This is if a solar pond is constructed and operated using locally available resources.

The use of local labour in the construction and operation of a solar pond can create employment opportunities in regional areas. Hence on a 'triple bottom line' evaluation, solar ponds in a well-chosen application can rate very well on economic, environmental and social criteria (Esquivel et al., 1993; Akbarzadeh et al., 2005).

7.4   SOLAR PONDS – WORLDWIDE

Since the 1950s a number of demonstration and a few industrial solar ponds have been constructed and operated around the world. This section presents a few examples of such solar ponds from around the world.

7.4.1   Solar ponds – Israel

Figure 7.4.1 shows the solar pond power station at Bet Ha Arava in Israel. Ormat constructed two solar ponds with a combined surface area of 250,000 m^2 near Bet Ha Arava north of the Dead Sea. These two solar ponds supplied required thermal energy to the Ormat power plant with a power production capacity of 5 MWe. Due to geopolitical reasons this solar pond power station was decommissioned in 1990. Solar pond technology is again gaining some interest for industrial process heating rather than power generation application (Tabor and Doron, 1990).

7.4.2   Solar ponds – Australia

Figure 7.4.2 shows the 50 m^2 experimental solar pond that was constructed in 1998 at the Bundoora East Campus of RMIT University located in Melbourne, Australia. The solar pond is circular in shape, 8 m in diameter and 2.5 m deep; it has an observation window to check water clarity. The overflow level control system maintains the solar pond water depth at 2.05 m from the bottom. The solar pond walls and floor is made from 0.2 m thick reinforced concrete and to protect the concrete floor and the inner surface of the concrete walls and steel reinforcement from corrosion due to salt, they are coated with a layer of epoxy. This solar pond is a partially in-ground type and has

Figure 7.4.1 The Dead Sea solar pond power station near Bet Ha Arava, Israel.

Figure 7.4.2 50 m² solar pond at RMIT University, Melbourne, Australia.

a wall around 1.2 m above the ground. As the pond wall is partially above the ground level, it allows using a gravity assisted overflow system to maintain the level of water in the pond. Salt lost by diffusion is replenished by adding solid sodium chloride salt to the bottom of the pond with a cylindrical salt charger (Andrews and Akbarzadeh, 2005; Leblanc et al., 2011; Yaakob et al., 2011).

The diffused salt that reaches the top convective zone must be removed to maintain the salinity gradient; at RMIT solar pond the top surface of the pond is continuously flushed with low salinity water, and this helps to remove the diffused salt. The flow rate of flushing should be kept at around twice the local yearly average evaporation rate. To reduce stirring of the top convective zone by natural wind, 0.6 m diameter floating rings made from high density polyethylene are spread over the top surface of the pond; these rings help reduce the amplitude of the waves formed due to natural wind. At present, the clarity of the water in this solar pond is maintained by keeping the pH in the range of 5 to 6.

Figure 7.4.3 shows the 3000 m² solar pond designed and constructed by RMIT University along with two industrial partners in early 2000 for supplying heat for salt drying process at Pyramid Salt Factory (Akbarzadeh et al., 2005). This pond has a total depth of 2.3 m, with a storage zone that is 0.8 m thick and a gradient zone of about 1.2 m. Initially this pond mainly supplied heat for the process of salt drying while later some amount of heat was also supplied to an aquaculture farm on the factory site. Saline ground water is used for surface flushing and the water that overflows is carried to a large evaporation pond. The location of the pond was around 200 m from Pyramid Hill's salt production plant. Unfortunately after the flash flooding in part of Victoria, Australia, the salt production facility was relocated and the solar pond is not in operation anymore.

7.4.3 Solar ponds – USA

In 1983 the University of Texas at El Paso along with Bruce Foods, Inc., constructed a 3700 m² solar 20 km northwest of El Paso city centre as shown in Figure 7.4.4 (Lu and Sandoval, 1993; Akbarzadeh et al., 2005). Through well-designed procedures for

Figure 7.4.3 Solar pond at Pyramid Hill, in northern Victoria, Australia; In operation (Top 2 photos); Photo taken in August 2012 – of out of operation Pyramid Hill solar pond (bottom).

Figure 7.4.4 Solar pond at the University of Texas at El Paso.

maintaining the gradient and clarity of the pond, a very high level of performance has been achieved, with storage zone temperatures remaining above 80°C year-round. This solar pond produced 120 kW maximum electrical power with an Ormat ORC engine in the summer of 1992. This solar pond operated for 16 years before it encountered problems with liner failure most likely due to high temperature deterioration of the polymer liners and was decommissioned at the end of 2003.

A 2000 m² experimental solar pond was constructed at the University of Illinois, USA to study and develop simple and cost effective construction methods (Newell et al., 1990).

At a 2000 m² solar pond in Miamisburg, Ohio, salt-gradient supplies the heat to warm-up a summer outdoor swimming pool and in winter a recreational building and the installation costs were only $35/m² (Sabetta, Pacetti et al. 1985).

7.4.4 Solar ponds – Tibet, China

Figure 7.4.5 shows the lithium carbonate (salt) solar ponds in Tibet, China. These ponds are constructed in a hilly barren land on the southwest of Zabuye Salt Lake. The solar pond has a top surface area of 3588 m² (78 m × 46 m), the depth is 4 m and the pond walls have a slope of 1.5:1. This is an in-ground solar pond, and after excavation

Figure 7.4.5 Photos clockwise from top left: (1) Zabuye salt lake; (2) Solar Pond in operation; (3) Lithium carbonate harvesting, and; (4) Lithium carbonate sheets (Nie, Bu et al. 2011).

the soil was properly compacted. The pond walls and floor have been covered with 8 cm thick steel reinforced concrete. The smooth concrete walls and the floor of this pond have been insulated with 0.5 mm thick sheet of Ethylene-Propylene- Diene-Monomer (S-801EPDM) to prevent brine leakage. This solar pond has an automatic temperature measurement system installed prior to the start of operation of the pond. Thirty ponds of similar size are been used to produce lithium carbonate at the same time in Zabuye (Nie et al., 2011).

7.4.5 Solar ponds – India

A 1200 m^2 pond was constructed at Central Salt and Marine Chemicals Research Institute (CSMCR) in Bhavnagar, Gujarat in 1970. This solar pond used magnesium chloride to create the salinity gradient. Magnesium chloride is a waste product from the process of making edible salt.

In 1980 a 100 m^2 experimental solar pond was constructed and operated for two years in Pondicherry. This pond used sodium chloride to create the salinity gradient and used low density polyethylene liners.

In 1980 another solar pond was constructed with a surface area of 1600 m^2 at CSMCR in Bhavnagar, Gujarat. This pond also used magnesium chloride to create the salinity gradient and had problems with the clarity of bittern.

In 1984 a 240 m^2 solar pond was constructed and operated for a long period at the Indian Institute of Science, Bangalore. Study of this pond has produced very useful performance data for a small solar pond in south India. It has also proven the technical and economic viability of small solar ponds.

Since the 1980s several small size solar ponds have been installed and operated for town water heating. A 400 m^2 solar pond was constructed to supply the hot water needs of a rural community at Masur on the west coast of India. A similar solar pond with 300 m^2 surface area was constructed to supply hot water to student hostels for an engineering college at Hubli in Karnataka.

Figure 7.4.6 shows the solar pond at Bhuj, India. The 6000 m^2 solar pond that was built at a dairy in Bhuj stood out in many regards. This was the first-ever solar pond in India to have connected itself to an industrial process, supplying heat to the Kutch Dairy. To reduce the construction cost of the solar pond the project developed a cost-effective, indigenous lining; it used locally mined clay and plastics. While the pond attained a record 99.8°C under stagnation, stability of the salinity gradient was maintained even at such elevated temperatures. With only one injection diffuser on one side of the pond, the desired salinity profile was achieved even at the farthest end.

Here an external heat exchanger is used to extract heat from the storage zone of this pond. Hot brine is withdrawn from the bottom of the pond and is pumped through a shell-and-tube heat exchanger where it heats the feedwater up to a temperature of 70°C. Further, this hot water was delivered to the Kutch Dairy plant to be used as pre-heated boiler feed water as well as for cleaning and washing. The entire exercise at the Bhuj solar pond successfully demonstrated the expediency of the technology by supplying 80,000 litres of hot water daily to the plant (Kumar and Kishore, 1999).

Figure 7.4.6 Solar pond in Bhuj, India.

7.5 SOLAR PONDS – APPLICATIONS

7.5.1 Heating

In the past solar ponds have been used for water heating and in some cases also for industrial process heating. Solar ponds have the highest potential in suitable applications and locations to supply low-temperature heat at competitive costs. The suitable heating applications of a solar pond are town water heating, salt drying, fruit drying, wood drying, hot water for the food industry, solution mining operations (Hull, 1989).

7.5.2 Aquaculture

Pyramid Hill solar pond has been used to supply low temperature heat for aquaculture for growing warm water fish and shrimps. Solar ponds are suitable where the desired supply temperature is low and a large amount of heat is required. Using fossil fuels for supplying the low temperature process heat is not sustainable; hence solar ponds should be used for heating aquaculture ponds to grow fish, shrimps, algae etc. The heat from the solar ponds can be used for controlling the temperature of the environment for growth control well as to supply other thermal energy needs of the plant.

This makes a solar pond a very economical and attractive technology (Akbarzadeh et al., 2005).

7.5.3 Desalination

Solar ponds can easily supply the large amount of heat that is usually required for thermal desalination processes. The low pressure (below atmosphere) thermal desalination systems like multiple effect evaporation (MEE) and multistage flash (MSF) can be easily coupled with solar ponds to produce fresh water with minimum environmental footprint. The waste brine from these desalination systems can be fed to the bottom of the pond to maintain the salinity gradient (Esquivel et al., 1993; Zhao et al., 2009).

7.5.4 Power production

There have been several large demonstration projects in past like the 5 MW solar pond power plant in Israel, a 15 kW power plant in Alice Springs, Australia and the 70 kW power plant at El Paso, USA (Akbarzadeh et al., 2005).

Organic Rankine Cycle (ORC) engines developed specifically to produce electric power from lower-temperature heat sources (80–90°C) have been used in these applications. However, their thermodynamic performance is well below the Carnot limit, so designs have low net thermal-to-electric energy conversion efficiencies (~7%), which adversely affects their economic viability.

With advancement in the low temperature heat engine technology, the electrical power production from low temperature heat stored in solar ponds will get economically competitive with other renewable energy technologies.

Consequently to date there has been very little commercial exploitation of solar ponds for generating electricity, despite the reduction in greenhouse gas emissions from burning primary fossil fuels that their utilization would yield.

7.6   FUTURE DIRECTIONS

The solar pond technology is now very well matured and has been successfully used on a trial basis around the world in the last half century for power production, industrial process heating, salt drying etc. The applications presently of greatest interest are industrial process heating, thermal desalination, aquaculture, mariculture, greenhouse heating, biogas production, and waste brine or other saline effluent management.

Lines of inquiry in solar pond science and technology that still require further investigation include alternative salts and new lining techniques that may increase stability and reduce construction costs; simpler more cost-effective methods to limit wind-driven mixing and thickening of the surface zone; increasing the thermal efficiency of solar ponds and maintaining high-temperature operation; and possibly the development of a new generation of solar ponds based on artificial solar pond liquids.

Globally, excellent sites for solar ponds are abundant and many solar pond applications may become economically viable as the prices of fossil fuels rise. Solar ponds as a renewable energy source capable of yielding environmental and social benefits can help make the transition to a truly sustainable energy system.

GLOSSARY

Density gradient:	The variation with depth of the density of the saline solution in a salt water solar pond (see 'salinity gradient').
Lower Convective Zone (LCZ): or Storage Zone	The bottom zone which is generally about 1 to 1.5 m thick and has near-saturated brine stores the incoming solar radiation in the form of heat.
Organic Rankine Cycle engine:	A heat engine based on the Rankine Cycle and using a low boiling point organic working fluid that can be used to generate electricity from low temperature heat (less than 100°C) as supplied by a solar pond.
Non-Convective Zone (NCZ) or Gradient zone:	The central zone, in a vertical direction, in a salt water solar pond in which there is a salinity gradient.
Salinity gradient:	The variation in salinity with depth used in a salt-water solar pond to create a density gradient that suppresses convection currents. In such a solar pond salinity is near saturation at the bottom of the pond and rises to near fresh water level in the top layer. Salt water solar ponds are also commonly called 'salinity gradient solar ponds'.
Solar pond:	A large-area collector of solar energy in the form of a shallow pond that stores heat for use for practical purposes. Designs include salt-water ponds, gel ponds, and ponds with covers. Incoming solar radiation is stored in the lower layer of the pond by suppressing the convection currents in the layer above that would otherwise lead to heat loss to the surroundings.
Utilisation coefficient or thermal efficiency:	(of a solar pond) on average for a given period is the total thermal output delivered to an application divided by the cumulative solar radiation incident upon the pond's surface over that period. Normally the period chosen to assess thermal efficiency would be a full year, after the solar pond had warmed up to its operating temperature range.
Upper Convective Zone (UCZ) or Top layer or Surface zone:	The upper layer of a salt water solar pond (also called the 'upper convective zone') comprising low salinity or fresh water that is continually added to the pond and removed at an overflow to take away salt that diffuses to the surface from the layers beneath. The top layer which is about 20 to 30 cm thick has very low and uniform salinity.

REFERENCES

Akbarzadeh, A., Andrews, J. and Golding, P. (2005) *Solar Pond Technologies: A review and Future Directions*. Advances in Solar Energy. Earthscan. London, UK, 16: 233–294.

Alagao, F.B., Akbarzadeh, A. and Johnson, P. (1993) The Design, Construction and Initial Operation of a closed cycle Salt Gradient Solar Pond. *In: Proceedings of 3rd International Conference on Progress in Solar Ponds*, May 23–27, El Paso Texas, USA.

Almanza, R. and Castaneda, R. (1993) How to test a liner clay for solar ponds. *In: Proceedings of 3rd International Conference on Progress in Solar Ponds*, May 23–27, El Paso Texas, USA.

Andrews, J. and Akbarzadeh, A. (2005) Enhancing the thermal efficiency of solar ponds by extracting heat from the gradient layer. *Solar Energy*, 78, 704–716.

Bryant, H.C. and Colbeck, I. (1977) A solar pond for London? *Solar Energy*, 19, 321–322.

Esquivel, P.A., Swift, A.H.P., McLean, T.J. and Golding, P. (1993) Solar Pond Economics for Industrial Process Heat, Baseload electricity and Water Desalting. *In: Proceedings of 3rd International Conference on Progress in Solar Ponds*, May 23–27, El Paso Texas, USA.

Gasulla, N., Yaakob, Y., Leblanc, J., Akbarzadeh, A. and Cortina, J.L. (2011) Brine clarity maintenance in salinity-gradient solar ponds. *Solar Energy*, 85, 2894–2902.

Hull, J. R., Nielsen, C.E. and Golding, P. (1989) *Salinity-gradient solar ponds*. Boca Raton, Fl, USA, CRC Press.

IUPAC (2007) NIST Standard Reference Database 106 IUPAC-NIST Solubility Database, NIST.

Jaefarzadeh, M.R. (2006) Heat extraction from a salinity-gradient solar pond using in pond heat exchanger. *Applied Thermal Engineering*, 26, 1858–1865.

Kumar, A. and Kishore, V.V.N. (1999) Construction and operational experience of a 6000 m² solar pond at Kutch, India. *Solar Energy*, 65, 237–249.

Leblanc, J., Akbarzadeh, A., Andrews, J., Lu, H. and Golding, P. (2011) Heat extraction methods from salinity-gradient solar ponds and introduction of a novel system of heat extraction for improved efficiency. *Solar Energy*, 85, 3103–3142.

Lu, H. and Sandoval, J. (1993) Experienences of clarity monitoring and maintanance at the El Paso solar pond. *In: Proceedings of 3rd International Conference on Progress in Solar Ponds*, May 23–27, El Paso Texas, USA.

Malik, N., Date, A., Leblanc, J., Akbarzadeh, A. and Meehan, B. (2011) Monitoring and maintaining the water clarity of salinity gradient solar ponds. *Solar Energy*, 85, 2987–2996.

Newell, T. A., Cowie, R.G., Upper, J.M., Smith, M.K. and Cler, G.L. (1990) Construction and operation activities at the University of Illinois Salt Gradient Solar Pond. *Solar Energy*, 45, 231–239.

Nie, Z., Bu, L., Zheng, M. and Huang, W. (2011) Experimental study of natural brine solar ponds in Tibet. *Solar Energy*, 85, 1537–1542.

Sabetta, F., Pacetti, M. and Principi, P. (1985) An internal heat extraction system for solar ponds. *Solar Energy*, 34, 297–302.

Tabor, H. (1980) Non-Convecting Solar Ponds. Philosophical Transactions of the Royal Society of London. *Series A, Mathematical and Physical Sciences*, 295, 423–433.

Tabor, H. and Doron, B. (1990) The Beith Ha'Arava 5 MW(e) Solar Pond Power Plant (SPPP) Progress report. *Solar Energy*, 45, 247–253.

Wang, J. and Seyed-Yagoobi, J. (1995) Effect of water turbidity on thermal performance of a salt-gradient solar pond. *Solar Energy*, 54, 301–308.

Wang, Y.F. and Akbarzadeh, A. (1982) A study on the transient behaviour of solar ponds. *Energy*, 7, 1005–1017.

Weinberger, H. (1964) The physics of the solar pond. *Solar Energy*, 8, 45–56.

Yaakob, Y., Date, A. and Akbarzadeh, A. (2011) Heat extraction from gradient layer using external heat exchangers to enhance the overall efficiency of solar ponds. *In: IEEE First Conference on Clean Energy and Technology (CET)*, 27–29 June 2011, Kuala Lumpur, Malaysia.

Zhao, Y., Akbarzadeh, A. and Andrews, J. (2009) Simultaneous desalination and power generation using solar energy. *Renewable Energy*, 34, 401–408.

Chapter 8

The solar thermal electrochemical production of energetic molecules: Step

Stuart Licht
Department of Chemistry, George Washington University, Washington, DC, USA

8.1 INTRODUCTION

Anthropogenic release of carbon dioxide and atmospheric carbon dioxide have reached record levels. One path towards CO_2 reduction is to utilize renewable energy to produce electricity. Another, less explored, path is to utilize renewable energy to directly produce societal staples such as metals, bleach, fuels, including carbonaceous fuels. Whereas solar-driven water splitting to generate hydrogen fuels has been extensively studied (Vayssieres 2009; Rajeshwar et al., 2008), there have been few studies of solar driven carbon dioxide splitting. "CO_2 *is a highly stable, noncombustible molecule, and its thermodynamic stability makes its activation energy demanding and challenging (Ohla et al., 2009).*" In search of a solution for climate change associated with increasing levels of atmospheric CO_2, the field of carbon dioxide splitting (solar or otherwise), while young, is growing rapidly, and as with water splitting, includes the study of photoelectrochemical, biomimetic, electrolytic, and thermal pathways of carbon dioxide splitting (Graves et al., 2011; Barber 2009). Recently we introduced a global process for the Solar Thermal Electrochemical Production (STEP) of energetic molecules, including CO_2 splitting (Licht 2009; Licht et al., 2010; Licht 2011) as well as the solar production of metals, fuels, bleach and other staples (Licht 2009, Licht et al., 2010a; Licht et al., 2010b; Licht and Wang 2010; Licht 2011; Licht et al., 2011b; Licht, and Wu. 2011; Licht et al., 2011a).

The direct thermal splitting of CO_2 requires excessive temperatures to drive any significant dissociation. As a result, lower temperature thermochemical processes using coupled reactions have recently been studied (Stamatiou et al., 2010; Venstrom and Davidson 2011; Chueh and Haile, 2010; Miller et al., 2008). The coupling of multiple reactions steps decreases the system efficiency. To date, such challenges, and the associated efficiency losses, have been an impediment to the implementation of the related, extensively studied field of thermochemical splitting of water (Rajeshwar et al., 2008). Photoelectrochemistry probes the energetics of illuminated semiconductors in an electrolyte, and provides an alternative path to solar fuel formation. Photoelectrochemical solar cells (PECs) can convert solar energy to electricity, (Licht, 1987; Licht and Peramunage, 1990; Oregan and Gratzel, 1991; Licht, 1998; Licht, 2002) and with inclusion of an electrochemical storage couple, have the capability for internal energy storage, to provide a level output despite variations in sunlight (Licht et al., 1987; Licht et al., 1999). Solar to photoelectrochemical energy can also be stored externally in chemical form, when it is used to drive the formation of energetically rich chemicals. Photochemical, and photoelectrochemical, splitting of carbon dioxide (Yan et al., 2011; Zhou et al., 2011; Richardson, Holland, Carpenter 2011; Barton et al., 2008; Kaneco et al., 2009; Pan and Chen, 2007) have demonstrated selective production of specific fuel products. Such systems function at low current density and efficiencies of ~1 percent, and as with photoelectrochemical water splitting face stability and bandgap challenges related to effective operation with visible light (Licht, 2002; Murphy, 2008; Currao, 2007).

The electrically driven (nonsolar) electrolysis of dissolved carbon dioxide is under investigation at or near room temperature in aqueous, non-aqueous and PEM media (Narayanan, et al., 2011; Delacourt and Newman 2010; Dufek et al., 2011; Gangeri et al., 2009; Innocent et al., 2008; Wang et al., 2009; Chu et al., 2008; Yano et al., 2007; Hori et al., 2005; Ogura et al., 2004). These are constrained by the thermodynamic and kinetic challenges associated with ambient temperature, endothermic processes, of a high electrolysis potential, large overpotential, low rate and low electrolysis efficiency. High temperature, solid oxide electrolysis of carbon dioxide dates back to suggestions from 1960 to use such cells to renew air for a space habitat, (Martin, 1965; Chandler et al., 1966; Erstfield, 1979; Stancati et al., 1981; Richter, 1981; Mizusaki et al., 1992; Tao et al., 2004; Green et al., 2008) and the sustainable rate of the solid oxide reduction of carbon dioxide is improving rapidly (Meyers et al., 2011; Kim-Lohsoontorn et al., 2011; Ebbesen et al., 2010; Jensen et al., 2010; Fu et al., 2010; Stoots et al., 2010; Fu et al., 2010). Molten carbonate, rather solid oxide, fuel cells running in the reverse mode had also been studied to renew air in 2002 (Lueck et al., 2002). In a manner analogous to our 2002 high temperature solar water splitting studies and described below (Licht, 2002; Licht, 2003; Licht et al., 2003; Licht, 2005), we showed in 2009 that molten carbonate cells are particularly effective for the solar driven electrolysis of carbon dioxide, (Licht, 2009; Licht, Wang et al., 2010a; Licht et al., 2011a; Licht 2011) and also CO_2-free iron metal production (Licht and Wang, 2010; Licht et al., 2011b; Licht, 2011).

Light-driven water splitting was originally demonstrated with TiO_2 (a semiconductor with a bandgap, $E_g > 3.0$ eV) (Fujishima and Honda, 1972). However, only a small fraction of sunlight has sufficient energy to drive TiO_2 photoexcitation. Studies had sought to tune (lower) the semiconductor bandgap to provide a better match to the electrolysis potential (Zou et al., 2001). In 2000, we used external multiple bandgap PVs (photovoltaics) to generate H_2 by splitting water at 18% solar energy conversion efficiency (Licht et al., 2000; Licht, 2001). However, that room temperature process does not take advantage of additional, available thermal energy.

An alternative to tuning a seminconductor bandgap to provide a better match to the solar spectrum, is an approach to tune (lower) the electrolysis potential (Licht et al., 2003; Licht, 2005). In 2002, we introduced a photo-electrochemical thermal water splitting theory, (Licht, 2002) which was verified by experiment in 2003, for H_2 generation at over 30% solar energy conversion efficiency, and providing the first experimental demonstration that a semiconductor, such as Si ($E_g = 1.1$ eV), with bandgap lower than the standard water-splitting potential ($E^\circ_{H_2O}(25^\circ C) = 1.23$ V), can directly drive hydrogen formation (Licht et al., 2003; Licht, 2005). With increasing temperature, the quantitative decrease in the electrochemical potential to split water to hydrogen and oxygen had been well known by the 1950s (deBethune and Licht, 1959;

Chase, 1998). In 1976 Wentworth and Chen wrote about "simple thermal decomposition reactions for storage of solar energy," with the limitation that the products of the reaction must be separated to prevent back reaction (and without any electrochemical component), (Wentworth, 1976) and as early as 1980 it was noted that thermal energy could decrease the necessary energy for the generation of H_2 by electrolysis (Bockris, 1980). However, the process combines elements of solid state physics, insolation and electrochemical theory, complicating rigorous theoretical support of the process. Our photo-electrochemical thermal water-splitting model for solar/H_2 by this process, was the first derivation of bandgap restricted, thermal enhanced, high solar water-splitting efficiencies. The model, predicting solar energy conversion efficiencies that exceed those of conventional photovoltaics was initially derived for AM(Air Mass)1.5, terrestrial insolation, and later expanded to include sunlight above the atmosphere (AM0 insolation) (Licht, 2002; Licht, 2003). The experimental accomplishment followed, and established that the water-splitting potential can be specifically tuned to match efficient photo-absorbers, (Licht et al., 2003; Licht, 2005) eliminating the challenge of tuning (varying) the semiconductor bandgap, and which can lead to over 30% solar to chemical energy conversion efficiencies. Our early process was specific to H_2 and did not incorporate the additional temperature enhancement of excess super-band gap energy and concentration enhancement of excess reactant to further decrease the electrolysis potential, in our contemporary STEP process.

8.2 SOLAR THERMAL ELECTROCHEMICAL PRODUCTION OF ENERGETIC MOLECULES: AN OVERVIEW

8.2.1 STEP theoretical background

A single, small band gap junction, such as in a silicon PV, cannot generate the minimum photopotential required to drive many room temperature electrolysis reactions, as shown in the left of Scheme 8.2.1. The advancement of such studies had focused on tuning semiconductor bandgaps (Zou et al., 2001) to provide a better match to the electrochemical potential (specifically, the water-splitting potential), or by utilizing more complex, multiple bandgap structures using multiple photon excitation (Licht et al., 2000; Licht 2001). Either of these structures are not capable of excitation beyond the bandedge and cannot make use of longer wavelength sunlight. Photovoltaics are limited to super-bandgap sunlight, $h\nu > E_g$, precluding use of long wavelength radiation, $h\nu < E_g$. STEP instead directs this IR sunlight to heat electrochemical reactions, and uses visible sunlight to generate electronic charge to drive these electrolyses.

Rather than tuning the bandgap to provide a better energetic match to the electrolysis potential, the STEP process instead tunes the redox potential to match the bandgap. The right side of Scheme 8.2.1 presents the energy diagram of a STEP process. The high temperature pathway decreases the free energy requirements for processes whose electrolysis potential decreases with increasing temperature. STEP uses solar energy to drive, otherwise energetically forbidden, pathways of charge transfer. The process combines elements of solid state physics, insolation (solar illumination) and high temperature electrochemical energy conversion. Kinetics improve, and endothermic thermodynamic potentials decrease, with increasing temperature. The result is a

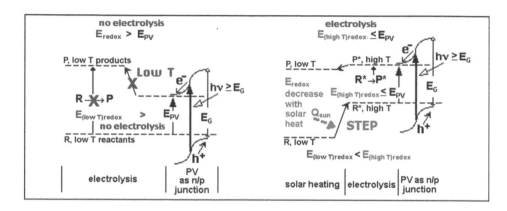

Scheme 8.2.1 Top: Comparison of PV and **STEP** solar driven electrolysis energy diagrams. STEP uses sunlight to drive otherwise energetically forbidden pathways of charge transfer. The energy of photodriven charge transfer is insufficient (left) to drive (unheated) electrolysis, but is sufficient (right) to drive endothermic electrolysis in the solar heated synergestic process. The process uses both visible & thermal solar energy for higher efficiency; thermal energy decreases the electrolysis potential forming an energetically allowed pathway to drive electrochemical charge transfer. Modified with permission from Licht 2009.

synergy, making use of the full spectrum of sunlight, and capturing more solar energy. **STEP** is intrinsically more efficient than other solar energy conversion processes, as it utilizes not only the visible sunlight used to drive PVs, but also utilizes the previously detrimental (due to PV thermal degradation) thermal component of sunlight, for the electrolytic formation of chemicals.

The two bases for improved efficiencies using the STEP process are: (i) excess heat, such as unused heat in solar cells, can be used to increase the temperature of an electrolysis cell, such as for electrolytic CO_2 splitting, while (ii) the product to reactant ratio can be increased to favor the kintetic and energetic formation of reactants. With increasing temperature, the quantitative decrease in the electrochemical potential to drive a variety of electrochemical syntheses is well known, substantially decreasing the electronic energy (the electrolysis potential) required to form energetic products. The extent of the decrease in the electrolysis potential, E_{redox}, may be tuned by choosing the constituents and temperature of the electrolysis. The process distinguishes radiation that is intrinsically energy sufficient to drive **PV** charge transfer, and applies all excess solar thermal energy to heat the electrolysis reaction chamber.

Scheme 8.2.2 summarizes the charge, heat and molecular flow for the **STEP** process; the high temperature pathway decreases the potential required to drive endothermic electrolyses, and also facilitates the kinetics of charge transfer (i.e., decreases overpotential losses), which arise during electrolysis. This process consists of (i) sunlight harvesting and concentration, (ii) photovoltaic charge transfer driven by super-bandgap energy, (iii) transfer of sub-bandgap and excess super-bandgap radiation to heat the electrolysis chamber, (iv) high temperature, low energy electrolysis forming energy rich products, and (v) cycle completion by pre-heating of the electrolysis reactant through heat exchange with the energetic electrolysis products.

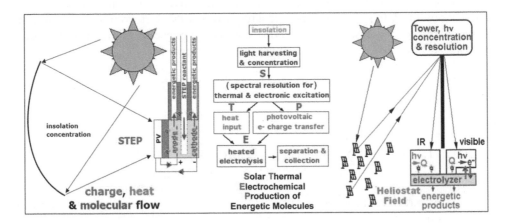

Scheme 8.2.2   Global use of sunlight to drive the formation of energy rich molecules. **Left:** Charge, & heat flow in **STEP**: heat flow (yellow arrows), electron flow (blue), & reagent flow (green). **Right:** Beam splitters redirect sub-bandgap sunlight away from the **PV** onto the electrolyzer. Modified with permission from Licht 2009.

As indicated on the right side of Scheme 8.2.2, the light harvesting can use various optical configurations; e.g. in lieu of parabolic, or Fresnel, concentrators, a heliostat/solar tower with secondary optics can achieve higher process temperatures (>1000°C) with concentrations of ~2000 suns. Beam splitters can redirect sub-bandgap radiation away from the PV (minimzing PV heating) for a direct heat exchange with the electrolyzer.

Solar heating can decrease the energy to drive a range of electrolyses. Such processes can be determined using available entropy, S, and enthalpy, H, and free-energy, G, data, (Chase, 1998) and are identified by their negative isothermal temperature coefficient of the cell potential (deBethune and 1959). This coefficient $(dE/dT)_{isoth}$ is the derivative of the electromotive force of the isothermal cell:

$$(dE/dT)_{isoth} = \Delta S/nF = (\Delta H - \Delta G)/nFT \tag{8.2.1}$$

The starting process of modeling any STEP process is the conventional expression of a generalized electrochemical process, in a cell which drives an n-electron charge transfer electrolysis reaction, comprising "x" reactants, R_i, with stoichiometric coefficients r_i, and yielding "y" products, C_i, with stoichiometric coefficients c_i. n-electron refers to the number of electrons gained to form the cathode products and lost to form the anode products in the electrolysis reaction.

Electrode 1 | Electrolyte | Electrode 2

Using the convention of $E = E_{cathode} - E_{anode}$ to describe the positive potential necessary to drive a non-spontaneous process, by transfer of n electrons in the electrolysis reaction:

$$\text{n-electron transfer electrolysis reaction:} \sum_{i=1}^{x} r_i R_i \rightarrow \sum_{i=1}^{y} c_i C_i \tag{8.2.2}$$

At any electrolysis temperature, T_{STEP}, and at unit activity, the reaction has electrochemical potential, $E°_T$. This may be calculated from consistent, compiled unit activity thermochemical data sets, such as the NIST condensed phase and fluid properties data sets, (Chase 1998) as:

$$E_T^\circ = -\Delta G^\circ(T = T_{STEP})/nF; \quad E_{ambient}^\circ \equiv E_T^\circ(T_{ambient});$$
$$\text{here } T_{ambient} = 298.15K = 25°C,$$

and:

$$\Delta G^\circ(T = T_{STEP}) = \sum_{i=1}^{y} c_i(H^\circ(C_i, T) - TS^\circ(C_i, T))$$

$$-\sum_{i=1}^{x} r_i(H^\circ(R_i, T) - TS^\circ(R_i, T)) \tag{8.2.3}$$

Compiled thermochemical data are often based on different reference states, while a consistent reference state is needed to understand electrolysis limiting processes, including water (Light et al., 2005; Licht, 1987). This challenge is overcome by modification of the unit activity (a = 1) consistent calculated electrolysis potential to determine the potential at other reagent and product relative activities via the Nernst equation (Licht, 1985; Licht et al., 1991). Electrolysis provides control of the relative amounts of reactant and generated product in a system. A substantial activity differential can also drive **STEP** improvement at elevated temperature, and will be derived.

The potential variation with activity, a, of the reaction: $\sum_{i=1}^{x} r_i R_i \rightarrow \sum_{i=1}^{y} c_i C_i$, is given by:

$$E_{T,a} = E_T^\circ - \left(\frac{RT}{nF}\right) \cdot \ln\left(\frac{\prod_{i=1}^{x} a(R_i)^{r_i}}{\prod_{i=1}^{y} a(C_i)^{c_i}}\right) \tag{8.2.4}$$

Electrolysis systems with a negative isothermal temperature coefficient tend to cool as the electrolysis products are generated. Specifically in endothermic electrolytic processes, the Equation 8.2.4 free-energy electrolysis potential, E_T, is less than the enthalpy based potential. This latter value is the potential at which the system temperature would remain constant during electrolysis. This thermoneutral potential, E_{tn}, is given by:

$$E_{tn}(T_{STEP}) = -\frac{\Delta H(T)}{nF};$$

$$\Delta H(T_{STEP}) = \sum_{i=1}^{b} c_i H(C_i, T_{STEP}) - \sum_{i=1}^{a} r_i H(R_i, T_{STEP}) \tag{8.2.5}$$

Two general STEP implementations are being explored. Both can provide the thermoneutral energy to sustain a variety of electrolyses. The thermoneutral potential, determined from the enthalpy of a reaction, describes the energy required to sustain an electrochemical process without cooling. For example, the thermoneutral potential we have calculated and reported for CO_2 splitting to CO and O_2 at unit activities, from Equation 8.2.5, is 1.46($\pm$0.01) V over the temperature range of 25–1400°C. As represented in Scheme 8.2.3 on the left, the standard electrolysis potential at room

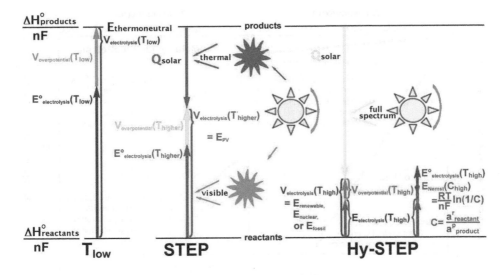

Scheme 8.2.3 Comparison of solar energy utilization in STEP and Hy-STEP implementations of the solar thermal electrochemical production of energetic molecules. Modified with permission from Licht 2011.

temperature, $E°$, can comprise a significant fraction of the thermoneutral potential. The first STEP mode, energetically represented next to the room temperature process in the scheme, separates sunlight into thermal and visible radiation. The solar visible generates electronic charge which drives electrolysis charge transfer. The solar thermal component heats the electrolysis and decreases both the $E°$ at this higher T, and the overpotential. The second mode, termed Hy-STEP (on the right) from "hybrid-STEP", does not separate sunlight, and instead directs all sunlight to heating the electrolysis, generating the highest T and smallest E, while the electrical energy for electrolysis is generated by a separate source (such as by photovoltaic, solar thermal electric, wind turbine, hydro, nuclear or fossil fuel generated electronic charge). As shown on the right side, high relative concentrations of the electrolysis reactant (such as CO_2 or iron oxide will further decrease the electrolysis potential).

8.2.2 STEP solar to chemical energy conversion efficiency

The Hy-STEP mode is being studied outdoors with wind or solar CPV generated electricity to drive $E_{electrolysis}$. The STEP mode is experimentally more complex and is presently studied indoors under solar simulator illumination. Determination of the efficiency of Hy-STEP with solar electric is straightforward in the domain in which $E_{electrolysis} < E_{thermoneutral}$ and the coulombic efficiency is high. Solar thermal energy is collected at an efficiency of $\eta_{thermal}$ to decrease the energy from $E_{thermoneutral}$ to $E_{electrolysis}$, and then electrolysis is driven at a solar electric energy efficiency of $\eta_{solar\text{-}electric}$:

$$\eta_{Hy\text{-}STEPsolar} = (\eta_{thermal} \cdot (E_{thermoneutral}\text{-}E_{electrolysis})$$
$$+ \eta_{solar\text{-}electric} \cdot E_{electrolysis})/E_{thermoneutral} \qquad (8.2.6)$$

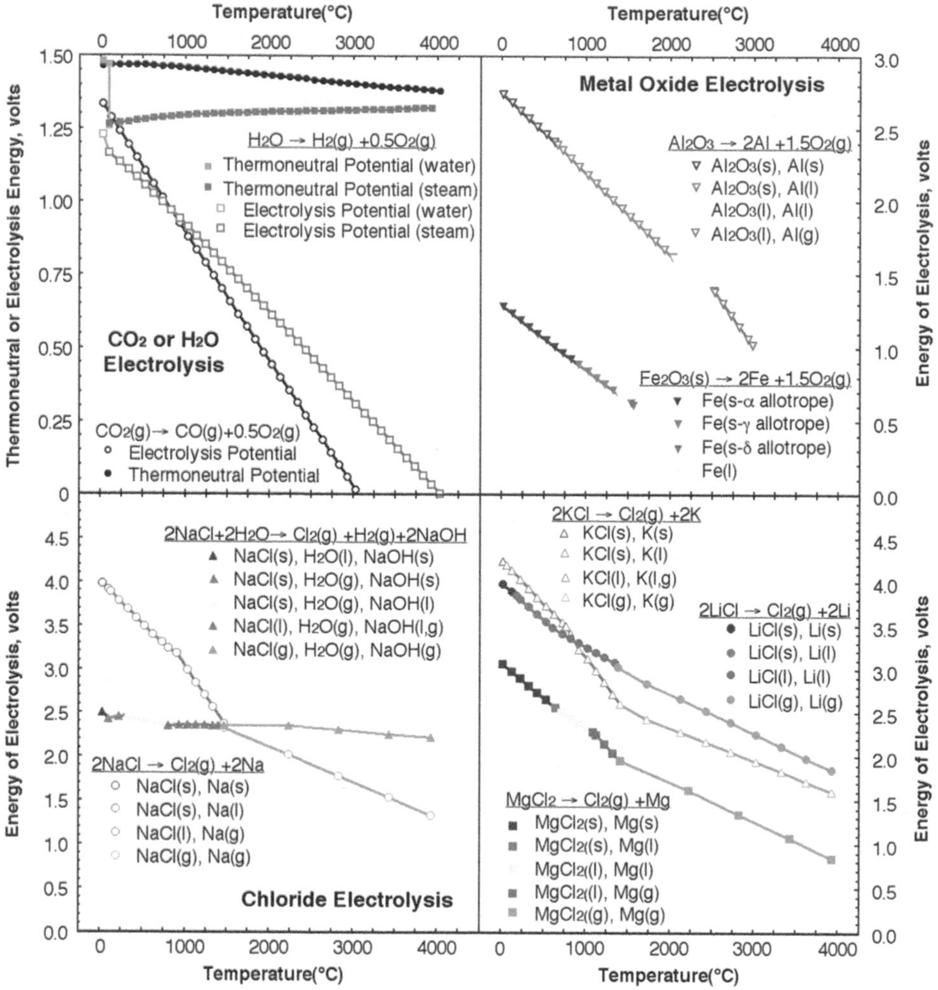

Figure 8.2.1 The calculated potential to electrolyze selected oxides (top) and chlorides (bottom). The indicated decrease in electrolysis energy, with increase in temperature, provides energy savings in the **STEP** process in which high temperature is provided by excess solar heat. Energies of electrolysis are calculated from Equation 8.2.3, with consistent thermochemical values at unit activity using NIST gas and condensed phase Shomate equations (Chase, 1998). Note with water excluded, the chloride electrolysis decreases (in the lower left of the figure). All other indicated electrolysis potentials, including that of water or carbon dioxide, decrease with increasing temperature. Thermoneutral potentials are calculated with Equation 8.2.5. Modified with permission from Licht 2009.

$\eta_{thermal}$ is higher than $\eta_{solar-electric}$, and gains in efficiency occur in Equation 8.2.6 in the limit as $E_{electrolysis}$ approaches 0. $E_{electrolysis} = 0$ is equivalent to thermochemical, rather than electrolytic, production. As seen in Figure 8.2.1, at unit activity $E°_{CO2/CO}$ does not approach 0 until 3000°C. Material constraints inhibit approach to this higher temperature, while electrolysis also provides the advantage of spontaneous product seperation.

At lower temperature, small values of $E_{electrolysis}$ can occur at higher reactant and lower product activities, as described in Equation 8.2.4. In the present configuration sunlight is concentrated at 75% solar to thermal efficiency, heating the electrolysis to 950°C, which decreases the high current density CO_2 splitting potential to 0.9 V, and the electrolysis charge is provided by CPV at 37% solar to electric efficiency. The solar to chemical energy conversion efficiency is in accordance with Equation 8.2.6:

$$\eta_{Hy\text{-}STEPsolar} = (75\% \cdot (1.46\,\text{V} - 0.90\,\text{V}) + 37\% \cdot 0.90\,\text{V})/1.46\,\text{V} = 52\% \quad (8.2.7)$$

A relatively high concentration of reactants lowers the voltage of electrolysis via the Nernst term in Equation 8.2.4. With appropriate choice of high temperature electrolyte, this effect can be dramatic, for example both in STEP iron and in comparing the benefits of the molten carbonate to solid oxide (gas phase) reactants for STEP CO_2 electrolytic reduction, sequestration and fuel formation. Fe(III) (as found in the common iron ore, hematite) is nearly insoluble in sodium carbonate, while it is soluble to over 10 m (molal) in lithium carbonate, (Licht et al., 2011b; Licht 2011) and as discussed in Section 8.2.3, molten carbonate electrolyzer provides 10^3 to 10^6 times higher concentration of reactant at the cathode surface than a solid oxide electrolyzer.

In practice, for STEP iron or carbon capture, we simultaneously drive lithium carbonate electrolysis cells together in series, at the CPV maximum power point (Figure 8.2.2). Specifically, a Spectrolab CDO-100-C1MJ concentrator solar cell is used to generate 2.7 V at maximum power point, with solar to electrical energy efficiencies of 37% under 500 suns illumination. As seen in Figure 8.2.2, at maximum power, the 0.99 cm^2 cell generates 1.3 A at 100 suns, and when masked to 0.2 cm^2 area generates 1.4 A at 500 suns. Electrolysis electrode surface areas were chosen to match the solar cell generated power. At 950°C at 0.9 V, the electrolysis cells generate carbon monoxide at 1.3 to 1.5 A (the electrolysis current stability is shown at the bottom of Figure 8.2.2).

In accord with Equation 8.2.6 and Scheme 8.2.3, Hy-STEP efficiency improves with temperature increase to decrease overpotential and $E_{electrolysis}$, and with increase in the relative reactant activity. Higher solar efficiencies will be expected, both with more effective carbonate electrocatalysts (as morphologies with higher effective surface area and lower overpotential) are developed, and as also as PV efficiencies increase. Increases in solar to electric (both PV, CPV and solar thermal-electric) efficiencies continue to be reported, and will improve Equation 8.2.7 efficiency. For example, multijunction CPV have been reported improved to $\eta_{PV} = 40.7\%$ (King et al., 2007; Green et al., 2011).

Engineering refinements will improve some aspects, and decrease other aspects, of the system efficiency. Preheating the CO_2, by circulating it as a coolant under the CPV (as we currently do in the indoor STEP experiment, but not outdoor, Hy-STEP experiments) will improve the system efficiency. In the present configuration outgoing CO and O_2 gases at the cathode and anode heat the incoming CO_2. Isolation of the electrolysis products will require heat exchangers with accompanying radiative heat losses, and for electrolyses in which there are side reactions or product recombination losses, $\eta_{Hy\text{-}STEPsolar}$ will decrease proportional to the decrease in coulombic efficiency. At present, wind turbine generated electricity is more cost effective than solar-electric, and we have demonstrated a Hy-STEP process with wind–electric, for CO_2 free production of iron (delineated in Section 8.3.3). Addition of long-term (overnight) molten

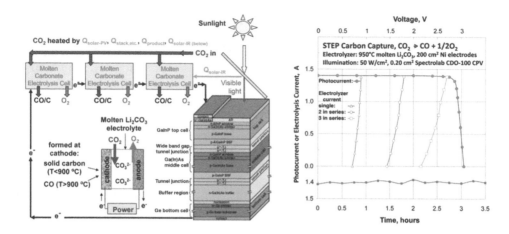

Figure 8.2.2 Left: STEP carbon capture in which three molten carbonate electrolysis in series are driven by a concentrator photovoltaic. Sunlight is split into two spectral regions; visible drives the CPV and thermal heats the electrolysis cell. In Hy-STEP (not shown) sunlight is not split and the full spectrum heats the electrolysis cell, and electronic charge is generated separately by solar, wind, or other source. Right: The maximum power point photovoltage of one Spectrolab CPV is sufficient to drive three in series carbon dioxide splitting 950°C molten Li_2CO_3 electrolysis cells. Top: Photocurrent at 500 suns (masked (0.20 cm²) Spectrolab CDO-100 CPV, or electrolysis current, versus voltage; electrolysis current is shown of one, two or three series 950°C Li_2CO_3 electrolysis cells with 200 cm² Ni electrodes. Three in series electrolysis cells provide a power match at the 2.7V maximum power point of the CPV at 950°C; similarly (not shown), two 750°C Li_2CO_3 electrolysis cells in series provide a power match at 2.7V to the CPV. Bottom: Stable carbon capture (with 200 cm² "aged" Ni electrodes at 750°C; fresh electrodes (not shown) exhibit an initial fluctuation as carbon forms at the cathode and Ni oxide layer forms on the anode. The rate of solid carbon deposition gradually increases as the cathode surface area slowly increases in time. Modified with permission from Licht et al. 2010a.

salt-insulated storage will permit continuous operation of the STEP process. Both STEP implementations provide a basis for practical, high solar efficiencies.

Components for STEP CO_2 capture and conversion to solid carbon are represented on the left side of Figure 8.2.2, and are detailed in Licht et al., 2010a; Licht et al., 2010; Licht et al., 2011b; Licht et al., 2010b; Licht et al., 2011a;. A 2.7 V CPV photopotential drives three in series electrolyses at 950°C. Fundamental details of the heat balance are provided in Licht et al., 2010a. The CPV has an experimental solar efficiency of 37%, and the 63% of insolation not converted to electricity comprises a significant heat source. The challenge is to direct a substantial fraction of this heat to the electrolysis. An example of this challenge is in the first stage of heating, in which higher temperatures increases CO_2 preheat, but diminishes the CPV power. Heating of the reactant CO_2 is a three tier process in the current configuration: the preheating of room temperature CO_2 consists of either (1a) flow-through a heat exchange fixed to the back of the concentrator solar cell and/or (1b) preheating to simulate CO_2 extracted from an available heat source such as a hot smoke (flue) stack, (2) secondary heating consists of passing this CO_2 through a heat exchange with the outgoing products, (3) tertiary heat is applied through concentrated, split solar thermal energy (Figure 8.2.2).

An upper limit to the energy required to maintain a constant system temperature is given in the case in which neither solar IR, excess solar visible, nor heat exchange from the environment or products would be applied to the system. When an 0.90 V electrolysis occurs, an additional 0.56 V, over $E_{tn} = 1.46V$, is required to maintain a constant system temperature. Hence, in the case of three electrolyses in series, as in Figure 8.2.2, an additional $3 \times 0.56\,V = 1.68\,V$ will maintain constant temperature. This is less than the 63% of the solar energy (equivalent to 4.6 V) not used in generating the 2.7 V of maximum power point voltage of electronic charge from the CPV in this experiment. Heating requirements are even less, when the reactant activity is maintained at a level that is higher than the product activity. For example, this is accomplished when products are continuously removed to ensure that the partial pressure of the products is lower than that of the CO_2. This lowers the total heat required for temperature neutrality to below that of the unit activity thermoneutral potential 1.46 V.

The STEP effective solar energy conversion efficiency, η_{STEP}, is constrained by both photovoltaic and thermal boost conversion efficiencies, η_{PV} and $\eta_{thermal\text{-}boost}$ (Licht et al., 2011a). Here, the CPV sustains a conversion efficiency of $\eta_{PV} = 37.0\%$. In the system, passage of electrolysis current requires an additional, combined (ohmic, & anodic + cathodic over-) potential above the thermodynamic potential. However, mobility and kinetics improve at higher temperature to decrease this overpotential. The generated CO contains an increase in oxidation potential compared to carbon dioxide at room temperature ($E_{CO2/CO}(25°C) = 1.33\,V$ for $CO_2 \rightarrow CO + 1/2\,O_2$ in Figure 8.2.1), an increase of 0.43 V compared to the 0.90 V used to generate the CO. The electrolysis efficiency compares the stored potential to the applied potential, $\eta_{thermal\text{-}boost} = E°_{electrolysis}(25°C)/V_{electrolysis}(T)$ (Licht et al., 2010a). Given a stable temperature electrolysis environment, the experimental STEP solar to CO carbon capture and conversion efficiency is the product of this relative gain in energy and the electronic solar efficiency:

$$\eta_{STEP} = \eta_{PV} \cdot \eta_{thermal\text{-}boost} = 37.0\% \cdot (1.33\,V/0.90\,V) = 54.7\% \qquad (8.2.8)$$

Ohmic and overpotential losses are already included in the measured electrolysis potential. This 54.7% STEP solar conversion efficiency is an upper limit of the present experiment, and as with the Hy-STEP mode, improvements are expected in electrocatalysis and CPV efficiency. Additional losses will occur when beamsplitter and secondary concentrator optics losses, and thermal systems matching are incorporated, but serves to demonstrate the synergy of this solar/photo/electrochemical/thermal process, leads to energy efficiency higher than that for solar generated electricity, (King et al., 2007; Green et al., 2011) or for photochemical, (Miller et al., 2008) photoelectrochemical, (Licht, 2002; Barton et al., 2008) solar thermal, (Woolerton et al., 2010) or other CO_2 reduction processes (Benson et al., 2009).

The CPV does not need, nor function with, sunlight of energy less than that of the 0.67 eV bandgap of the multi-junction Ge bottom layer. From our previous calculations, this thermal energy comprises 10% of AM1.5 insolation, which will be further diminished by the solar thermal absorption efficiency and heat exchange to the electrolysis efficiency, (Licht, 2003) and under $0.5\,MW\,m^{-2}$ of incident sunlight (500 suns illumination), yields $\sim 50\,kW\,m^{-2}$, which may be split off as thermal energy towards heating the electrolysis cell without decreasing the CPV electronic power. The CPV, while efficient, utilizes less than half of the super-bandgap ($h\nu > 0.67\,eV$) sunlight. A portion of this $> \sim 250\,kW\,m^{-2}$ available energy, is extracted through heat exchange at the backside of the CPV. Another useful source for consideration as supplemental heat is industrial exhaust. The temperature of industrial flue stacks varies widely, with fossil fuel source and application, and ranges up to $650°C$ for an open circuit gas turbine. The efficiency of thermal energy transfer will limit use of this available heat.

A lower limit to the STEP efficiency is determined when no heat is recovered, either from the CPV or remaining solar IR, and when heat is not recovered via heat exchange from the electrolysis products, and when an external heat source is used to maintain a constant electrolysis temperature. In this case, the difference between the electrolysis potential and the thermoneutral potential represents the enthalpy required to keep the system from cooling. In this case, our 0.9 V electrolysis occurs at an efficiency of $(0.90\,V/1.46\,V) \cdot 54.7\% = 34\%$. While the STEP energy analysis, detailed in Section 8.4.2 for example for CO_2 to CO splitting, is more complex than that of the Hy-STEP mode, more solar thermal energy is available including a PV's unused or waste heat to drive the process and to improve the solar to chemical energy conversion efficiency. We determine the STEP solar efficiency over the range from inclusion of no solar thermal heat (based on the enthalpy, rather than free energy, of reaction) to the case where the solar thermal heat is sufficient to sustain the reaction (based on the free energy of reaction). This determines the efficiency range, as chemical flow out to the solar flow in (as measured by the increase in chemical energy of the products compared to the reactants), from 34% to over 50%.

8.2.3 Identification of STEP consistent endothermic processes

The electrochemical driving force for a variety of chemicals of widespread usewill be shown to significantly decrease with increasing temperature. As calculated and summarized in the top left of Figure 8.2.1, the electrochemical driving force for electrolysis of either carbon dioxide or water, significantly decreases with increasing temperature. The ability to remove CO_2 from exhaust stacks or atmospheric sources, provides a response to linked environmental impacts, including global warming due to anthropogenic CO_2 emission.From the known thermochemical data for CO_2, CO and O_2, and in accord with Equation 8.2.1, CO_2 splitting can be described by:

$$CO_2\,(g) \rightarrow CO\,(g) + 1/2O_2\,(g);$$
$$E°_{CO_2\text{-split}} = (G°_{CO} + 0.5G°_{O_2} - G°_{CO_2})/2F; \quad E°\,(25°C) = 1.333\,V \qquad (8.2.9)$$

As an example of the solar energy efficiency gains, this progress report focuses on CO_2 splitting potentials, and provides examples of other useful STEP processes. As seen in Figure 8.2.1, CO_2 splitting potentials decrease more rapidly with temperature than those for water splitting, signifying that the STEP process may be readily applied to CO_2 electrolysis. Efficient, renewable, non-fossil fuel energy-rich carbon sources are needed, and the product of Equation 8.2.9, carbon monoxide is a significant industrial gas with a myriad of uses, including the bulk manufacturing of hydrocarbon fuels, acetic acid and aldehydes (and detergent precursors), and for use in industrial nickel purification (Elschenbroich and Salzer, 1992). To alleviate challenges of fossil fuel resource depletion, CO is an important syngas component and methanol is formed through the reaction with H_2. The ability to remove CO_2 from exhaust stacks or atmospheric sources, also limits CO_2 emission. Based on our original analogous experimental photo-thermal electrochemical water electrolysis design, (Licht et al., 2003; Licht, 2005) the first CO_2 STEP process consists of solar driven and solar thermal assisted CO_2 electrolysis. In particular, in a molten carbonate bath electrolysis cell, fed by CO_2.

$$\text{cathode: } 2CO_2 \text{ (g)} + 2e- \rightarrow CO_3^= \text{ (molten)} + CO \text{ (g)}$$

$$\text{anode: } CO_3^= \text{ (molten)} \rightarrow CO_2 \text{ (g)} + 1/2O_2 \text{ (g)} + 2e-$$

$$\text{cell: } CO_2 \text{ (g)} \rightarrow CO \text{ (g)} + 1/2O_2 \text{ (g)} \tag{8.2.10}$$

Molten alkali carbonate electrolyte fuel cells typically operate at 650°C. Li, Na or K cation variation can affect charge mobility and operational temperatures. Sintered nickel often serves as the anode, porous lithium doped nickel oxide often as the cathode, while the electrolyte is suspended in a porous, insulating, chemically inert $LiAlO_2$ ceramic matrix (Sunmacher, 2007).

Solar thermal energy can be used to favor the formation of products for electrolyses characterized by a negative isothermal temperature coefficient, but will not improve the efficiency of heat neutral or exothermic reactions. An example of this restriction occurs for the electrolysis reaction currently used by industry to generate chlorine. During 2008, the generation of chlorine gas (principally for use as bleach and in the chlor-alkali industry) consumed approximately 1% of the world's electricity, (Pellegrino, 2000) prepared in accord with the industrial electrolytic process:

$$2NaCl + 2H_2O \rightarrow Cl_2 + H_2 + 2NaOH; \quad E° \text{ (25°C)} = 2.502 \text{ V} \tag{8.2.11}$$

In the lower left portion of Figure 8.2.1, the calculated electrolysis potential for this industrial chlor-alkali reaction exhibits little variation with temperature, and hence the conventional generation of chlorine by electrolysis would not benefit from the inclusion of solar heating. This potential is relatively invariant, despite a number of phase changes of the components (indicated on the figure and which include the melting of NaOH or NaCl). However, as seen in the figure, the calculated potential for the anhydrous electrolysis of chloride salts is endothermic, including the electrolyses to generate not only chlorine, but also metallic lithium, sodium and magnesium, and can be greatly improved through the **STEP** process:

$$MCl_n \rightarrow n/2Cl_2 + M;$$
$$E°_{MCl\text{-split}} \text{ (25°C)} = 3.98\text{V-M} = Na, 4.24\text{V-K}, 3.98\text{V-Li}, 3.07\text{V-Mg} \tag{8.2.12}$$

The calculated decreases for the anhydrous chloride electrolysis potentials are in the order of volts per 1000°C temperature change. For example, from 25°C up to the

$MgCl_2$ boiling point of $1412°C$, the $MgCl_2$ electrolysis potential decreases from 3.07 V to 1.86 V. This decrease provides a theoretical basis for significant, non-CO_2 emitting, non-fossil fuel consuming processes for the generation of chlorine and magnesium, to be delineated in Section 8.3.4, and occurring at high solar efficiency analogous to the similar CO_2 **STEP** process.

In Section 8.3.2 the **STEP** process will be derived for the efficient solar removal/recycling of CO_2. In addition, thermodynamic calculation of metal and chloride electrolysis rest potentials identifies electrolytic processes which are consistent with endothermic processes for the formation of iron, chlorine, aluminum, lithium, sodium and magnesium, via CO_2–free pathways. As shown, the conversion and replacement of the conventional, aqueous, industrial alkali-chlor process, with an anhydrous electrosynthesis, results in a redox potential with a calculated decrease of 1.1 V from $25°C$ to $1000°C$.

As seen in the top right of Figure 8.2.1, the calculated electrochemical reduction of metal oxides can exhibit a sharp, smooth decrease in redox potential over a wide range of phase changes. These endothermic process provide an opportunity for the replacement of conventional industrial processes by the **STEP** formation of these metals. In 2008, industrial electrolytic processes consumed ~5% of the world's electricity, including for aluminum (3%), chlorine (1%), and lithium, magnesium and sodium production. This 5% of the global 19×10^{12} kWh of electrical production, is equivalent to the emission of 6×10^8 metric tons of CO_2 (Pellegrino, 2000). The iron and steel industry accounts for a quarter of industrial direct CO_2 emissions. Currently, iron is predominantly formed through the reduction of hematite with carbon, emitting CO_2:

$$Fe_2O_3 + 3C + 3/2O_2 \rightarrow 2Fe + 3CO_2 \qquad (8.2.13)$$

A non-CO_2 emitting alternative is provided by the **STEP** driven electrolysis of Fe_2O_3:

$$Fe_2O_3 \rightarrow 2Fe + 3/2O_2 \qquad E° = 1.28\,V \qquad (8.2.14)$$

As seen in the top right of Figure 8.2.1, the calculated iron-generating electrolysis potential drops 0.5 V (a 38% drop) from $25°C$ to $1000°C$, and as with the CO_2 analogue, will be expected to decrease more rapidly with non-unit activity conditions, as will be delineated in a future study. Conventional industrial processes for these metals and chlorine, along with CO_2 emitted from power and transportation, are responsible for the majority of anthropogenic CO_2 release. The **STEP** process, to efficiently recover carbon dioxide and in lieu of these industrial processes, can provide a transition beyond the fossil fuel-electric grid economy.

The top left of Figure 8.2.1 includes calculated thermoneutral potentials for CO_2 and water-splitting reactions. At ambient temperature, the difference between E_{th} and E_T does not indicate an additional heat requirement for electrolysis, as this heat is available via heat exchange with the ambient environment. At ambient temperature, $E_{tn} - E_T$ for CO_2 or water is respectively 0.13 and 0.25 V, is calculated (not shown) as 0.15 ± 0.1 V for Al_2O_3 and Fe_2O_3, and 0.28 ± 0.3 V for each of the chlorides.

We find that molten electrolytes present several fundamental advantages compared to solid oxides for CO_2 electrolysis. (i) Molten carbonate electrolyzer provides 10^3 to 10^6 times higher concentration of reactant at the cathode surface than a solid oxide electrolyzer. Solid oxides utilize gas phase reactants, whereas carbonates utilize molten phase reactants. Molten carbonate contains 2×10^{-2} mol reducible tetravalent carbon/cm^3. The density of reducible tetravalent carbon sites in the gas phase is considerably lower. Air contains 0.03% CO_2, equivalent to only 1×10^{-8} mol of tetravalent carbon/cm^3, and flue gas (typically) contains 10–15% CO_2, equivalent to 2×10^{-5} mol reducible C(IV)/cm^3. Carbonate's higher concentration of active, reducible tetravalent carbon sites, logarithmically decreases the electrolysis potential, and can facilitate charge transfer at low electrolysis potentials. (ii) Molten carbonates can directly absorb atmospheric CO_2, whereas solid oxides require an energy consuming pre-concentration process. (iii) Molten carbonates electrolyses are compatible with both solid and gas phase products. (iv) Molten processes have an intrinsic thermal buffer not found in gas phase systems. Sunlight intensity varies over a 24-hour cycle, and more frequently with variations in cloud cover. This disruption to other solar energy conversion processes is not necessary in molten salt processes. For example as discussed in Section 8.4.3, the thermal buffer capacity of molten salts has been effective for solar to electric power towers to operate 24/7. These towers concentrate solar thermal energy to heat molten salts, which circulate and via heat exchange boil water to drive conventional mechanical turbines.

8.3   DEMONSTRATED STEP PROCESSES

8.3.1   STEP hydrogen

STEP occurs at both higher electrolysis and higher solar conversion efficiencies than conventional room temperature photovoltaic (PV) generation of hydrogen. Experimentally, we demonstrated a sharp decrease in the water splitting potential in an unusual molten sodium hydroxide medium, Figure 8.3.1, and as shown in Figure 8.3.2, three series connected Si CPVs efficiently driving two series molten hydroxide water-splitting cells at 500°C to generate hydrogen (Licht et al., 2003; Licht, 2005).

Recently we have considered the economic viability of solar hydrogen fuel production. That study provided evidence that the STEP system is an economically viable solution for the production of hydrogen (Licht et al., 2003; Licht, 2005).

8.3.2   STEP carbon capture

In this process carbon dioxide is captured directly, without the need to pre-concentrate dilute CO_2, using a high temperature electrolysis cell powered by sunlight in a single step. Solar thermal energy decreases the energy required for the endothermic conversion of carbon dioxide and kinetically facilitates electrochemical reduction, while solar visible generates electronic charge to drive the electrolysis. CO_2 can be captured as solid carbon and stored, or used as carbon monoxide to feed chemical or synthetic fuel production. Thermodynamic calculations are used to determine, and then demonstrate, a specific low energy, molten carbonate salt pathway for carbon capture.

Prior investigations of the electrochemistry of carbonates in molten salts tended to focus on reactions of interest to fuel cells, (Sunmacher, 2007) rather than the (reverse) electrolysis reactions of relevance to the STEP reduction of carbon dioxide, typically in

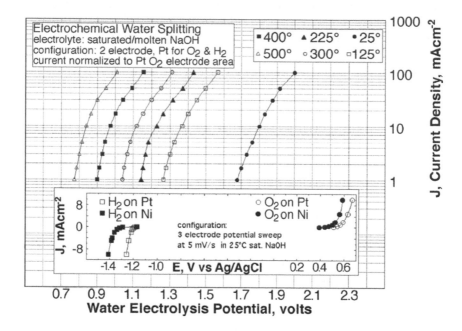

Figure 8.3.1 VH$_2$O, measured in aq.saturated or molten NaOH, at 1 atm. Steam is injected in the molten electrolyte. O$_2$ anode is 0.6 cm^2 Pt foil. IR and polarization losses are minimized by sandwiching 5 mm from each side of the anode, oversized Pt gauze cathode. Inset: At 25°C, 3 electrode values comparing Ni and Pt working electrodes and with a Pt gauze counterelectrode at 5 mV/s. Modified with permission from Licht 2011.

alkali carbonate mixtures. Such mixtures substantially lower the melting point compared to the pure salts, and would provide the thermodynamic maximum voltage for fuel cells. However, the electrolysis process is maximized in the opposite temperature domain of fuel cells, that is at elevated temperatures which decrease the energy of electrolysis, as schematically delineated in Scheme 8.2.1. These conditions provide a new opportunity for effective CO$_2$ capture.

CO$_2$ electrolysis splitting potentials are calculated from the thermodynamic free energy components of the reactants and products (Licht, 2009; Licht et al., 2010a; Chase, 1998) as $E = -\Delta G$(reaction)/nF, where n = 4 or 2 for the respective conversion of CO$_2$ to the solid carbon or carbon monoxide products. As calculated using the available thermochemical enthalpy and entropy of the starting components, and as summarized in the left side of Figure 8.3.3, molten Li$_2$CO$_3$, via a Li$_2$O intermediate, provides a preferred, low energy route compared to Na$_2$CO$_3$ or K$_2$CO$_3$ (via Na$_2$O or K$_2$O), for the conversion of CO$_2$. High temperature is advantageous as it decreases the free energy energy necessary to drive the STEP enodthermic process. The carbonates, Li$_2$CO$_3$, Na$_2$CO$_3$ and K$_2$CO$_3$, have respective melting points of 723°C, 851°C and 891°C. Molten Li$_2$CO$_3$ not only requires lower thermodynamic electrolysis energy, but in addition has higher conductivity (6 S cm^{-1}) than that of Na$_2$CO$_3$ (3 S cm^{-1}) or K$_2$CO$_3$ (2 S cm^{-1}) near the melting point (Zhang and

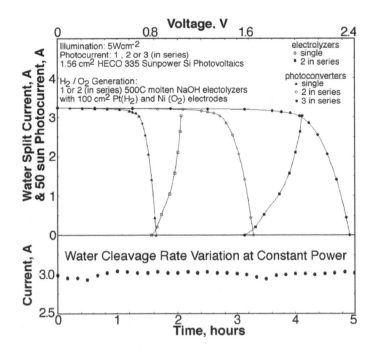

Figure 8.3.2 Photovoltaic and electrolysis charge transfer of STEP hydrogen using Si CPV's driving molten NaOH water electrolysis. Photocurrent is shown for 1, 2 or 3 1.561 cm² HECO 335 Sunpower Si photovoltaics in series at 50 suns. The CPV's drive 500°C molten NaOH steam electrolysis using Pt gauze electrodes. Left inset: electrolysis current stability. Modified with permission from Licht 2011.

Wang, 2006). Higher conductivity is desired as it leads to lower electrolysis ohmic losses. Low carbonate melting points are achieved by a eutectic mix of alkali carbonates (T_{mp} $Li_{1.07}Na_{0.93}CO_3$: 499°C; $Li_{0.85}Na_{0.61}K_{0.54}CO_3$: 393°C). Mass transport is also improved at higher temperature; the conductivity increases from 0.9 to 2.1 S cm^{-1} with temperature increase from 650°C to 875°C for a 1:1:1 by mass mixture of the three alkali carbonates (Kojima et al., 2008).

In 2009 we showed that molten carbonate electrolyzers can provide an effective media for solar splitting of CO_2 at high conversion efficiency. In 2010 Kaplan, et al., and our group separately reported that molten lithiated carbonates provide a particularly effective medium for the electrolytsis reduction of carbon dioxide (Licht et al., 2010a; Kaplan et al., 2010). As we show in the photograph in Figure 8.3.3, at 750°C, carbon dioxide is captured in molten lithium carbonate electrolyte as solid carbon by reduction at the cathode at low electrolysis potential. It is seen in the cyclic voltammetry, CV, that a solid carbon peak that is observed at 750°C is not evident at 950°C. At temperatures less than ~900°C in the molten electrolyte, solid carbon is the preferred CO_2 splitting product, while carbon monoxide is the preferred product at higher temperature. As seen in the main portion of the figure, the electrolysis potential is <1.2 V at either 0.1 or 0.5 A/cm², respectively at 750 or 850°C. Hence,

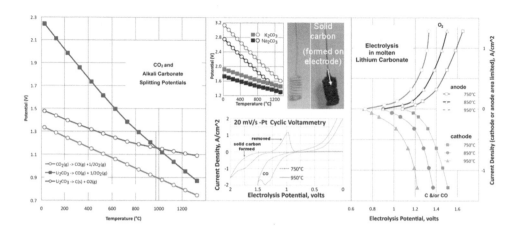

Figure 8.3.3 The calculated (left) and measured (right) electrolysis of CO_2 in molten carbonate. Left: The calculated thermodynamic electrolysis potential for carbon capture and conversion in Li_2CO_3 (main figure), or Na_2CO_3 or K_2CO_3 (left middle); squares refer to M_2CO_3 to $C + M_2O + O_2$ and circles to a M_2CO_3 to $CO + M_2O + 1/2O_2$. To the left of the vertical brown line, solid carbon is the thermodynamically preferred (lower energy) product. To the right of the vertical line, CO is preferred. Carbon dioxide fed into the electrolysis chamber is converted to solid carbon in a single step. Photographs: coiled platinum cathode before (left), and after (right), CO_2 splitting to solid carbon at 750°C in molten carbonate with a Ni anode. Right: The electrolysis full cell potential is measured, under anode or cathode limiting conditions, at a platinum electrode for a range of stable anodic and cathodic current densitites in molten Li_2CO_3. Lower midde: cathode size restricted full cell cyclic voltammetry, CV, of Pt electrodes in molten Li_2CO_3. Modified with permission from Licht et al. 2010a.

the electrolysis energy required at these elevated, molten temperatures is less than the minimum energy required to split CO_2 to CO at 25°C:

$$CO_2 \rightarrow CO + 1/2\ O_2 \quad E°(T = 25°C) = 1.33\ V \tag{8.3.1}$$

The observed experimental carbon capture correlates with:

$$Li_2CO_3\ (molten) \rightarrow C\ (solid) + Li_2O\ (dissolved) + O_2\ (gas) \tag{8.3.2A}$$

$$Li_2CO_3\ (molten) \rightarrow CO\ (gas) + Li_2O\ (dissolved) + 1/2\ O_2\ (gas) \tag{8.3.2B}$$

When CO_2 is bubbled in, a rapid reaction back to the original lithium carbonate is strongly favored:

$$Li_2O\ (dissolved) + CO_2\ (gas) \rightarrow Li_2CO_3\ (molten) \tag{8.3.3A}$$

$$Li_2CO_3 \rightarrow Li_2O + CO_2 \tag{8.3.3B}$$

In the presence of carbon dioxide, reaction (8.3.3A) is strongly favored (exothermic), and the rapid reaction back to the original lithium carbonate occurs while CO_2 is bubbled into molten lithium carbonate containing the lithium oxide.

The carbon capture reaction in molten carbonate, combines Equations 8.3.2 and 8.3.3:

$$CO_2 \text{ (gas)} \rightarrow C \text{ (solid)} + O_2 \text{ (gas)} \qquad T \leq 900°C \qquad (8.3.4A)$$

$$CO_2 \text{ (gas)} \rightarrow CO \text{ (gas)} + 1/2 O_2 \text{ (gas)} \quad T \geq 950°C \qquad (8.3.4B)$$

The electrolysis of carbon capture in molten carbonates can occur at lower experimental electrolysis potentials than the unit activity potentials calculated in Figure 8.3.3. A constant influx of carbon dioxide to the cell maintains a low concentration of Li_2O, in accord with reaction 23. The activity ratio, Θ, of the carbonate reactant to the oxide product in the electrolysis chamber, when high, decreases the cell potentials with the Nernst concentration variation of the potential in accord with Equation 8.3.2, as:

$$E_{CO2/X}(T) = E°_{CO2/X}(T) - 0.0592 \text{ V} \cdot T(K)/(n \cdot 298 \text{ K}) \cdot \log(\Theta);$$
$$n = 4 \text{ or } 2, \text{ for } X = C_{solid} \text{ or CO product} \qquad (8.3.5)$$

For example, from Equation 8.3.5, the expected cell potential at 950°C for the reduction to the CO product is $E_{CO2/CO} = 1.17\,V - (0.243\,V/2) \cdot 4 = 0.68\,V$, with a high $\Theta = 10,000$ carbonate/oxide ratio in the electrolysis chamber. As seen in the Figure 8.3.3 photograph, CO_2 is captured in 750°C Li_2CO_3 as solid carbon by reduction at the cathode at low electrolysis potential. The carbon formed in the electrolysis in molten Li_2CO_3 at 750°C is in quantitative accord with the 4 e- reduction of Equation 8.3.2, as determined by (i) mass, at constant 1.25 A for both 0.05 and 0.5 A/cm² (large and small electrode) electrolyses (the carbon is washed in a sonicator, and dried at 90°C), by (ii) ignition (furnace combustion at 950°C) and by (iii) volumetric analysis in which KIO_3 is added to the carbon, converted to CO_2 and I_2 in hot phosphoric acid ($5C + 4KIO_3 + 4H_3PO_4 \rightarrow 5CO_2 + 2I_2 + 2H_2O + 4KH_2PO_4$), the liberated I_2 is dissolved in 0.05 M KI and titrated with thiosulfate using a starch indicator. We also observe the transition to the carbon monoxide product with increasing temperature. Specifically, while at 750°C the molar ratio of solid carbon to CO-gas formed is 20:1, at 850° in molten Li_2CO_3, the product ratio is a 2:1, at 900°C, the ratio is 0.5:1, and at 950°C the gas is the sole product. Hence, in accord with Figure 8.2.2, switching between the C or CO product is temperature programmable.

We have replaced Pt, with Ni, nickel alloys (inconel and monel), Ti and carbon, and each are effective carbon capture cathode materials. Solid carbon deposits on each of these cathodes at similar overpotential in 750°C molten Li_2CO_3. For the anode, both platinum and nickel are effective, while titanium corrodes under anodic bias in molten Li_2CO_3. As seen in the right side of Figure 8.3.3, electrolysis anodic overpotentials in Li_2CO_3 electrolysis are comparable, but larger than cathodic overpotentials, and current densities of over 1 A cm⁻² can be sustained. Unlike other fuel cells, carbonate fuel cells are resistant to poisoning effects, (Sunmacher, 2007) and are effective with a wide range of fuels, and this appears to be the same for the case in the reverse mode (to capture carbon, rather than to generate electricity). Molten Li_2CO_3 remains transparent and sustains stable electrolysis currents after extended (hours/days) carbon capture over a wide range of electrolysis current densities and temperatures.

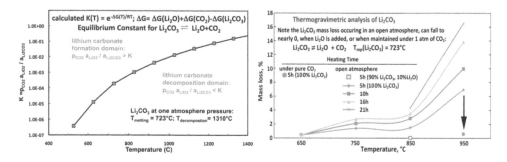

Figure 8.3.4 Left: Species stability in the lithium carbonate, lithium oxide, carbon dioxide system, as calculated from Li_2CO_3, Li_2O, and CO_2 thermochemical data. Right: Thermogravimetric analysis of lithium carbonate. The measured mass loss in time of Li_2CO_3. Not shown: The Li_2CO_3 mass loss rate also decreases with an increasing ratio of Li_2CO_3 mass to the surface area of the molten salt exposed to the atmosphere. This increased ratio, may increase the released partial pressure of CO_2 above the surface, increase the rate of the back reaction ($Li_2O + CO_2 \rightarrow Li_2CO_3$), and therefore result in the observed decreased mass loss. Hence, under an open atmosphere at 950°C, the mass loss after 5 hours falls from 7% to 4.7%, when the starting mass of pure Li_2CO_3 in the crucible is increased from 20 to 50 g. Under these latter conditions (open atmosphere, 950°C, 50 g total electrolyte), but using a 95% Li_2CO_3, 5% Li_2O mix, the rate of mass loss is only 2.3%. Modified with permission from Licht et al. 2011a.

As delineated in Section 8.2.3, in practice, either STEP or Hy-STEP modes are useful for efficient solar carbon capture. CO_2 added to the cell is split at 50% solar to chemical energy conversion efficiency by series coupled lithium carbonate electrolysis cells driven at maximum power point by an efficient CPC. Experimentally, we observe the facile reaction of CO_2 and Li_2O in molten Li_2CO_3. We can also calculate the thermodynamic equilibrium conditions between the species in the system, Equation 8.2.3B. Using the known thermochemistry of Li_2O, CO_2 and Li_2CO_3, (Chase, 1998) we calculate the reaction free-energy of Equation 8.2.1, and from this calculate the thermodynamic equilibrium constant as a function of temperature. From this equilibrium constant, the area above the curve on the left side of Figure 8.3.4 presents the wide domain (above the curve) in which Li_2CO_3 dominates, that is where excess CO_2 reacts with Li_2O such that $p_{CO2} \cdot a_{Li2O} < a_{Li2CO3}$. This is experimentally verified when we dissolve Li_2O in molten Li_2CO_3, and inject CO_2 (gas). Through the measured mass gain, we observe the rapid reaction to Li_2CO_3. Hence, CO_2 is flowed into a solution of 5% by weight Li_2O in molten Li_2CO_3 at 750°C, the rate of mass gain is only limited by the flow rate of CO_2 into the cell (using an Omega FMA 5508 mass flow controller) to react one equivalent of CO_2 per dissolved Li_2O. As seen in the measured thermogravimetric analysis on the right side of Figure 8.3.4, the mass loss in time is low in lithium carbonate heated in an open atmosphere (~0.03% CO_2) up to 850°C, but accelerates when heated to 950°C. However the 950°C mass loss falls to nearly zero, when heated under pure (1 atm) CO_2. Also in accord with Equation 8.2.1 added Li_2O shifts the equilibrium to the left. As seen in the figure in an open atmosphere, there is no mass loss in a 10% Li_2O, 90% Li_2CO_3 at 850°C, and the Li_2O containing electrolyte absorbs CO_2 (gains mass) at 750°C to provide for the direct carbon capture of atmospheric CO_2, without a CO_2 pre-concentration stage. This consists of the absorption of atmospheric CO_2 (in molten Li_2CO_3 containing Li_2O, to form Li_2CO_3), combined with a facile rate of CO_2 splitting due to the high carbonate concentration, compared to the atmospheric concentration of CO_2, and the continuity of the steady-state of concentration Li_2O, as Li_2CO_3 is electrolyzed in Equation 8.3.2.

8.3.3 STEP iron

A fundamental change in the understanding of iron oxide thermochemistry can open a facile, new CO_2-free, route to iron production. Along with control of fire, iron production is one of the founding technological pillars of civilization, but is a major source of CO_2 emissions. In industry, iron is still produced by the carbothermal greenhouse gas intensive reduction of iron oxide by carbon-coke, and a carbon dioxide free process to form this staple is needed.

The earliest attempt at electrowinning iron (the formation of iron by electrolysis) from carbonate appears to have been in 1944 in the unsuccessful attempt to electrodeposit iron from a sodium carbonate, peroxide, metaborate mix at 450–500°C, which deposited sodium and magnetite (iron oxide), rather than iron (Andrieux and Weiss, 1944; Haarberg et al., 2007). Other attempts (Haarberg et al., 2007) have focused on iron electrodepostion from molten mixed halide electrolytes, which has not provided a successful route to form iron, (Wang et al., 2008; Li et al., 2009), or aqueous iron electrowinning (Yuan et al., 2009; Palmaer and Brinell, 1913; Eustis, 1922; Mostad et al., 2008) that is hindered by the high thermodynamic potential ($E° = 1.28$ V) and diminished kinetics at low temperature.

We present a novel route to generate iron metal by the electrolysis of dissolved iron oxide salts in molten carbonate electrolytes, unexpected due to the reported insolubility of iron oxide in carbonates. We report high solubility of lithiated iron oxides, and facile charge transfer that produces the staple iron at high rate and low electrolysis energy, and can be driven by conventional electrical sources, but is also demonstrated with STEP processes that decrease or eliminate a major global source of greenhouse gas emissions (Licht, 2009; Licht et al., 2010a; Licht and Wu, 2011).

As recently as 1999, the solubility of ferric oxide, Fe_2O_3, in 650°C molten carbonate was reported as very low, a $10^{-4.4}$ mole fraction in lithium/potassium carbonate mixtures, and was reported as invariant of the fraction of Li_2CO_3 and K_2CO_3 (Qingeng et al., 1999). Low solubility, of interest to the optimization of molten carbonate fuel cells, had likely discouraged research into the electrowinning of iron metal from ferric oxide in molten lithium carbonate. Rather than the prior part per million reported solubility, we find higher Fe(III) solubilities, in the order of 50% in carbonates at 950°C. The CV of a molten Fe_2O_3 Li_2CO_3 mixture presented in Figure 8.3.5, and exhibits a reduction peak at -0.8 V, on Pt (gold curve); which is more pronounced at an iron electrode (light gold curve). At constant current, iron is clearly deposited. The cooled deposited product contains pure iron metal and trapped salt, and changes to rust color with exposure to water (Figure 8.3.5 photograph). The net electrolysis is the redox reaction of ferric oxide to iron metal and O_2, Equation 8.2.14. The deposit is washed, dried, and is observed to be reflective, grey metallic, responds to an external magnetic field, and consists of dendritic iron crystals.

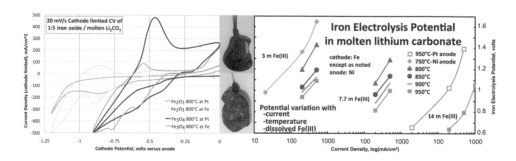

Figure 8.3.5   Middle: Photographs of electrolysis products from 20% Fe_2O_3 or Fe_3O_4 by mass in 800°C Li_2CO_3: following extended 0.5A electrolysis at a coiled wire (Pt or Fe) cathode with a Ni anode. Left: cathode restricted CV in Li_2CO_3, containing 1:5 by weight of either Fe_2O_3 or Fe_3O_4. Right: The measured iron electrolysis potentials in molten Li_2CO_3, as a function of the temperature, current density, and the concentration of dissolved Fe(III). Modified with permission from Licht and Wang 2010.

The two principal natural ores of iron are hematite (Fe_2O_3) and the mixed valence $Fe^{2+/3+}$ magnetite (Fe_3O_4). We observe that, Fe_3O_4 is also highly soluble in molten Li_2CO_3, and may also be reduced to iron with the net electrolysis reaction:

$$Fe_3O_4 \rightarrow 3Fe + 2O_2 \quad E° = 1.32\,V, E_{thermoneutral} = 1.45\,V \tag{8.3.6}$$

Fe_3O_4 electrolysis potentials run parallel, but ~0.06 V higher, than those of Fe_2O_3 in Figure 8.2.1. The processes are each endothermic; the required electrolysis potential decreases with increasing temperature. For Fe_3O_4 in Figure 8.3.5, unlike the single peak evident for Fe_2O_3, two reduction peaks appear in the CV at 800°C. Following the initial cathodic sweep (indicated by the left arrow), the CV exhibits two reduction peaks, again more pronounced at an iron electrode (grey curve), which appear to be consistent with the respective reductions of Fe^{2+} and Fe^{3+}. In either Fe_2O_3, or Fe_3O_4, the reduction occurs at a potential before we observe any reduction of the molten Li_2CO_3 electrolyte, and at constant current, iron is deposited. Following 1 hour of electrolysis at either 200 or 20 mA/cm² of iron deposition, as seen in the Figure 8.3.5 photographs, and as with the Fe_2O_3 case, the extracted cooled electrode, following extended electrolysis and iron formation, contains trapped electrolyte. Following washing, the product weight is consistent with the eight electron per Fe_3O_4 coulombic reduction to iron.

The solid products of the solid reaction of Fe_2O_3 and Li_2CO_3 had been characterized (Collongues and Chaudron, 1950; Wijayasinghe et al., 2003). We prepare and probe the solubility of lithiated iron oxide salts in molten carbonates, and report high Fe(III) solubilities, in the order of 50% in molten carbonates, are achieved via the reaction of Li_2O with Fe_2O_3, yielding an effective method for CO_2 free iron production.

Lithium oxide, as well as Fe_2O_3 or Fe_3O_4, each have melting points above 1460°C. Li_2O dissolves in 400–1000°C molten carbonates. We find the solubility of Li_2O in molten Li_2CO_3 increases from 9 to 14 m from 750° to 950°C. Following preparation of specific iron oxide salts, we add them to molten alkali carbonate. The resultant

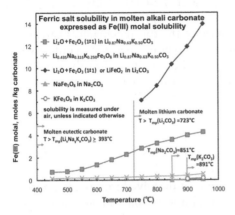
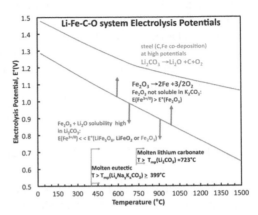

Figure 8.3.6 Left: Measured ferric oxides solubilities in alkali molten carbonates. Right: Calculated unit activity electrolysis potentials of $LiFe_5O_8$, Fe_2O_3 or Li_2CO_3. Vertical arrows indicate Nernstian shifts at high or low Fe(III). Modified with permission from Licht and Wang 2010.

Fe(III) solubility is similar when either $LiFeO_2$, or $LiFeO_2$ as $Fe_2O_3 + Li_2O$, is added to the Li_2CO_3. As seen in the left side of Figure 8.3.6, the solubility of $LiFeO_2$ is over 12 m above 900C° in Li_2CO_3.

Solid reaction of Fe_2O_3 and Na_2CO_3 produces both $NaFeO_2$ and $NaFe_5O_8$ products (Lykasov and Pavlovskaya, 2003). As seen in Figure 8.3.6, our unlike Li_2CO_3, measurements in either molten Na_2CO_3 or K_2CO_3, exhibit <<1 wt% iron oxide solubility, even at 950°C. However the solubility of $(Li_2O + Fe_2O_3)$ is high in the alkali carbonate eutectic, $Li_{0.87}Na_{0.63}K_{0.50}CO_3$, and is approximately proportional to the Li fraction in the pure Li_2CO_3 electrolyte. Solubility of this lithiated ferric oxide in the $Li_xNa_yK_zCO_3$ mixes provides an alternative molten media for iron production, which compared to pure lithium carbonate, has the disadvantage of lower conductivity, (Licht and Wang, 2010) but the advantage of even greater availability, and a wider operating temperature domain range (extending several hundred degrees lower than the pure lithium system).

Fe_2O_3 or $LiFe_5O_8$ dissolves rapidly in molten Li_2CO_3, but reacts with the molten carbonate as evident in a mass loss, which evolves one equivalent of CO_2 per Fe_2O_3, to form a steady state concentration of $LiFeO_2$ in accord with the reaction of Equation 8.3.7 (but occurring in molten carbonate) (Licht et al., 2011b). However, 1 equivalent of Li_2O and 1 equivalent of Fe_2O_3, or $LiFeO_2$, dissolves without the reactive formation of CO_2. This is significant for the electrolysis of Fe_2O_3 in molten carbonate. As $LiFeO_2$ is reduced Li_2O is released, Equation 8.3.8, facilitating the continued dissolution of Fe_2O_3 without CO_2 release or change in the electrolyte, More concisely, iron production via hematite in Li_2CO_3 is given by I and II:

I dissolution in molten carbonate: $Fe_2O_3 + Li_2O \rightarrow 2LiFeO_2$ (8.3.7)

II electrolysis, Li_2O regeneration: $2LiFeO_2 \rightarrow 2Fe + Li_2O + 3/2\,O_2$ (8.3.8)

Iron Production, Li_2O unchanged(I + II): $Fe_2O_3 \rightarrow 2Fe + 3/2\,O_2$ (8.3.9)

As indicated in Figure 8.3.4, a molar excess, of greater than 1:1 of Li_2O to Fe_2O_3 in molten Li_2CO_3, will further inhibit the Equation 8.2.1 disproportionation of lithium carbonate. The right side of Figure 8.3.6 summarizes the thermochemical calculated potentials constraining iron production in molten carbonate. Thermodynamically it is seen that at higher potential, steel (iron containing carbon) may be directly formed via the concurrent reduction of CO_2, which we observe in the Li_2CO_3 at higher electrolysis potential, as $Li_2CO_3 \rightarrow C + Li_2O + O_2$, followed by carbonate regeneration via Equation 8.2.3, to yield by electrolysis in molten carbonate:

Steel Production: $Fe_2O_3 + 2xCO_2 \rightarrow 2FeC_x + (3/2 + 2x)O_2$ (8.3.10)

From the kinetic perspective, a higher concentration of dissolved iron oxide improves mass transport, decreases the cathode overpotential and permits higher steady-state current densities of iron production, and will also substantially decrease the thermodynamic energy needed for the reduction to iron metal. In the electrolyte Fe(III) originates from dissolved ferric oxides, such as $LiFeO_2$ or $LiFe_5O_8$. The potential for the $3e^-$ reduction to iron varies in accord with the general Nerstian expression, for a concentration [Fe(III)], at activity coefficient, α:

$$E_{Fe(III/0)} = E^\circ_{Fe(III/0)} + (RT/nF)\log(\alpha_{Fe(III)}[Fe(III)])^{1/3}$$ (8.3.11)

This decrease in electrolysis potential is accentuated by high temperature and is a $\sim$0.1 V per decade 10 increase in Fe(III) concentration at 950°C. Higher activity coefficient, $\alpha_{Fe(III)} > 1$, would further decrease the thermodynamic potential to produce iron. The measured electrolysis potential is presented on the right of Figure 8.3.5 for dissolved Fe(III) in molten Li_2CO_3, and is low. For example 0.8 V sustains a current density of 500 mA cm^{-2} in 14 m Fe(III) in Li_2CO_3 at 950°C. Higher temperature, and higher concentration, lowers the electrolysis voltage, which can be considerably less than the room potential required to convert Fe_2O_3 to iron and oxygen. When an external source of heat, such as solar thermal, is available then the energy savings over room temperature iron electrolysis are considerable.

Electrolyte stability is regulated through control of the CO_2 pressure and/or by dissolution of excess Li_2O. Electrolyte mass change was measured in 7 m $LiFeO_2$ & 3.5 m Li_2O in molten Li_2CO_3 after 5 hours. Under argon there is a 1, 5 or 7 wt% loss respectively at 750°C, 850°C or 950°C), through CO_2 evolution. Little loss occurs under air (0.03% CO_2), while under pure CO_2 the electrolyte gains 2–3 wt% (external CO_2 reacts with dissolved Li_2O to form Li_2CO_3).

The endothermic nature of the new synthesis route, that is the decrease in iron electrolysis potential with increasing temperature, provides a low free energy opportunity for the STEP process. In this process, solar thermal provides heat to decrease the iron electrolysis potential, Figure 8.3.5, and solar visible generates electronic charge to drive the electrolysis. A low energy route for the carbon dioxide free formation of iron metal from iron ores is accomplished by the synergistic use of both visible and infrared sunlight. This provides high solar energy conversion efficiencies, Figure 8.2.2, when applied to Equations 8.2.14 and (8.3.6) 20 in a molten carbonate electrolyte. We again use a 37% solar energy conversion efficient concentrator photovoltaic (CPV) as

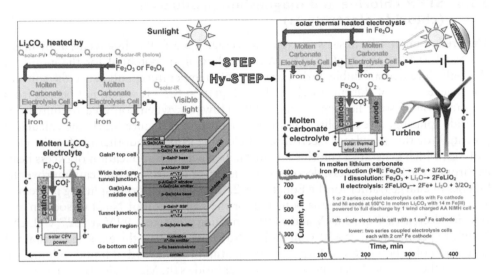

Figure 8.3.7 STEP and (wind) Hy-STEP iron. Left: STEP iron production in which two molten carbonate electrolysis in series are driven by a concentrator photovoltaic. The 2.7V maximum power of the CPV can drive either two 1.35V iron electrolyses at 800°C (schematically represented), or three 0.9V iron electrolyses at 950°C. At 0.9V, rather than at E°(25°C) = 1.28V, there is a considerably energy savings, achieved through the application of external heat, including solar thermal, to the system. Right: The Hy-STEP solar thermal/wind production of CO_2 free iron. Concentrated sunlight heats, and wind energy drives electronic transfer into the electrolysis chamber. The required wind powered electrolysis energy is diminished by the high temperature and the high solubility of iron oxide. Bottom: Iron is produced at high current density and low energy at an iron cathode and with a Ni anode in 14 m Fe_2O_3 + 14 m Li_2O dissolved in molten Li_2CO_3. Modified with permission from Licht et al. 2011b.

a convenient power source to drive the low electrolysis energy iron deposition without CO_2 formation in Li_2CO_3, (Licht, 2009) as schematically represented in Figure 8.3.7.

A solar/wind **Hybrid Solar Thermal Electrochemical Production** iron electrolysis process is also demonstrated (Licht et al., 2011b). In lieu of solar electric, electronic energy can be provided by alternative renewables, such as wind. As shown on the right side of Figure 8.3.7, in this Hy-STEP example, the electronic energy is driven by a wind turbine and concentrated sunlight is only used to provide heat to decrease the energy required for iron splitting. In this process, sunlight is concentrated to provide effective heating, but is not split into separate spectral regions as in our alternative implmentation. Hy-STEP iron production is measured with a 31.5″ × 44.5″ Fresnel lens (Edmund Optics) which concentrates sunlight to provide temperatures of over 950°C, and a Sunforce-44444 400 W wind turbine provides electronic charge, charging series nickel metal hydride, MH, cells at 1.5 V). Each MH cell, provides a constant discharge potential of 1.0–1.3 V, which are each used to drive one or two series connected iron electrolysis cells as indicated in the right side of Figure 8.3.7, containing 14 m Fe(III) molten Li_2CO_3 electrolysis cells. Electrolysis current is included in the lower right of Figure 8.3.7. Iron metal is produced. Steel (iron containing carbon) may be directly formed via the concurrent reduction of CO_2, as will be delineated in an expanded study.

8.3.4 STEP chlorine and magnesium production (chloride electrolysis)

The predominant salts in seawater (global average $3.5 \pm 0.4\%$ dissolved salt by mass) are NaCl (0.5 M) and $MgCl_2$ (0.05 M). The electrolysis potential for the industrial chlor-alkali reaction exhibits little variation with temperature, and hence the conventional generation of chlorine by electrolysis, Equation 8.2.11, would not benefit from the inclusion of solar heating (Licht, 2009). However, when confined to anhydrous chloride splitting, as exemplified in the lower portion of Figure 8.2.1, the calculated potential for the anhydrous electrolysis of chloride salts is endothermic for the electrolyses, which generate a chlorine and metal product. The application of excess heat, as through the STEP process, decreases the energy of electrolysis and can improve the kinetics of charge tranfer for the Equation 8.2.12 range of chloride splitting processes. The thermodynamic electrolysis potential for the conversion of NaCl to sodium and chlorine decreases, from 3.24 V at the 801°C melting point, to 2.99 V at 1027°C (Licht, 2009). Experimentally, at 850°C in molten NaCl, we observe the expected, sustained generation of yellow-green chlorine gas at a platinum anode and of liquid sodium (mp 98 °C) at the cathode. Electrolysis of a second chloride salt, $MgCl_2$, is also of particular interest. The magnesium, as well as the chlorine, electrolysis products are significant global commodities. Magnesium metal, the third most commonly used metal, is generally produced by the reduction of calcium magnesium carbonates by ferrosilicons at high temperature, (Li and Xie, 2005) which releases substantial levels of carbon dioxide contributing to the anthropogenic greenhouse effect. However, traditionally, magnesium has also been produced by the electrolysis of magnesium chloride, using steel cathodes and graphite anodes, and alternative materials have been invesitgated (Demirci and Karakaya, 2008).

Of significance here to the STEP process is the highly endothermic nature of anhydrous chloride electrolysis, such as for $MgCl_2$ electrolysis, in which solar heat will also decrease the energy (voltage) needed for the electrolysis. The rest potential for electrolysis of magnesium chloride decreases from 3.1 V, at room temperature, to 2.5 V at the 714°C melting point. As seen in Figure 8.3.8, the calculated thermodynamic potential for the electrolysis of magnesium chloride continues to decrease with increasing temperature, to ~2.3 V at 1000°C. The 3.1 V energy stored in the magnesium and chlorine room temperature products, when formed at 2.3 V, provide an energy savings of 35%, if sufficient heat applied to the process can sustain this lower formation potential. Figure 8.3.8 also includes the experimental decrease in the $MgCl_2$ electrolysis potential with increasing temperature in the lower right portion. In the top portion of the figure, the concurrent shift in the cyclic voltammogram is evident, decreasing the potential peak of magnesium formation, with increasing temperature from 750°C to 950°C. Sustained electrolysis and generation of chlorine at the anode and magnesium at the cathode (Figure 8.3.8, photo inset) is evident at platinum electrodes. The measured potential during constant current electrolysis at 750°C in molten $MgCl_2$ at the electrodes is included in the figure.

In the magnesium chloride electrolysis cell, nickel electrodes yield similar results to platinum, and can readily be used to form larger electrodes. The nickel anode sustains extended chlorine evolution without evident deterioration; the nickel cathode may slowly alloy with deposited magnesium. The magnesium product forms both as

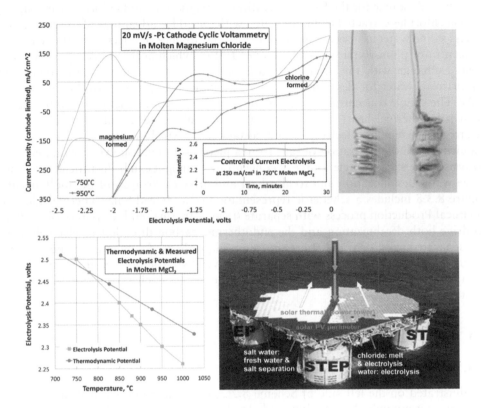

Figure 8.3.8   Photograph lower left: coiled platinum before (left), and after (right), $MgCl_2$ electrolysis forming Mg metal on the cathode (shown) and evolving chlorine gas on the anode. Main figure: cathode size restriced cyclic voltammetry of Pt electrodes in molten $MgCl_2$. Inset: The measured full cell potential during constant current electrolysis at 750°C in molten $MgCl_2$. Lower right: Thermodynamic and measured electrolysis potentials in molten $MgCl_2$ as a function of temperature. Electrolysis potentials are calculated from the thermodynamic free energies components of the reactants and products as $E = -\Delta G(reaction)/2F$. Measured electrolysis potentials are stable values on Pt at 0.250A/cm^2 cathode (Licht et al., 2011a). Lower right: A schematic representation of a separate (i) solar thermal and (ii) photovoltaic field to drive both water purification, hydrogen generation, and the endothermic electrolysis of the separated salts to useful products. Modified with permission from Licht et al. 2011a.

the solid and liquid (Mg mp 649°C). The liquid magnesium is less dense than the electrolyte, floats upwards, and eventually needs to be separated and removed to prevent an inter-electrode short, or to prevent a reaction with chlorine that is evolved at the anode. In a scaled-up cell configuration (not shown in Figure 8.3.8, a larger Ni cathode (200 cm^2 cylindrical nickel sheet (McMaster 9707K35) was employed, sandwiched between two coupled cylindrical Ni sheet anodes (total 200 cm^2, of area across from the cathode) in a 250 ml alumina (Adavalue) crucible, and sustains multi-amp large ampere currents. The potential at constant current is initially stable, but this cell configuration leads to electrical shorts, unless liquid magnesium is removed.

One salt source for the STEP generation of magnesium and chlorine from $MgCl_2$ are via chlorides extracted from salt water, with the added advantage of the generation of less saline water as a secondary product. In the absence of effective heat exchanger, concentrator photovoltaics heat up to over 100°C, which decreases cell performance. Heat exchange with the (non-illuminated side of) concentrator photovoltaics can vaporize seawater for desalinization and simultaneously prevent overheating of the CPV. The simple concentrator STEP mode (coupling super-bandgap electronic charge with solar thermal heat) is applicable when sunlight is sufficient to both generate electronic current for electrolysis and sustain the electrolysis temperature. In cases requiring both the separation of salts from aqueous solution followed by molten electrolysis of the salts, a single source of concentrated sunlight can be insufficient, to both drive water desalinization and to also heat and drive electrolysis of the molten salts. Figure 8.3.8 includes a schematic representation of a Hybrid-Solar Thermal Electrochemical Production process with separate (i) solar thermal and (ii) photovoltaic field to drive both desalinization and the endothermic carbon dioxide-free electrolysis of the separated salts, or water splitting, to useful products. As illustrated, the separate thermal and electronic sources may each be driven by insolation, or alternatively, can be (i) solar thermal and (ii) (not illustrated) wind, water, nuclear or geothermal driven electronic transfer.

8.4 STEP CONSTRAINTS

8.4.1 STEP limiting equations

As illustrated on the left side of Scheme 8.2.2, the ideal **STEP** electrolysis potential incorporates not only the enthalpy needed to heat the reactants to T_{STEP} from $T_{ambient}$, but also the heat recovered via heat exchange of the products with the inflowing reactant. In this derivation it is convenient to describe this combined heat in units of voltage via the conversion factor nF:

$$Q_T \equiv \sum_i H_i(R_i, T_{STEP}) - \sum_i H_i(R_i, T_{ambient}) - \sum_i H_i(C_i, T_{STEP}) + \sum_i H_i(C_i, T_{ambient});$$

$$E_Q(V) = -Q_T(J/mol)/nF \tag{8.4.1}$$

The energy for the process, incorporates E_T, E_Q, and the non-unit activities, via inclusion of Equation 8.4.1 into Equation 8.2.4, and is termed the **STEP** potential, E_{STEP}:

$$E_{STEP}(T, a) = [-\Delta G°(T) - Q_T - RT \cdot \ln(\prod_{i=1}^{x} a(R_i)^{ri} / \prod_{i=1}^{y} a(P_i)^{pi})]/nF;$$

$$E_{STEP}°(a = 1) = E_T° + E_Q \tag{8.4.2}$$

In a pragmatic electrolysis system, product(s) can be be drawn off at activities that are less than that of the reactant(s). This leads to large activity effects in Equation 8.4.2 at higher temperature, (Licht, 2009; Licht et al., 2010a; Licht and Wang, 2010; Licht et al., 2011b; Licht et al., 2011a; Licht, 2002; Licht, 2003; Licht et al., 2003; Licht, 2005) as the RT/nF potential slope increases with T (e.g. increasing 3-fold from 0.0592 V/n at 25°C to 0.183 V/n at 650°C).

The **STEP** factor, A_{STEP} is the extent of improvement in carrying out a solar driven electrolysis process at T_{STEP}, rather than at $T_{ambient}$. For example, when applying the same solar energy, to electronically drive the electrochemical splitting of a molecule which requires only two thirds the electrolysis potential at a higher temperature, then $A_{STEP} = (2/3)^{-1} = 1.5$. In general, the factor is given by:

$$A_{STEP} = E_{STEP}(T_{ambient}, a)/E_{STEP}(T_{STEP}, a); \quad e.g. T_{ambient} = 298K \quad (8.4.3)$$

The **STEP** solar efficiency, η_{STEP}, is constrained by both photovoltaic and electrolysis conversion efficiencies, η_{PV} and $\eta_{electrolysis}$, and the **STEP** factor. In the operational process, passage of electrolysis current requires an additional, combined (anodic and cathodic) overpotential above the thermodynamic potential; that is $V_{redox} = (1 + z)E_{redox}$, Mobility and kinetics improve at higher temperature and $\xi(T > T_{ambient}) < \xi(T_{ambient})$ (Light, 1987; Sunmacher, 2007). Hence, a lower limit of $\eta_{STEP}(V_T)$ is given by $\eta_{STEP-ideal}(E_T)$. At $T_{ambient}$, $A_{STEP} = 1$, yielding $\eta_{STEP}(T_{ambient}) = \eta_{PV} \cdot \eta_{electrolysis}$. η_{STEP} is additionally limited by entropy and black body constraints on maximum solar energy conversion efficiency. Consideration of a black body source emitted at the sun's surface temperature and collected at ambient earth temperature, limits solar conversion to 0.933 when radiative losses are considered, (Solanki and Beaucarne, 2007) which is further limited to $\eta_{PV} < \eta_{limit} = 0.868$ when the entropy limits of perfect energy conversion are included (Luque and Marti 2003). These constraints on $\eta_{STEP-ideal}$ and the maximum value of solar conversion, are imposed to yield the solar chemical conversion efficiency, η_{STEP}:

$$\eta_{STEP-ideal}(T, a) = \eta_{PV} \cdot \eta_{electrolysis} \cdot A_{STEP}(T, a)$$

$$\eta_{STEP}(T, a) \cong \eta_{PV} \cdot \eta_{electrolysis}(T_{ambient}, a) \cdot A_{STEP}(T, a); \quad (\eta_{STEP} < 0.868) \quad (8.4.4)$$

As calculated from Equation 8.2.3 and the thermochemical component data (Chase, 1998) and as presented in Figure 8.2.1, the electrochemical driving force for a variety of chemicals of widespread use by society, including aluminium, iron, magnesium and chlorine, significantly decreases with increasing temperature.

8.4.2 Predicted **STEP** efficiencies for solar splitting of CO_2

The global community is increasingly aware of the climate consequences of elevated greenhouse gases. A solution to rising carbon dioxide levels is needed, yet carbon dioxide is a highly stable, noncombustible molecule, and its thermodynamic stability makes its activation energy demanding and challenging. The most challenging stage in converting CO_2 to useful products and fuels is the initial activation of CO_2 for which energy is required. It is obvious that using traditional fossil fuels as the energy source would completely defeat the goal of mitigating greenhouse gases. A preferred route is to recycle and reuse the CO_2 and provide a useful carbon resource. We limit the non-unit activity examples of CO_2 mitigation in Equation 8.3.1 to the case when CO and O_2 are present as electrolysis products, which yields $a_{O2} = 0.5 a_{CO}$, and upon substitution into Equation 8.4.2:

$$E_{STEP}(T, a) = E^{\circ}_{STEP}(T) - (RT/2F) \cdot \ln(N);$$

$$E^{\circ}(25°C) = 1.333 \, V; \quad N = \sqrt{2} a_{CO_2} a_{CO}^{-3/2} \quad (8.4.5)$$

The example of $E_{STEP}(T, a \neq 1)$ on the left side of Figure 8.4.1 is derived when $N = 100$, and results in a substantial drop in the energy to split CO_2 due to the discussed influence of $RT/2F$. Note that at high temperature conditions in the figure, $E_{STEP} < 0$ occurs, denoting the state in which the reactants are spontaneously formed (without an applied potential). This could lead to the direct thermochemical generation of products, but imposes substantial experimental challenges. To date, analogous direct water-splitting attempts are highly inefficient due to the twin challenges of high temperature material constraints and the difficulty in product separation to prevent back reaction upon cooling (Kogan, 1998). The **STEP** process avoids this back reaction through the separation of products, which spontaneously occurs in the electrochemical, rather than chemical, generation of products at separate anode and cathode electrodes.

The differential heat required for CO_2 splitting, E_Q, and the potential at unit activity, E°_{STEP}, are calculated and presented in the top of Figure 8.4.1. E_Q has also been calculated and is included. E_Q is small (comprising tens of millivolts or less) over the entire temperature range. Hence from Equation 8.4.2, E°_{STEP} does not differ significantly from the values presented for E°_T for CO_2 in Figure 8.2.2. $E_{CO2split}(25°C)$ yields $A_{STEP}(T) = 1.333 \, V/E^\circ_{STEP}(T)$ with unit activity, and $A_{STEP}(T) = 1.197 \, V/E_{STEP}(T)$ for the $N = 100$ case. Large resultant **STEP** factors are evident in the left of Figure 8.4.1. This generates substantial values of solar to chemical energy conversion efficiency for the **STEP** CO_2 splitting to CO and O_2.

A **STEP** process operating in the $\eta_{PV} \cdot \eta_{electrolysis}$ range of 0.20 to 0.40 includes the range of contemporary 25 to 45% efficient concentrator photovoltaics, (King et al., 2007; Green et al., 2011) and electrolysis efficiency range of 80 to 90%. From these, the CO_2 solar splitting efficiencies are derived from Equations 8.4.4 and 8.4.5, and are summarized on the right side of Figure 8.4.1. The small values of $E_{STEP}(T)$ at

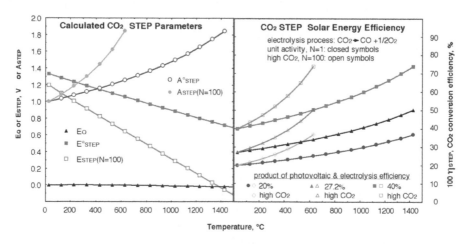

Figure 8.4.1 Top: Calculated **STEP** parameters for the solar conversion of CO_2. Bottom: Solar to chemical conversion efficiencies calculated through Equation 8.4.4 for the conversion of CO_2 to CO and O_2. In the case in which the product of the photovoltaic and electrolysis efficiency is 27.2% ($\eta_{PV} \cdot \eta_{electrolysis} = 0.272$), the **STEP** conversion efficiency at unit activity is 35%, at the 650°C temperature consistent with molten carbonate electrolysis, rising to 40% at the temperature consistent with solid oxide electrolysis (1000°C). Non-unit activity calculations presented are for the case of $\sqrt{2} \, a_{CO_2} a_{CO}^{-3/2} = 100$. A solar conversion efficiency of 50% is seen at 650°C when $N = 100$ (the case of a cell with 1 bar of CO_2 and ~58 mbar CO). Modified with permission from Licht 2009.

higher T, generate large **STEP** factors, and result in high solar to chemical energy conversion efficiencies for the splitting of CO_2 to CO and O_2. As one intermediate example from Equation 8.4.5, we take the case of an electrolysis efficiency of 80% and a 34% efficient photovoltaic ($\eta_{PV} \cdot \eta_{electrolysis} = 0.272$). This will drive **STEP** solar CO_2 splitting at molten carbonate temperatures (650°C) at a solar conversion efficiency of 35% in the unit activity case, and at 50% when N = 100 (the case of a cell with 1 bar of CO_2 and ~58 mbar CO).

8.4.3   Scaleability of STEP processes

STEP can be used to remove and convert carbon dioxide. As with water splitting, the electrolysis potential required for CO_2 splitting falls rapidly with increasing temperature (Figure 8.2.1), and we have shown here (Figure 8.2.2) that a photovoltaic, converting solar to electronic energy at 37% efficiency and 2.7 V, may be used to drive three CO_2 splitting lithium carbonate electrolysis cells, each operating at 0.9 V, and each generating a 2 electron CO product. The energy of the CO product is 1.3 V (Equation 8.2.1), even though generated by electrolysis at only 0.9V due to synergistic use of solar thermal energy. As seen in Figure 8.2.5, at lower temperature (750°C, rather than 950°C), carbon, rather than CO, is the preferred product, and this four electron reduction approaches 100% Faradaic efficiency.

The CO_2 STEP process consists of solar-driven and solar thermal assisted CO_2 electrolysis. Industrial environments provide opportunities to further enhance efficiencies; for example fossil-fueled burner exhaust provides a source of relatively concentrated, hot CO_2. The product carbon may be stored or used, and the higher temperature product carbon monoxide can be used to form a myriad of industrially relevant products including conversion to hydrocarbon fuels with hydrogen (which is generated by STEP water splitting in Section 8.3.1), such as smaller alkanes, dimethyl ether, or the Fischer Tropsch generated middle-distillate range fuels of C11–C18 hydrocarbons including synthetic jet, kerosene and diesel fuels (Andrews and Logan, 2008). Both STEP and Hy-STEP represent new solar energy conversion processes to produce energetic molecules. Individual components used in the process are rapidly maturing technologies including wind electric, (Barbier, 2010) molten carbonate fuel cells (Sunmacher, 2007), and solar thermal technologies (BrightSource, 2012; AREVA, 2012; Siemens, 2011; Solar Reserve, 2012; Amonix, 2012; Energy Innovations, 2012; Pitz-Paul, 2007).

It is of interest whether material resources are sufficient to expand the process to substantially impact (decrease) atmospheric levels of carbon dioxide. The buildup of atmospheric CO_2 levels from a 280 to 392 ppm occurring over the industrial revolution comprises an increase of 1.9×10^{16} mole (8.2×10^{11} metric tons) of CO_2,(Tans, 2009) and will take a comparable effort to remove. It would be preferable if this effort results in useable, rather than sequestered, resources. We calculate below a scaled up STEP capture process can remove and convert all excess atmospheric CO_2 to carbon.

In STEP, $6 \, kWh \, m^{-2}$ of sunlight per day, at 500 suns on $1 \, m^2$ of 38% efficient CPV, will generate 420 kAh at 2.7 V to drive three series-connected molten carbonate electrolysis cells to CO, or two series-connected molten carbonate electrolysis cells to form solid carbon. This will capture 7.8×10^3 moles of CO_2 day^{-1} to form solid carbon (based on $420 \, kAh \cdot 2$ series cells/4 Faraday mol^{-1} CO_2). The CO_2 consumed per day is three fold higher to form the carbon monoxide product (based on 3 series cells and

$2\,F\,mol^{-1}\,CO_2$) in lieu of solid carbon. The material resources to decrease atmospheric carbon dioxide concentrations with STEP carbon capture, appear to be reasonable. From the daily conversion rate of 7.8×10^3 moles of CO_2 per square meter of CPV, the capture process, scaled to $700\,km^2$ of CPV operating for 10 years can remove and convert all the increase of 1.9×10^{16} mole of atmospheric CO_2 to solid carbon. A larger current density at the electrolysis electrodes, will increase the required voltage and would increase the required area of CPVs. While the STEP product (chemicals, rather than electricity) is different than contemporary concentrated solar power (CSP) systems, components including a tracker for effective solar concentration are similar (although an electrochemical reactor replaces the mechanical turbine). A variety of CSP installations, which include molten salt heat storage, are being commercialized, and costs are decreasing. STEP provides higher solar energy conversion efficiencies than CSP, and secondary losses can be lower (for example, there are no grid-related transmission losses). Contemporary concentrators, such as based on plastic Fresnel or flat mirror technologies, are relatively inexpensive, but may become a growing fraction of cost as concentration increases (Pitz-Paal et al., 2007). A greater degree of solar concentration, for example 2000 suns, rather than 500 suns, will proportionally decrease the quantity of required CPV to $175\,km^2$, while the concentrator area will remain the same at $350{,}000\,km^2$, equivalent to 4% of the area of the Sahara desert (which averages $\sim 6\,kWh\,m^{-2}$ of sunlight per day), to remove anthropogenic carbon dioxide in ten years.

A related resource question is whether there is sufficient lithium carbonate, as an electrolyte of choice for the STEP carbon capture process, to decrease atmospheric levels of carbon dioxide. $700\,km^2$ of CPV plant will generate 5×10^{13} A of electrolysis current, and require ~ 2 million metric tonnes of lithium carbonate, as calculated from a 2 kg/l density of lithium carbonate, and assuming that improved, rather than flat, morphology electrodes will operate at $5\,A/cm^2$ $(1{,}000\,km^2)$ in a cell of 1 mm thick. Thicker, or lower current density, cells will require proportionally more lithium carbonate. Fifty, rather than ten, years to return the atmosphere to pre-industrial carbon dioxide levels will require proportionally less lithium carbonate. These values are viable within the current production of lithium carbonate. Lithium carbonate availability as a global resource has been under recent scrutiny to meet the growing lithium battery market. It has been estimated that the current global annual production of 0.13 million tonnes of LCE (lithium carbonate equivalents) will increase to 0.24 million tonnes by 2015 (Tahil, 2008). Potassium carbonate is substantially more available, but as noted in the main portion of the paper can require higher carbon capture electrolysis potentials than lithium carbonate. An additional modified barium carbonate STEP electrolyte has been introduced (Licht et al., 2013), and STEP mechanisms continue to be probed (Cui and Licht, 2013), and the portfolio of new STEP processes and products, such as STEP Cement (Licht, 2012) and STEP Water Treatment (Wang et al., 2012; Wang et al., 2103), continues to expand.

8.5 CONCLUSIONS

To ameliorate the consequences of rising atmospheric carbon dioxide levels and its effect on global climate change, there is a drive to replace conventional fossil fuel driven electrical production by renewable energy driven electrical production. In addition to the replacement of the fossil fuel economy by a renewable electrical economy, we suggest that a renewable chemical economy is also warranted. Solar energy can be efficiently used, as demonstrated with the STEP process, to directly and efficiently form the chemicals needed by society without carbon dioxide emissions. Iron, a basic commodity, currently accounts for the release of one quarter of worldwide CO_2 emissions by industry, which may be eliminated by replacement with the STEP iron process. The unexpected solubility of iron oxides in lithium carbonate electrolytes, coupled with facile charge transfer and a sharp decrease in iron electrolysis potentials with increasing temperature, provides a new route for iron production. Iron is formed without an extensive release of CO_2 in a process compatible with the predominant naturally occurring iron oxide ores, hematite, Fe_2O_3, and magnetite, Fe_3O_4. STEP can also be used in direct carbon capture, and the efficient solar generation of hydrogen and other fuels.

In addition to the removal of CO_2, the STEP process is shown to be consistent with the efficient solar generation of a variety of metals, as well as chlorine via endothermic electrolyses. Commodity production and fuel consumption processes are responsible for the majority of anthropogenic CO_2 release, and their replacement by STEP processes provide a path to end the root cause of anthropogenic global warming, as a transition beyond the fossil fuel, electrical or hydrogen economy, to a renewable chemical economy based on the direct formulation of the materials needed by society. An expanded understanding of electrocatalysis and materials will advance the efficient electrolysis of STEP's growing porfolio of energetic products.

REFERENCES

Abanades, S. and Chambon, M. (2010) CO_2 Dissociation and upgrading from two-step solar thermochemical processes based on ZnO/Zn and SnO_2/SnO redox pairs. *Energy Fuels*, 24, 6677–6674.

Andrews, A. and Logan, J. (2008) Fischer-Tropsch fuels from coal, natural gas, and biomass: background and policy. *Congressional Research Service Report for Congress*. RL34133, (March 27, 2008); available at: http://assets.opencrs.com/rpts/RL34133_20080327.pdf.

Amonix (2012) at: http://www. amonix.com/

Andrieux, L. and Weiss, G. (1944) Productions of electrolysis of molten salts with an iron anode. *Comptes Rendu* 217 615.

AREVA (2012) at: http://www.areva.com/EN/solar-220/areva-solar.html

Barber, J. (2009) Photosynthetic energy conversion: natural and artificial. *Chemical Society Reviews*, 38, 185–196.

Barbier, E. (2010) How is the global green new deal going?. *Nature*, 464, 832–833.

Barton, E.E., Rampulla, D.M. and Bocarsly, A.B. (2008) Selective solar-driven reduction of CO_2 to methanol using a catalyzed p-GaP based photoelectrochemical cell. *Journal of the American Chemical Society*, 130, 6342–6344.

Benson, E., Kubiak, C.P., Sathrum, A.J. and Smieja, J.M. (2009) Electrocatalytic and homogeneous approaches to conversion of CO_2 to liquid fuels. *Chemical Society Reviews*, 38, 89–99.

Bockris, J. O'M. (1980) *Energy Options*. Halsted Press New York.

BrightSource (2012) at: http://brightsourceenergy.com

Chandler, H., Pollara, F., Elikan, L., Archer, D. and Zahradnik, R. (1966) Aerospace life support. *AICHE Chemical Engineering Progress Series*, 62, 38–42.

Chase, M.W. (1998) NIST-JANAF Thermochemical Tables. 4th Edition. *Journal of Physical and Chemical Reference Data*, 9, 1; data available at: http://webbook.nist.gov/chemistry/form-ser.html

Chu, D., Qin, G., Yuan, X., Xu, M., Zheng, P. and Lu, J. (2008) Fixation of CO_2 by electro-catalytic reduction and electropolymerization in ionic liquid-H_2O solution. *Chemistry and Sustainability*, 1, 205–209.

Chueh, W. and Haile, S. (2010) A Thermochemical study of ceria: Exploiting and old materials for new modes of energy conversion and CO_2 mitigation. *Philosophical Transactions of the Royal Society A*, 368, 3269–3294.

Collongues, R. and Chaudron, G. (1950) Sur la preparation des ferrites de lithium. *Comptes Rendus*, 124, 143–145.

Cui, B. and Licht, S. (2013) Critical STEP advances for sustainable iron production *Green Chemistry* 15(4), 881–884, with 16 page online supplementary information.

Currao, A. (2007) Photoelectrochemical water splitting. *Chimia*, 61, 815–819.

deBethune, A.J. and Licht, S. (1959) The temperature coefficients of electrode potentials. *Journal of the Electrochemical Society*, 106, 616–625.

Delacourt, C. and Newman, J. (2010) Mathematical modeling of CO_2 reduction to CO in aqueous electrolytes. *Journal of the Electrochemical Society*, 157, B1911–B1926.

Demirci, G. and Karakaya, I. (2008) Electrolytic magnesium production and its hydrodynamics by using an Mg–Pb alloy cathode. *Journal of Alloys and Compounds*, 465, 255–260.

Dufek, E., Lister, T. and McIlwain, M. (2011) Bench-scale electrochemical system for generation of CO and syn-gas. *Journa. Applied Electrochemestry*, 41, 623–631.

Ebbesen, S., Graves, C., Hausch, A., Jensen, S.H. and Mogensen, M. (2010) Poisoning of solid oxide electrolysis cells by impurities. *Journal of the Electrochemical Society*, 157, B1419–B1429.

Elschenbroich, C. and Salzer, A. (1992) *Organometallics*. 2nd Ed., Wiley-VCH, Weinheim, Germany.

Energy Innovations (2012) at: http://www.energyinnovations.com/

Erstfeld, T. (1979) Carbon dioxide electrolysis using a ceramic electrolyte for space processing. AIAA Paper 79-1375, 1.

Eustis, F. A. (1922) Electrolytic iron from sulfide ores. *Chemical and Metallurgical Engineering*, 27, 684.

Fu, Q., Mabilat, C., Zahid, M., Brisse, A. and Gautier, L. (2010) Syngas production via high-temperature steam/CO_2 co-electrolysis: an economic assessment. *Energy and Environmental Science*, 3, 1382–1397.

Fujishima, A. and Honda, K. (1972) Electrochemical photolysis of water at a semiconductor electrode. *Nature*, 238, 37–38.

Gangeri, M., Perathoner, S., Caudo, S., Centi, G., Amadou, J., Begin, D., Pham-Huu, C., Ledoux, M.J., Tessonnier, J.P., Su, D.S. and Schologl, R. (2009) Fe and Pt carbon nanotubes for the electrocatalytic conversion of carbon dioxide to oxygenates. *Catalysis Today*, 143, 57–63.

Graves, C., Ebbsen, S., Mogensen, M. and Lackner, K.S. (2011) Sustainable hydrocarbon fuels by recycling CO_2 and H_2O with renewable or nuclear energy. *Renewable and Sustainable Energy Reviews*, 15, 1–23.

Green, M., Emery, K., Hishikawa, Y. and Warta, W. (2011) Solar cell efficiency tables. *Progress in Photovoltaics: Research and Applications*, 19, 84–92.

Green, R. Liu, C. and Adler, S. (2008) Carbon dioxide reduction on gadolinia-doped ceria cathodes. *Solid State Ionics*, 179, 647–660.

Haarberg, G.M., Kvalheim, E., Rolseth, S., Murakami, T., Pietrzykd, S. and Wange, S. (2007) Electrodeposition of iron from molten mixed chloride/fluoride electrolytes. *ECS Transactions* 3, 341–345.

Hori, Y., Konishi, H., Futamura, T., Murata, A., Koga, O., Sakurai, H. and Oguma, H. (2005) Deactivation of copper electrode in electrochemical reduction of CO_2. *Electrochimica Acta*, 50, 5354–5369.

Innocent, B., Liaigre, D., Pasquier, D., Ropital, F., Leger, J.M. and Kokoh, K.B. (2009) Electro-reduction of carbon dioxide to formate on lead electrode in aqueous medium. *Journal of Applied Electrochemistry*, 39, 227–232.

Jensen, S., Sun, X., Ebbesen, S., Knibbe, R. and Mogensen, M. (2010) Hydrogen and synthetic fuel production using pressurized solid oxide electrolysis cells. *International Journal of Hydrogen Energy*, 35, 9544–9549.

Kaneco, S., Ueno, Y., Katsumata, H, Suzuki, T. and Ohta, K. (2009) Photoelectrochemical reduction of CO_2 at p-InP electrode in copper particle-suspended methanol. *Chemical Engineering Journal*, 148, 57–62.

Kaplan, V., Wachtel, E., Gartsman, K., Feldman, Y. and Lubomirsky, I. (2010) Conversion of CO_2 to CO by electrolysis of molten lithium carbonate. *Journal of The Electrochemical Society*, 157, B552–B556.

Kim-Lohsoontorn, P., Laosiripojana, N. and Bae, J. (2011) Electrochemical performance of solid oxide electrolysis cell electrodes under high-temperature coelectrolysis of steam and carbon dioxide. *Current Applied Physics*, 11, S223–S228.

King, R.R., Law, D.C., Edmonson, K.M., Fetzer, C.M., Kinsey, G.S., Yoon, H., Sherif, R.A. and Karam N.H. (2007) 40% efficient metamorphic GaInP/GaInAs/Ge multijunction solar cells. *Applied Physics Letters*, 90, 183516.

Kogan, A. (1998) Direct solar thermal splitting of water and on-site separation of the products—II. Experimental feasibility study. *International Journal of Hydrogen Energy*, 23, 89–98.

Kojima, T., Miyazaki, Y., Nomura, K. and Tanimoto, K. (2008) Density, surface tension, and electrical conductivity of ternary molten carbonate system Li_2CO_3–Na_2CO_3–K_2CO_3 and methods for their estimation. *Journal of The Electrochemical Society*, 155, F150–F156.

Li, G.M., Wang, D.H. and Chen, Z. (2009) Direct reduction of solid Fe2O3 in molten CaCl2 by potentially green process. *Journal of Materials Science and Technology* 25, 767–771.

Li, H.Q. and Xie, S.S. (2005) Research on development of dolomite-ferrosilicon thermal reduction Process of magnesium production. *Journal of Rare Earths*, 23, 606–610.

Licht, S. (1985) pH measurement in concentrated alkaline solutions. *Analytical Chemistry*, 57, 514–519.

Licht, S. (1987) A description of energy conversion in photoelectrochemical solar cells. *Nature*, 330, 148–151.

Licht, S., Hodes, G., Tenne, R. and Manassen, J. (1987) A light variation insensitive high efficiency solar cell. *Nature*, 326, 863–864.

Light, T.S. and Licht, S. (1987) Conductivity and resistivity of water from the melting through critical points. *Analytical Chemistry* 59, 2327–2330.

Licht, S. and Peramunage, D. (1990) Efficient photoelectrochemical solar cells from electrolyte modification. *Nature*, 345, 330–330.

Licht, S., Longo, K., Peramunage, D. and Forouzan, F. (1991) Conductometric analysis of the second acid dissociation constant of H2S in highly concentrated aqueous media *Journal of Electroanalytical Chemistry and Interfacial Electrochemistry*, 318, 119–129.

Licht, S. (1986) Combined solution effects yield stable thin-film cadmium selenide telluride/ polysulfide photoelectrochemical solar cells. *Journal of Physical Chemistry*, 90, 1096–1099.

Licht, S., Wang, B., Soga, T. and Umeno, M. (1999) Light invariant, efficient, multiple bandgap AlGaAs/Si/metal hydride solar cell. *Applied Physics Letter*, 74, 4055.

Licht, S., Wang, B., Mukerji, S., Soga, T., Umeno, M. and Tributsch, H. (2000) Efficient solar water splitting; Exemplified by RuO_2 catalyzed AlGaAs/Si photohydrolysis. *Journal of Physical Chemistry*, 104, 8920–8924.

Licht, S. (2001) Multiple bandgap semiconductor/electrolyte solar energy conversion. *Journal of Physical Chemistry B*, 105, 6281–6294.

Licht, S. (2002) Efficient solar generation of hydrogen fuel – a fundamental analysis. *Electrochemistry Communications*, 4, 790–795.

Licht, S. Ed. (2002) *Semiconductor Electrodes and Photoelectrochemistry*. Weinheim: Wiley-VCH.

Licht, S. (2003) Solar water splitting to generate hydrogen fuel: photothermal electrochemical analysis. *Journal of Physical Chemistry B*, 107, 4253–4260.

Licht, S., Halperin, L., Kalina, M., Zidman, M. and Halperin, N. (2003) Electrochemical potential tuned solar water splitting. *Chemical Communication*, 2003, 3006–3007.

Licht, S. (2005) Thermochemical solar hydrogen generation. *Chemical Communication*, 2005, 4635–4646

Light, T.S., Licht, S., Bevilacqua, A.C. and Morash, K.R. (2005) The fundamental conductivity and resistivity of water. *Electrochemal and Solid-State Letters* 8, E16–E19.

Licht, S. (2009) STEP (solar thermal electrochemical photo) generation of energetic molecules: A solar chemical process to end anthropogenic global warming. *Journal of Physical Chemistry C*, 113, 16283–16292.

Licht, S. and Wang, B. (2010) High solubility pathway to the carbon dioxide free production of iron. *Chemical Communications*, 46, 7004–7006.

Licht, S., Wang B., Ghosh, S., Ayub, H., Jiang, D. and Ganley, J. (2010a) A new solar carbon capture process: solar thermal electrochemical photo (STEP) carbon capture. *Journal of Physical Chemistry Letters*, 1, 2363–2368.

Licht, S., Chityat, O., Bergmann, H., Dick, A., Ayub, H. and Ghosh, S. (2010b) Efficient STEP (Solar thermal electrochemical photo) production of hydrogen – an economic assessment. *International Journal of Hydrogen Energy*, 35, 10867–10882.

Licht, S. (2011) Efficient solar-driven synthesis, carbon capture, and desalinization, STEP: solar thermal electrochemical production of fuels, metals, bleach. *Advanced Materials*, 20, 1–21.

Licht, S. and Wu, H. (2011) STEP – STEP iron, a chemistry of iron formation without CO_2 emission: Molten carbonate solubility and electrochemistry of iron ore impurities. *Journal of Physical Chemistry C*, 115, 25138–25147.

Licht, S., Wang, B. and Wu, H. (2011a) STEP – A solar chemical process to end anthropogenic global warming II: experimental results. *Journal of Physical Chemistry C*, 115, 11803–11821.

Licht, S., Wu, H., Zhang, Z. and Ayub, H. (2011b) Chemical mechanism of the high solubility pathway for the carbon dioxide free production of iron. *Chemical Communications*, 47, 3081–3083.

Licht, S., Wu, H., Hettige, C., Wang, B., Lau, J., Asercion, J., Stuart, J. (2012) STEP Cement: Solar Thermal Electrochemical Production of CaO without CO2 emission. *Chemical Communications* 48, 6019–6021, with online 20 page supplement.

Licht, S. Cui, B., Wang, B. (2013) STEP Carbon Capture: the barium advantage *Journal of CO_2 Utilization* 2, 58–63, with 12 page online supplementary information.

Lueck, D., Buttner, W. and Surma, J. (2002) Space habitat carbon dioxide electrolysis to oxygen. *Fluid System Technologies*, at: http://rtreport.ksc.nasa.gov/techreports/2002report/600%20Fluid%20Systems/609.html.

Luque, A. and Marti, A. (2003) *Handbook of Photovoltaic Sci. & Eng.*, Eds. A. Luque, S. Haegedus, S. Weinheim:Wiley-VCH, 113.

Lykasov, A. and Pavlovskaya, M. (2003) Phase equilibria in the Fe–Na–O system between 1100 and 1300 K. *Inorganic Materials*, 39, 1088–1091.

Martin, N. (1965) Carbon dioxide reduction systems. *AIAA Fourth Manned Space Flight Meeting*, October 11–13, 1965, St. Louis, Missouri, USA.

Miller, J., Allendorf, M., Diver, R., Evans, L.R., Siegel, N.P. and Stuecker. J.N. (2008) Metal oxide composites and structures for ultra-high temperature solar thermochemical cycles. *Journal of Materials Science*, 43, 4714–4728.

Mizusaki, J., Tagawa, H., Miyaki, Y., Yamauchi, S., Koshiro, I. and Hirano, K. (1992) Kinetics of the electrode reaction at the $CO\text{-}CO_2$, porous Pt/stabilized zirconia interface. *Solid State Ionics*, 53–56, 126–134.

Mostad, E., Rolseth, S. and Thonstad, S. (2008) Electrowinning of Iron from Sulphate Solutions. *Hydrometallurgy*, 90, 213–220.

Moyer, C., Sullivan, N., Zhu, H. and Kee, R.J. (2011) Polarization characteristics and chemistry in reversible tubular solid-oxide cells operating on mixtures of H_2, CO, H_2O, and CO_2. *Journal of The Electrochemical Society*, 158, B117–B131.

Murphy, A.B. (2008) Does carbon doping of TiO_2 allow water splitting in visible light? Comments on "Nanotube enhanced photoresponse of carbon modified (CM)-n-TiO_2 for efficient water splitting". *Solar Energy Materials and Solar Cells*, 92, 363–367.

Narayanan, S.R., Haines, B., Soler, J. and Valdez, T.I. (2011) Electrochemical conversion of carbon dioxide to formate in alkaline polymer electrolyte membrane cells. *Journal of The Electrochemical Society*, 159, F353–F359.

Ohla, O. Surya, P., Licht, S. and Jackson, N. (2009) *Reversing global warming: chemical recycling and utilization of CO_2*. Report of the National Science Foundation sponsored 7-2008 Workshop, 17 pages; full report available at: http://www.usc.edu/dept/chemistry/loker/ReversingGlobalWarming.pdf

Ogura, K., Yano, H. and Tanaka, T. (2004) Selective formation of ethylene from CO2 by catalytic electrolysis at a three-phase interface. *Catalysis Today*, 98, 515–521.

Oregan, B. and Gratzel, M. (1991) A low-cost, high-efficiency solar cell based on dye-sensitized colloidal TiO2. *Nature* 353, 737–740.

Palmaer, W. and Brinell, J. A., (1913) Iron sheets and tubes. *Metallurgical and Chemical Engineering*, 11, 197–203.

Pan, P. and Chen, Y. (2007) Photocatalytic reduction of carbon dioxide on NiO/InTaO$_4$ under visible light irradiation. *Catalysis Communications*, 8, 1546–1549.

Pellegrino, J.L. (2000) *Energy & Environmental Profile of the U.S. Chemical Industry*. available online at: http://www1.eere.energy.gov/industry/chemicals/tools_profile.html.

Pitz-Paal, R., (2007) High temperature solar concentrators. *Solar Energy Conversion and Photoenergy Systems*, Eds. Galvez, J. B.; Rodriguez, S. M. Oxford: EOLSS Publishers

Qingeng, L., Borum, F., Petrushina, I. and , Bjerrum, N.J. (1999) Complex formation upon dissolution of metal oxides in molten alkali carbonates. *Journal of The Electrochemical Society*, 146, 2449–2454.

Rajeshwar, K., McConnell, R. and Licht, S. Eds. (2008) *The Solar Generation of Hydrogen: Towards a Renewable Energy Future*. Springer, New York, USA.

Richardson, R., Holland, E. and Carpenter, B. (2011) A renewable amine for photochemical reduction of CO_2. *Nature Chemistry*, 3, 301–303.

Richter, R. (1981) Basic investigation into the production of oxygen in a solid electrolyte process. *American Institute of Aeronautics and Astronautics, 16th Thermophysics Conference*, June 23–25, 1981, Palo Alto, California, USA.

Siemens (2011) at: http://www.siemens.com/press/pool/de/pressemitteilungen/2011/renewable_energy/ERE201102037e.pdf.

Solanki, C.S. and Beaucarne, G. (2007) Advanced solar cell concepts. *Energy for Sustainable Developent*, 11, 17–23.

SolarReserve (2012) at: http://www.solarreserve.com/.

Stamatiou, A., Loutzenhiser, P.G. and Steinfeld, A. (2010) Solar syngas production from H_2O and CO_2 via two-Step thermochemical cycles based on Zn/ZnO and FeO/Fe$_3$O$_4$ redox reactions: kinetic analysis. *Energy Fuels*, 24, 2716–2722.

Stancati, M., Niehoff, J., Wells, W. and Ash, R.L. (1979) In situ propellant production – A new potential for round-trip spacecraft. *AIAA Conference on Advanced Technology for Future Space Systems*, May 1979, NASA Langley Research Center AIAA Paper No. 79-0906.

Stoots, C.M., O'Brien, J.E., Condie, K.G. and Hartvigsen, J.J. (2010) High-temperature electrolysis for large-scale hydrogen production from nuclear energy – Experimental investigations. *International Journal of Hydrogen Energy*, 35, 4861–4870.

Sunmacher, K. (2007) *Molten carbonate fuel cells*. Weinheim: Wiley-VCH, Germany.

Tahil, W. (2008) *The trouble with lithium 2; Under the microscope*, Martainsville, France: Meridan International Research. at: http://www.meridian-int-res.com/Projects/Lithium_Microscope.pdf.

Tans, P. (2009) An accounting of the observed increase in oceanic and atmospheric CO_2 and an outlook for the future. *Oceanography*, 22, 26–35.

Tao, G., Sridhar, K. and Chan, C. (2004) Study of carbon dioxide electrolysis at electrode/electrolyte interface: Part I. Pt/YSZ interface. *Solid State Ionics*, 175, 615–619.

Tao, G., Sridhar, K. and Chan, C. (2004) Study of carbon dioxide electrolysis at electrode/electrolyte interface: Part II. Pt-YSZ cermet/YSZ interface. *Solid State Ionics*, 175, 621–624.

Vayssieres, L., Ed. (2009) *On Solar Hydrogen & Nanotechnology*. Weinheim: John Wiley and Sons, USA.

Venstrom, L.J. and Davidson, J.H. (2011) Splitting water and carbon dioxide via the heterogeneous oxidation of zinc vapor: Thermodynamic considerations. *Journal of Solar Energy Engineering, 133*, 011017.

Wang, A., Liu, W., Cheng, S., Xing, D., Zhou, J. and Logan, B.E. (2009) Source of methane and methods to control its formation in single chamber microbial electrolysis cells. *International Journal of Hydrogen Energy*, 34, 3653–3658.

Wang, S., Haarberg, G.M. and Kvalheim, E. (2008) Electrochemical behavior of dissolved Fe_2O_3 in molten CaCl2-KF. *Journal of Iron and Steel Research, International*, 15, 48–51.

Wang, B., Wu, H., Zhang, G., Licht, S. (2012) STEP wastewater treatment: Solar Thermal Electrochemical Pollutant Oxidation. *ChemSusChem*. 5, 2000–2010.

Wang, B., Wu, H., Hu, Y., Licht, S. (2013) STEP Pollutant to Solar Hydrogen: Solar driven thermal electrochemical wastewater treatment with synergetic production of hydrogen *Electrochemical Science Letters* 2, H34-H36.

Wentworth, W. and Chen, E. (1976) Simple thermal-decomposition reactions for storage of solar thermal energy. *Solar Energy*, 18, 205–214.

Woolerton, Y., Sheard, S., Reisner, E., Pierce, E., Ragsdale, S.W. and Armstrong F.A. (2010) Efficient and clean photoreduction of CO_2 to CO by enzyme-modified TiO_2 Nanoparticles using visible light. *Journal of the American Chemical Society*, 132, 2132–2133.

Wijayasinghe, A., Bergman, B., and Lagergren, C. (2003) LiFeO2-LiCoO2-NiO cathodes for molten carbonate fuel cells. Journal of the *Electrochemical Society*, 150, A558–A564.

Yan, S., Wan, L., Li, Z. and Zou, Z. (2011) Facile temperature-controlled synthesis of hexagonal Zn2GeO4 nanorods with different aspect ratios toward improved photocatalytic activity for overall water splitting and photoreduction of CO_2. *Chemical Communications*, 47, 5632–5634.

Yano, J., Morita, T., Shimano, K., Nagami, Y. and Yamasaki, S. (2007) Selective ethylene formation by pulse-mode electrochemical reduction of carbon dioxide using copper and copper-oxide electrodes. *Journal of Solid State Electrochemistry*, 11, 554–557.

Yuan, B.Y., Kongstein, O.E. and Haarberg, G.M. (2009) Electrowinning of iron in aqueous alkaline solution using a rotating cathode. *Journal of The Electrochemical Society, 156*, D64–D69.

Zhang, Z. and Wang, Z., Eds. (2006) *Principles and applications of molten salt electrochemistry*. Chemical Industry Press, Beijing, China.

Zhou, H., Fan, T. and Zhang, D. (2011) An insight into artificial leaves for sustainable energy inspired by natural photosynthesis. *ChemCatChem*, 3, 513–528.

Zou, Z., Ye, Y., Sayama, K. and Arakawa, H. (2001) Direct splitting of water under visible light irradiation with an oxide semiconductor photocatalyst. *Nature*, 414, 625–627.

Chapter 9

Solar hydrogen production and CO$_2$ recycling

Zhaolin Wang[1] *& Greg F. Naterer*[2]

[1]*Faculty of Engineering and Applied Science, University of Ontario Institute of Technology, Oshawa, Ontario, Canada*
[2]*Faculty of Engineering and Applied Science, Memorial University of Newfoundland, St. John's, Newfoundland and Labrador, Canada*

9.1 SUSTAINABLE FUELS WITH SOLAR-BASED HYROGEN PRODUCTION AND CARBON DIOXIDE RECYCLING

The need for energy will continue to increase rapidly as the world aims to improve its living quality. As the usable fossil fuel resources are diminishing due to the ongoing depletion, the scenario of using carbon-based fuels is unsustainable. It was estimated by the International Energy Agency (IEA) and the US Energy Information Administration (EIA) that the global primary energy demand will be still met primarily by fossil fuels in the medium term before 2035 (IEA, 2011, 2012; EIA, 2012). In addition, the usage of fossil fuels generates toxic pollutants threatening the ecobalance and our health. The CO$_2$ emissions induce undesirable climate effects. Therefore, clean energy alternatives and carbon recycling are needed for solving the future energy sustainability problems.

Among several clean energy substitutes such as nuclear, hydroelectric, geothermal, and solar, nuclear fission energy has been well proven as a technology widely used in the world. However, nuclear waste disposal challenges and unpredictable accidents such as Three Mile Island, Chernobyl, and Fukushima are often cited by the public (Bodansky 2001; Grady, 2011; Makhijani, 2009; Pearce, 2008). As to hydroelectric power, it is limited by the availability of waterways. River dams may also cause unpredictable influences on the aquatic ecosystems, fisheries, and river transport. Regarding deep underground geothermal electricity, the usable energy is often located kilometres below the surface and there are concerns about seismic impacts. In comparison, solar energy is a safe, clean and unlimited resource (Roeb et al., 2010). However, the availability of sunlight on the earth's surface is intermittent because it is not available at night and on rainy and cloudy days. The energy distributed on various areas may differ significantly. Therefore, if the solar energy captured at daytime or regions rich in sunlight can be stored, then the intermittency issue would be resolved. The usage of solar energy for hydrogen production is a good option for solar energy storage, because hydrogen has a much higher mass energy density than most current fossil fuels (Envestra, 2010; EVWorld, 2010; ForestBioEnergy, 2010). The only product of hydrogen combustion is water vapour, which can be recycled to produce hydrogen with solar energy. It was reported that the efficiency of a hydrogen internal combustion engine could be 10–40% higher than a gasoline engine. The hybrid electric motor and fuel cell vehicle could even be 2 to 3 times more efficient than an internal gasoline combustion engine (Berger,

2006). Therefore, solar-based hydrogen production with water is widely viewed as a very promising option for a sustainable future.

Even in the conventional fossil fuel industry, hydrogen has a major role in the upgrading of petroleum products. Also, hydrogen is a necessity for the production of fertilizers in the agricultural industry. Currently, oil upgrading and fertilizer production account for about 50% and 40% of the hydrogen consumption, respectively (Dalcor, 2005; Freedonia, 2010; Kramer, 2005). The rising need of hydrogen by modern agriculture and petroleum products will strongly advance the hydrogen economy (Forsberg, 2002; Naterer et al., 2008). However, the major hydrogen production methods of today are generally not "clean" because more than 95% of the global hydrogen is produced from fossil fuels, i.e., 48% from steam methane reforming (SMR), 30% from refinery/chemical off-gases, and 18% from coal gasification (IEA, 2010; NYSERDA, 2010). Water electrolysis accounts for less than 4%. Even this 4% is not fully clean because the electricity used for hydrogen production is not fully generated from clean fuels. The usage of fossil fuels to produce hydrogen generates large amounts of greenhouse gases. Therefore, the future hydrogen production pathway from solar-based water splitting is a promising solution.

Another option for reducing the depletion rate of our planet's fossil fuel reserves is to recycle the CO_2 emissions with solar energy. This may improve the renewability of limited fossil fuels and at the same time store the solar energy and minimize the impact of intermittency of the sunlight. In addition to the renewable benefit, the CO_2 recycling is also a safer measure compared with the geological sequestration of CO_2 into deep oceans or geological formations, because the sequestrated CO_2 has many unpredictable risks such as leakage of CO_2 back to the atmosphere and the change of ocean water properties (Yang, 2011; Little et al., 2010; Spicer, 2007). Even though care is taken to identify the appropriate geological areas for the storage of CO_2, there is always a likelihood of leakage due to different reasons such as an earthquake. The liquefied CO_2 on the basin of the ocean floor may have a much higher CO_2 concentration than normal levels. This makes it difficult for some ocean organisms to survive near the ocean basin and as a result, the whole ecosystem is disturbed. As for deep ground sequestration, the leaked CO_2 may mix with groundwater and consequently make the water toxic and unsuitable for human consumption. Although deep saline reservoirs potentially have a large capacity to store CO_2, high pressure CO_2 can significantly acidify the fluids in the reservoir and dissolve minerals such as calcium carbonate. As a consequence, the permeability is increased which could allow CO_2-rich fluids to escape the reservoir along new pathways and contaminate aquifers used for drinking water (Kharaka et al., 2006).

Since CO_2 alone is not a fuel and most fuels and organic compounds comprise hydrogen, so hydrogen is a necessity in the conversion of CO_2 into other useful fuels and organic compounds such as syngas, methanol, and dimethyl ether. However, nowadays more than 96% of the world's hydrogen is produced from fossil fuels through processes such as steam-methane reforming (SMR) and coal gasification with steam (IEA, 2012). As discussed previously, these hydrogen production processes are depleting the fossil fuel reserves and emitting large amounts of pollutants and greenhouse gases. Hence, to find renewable and low-cost methods of producing hydrogen in large capacities is also critical to the CO_2 recycling.

Engineers and scientists have been developing numerous methods for clean hydrogen production and CO_2 recycling. This chapter will examine the energy efficiencies, requirements, and thresholds of different technologies. The major water splitting and CO_2 recycling technologies will be compared and categorized according to the reaction mechanisms and engineering approaches. The engineering approaches would be the focus from the perspectives of apparatus, components, materials, equipment and layout of processes.

9.2   SOLAR-BASED HYDROGEN PRODUCTION WITH WATER SPLITTING METHODS

This section examines the technologies of utilizing solar energy for hydrogen production with water splitting, which are different from conventional fossil fuel-based methods. The conventional methods such as modified steam methane reforming and coal gasification suitable for the usage of solar energy, are still fossil fuel-based although the green house gas emissions are significantly reduced due to the usage of solar thermal energy to replace the combustion of an extra amount of methane or coal for the supply of reaction heat. Therefore, the solar-assisted fossil fuel-based conventional methods for hydrogen production won't be discussed in this chapter. Since the only feedstock to a water splitting cycle is water and the products are hydrogen and oxygen with no greenhouse gas emissions, so the hydrogen production with water splitting is the focus of this chapter.

9.2.1   Solar-to-hydrogen efficiency of water splitting processes

In water splitting methods, the only feedstock to the hydrogen production cycle is water, and the only products are hydrogen and oxygen. Since the combustion of hydrogen produces water again, the solar-based water splitting methods are fully renewable compared with the prior modified conventional fossil fuel-based methods for the usage of solar energy. This section will focus on these unconventional methods. Engineers and scientists have been developing numerous methods for the clean hydrogen production from water splitting with solar energy, such as thermolysis, thermochemical, water electrolysis, photoeletrolysis, photoeletrochemical, photochemical, photodissociation, photodecomposition, photolysis, photodegradation, photocatalytic, photobiological, and hybrid methods. This section aims to introduce and compare the reaction mechanisms and the major water splitting technologies and engineering approaches.

Different water splitting mechanisms and engineering approaches may indicate various hydrogen production efficiencies. Considering the variety of complex efficiency related factors, the following efficiency definition is adopted for a more consistent comparison:

$$\eta = \frac{m_P \cdot \Delta H^0_{Liq,298}}{I_S} \tag{9.2.1}$$

where η is the solar-to-hydrogen production efficiency, I_S is the total incident solar irradiance of the whole solar spectrum on the basis of the sunlight receiving area of the device for the reacting system, and its units are $J \cdot s^{-1}$. Also, m_P in Equation (9.2.1) is the hydrogen production rate in units of $mole \cdot s^{-1}$, and ΔH^{θ}_{Liq} is the standard enthalpy change of the following water splitting reaction at 298K (25°C) in units of $J \cdot mole^{-1}$ with a value of 285,800 J/mol:

$$H_2O \text{ (liquid)} = H_2 \text{ (gas)} + \tfrac{1}{2}O_2 \text{ (gas)} \quad \Delta H^0_{298} = 285,800 \, J/mol \tag{9.2.2}$$

If 25°C is also used as the reference temperature for the higher heating value (HHV), then the HHV is equal to the enthalpy change ΔH^{θ}_{Liq}. The product of m_P and ΔH_{Liq} indicates the theoretical minimum energy needed to split water into hydrogen and oxygen, or the maximum energy that can be recovered from hydrogen when hydrogen is used as a fuel. The reason to use liquid water at 25°C rather than gaseous water is because the starting state of the water used in industry for hydrogen production is mostly in the form of liquid at an ambient temperature, although in the hydrogen production reactor it could be in other forms and the temperature could be slightly different. Also, since the efficiency is discussed from the perspectives of production rather than usage, the actual energy input to the production may attract more engineering interest. If the reported values in past literature are based on ΔH_{Gas}, ΔG_{Gas} and ΔG_{Liq}, they will be converted to ΔH_{Liq} in this chapter for a consistent comparison.

The reason to use the whole spectrum in Equation (9.2.1) for the efficiency comparison is to make the devices working at different wavelengths to be more comparable, and also more convenient for the evaluation of the energy losses due to the non-use of other wavelengths. The efficiency is also influenced by the working wavelength of the devices and the sunlight absorbing material may work only for a specific range of wavelengths. Consequently, other wavelengths of the solar spectrum will be unused and the efficiency of sunlight usage is reduced. Different devices may work at different wavelengths of the solar spectrum, so the comparison is not made on the same basis when using the working wavelength of the devices to evaluate their performance. Different wavelengths correspond to different portions of solar irradiance. Table 9.2.1 shows the energy distribution of different wavelengths of the solar spectrum from past data (Thuillier et al., 2003). It can be concluded that if the hydrogen production device works only in the ultraviolet region, then it can only use a maximum of 10% of the total incident solar energy. By comparison, if it works in the infrared region, then potentially 50% of the total incident solar energy can be used. In addition, various devices and technologies have different solar irradiance tracking and capturing capabilities. As a result, the requirement of land area and dimensions of auxiliary equipment may be quite different.

Table 9.2.1   Energy distribution of the solar spectrum.

Irradiance	Wavelength	% of total	
Infrared	700–2,400 nm	49.4	91.7
Visible	400–700 nm	42.3	
Ultraviolet A	320–400 nm	6.3	8.3
Ultraviolet B	290–320 nm	1.5	
Ultraviolet C	200–290 nm	0.5	

Currently, some other efficiency definitions are also used to evaluate the solar-to-hydrogen technologies and equipment. The existence of multiple definitions arises from the difference of reaction mechanisms, which will be discussed in the following sections. A direct comparison for different solar-to-hydrogen technologies is challenging because a wide scope of technologies is involved and many technologies are still at an early stage of development. This section will provide some general comparisons on the basis of similar standards reported in literature.

9.2.2   Matching the temperature requirements of solar-based hydrogen production methods

In many unconventional hydrogen production methods, e.g., thermochemical water splitting and high temperature electrolysis, high temperature heat is needed (Hinkley et al., 2011; Monnerie et al., 2011; Corgnale et al., 2011; Summers et al., 2009; Xiao et al., 2012). Therefore, a large amount of solar irradiance must be concentrated to reach the temperature requirements and a large additional land area is needed to concentrate the solar irradiance. The high temperature requirements bring some significant engineering challenges. For example, it is still a challenge to find a coating material for the solar receivers to improve the absorptance and reduce the emittance at 600°C (Sergeant et al., 2010; Barshilia et al., 2006; Kennedy et al., 2005). Furthermore, it is challenging to select an appropriate working fluid and equipment material for a solar irradiance receiver and reactor. The working fluids include water, thermal oils, molten salts, steam, air, and other gases. Due to the good heat transfer performance and low melting points, thermal oils are widely used in the solar concentrating devices such as solar troughs. However, thermal oils are volatile and toxic, and may decompose at a high temperature, so the thermal oils are currently operated below 450°C (Moens et al., 2003, 2004; Eck et al., 2007; Wu et al., 2001), which is not sufficiently high to cover the temperature threshold of some thermochemical hydrogen production cycles (Xiao et al., 2012; Le Gal et al., 2010; Corgnale et al., 2011), which will be discussed in later sections.

The solar irradiance concentrating devices include solar troughs, lenses, parabolic dishes, heliostats, and reflection mirrors. Currently, a solar trough can concentrate more power than a lens and a parabolic dish, but it is challenging to reach up to 500°C even if it uses a molten salt as the working fluid (Herrmann et al., 2004), because the receiving area of the tubular irradiance receiver makes the relative number of suns fewer than that otherwise concentrated to a focal spot by a lens and dish. So the temperatures provided by solar troughs are not suitable for some thermochemical cycles with higher temperature requirements (Xiao et al., 2012; Le Gal et al., 2010).

A solar tower capable of concentrating thousands of suns as well as tens of megawatts of irradiance with heliostats or reflection mirrors can reach a temperature range of 500–1000°C (Schramek et al., 2009; Spelling 2009; Dersch et al., 2011). When utilizing a molten salt as the working fluid, the operating temperature of large lab-scale equipment can reach up to 900°C (Forsberg et al., 2007; Patel 2011; Dunn et al., 2012; Moore et al., 2010; Matsunami et al., 2000). Some recent small industrial scale construction projects utilizing molten salts operate at up to 650°C (Khan et al., 2004; Martín, 2007; NREL, 2010). A disadvantage of molten salts is their higher melting points than thermal oils and gases, which limits the heat transfer and storage performance. Therefore, utilizing gas as the working fluid of the solar tower is another option for obtaining a high operating temperature range of 700–1000°C (Schwarzbözl et al., 2006; Ahlbrink et al., 2009; Göttsche et al., 2010). Some operational small industrial scale solar thermal plants using air and other gases can operate at a temperature of 1,000°C. Currently, a high temperature up to 3,500°C can be obtained with solar concentrating furnaces of laboratory and pilot scale equipment (Haueter et al., 1999; Riveros-Rosas et al., 2010). This is a promising scenario for the high temperature hydrogen production cycles.

9.2.3 Thermolysis, thermal decomposition and thermochemical methods

The water in Equation (9.2.2) may take part in the reaction in the form of either liquid water or steam. Since the generation of steam is also from liquid water, then the liquid form for Equation (9.2.2) is more widely used. The changes of standard enthalpy of the water splitting at 298K are equal to the negative of the higher and lower heating values of hydrogen if the same reference temperature is adopted:

$$\Delta H_{Liq,298}^0 = -HHV = -285.8 \, kJ/mol = 2.97 \, eV/molecule \tag{9.2.3}$$

$$\Delta H_{Gas,298}^0 = -LHV = -241.8 \, kJ/mol = 2.52 \, eV/molecule \tag{9.2.4}$$

where the superscript "0" means the standard state, the subscript 298 means the temperature of 298K (25°C), and the subscripts Liq and Gas mean the liquid and gaseous states, respectively. Figure 9.2.1 shows the influence of the temperature on the standard enthalpy change of water splitting reaction for the gaseous form of water. It can be found that the reaction enthalpy at 5,000°C is only about 7% higher than at room temperature. This means the value of the reaction enthalpy change at room temperature can well represent the values under current engineering temperature ranges. Also, the value adopted in Equation (9.2.1) for the efficiency definition is reasonable.

However, the enthalpy change mainly gives the energy balance and requirements. The balance is not sufficient for examining the spontaneity of direct water splitting, which is reflected by the changes of the Gibbs free energy. The values of the standard Gibbs free energy of Reaction (9.2.2) at 298K are (Licht, 2005):

$$\Delta G_{Liq}^0 = 237.0 \, kJ/mol = 2.47 \, eV/molecule \tag{9.2.5}$$

$$\Delta G_{Gas}^0 = 228.4 \, kJ/mol = 2.38 \, eV/molecule \tag{9.2.6}$$

It can be found that the values of the Gibbs energy for liquid and gas forms differ by only 3.8%. So either value can be used to examine the spontaneity of water splitting. The large positive values of the Gibbs free energy in Equations (9.2.5) and (9.2.6) indicate that the direct decomposition of water is far from spontaneous except that the temperature is increased or a large amount of non-PV work such as electrical work is injected into the water molecules. To study the spontaneity, Figure 9.2.2 shows the standard Gibbs energy change of the water splitting reaction at different temperatures. It can be found that the transition temperature under standard conditions is about

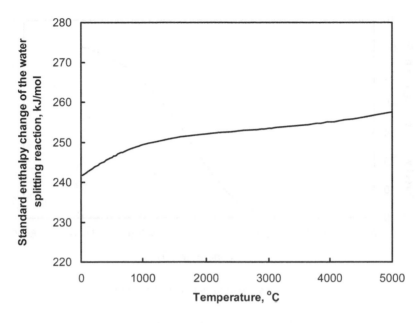

Figure 9.2.1   Standard enthalpy change of water splitting reaction vs. temperature.

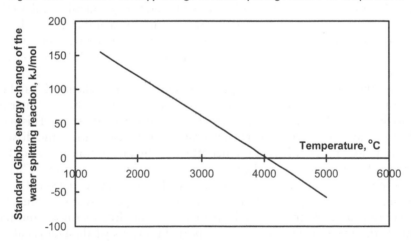

Figure 9.2.2   Standard Gibbs energy change of water splitting reaction vs. temperature.

4,000°C, at which the Gibbs energy change becomes negative and the water splitting becomes spontaneous.

From the Gibbs energy change in Figure 9.2.2, the water decomposition equilibrium constant can be calculated. Then according to the equilibrium constant, the water decomposition percentage can be estimated. Figure 9.2.3 shows the direct water decomposition percentage at different temperatures. It can be found that the below 2% of water directly split at 2,000°C, and if 40% of water is split, the temperature must be higher than 4,000°C. This is a very high temperature that would forego engineering

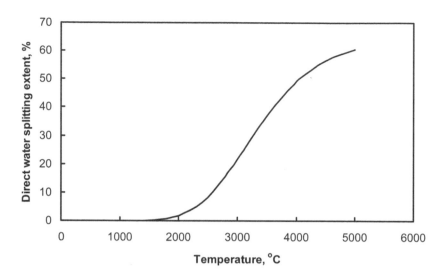

Figure 9.2.3 Direct water splitting extent at different temperatures.

practice in the near future, although currently a high temperature up to 3,500°C can be obtained with solar concentrating furnaces of laboratory and pilot scales (Haueter et al., 1999; Riveros-Rosas et al., 2010). At this high temperature, it is challenging for refractory materials and construction of equipment. Moreover, the direct thermolysis product is a gas mixture of hydrogen and oxygen, which has a considerable explosion risk.

To lower the water themolysis temperature, some auxiliary chemicals can be introduced to form at least one intermediate in an association step and then the intermediate releases hydrogen and/or oxygen separately in other dissociation steps. This will form an "indirect" water splitting cycle. The integration of the association and dissociation steps forms a closed cycle wherein the net effect of the integration is that only water is decomposed and the auxiliary chemicals are recycled inside the cycle. This type of water decomposition is often termed as a "thermochemical" cycle. If a small portion of energy is supplied to the cycle as electricity, it is then a type of hybrid thermochemical cycle rather than a purely thermal cycle. About three hundred thermochemical cycles, either purely thermal or hybrid, have been reported previously (Abanades et al., 2006).

For fully thermal hydrogen production, two-step water splitting cycles based on metal redox reactions are leading examples. These cycles usually consist of an endothermic reduction reaction where oxygen is produced from a metal oxide, and a hydrolysis reaction where hydrogen is produced (Xiao et al., 2012; Le Gal et al., 2010):

$$M_xO_y \rightarrow M_xO_{y-1} + \tfrac{1}{2}O_2 \tag{9.2.7}$$

$$M_xO_{y-1} + H_2O \rightarrow M_xO_y + H_2 \tag{9.2.8}$$

where M denotes a metal, and the subscripts x and y mean the numbers of the metal and oxygen atoms in a metal oxide molecule. Zinc is a metal example currently under active investigation (Haltiwanger et al., 2010; Steinfeld 2002; Melchior, 2009):

$$ZnO = Zn + \tfrac{1}{2}O_2 \tag{9.2.9}$$

$$Zn + H_2O = ZnO + H_2 \tag{9.2.10}$$

The redox pairs of metal oxides that have been reported include Fe_3O_4/FeO, TiO_2/TiO_x, Mn_3O_4/MnO, CeO_2/Ce_2O_3, Co_3O_4/CoO, Nb_2O_5/NbO_2, In_2O_3/In, WO_3/W, and CdO/Cd, among others (Xiao et al., 2012; Le Gal et al., 2010; Haltiwanger et al., 2010; Steinfeld, 2002; Charvin et al., 2007). A significant advantage of using these cycles for hydrogen production is that only two chemical reactions are involved, which reduces the challenges of the system integration that otherwise occurs for three or more chemical reactions. However, the temperature for the oxygen production reaction is usually in the range of 1,500–2,500°C. This is a major challenge for equipment materials, which at the same time are expected to have the characteristics of high solar absorptance, low thermal emittance, corrosion resistance, and thermal stability.

Another leading example of fully thermal cycles is the sulfur–iodine (S-I) cycle, which was first investigated at General Atomics in 1970s (Schultz, 2003; Riccardi et al., 2011; Khan, 2004). An advantage of the S-I cycle over the metal oxide redox cycles is its lower temperature requirement of 850°C. The S-I cycle has been scaled up from proof-of-principle tests to a larger engineering scale by the Japan Atomic Energy Agency (JAEA) (Kubo et al., 2004; Lewis et al., 2006; Sakurai et al., 2000; Terada et al., 2007). The scale of the S-I cycle at JAEA can reach 0.065 kg/day of hydrogen production at present. Commissariat à l'énergie atomique (CEA, (Anzieu et al., 2006)) and the Sandia National Laboratory (SNL, (Moore et al., 2007)) are also active developers of the S-I cycle. There are several types of S-I cycles. Table 9.2.2 shows a typical three-step purely thermal cycle that is commonly studied (Riccardi et al., 2011; Khan et al., 2004; Kubo et al., 2004; T-Raissi et al., 2003). The temperatures for each step of the S-I cycle adopted by different researchers have some differences, depending on the reactor technology. However, these differences are not significant enough to have major differences in the S-I cycle.

The presence of iodine-based chemicals in the S-I cycle brings some significant engineering challenges. For example, great precaution must be taken to process the mixture of combustible H_2 and I_2 at 450°C. Also, the separation of HI, H_2, and I_2 is a complex multiple-stage process and the distillation of azeotropic HI would significantly enhance the energy cost of the cycle (Elder et al., 2005; Stewart 2005; Guo et al., 2011). To avoid these challenges, another sulfur-based thermochemical cycle named as a "hybrid sulfur cycle" or "Westinghouse cycle" has attracted attention (Riccardi et al., 2011; Hinkley et al., 2011; Monnerie et al., 2011; Corgnale et al., 2011; Roeb et al., 2010). As shown in Table 9.2.2, in the hybrid sulfur cycle, hydrogen is produced from the electrolysis of an aqueous solution of SO_2 and the operating temperature is about 120°C, which is significantly lower than the HI decomposition temperature of 450°C for the H_2 production in the S-I cycle.

Both the S-I and the hybrid sulfur cycles require an input temperature of 850°C, which still brings many high temperature-related challenges, although the temperature is significantly lower than redox cycles with metal oxide pairs. In recent years, the

Table 9.2.2 Major chemical processes of a fully thermal S-I cycle and hybrid sulfur cycle.

Step	Process and heat flow	Major reaction
Processes in an S-I thermochemical cycle		
I	Hydrolysis step (exothermic)	$I_2(l+g) + SO_2(g) + 2H_2O(g) = 2HI(g) + H_2SO_4(l) + Q$, at $120°C$
$2^{(a)}$	Oxygen production step (endothermic)	$H_2SO_4(g) + Q = SO_2(g) + H_2O(g) + 0.5O_2(g)$, at $800-1000°C$
3	Hydrogen production step (endothermic)	$2HI(g) + Q = I_2(g) + H_2(g)$, at $450°C$
Major chemical processes of a hybrid sulfur cycle		
A	Hydrogen production step (electrolytic)	$SO_2(aq) + 2H_2O(l) + V_E = H_2SO_4(aq) + H_2(g)$ at $80-120°C$
$B^{(a)}$	Oxygen production step (endothermic)	$H_2SO_4(g) + Q = SO_2(g) + H_2O(g) + 0.5O_2(g)$, at $850°C$

Symbols: aq – aqueous, g – gas, l – liquid, Q – heat, V_E – electricity
(a) This step can be divided $H_2SO_4(aq) + Q = SO_3(g) + H_2O(g)$, at $300-450°C$
into two separate steps: $SO_3(g) + Q = SO_2(g) + 1/2O_2(g)$, at $800-1000°C$

copper-chlorine (Cu-Cl) hybrid cycle has gained major attraction due to its lower temperature requirement of $530°C$, which can be accommodated by more technologies of solar thermal energy (Khan et al., 2004; Xu et al., 2012; Litwin et al., 2010). The Cu-Cl cycle also has several variations with various numbers of steps from 2 to 5 depending on reaction conditions [Lewis, 2008; Wang et al., 2008, 2009]. The cycle with 4 steps shown in Table 9.2.3 is a typical hybrid Cu-Cl cycle. The energy structure and heat requirements are also shown in Table 9.2.3, so as to provide a basis for the efficiency evaluation of the hybrid cycle. A solarium laboratory apparatus that can absorb a maximum of 50 kW solar irradiance and temperature of $800°C$ is under development for the study of a solar-based Cu-Cl cycle and other photochemical processes at the University of Ontario Institute of Technology (UOIT). The scale-up of the cycle from proof-of-principle to a larger engineering scale of 3 kg/day is also in progress at UOIT (Wang et al., 2008, 2009; Naterer et al., 2008) in collaboration with partners that include Atomic Energy of Canada Limited (AECL).

As suggested in Tables 9.2.2 and 9.2.3, more steps are needed by the fully thermal S-I and hybrid Cu-Cl cycles than the hybrid metal redox cycles to complete the water decomposition in a closed loop. Multiple chemical reactors and auxiliary equipment for chemical reactions and heat transfer are needed. This may increase the capital cost of the equipment and operating cost. Therefore, the S-I and Cu-Cl thermochemical cycles are more appropriate for large scale hydrogen production to offset the costs arising from multiple chemical processes.

Since heat is the major form of energy input to thermochemical cycles, the energy loss in the conversion of heat to electricity is then avoided. This indicates a great potential to improve the overall thermal efficiency for hydrogen production. For example, it was estimated that the efficiency could reach 40–56% by Zn/ZnO cycles (Xiao et al., 2012; Haltiwanger et al., 2010; Melchior, 2009; Schunk et al., 2009), 39–45% by Fe_3O_4/FeO cycles (Xiao et al., 2012; Charvin et al., 2008), 35–46% by hybrid sufur

Table 9.2.3 Major chemical processes of a hybrid copper-chlorine cycle and energy distribution.

Step	Process and heat flow	Major reaction
	Processes in Cu-Cl thermochemical cycle	
I	Electrolytic hydrogen production	$2CuCl$ (s) $+ 2HCl$ (aq) $+V_E = H_2$ (g) $+ 2CuCl_2$ (aq) in aqueous solution, at 30~100°C
II	Drying of cupric chloride (endothermic)	$CuCl_2$ (aq) $+ n_f H_2O$ (l) $+ Q = CuCl_2 \cdot n_h H_2O$ (s) $+ (n_f - n_h)$ H_2O, where $n_f > 7.5$, $n_h = 0$~4, depending on temperature. Below 80°C, crystallization; at 100~200°C, spray drying.
III	Hydrolysis of cupric chloride (endothermic)	$2CuCl_2 \cdot n_h H_2O$ (s) $+ H_2O$ (g) $+ Q = CuOCuCl_2$ (s) $+ 2HCl$ (g) $+ n_h H_2O$ (g), at 400°C
IV	Oxygen production (endothermic)	$CuOCuCl_2$ (s) $+ Q = 2CuCl$ (molten) $+ 0.5O_2$ (g), at 530°C

Symbols: aq – aqueous, g – gas, l – liquid, n_f – number of free water, n_h – number of hydrated water, Q – heat, s – solid, V_E – electricity

Energy distribution: In the total energy input, thermal energy and electricity occupy 70–90% and 10–30%, respectively.

Heat requirements for various hydrogen production scales

H_2 production rate, tonnes/day	0.001 (1 kg/day)	1	50	100	200
Heat requirement, MW$_{th}$	0.00263 (2.63 kW$_{th}$)	2.63	132	263	525

cycles (Hinkley et al., 2011; Monnerie et al., 2011; Corgnale et al., 2011; Summers et al., 2009), and 40–60% by S-I and Cu-Cl cycles. These efficiencies have the potential to compete with current steam methane reforming (Lewis, 2008; Wang et al., 2008, 2009). Another advantage of thermochemical cycles is that water decomposition may utilize separate facilities that are independent of the capturing and processing of solar thermal energy. Therefore, the design and maintenance of the hydrogen production and solar thermal energy facilities can be separately performed. The solar thermal energy plant can be designed in a compact fashion that mainly aims at efficiently capturing and concentrating the solar irradiance, wherein a solar tracking system can be readily utilized.

If the captured solar thermal energy needs to be transported by a heat transfer fluid over a distance from a solar tower to the thermochemcial hydrogen production cycle, the heat losses must be controlled to below 30% so as to compete with water electrolysis and steam methane reforming. The pipeline diameter (including the thermal insulation) for the heat transport between solar thermal and thermochemical hydrogen production plants must be large, either utilizing molten salt or pressurized helium as the heat transfer fluid when the heat transport lies in the range of 100–700 MW$_{th}$ which corresponds to 40–200 tonnes of hydrogen production per day. A long distance (>10 km), heat transport is not suggested.

9.2.4 Water electrolysis

The energy input for Equation (9.2.2) could also be in the form of electricity. This means the solar energy must be converted to electricity and the corresponding facilities should then be designed for the distribution of electrical current. Regarding the usage of electricity as the major energy input, the water molecule is split by an imposed electric potential (Bockris et al., 1983):

$$2H_2O(l) = O_2\ (g) + 4H^+\ (aq) + 4e^-, \Phi_A = 1.23\,V \text{ on surface of anode (oxidation)}$$
(9.2.11)

$$2H^+\ (aq) + 2e^- = H_2\ (g), \Phi_C = 0.00\,V \text{ on surface of cathode (reduction)} \quad (9.2.12)$$

where 1.23 V is the standard potential of the anode that indicates the theoretical minimum requirement. If viewed from the level of molecules, water electrolysis, photoelectrolysis, and photoelectrochemical methods can be categorized as the same type. However, their engineering approaches may be very different, which will be discussed in detail in the following sections.

As to splitting of water molecules with electricity as indicated by Equations (9.2.11) and (9.2.12), an electrode efficiency defined on the basis of the gap between the actual potential bias and the theoretical minimum value of 1.23 V is often used to assess the performance of the electrode (Licht, 2005; Bockris et al., 1983). This will not be discussed in detail in this chapter.

In water electrolysis, water is split with an electric current to produce hydrogen. Direct current (DC) passes through two electrodes immersed in water, i.e., anode and cathode, and hydrogen is produced on the surface of the cathode when the electric potential is sufficiently high. The electrodes can be shaped to rods or plates, and the reactions taking place on the surface of the electrodes are shown in Equations (9.2.11) and (9.2.12). In order to avoid confusion with other terminology such as photoeletrolysis and photoelctrochemical methods, this paper suggests that the terminology "water electrolysis" is only used when the electricity is fully obtained from an external power generated from photovoltaic panels or turbines driven by solar-generated steam or other gases. Therefore, the electrolyzer and water do not receive the sunlight for water splitting. This categorization considered the engineering flexibility and engineering practicality for the integration of independent power sources and various electrolyzers.

As shown in Figure 9.2.4, the basic components of the hydrogen production unit include two electrodes (anode and cathode) and one external power source. It can be found that the overall efficiency of the hydrogen production depends on the electricity-to-hydrogen efficiency of the electrolyzer, and the solar-to-electricity conversion efficiency. Table 9.2.4 shows the power consumption and efficiency of different industrial electrolysis systems in the U.S. (Ivy, 2004), Europe (European Commision, 2001) and China (CSPCS, 2009).

Since conventional electrolysis is mature technology and the electricity-to-hydrogen efficiency of a commercially available electrolyzer lies in the range of 50%–80% either using alkaline or polymer electrolyte membrane electrolyzers, the electricity generation dominates the overall efficiency of the hydrogen production. Currently, the power generation efficiency with photovoltaic panels is about 10–20% (van Helden et al., 2004; Yamada et al., 2011; Hanna et al., 2006). Therefore, the maximum overall efficiency of hydrogen production is below 16% (Khaselev et al., 2001). It is anticipated that the power generation efficiency of photovoltaic panels can be enhanced in the future by use of new materials to accommodate more irradiation of the solar spectrum.

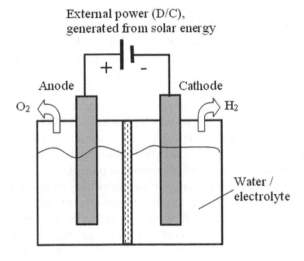

Figure 9.2.4 Components of a solar-based electrolyzer.

Table 9.2.4 Power consumption and efficiency of different industrial electrolyzers.

Country	Electrolyzer model or type	Electricity required for system, $kWh/m^3(H_2)$	Electricity required for electrolyzer only, $kWh/m^3(H_2)$	High heating value efficiency of electrolysis system (electricity-to-H_2), %
US[a]	Stuart: IMET 1000	4.8	4.2	73.9
	Teledyne: EC-750	5.6	N/A	63.3
	Proton: HOGEN 380	6.3	4.3	56.3
	Norsk Hydro: Atmospheric Type No. 5040 (5150 Amp DC)	4.8	N/A	73.9
	Avalence: Hydrofiller 175	5.4	N/A	65.7
Europe[b]	Membrane	8.8	N/A	40.1
	Amalgam	11.3	N/A	31.4
	Diaphragm	9.4	N/A	37.7
China[b]	Membrane	8.9	N/A	39.8

(a) The electrolyte is an aqueous solution of KOH and the electrolysis system is specifically designed for hydrogen production from water electrolysis.
(b) The electrolyte is aqueous solution of NaCl and the electrolysis system also produces Cl_2 and NaOH as commercial products.

The external electricity can also be generated from a solar thermal plant that uses a sunlight concentrating device to generate high temperature fluid and then use the heat captured by the fluid to generate electricity. The concentrating devices and working fluids have been discussed in the former section regarding thermochemical cycles. Currently, thermal oils are usually used to generate steam to drive a steam turbine

(Moens et al., 2003, 2004; Eck et al., 2007; Wu et al., 2001), and molten salts are planned to be used for gaining a higher temperature than thermal oils (Forsberg et al., 2007; Patel, 2011; Dunn et al., 2012; Moore et al., 2010), and air and other gases are used to drive a gas turbine (Schwarzbözl et al., 2006; Ahlbrink et al., 2009; Göttsche et al., 2010). The currently operational solar thermal plant can generate the working fluid at more than 500°C and reach 1,000°C, so the conversion efficiency from a working fluid to electricity is in the range of 30–60%. Currently operational solar thermal plants show that the solar radiation capturing efficiency of solar concentrating devices with a tracking system is usually higher than 70% (European Commission, 2010; Schmitz, 2009; Taggart, 2008). Therefore, the integration of an electrolyzer and solar thermal power plant can deliver higher hydrogen production efficiency (15%–56%) than an electrolyzer and photovoltaic panels.

In addition to the engineering maturity and flexibility of integration with various solar power generation technologies, another major advantage of water electrolysis is that the hydrogen production can still operate at nights or days by using power when sunlight is not available. The power could be either generated from the stored solar thermal energy for the use at nights and undesirable weather conditions, or directly from the power grid. In a concentrated solar power plant utilizing solar troughs or solar towers, a large amount of solar energy can be stored in thermal oils or molten salts for the times when sunlight is not available (Moens et al., 2003, 2004; Wu et al., 2001; Herrmann et al., 2004; Patel et al., 2011; Dunn et al., 2012). As for electricity from the power grid, even if the energy sources on the power grid may not be "clean", the impacts of unpredicted weather conditions can be minimized.

Since water electrolysis utilizes an external power source, the design of the power generation plant does not need to consider the location of the solar power plant. It can be designed in a compact way that aims at efficient sunlight capturing. A solar tracking system can be readily utilized for collecting or concentrating the sunlight. In addition, unpredictable mutual safety impacts are minimized and the distance between the solar thermal power plant and the facilities of electrolyzer are flexible. The electrolytic facilities do not need to occupy the space where sunlight is more suitable for the power generation.

9.2.5 Photoelectrolysis and photoelectrochemical water splitting

As to the water electrolysis presented in the former section, whether the electricity is fully obtained from an external source does not cause a significant difference from the perspectives of the chemical reaction mechanisms on the electrode surface. For example, the electricity can be generated by an electrode on its own if the electrode is made of materials that can create electric potential due to its exposure to sunlight (Licht, 2005). The materials could be either n-type or p-type semiconductors. However, from the aspect of an engineering and equipment setup, the difference is significant. In this section, it is suggested that "photoelectrolysis" and "photoelectrochemical" water splitting are not categorized as "water electrolysis". Instead, they are adopted when at least one light absorbing electrode is needed, and only a part or no electricity for the reduction or oxidization reaction on the electrode is obtained from external power sources.

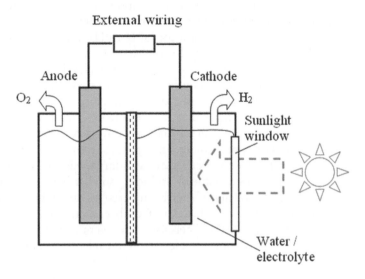

Figure 9.2.5 Components of a photoelectrolysis or photoelectrochemical unit.

Figure 9.2.5 shows the basic components of a photoelectrolysis or photoelectrochemical (PEC) hydrogen production unit, including a sunlight absorbing electrode (typically made or coated with semiconductor) and a counter electrode (typically metal) immersed in an electrolyte. The sunlight absorbing electrode must be arranged to face the sunlight window to capture sufficient solar radiation to generate electric potential. The sunlight absorbing electrode should be wired either externally or internally with the other counter electrode so as to form a closed circuit. The reaction mechanism is described in Equations 9.2.11 and 9.2.12. It can be found that the sunlight absorbing material has a dominating role in determining the hydrogen production efficiency. The band gap, i.e., the potential, created by the electrode material must exceed the bottom theoretical limit of 1.23 eV to split water molecules, plus overcoming the electric resistance of the closed circuit. It was also reported that the materials for the sunlight absorbing electrode are very likely subject to electrochemical corrosion. Therefore, many studies are conducted to the development of new anti-corrosion and high efficiency materials (Grätzel, 2003). Another option to improve the performance of photoelectrolysis is to utilize external power to boost the electrode potential. This is a type of hybrid system of water electrolysis and photoelectrolysis. In the hybrid system, the electrode must be a sunlight absorbing material, so this section suggests that the system is still viewed as photoelectrolysis rather than water electrolysis due to the distinct sunlight capturing and electricity generation patterns.

Recently, there is an alternative design of a PEC that makes the entire device a microparticle, nanoparticle, or nanofibre (Vayssieres, 2009; Solarska et al., 2012; Yang, 2011; Grimes et al., 2007). The device includes the mini-cathode, mini-anode and photovoltaic components sandwiched together in single particles that are suspended in an aqueous solution. Hydrogen is evolved at the cathode and oxygen at the anode. The reaction byproducts, OH^- and H^+ recombine in the solution completing the cycle. No external wiring is required as all components are internally connected. Short electron pathways and large surface areas result in improved efficiencies. The major drawback of this design is the generation of both H_2 and O_2 at the same location. Therefore, hydrogen is not produced separately in the reactor, although each hydrogen molecule is formed separately if viewed from the level of a single molecule. The mixture of hydrogen and oxygen is quite sensitive to sparks that may lead to an explosion. The separation of hydrogen and oxygen is a major engineering challenge.

It is difficult to directly compare the photoelectrolysis and water electrolysis because the range of technologies is very broad, so only a few general conclusions are made here on the basis of approximate similarity. Currently, the band gaps that can be provided by photoelectrode materials are still large, e.g., greater than 3.2 eV, although there are some materials capable of providing smaller band gaps. This makes the solar irradiance of larger wavelengths such as infrared less available, hence the sunlight usage efficiency is low (Walter et al., 2010; Conibeer et al., 2007; Currao, 2007; Prakasam, 2008; Grimes et al., 2007). Therefore, the overall solar-to-hydrogen efficiency of current photoelectrolysis or photoelectrochemical hydrogen production units rarely reaches higher than 16% (Solarska et al., 2012; Yang et al., 2011; Prakasam 2008; Licht et al., 2000; Mohapatra et al., 2007). Also, in comparison with water electrolysis, it is more challenging for the photoelectrolysis or photoelectrochemical unit to efficiently track the sun because of the structure and operating complexity of the equipment to simultaneously process hydrogen, oxygen, water, and sunlight window. For example, the contact between an electrode and water may be changed when the equipment is tilted for efficient sunlight tracking. In addition, the auxiliary components of the system may occupy a large portion of the sunlight projection area or shade the sunlight in the sunlight tracking operation.

The lack of a combination of a stable, efficient light absorption system consisting of suitable photoelectrodes and light windows partly accounts for the low efficiency. No reliance on the external power source may bring some advantages, including simplicity of system design because of the elimination of the auxiliary components required by the electrolyzer, and potentially a large photoanode and photocathode surface with nanosize materials (Yang et al., 2011; Walter et al., 2010; Conibeer et al., 2007; Currao, 2007). Also, photoelectrolysis or photoelectrochemical hydrogen production plants can be more readily distributed in hydrogen fueling stations or remote geographic areas to avoid building an expensive power transmission and distribution grid for otherwise using water electrolysis.

9.2.6 Photochemical, photocatalytic, photodissociation, photodecomposition, and photolysis

In addition to using concentrated solar thermal energy or electricity, there is another way to use the irradiance for the water splitting. As shown in Equations 9.2.3 and 9.2.5, for a single water molecule, if the photons are directly trapped by some auxiliary substances (sensitizers and catalysts) to activate the electrons to a higher energy state, then the water molecules can capture the activated electrons from auxiliary substances. As a result, the water molecules are activated to a high energy state, preparing for further formation of hydrogen and oxygen atoms (Hagiwara et al., 2006). This series of photochemical reactions are described as follows:

$$4S + 4h\gamma = 4S^* \tag{9.2.13}$$

$$4S^* + 4C_R = 4S^+ + 4(C_R^-)^* \tag{9.2.14}$$

$$4(C_R^-)^* + 4H^+ = 4C_R + 4H^* \tag{9.2.15}$$

$$4S^+ + 4C_{OX} = 4S + 4C_{OX}^+ \tag{9.2.16}$$

$$4C_{OX}^+ + 2H_2O = 4C_{OX} + 2O^* + 4H^+ \tag{9.2.17}$$

$$4H^* = 2H_2 \tag{9.2.18}$$

$$2O^* = O_2 \tag{9.2.19}$$

where h is the Planck constant (6.626×10^{-34} J·s), γ is the photon frequency, and their product means a photon and its energy. Also, S means sensitizer and C_R and C_{OX} indicate the catalysts for the reduction and oxidization reactions, respectively. The superscript asterisk means the activated state. If viewed from the reaction mechanism shown in Equations 9.2.13–9.2.19, photochemical, photodissociation, photodecomposition, photodegradation, photocatalytic, and photolysis can be categorized as the same type. The reaction mechanism may suggest very different engineering approaches, compared with the heat and electric potential driven water spitting, which will be discussed in later sections.

In this type of water splitting processes, the performance of the sensitizers and catalysts is often assessed with photon-use efficiency, which is defined on the basis of the absorbed photons (Melis 2004), which is often adopted for describing the efficiency of the photocatalytic reactions. Another useful efficiency is called "quantum efficiency", which is defined as the ratio of the number of charge carriers collected by a solar cell to the number of photons illuminating on the solar cell (Park et al., 2009). This efficiency is often adopted to evaluate the yield of incident photon to charge carriers for photovoltaic panels. Since photon energy varies with wavelength, consequently the quantum efficiency may vary for different wavelengths of light. There are also definitions on the values of enthalpy and Gibbs free energy (Rajeshwar et al., 2008) but in units of ev/molecule from the molecular level. Their values are shown in Equations 9.2.3 and 9.2.5, i.e., 2.97 ev/molecule and 2.47 ev/molecule for energy balance and spontaineity threshhold, respectively.

As discussed in the previous sections, an electrode must be utilized to create sufficient potential and sunlight must be converted to electric current in the water electrolysis and photoelectrolysis. If an electrode is not needed, then the terminologies "photochemical", "photocatalytic", "photodissociation", "photodecomposition", and "photolysis" can be regarded to have a common engineering setup and similar reaction mechanisms that do not utilize the electric potential to break the chemical bond of hydrogen and oxygen. If there is no electrode, the energy of the photons must be absorbed and stored in some intermediate reagents and then delivered by the reagent to water molecules. Water is transparent to a large portion of the photons in the terrestrial solar spectrum and the photons cannot be directly utilized to break the hydrogen and oxygen bond. Firstly, a reagent is needed that must have the ability to serve as a photon sensitizer to absorb photons and use the photons to activate

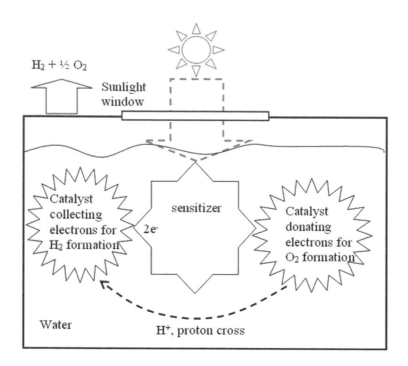

Figure 9.2.6 Components of a photochemical unit.

the electrons to a higher energy state. Secondly, two electrons are needed to form a hydrogen molecule. However, the energy carried by a single photon in the solar spectrum can activate only one electron for the reduction of a proton. Therefore, another reagent (catalyst) is needed to accumulate the activated electrons for the formation of hydrogen molecules. In order to split water molecules rather than other substances, the catalyst, or again one more catalyst, must also have the ability to capture electrons from the negative valence oxygen atom of the water molecule (Kudo, 2007). As to the formation of an oxygen molecule, four electrons must be extracted by the catalyst to prepare for the oxidization process. The processes and reactions taking place on the sensitizers and catalysts are shown in Equations 9.2.13–9.2.19.

Figure 9.2.6 shows the most basic components needed by a photochemical hydrogen production cell, including a sunlight window, sensitizer and at least one catalyst. It can be found that a sensitizer and catalyst are preferably distributed uniformly in the water so as to intercept more sunlight and approach sufficient contact with water. As with particle-based PEC, hydrogen is not produced separately, leading to similar challenges. It can also be found that the preparation and performance of the sensitizer and catalyst may greatly influence the efficiency and economics. For example, if the lifetime of the catalyst or sensitizer is not long, then it must be replenished frequently. This may increase the operating cost and generate a waste stream, which then may not be a strictly clean hydrogen production technology. There are many engineers and scientists focusing on the development of low-cost, highly-efficient, and long-lifetime sensitizers and catalysts (Grätzel, 2003; Hagiwara et al., 2006). By improving the technology of manufacturing the sensitizer and catalyst, for example with nanotechnology, then photochemical hydrogen production has the potential to be improved significantly.

Recent advancement in the creation of supramolecular catalysts has combined the sensitizer and the catalyst into a single unit. These units are designed to be either to be a Hydrogen Evolving Reaction (HER) (Vayssieres, 2009) or Oxygen Evolving Reaction (OER) (Crabtree, 2010). Each of these reactions operates as half cells. The HER requires an influx of electrons and light, then reduces water to produce hydrogen gas and OH^- ions. The OER requires light and oxidizes water to produce oxygen gas, H^+ ions, and an excess of electrons. By coupling the two half cells together using electrodes and a proton exchange membrane (PEM), a complete reactor can be built (Zamfirescu et al., 2011). Alternatively, the OER could instead use light to oxidize the OH^- produced from the HER to produce oxygen gas, water and an excess of electrons. Overall, this is a lower energy pathway than the OER presented above, but there are larger challenges in finding a suitable membrane. Even though electrodes are present, this is not classified as electrolysis or photoelectrolysis as the reactions do not occur at the electrodes but at the catalysts. The electrodes are only present to complete the electron circuit. Unlike the previously described method, the hydrogen and oxygen are produced separately, creating similar engineering challenges to those discussed in photoelectrolysis.

The advantages of photoelectrolysis over conventional electrolysis are also expected for photochemical water splitting, such as the elimination of a power source and auxiliary components of the electrolyzer. In addition, photochemical processes can be implemented in homogeneous catalytic compounds promoting the HER (Wang et al., 2011). Hence, the processing of catalysts can be greatly simplified to the processing of a fluid in engineering. Also, greater tunability is possible with modular architectures and precise details of molecular scale transformations are more accessible for the research (Teets et al., 2011). By comparison, the photonelectrode cannot be homogeneous even if the size is at the nanoscale, otherwise the band gap won't be satisfied and the photovoltage cannot be created.

Similar to photoelectrolysis, it is challenging for the photochemical unit to efficiently make use of the solar irradiance of all wavelengths due to the wavelength selectivity of catalysts (Maeda et al., 2006, 2010; Li et al., 2011). The solar-to-hydrogen efficiency of an operational small pilot-scale photocatalytic hydrogen production demonstration of 1.88 liters per hour is even below 1% (Jing et al., 2010). Therefore, the overall efficiency of current photochemical hydrogen production units rarely reaches 10%, although the quantum efficiency at a specific wavelength could reach 56% (Li et al., 2011; Maeda, 2010, 2011; Kudo, 2009). Considering the sunlight tracking challenges due to the structure and operating complexity of the equipment to simultaneously process hydrogen, oxygen, water, and a sunlight window, it can be concluded that much further research and development is needed towards commercialization of the photocatalytic hydrogen production.

9.2.7 Hybrid and other hydrogen production methods

Two or more of the technologies presented in the previous sections can be combined together for the production of hydrogen, for a hybrid production technology possessing

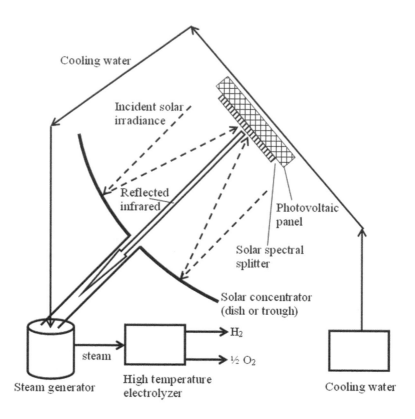

Figure 9.2.7 Hybrid use of concentrator photovoltaic panel and high temperature electrolysis.

the potential to deliver a higher energy and hydrogen production efficiency. A reported hybrid technology is a concentrator photovoltaic (CPV) utilizing water-cooled multi-sun photovoltaic panels to receive tens or hundreds of suns to generate electricity with visible light at a much higher efficiency than one-sun PV panels, and at the same time generate steam with concentrated infrared for high temperature electrolysis (HTE) (Lasich, 1999). As shown in Figure 9.2.7, the sunlight is concentrated by a parabolic dish or trough, and the water is heated by cooling the PV panels and evaporated by receiving the reflected infrared radiation. At the focal spot, a spectral splitter reflects infrared radiation but allows for the transmission of visible light to high-efficiency solar photovoltaic cells behind the splitter. The reflected infrared radiation is conducted to the steam generator for the HTE process.

The electricity for the electrolysis is generated from the concentrated visible light. It was reported that this type of system could provide 40% for the solar-to-hydrogen efficiency in the near term (McConnell et al., 2005, 2006; Thompson, 2005). A major challenge of this technology is that the hydrogen and oxygen are produced in a mixture, which may have drawbacks for the separation in the scale-up of the system.

9.3   SOLAR-BASED CO_2 RECYCLING WITH HYDROGEN

As presented previously, solar-based hydrogen production is not only a substitute for fossil fuels in the future, it is also a necessity for CO_2 recycling and hydrogenation in synfuel production. With the addition of hydrogen to CO_2, methanol and its derivatives can be produced. This will significantly increase the sustainability of our limited fossil fuel resources. Currently, there are several methods of adding hydrogen to CO_2. A technology is methane-assisted processes utilizing the hydrogen in methane to convert CO_2 to carbon-based fuels, for example (Von Zedtwitz-Nikulshyn, 2009),

$$CH_4 \text{ (g)} + CO_2 \text{ (g)} = 2H_2 \text{ (g)} + 2CO \text{ (g)} \tag{9.3.1}$$

The CO_2 present in Equation (9.3.1) can also be captured by a CaO-based cycle and then used in the following reaction to synthesize fuels:

$$CaO \text{ (s)} + CO_2 \text{ (g)} = CaCO_3 \text{ (g)} \tag{9.3.2}$$

$$CaCO_3 \text{ (s)} + CH_4 \text{ (g)} = CaO \text{ (g)} + 2CO \text{ (g)} + 2H_2 \text{ (g)} \tag{9.3.3}$$

However, this technology is not strictly renewable because methane is used in the processes. So it won't be further discussed in detail in this section.

A renewable option is to use H_2 and CO_2 to synthesize methanol catalytically (Fornero et al., 2011; Olah et al., 2009):

$$3H_2 \text{ (g)} + CO_2 \text{ (g)} = CH_3OH \text{ (g)} + H_2O \text{ (g)}, 260°C, \Delta H° = -49.7 \text{ kJ/mol} \tag{9.3.4}$$

The enthalpy change of the preheating process of the reactants is:

$$[3H_2 \text{ (g)} + CO_2 \text{ (g)}] \text{ of } 20°C = [3H_2 \text{ (g)} + CO_2 \text{ (g)}] \text{ of } 260°C, \Delta H° = 30.6 \text{ kJ/mol} \tag{9.3.5}$$

The enthalpy change of Reaction 9.3.4 is a negative value, indicating an exothermic reaction. As the preheating of 3 moles of H_2 and 1 mole of CO_2 from 20°C to 260°C requires 30.6 kJ, which is smaller than the heat released from Reaction 9.3.4, the methanol production process can be assumed as a self-sustained process if the heat losses and the heat recovered from the cooling of the products can offset the electricity requirement. Then it can be approximated that the energy consumption for the CO_2 recycling is mainly established by the H_2 production and CO_2 capture, which will be examined in the following sections.

Table 9.3.1 lists the energy requirements of hydrogen production with the solar-based conventional water electrolysis and hybrid Cu-Cl thermochemical cycle. The energy requirement of the hybrid Cu-Cl cycle for hydrogen production is 222 MJ/kg H_2 of solar thermal energy and 32 MJ/kg of solar electricity (Wang et al., 2010).

The energy requirements of CO_2 recycling for synfuel production include the capture and purification of CO_2 from industrial emissions and ambient air. A challenge in the industrial design of CO_2 recycling is the major energy requirements of CO_2 capture, which corresponds to a high energy cost. The energy requirements are influenced

Table 9.3.1 Energy requirements of electrolysis and hybrid thermochemical Cu-Cl cycle.

H₂ production method	Solar thermal energy-to hydrogen efficiency %	Energy requirement	
		thermal MJ_{th}/kg $(MJ_{th}/kmol)$	electricity MJ_e/kg
Water electrolysis	40%	0 (0)	161
Cu-Cl cycle	50%	222 (444)	32

by many factors. They have been actively studied by others for the capture of carbon dioxide from flue gases or air (Finkenrath, 2011; Von Zedtwitz-Nikulshyna, 2009; Stolaroff, 2006). The focus of this section is to examine the heat requirements by sorption processes. The energy requirements include thermal energy for CO_2 release from sorbents and electricity for capturing and transporting emissions comprising CO_2 and other gases to CO_2 absorption equipment. As the electricity requirement is influenced by many parameters such as the emission composition, distance between the solar thermal energy capture site and CO_2 capturing plant, and the flow type for CO_2 absorption and desorption processes, this section will focus on the thermal energy requirement of a CO_2 capture method, which is characterized by both quality and quantity influencing the feasibility of linking the CO_2 capture process with a nuclear reactor.

Table 9.3.2 summarizes the heat requirements of CO_2 absorption and desorption processes of some CO_2 capture cycles currently under active investigation, including Na_2CO_3-based (Nikulshina et al., 2008; Liang et al., 2004; Lee et al., 2008), K_2CO_3-based (Zhao et al., 2010; Lee et al., 2004, 2008), CaO-based (Blamey et al., 2010; MacKenzie et al., 2007; Salvador et al., 2003), CaO-NaOH-based (Mahmoudkhani et al., 2009; Siriwardane, 2007), and MEA-based methods (Han et al., 2011; Yeh et al., 2001). The adsorption heat is evaluated with data of the National Institute of Standards and Technology (NIST, 2012), and other investigators (Han et al., 2011; Yeh et al., 2001). The table shows that all adsorption processes are exothermic but occur below 100°C, so heat is not easily recovered. As for the CO_2 desorption processes, they are all endothermic. Also, CaO-based and NaOH-CaO-based cycles may directly provide high purity CO_2 because of no need of separation of CO_2 and water vapour. However, these two processes have a temperature threshold of 900°C, which can only be satisfied by a large intensity of concentrated solar irradiance. The temperature requirements of CO_2 desorption processes of other cycles are below 200°C, which can be satisfied by current industrial solar concentrators. This provides a good flexibility for the linkage of a solar thermal power plant and a CO_2 capture plant.

The enthalpy changes of different desorption processes may differ significantly in the range of 135–180 kJ/mol CO_2, as shown in Table 9.3.2. Except for CaO-based and NaOH-CaO-based cycles, other carbon capture cycles require a thermal energy range of 135–165 kJ/mol CO_2 below 200°C. To have an approximation of the CO_2 capture capacity with off-peak hours of a nuclear plant, the thermal efficiency of the CO_2

Table 9.3.2 Heat requirements of some typical CO$_2$ capture cycles.

Cycle	Function	Processes	T, °C	Process enthalpy $\Delta H^{(a)}$, kJ/mol
Na$_2$CO$_3$-based	Absorption	Na$_2$CO$_3$ (s) + H$_2$O (g) + CO$_2$ (g) = 2NaHCO$_3$ (s)	20–60	−135.5
	Desorption	2NaHCO$_3$ (s) = Na$_2$CO$_3$ (s) + H$_2$O (g) + CO$_2$ (g)	120–180	135.5
K$_2$CO$_3$-based	Absorption	K$_2$CO$_3$ (s) + H$_2$O (g) + CO$_2$ (g) = 2KHCO$_3$ (s)	20–60	−140.9
	Desorption	2KHCO$_3$ (s) = K$_2$CO$_3$ (s) + H$_2$O (g) + CO$_2$ (g)	120–180	140.9
CaO-based	Regeneration	CaO (s) + H$_2$O (l) = Ca (OH)$_2$ (s)	100	−65.3
	Absorption	Ca (OH)$_2$ (aq) + CO$_2$ (g) = CaCO$_3$ (s) + H$_2$O (g)	100	−69.8
	Desorption	CaCO$_3$ (s) = CaO (s) + CO$_2$ (g)	900	179.2
CaO-NaOH-based	Absorption	2NaOH (s) + CO$_2$ (g) = Na$_2$CO$_3$ (s) + H$_2$O (g)	20–60	−127.2
	Precipitation	Na$_2$CO$_3$ (s) + Ca (OH)$_2$ (aq) = CaCO$_3$ (s) + 2NaOH (aq)	20–60	−5.3
	Desorption	CaCO$_3$ (s) = CaO (s) + CO$_2$ (g)	900	179.2
	Alkalization	CaO (s) + H$_2$O (l) = Ca (OH)$_2$ (s)	100	−65.3
MEA-based[b]	Absorption	RNH$_2$ + H$_2$O (l) + CO$_2$ (g) = RNH$_3^+$ + HCO$_3^-$	38	−72.0
	Desorption	RNH$_3^+$ + HCO$_3^-$ = RNH$_2$ + H$_2$O (g) + CO$_2$ (g)	120	165.0

(a) ΔH is the process enthalpy change. A positive value for ΔH means endothermic (requiring heat), otherwise exothermic (releasing heat).
(b) MEA, also ETA, is monoethanolamine, which is often denoted by RNH$_2$, where R is "OH (CH$_2$)$_2$" [Ali 2004].

capture from flue gases rather than air is assumed to be 50%, which is an average value of the 40%–60% range reported by investigators for various CO$_2$ capture methods (Tzimas, 2009; Von Zedtwitz-Nikulshyna, 2009; David et al., 2000). Therefore, the thermal energy for CO$_2$ capture from flue gases is approximately 270–360 kJ/mol CO$_2$.

Other energy requirements in the CO$_2$ capture process include at least three portions: (i) work to transport the flue gas to the CO$_2$ capture process for the separation of CO$_2$ and other gases; (ii) work to compress the concentrated CO$_2$ to the reservoir pressure, and (iii) work to move the compressed CO$_2$ into a distant storage location including a storage tank or geologic formation. It can be shown that the lower bound of the total work with ideal Second-Law efficiencies for these three portions is about 9, 13 and 2 kJ/mol CO$_2$, respectively, assuming the flue gas comprises 78% N$_2$ from the atmosphere, 15% CO$_2$ from the oxidation of the carbon in the hydrocarbon, 7% steam, reservoir pressure of 70 bars, and the ground water depth is only 2 km. Assuming further the isothermal compression efficiency is 65%, then the total electricity requirement to complete the above three steps is approximately 37 kJ/mol CO$_2$ (House et al., 2009). Taking the value of 45% as the conversion efficiency for the solar thermal energy conversion to electricity, then the primary solar thermal energy is about 82 kJ/mol CO$_2$. Consequently, the total thermal energy requirement for CO$_2$ capture and storage lies in the range of 352–442 kJ/mol CO$_2$.

Table 9.3.3 summarizes the energy requirements of the H$_2$ production, CO$_2$ capture and compression for methanol synthesis for the production of 1 mole of methanol with Reaction 9.3.4. Figure 9.3.1 shows the percentage of CO$_2$ capture in the synthesis of methanol production based on the data of Table 9.3.3. It can be concluded that a key to CO$_2$ recycling is an economic hydrogen source, because the energy required

Table 9.3.3 Distribution of energy requirements of methanol production.

H_2 production: equivalent thermal energy requirement for 3 moles of H_2		CCS: equivalent thermal energy requirement for 1 mole of CO_2	
Method	kJ	method	kJ
Cu-Cl cycle	1759	Na_2CO_3-based	353.0
	1759	K_2CO_3-based	363.8
	1759	CaO-based	440.4
	1759	MEA-based	412.0
Water electrolysis	2147	Na_2CO_3-based	353.0
	2147	K_2CO_3-based	363.8
	2147	CaO-based	440.4
	2147	MEA-based	412.0

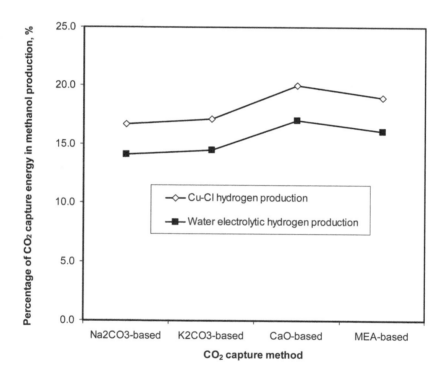

Figure 9.3.1 Percentage of CO_2 capture energy in the synthesis of methanol production.

by hydrogen preparation accounts for the majority of energy cost in the methanol production process.

As Reaction 9.3.4 can be self-sustained and the solar reactors for hydrogen production have been presented in the hydrogen production section of the chapter, this section will focus on the reactors for CO_2 capture, particularly the extraction of solid carbonates from the aqueous solution and the release of CO_2 from the carbonates or sorbents. As the carbonation and calcination operations are very mature in industry, the major modifications are conducted for the usage of solar thermal energy. A solar-based spray carbonator and fluidized bed calcinator were tested at a large laboratory scale (Nikulshina et al., 2006, 2009). Both reactors have a transparent section allowing for the direct heating of the solid reactants. As indicated in Table 9.3.2, solid and gas must be processed in the same reactor. Sometimes the aqueous solution of the solid is used for the CO_2 capture, so the heat transfer is a multiphase process and preferable that the solid can be directly heated by the solar irradiation due to the poor heat transfer performance of solid particles. This explains why the reported reactors are transparent, which are different from solar hydrogen production reactors, where only gases are processed.

9.4 SUMMARY

This chapter presented scenarios of using solar-based hydrogen and CO_2 recycling to provide a sustainable solution to the increasing demand of clean energy and ongoing depletion of conventional fossil fuels. The intermittency issue of solar energy can also be significantly addressed with the usage of solar-based hydrogen as well as synfuels produced from recycled CO_2 and H_2. Then this chapter examined the solar-to-hydrogen reaction mechanisms and technologies including thermochemical cycles utilizing solar thermal energy to split water molecules, conventional electrolysis utilizing solar-generated electricity to split water molecules, and photochemical processes utilizing photon-activated electrons of auxiliary reagents (sensitizer and catalyst) to activate and split water molecules. This chapter also examined and suggested some categorization criteria for technologies from the reaction mechanisms and engineering approaches, particularly the latter.

The basic components for the hydrogen production apparatus and major advantages and challenges of the technologies were also examined. It was concluded that conventional water electrolysis powered by solar generated electricity is more mature than other technologies, but the solar-to-hydrogen efficiency is currently below 16% due to the energy loss in the conversion of solar irradiation to electricity. Thus, its efficiency improvement is mainly determined by the increase of solar-to-electricity conversion efficiency, which is a maximum of about 20% for currently operating PV panels and solar thermal plants. A hybrid method involving a high temperature electrolysis and spectrum splitter may utilize more heat than conventional electrolysis. Consequently, the hydrogen production efficiency is greatly increased to 40–50%, but it is challenging to find appropriate materials for the electrodes and electrolyte that must withstand high temperature steam, hydrogen and oxygen.

Thermochemical cycles benefit from large production scales in order to minimize the energy losses arising from high temperature requirements and multiple auxiliary processes for an integrated operation of the thermochemical cycle. A high temperature of 1,500–2,500°C required by the metal oxide redox pair cycles may keep the cycles from being utilized in the short term, although the cycles usually have only two chemical reactions. The solar-to-hydrogen efficiency of thermochemical cycles was estimated to be in the range of 40–60%, which is higher than conventional electrolysis because the cycles use thermal energy as the major energy input with no energy loss due to the conversion of thermal energy to electricity.

The solar-to-hydrogen efficiencies of both photoelectrochemical and photochemical technologies are significantly limited by the activity and wavelength range of the photoelectrodes and photocatalysts. Even if the quantum efficiency reaches as high as 56%, the solar-to-hydrogen efficiency is still below 16% and 10% for the existing photoelectrochemical and photochemical cells, respectively.

Both thermochemical cycles and water electrolysis may need additional hydrogen distribution systems. By comparison, photoelectrochemical and photochemical technologies are more suitable at hydrogen fueling stations because fewer processes are needed. Therefore, they are more suitable for serving as hydrogen fueling stations with no need of extra hydrogen distribution systems. The extension of the working wavelength of the materials for a solar PV panel, photoelecltrodes, and photo catalysts, to visible light (400–700 nm) and the infrared (700–2400 nm) range is a useful future research direction to improve the solar-to-hydrogen efficiency, as these two spectral ranges occupy more than 90% of the total solar irradiance.

This chapter also examined the energy requirements of synfuel production from captured CO_2 and hydrogen. It was found that the synfuel production reaction is self-sustainable, so it was concluded that the energy requirements of synfuel production are mainly determined by the hydrogen production and CO_2 capture. The hydrogen production energy consumption is about 5–7 times the level for CO_2 capture. The CO_2 capture methods such as Na_2CO_3-based, K_2CO_3-based, CaO-based, and MEA-based processes were examined. A challenge of the solar-based carbonation and calcination reactors for CO_2 absorption and release is to provide more efficient heat transfer from the sunlight to solid particles. Improved multiphase flow reactors are needed because gas and solid are present simultaneously.

Nomenclature

A Area, m^2
C_{OX} Catalyst for oxidization reaction
C_R Catalyst for reduction reaction
I_S Total incident solar irradiance of the solar spectrum, $J \cdot s^{-1}$
h Planck constant, $6.626 \times 10^{-34} J \cdot s$
m_P Hydrogen production per unit time, $mole \cdot s^{-1}$
S Sensitizer
T Temperature, K

Greek

η Efficiency
γ Photon frequency, Hz

REFERENCES

Abanades, S., Charvin, P., Flamant, G. and Neveu, P. (2006) Screening of water splitting thermo-chemical cycles potentially attractive for hydrogen production by concentrated solar energy. *Energy*, 31, 2805–2822.

Ahlbrink, N., Belhomme, B. and Pitz-Paal, R. (2009) Modeling and simulation of a solar tower power plant with open volumetric air receiver. *Proceedings 7th Modelica Conference*, Sep. 20–22, 2009, Como, Italy. pp. 685–693.

Ali, C. F. (2004) CO$_2$ Capture with MEA: integrating the absorption process and steam cycle of an existing coal-fired power plant. *MSc Thesis*. Department of Chemical Engineering. University of Waterloo, Ontario, Canada.

Anzieu, P., Carles, P., Le Duigou, A., Vitart, X. and Lemort, F. (2006) The sulfur–iodine and other thermochemical process studies at CEA. *International Journal of Nuclear Hydrogen Production and Applications*, 1, 144–153.

Barshilia, H. C., Selvakumar, N., Rajam, K. S., Rao, S., Muraleedharan, K. and Biswas, A. (2006) TiAlN/TiAlON/Si$_3$N$_4$ tandem absorber for high temperature solar selective applications. Applied Physics Letters, 89, 191909–191909-3.

Berger, E. (2006) BMW hydrogen near zero emission vehicle development. *BMW CleanEnergy. CARB ZEV Technology Symposium*. pp. 6–13. September 2006.

Blamey, J, Anthony, E. J., Wang, J. and Fenell, P.S. (2010) The calcium looping cycle for large-scale CO$_2$ capture. *Progress in Energy and Combustion Science*, 36, 260–279.

Bodansky, D. (2001) The environmental paradox of nuclear power. *Environmental Practice*, 3, 86–88.

Böhmer, M., Langnickel, U. and Sanchez, M. (1991) Solar steam reforming of methane. *Solar Energy Materials*, 24, 441–448.

Bockris, J. O., Szklarczyk, M., Aliasgar, Q. (1983) Photo-assisted electrolysis cell with p-silicon and n-silicon electrodes. *United States Patent 4501804*. Filing Date: 08/08/1983

Charvin, P., Abanades, S., Flamant, G. and Lemort, F. (2007) Two-step water splitting thermo-chemical cycle based on iron oxide redox pair for solar hydrogen production. *Energy*, 32, 1124–1133.

Charvin, P., Stéphane, A., Florent, L. and Gilles, F. (2008) Analysis of solar chemical processes for hydrogen production from water splitting thermochemical cycles. *Energy Conversion and Management*, 49, 1547–1556.

Conibeer, G. J. and Richards, B.S. (2007) A comparison of PV/electrolyser and photoelectrolytic technologies for use in solar to hydrogen energy storage systems. *International Journal of Hydrogen Energy*, 32, 2703–2711.

Corgnale, C. and Summers, W. A. (2011) Solar hydrogen production by the Hybrid Sulfur process. *International Journal of Hydrogen Energy*, 36, 11604–11619.

Crabtree R. H. (2010) Energy Production and Storage: Inorganic Chemical Strategies for a Warming World. John Wiley & Sons. 2010. West Sussex, UK. ISBN 978-0-470-74986-9. pp. 35–52.

CSPCS. (2009) Clean production reference for the alkali industry. *China State Petroleum Clean Production Standards*. No.Guofa 16, 2009. (in Chinese).

Currao, A. (2007). Photoelectrochemical water splitting. *Chimia*, 61, 815–819.

Dahl, J., Buechler, K., Finley, R., Stanislaus, T., Weimer, A., Lewandowski, A., Bingham, C., Smeets, A. and Schneider, A. (2002) Rapid solar-thermal dissociation of natural gas in an aerosol flow reactor. *Proceedings of the 2002 U.S. DOE Hydrogen Program Review*. NREL/CP-610-32405

Dalcor Consultants Ltd. (2005) *Canadian Hydrogen Inventory*. West Vancouver, BC, June, 2005.

David, J. and Herzog, H. (2000) The cost of carbon capture. *The 5th International Conference on Greenhouse Gas Control Technologies*, August 13–16, 2000, Cairns, Australia.

De Falco, M. annd Piemonte, V. (2010) Solar enriched methane production: Assessment of plant potentialities and applications. *Applied Technologies and Innovations* 1 1–8.

Dersch, J., Schwarzbözl, P. and Richert, T. (2011) Annual yield analysis of solar tower power plants with GREENIUS. *Journal of Solar Energy Engineering*, 133, 031017-1~031017-9.

Dunn, R. I., Hearps, P. J. and Wright, M. N. (2012) Molten-salt power towers: newly commercial concentrating solar storage. *Proceedings of the IEEE*, 100, 504–515.

Eck, M. and Hennecke, K. (2007) Heat transfer fluids for future parabolic trough solar thermal power plants. *Proceedings of ISES World Congress 2007 (Vol. I–Vol. V)*. Eds: Goswami D. Y., Zhao Y. Publisher: Springer. 2009, 5, 1806–1812.

EIA (US Energy Information Administration). (2008) *World Energy Overview: 1996–2006*. Report Released: June–December 2008. http://www.eia.gov/iea/overview.html. Accessed on March 10, 2012.

Elder, R. H., Priestman, G. H., Ewan, B. C. and Allen, R.W.K. (2005) The separation of HIx in the sulphur-iodine thermochemical cycle for sustainable hydrogen production. *Process Safety and Environmental Protection*, 83, 343–350.

Envestra Limited. (2010) About Natural Gas. http://www.natural-gas.com.au/about/references.html. Accessed on December 8, 2010.

European Commission. (2009) Joint Research Centre (JRC). Institute for Energy. EUR 24125 EN – 2009. ISBN 978-92-79-14612-1. Luxembourg: Publications Office of the European Union.

European Commission. Solar hybrid gas turbine electric power system. Solgate Progress Report. EUR 21615. http://ec.europa.eu/research/energy/pdf/solgate_en.pdf. Accessed on May 7th, 2010.

EVWorld. (2010) Energy Content of Fuels. http://www.evworld.com/library/energy_numbers.pdf Accessed on December 8, 2010.

ForestBioEnergy. (2010) Sustainable forestry for bioenergy and bio-based products. Energy Basics. Fact Sheet 5.8, pp. 189–191. http://www.forestbioenergy.net/training-materials/fact-sheets/module-5-fact-sheets/fact-sheet-5-8-energy-basics Accessed on December 20, 2010.

Finkenrath, M. (2011) *Cost and performance of carbon dioxide capture from power generation*. http://www.iea.org/papers/2011/costperf_ccs_powergen.pdf. Accessed on April 11, 2012.

Fornero, E. L., Chiavassa, D. L., Bonivardi, A. L. and Baltanas, M.A. (2011) CO_2 capture via catalytic hydrogenation to methanol: thermodynamic limit vs. 'kinetic limit'. *Catalysis Today*, 172, 158–165.

Forsberg, C. W. (2002) Hydrogen, electricity, and nuclear power. *Nuclear news*, pp. 30–31. Sept. 2002.

Forsberg, C. W., Peterson, P. F. and Zhao, H. (2007) High-temperature liquid-fluoride-salt closed-Brayton-cycle solar power towers. *Journal of Solar Energy Engineering*, 129, 141–146.

Freedonia Group. (2010) World Hydrogen Industry Study with Forecasts for 2013 & 2018. Study# 2605. February 2010.

Giaconia, A., Labach, I., Caputo, G. and Sau, S. (2010) Experimental and Theoretical Studies of Solar Steam Reforming Assisted by Molten Salts. *18th World Hydrogen Energy Conference 2010 – WHEC 2010. Parallel Sessions Book 3: Hydrogen Production Technologies – Part 2. Proceedings of the WHEC*. May 16–21, 2010, Essen, Germany.

Göttsche, J., Hoffschmidt, B., Schmitz, S., Sauerborn, N., Buck, R., Teufel, E., Badstubner, K., Ifland, D. and Rebholz, C. (2010) Solar concentrating systems using small mirror arrays. *Journal of Solar Energy Engineering*, 132, 011003-1–011003-4.

Grady, D. (2011) Precautions should limit health problems from nuclear plant's radiation. *The New York Times-Asia Pacific*. March 15, 2011. http://www.nytimes.com/2011/03/16/world/asia/16health.html. Accessed on April 12, 2012.

Grätzel, M. (2003) Dye-sensitized solar cells. *Journal of Photochemistry and Photobiology C: Photochemistry Reviews*, 4, 145–153 .

Grimes, C. A. and Varghese, O. K. (2007) *Light, water, hydrogen: the solar generation of hydrogen by water photoelectrolysis*. Publisher: Springer. 1st edition, December 4, 2007. pp. 115–157.

Guo, H., Kasahara, S., Onuki, K., Zhang, P. and Xu, J. (2011) Simulation study on the distillation of hyper-Pseudoazeotropic $HI–I_2–H_2O$ mixture. *Industry and Engineering Chemistry Research*, 50, 11644–11656.

Hagiwara, H., Ono, N., Inoue, T., Matsumoto, H. and Ishihara, T. (2006) Dye-sensitizer effects on a Pt/KTa(Zr)O$_3$ catalyst for the photocatalytic splitting of water. *Angewandte Chemie International Edition*, 45, 1420–2422.

Haltiwanger, J. F., Davidson, J. H. and Wilson, E. J. (2010) Renewable hydrogen from the Zn/ZnO solar thermo chemical cycle: A cost and policy analysis. *Journal of Solar Energy Engineering*, 132, 041011-1–041011-8.

Han, B., Zhou, C., Wu, J., Tempel, D. and Cheng, H. (2011) Understanding CO_2 capture mechanisms in aqueous monoethanolamine via first principles simulations. *The Journal of Physical Chemistry Letters*, 2, 522–526.

Hanna, M. C. and Nozik, A. J. (2006) *Solar conversion efficiency of photovoltaic and photoelectrolysis cells with carrier multiplication absorbers. Journal of Applied Physics*, 100, 074510-1–074510-8.

Haueter, P., Seitz, T. and Steinfeld, A. (1999) A new high-flux solar furnace for high-temperature thermochemical research. *Journal of Solar Energy engineering*, 121, 77–80.

Herring. S. (2006) *Laboratory-Scale High Temperature Electrolysis System*. 2006 DOE Hydrogen, Fuel Cells & Infrastructure Technologies Program Review Washington DC, May 17, 2006.

Herrmann, U., Kelly, B. and Price, H. (2004) Two-Tank molten salt storage for parabolic trough solar power plants. *Energy*, 29, 883–893.

Hinkley, J. T., O'Brien, J. A., Fell, C. J. and Lindquist, S. (2011) Prospects for solar only operation of the hybrid sulphur cycle for hydrogen production. *International Journal of Hydrogen Energy*, 36, 11596–11603.

Hirsch, D., Epstein, M. and Steinfeld, A. (2001) The solar thermal decarbonization of natural gas. *International Journal of Hydrogen Energy*, 26, 1023–1033.

House, K. Z., Harvey, C. F., Aziz, M. J. and Schrag, D.P. (2009) The energy penalty of postcombustion CO2 capture & storage and its implications for retrofitting the U.S. installed base. *Energy and Environmental Science*, 2, 193–205.

IEA (International Energy Agency). (2011) *World Energy Outlook 2011*. Released on November 9, 2011. http://www.worldenergyoutlook.org/docs/weo2011/key_graphs.pdf. Accessed on March 20, 2012.

IEA (International Energy Agency). (2012) Hydrogen Production & Distribution. *IEA Energy Technology Essentials*. http://www.iea.org/techno/essentials5.pdf. Accessed March 20, 2012.

IEA (International Energy Agency). (2010) Hydrogen production & distribution. *IEA Energy Technology Essentials* http://www.iea.org/techno/essentials5.pdf. Accessed on December 20, 2010.

Ivy, J. (2004) Summary of electrolytic hydrogen production milestone completion report. *Technical report of National Renewable Energy Laboratory*. September 2004. NREL/MP-560-36734.

Jing, D., Guo, L., Zhao, L., Zhang, X., Liu, H., Li, M., Shen, S., Liu, G., Hu, X., Zhang, X., Ma, L. and Guo, P. (2010) Efficient solar hydrogen production by photocatalytic water splitting: From fundamental study to pilot demonstration. *International Journal of Hydrogen Energy*, 35, 7087–7097.

Kennedy, C. E. and Price, H. (2005) Progress in development of high-temperature solar-selective coating. *Proceedings of ISEC 2005. Paper No. ISEC2005-76039. 2005 International Solar Energy Conference*. August 6–12, 2005, Orlando, Florida USA.

Khan, M. A., Chen, Y. and Boehm, R. (2004) Process analysis and simulation of the solar thermochemical hydrogen generation. *ASME 2004 International Mechanical Engineering Congress and Exposition (IMECE2004)*. Paper no. IMECE2004-59647. pp. 195–202. November 13–19, 2004, Anaheim, California, USA.

Kharaka, Y. K., Cole, D. R., Hovorka, S. D., Gunter, W.D., Knauss, K.G. and Freifeld, B.M. (2006) Gas-water-rock interactions in the Frio Formation following CO_2

injection: implications for the storage of greenhouse gases in sedimentary basins. *Geology*, 34, 577–580.

Khaselev, O., Bansal, A. and Turner, J. A. (2001) High-efficiency integrated multijunction photovoltaic/electrolysis systems for hydrogen production. *International Journal of Hydrogen Energy*, 26, 127–132.

Kramer, D. A. (2005) Nitrogen (Fixed)—Ammonia, U.S. Geological Survey, Mineral Commodity Summaries. Released on January 2005. pp. 116–117. http://minerals.usgs.gov/minerals/pubs/commodity/nitrogen/nitromcs05.pdf. Accessed on January 20, 2011.

Kubo, S, Kasahara, S., Okuda, H., Terada, A., Tanaka, N., Inaba, Y., Ohashi, H., Inagaki, Y., Onuki, K. And Hino, R. (2004) A pilot test plan of the thermochemical water-splitting iodine–sulfur process. *Nuclear Engineering and Design*, 233, 355–362.

Kudo, A. (2007) Photocatalysis and solar hydrogen production. *Pure and Applied Chemistry*, 79, 1917–1927.

Kudo, A. and Miseki, Y. (2009) Heterogeneous photocatalyst materials for water splitting. *Chemical Society Review*, 38, 253–278.

Lasich, J. B. (1999) Production of hydrogen from solar radiation at high efficiency. *US Patent No. 5,973,825*. October 26, 1999.

Lee, S. C., Choi, B. Y., Lee, S. J., Jung, S.Y., Ryu, C.K. and Kim, J.C. (2004) CO_2 absorption and regeneration using Na and K based sorbents. *Studies in Surface Science and Catalysis*, 153, 527–530.

Lee, J. B., Ryu, C. K., Baek, J., Lee, J.H., Eom, T.H. and Kim, S.H. (2008) Sodium-based dry regenerable sorbent for carbon dioxide capture from power plant flue gas. *Industrial and Engineering Chemistry Research*, 47, 4465–4472.

Le Gal, A., Abanades, S. and Flamant, G. (2010) Development of mixed metal oxides for thermochemical hydrogen production from solar water-splitting. *Parallel Sessions Book 2: Hydrogen Production Technologies – Part 1. Proceedings of the WHEC*. 18th World Hydrogen Energy Conference 2010 – WHEC, May 16–21, 2010, Essen, Germany.

Lewis, M. A. (2008) Update on the Cu-Cl cycle R& D effort. Workshop of the ORF Hydrogen Project at AECL Chalk River Laboratories. pp. 25–33. Chalk River, Ontario, October 17, 2008.

Lewis, M. A. and Taylor, A. (2006) High temperature thermochemical processes. *DOE Hydrogen Program, Annual Progress Report*, pp. 182–185, Washington DC, 2006.

Li, Q., Guo, B., Yu, J., Ran, J., Zhang, B., Yan, H. and Gong, J.R. (2011) Highly efficient visible-light-driven photocatalytic hydrogen production of CdS-cluster-decorated graphene nanosheets. *Journal of the American Chemical Society* 133 10878–10884.

Liang, Y. and Harrison, D. P. (2004) Carbon Dioxide Capture Using Dry Sodium-Based Sorbents. *Energy and Fuels*, 18, 569–575.

Licht, S. (2005) Solar water splitting to generate hydrogen fuel – a photothermal electrochemical analysis. *International Journal of Hydrogen Energy*, 30, 459–470.

Licht, S., Wang B., Mukerji S., T. Soga, Umeno M. and Tributsch H. (2000) Efficient solar water splitting, exemplified by RuO_2-catalyzed AlGaAs/Si photoelectrolysis. *Journal of Physical Chemistry, B, 104*, 8920–8924.

Little, M. and Jackson, R.B. (2010) Potential impacts of leakage from deep CO_2 geosequestration on overlying freshwater aquifers. *Environmental Science and Technology*, 44, 9225–9232.

Litwin, R. Z. and Pinkowski, S.M. (2010) Solar power for thermochemical production of hydrogen. *US patent*. Patent No.: US7,726,127 B2, June 1, 2010.

Maeda, K. (2011) Photocatalytic water splitting using semiconductor particles: history and recent developments. *Journal of Photochemistry and Photobiology C: Photochemistry Reviews* 12, 237–268.

Maeda, K. and Domen, K. (2010) Photocatalytic water splitting: recent progress and future challenges. *Journal of Physical Chemistry Letters*, 1, 2655–2661.

Maeda, K., Teramura, K., Lu, D., Takata, T., Saito, N., Inoue, Y. and Domen, K. (2006) Photocatalyst releasing hydrogen from water. *Nature*, 440, 295.

Maegami, Y., Iguchi, F. and Yugami H. (2011) Efficient solar methane reforming using spectrally controlled thermal radiation produced by concentrated solar radiation. *Proceedings of the ASME 5th International Conference on Energy Sustainability (ES2011)*. Paper No. ES2011-54612. August 7–10, 2011, Washington, DC, USA.

MacKenzie, A., Granatstein, D. L., Anthony, E. J. and Abanades, J.C. (2007) Economics of CO_2 capture using the calcium cycle with a pressurized fluidized bed combustor. *Energy Fuels*, 21, 920–926.

Mahmoudkhani, M. and Keith, D. W. (2009) Low-energy sodium hydroxide recovery for CO_2 capture from atmospheric air—thermodynamic analysis. *International Journal of Greenhouse Gas Control*, 3, 376–384.

Makhijani, A. (2009) Nuclear power and CO_2 emission reductions comments on radioactive waste management and relative costs of options. *IEER (Institute of Energy and Environmental Research)*. http://www.ieer.org/carbonfree/NuclearPower_wastes_and_CO2_cost_reduction_considerations.pdf. Accessed on April 20th, 2010.

Martín, J. C. (2007) Solar Tres – First commercial molten salt central receiver. *NREL CSP Technology Workshop*. March 7, 2007, Denver, US.

Matsunami, J., Yoshida, S. and Oku, Y. (2000) Coal gasification with CO_2 in molten salt for solar thermal/chemical energy conversion. *Energy*, 25, 71–79.

McConnell, R., Symko-Davies, M. and Friedman, D. (2006) Multijunction photovoltaic technologies for high performance concentrators. *Proceedings of the 2006 IEEE 4th World Conference on Photovoltaic Energy Conversion*. May 2006, Waikoloa, Hawaii.

McConnell, R. D., Lasich, J. B. and Elam, C. (2005) A hybrid solar concentrator PV system for the electrolytic production of hydrogen. *Proceedings of the 20th European Photovoltaic Solar Energy Conference and Exhibition*. June 6–11, 2005, Barcelona, Spain.

Melchior, T. (2009) H_2 Produciton by the two-step water-splitting thermochemical cycle based on Zn/ZnO redox reactions. *PhD thesis*. ETH ZURICH (Swiss Federal Institute of Technology Zurich), Switzerland.

Melis, T. (2004) Maximizing Photosynthetic Efficiencies and Hydrogen Production in Microalgal Cultures. UC Berkeley, May 24, 2004. http://www.hydrogen.energy.gov/pdfs/review04/hpd_9_melis.pdf. Accessed on March 20, 2011.

Moens, L., Blake, D. M., Rudnicki, D. L. and Hale, M.J. (2003) Advanced thermal storage fluids for solar parabolic trough systems. *Journal of Solar Energy Engineering*, 125, 112–116.

Moens, L. and Blake, D. M. (2004) Advanced heat transfer and thermal storage fluids. *2004 DOE Solar Energy Technologies Program Review Meeting*. October 25–28, 2004. Denver, Colorado.

Mohapatra, S. K., Misra, M., Mahajan, V. K. and Raja, K.S. (2007) Design of a highly efficient photoelectrolytic cell for hydrogen generation by water splitting: application of TiO_2-xCx nanotubes as a photoanode and Pt/TiO_2 nanotubes as a cathode. *Journal of Physical Chemistry. C*, 111, 8677–8685.

Monnerie, N., Schmitz, M., Roeb, M., Quantius, D., Graf, D., Sattler, C. and De Lorenzo, D. (2011) Potential of hybridisation of the thermochemical hybrid-sulphur cycle for the production of hydrogen by using nuclear and solar energy in the same plant. *International Journal of Nuclear Hydrogen Production and Applications*, 2, 178–201.

Moore, R. and Parma, E. (2007) A laboratory-scale sulfuric acid decomposition apparatus for use in hydrogen production cycles. *American nuclear society annual meeting. Boston, Massachusetts*. June 24–28, 2007.

Moore, R., Vernon, M., Ho, C. K., Siegel, N.P. and Kolb, G.J. (2010) Design considerations for concentrating solar power tower systems employing molten salt. *Sandia Report SAND2010-6978*. Available to the public from U.S. Department of Commerce National Technical

Information Service. 5285 Port Royal Rd. Springfield, VA 22161. Printed September 2010.

Naterer, G.F., Fowler M., Cotton J. and Gabriel, K. (2008) Synergistic roles of off-peak electrolysis and thermo chemical production of hydrogen from nuclear energy in Canada. *International Journal of hydrogen energy* 33, 6849–6857.

Naterer, G, Gabriel, K. and Wang, Z. L. (2008) Thermochemical hydrogen production with a copper–chlorine cycle. I: oxygen release from copper oxychloride decomposition. *International Journal of Hydrogen Energy*, 33, 5439–5450.

Nikulshina, V., Hirsch, D., Mazzotti, M. and Steinfeld, A. (2006) CO_2 capture from air and co-production of H2 via the $Ca(OH)_2$–$CaCO_3$ cycle using concentrated solar power–Thermodynamic analysis. *Energy*, 31, 1379–1389.

Nikulshina, V., Gebald, C. and Steinfeld, A. (2009) CO_2 capture from atmospheric air via consecutive CaO-carbonation and $CaCO_3$-calcination cycles in a fluidized-bed solar reactor *Chemical Engineering Journal*, 146, 244–248.

Nikulshina, V., Ayesa, N., Gálvez, M. E. and Steinfeld, A. (2008) Feasibility of Na-based thermochemical cycles for the capture of CO_2 from air—Thermodynamic and thermogravimetric analyses. *Chemical Engineering Journal*, 140, 62–70.

NREL (National Renewable Energy Lab). (2010) Concentrating solar power: energy from mirrors. *Technical report-DOE/GO-102001-1147.* http://www.nrel.gov/docs/fy01osti/28751.pdf. Accessed on April 20th, 2010.

NYSERDA (New York State Energy Research and Development Authority). (2010) Hydrogen fact sheet: hydrogen production – steam methane reforming (SMR). *Technical report-Clean Energy Initiative.* http://www.getenergysmart.org/files/hydrogeneducation/6hydrogenproductionsteammethanereforming.pdf. Accessed on May 1st, 2010.

NIST (National Institute of Standards and Technology). (2012) *Chemistry WebBook.* http://webbook.nist.gov/chemistry/. Accessed on February 20, 2012.

Olah, G. A., Goeppert, A. and Prakash G. K. S. (2009) Chemical recycling of carbon dioxide to methanol and dimethyl ether: from greenhouse gas to renewable, environmentally carbon neutral fuels and synthetic hydrocarbons. *Journal of Organic Chemistry*, 74, 487–498.

Park, S. H., Roy, A., Beaupré, S., Cho, S., Coates, N., Moon, J.S., Moses, D., Leclerc, M., Lee, K. and Heeger, A.J. (2009) Bulk heterojunction solar cells with internal quantum efficiency approaching 100%. *Nature Photonics*, 3, 297–302.

Patel, S. (2011) Spanish power tower supplies 24 hours of electricity. *Power-Business and Technology for the Global Generation Industry.* September 1, 2011.

Pearce, J. M. (2008) Thermodynamic limitations to nuclear energy deployment as a greenhouse gas mitigation technology. *International Journal of Nuclear Governance, Economy and Ecology*, 2, 113–130.

Perkins, C. M., Woodruff, B. and Andrews, L. (2008) Synthesis gas production by rapid solar thermal gasification of corn stover. NREL/CD-550-42709. *14th Biennial CSP SolarPACES (Solar Power and Chemical Energy Systems) Symposium*, 4–7 March 2008, Las Vegas, Nevada.

Petrasch, J. and Steinfeld, A. (2007) Dynamics of a solar thermochemical reactor for steam-reforming of methane. *Chemical Engineering Science*, 62, 4214–4228.

Piatkowski, N. and Steinfeld, A. (2008) Solar-driven coal gasification in a thermally irradiated packed-bed Reactor. *Energy and Fuels*, 22, 2043–2052 .

Piatkowski, N., Wieckert, C. and Steinfeld, A. (2008) Experimental investigation of a packed-bed solar reactor for the steam-gasification of biomass charcoal. ES2008-54118. *Proceedings of ES2008 Energy Sustainability.* August 10–14, 2008, Jacksonville, Florida, USA.

Piatkowski, N., Wieckert, C. and Weimer A. W. (2011) Solar-driven gasification of carbonaceous feedstock—a review. *Energy and Environmental Science*, 4, 73–82.

Prakasam, H. E. (2008) Towards highly efficient water photoelectrolysis. *PhD thesis*. The Graduate School Department of Electrical Engineering. The Pennsylvania State University. May 2008. pp. 203–206.

Rajeshwar, K., McConnell, R. and Licht, S. (2008) *Solar Hydrogen Generation-Toward a Renewable Energy Future*. pp. 178–183. Boston, USA, ISBN: 978-0-387-72809-4, Springer Science + Business Media, LLC, 2008.

Riccardi J., Massimo S., Fastelli I. and Smitkova M. (2011) Modelling of Westinghouse and sulphur-iodine water splitting cycles for hydrogen production. *International Journal of Energy and Environmental Engineering*, 2, 49–62.

Riveros-Rosas, D., Herrera-Vázquez, J., Pérez-Rábago, C. A., Arancibia-Bulnes, C.A., Vazquez-Montiel, S., Sanchez-Gonzalez, M., Granados-Agustin, F., Jaramillo, O.A. and Estrada, C.A. (2010) Optical design of a high radiative flux solar furnace for Mexico. *Solar Energy*, 84, 792–800.

Rodat, S., Abanades, S. and Flamant, G. (2009) High-Temperature Solar Methane Dissociation in a Multitubular Cavity-Type Reactor in the Temperature Range 1823–2073 K. Energy and Fuels, 23, 2666–2674.

Rodat, S., Abanades, S., Grivei, E., Patrianakos, G., Zygogianni, A., Konstandopoulos, A.G. and Flamant, G. (2011) Characterisation of carbon blacks produced by solar thermal dissociation of methane. *Carbon* 49, 3084–3091.

Roeb, M. and Müller-Steinhagen, H. (2010) Concentrating on solar electricity and fuels. *Science*, 329, 773–774.

Roeb, M., Thomey, D., Graf, D., Sattler, C., Poitou, S., Pra, F., Tochon, P., Mansilla, C., Robin, J.C., Le Naour, F., Allen, R.W.K., Elder, R., Atkin, I., Karagiannakis, G., Agrafiotis, C., Konstandopoulos, A.G., Musella, M., Haehner, P., Giaconia, A., Sau, S., Tarquini, P., Haussener, S., Steinfeld, A., Martinez, S., Canadas, I., Orden, A., Ferrato, M., Hinkley, J., Lahoda, E. and Wong, B. (2010) HycycleS – a project on solar and nuclear hydrogen production by sulphurbased thermochemical cycles. *18th World Hydrogen Energy Conference 2010 – WHEC 2010, Parallel Sessions Book 2: Hydrogen Production Technologies- Part 1. Proceedings of the WHEC*. pp. 267–274. May 16–21, 2010, Essen, Germany.

Sakurai, M., Nakajima, H., Amir, R., Onuki, K. and Shimizu, S. (2000) Experimental study on side-reaction occurrence condition in the iodine-sulfur thermochemical hydrogen production process. *International Journal of Hydrogen Energy*, 23, 613–619.

Salvador, C., Lu, D., Anthony, E. J. and Abanades, J.C. (2003) Enhancement of CaO for CO_2 capture in an FBC environment. *Chemical Engineering Journal*, 96, 187–195.

Sattler, C., Roeb, M. and Houaijia, A. (2011) Solar heat and power for hydrogen production via high temperature electrolysis. *2nd-RelHy International Workshop on High Temperature Water Electrolysis Towards Large Scale Demonstration*. July 5–7, 2011, Imperial College, London, UK.

Schmitz, M. (2009) Salt-free solar: CSP tower using air. *Renewable Energy World Magazine*. http://www.renewableenergyworld.com/rea/news/article/2009/03/salt-free-solar-csp-tower-using-air. Accessed on May 20th, 2010.

Schramek, P., Mills, D. R. and Stein, W. (2009) Design of the heliostat field of the CSIRO solar tower. *Journal of Solar Energy Engineering*, 131, 024505-1–024505-6.

Schultz, K. (2003) Thermochemical production of hydrogen from solar and nuclear energy. *Technical report for the Stanford global climate and energy project*. General Atomics, Sandiego, CA, 2003.

Schunk, L. O., Lipiński, W. and Steinfeld, A. (2009) Heat transfer model of a solar receive reactor for the thermal dissociation of ZnO-Experimental validation at 10 kW and scale-up to 1 MW. *Chemical Engineering Journal*, 150, 502–508.

Schwarzbözl, P., Buck, R., Sugarmen, C., Ring, A., Crespo, M.J.M., Altwegg, P. and Enrile, J. (2006) Solar gas turbine systems: Design, cost and perspectives. *Solar Energy*, 80, 1231–1240.

Sergeant, N. P., Agrawal, M. and Peumans, P. (2010) High performance solar-selective absorbers using coated sub-wavelength gratings. *Optics Express*, 18, 5525–5540.

Siriwardane, R. V., Robinson, C., Shen, M. and Simonyi, T. (2007) Novel regenerable sodium-based sorbents for CO_2 capture at warm gas temperatures. *Energy and Fuels*, 21, 2088–2097.

Solarska, R., Jurczakowski, R. and Augustynski, J. (2012) A highly stable, efficient visible-light driven water photoelectrolysis system using a nanocrystalline WO3 photoanode and a methane sulfuric acid electrolyte. *Nanoscale*, 4, 1553–1556.

Stolaroff, J. K. (2006) Capturing CO_2 from ambient air: a feasibility assessment. *PhD thesis*. August, 2006. Carnegie Mellon University, Pittsburgh, PA, US.

Spelling, J. (2009) Thermo-economic optimisation of solar tower thermal power plants. *22nd International Conference on Efficiency, Cost, Optimisation, Simulation and Environmental Impact of Energy Systems (ECOS 2009)*. August 31–September 3, 2009. Foz do Iguaçú – BRAZIL.

Spicer, J. I., Raffo, A. and Widdicombe, S. (2007) Influence of CO_2-related seawater acidification on extracellular acid–base balance in the velvet swimming crab Necora puber. *Marine Biology*, 151, 1117–1125.

Steinfeld, A. (2002) Solar hydrogen production via a two-step water-splitting thermochemical cycle based on Zn/ZnO redox reactions. *International Journal of Hydrogen Energy*, 27, 611–619.

Stewart, F.F. (2005) Evaluation and characterization of membranes for $HI/H_2O/I_2$ water separation for the S-I cycle FY 2005 Report for Project: sulfur cycles – acid concentration membranes nuclear hydrogen initiative. *Idaho National Laboratory Technical Report*. September 2005. Document No.: INL/EXT-05-00723.

Summers, W.A. (2009) Hybrid sulfur thermochemical cycle. *DOE Hydrogen Program 2009 Annual Merit Review*. Report No.SRNS-ST1-2009-00223. Released on May 19, 2009.

Taggart, S. (2008) Hot stuff : CSP and the power tower. *Renewable Energy Focus*. pp. 51–54. http://www.trec-uk.org.uk/articles/REF/ref_0903_pg51_55.pdf. Accessed on May 7th, 2010.

Teets, T.S. and Nocera, D.G. (2011) Photocatalytic hydrogen production. *Chemical Communications*, 47, 9268–9274.

Terada, A., Iwatsuki, J., Ishikura, S., Noguchi, H., Kubo, S., Okuda, H., Kasahara, S., Tanaka, N., Ota, H., Onuki, K. and Hino, R. (2007) Development of hydrogen production technology by thermochemical water splitting IS process pilot test plan. *Journal of Nuclear Science and Technology*, 44, 477–482.

Thompson, J.R. (2005) Cost analysis of a concentrator photovoltaic hydrogen production system. *International Conference on Solar Concentrators for the Generation of Electricity or Hydrogen*. 1–5 May 2005, Scottsdale, Arizona.

Thuillier, G., Hersé, M., Labs, D., Foujols, T., Peetermans, W., Gillotay, D., Simon, P.C. and Mandel, H. (2003) The solar spectral irradiance from 200 to 2400 nm as measured by the SOLSPEC spectrometer from the atlas and EURECA missions. *Solar Physics*, 214, 1–22.

Tzimas, E. (2009) *The Cost of Carbon Capture and Storage Demonstration Projects in Europe*.

T-Raissi, A. and Paster, M. (2003) Analysis of solar thermochemical water-splitting cycles for hydrogen production. *Hydrogen, Fuel Cells, and Infrastructure Technologies. FY 2003 Progress Report*. http://www.fsec.ucf.edu/en/research/hydrogen/analysis/documents/FY03_ProgressReport.pdf. Accessed on March 10, 2011.

van Helden, W.G.J., van Zolingen, R. J.Ch. and Zondag, H.A. (2004) PV thermal systems: PV panels supplying renewable electricity and heat. *Progress in photovoltaics: Research and Applications*, 12, 415–426.

Von Zedtwitz-Nikulshyna. (2009) CO_2 capture from atmospheric air via solar driven carbonation-calcination cycles. *PhD thesis*. Department of mechanical engineering. ETH Zurich.

Vayssieres, L. (2009) Nanoparticle-assembled catalysts for photochemical water splitting. pp. 507–521 and pp. 589–622. Singapore, ISBN 978-0-470-82397-2, John Wiley & Sons, 2009.

Wang, Z., Naterer, G.F. and Gabriel, K. (2008) Multiphase reactor scale-up for Cu–Cl thermochemical hydrogen production. *International Journal of Hydrogen Energy*, 33, 6934–6946.

Wang, Z.L., Naterer, G.F. and Gabriel, K.S. (2009) Comparison of different copper-chlorine thermochemical cycles for hydrogen production. *International Journal of Hydrogen Energy*, 34, 3267–3276.

Walter, M.G., Warren, E.L., McKone, J. R., Boettcher, S.W., Mi, Q., Santori, E.A. and Lewis, N.S. (2010) Solar water splitting cells. *Chemical Reviews*, 110, 6446–6473.

Wang, Z.L. and Naterer, G.F. (2010) Greenhouse gas reduction in oil sands upgrading and extraction operations with thermochemical hydrogen production. *International Journal of Hydrogen Energy*, 35, 11816–11828.

Wang, X., Goeb, S., Ji, Z., Pogulaichenko, N.A. and Castellano, F.N. (2011) Homogeneous photocatalytic hydrogen production using π-conjugated platinum(II) arylacetylide sensitizers. *Inorganic Chemistry*, 50, 705–707.

Weimer A. W., Dahl, J. and Buechler, K. (2001) Thermal dissociation of mehane using a solar coupled aerosol reactor. *Proceedings of the 2001 DOE Hydrogen Program Review*. NREL/CP-570-30535.

Wu, B. and Reddy, R.G. (2001) Novel ionic liquid thermal storage for solar thermal electric power systems. *Proceedings of Solar Forum 2001. Solar Energy: The Power to Choose*. April 21–25, 2001, Washington, DC.

Xiao, L., Wu, S.Y. and Li, Y.R. (2012) Advances in solar hydrogen production via two-step water-splitting thermochemical cycles based on metal redox reactions. *Renewable Energy*, 41, 1–12.

Xu, R. and Wiesner, T.F. (2012) Dynamic model of a solar thermochemical water-splitting reactor with integrated energy collection and storage. *International Journal of Hydrogen Energy*, 37, 2210–2223.

Xue, E., O'Keeffe, M. and Ross, J.R.H. (1996) Water-gas shift conversion using a feed with a low steam to carbon monoxide ratio and containing sulphur. *Catalysis Today*, 30, 107–118.

Yamada, N. and Ijiro, T. (2011) Design of wavelength selective concentrator for micro PV/TPV systems using evolutionary algorithm. *Optics Express*, 19, 13140–13149.

Yang, L. and Leung, W.W. (2011) Application of a bilayer TiO_2 nanofiber photoanode for optimization of dye-sensitized solar cells. *Advanced Materials*, 23, 4559–4562.

Yeh, J. T. and Pennline, H.W. (2001) Study of CO_2 absorption and desorption in a packed column, Energy and Fuels 15, 274–278.

Yang, Y. (2011) Statistical methods for integrating multiple CO_2 leak detection techniques at geologic sequestration sites. *PhD thesis*. Civil and Environmental Engineering. Carnegie Mellon University Pittsburgh, PA, US. May, 2011.

Zamfirescu, C. and Dincer, I., Naterer, G. F. (2011) Analysis of a photochemical water splitting reactor with supramolecular catalysts and a proton exchange membrane. *International Journal of Hydrogen Energy*, 36, 11273–11281.

Zedtwitz, P. v., Petrasch, J., Trommer, D. and Steinfeld, A. (2006) Hydrogen production via the solar thermal decarbonization of fossil fuels. *Solar Energy*, 80, 1333–1337.

Zedtwitz, P. v. and Steinfeld, A. (2003) The solar thermal gasification of coal – energy conversion efficiency and CO2 mitigation potential. *Energy*, 28, 441–456.

Zhao, C., Chen, X. and Zhao, C. (2010) Multiple-cycles behavior of K_2CO_3/Al_2O_3 for CO_2 Capture in a fluidized-bed reactor. *Energy and Fuels*, 24, 1009–1012.

Chapter 10

Photoelectrochemical cells for hydrogen production from solar energy

Tania Lopes, Luisa Andrade & Adelio Mendes
Laboratório de Engenharia de Processos, Ambiente e Energia (LEPAE), Faculdade de
Engenharia da Universidade do Porto, Rua Roberto Frias, Porto, Portugal

10.1 INTRODUCTION

The awareness concerning carbon dioxide emissions and the depletion of fossil fuel reserves motivates the development of innovative processes to take advantage from renewable energy sources (Grätzel, 2005). The world power consumption is currently about 13 TW and it is expected to increase up to 23 TW by 2050 (Dobran, 2010). With approximately 120 PW of solar energy continuously striking the earth at any given moment, the challenge in converting sunlight into electricity via photovoltaic (PV) cells is to reduce the cost per watt of delivered solar electricity (Krol et al., 2008 and Nathan, 2005), which is already approximately 0.65 €/Wp for crystalline silicon modules. The solar PV technology has greatly evolved in the last decade and it is now a well-established way to convert solar energy into electric energy, which accounts presently more than 21 GW installed worldwide (The Energy Report, 2011). Nevertheless, this technology only works on a daily basis and it largely depends on the amount of solar radiation available. Thus, an effective method to store energy for later dispatch is still needed (Trieb, 2005). A practical way to convert sunlight into a storable energy form is using a photoelectrochemical (PEC) cell that splits water into hydrogen and oxygen by light-induced electrochemical processes (Grimes et al., 2008).

Hydrogen production via photoelectrochemical water-splitting is a thriving alternative that combines photovoltaic cells with an electrolysis system (Bard and Fox, 1995; Khaselev and Turner, 1998). The major advantage is that solar harvesting, conversion and storage are combined in a single integrated system (Nathan, 2005). The hydrogen generated by this process has the potential to be a sustainable carbon-neutral fuel since it is produced from a renewable source and it can be stored or transformed into other chemicals such as methanol or methane (Grimes, 2008; Zerta, 2008).

10.2 PHOTOELECTROCHEMICAL CELLS SYSTEMS OVERVIEW

10.2.1 Solar water-splitting arrangements

Converting sunlight into hydrogen and oxygen through water-splitting can be accomplished via different technologies, as sketched in Figure 10.2.1. More specifically, via three general types of devices: *i*) composed devices – photovoltaic (PV) cell associated with an electrolyzer or photovoltaic (PV) cell associated with a PEC cell; *ii*) stand

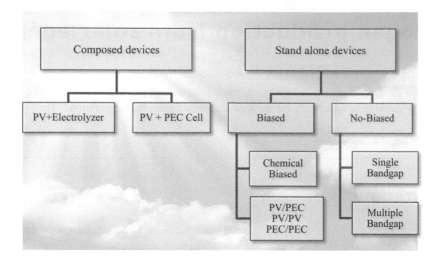

Figure 10.2.1 Solar water splitting based on PEC cells, PV cells or combined arrangements systems. The red line represents the "holy grail" of the PEC system.

alone devices – semiconductor-liquid junction (SCLJ) photoelectrochemical cell (Krol et al., 2008; Grimes et al., 2008; Aruchamy et al., 1982; Minggu et al., 2010); and *iii*) by thermochemical cycles (Coelho et al, 2010). The third technology will not be considered in this chapter.

10.2.1.1 Composed devices

Up to today, no semiconductor photoelectrode is able to efficiently perform alone water-splitting and thus an extra bias must be supplied. The most developed technology is the PV device/electrolyzer arrangement where the photovoltaic cells are silicon based, achieving maximum efficiencies of 15%, and efficiencies of electrolyzers is often around 75%[1]. For instance, combining commercial 12% efficiency PV modules with a water electrolysis unit operating with an energy conversion efficiency of approximately 65% (output voltage of 1.9 V) results in a solar-to-hydrogen efficiency of about 7.8% (Kruse et al., 2002; Bansal et al., 1999; Bilgen, 2001). Only by combining optimized PV technologies it is possible to achieve higher solar-to-hydrogen conversion efficiencies (Green et al., 2008; Conibeer and Richards, 2007).

Even if biasing an electrolyzer with a separate set of solar cells is very attractive from an efficiency point of view, the fact of involving two separate devices complicates the system and increases the cost (Krol et al., 2008), besides being more energy dissipative. Furthermore, in a system PV + electrolyzer at least four silicon PV cells connected in series are required to generate the desired voltage for water-splitting, which under

---

[1]The efficiency of an electrolyzer is defined as $\eta_Z = E°/V$, where $E°$ is the thermodynamic cell potential (1.23 V for water electrolysis) and V is the voltage applied to the cell under operating conditions.

unfavorable climatic conditions such as partial shading, haze or cloudiness may ultimately interrupt the photoelectrolysis. Similar problems are observed when a PV cell is used as external bias for a PEC cell to promote water-splitting. In this case, the electric current generated by the PV cell goes directly to the PEC cell instead of feeding an electrolyzer, resulting in a cheaper but not necessarily more efficient embodiment (Grimes et al., 2008; Minggu et al., 2010). Still, two separated parts must be considered when estimating the initial and operating costs. Moreover, the available area for solar exposure must be substantially increased since both PV and PEC cells have to be illuminated (Minggu et al., 2010; Conibeer and Richards, 2007). In this sense, a more effective approach would be to "merge" the PV cell with an electrolyzer to make photoelectrochemical devices with a semiconductor-liquid junction. Thus, several efforts have been made to design a monolithic system to partially avoid the previously mentioned technological and economic drawbacks (Minggu et al., 2010).

10.2.1.2 Single devices

Single water-splitting devices (Figure 10.2.1) can be divided into **biased** and **zero biased** systems. Concerning **biased** systems, there are chemically biased photo-assisted photoelectrolysis cells and tandem devices (Grimes et al., 2008; Minggu et al., 2010). In the first case, the bias is achieved using two different electrolytes (e.g. acid and basic electrolytes) placed in two separated half-cells. However, this configuration is not self-sufficient, relying not only on sunlight but also on additional input of chemicals to stabilize the electrolyte solutions (Minggu et al., 2010). In the tandem approach, the cell is normally characterized by layered stacked or hybrid structures involving several different semiconductor films placed on top of each other. In this configuration, at least one of the substructures must work as a bias source. These internal biased photoelectrode tandem structures can be subdivided into: *i*) PV/PEC (Miller et al., 2005); *ii*) PV/PV (Khaselev et al., 2001) and; *iii*) PEC/PEC (Grätzel and Augustynski, 2001).

The use of PV/PEC systems has an advantage over PV/PV systems because the PEC face (layer) can replace the face conductor grids that partially obscure the PV layer. Consequently, PEC panels are able to reduce some cost components and improve photon capture of PV layer (James et al., 2009). Recently a new multiphoton combination of a PEC cell and two dye cells (tandem arrangement) was proposed (Brillet, 2010). Three different architectures were suggested. The authors found that the "trilevel" tandem architecture (hematite/squaraine dye/black dye) produces the highest operating current density. The expected highest solar-to-hydrogen efficiency was about 1.36%. However, this value is far below the expected 3.3% that should be possible with the nanostructured hematite photoanodes used (Brillet, 2010).

Photoelectrochemical devices with **no additional bias** represent a prospective pathway to overcome the complexity of biased systems. No-bias photoelectrochemical devices comprise single and multiple photo-system arrangements. The possible arrangements of single photo-system are:

i)   n-type semiconductor photoanode and a metal counter-electrode (Figure 10.2.2a) (Kay et al., 2006);

ii) p-type semiconductor photocathode and a metal counter-electrode (Figure 10.2.2b) (Chandra et al., 1985);

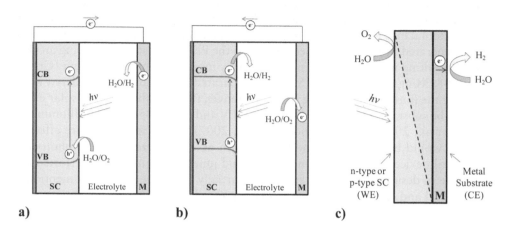

Figure 10.2.2 No-biased single photo-system configurations for solar water splitting (SC – semiconductor; M – metal; WE – working electrode; CE – counter-electrode) – (adapted from Minggu et al., 2010): a) n-type semiconductor photoanode and a metal counter-electrode; b) p-type semiconductor photocathode and a metal counter-electrode; and c) monolithic configuration.

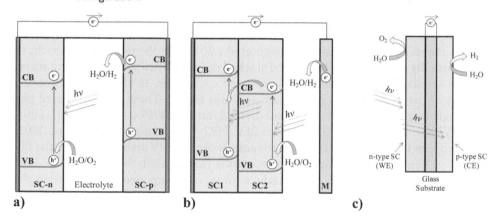

Figure 10.2.3 Different no-biased multiple photo-system configurations for solar water splitting (SC – semiconductor; M – metal; WE – working electrode; CE – counter-electrode) – (adapted fromMinggu et al., 2010): a) n- and p-type semiconductors wired; b) n- and p-type semiconductors linked by an ohmic contact; and c) hybrid systems.

iii) monolithic configuration – bipolar system and a layered metal counter-electrode (Figure 10.2.2c).

Concerning multiple photo-system arrangements, the following configurations can be identified: (Minggu et al., 2010)

i) n- and p-type semiconductors (acting as photoanode and photocathode, respectively) (Nozik, 1976), wired (Figure 10.2.3a) or linked by an ohmic contact (Figure 10.2.3b);

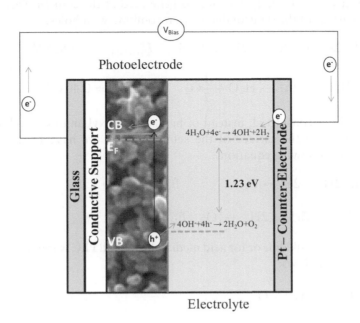

Figure 10.2.4 Schematic representation of the relevant processes involved in the photo-hydrolysis of water.

ii)   hybrid systems with layered structures involving several different semiconductor films stacked on top of each other (Figure 10.2.3c).

All the presented routes to convert solar energy into hydrogen show advantages and disadvantages. Nevertheless, there is a consensus that the single photo-system is the holy grail of PEC technology in terms of simplicity, packaging and overall system costs (Krol et al., 2008).

10.2.2 Working principles of photoelectrochemical cells for water-splitting

The principle of converting sunlight into hydrogen by water photoelectrolysis using a single photon-system, taking as an example a n-type semiconductor in an alkaline media, is illustrated in Figure 10.2.4.

The single-photon PEC system for water-splitting is composed of a semiconductor photoelectrode that absorbs photons with sufficient energy to inject electrons from the valence to the conduction band, creating electron-hole pairs – Equation 10.2.1.

$$2h\upsilon \rightarrow 2e^- + 2h^+ \quad \text{Photon-induced electron-hole pair generation} \qquad (10.2.1)$$

As sketched in Figure 10.2.4, the excited electrons percolate through the semiconductor layer reaching the counter-electrode, via the external circuit, to promote water reduction at its surface – Equation 10.2.2 – while holes oxidize water in the semiconductor surface – Equation 10.2.3 (Krol et al., 2008; Archer and Nozik, 2008). The cycle is closed when the electrolyte anions generated at the counter-electrode diffuse back to the surface of the semiconductor to recombine with holes.

$$\text{Cathode: } 2H_2O + 2e^- \rightarrow 2H_2 + 2OH^- \quad E^\circ_{H_2O/H_2} = -0.828\,V \quad (10.2.2)$$

$$\text{Anode: } 2OH^- + 2h^+ \rightarrow H_2O + \frac{1}{2}O_2 \quad E^\circ_{O_2/OH^-} = 0.401\,V \quad (10.2.3)$$

If an acid media is considered, instead of having hydroxyl anions traveling from the counter electrode to the surface of the semiconductor we have hydrogen ions, as described by the following equations:

$$\text{Cathode: } 2H^+ + 2e^- \rightarrow H_2 \quad\quad E^\circ_{H^+/H_2} = 0.0\,V \quad (10.2.4)$$

$$\text{Anode: } H_2O + 2h^+ \rightarrow 2H^+ + \tfrac{1}{2}O_2 \quad E^\circ_{H_2O/O_2} = 1.23\,V \quad (10.2.5)$$

In both cases, i.e. for alkaline or for acid media, the overall PEC water-splitting reaction can be written as follows:

$$H_2O + 2\,h\nu \rightarrow H_2 + \frac{1}{2}O_2 \quad (10.2.6)$$

A similar phenomenon occurs when a p-type semiconductor is used. Nevertheless, for this case the dominant (or the majority) charge carrier is holes, which will travel through the external circuit towards the metal counter-electrode, working now as the anode. On the other hand, electrons travel to the surface of the semiconductor in contact with the electrolyte to reduce water (Nozik, 1978).

The minimum potential of $-1.23\,V$ at $25°C$ is needed to electrolyze water. The negative sign identifies the process as not being spontaneous and so the reaction cannot occur without additional energy from an external electrical power source. This value is obtained from the following relation:

$$\Delta E^\circ = -\frac{\Delta G^\circ}{n_e F} \quad (10.2.7)$$

ΔG° is the standard Gibbs free energy change ($+237\,kJ \cdot mol^{-1}$), representing a thermodynamic minimum for splitting water into the gaseous hydrogen and oxygen at $25°C$ and 1 bar; ΔE° is the electric standard potential of the reaction.

For a direct photoelectrochemical water-split using a single-photon system, several key criteria must be simultaneously fulfilled:

i) The semiconductor system must generate sufficient voltage upon irradiation to split water;
ii) The bulk bandgap must make efficient use of the solar spectrum;
iii) The band-edge potentials at the surface must straddle the hydrogen and oxygen redox potentials according to the half-reactions described in Equations 10.2.2– 10.2.5;
iv) Low overpotentials;
v) The system must exhibit long-term stability in aqueous electrolytes;

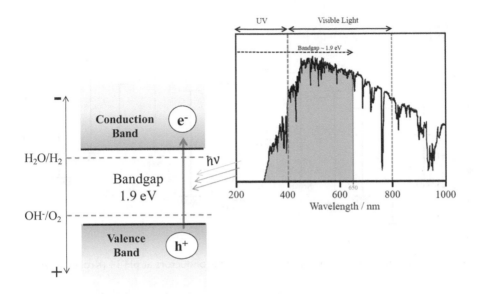

Figure 10.2.5 Schematic illustration of a semiconductor with a hypothetically ideal bandgap of 1.9 eV. Right: Intensity of sunlight vs. wavelength for AM1.5 conditions. The grey area represents the part of the spectrum that can be absorbed by a semiconductor with a bandgap of 1.9 eV.

vi) The charge transfer from the surface of the semiconductor to the solution must be selective for water-splitting and exhibit low kinetic overpotentials (Chen et al., 2010);

vii) The material must be sufficiently abundant, harmless and cost-effective.

Since overpotentials are required at various points in the system to ensure sufficiently fast reaction kinetics, i.e. related to the electrochemical reaction kinetics at anode and cathode and charge transfer (inside the electrodes and in the electrolyte), the minimum bandgap required to split water is at least 1.9 eV. This value also imposes that the semiconductor is able to absorb light for wavelengths lower than 650 nm, as show in Figure 10.2.5 (Krol et al., 2008).

Despite the research efforts to date no single semiconducting material has been found that will fulfill all the requirements needed to generate standalone devices for solar hydrogen production from water-splitting (Krol et al., 2008; Sivula et al., 2011).

10.2.3 Materials overview

The keystone in water photoelectrolysis is the development of an efficient, robust, reliable, cost-effective, and stable photoelectrode system (Grimes et al., 2008). The first material recognized to split water under UV light was TiO_2, reported by Fujishima and Honda in 1971 (Fujishima and Honda, 1972). Thenceforward, extensive efforts have been made to find a suitable material for efficient photoelectrodes. Thus, during the last three decades, different types of semiconductors were studied such as metal

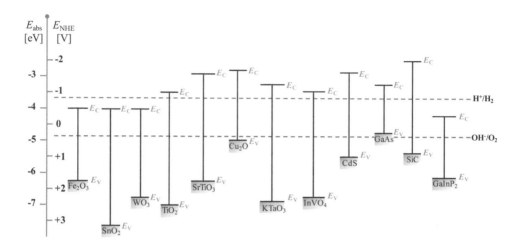

Figure 10.2.6 Energy band positions for various semiconductors at pH 14 (Krol et al., 2008).

oxide (e.g. Fe_2O_3, $SrTiO_3$, TiO_2, WO_3, $BiVO_4$, Cu_2O, etc.) and non metal oxide semiconductors (e.g. GaAs, CdS, InP, etc.) (Grimes et al., 2008). The photocorrosion stability of photoanode or photocathode, its wavelength response (bandgap) and current-voltage characteristic are important factors that determine the semiconductor performance in water-splitting – Figure 10.2.6.

10.2.3.1 Metal oxide semiconductor

Concerning the metal-oxide semiconductors, only a few are able to fulfill the bandgap and band edge requirements for operating at zero bias voltage; $SrTiO_3$, $KTaO_3$ and ZrO_2 are among them but only $SrTiO_3$ has been in fact studied (Memming, 2001). $SrTiO_3$ generates hydrogen without any additional bias even if with a barely small efficiency, less than 1%, which was ascribed to its large bandgap energy (3.4–3.5 eV) (Mavroides et al., 1976).

Preparing a semiconductor oxide that meets all criteria needed to achieve efficient water-splitting is a great challenge and the most frequently studied photoelectrode materials are TiO_2, WO_3, Fe_2O_3, $BiVO_4$ SnO_2 and Cu_2O and their modifications.

The well known titanium dioxide (TiO_2), with different crystalline structures (anatase mostly but also rutile) and arrangements as single crystal, polycrystalline or thin films, have been largely investigated (Nowotny et al., 2007a; Nowotny et al., 2007b; Nowotny et al., 2007c; Nowotny et al., 2007d; Grimes et al., 2008) mainly due to its excellent stability over a wide range of pH and applied potentials, low cost and abundance (Nowotny et al., 2007a). As a major drawback, TiO_2 only absorbs in the UV light spectrum due to its large bandgap of approximately 3.2 eV. Several attempts have been made to extend TiO_2 spectral response into the visible light by doping it with aliovalent ions, such as W, Ta, Nb, Zn, In, Li and Ge (Karakitsou et al., 1993), Pb (Rahman et al., 1999), Mo and Cr (Wilke and Breuer, 1999), Cr (Bak et al., 2002), C (Khan et al., 2002) and N (Nakamura et al., 2004; Babu et al., 2012). Wilke and co-workers showed the important effect of doping TiO_2 with Mo and Cr ions on the decrease of TiO_2 bandgap (Wilke and Breuer, 1999). An impressive reduction on the bandgap of TiO_2 photoanodes was achieved also by Khan et al. (2002) with carbon incorporation into the TiO_2-x lattice during heating in a natural gas flame (Khan et al., 2002). Nevertheless, the studies concerning the doping effect of TiO_2 semiconducting material do not provide clear conclusions, since the dopant that may have a positive effect on the bandgap (E_g) reduction, and thereby increasing the light absorption, has a negative effect in the energy conversion efficiency (ECE). Moreover, the reduction of E_g should be followed by changing other relevant functional properties. Finally, the procedures used to incorporate the dopants are often arbitrarily selected. Thus, without a solid knowledge of the time and the temperature required to incorporate the dopants, it is truly difficult to replicate the TiO_2 material and to obtain a homogeneous distribution in the semiconductor (Bak et al., 2002).

The same holds for tin dioxide (SnO_2) semiconductor; it has also a large bandgap in the range of 3.1–3.3 eV that makes this material able to absorb only the UV solar spectrum (Grimes et al., 2008). Nevertheless, an n-type single crystal of SnO_2 doped with Sb was investigated by Wrighton and co-workers for H_2 and O_2 production with an applied bias of 0.5 V under UV light illumination (Wrighton et al., 1976). However, comparing TiO_2 with SnO_2, the latter requires a slightly higher potential to achieve the photocurrent onset (Grimes et al., 2008).

Tungsten trioxide (WO_3) is an interesting semiconductor since it has an attractive bandgap of 2.5–2.7 eV (Grimes et al., 2008; Santato et al., 2001; Butler, 1977). Theoretically, a bandgap of 2.7 eV allows use of 12% of the AM 1.0 solar spectrum, a very high value compared to the barely 4% achieved with TiO_2 (Grimes et al., 2008; Butler, 1977). Although this material had been widely studied by Deb in 1972, it was Hodes in 1976 that first recognized it as an active visible-light driven photoanode for water-splitting (Hodes et al., 1976). This material shows good stability in water for pH < 4 and a favorable energy band edge for oxygen evolution (Butler, 1977). Nevertheless, the minority carrier (hole) diffusion length plays a limiting role in the photoresponse of tungsten trioxide photoanodes due to the indirect bandgap transition (Solarska et al., 2012). Usually, WO_3 photoanode is used as thin films and can be found either in the crystalline or in the amorphous forms. To obtain efficient WO_3 photoanodes, a highly crystalline structure is desirable since it minimizes the imperfections and the surface contaminations which may lead to a charge trapping and carrier recombination (Meda et al., 2010). Recently, a nanostructured WO_3 photoanode has been described capable of producing a photocurrent of about $3\,mA\,cm^{-2}$ in 3 M CH_3HSO_3 (AM 1.5 G) (Solarska et al., 2012).

Alternatively, small bandgap materials can be considered as a starting point in the research into single-photon systems, such as Fe_2O_3 or $BiVO_4$ (Krol et al., 2008; Kay et al., 2006; Luo et al., 2008; Long and Cai, 2008; Liang et al., 2008; Khan and Akikusa, 1999). Particular attention has been given to hematite (α-Fe_2O_3) and it has actually been considered a material with great potential for PEC applications. Hematite is one of the most abundant and inexpensive oxide semiconductors with an interesting bandgap of 1.9–2.3 eV, is a non-toxic material and is stable in water (Satsangi et al., 2010). As a drawback, pure-phase α-Fe_2O_3 has intrinsically poor charge carrier transportation, which limits its quantum efficiency. Moreover, it has poor oxygen evolution reaction (OER) kinetics and the band edges are not well positioned to directly carry out the reduction of water. Intensive research efforts have been conducted to improve hematite's intrinsic electronic properties by varying the deposition method or by doping the photoelectrode with Si, Ti, Pt, Mo and Cr, among other atoms (Kay et al., 2006; Satsangi et al., 2010; Glasscock et al., 2007; Hu et al., 2008; Kleiman-Shwarsctein et al., 2008; Brillet et al., 2010). In an ideal hematite photoanode, the onset is just anodic of the flat band potential with a photocurrent plateau of 12.6 mA cm^{-2} (Tilley et al., 2010). However, until the present, no research group has been successful in splitting water by means an α-Fe$_2$O$_3$ photoanode without assistance of an external bias voltage (Krol et al., 2008).

Another interesting material is BiVO$_4$ semiconductor, with a bandgap of 2.4–2.5 eV and a reasonable band edge alignment with respect to water redox potentials; in fact it has the ability to carry out the water photosplitting reaction (Sayama et al., 2006). Moreover, it has been reported that BiVO$_4$ is able to show both semiconducting properties, n- and p-types, (Vinke et al., 1992) as well as a high photon-to-current conversion efficiencies (>40%) at 420 nm (Memming, 2001; Karakitsou and Verykios, 1993). Nevertheless, further improvements on its fundamental electronic structure and stability are still needed (Sayama et al., 2006).

Over the past few years several efforts have been made in order to find an efficient harvesting semiconductor under visible light. Cuprous oxide (Cu$_2$O), which works as a photocathode, has an interesting bandgap of 2.0–2.1 eV (Hara et al., 1998). Theoretical calculations indicate that Cu$_2$O can produce up to 14.7 mA cm^{-2}, corresponding to a light-to-hydrogen conversion efficiency of 18% based on the AM 1.5 spectrum (Paracchino et al., 2011). For solar water-splitting purposes, Cu$_2$O has favorable energy band positions; the conduction band is located 0.7 V negative of the hydrogen evolution potential with the valence band lying just positive of the oxygen evolution potential (Paracchino et al., 2011). Since no overpotential is available for oxygen evolution, the reduction band edge is close to the water reduction potential, the p-type Cu$_2$O can drive half of the water-splitting reaction but an external bias must be applied to conduct the other half reaction (water oxidation) (Paracchino et al., 2011). However, the limiting factor of this material is the poor stability in aqueous solutions, since the redox potentials for the reduction and oxidation of monovalent copper oxide lie within the bandgap (Hara et al., 1998; Paracchino et al., 2011). The corrosion sensitivity issue of cuprous oxide under illumination can be addressed by depositing very thin protective layers by e.g. atomic layer deposition (ALD). In fact, using this methodology, Grätzel and co-workers have designed, up to now, the best performing oxide photoelectrode, 7.6 mA cm^{-2} at 0 V$_{RHE}$, using Cu$_2$O electrodes protected with nanolayers of Al-doped zinc oxide and titanium dioxide activated for hydrogen evolution with electrodeposited platinum nanoparticles, i.e. Cu$_2$O was coated with layers of n-type oxides with structure $5 \times (4$ nm ZnO/0:17 nm Al$_2$O$_3$)/11 nm TiO$_2$ (Paracchino et al., 2011).

10.2.3.2 Non-oxide semiconductor

Non-oxide semiconductors (p-type and/or n-type photoelectrodes) are known to efficiently harvest sunlight, converting it into electricity: amorphous, polycrystalline and crystalline silicon (a-Si, p-Si and c-Si), gallium arsenide (GaAs), cadmium telluride (CdTe), gallium phosphide (GaP), indium phosphide (InP), copper indium diselenide (CIS), copper indium gallium diselenide (CIGS) and gallium indium phosphide

Table 10.2.1 Non-oxide n-type semiconducting materials with small bandgap (Grimes et al., 2008).

Semiconductor	Bandgap
CdSe	1.7 eV
CdTe	1.4 eV
GaP	2.24 eV
GaAs	1.35 eV
InP	1.35 eV
MoS_2	1.75 eV
$MoSe_2$	1.5 eV

($GaInP_2$) (Grimes et al. 2008). As mentioned, most of the oxide semiconductors studied for water-splitting show bandgap and stability issues, thus, particular attention has been given to non-oxide materials since they have smaller bandgaps enabling the capture of a larger portion of the solar spectrum energy.

Cadmium sulfide (CdS) has well positioned band edges to efficiently reduce and oxidize water with an optical absorption of 520 nm (corresponding to a bandgap of 2.4 eV). However, it suffers from anodic photodecomposition by the photogenerated holes (Grimes et al., 2008). Similar problems of photocorrosion can be found with other n-type non-oxide semiconductors as the ones presented in Table 10.2.1.

Considering now p-type non-oxide materials, they usually show stable behavior against cathodic photodecomposition since the photoelectrons in excess migrate towards the semiconductor/electrolyte interface such as the p-Si and p-GaP (Grimes et al., 2008). However, the flat band position of these materials is unfavorable for the H_2O/O_2 redox and a large bias voltage must be applied (Nozik, 1978).

One approach to overcome stability issues is covering the unstable photoelectrodes with thin films of stable wide bandgap semiconductors with suitable band edges or with thin metal films (e.g. by chemical vapor deposition (CVD), sputtering or atomic layer deposition) (Nozik, 1978). Although single-photon systems seem to be the preferable route to produce hydrogen from solar energy, either with oxide or non-oxide materials, significant improvements should be accomplished on electronic structure and stability in order to be used alone in PEC systems – Figure 10.2.2a and 2b.

Clearly, there are three routes that may result in a high efficient system for water-splitting and without need of an additional bias. These are illustrated in Figures 10.2.2c and 10.2.3a and 10.2.3c. All strategies share the feature of having two semiconductors with different bandgaps. This provides a mechanism by which a single electron is photoexcited twice and, correspondingly generates a larger bias from light. It has been calculated that this type of system could realistically achieve a solar-to-hydrogen conversion efficiency of 21.6% (Bolton et al., 1985).

The most compelling approach is however illustrated in Figure 10.2.2c and Figure 10.2.3a. Here, various combinations of n-type and p-type semiconductors, oxide and non-oxide, such as n-TiO_2/p-GaP, n-$SrTiO_3$/p-GaP, n-Fe_2O_3/p -Fe_2O_3 have been used to eliminate the bias needed for water-splitting – Table 10.2.2. Because of the low performance of the individual electrodes in these dual-photoelectrode devices, the resulting overall efficiency is low.

Table 10.2.2 Examples of n-p photoelectrochemical cells for water splitting (FIGURE 10.2.3a).

n-SC/p-SC	Electrolyte	Energy conversion Efficiency, η	Reference
n-TiO$_2$/p-GaP	0.2 M H$_2$SO$_4$	0.25%	Nozik, 1976
n-SrTiO$_3$/p-GaP	I M NaOH	0.67%	Ohashi et al., 1977
n-Fe$_2$O$_3$/p-Fe$_2$O$_3$	0.1 M H$_2$SO$_4$	0.10%	Ingler et al., 2006

10.2.4 Stability issues – photocorrosion

In photoelectrochemical cells, stability issues are one of the major problems to be solved (Krol et al., 2008). Usually, when a semiconductor electrode is placed in contact with an electrolyte solution some reactions may occur, for instance ionic oxidation or reduction of the semiconductor with simultaneous reduction or oxidation of a component (Gadgil, 1990). The electrolytic reduction of a semiconductor is often associated with the electrons in the valence band, while the electrolytic oxidation reaction is related to holes in the conduction band as electronic reactants (Gadgil, 1990). Following the Gerischer's derivations, it is possible to formulate the simplest type of decomposition reaction involving a binary semiconductor MX and the solvation (complexing) of the elements (labeled hereafter as "solv") as stated next (Memming, 2001):

$$MX + ze^- + solv \rightarrow M + X_{solv}^{z-} \tag{10.2.8}$$

for a cathodic reaction, and

$$MX + zh^+ + solv \rightarrow M_{solv}^{z+} + X \tag{10.2.9}$$

for an anodic reaction. Using H$^+$/H$_2$ standard potentials as reference, the corresponding reaction for hydrogen may be written as:

$$\frac{1}{2}zH_2 + solv \rightarrow zH_{solv}^+ + ze^- \tag{10.2.10}$$

The addition of Equation 10.2.10 to Equation 10.2.8 or to Equation 10.2.9 yields the corresponding equations for the free energy values, $_n\Delta G_{sH}$ and $_p\Delta G_{sH}$, respectively. The decomposition potentials equations are:

$$_pE_{decomp} = {_p\Delta G_{sH/z}} \tag{10.2.11}$$

for the oxidation, and

$$_nE_{decomp} = -{_n\Delta G_{sH/z}} \tag{10.2.12}$$

for the reduction of the semiconductor (Memming, 2001).

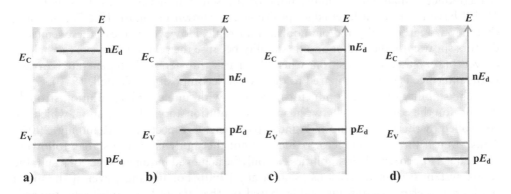

Figure 10.2.7 Relative positions of decomposition Fermi Levels of a semiconductor with respect to its band edges: a) cathodically and anodically stable, b) cathodically and anodically unstable, c) cathodically stable but anodically unstable and d) anodically stable but catodically unstable. Adapted from ref (Gerischer, 1977).

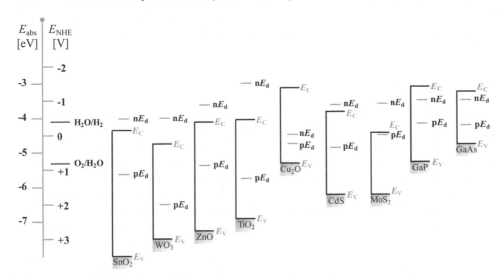

Figure 10.2.8 Positions of band edges and decomposition Fermi levels for different oxide and non-oxide semiconductors at pH 7. Adapted from (Memming, 2001).

The energy positions of the electron-induced potential $_nE_d$ and the hole-induced corrosion value $_pE_d$ can be plotted with respect to the band edges E_c and E_v, as shown in Figure 10.2.7. In fact, the criterion for thermodynamic stability of the semiconductor is:

$$_pE_d > E_{redox} >_n E_d \tag{10.2.13}$$

Figure 10.2.8 shows some decomposition potentials for various semiconductors used to carry out the water-splitting reaction under solar radiation. However, these diagrams show some practical limitations since, besides the thermodynamic stability, the reaction kinetics may also play an important role in the stability definition of a given semiconductor material. Actually, none of the semiconductors presented in Figure 10.2.8 have their Fermi level edges positioned as shown in Figure 10.2.7a, meaning that they are cathodically and/or anodically unstable. However, some of these semiconductors show cathodic and/or anodic stability because the reaction kinetics can help in preventing crystal decompositions where the charge transfer of photogenerated carriers in the interface is faster compared to the crystal decomposition (Gadgil, 1970).

The stability of a semiconductor in contact with an electrolyte solution strongly depends on the competition between anodic dissolution and redox reaction, which are controlled by thermodynamic and kinetic parameters, respectively (Memming, 2001). Thus, even if the semiconductor oxides are not thermodynamically stable, following Gerischer's approach, their stabilities can only be achieved in the presence of a suitable redox system for kinetic reasons (Sinn et al., 1990). For instance, even if the metal oxides are thermodynamically stable, based on the $_nE_d$ and $_pE_d$ positions, towards cathodic photocorrosion, most of them are unstable towards anodic photocorrosion. Nevertheless, there are some n-type semiconductors, such as α-Fe_2O_3 and TiO_2, which are sufficiently stable in aqueous electrolytes because their decomposition is controlled by very slow corrosion reaction kinetics (Krol and Schoonman, 2008). As suggested by Krol, if the material has a tendency for photodecomposition, this may be prevented by adding a suitable co-catalyst to favor the water oxidation route (Krol and Schoonman, 2008).

10.2.5 PEC reactors

A photoelectrochemical cell combines the harvesting of solar energy and the electrolysis of water process in a single device. Thus, when a semiconductor with the ideal set of properties is immersed in an aqueous electrolyte and illuminated, the corresponding photon energy is directly used to split water into hydrogen and oxygen (i.e. in a chemical energy). The basic setup for water-splitting comprises two electrodes immersed in an aqueous electrolyte solution, where one or both electrodes are photoactive. The electrolyte container must be transparent, or have at least a transparent window for allowing light to strike the photoelectrode; then, water-splitting occurs when the energetic requirements are met (Minggu et al., 2010). In a laboratory setup, to measure the PEC cell efficiency, a three-electrode configuration is normally used, the third electrode being a reference one. However, to simulate a real PEC cell application the two-electrode configuration is preferable. The more common PEC cell design is the conventional electrochemical cell used for corrosion studies, with an optically transparent window, as reported by Chen et al. in 2010 (Chen et al., 2010). The optically transparent window is very important for PEC cells to work properly, for instance, a normal soda lime glass cuts off the transmission for wavelengths lower than 350 nm, while a quartz window will normally have a transmittance higher than 90% from 250 nm. Nevertheless, a cheaper material can be used with similar performance; fused silica (amorphous silica) allows transmission values higher than 90% and shows an excellent stability in both acid and alkaline aqueous solutions (except for fluoridric acid) (Krol and Schoonman, 2008). Normally, the research laboratories in water photosplitting usually manufacture their own PEC cells, e.g. simple cubic or cylindrical open vessels, closed vessels equipped with an ion exchange membrane separating hydrogen from oxygen evolutions, H-type PEC cells, sandwich assembly,

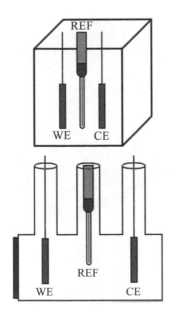

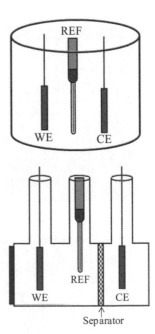

Figure 10.2.9 Example of PEC cells for water splitting with different designs, in a three-electrode configuration.

among other more complex cells, such as the ones that allow tandem configurations (PV + PEC system in a single embodiment) – Figure 10.2.9 (Minggu et al., 2010). Presently there are some PEC cells available commercially, e.g. Pine Research Instruments Company (Pine Instrument Company); however, these cells are normally limited in what concerns innovative configurations and characterization methods (Krol and Grätzel, 2012). The best option is still the PEC cells in-house designed and built.

Figure 10.2.10 shows an example of a versatile PEC cell designed by the authors; for small (Figure 10.2.10a) and for large photoanodes (Figure 10.2.10b), up to $10 \times 10 \, cm^2$. The developed cells, named *Portocell*, have two removable windows (front and back) screwed to a transparent acrylic part. The small cell has a mask that allows an illumination area of $4 \, cm^2$, crossing a synthetic quartz window (Robson Scientific, England), which is pressed against an o-ring by means of five screws. After assembling the black and the transparent acrylic part, the cell is then filled with the appropriate electrolyte solution where both electrodes are immersed. An acrylic cap can then be screwed on the top of the cell, allowing a reference electrode to be connected, if a three electrode configuration is desired, or just permitting the evolution of the electrolysis decomposition gases. *Portocell* permits back and front illumination and allows one to place a separator between the electrodes to avoid gas (hydrogen and oxygen) mixture. This separator can be a Nafion® membrane that allows just protons to permeate or a porous hydrophobic Teflon® membrane, which exhibits a very small ionic transfer resistance and prevents hydrogen and oxygen gas bubbles to mix (Mendes, et al., under protection).

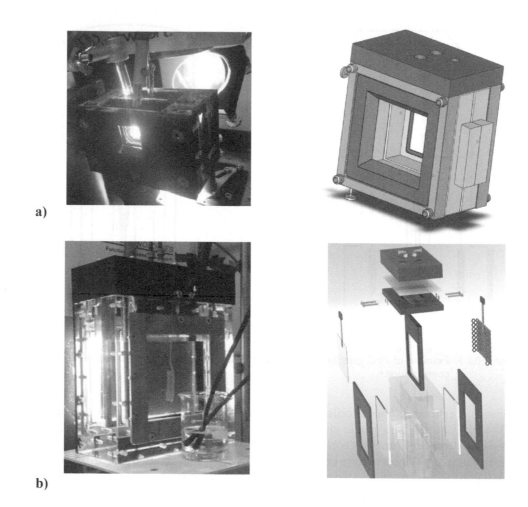

Figure 10.2.10 Photoelectrochemical cells designed by the authors for small scale photoelectrodes (a) and for large scale photoelectrodes (b) – (Lopes et al., 2012).

In the upper part of the larger *Portocell*, a membrane separates the electrolyte vessel and the gas collecting chamber. This membrane, of porous Teflon®, allows the gases to permeate but prevents the electrolyte to cross and consequently, in outdoor applications, the cell can be tilted to maximize the sunlight harvesting without fearing electrolyte leaks. Moreover, it avoids the use of a complex electrolyte feeding control system in continuous operating cells since it is just necessary to feed anode and cathode sides and the exit stream leaves from the other side without leaking to the gas collecting chamber (Mendes, et al., under protection).

10.2.5.1 The electrolyte

The electrolyte chemistry is a key factor that can dramatically influence the photoresponse of a PEC cell (Archer and Nozik, 2008). In electrochemical cells the electrolyte consists of a solvent with active species to be reduced or oxidized, depending if it is an alkaline or an acid media. Nevertheless, pure water is not conductive and thus supporting ions must be added to ensure the desired charge transfer (Krol et al., 2008). The photoactive semiconductor immersed in a redox electrolyte is greatly affected by the solution properties, redox level and stability, interfacial kinetics (adsorption), viscosity, conductivity, ionic activity and transparency within the crucial wavelength region (Archer and Nozik, 2008). As mentioned, the choice of a suitable electrolyte solution is very important, mainly in what concerns the redox couple selection. It should improve the charge-transfer kinetics, the photoelectrode stability, and also should help in preventing undesirable phenomena such as surface recombination and trapping (Archer and Nozik, 2008). Also, the electrolyte concentration should be sufficiently high to avoid large ohmic voltage losses. As reported by Roel and co-authors, the voltage drop is given by $V_{loss} = I \times R_E$, where I is the total current flowing between the working electrode and the counter-electrode, and R_E is the electrolyte resistance (Krol and Grätzel, 2012). The electrolyte conductivity strongly depends on the type of ions and the corresponding concentration value. It is important to add that there is not a linear relation between conductivity and ion concentration due to incomplete dissociation of anions and cations and/or ion-solvent interactions. Moreover, deviations from linearity can occur for concentrations above $1 \, mmol \, L^{-1}$; at high concentrations, i.e. $>1 \, M$, the formation of ion-pairs can result in a decrease of the conductivity with the concentration increase. This behavior explains why the conductivity starts to decrease for concentrations higher than $\sim 6 \, M$ of KOH aqueous solutions. To avoid large ohmic losses it is then important to guarantee concentrations of at least $0.5 \, M$ (Krol and Grätzel, 2012).

Usually, acid electrolytes such as aqueous H_2SO_4 or HCl solution ($0.5–1 \, M$) are often used with WO_3 and TiO_2 photoelectrodes (Krol and Grätzel, 2012). Aqueous NaCl solutions are also used with WO_3 photoanodes, simulating sea water conditions (Alexander and Augustynski, 2010). For electrodes such as α-Fe_2O_3, where alkaline or neutral electrolyte solutions are preferable, concentrations of $0.5–1 \, M$ NaOH or KOH should be used. Metal oxide semiconductors that are only stable in fairly neutral environments such as $BiVO_4$, $0.5 \, M$ Na_2SO_4 or K_2SO_4 solutions should be used and the electrolyte should be buffered (KH_2PO_4/K_2HPO_4) to prevent local pH fluctuations (Krol and Grätzel, 2012).

In photoelectrochemical cells bubbles usually get stacked in the electrode surface which can generate excessive noise on the photocurrent signal and thus it is necessary to remove them. This can be done at lab level flashing using a nitrogen or argon stream or simply by using a magnetic stir bar for stirring the electrolyte (Krol and Grätzel, 2012). Moreover, by using these procedures, the back-reaction of dissolved hydrogen and oxygen again to water can also be prevented ensuring that the redox potentials do not change over time. The *Portocell* with the integrated electrolyte recirculation systems helps to remove the stacked bubbles, being a solution applicable for both laboratory and industrial contexts.

10.2.5.2 The counter-electrode

As stated before, the preferable configuration is the single-photon system, in which the photoactive semiconductor works as working-electrode and a metallic material as

Table 10.2.3 Common reference electrodes for PEC research overview (Peterson, 2012).

Reference Electrode	Filling Solution	Potential (vs. SHE)
Reversible Hydrogen Electrode (RHE)	The electrode is in the actual electrolyte solution and not separated by a salt bridge	$E_0 = 0.0 + 0.059 \times pH$
Standard Hydrogen Electrode (SHE) (= normal hydrogen)	Acid solution with activity equal to 1 $[H^+] = 1.18\,M$	$E_0 = 0.0$
Calomel (Hg/Hg_2Cl_2)	0.1 M KCl	0.334
	1 M KCl (**NCE**)	0.281
	3.5 M KCl	0.250
	Saturated KCl (**SCE**)	0.242
	Saturated NaCl (**SSCE**)	0.236
Silver/Silver chloride (Ag/AgCl)	0.1 M KCl	0.288
	1 M KCl	0.237
	3 M KCl	0.210
	3.5 M KCl	0.205
	Saturated KCl	0.198
	3 M NaCl	0.209
	Saturated NaCl	0.197
	SeaWater	0.25

counter-electrode. The reaction at the counter-electrode should be as fast as possible and should have a high catalytic activity in order to prevent performance limitations (Krol and Grätzel, 2012). Usually, platinum is used as counter-electrode; this material presents good stability over a wide range of electrolytes and pH, as well as showing low overpotentials for hydrogen evolution (~0.1 V). To avoid inhomogeneous current densities at the working electrode, the counter-electrode should face it symmetrically; this is critical for electrolytes concentration lower than 0.5 M (Krol and Grätzel, 2012). Moreover, the counter-electrode area should be twice as large than the photoelectrode area (Krol and Grätzel, 2012). In PEC systems, a compromise must be maintained between the working electrode, the counter electrode and the electrolyte solution in order to ensure low overpotentials, fast charge transport and efficient light absorption.

10.2.5.3 The reference-electrode

In single photon-system PEC cells, the applied potential is an important parameter when studying the properties of the photoelectrode (photoanode or photocathode). The three-electrode configuration allows one to measure the applied potential with respect to a fixed reference electrode, allowing to turn visible the independent response of the working electrode to any change in the applied potential. The same cannot be hold for the metal counter-electrode since its overpotential at the interface with the electrolyte is usually unknown and varies with the amount of current flowing through the cell according to the Butler Volmer relation. Table 10.2.3 shows an overview considering the reference electrodes most used in PEC research applications.

As exemplified in Table 10.2.3 there are several choices for the reference electrode. However, the most commonly used reference electrode is the silver/silver chloride.

In water-splitting studies the applied potential is reported against RHE, thus the potential measured with the Ag/AgCl electrode must be converted into RHE scale using the following expression:

$$E_{RHE} = E_{Ag/AgCl} + E^0_{Ag/AgCl \text{ vs. SHE}} + 0.059 \times pH \tag{10.2.14}$$

where the $E^0_{Ag/AgCl \text{ vs. SHE}}$ is the potential of the Ag/AgCl reference electrode with respect to the SHE, see Table 10.2.3 (Peterson, 2012).

There are experimental reasons for the choice of a reference electrode; one important selection parameter is its stability on the electrolyte solution where it is immersed as well as the operating temperature (Peterson, 2012). All the reference electrodes are very sensitive and so it is crucial their good maintenance; the lifetime of a reference electrode is 2–3 years with a daily basis usage. The feasibility of a reference electrode can be checked using three identical reference electrodes and by observing potential differences at every two weeks and confirming if the deviation between any two individual electrodes is less than ±3 mV (Krol and Grätzel, 2012).

10.3 ELECTROCHEMICAL IMPENDANCE SPECTROSCOPY

Several photoelectrochemical techniques have been used to characterize photoelectrodes with the goal of understanding its performance and limitations. This kind of measurement is usually performed under steady-state and includes simulated sunlight measurements, as photocurrent-voltage, wavelength-dependent measurements, photocurrent action spectra and quantum efficiencies. Nevertheless, more detailed properties cannot be extracted from steady-state measurements and so dynamic techniques should be considered to identify performance-limiting steps or to determine certain materials properties. These techniques allow the interpretation of the charge transfer kinetics, mainly characterized by diffusion coefficients and lifetime of the different charge carriers. One of the most powerful characterization techniques of photoelectrochemical cells involving transient probing is Electrochemical Impedance Spectroscopy (EIS).

EIS is a dynamic technique that has many advantages, not only because it is user-friendly, but also because of to its sensitivity and ability to separate different complex processes, such as those occurring in a photoelectrochemical system (Bisquert, 2002; Bisquert, 2003). The foundations of EIS began in the 19th century with the controversial but extraordinary work of Oliver Heaviside (Heaviside, 2012), where he defined the terms "impedance", "reactance" and "admittance". At the end of the 19th century, Warburg derived the impedance function for a diffusional process in a remarkable work where he extended the concept of impedance to electrochemical systems (Macdonald, 2006). In the early 20th century, EIS experiments were performed mainly for capacitance measurements of ideally polarizable electrodes, e.g. mercury, using reactive bridges at relatively high frequencies. Even though these studies yielded important information about the double layer behavior, complete impedance studies including the low frequency range were only possible at the 1940s with the invention of electronic potentiostats. Three decades later, the frequency response analyzer (FRA) was developed, allowing one to probe electrochemical interfaces at sub-millihertz range

(Chang and Park, 2010). Since then, EIS has been widely used in electrochemistry, photoelectrochemistry and corrosion studies. However, this method is viable only for a stable and reversible system in equilibrium, as the system's linearity, stability and causality must be ensured (Macdonald, 2006).

10.3.1 Fundamentals

EIS is a technique widely used for characterizing the electrical behavior of systems in which the overall performance is determined by a number of strongly coupled processes, each proceeding at a different rate. The most common and standard procedure in impedance measurements consists of applying a small voltage sinusoidal perturbation and monitoring the resulting current response of the system at the corresponding frequency. An EIS measurement can be performed under any bias illumination and at any working condition of the solar cell. Nevertheless, the single-frequency voltage perturbation is usually done in open-circuit conditions with a modulation signal of magnitude V_0:

$$V(t) = V_{OC} + V_0\cos(\omega t) \tag{10.3.1}$$

The response in current has the same period as the voltage perturbation but will be phase-shifted by ϕ:

$$I(t) = I_{OC} + I_0\cos(\omega t - \phi) \tag{10.3.2}$$

V_0 and I_0 are the amplitudes of the voltage and current signals, respectively, and $\omega(=2\pi f)$ is the radial frequency in radians per second; the open-circuit current I_{OC} is zero – Figure 10.3.1.

Similar to resistance, impedance is a measure of the ability of a system to impede the flow of electrical current. Thus, impedance is the ratio of a time-dependent voltage and a time-dependent current as defined by Equations 10.3.1 and 10.3.2:

$$Z = \frac{V_0 \cos(\omega t)}{I_0 \cos(\omega t - \phi)} = Z_0 \frac{\cos(\omega t)}{\cos(\omega t - \phi)} \tag{10.3.3}$$

The impedance is therefore expressed in terms of a magnitude, Z_0, and a phase shift ϕ. Applying complex notation, the impedance response of a system can be described in terms of real and imaginary components, as follows (Barsoukov and Macdonald, 2005):

$$Z = Z_0 \frac{\exp(j\omega t)}{\exp(j\omega t - j\phi)} = Z_0(\cos\phi + j\sin\phi) \tag{10.3.4}$$

Knowing the values of $Z_{Real} = Z_0\cos\phi$, $Z_{Imag} = Z_0\sin\phi$ and the phase angle ϕ, Bode and Nyquist diagrams can be plotted. The Bode diagram is the representation of the symmetric of the phase angle ϕ vs frequency. This is a semi-logarithmic plot because both the impedance and the frequency often span orders of magnitude. Bode plots explicitly show the frequency-dependence of the impedance of the device under test. A Nyquist plot is the representation of the imaginary impedance, $-Z_{Imag}$, which is

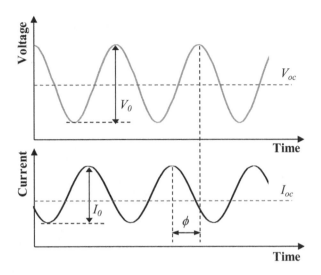

Figure 10.3.1 Sinusoidal voltage perturbation and resulting sinusoidal current response, phase-shifted by ϕ. V_0 – amplitude of the voltage signal; I_0 – amplitude of the current signal; V_{oc} – open-circuit voltage; I_{oc} – open-circuit current (Andrade et al., 2010).

indicative of the capacitive and inductive character of the cell, vs the real impedance of the cell, Z_{Real}. Nyquist plots have the advantage that activation-controlled processes with distinct time-constants show up as unique impedance arcs and the shape of the curve provides insight into possible mechanisms or governing phenomena. However, this format of representing impedance data has the disadvantage that the frequency-dependence is implicit; therefore, the AC frequency of selected data points should be indicated. Because both data formats have their advantages, it is usually best to present both Bode and Nyquist plots – Figure 10.3.2.

It is important to note that impedance analysis is based on the assumption that the system under study behaves linearly. Since linear systems typically exhibit features and properties that are much simpler than the general nonlinear cases, the analysis becomes less complex. A system is linear if it complies with both homogeneity and additivity principles, which state that: i) when a perturbation is imposed to a system, the response will be proportional and of the same type as the input signal (for instance, if a tensile strength applied to a sample increases twofold, the corresponding strain will double); ii) if the perturbation imposed on a system consists of the weighted sum of several signals, then, the output is simply the weighted sum of the system's responses to each input signal. Mathematically, let us consider $y_1(t)$ the response of a continuous time system $x_1(t)$ and $y_2(t)$ the output corresponding to the input $x_2(t)$. Then the system is linear if:

i)   *Principle of homogeneity*: the response to $a \cdot x_1(t)$ is $a \cdot y_1(t)$
ii)   *Principle of additivity*: the response to $x_1(t) + x_2(t)$ is $y_1(t) + y_2(t)$

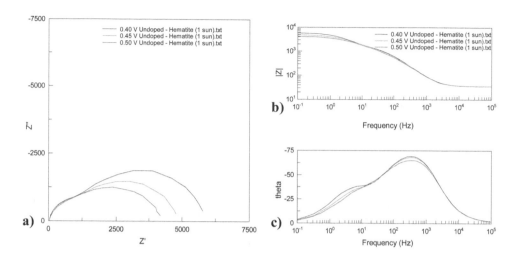

Figure 10.3.2   Graphical representation of the AC impedance of a PEC cell in 3-electrodesconfiguration: (a) Nyquistdiagram; (b) impedance Bode diagram; (c) phase Bode diagram.

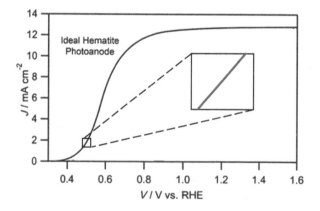

Figure 10.3.3   Current versus voltage curve showing pseudo-linearity. (Current-voltage characteristic is a steady-state technique that determines the performance response of a photoelectrode in the dark and under different light conditions. The I-V characteristic applied for water splitting is usually performed in a three electrodes configuration (being the third one is the reference electrode, usually Ag/AgCl)).

Attempting for the system under study, i.e. a photoelectrochemical cell for water-splitting and its $I-V$ characteristic shown in Figure 10.3.3, it is clear that the response to a voltage input signal is not linear. The way to circumvent this situation is to consider only a small portion of the cell's current *versus* voltage curve, which appears to be linear – Figure 10.3.3. In practice, for EIS measurements a small voltage perturbation (1–20 mV) is applied to the cell, ensuring that the response is in the pseudo-linear range (Conway et al., 2002).

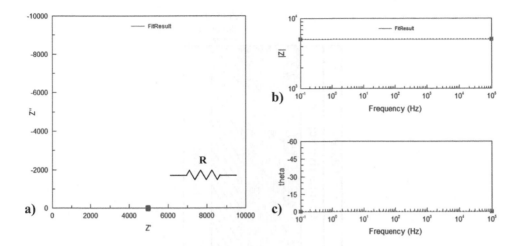

Figure 10.3.4 Graphical representation of the AC impedance of a resistor ($R = 5\,k\Omega$): (a) Nyquist diagram; (b) impedance Bode diagram; (c) phase Bode diagram.

10.3.2 Electrical analogues

EIS data usually represents the electrochemical systems as an electronic circuit, which may consist of resistors, capacitors, inductors and more complex elements, assembled in series or in parallel. Equivalent electrical analogues are a useful tool for the interpretation of experimental results, by fitting the experimental data to specific arrangements of electrical elements. This can provide relevant information concerning reaction kinetics, ohmic conduction processes and even mass transfer phenomena occurring in electrochemical systems. The different circuit elements in alternating current (AC) are briefly described hereafter. However, complementary knowledge about standard circuit elements is strongly encouraged (Barsoukov and Macdonald, 2005).

10.3.2.1 Ohmic resistance

The equivalent analogue for an ohmic conduction process is a simple resistor, which according to the Ohm's law represents the resistance to electric charge transfer. For a sinusoidal perturbation the impedance of a resistor Z_R in the complex plane is simply defined as:

$$Z_R = R \tag{10.3.5}$$

For the case of a simple resistor, the correspondent Nyquist plot is just a point in the real axis with value R with no imaginary component and independent of frequency – Figure 10.3.4.

10.3.2.2 Double layer capacitance

An electrical double layer exists at the interface between an electrode and its surrounding electrolyte as shown in Figure 10.3.5. This double layer is formed due to the charge

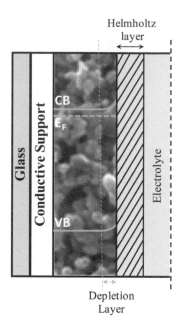

Figure 10.3.5 Schematic representation for an n-type semiconductor of: depletion layer zone and the electrical double layer (Helmholtz layer).

separation that occurs across the interface: an excess of ions of opposite charge to that on the electrode will be found at the electrolyte phase boundary. A simple way of understanding the double layer behavior is to imagine that ions at each side of the interface approach the electrode surface as closely as possible, originating two parallel layers of equal and opposite charge, one on the electrode side and the other on the electrolyte side – Figure 10.3.5. This double-layer will act as a charge storage (Bard and Faulkner, 2001), i.e. a capacitor with an impedance response defined as follows:

$$Z_C = \frac{1}{j\omega C} \tag{10.3.6}$$

In real cells, formed by nanoporous semiconductors, the double layer capacitor does not behave ideally. Instead it acts like a constant phase element (CPE), a non-ideal capacitance with a non-uniform distribution of current in the heterogeneous material. In this case, the impedance of the double layer capacitance is defined as:

$$Z_C = \frac{1}{j\omega C^{n_z}} \tag{10.3.7}$$

where n_z $(0 < n_z < 1)$ is an empirical constant with no real physical meaning; for ideal capacitors $n_z = 1$ (Barsoukov and Macdonald, 2005).

For the case of a simple capacitor, the corresponding Nyquist plot is just a vertical line coincident with the imaginary axis and thus with no real component –

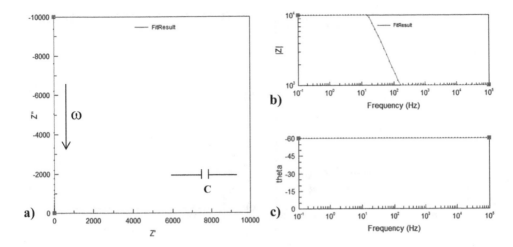

Figure 10.3.6 Graphical representation of the AC impedance of a capacitor: (a) Nyquist diagram; (b) impedance Bode diagram; and (c) phase Bode diagram. ω is the radial frequency; $C = 1\,\mu F$.

Figure 10.3.6. As frequency decreases, the imaginary component of the impedance dominates the response of the circuit.

Components of an electrical circuit can be connected in series or in parallel. The net impedance of a system with elements assembled in series is the sum of the impedances of the circuit elements:

$$Z_{Series} = Z_1 + \cdots + Z_i \tag{10.3.8}$$

i being the number of circuit elements. Thus, if a capacitor is assembled in series with a resistor, the net impedance is given by:

$$Z = R + \frac{1}{j\omega C} \tag{10.3.9}$$

The corresponding Nyquist diagram is a vertical line displaced with the real component given by the value of the resistor R. As frequency decreases, the imaginary component of the impedance dominates the response of the circuit – Figure 10.3.7.

10.3.2.3 Electrochemical reaction

The impedance behavior of an electrochemical reaction can be well represented by a parallel assembling of a resistor, R, and a capacitor, C. While the resistor models the kinetics of the electrochemical reaction, the capacitor describes the charge separation across the interface (O'Hayre et al., 2006). The net impedance of a system with elements assembled in parallel is given by:

$$\frac{1}{Z_{parallel}} = \frac{1}{Z_1} + \cdots + \frac{1}{Z_n} \tag{10.3.10}$$

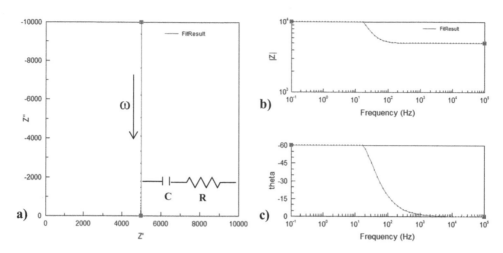

Figure 10.3.7 Graphical representation of the AC impedance of a capacitor in series with a resistor: (a) Nyquist diagram; (b) impedance Bode diagram; and (c) phase Bode diagram. ω is the radial frequency; $C = 1\,\mu F$ and $R = 5\,k\Omega$.

i being the number of circuit elements. Thus, for an electrochemical reaction the total impedance may be defined as:

$$Z = \frac{1}{1/R + j\omega C} \tag{10.3.11}$$

The corresponding Nyquist diagram, presented in Figure 10.3.8, shows a semicircle with diameter R. The extension of the semicircle, therefore, provides useful information concerning the reaction kinetics of the system: facile reaction kinetics will show a small diameter, while a blocking electrode will be characterized by a huge semicircle. Finally, the time constant of the reaction kinetics, τ, is given by:

$$\omega_{max} = \frac{1}{\tau} = \frac{1}{RC} \tag{10.3.12}$$

where ω_{max} is the radial frequency at the semicircle maximum (Figure 10.3.8). The high-frequency intercept of the semicircle is zero, while the low-frequency intercept of the impedance semicircle is R.

10.3.3   EIS analysis of PEC cells for water-splitting

Photoelectrochemical cells for water-splitting have been extensively characterized by the well-known Mott-Schottky relation, which allows obtaining the flat-band potential and the donor density by plotting the inverse square route of the space charge capacitance as a function of the applied potential. The capacitance is usually determined by fitting the experimental data to a simple resistor-capacitor (RC) electrical analogue assembled in parallel – Figure 10.3.9a. In fact, C_1 in Figure 10.3.9a corresponds

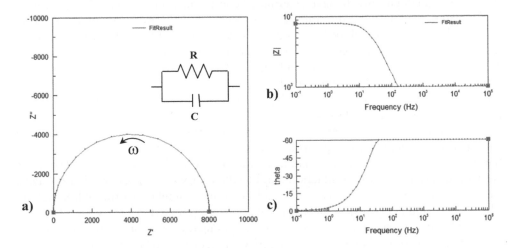

Figure 10.3.8 Circuit diagram and Nyquist plot representing the impedance behavior of an electrochemical reaction: (a) Nyquist diagram; (b) impedance Bode diagram; and (c) phase Bode
diagram. ω is the radial frequency; $Z_R = 8\,k\Omega$ and $Z_C = 1\,\mu F$.

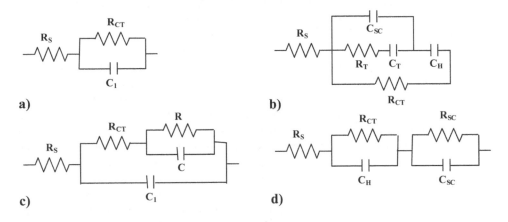

Figure 10.3.9 Proposed electric models to describe PEC cells behavior: (a) simple RC element used
to calculate the flat band potential and the donor density from Mott-Schoktty equation,
applicable at high-frequency; (b) electric model taking into account trap states (interface states at low-potentials and near the flat band potential and deep donor levels far
from the flat-band potential); (c) electric model describing electron transfer from the
conduction band to the redox system via two steps: electron trapping in the surface
states and posterior tunneling to the redox system; and (d) typical response of hematite
photoelectrodes prepared by APCVD.

to a series connection of two capacitances, the depletion layer of the semiconductor
capacitance, C_{SC}, and the Helmholtz layer capacitance in the electrolyte side, C_H.
The latter is negligible in the most cases, allowing isolating C_{SC} to be used in the
Mott-Schottky analysis. Different ranges of frequencies were considered to validate

the referred analogue and all have determined consistent values of flat-band potential, in the range of 0.4–0.6 V_{RHE}, and of donor density, $10^{17} cm^{-3}$ for undoped structures and $10^{21} cm^{-3}$ for doped materials (Sivula et al., 2011). Nevertheless, several discrepancies were identified for oxide semiconductors electrodes from the typical Mott-Schottky behavior: the near and far flat-band potential, ascribed to localized interface states and to deep donor levels, respectively. To account for this, more complete models were developed considering these intrabandgap states – Figure 10.3.9b (Leduc and Ahmed, 1988; Horowitz, 1983; Goodman, 1963).

Considering an n-type semiconductor under forward bias, there are two main charge transfer routes involving surface states: trapping electrons from the conduction band or holes from the valence band, acting as a recombination center. Thus, the electron transfer from the conduction band to the redox system occurs via two steps: electron trapping in the surface states and posterior tunneling to the redox system. The corresponding equivalent electrical analogue is presented in Figure 10.3.9c. This analysis has also been extensively used for determining the injection of minority carriers and subsequent recombination with majority carriers via surface states (Gomes and Vanmaekelbergh, 1996). The physical meaning of the simple elements depends upon the details of the entire mechanism.

More recently, the electrical analogue presented in Figure 10.3.9d was proposed to describe the impedance response of a nanostructured hematite photoanode prepared by atomic pressure chemical vapor deposition (APCVD) (Le Formal et al., 2011). In this model the element R_S is the series resistance, which includes the sheet resistance of the TCO glass substrate and the external contacts resistance of the cell (e.g. wire connections). Then, two RC elements in series are considered representing the semiconductor bulk and the surface phenomena, respectively (Le Formal et al., 2011). Accordingly, and bearing in mind that the electronic processes in bulk are generally faster than the charge transfer processes or diffusion of ions in solution, the low-frequency response was assigned to the semiconductor-electrolyte charge transfer resistance, R_{CT}, together with the C_H (Sivula et al., 2011; Le Formal et al., 2011). The faster electronic processes occur in the semiconductor (high-frequency range, 0.1 Hz to 100 Hz); they are ascribed to the resistance on the depletion layer, R_{SC}, and the space charge capacitance, C_{SC}, similarly to the simple RC circuit shown in Figure 10.3.9a. Moreover, this RC element is the combination of different resistances and capacitances related to transport in the semiconductor layer, charge diffusion in the space charge layer and surface trap charging by electrons and holes (Le Formal et al., 2011).

10.4 FUNDAMENTALS IN ELECTROCHEMISTRY APPLIED TO PHOTOELECTROCHEMICAL CELLS

Edmond Becquerel, in 1839, discovered the photoelectric effect while he was experimenting with an electrolytic cell made up of two metal electrodes (Becquerel, 1839). This phenomenon was not completely understood until 1954, when Brattain and Garret demonstrated how electrochemical reactions occurring at germanium electrodes can be influenced by changing the germanium semiconducting properties and also by light excitation (Archer and Nozik, 2008). This pioneer work was followed up by several other investigations on semiconductors between 1954 and 1970 (Nozik and

Memming, 1996). These studies established the first models for the charge distribution, kinetics and energetic of charge transfer mainly across the semiconductor-liquid interface. However, it was only in the seventies that the potential application of photoelectrochemical systems for solar energy conversion and storage was truly recognized (Fujishima and Honda, 1972b). Given the enormous efforts devoted to understanding the physics behind semiconductors, it might be expected that this is a mature field. Nevertheless, most of this knowledge pertains to bulk and transport phenomena and for the above mentioned applications; it is the surface (photo)chemistry that governs the interfacial transfer and energy transduction, and here we are still evolving our understanding (Archer and Nozik, 2008). Despite these difficulties, the progress that has been made, mainly through the phenomenological understanding of the processes occurring in the semiconductor surface, and more concretely at the semiconductor-electrolyte interface, was indeed shown to be of great importance (Butler, 1977; Gärtner, 1959; Henry et al., 1978; Reiss, 1978; Wilson, 1977). This section targets a fundamental understanding of semiconductor physics, by describing the different type of semiconductors and the main equations governing the charge carrier distribution. The semiconductor/electrolyte interface will be described in terms of potential distribution. It is then described as an integrated phenomenological model for photoelectrochemical cells for water-splitting. The continuity and transport-governing equations are defined for the several mobile species involved in the phenomena occurring in the different regions of the PEC cell (Andrade et al., 2010; Andrade et al., 2011).

10.4.1 Semiconductor energy

Solid materials can be categorized as conductors, insulators or semiconductors, depending on their ability to transport electrical current. A conductor carries electrical current, whereas an insulator cannot carry current. Between these two materials are the semiconductors. Electrons in semiconductors can have energies only within certain bands between the energy of the ground state, corresponding to electrons tightly bound to the atomic nuclei of the material, and the free electron energy, which is the energy required for an electron to escape. Between these two bands there is a bandgap with energy E_g, which greatly determines the properties of the material (Andrade et al., 2010; Andrade et al., 2011). The filled energy states that correspond to the valence band are localized in the energy range below the gap and the empty energy states corresponding to the conduction band are localized in the energy range above the gap (Memming, 2001).

10.4.1.1 Intrinsic semiconductor

The Fermi-Dirac function $f(E)$ gives the probability that a single-particle state of energy E would be occupied by an electron at thermodynamic equilibrium (at a constant temperature with no external injection or generation of carriers). The Fermi-Dirac distribution function, also called the Fermi function, is given by:

$$f(E) = \frac{1}{1 + \exp\left(\frac{E - E_F}{kT}\right)} \tag{10.4.1}$$

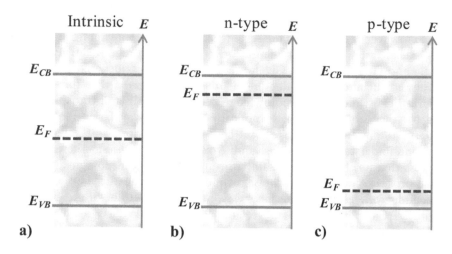

Figure 10.4.1 Schematic representation of semiconductor energy band levels.

where k is the Boltzman constant. The system is characterized by its temperature, T, and its Fermi level, E_F, defined as the energy level at which the probability to be occupied by an electron is 0.5.

For an intrinsic semiconductor (undoped), the Fermi Level lies at the mid-point of the bandgap – Figure 10.4.1a – and the energy of its Fermi Level can be calculated as follows (Nozik and Memming, 1996):

$$E_F = \frac{1}{2}(E_V + E_C) + \frac{1}{2}\left(kT\ln\frac{N_V}{N_C}\right) \tag{10.4.2}$$

where E_V and E_C are, respectively, the energy levels of the valence and the conduction band edges and T is the absolute temperature. N_C and N_V are the density-of-states at the conduction and valence bands, respectively, and given by (Memming, 2001):

$$N_C = 2\frac{\left(2\pi m_e^* kT\right)^{3/2}}{h^3} \tag{10.4.3}$$

$$N_V = 2\frac{\left(2\pi m_h^* kT\right)^{3/2}}{h^3} \tag{10.4.4}$$

where h is Plank's constant and m_h^* and m_e^* are the effective masses of holes and electrons, respectively. The electron and hole densities in an intrinsic semiconductor material in the conduction and valence bands, respectively, are given by:

$$n = N_C \exp\left(-\frac{E_C - E_F}{kT}\right) \tag{10.4.5}$$

$$p = N_V \exp\left(\frac{E_V - E_F}{kT}\right) \tag{10.4.6}$$

In an intrinsic semiconductor material the number of electrons in the conduction band equals the number of holes in the valence band: $n = p = n_i$. Considering Equations 10.4.5 and 10.4.6, the intrinsic carrier concentration is given by:

$$np = N_c N_v \exp\left(\frac{E_c - E_v}{kT}\right) = n_i^2 \qquad (10.4.7)$$

where n_i is the intrinsic carrier density, a material constant that depends on the semiconductor bandgap and temperature (Grimes et al., 2008). Typical carrier densities in intrinsic semiconductor materials range from 10^{15} to 10^{19} cm^{-3} (Archer and Nozik, 2008). Optical excitation perturbs this relation, and thus the relaxation back towards the equilibrium occurs through various scattering and recombination processes. Equation 10.4.7 is also valid for doped semiconductors and actually it is of great importance since knowing one carrier density, for instance n, the other (here p) can be computed.

10.4.1.2 Doped semiconductors

For doped structures the relative position of the band edges, as well as the Fermi level in the bulk of the semiconductor, depends on the doping level. When a semiconductor is doped with donors or acceptor atoms, then the corresponding energy levels are introduced within the so-called forbidden zone. Usually, the donor level is close to the conduction band giving rise to an n-type semiconductor – Figure 10.4.1b. On the other hand, the acceptor level is located near the valence band originating a p-type semiconductor – Figure 10.4.1c. By definition, a donor level is neutral if filled by an electron and positive if empty. In opposition, the acceptor level is neutral if empty and negative if filled by an electron (Nozik and Memming, 1996). The majority charge carriers for p-type and n-type semiconductors are holes and electrons, respectively.

By doping, a semiconductor remains electrically neutral and the Fermi Level must adjust itself in order to ensure the neutrality. Thus, for an n-type semiconductor:

$$n = N_D^+ + p \qquad (10.4.8)$$

in which N_D^+ is the density of ionized donors, related to the occupied donor density; N_D means the Fermi function:

$$N_D^+ = N_D \left[1 - \frac{1}{1 + \exp\left(\frac{E_D - E_F}{kT}\right)} \right] \qquad (10.4.9)$$

where E_D is the donor energy. The same happens for p-type semiconductor materials:

$$N_A^- = N_A \left[\frac{1}{1 + \exp\left(\frac{E_A - E_F}{kT}\right)} \right] \qquad (10.4.10)$$

where N_A^- is the density of ionized acceptors and E_A is the acceptor energy.

Table 10.4.1   Electron and hole densities in n-type and p-type semiconductor materials at room temperature (Würfel, 2005).

	n	p	E_F
n-type	$n \approx N_D$	$p = \dfrac{n_i^2}{n} = \dfrac{n_i^2}{N_D}$	$E_F = E_C - kT \ln \dfrac{N_C}{N_D}$
p-type	$n = \dfrac{n_i^2}{p} = \dfrac{n_i^2}{N_A}$	$p \approx N_A$	$E_F = E_V - kT \ln \dfrac{N_V}{N_A}$

Under room temperature and assuming that all donors and acceptors are completely ionized, the electron and the hole densities equal the donor and acceptor densities, respectively: $n = N_D$ and $p = N_A$. Table 10.4.1 summarizes the equations for obtaining electron and hole densities as well as their Fermi Level positions for an n-type and a p-type semiconductor, with either shallow donors or shallow acceptors, at room temperature.

10.4.1.3   Potential distribution at the semiconductor–electrolyte interface

A key feature of a semiconductor material in contact with a liquid solution is the formation of a built-in electric-field or space-charge region in the semiconductor side and a Helmholtz double-layer adjacent to the semiconductor, in the liquid side. The space charge layer is an important concept in solar conversion systems since it is linked to the efficient separation of photogenerated electrons and holes and by consequently preventing the recombination phenomena (Würfel, 2005). Those layers arise when a semiconductor is brought into contact with a second phase, both with different initial chemical potentials. Charges are transferred between them until equilibrium is established. The electrochemical potential of the liquid phase is determined by its redox potential; on the other hand, the redox potential of the semiconductor material is determined by the corresponding Fermi level.

Considering an n-type semiconductor, its Fermi level is typically higher than the redox potential of the electrolyte solution. Consequently, the electrons will be transferred from the semiconductor to the electrolyte. The Fermi level in the semiconductor moves "down" and the process stops when the Fermi level equals the redox potential in the electrolyte side. This movement is also followed by the conduction band edge, giving rise to a band bending as illustrated in Figure 10.4.2a and described by Equation 10.4.14. Similarly, for a p-type semiconductor a depletion layer also occurs when the region containing negative charges is depleted of holes since they are transferred to the electrolyte media until Fermi level and redox potential equilibration. This negative layer will be compensated by the positive charges adsorbed on the solid electrode surface in the electrolyte side – Figure 10.4.2b.

The potential distribution and the width of the space charge layer can be quantitatively described by the Poisson – Equation 10.4.11. This depends on the amount

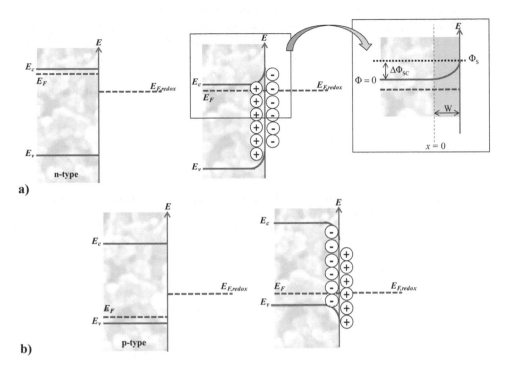

Figure 10.4.2 Schematic representation of the band bending effect for (a) an n-type semiconductor and (b) a p-type semiconductor.

of transferred charge to the surface of the semiconductor and the density of shallow donors in the semiconductor material, N_D.

$$\frac{d^2 \Delta \Phi}{dx^2} = -\frac{1}{\varepsilon \varepsilon_0} \rho(x) = -\frac{d\xi}{dx} \qquad (10.4.11)$$

ε and ε_0 are the dielectric constant and the permittivity of free space, respectively. ξ is the electric field and the total charge density ρ is given by (Krol and Grätzel, 2012):

$$\rho(x) = e[N_D^+ - n(x)] \qquad (10.4.12)$$

where x is the distance from the surface, N_D^+ is ionized donor density, given by the doping of the semiconducting material – Equation 10.4.9. The concentration of free electrons, $n(x)$ varies with distance – Equation 10.4.5.

Since the Fermi level is expected to be constant within the space charge region, the position of the conduction band energy $E_C(x)$ varies with the distance by the band bending effect, $\Phi(x)$, as follows:

$$n(x) = N_C \exp\left(-\frac{E_C(x) - E_F - e\Phi(x)}{kT}\right) = n_0 \exp\left(\frac{e\Phi(x)}{kT}\right) \qquad (10.4.13)$$

It is important to note that the energy of an electron is related to its potential by $E = -e\Phi$ and that the bulk potential is conveniently chosen to be zero. In Equation 10.4.13 n is the bulk concentration that must be equal to the ionized donors N_D^+ since the bulk is electrically neutral. Moreover, assuming that all the electrons are ionized, this leads to $N_D^+ = N_D = n_0$. This relation can be introduced in Equation 10.4.12 that combined with Equation 10.4.12 allows to calculate the total charge density at the point x:

$$\rho(x) = eN_D\left[1 - \exp\left(\frac{e\Phi(x)}{kT}\right)\right] \tag{10.4.14}$$

However, Equations 10.4.3 and 10.4.7 have an implicit dependence on the band bending $\Phi(x)$ parameter (Krol and Grätzel, 2012). An alternative path to overcome this difficulty is considering the derivative of the square of the electric field, ξ^2 (Krol and Grätzel, 2012):

$$\frac{d(\xi^2)}{dx} = 2\xi\frac{d\xi}{dx} = 2\xi\frac{\rho(x)}{\varepsilon\varepsilon_0} = -2\frac{\rho(x)}{\varepsilon\varepsilon_0}\frac{d\Phi}{dx} \tag{10.4.15}$$

Since a one-to-one correspondence between Φ and x can now be identified and by changing the independent variable from x to Φ, Equation 10.4.16 can be written:

$$\xi^2 = \int_0^\Phi (x) - 2\frac{\rho(\Phi)}{\varepsilon\varepsilon_0}d\Phi \tag{10.4.16}$$

The charge accumulated between the neutral bulk and up to the beginning of the depletion layer, at point $x = a$ (see Figure 10.4.3), is related to the electric field by the integral form of Gauss' law:

$$\xi = \frac{Q}{\varepsilon\varepsilon_0 A} \tag{10.4.17}$$

where the semiconductor area is represented by A. Combining both Equations 10.4.16 and 10.4.17 one obtain:

$$Q = \sqrt{-2\varepsilon\varepsilon_0 A^2 \int_0^{\Phi(x)} \rho(\Phi)d\Phi} \tag{10.4.18}$$

This equation is of general application and can be used with non-homogeneous doping profiles and deep donors/acceptors (Krol and Grätzel, 2012). Now, it is possible to calculate the total charge accumulated in the space charge region using

Equations 10.4.14 and 10.4.18 and integrating between $\Phi = 0$ and $\Phi = \Phi_S = -\Phi_{SC}$ (see Figure 10.4.2), Equation 10.4.19 is obtained (Krol and Grätzel, 2012):

$$Q_{SC} = \sqrt{-2\varepsilon\varepsilon_0 A^2 \int\limits_0^{-\Phi_{SC}} eN_D \left(1 - \exp\left(\frac{e\Phi}{kT}\right)\right)}$$
$$= \sqrt{2\varepsilon\varepsilon_0 eN_D A^2 \left(\Phi_{SC} + \frac{kT}{e}\exp\left(-\frac{e\Phi_{SC}}{kT}\right) - \frac{kT}{e}\right)} \tag{10.4.19}$$

Assuming that for a PEC cell under normal operating conditions the potential drop across the space charge is at least 0.1 V, Equation 10.4.19 can be simplified to:

$$Q_{SC} = \sqrt{2\varepsilon\varepsilon_0 eN_D A^2 \left(\Phi_{SC} - \frac{kT}{e}\right)} \tag{10.4.20}$$

The accumulated total charge, Q_{SC}, is given by the product of the amount of charges N_D and the corresponding volume $(A \cdot W)$. This allows the relation between the accumulated total charge, Q_{SC}, to be obtained and the depletion layer thickness, W:

$$Q_{SC} = eN_D AW \tag{10.4.21}$$

and so the space charge layer width can be expressed as:

$$W = \sqrt{\frac{2\varepsilon\varepsilon_0}{eN_D}\left(\Phi_{SC} - \frac{kT}{e}\right)} \tag{10.4.22}$$

Even though these equations have been derived for an n-type semiconductor, an analogous approach can be followed for a p-type material, where N_D is replaced by the shallow acceptor density, N_A. Typically, values of space charge layer width range from 5 to 500 nm. It is notable that the accumulated charge in the depletion layer is compensated by an opposite charge electrolyte layer accumulated at the surface of the semiconductor (i.e. trapped electrons) (Krol and Grätzel, 2012).

In fact, the dopant density and the depletion layer width can be experimentally determined by capacity measurements. The differential capacity of the space charge layer, C_{SC}, under certain conditions can be obtained from the total impedance of the system by differentiating Equation 10.4.20 with respect to Φ_{SC}:

$$\frac{1}{C_{SC}^2} = \left(\frac{dQ_{SC}}{d\Phi_{SC}}\right)^{-2} = \frac{2}{\varepsilon\varepsilon_0 eN_D A^2}\left(\Phi_{SC} - \frac{kT}{e}\right) \tag{10.4.23}$$

This is the so-called Mott-Schottky equation, only valid for a space charge region depleted of the majority charge carriers with respect to the bulk density. The thickness of the space charge layer decreases as the doping concentration increases (Krol and Grätzel, 2012). Plotting $1/C_{SC}^2$ as a function of the applied potential Φ_A the donor

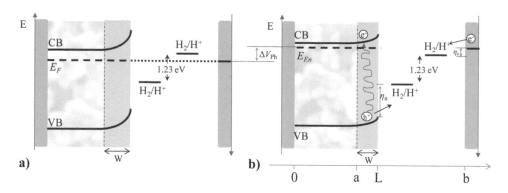

Figure 10.4.3 Energetic diagram of a PEC cell under dark (a) and illumination (b) conditions.

density, N_D, can be obtained from the slope of Equation 10.4.23 while, with extrapolation for $1/C_{SC}^2 = 0$ it is possible to obtain the flat band potential, Φ_{FB}, using the following relation:

$$\Phi_{SC} = \Phi_{Appl.} - \Phi_{FB} \qquad (10.4.24)$$

10.4.2 Continuity and kinetic equations

The net charge transfer from the semiconductor to the redox species in the electrolyte solution occurs via majority charge carriers in dark conditions and, via minority charge carriers under illumination (Nozik and Memming, 1996). In the present section, we will focus our attention in the light-induced process of the PEC cells. Figure 10.4.3 shows the energy diagram for a PEC cell in the dark and under illumination conditions. The efficient separation of the photogenerated electron-hole pairs in the semiconductor highly depends on the electrical field formed at the semiconductor/electrolyte interface. As mentioned, this field-induced charge separation is a key parameter to high energy-conversion efficiencies. The strength of the electric field depends on the doping level of the semiconductor and on other energy features of the systems (Nozik and Memming, 1996). At an n-type semiconductor/electrolyte interface, the electric field drives the minority carriers (i.e. holes) towards the solid/liquid interface and drives the majority carriers (i.e. electrons) into the bulk of the semiconductor. The energy storage process is completed if the holes undergo interfacial charge transfer to the donors in the electrolyte prior to any recombination events (Nozik and Memming, 1996). These recombination events can occur by way of: bulk recombination via bandgap states or direct electron loss to holes in the valence band; photocorrosion of the semiconductor; and dissolution reactions (Nozik and Memming, 1996). The last two processes are responsible for the degradation of the electrode and for the consequent stability issues.

Optical excitation of a PEC cell leads to the perturbation of the equilibrium attained when the semiconductor is placed into contact with an electrolyte solution under dark conditions. Charge transport and relaxation rates of the carrier populations to re-establish the former equilibrium can be described using the continuity equation (Andrade et al., 2011):

$$-\frac{\partial j_i}{\partial x} + G_i(x) - R_i(x) = \frac{\partial n_i}{\partial t}, \quad i = e^-, h^+ \quad \text{and} \quad OH^- \qquad (10.4.25)$$

The first term on the left-hand side of the equation represents the carrier flux defined by the respective transport equation. The second and third terms are the generation and relaxation rates of species i, respectively. Both reaction kinetics, are characterized by a time constant. The term on the right-hand side of the equation corresponds to the carrier concentration history of species i.

Equation 10.4.25 will be applied to the different PEC system components and for the different species that exist on each region: semiconductor bulk, depletion layer or electrolyte solution. For simplicity, a flat structure of semiconductor material stable at an alkaline media was considered.

a)   *Electrons in the depletion layer $(a < x < L)$*

In the depletion layer there are two contributions for the electron flux, j_e^-: the diffusive transport of electrons and the convection transport driven by a macroscopic electric field. Equation 10.4.26 relates the electron flux at any position x with the gradient of electrons concentration across the depletion layer, n_e^-, by means of the electron diffusion coefficient, D_e^-, the electron mobility, μ_e^-, and the macroscopic electric field, ξ:

$$j_{e^-} = -D_{e^-}\frac{\partial n_{e^-}}{\partial x} - \mu_{e^-} n_{e^-} \xi \qquad (10.4.26)$$

The generation term for electrons in Equation 10.4.25 is given by the Beer Lambert equation, which relates the absorption of light to the properties of the material through which the light is travelling (Andrade et al., 2011):

$$G_{e^-} = \eta_{inj}\alpha(\lambda)I_0^* e^{-\alpha(\lambda)(1-x)} \qquad (10.4.27)$$

which assumes that each photon, with energy $h\nu \geq E_g$, absorbed by the semiconductor results in an injected electron into its conduction band. $\alpha(\lambda)$ is the wavelength-dependent absorption coefficient, which is a material property; I_0^* is the incident photon flux, corrected for reflection losses of the TCO glass; and η_{inj} is the electron injection efficiency. The expression $(1 - x)$ in the exponential term indicates that front illumination of the semiconductor is considered. The electron recombination term is neglected, it being assumed that in the depletion layer electrons and holes are efficiently separated (Reichman, 1980).

The continuity equation for electrons in the depletion layer between $x = a$ and $x = L$ can be written as:

$$\frac{\partial n_{e^-}}{\partial t} = D_{e^-}\frac{\partial^2 n_{e^-}}{\partial x^2} + \mu_{e^-}\frac{\partial(n_{e^-}\xi)}{\partial x} + \eta_{inj}\alpha(\lambda)I_0^* e^{-\alpha(\lambda)(1-x)} \qquad (10.4.28)$$

b)   *Electrons in the semiconductor bulk ($0 < x < a$)*

In the semiconductor bulk the effect of the macroscopic electric field is negligible. Thus, the charge transport within the semiconductor takes place only via diffusion; the flux equation only considers the transport driven by the concentration gradient as follows:

$$j_{e^-} = -D_{e^-} \frac{\partial n_{e^-}}{\partial x} \tag{10.4.29}$$

In this region, outside the depletion layer zone, semiconductor bulk recombination of conduction band electrons with holes in the valence band is taken into account, since the contribution of the macroscopic electric field responsible for charge separation is negligible. The electron recombination is given by a first-order reaction based on the electron density excess regarding its equilibrium value, $\Delta n_{e^-} = n_{e^-}(x,t) - n_{eq}$:

$$R_{e^-} = k_{e^-} \Delta n_{e^-} \tag{10.4.30}$$

where $k_{e^-} = 1/\tau_{e^-}$ is the first-order reaction rate constant, τ_{e^-} is the electron lifetime and n_{eq} is the dark equilibrium electron density, which can be calculated using Equation 10.4.5. The density-of-states of the conduction band in Equation 10.4.5 can be determined from Equation 10.4.3.

The continuity equation for electrons in the bulk semiconductor is written as:

$$\frac{\partial n_{e^-}}{\partial t} = D_{e^-} \frac{\partial^2 n_{e^-}}{\partial x^2} + \eta_{inj}\alpha(\lambda)I_0^* e^{-\alpha(\lambda)(1-x)} - \frac{(n_{e^-} - n_{eq})}{\tau_{e^-}} \tag{10.4.31}$$

c)   *Holes in the semiconductor depletion layer ($a < x < L$)*

The holes produced in the depletion layer region are perhaps the only ones capable of reacting with the hydroxyl ions in the semiconductor surface adjacent to the electrolyte solution. The continuity equation for the holes is similar to the one written for electrons in the depletion layer:

$$\frac{\partial n_{h^+}}{\partial t} = D_{h^+} \frac{\partial^2 n_{h^+}}{\partial x^2} - \mu_{h^+} \frac{\partial(n_{h^+}\xi)}{\partial x} + \eta_{inj}\alpha(\lambda)I_0^* e^{-\alpha(\lambda)(1-x)} \tag{10.4.32}$$

When the semiconductor absorbs solar radiation with enough energy to excite an electron from the valence band to the conduction band, an electron-hole pair is generated. Thus, holes and electrons are generated at the same rate:

$$G_{h^+} = G_{e^-} \tag{10.4.33}$$

and then the generation terms are equal for the electrons and holes species.

d)   *Holes in the semiconductor bulk ($0 < x < a$)*

Similar to what happens with the electrons in the semiconductor bulk, the macroscopic electric field is negligible and the transport of holes to the surface of the semiconductor takes place via diffusion. As mentioned the generation term is equal for both electrons and holes $G_{h+} = G_{e-}$. On the other hand, if a hole in the valence band reacts with an electron from the conduction band, a photon is produced – exactly the reverse process of absorption, i.e. radiative recombination takes place. The recombination of a hole with a conduction band electron occurs at the same rate of the spontaneous transition of an electron from the conduction band to an unoccupied state in the valence band:

$$R_{h+} = R_{e-} \rightarrow \frac{(n_{h+} - n_{eq})}{\tau_{h+}} = \frac{(n_{e-} - n_{eq})}{\tau_{e-}} \tag{10.4.34}$$

The continuity equation for holes can then be written:

$$D_{h+} \frac{\partial^2 n_{h+}}{\partial x^2} + \eta_{inj} I_0^* \alpha e^{-\alpha x} - \frac{\Delta n_{e-}}{\tau_{e-}} = \frac{\partial n_{h+}}{\partial t} \tag{10.4.35}$$

e) *Hydroxyl ions in the electrolyte solution (L < x < b)*

Hydroxyl ions are formed in the counter-electrode by reduction of water to hydrogen gas – Equation 10.2.4. Then OH^- ions diffuse to the semiconductor surface where they react with holes to produce oxygen gas – Equation 10.2.5. Only electrolyte exists in the region $L < x < b$, so no ions are generated or lost by recombination with holes. Accordingly, the respective continuity equation for OH^- ions, applied to this electrolyte region, can be written as:

$$\frac{\partial n_{OH^-}}{\partial t} = D_{OH^-} \frac{\partial^2 n_{OH^-}}{\partial x^2} \tag{10.4.36}$$

The previous equations describe the charge transport within a PEC cell under operating light conditions. With the appropriate boundary and initial conditions and the adequate characteristic parameters of the system, those equations can be used to model a PEC cell (Reiss, 1978).

10.4.2.1 Energy balance

In water photoelectrolysis a net chemical reaction occurs by means of water decomposition into hydrogen and oxygen by light induced processes. For photoelectrochemical cells the energy balance behind these reactions was written by Nozik in 1976 as follow (Andrade et al., 2011):

$$E_{Bias} + E_g - \Phi(x) - \Delta E_F = \frac{\Delta G^o}{n_e F} + \eta_a + \eta_c + \eta_{ohmic} \tag{10.4.37}$$

the term E_{Bias} is added when an external bias applied to the PEC cell is required. E_g is the bandgap energy and ΔE_F, defined as the difference between the Fermi level and majority carrier band edge of the semiconductor. The free energy change per electron for the overall reaction taken as $1.23\,eV$ for water decomposition is here represented by $\Delta G^o/n_e F$. η_a and η_c are the overpotentials at the anode and cathode, respectively.

The η_{ohmic} term decribes the ohmic loss through the electrode area A and may be written in terms of an ohmic resistance, R_{ohmic} (Nozik, 1978):

$$\eta_{\text{ohmic}} = J_{\text{cell}} A R_{\text{ohmic}} \tag{10.4.38}$$

The left term in Equation 10.4.38 represents the net photon energy necessary to generate an electron-hole pair in the semiconductor and that has to equal the electrochemical work described by the right hand of the same equation.

10.4.2.2   Metal–liquid interface

In a PEC cell for water-splitting electrons return from the external circuit to the counter-electrode in order to reduce water into hydrogen. This reaction occurs in the surface of a metal electrode, typically a platinum wire. The interface electrolyte metal electrode can be treated as an electrochemical half-cell and described using the well-known Butler-Volmer equation:

$$j = j_0 \left[\frac{C_R^*}{C_R^{0*}} \exp\left(\frac{\beta z e \eta_c}{kT} \right) - \frac{C_P^*}{C_P^{0*}} \exp\left(\frac{-(1-\beta) z e \eta_c}{kT} \right) \right] \tag{10.4.39}$$

where η is the overvoltage, z is the number of electrons transferred in the electrochemical reaction, C_R^* and C_P^* are the surface concentrations of the reactants and products, j_0 is the exchange current density (Aruchamy et al., 1982). Basically, Butler-Volmer equation states that the current produced by an electrochemical reaction increases exponentially with the activation overvoltage. Taken as an example the electrolyte/platinum interface in a PEC cell system for water-splitting, Equation 10.4.38 becomes:

$$j = j_0 \left[\frac{n_{\text{H}_2\text{O}}(b)}{n_{\text{H}_2\text{O}}^{ref}(b)} \exp\left(\frac{\beta z q \eta_{Pt}}{k_B T} \right) - \frac{n_{\text{H}_2}(b) n_{\text{OH}^-}^2(b)}{n_{\text{H}_2}^{ref}(b)(n_{\text{OH}^-}^{ref})^2(b)} \exp\left(\frac{-(1-\beta) z q \eta_{Pt}}{k_B T} \right) \right] \tag{10.4.40}$$

The overpotential may be regarded as the extra voltage needed to reduce the energy barrier of the rate-determining step to a value such that the electrode reaction proceeds at a desired rate. Thus, this equation tells us that if we want more current we have to pay a price in terms of voltage lost – overvoltage.

10.4.2.3   PEC cell photoresponse

The photoresponse of the PEC cell will be determined by the behavior of photogenerated electron-hole pairs and thus the physical properties of the semiconductor. Thus, the photocurrent flowing through the interface, under illumination, was derived by Gärtner, for an n-type semiconductor, as follows:

$$j_G = j_0 + e I_0^* \left[1 - \frac{\exp(-\beta W)}{1 + \beta L_p} \right] \tag{10.4.41}$$

where I_0^*, is the incident photon flux, β is the absorption coefficient (assuming monochromatic illumination), W is the depletion layer width, L_p is the hole diffusion length and j_0 is the exchange current density. The Gärtners's model assumes that there is no recombination in the space charge region and at the interface (O'Hayre et al., 2006).

In 1989 Reichman improved this model introducing more appropriated boundary conditions for the total valence band photocurrent in an n-type semiconductor (Krol and Schoonman, 2008):

$$j_p = \frac{j_G - j_0 \exp\left(-\frac{e\eta}{kT}\right)}{1 + \frac{j_0}{j_p^0}\exp\left(-\frac{e\eta}{kT}\right)} \tag{10.4.42}$$

where j_p^0 is the hole transfer rate at the semiconductor/electrolyte interface, and j_0 is the saturation current density, i.e. the hole current in the valence band at $x = W$ when $I_0 = 0$. η is the overvoltage, which is defined as the difference between the applied potential and the open-circuit potential under illumination. The Reichman derivation is of special interest since it includes the possibility of recombination in the space charge region, and it can be used to model the effect of the slow hole transfer kinetics that is often observed in oxides such as α-Fe_2O_3 (Reichman, 1980). These models, which describe only the minority carriers process, cannot be used under near-flatband conditions since at this point also the majority charge carriers contribute to the overall current.

10.5 PEC CELLS BOTTLENECKS AND FUTURE PROSPECTS

In past few years, photoelectrochemical technology for hydrogen production from solar energy has been growing fast, not only due to the opportunity of having a storable form of energy, but also because this kind of energy is totally produced in a clean and environmentally friendly way. Photoelectrochemical hydrogen production is not yet a commercial solution for our energy problems; in fact for considering it as an alternative there are some key challenges that should be addressed. Indeed material improvements are strongly envisaged, closely followed by the implementation of robust and cost-effective systems. Thus, an effective way would be to use a tandem system where the voltage difference to split water is given by a photovoltaic arrangement. In fact, with materials capable of producing photocurrents of $8\,mA\,cm^{-2}$ at $100\,mW\,cm^{-2}$ with a 0.8 V voltage difference would open the opportunity of exploitation PEC systems with a cost/m^2 of around \$80, excluding the PV bias cost with lifetimes higher than 10 years (Krol and Grätzel, 2012). Thus, it is expected that this process for producing a chemical fuel would be affordable with no big fluctuations in the prices, as often happens with the oil market.

Nomenclature

A – Cell area, m^2

b – Cell thickness, m

C_i – Concentration of species i, $mol \cdot m^{-3}$

D_i – Diffusion coefficient of species i, $m^2 \cdot s^{-1}$

D_i^b – Diffusion coefficient of species i in the electrolyte bulk, $m^2 \cdot s^{-1}$

D_i^L – Diffusion coefficient of species i in the electrolyte in the pores, $m^2 \cdot s^{-1}$

D_{ref} – Reference diffusion coefficient, $m^2 \cdot s^{-1}$

e – Elementary charge, $1.60217646 \times 10^{-19} C$

E_A – Acceptor energy, V

E_{Bias} – External bias potential, V

E_C – Conduction band potential, V

E_D – Donor energy, V

$_pE_d$ – Hole-induced corrosion value, V

$_nE_d$ – Electron-induced corrosion value, V

E_{fb} – Flat band potential, V

E_g – Semiconductor bandgap, V

E_F – Fermi level of the semiconductor, V

E_{H_2O/H_2} – Water reduction potential in alkaline electrolyte, V

E_{O_2/OH^-} – Water oxidation potential in alkaline electrolyte, V

E_{H^+/H_2} – Water redution potential in acid electrolyte, V

E_{H_2O/O_2} – Water oxidation potential in acid electrolyte, V

E_{Redox} – Redox potential, V

E_{VB} – Valence band potential, V

F – Faraday constant, $9.6485339 \times 10^4 \ C \cdot mol^{-1}$

G_i – Generation rate of species i, $m^{-3} \cdot s^{-1}$

h – Planck constant, $6.6260693(11) \times 10^{-34} \ J \cdot s$

I – Current, A

I_0^* – Incident photon flux corrected for reflection losses, $m^{-2} \cdot s^{-1}$

I_0 – Amplitude of the current signal, A

I_{OC} – Open-circuit current, A

$I(t)$ – Current response, A

j_i – Current density of species i, $s^{-1} \ m^{-2}$

J_{cell} – Net current density, $A \ m^{-2}$

J_0 – Exchange current density at Pt electrode, $A \cdot m^{-2}$

k – Boltzman constant, $1.3806503 \times 10^{-23} \ J \cdot K^{-1}$

k_{e^-} – Back reaction rate constant, s^{-1}

L – Thickness of the semiconductor, m

$m_{e^-}^*$ – Effective electron mass, kg

n – Electron density, m^{-3}

n_0 – Electron bulk concentration, m^3

n_Z – Empirical constante

n_e – Moles of electrons, mol

n_{eq} – Dark equilibrium electron density, m^{-3}

n_i – Density of species i, m^{-3}

n_i^b – Density of species i in the electrolyte bulk, m^{-3}

n_{int} – Intrinsic carrier concentration, m^{-3}

n_{ref} – Reference particle density, m^{-3}

Q_{SC} – Space charge below the semiconductor surface, $A \ s \ cm^{-2}$

N_A – Occupied acceptor density, m^{-3}

N_A^+ – Density of ionized acceptors, m^{-3}

N_C – Effective density of states in the conduction band, m^{-3}

N_D – Occupied donor density, m^{-3}

N_D^+ – Density of ionized donors, m^{-3}

N_V – Effective density of states in the valence band, m^{-3}

R_i – Recombination rate of species i, m^{-3}·s^{-1}

R_E – Electrolyte resistance, Ω

R_{Ohmic} – Ohmic resistance, Ω

T – Absolute temperature, K

t – Time, s

V_{OC} – Open-circuit voltage, V

V_{loss} – Ohmic voltage loss, V

$V(t)$ – Voltage perturbation, V

V_0 – Amplitude of the voltage signal, V

W – Space charge thickness, m

x – Coordinate, m

z – Number of electrons transferred in the reaction

Z – Impedance, Ω

Z_C – Impedance of capacitor, Ω

Z_{Imag} – Imaginary Part of impedance, Ω

Z_{Real} – Real Part of impedance, Ω

Z_R – Impedance of a resistor R, Ω

Z_0 – Ratio of the amplitude of the voltage signal and the amplitude of the current collector

Greek symbols

$\alpha(\lambda)$ – Wavelength-dependent absorption coefficient, m^{-1}

β – Transfer coefficient

ε – Dielectric constant

ε_0 – Permittivity of free space, 8.85419×10^{-12} F m^{-1}

ξ – Macroscopic electric field, V·m^{-1}

ϕ – Phase angle, $^\circ$

Φ – Electrostatic potential, V

$\Phi(x)$ – Band Bending, V

Φ_{FB} – Flat Band potential, V

η – $E_{Bias} - V_{OC}$, V

η_a – Overpotential at the anode, V

η_c – Overpotential at the cathode, V

η_{inj} – Electron injection efficiency

η_{Ohmic} – Ohmic overpotential, V

η_{Pt} – Overpotential at Pt electrode (cathode), V

λ – Wavelength, m

μ_i – Mobility of species i, m^2·V^{-1}s^{-1}

$\rho(x)$ – Total charge density, m^{-3}

τ – Time constant of the reaction kinetics, s

τ_n – Carrier lifetime, s

ΔE_F – Variation of the Fermi level potential, V

ΔG – Free energy for the overall cell reaction, J

ΔE – Electric potential of the reaction, V mol^{-1}

Superscripts

^{o} – Standard

Subscripts

d – Decomposition
e− – Electrons
FB – Flat Band
h+ – Holes
S – Series resistance
SC – Space charge

Abbreviations

ALD – Atomic layer deposition
CB – Conduction Band
CE – Counter-Electrode
CVD – Chemical vapor deposition
EIS – Electrochemical impedance spectroscopy
IPCE – Incident photon to current efficiency
OER – Oxygen evolution reaction
ref – reference
REF – Reference electrode
RHE – Reversible Hydrogen Electrode
SC – Semiconductor
SCLJ – Semiconductor-liquid junction
SHE – Standard Hydrogen Electrode
Solv – Solvation of the elements
PEC – Photoelectrochemical
PV – Photovoltaic
TCO – Transparent conductor oxide
VB – Valence Band
WE – Working electrode

REFERENCES

Alexander, B.D. and Augustynski, J. (2010) Nanostructured thin-Film WO₃ photoanodes for solar water and sea-water splitting. *Solar Hydrogen & Nanotechnology*. John Wiley & Sons, Ltd.

Andrade, L., Ribeiro, H.A. and Mendes, A. (2010) Dye-sensitized solar cells: an overview. *Energy Production and Storage: Inorganic Chemical Strategies for a Warming World*, R.H. Crabtree, Ed, John Wiley & Sons, Ltd: Chichester, UK.

Andrade, L., Lopes, T., Ribeiro, H. A. and Mendes, A. (2011) Transient phenomenological modeling of photoelectrochemical cells for water splitting – Application to undoped hematite electrodes. *International Journal of Hydrogen Energy*, 36, 175–188.

Archer, M.D. and Nozik, A.J. (2008) Nanostructured and photoelectrochemical systems for solar photon conversion. *Series on Photoconversion of Solar Energy*, Imperial College Press, London.

Aruchamy, A., Aravamudan, G. and Subba Rao, G. (1982) Semiconductor based photoelectro-chemical cells for solar energy conversion – An overview. *Bulletin of Materials Science*, 4, 483–526.

Babu, V.J., Kumar, M.K., Nair, A.S., Khen, T.L., Allakhverdiev, S.I. and Ramakrishna, S. (2012) Visible light photocatalytic water splitting for hydrogen production from N-TiO$_2$ rice grain shaped electrospun nanostructures. *International Journal of Hydrogen Energy*, 37, 8897–8904.

Bak, T., Nowotny, J., Rekas, M. and Sorrell, C.C. (2002) Photo-electrochemical hydrogen generation from water using solar energy. Materials-related aspects. *International Journal of Hydrogen Energy*, 27, 991–1022.

Bansal, A., Beach, J., Collins, R., Khaselev, O. and Turner, J.A. (1999) Photoelectrochemical based direct conversion systems for hydrogen production. *In: Proceedings of the 1999 Hydrogen Program Review*. US. Department of Energy National Renewable Energy Laboratory NREL/CP-570-26938.

Bard, A.J. and Fox, M.A. (1995) Artificial photosynthesis: Solar splitting of water to hydrogen and oxygen. *Accounts of Chemical Research*, 28, 141–145.

Bard, A.J. and Faulkner, L.R. (2001) *Electrochemical Methods Fundamentals and Applications*. 2nd ed., John Wiley & Sons, New York.

Barsoukov, E. and Macdonald, J.R. (2005) *Impedance Spectroscopy: Theory, Experiment, and Applications*. 2nd ed., Wiley-Interscience, Hoboken, N.J.

Becquerel, A.E. (1839) Mémoire sur les effets électriques produits sous l'influence des rayons solaires. *Comptes Rendus des Séances Hebdomadaires*, 9, 561–567.

Bilgen, E. (2001) Solar hydrogen from photovoltaic-electrolyzer systems. *Energy Conversion and Management*, 42, 1047–1057.

Bisquert, J. (2002) Theory of the impedance of electron diffusion and recombination in a thin layer. *The Journal of Physical Chemistry B*, 106, 325–333.

Bisquert, J. (2003) Chemical capacitance of nanostructured semiconductors: its origin and significance for nanocomposite solar cells. *Physical Chemistry Chemical Physics*, 5, 5360–5364.

Bolton, J.R., Strickler, S.J. and Connolly, J.S. (1985) Limiting and realizable efficiencies of solar photolysis of water. *Nature*, 316, 495–500.

Brillet, J., Cornuz, M., Le Formal, F., Yum, J.-H., Grätzel, M. and Sivula, K. (2010) Examining architectures of photoanode-photovoltaic tandem cells for solar water splitting. *Journal of Materials Research*, 25, 8.

Brillet, J., Grätzel, M. and Sivula, K. (2010) Decoupling feature size and functionality in solution-processed, porous hematite electrodes for solar water splitting. *Nano Letters*, 10, 4155–4160.

Butler, M.A. (1977) Photoelectrolysis and physical properties of the semiconducting electrode WO$_3$. *Journal of Applied Physics*, 48, 1914–1920.

Chandra, N., Wheeler, B.L. and Bard, A.J. (1985) Semiconductor electrodes. 59. Photocurrent efficiencies at p-InP electrodes in aqueous solutions. *The Journal of Physical Chemistry*, 89, 5037–5040.

Chang, B.-Y. and Park, S.-M. (2010) Electrochemical impedance spectroscopy. *Annual Review of Analytical Chemistry*, 3, 207–229.

Chen, Z., Jaramillo T.F., Deutsch, T.G., Kleiman-Shwarscstein, A., Forman, A.J., Gaillard, N., Garland, R., Takanabe, K., Heske, C., Sunkara, M., McFarland, E.W., Domen, K., Miller, E. L., Turner, J.A. and Dinh, H.N. (2010) Accelerating materials development for photoelectrochemical hydrogen production: Standards for methods, definitions, and reporting protocols. *Journal of Materials Research*, 25, 3–16.

Coelho, B., Oliveira, A.C. and Mendes, A. (2010) Concentrated solar power for renewable electricity and hydrogen production from water—a review. *Energy & Environmental Science*, 3, 1398–1405.

Conibeer, G.J. and Richards, B.S. (2007) A comparison of PV/electrolyser and photoelectrolytic technologies for use in solar to hydrogen energy storage systems. *International Journal of Hydrogen Energy*, 32, 2703–2711.

Conway, B.E., Bockris, J.O'M., White, R.E. (2002) Modern Aspects of Electrochemistry No. 32, Kluwer Acamdemic Publishers, New York, Boston, Dordrecht, London, Moscow.

Dobran, F. (2010) Energy supply options for climate change mitigation and sustainable development. *In: Proceedings of XXI World Energy Congress.* 12–16 September, Montreal, Canada.

Fujishima, A. and Honda, K. (1972) Electrochemical photolysis of water at a semiconductor electrode. *Nature,* 238, 37–38.

Gadgil, P.N. (1990) Preparation of iron pyrite films for solar cells by metalorganic chemical vapor deposition. Simon Fraser University.

Gärtner, W.W. (1959) Depletion-layer photoeffects in semiconductors. *Physical Review,* 116(1), 84–87.

Gerischer, H. (1977) On the stability of semiconductor electrodes against photodecomposition. *Journal of Electroanalytical Chemistry,* 82, 133–143.

Glasscock, J.A., Barnes, P.R.F., Plumb, I.C. and Savvides, N. (2007) Enhancement of photoelectrochemical hydrogen production from hematite thin films by the introduction of Ti and Si. *The Journal of Physical Chemistry C,* 111, 16477–16488.

Goodman, A.M. (1963) Metal–Semiconductor barrier height measurement by the differential capacitance method – one carrier system. *Journal of Applied Physics,* 34, 329–338.

Grätzel, M. (2005) Solar energy conversion by dye-sensitized photovoltaic cells. *Inorganic Chemistry,* 44, 6841–6851.

Grätzel, M. and Augustynski, J. (2001) *Tandem Cell for Water Cleavage by Visible Light.* Patent No. WO/2001/002624.

Green, M.A., Emery , K., Hishikawa, Y. and Warta, W. (2008) Short communication solar cell efficiency tables (Version 31). *Progress in Photovoltaics: Research and Applications,* 16, 61–67.

Grimes, C.A., Varghese, O.K. and Ranjan, S. (2008) Oxide semiconductors nano-crystalline tubular and porous systems. In: *Light, Water, Hydrogen.* pp. 257–369, Springer: New York.

Grimes, C.A., Varghese, O.K. and Ranjan, S. (2008) Non-oxide semiconductor nanostructures. In: *Light, Water, Hydrogen.* pp. 427–483, Springer: New York.

Grimes, C.A., Varghese, O.K. and Ranjan, S. (2008) The solar generation of hydrogen by water photoelectrolysis. In: *Light, Water, Hydrogen.* pp. 115–190, Springer: New York.

Grimes, C.A., Varghese, O.K. and Ranjan, S. (2008) Photoelectrolysis. In: *Light, Water, Hydrogen.* pp. 115–190, Springer: New York.

Grimes, C.A., Varghese, O.K., and Ranjan, S. (2008) Oxide semiconducting materials as photoanodes. In: *Light, Water, Hydrogen.* pp. 191–255, Springer: New York.

Gomes, W.P. and Vanmaekelbergh, D. (1996) Impedance spectroscopy at semiconductor electrodes: Review and recent developments. *Electrochimica Acta,* 41, 967–973.

Hara, M., Kondo , T., Komoda, M., Ikeda, S., N. Kondo, J., Domen, K., Shinohara, K. and Tanaka, A. (1998) Cu_2O as a photocatalyst for overall water splitting under visible light irradiation. *Chemical Communications,* 3, 357–358.

Heaviside, O. http://www-groups.dcs.st-and.ac.uk/~history/Mathematicians/Heaviside.html.

Henry, C.H., Logan, R.A. and Merritt, F.R. (1978) The effect of surface recombination on current in $Al_xGa_{1-x}As$ heterojunctions. *Journal of Applied Physics,* 49, 3530–3542.

Hodes, G., Cahen, D. and Manassen, J. (1976) Tungsten trioxide as a photoanode for a photoelectrochemical cell (PEC). *Nature,* 260, 312–313.

Horowitz, G. (1983) Capacitance-voltage measurements and flat-band potential determination on Zr-doped α-Fe_2O_3 single-crystal electrodes. *Journal of Electroanalytical Chemistry and Interfacial Electrochemistry,* 159, 421–436.

Hu, Y.-S., Kleiman-Shwarsctein, A., Forman, A.J., Hazen, D., Park, J.-N. and McFarland, E.W. (2008) Pt-Doped α-Fe_2O_3 thin films active for photoelectrochemical water splitting. *Chemistry of Materials,* 20, 3803–3805.

Ingler, W.B. and Khan, S.U.M. (2006) A self-driven p/n-Fe$_2$O$_3$ tandem photoelectrochemical cell for water splitting. *Vol. 9. ETATS-UNIS: Institute of Electrical and Electronics Engineers*, Pennington, NJ, USA.

James, B.D., Baum, G.N., Perez, J. and Baum, K.N. (2009) *Technoeconomic analysis of photoelectrochemical (PEC) hydrogen production*. U.D. Report, Editor. Directed Technologies Inc.: Virginia, USA.

Karakitsou, K.E. and Verykios, X.E. (1993) Effects of altervalent cation doping of titania on its performance as a photocatalyst for water cleavage. *The Journal of Physical Chemistry*, 97(6), 1184–1189.

Kay, A., Cesar, I. and Grätzel, M. (2006) New Benchmark for Water Photooxidation by Nanostructured α-Fe$_2$O$_3$ Films. *Journal of the American Chemical Society*, 128, 15714–15721.

Khan, S.U.M., Al-Shahry, M. and Ingler, W.B. (2002) Efficient photochemical water splitting by a chemically modified n-TiO$_2$. *Science*, 297, 2243–2245.

Khan, S.U.M. and Akikusa, J. (1999) Photoelectrochemical splitting of water at nanocrystalline n-Fe$_2$O$_3$ thin-film electrodes. *The Journal of Physical Chemistry B*, 103, 7184–7189.

Khaselev, O. and Turner, J.A. (1998) A monolithic photovoltaic-photoelectrochemical device for hydrogen production via water splitting. *Science*, 280, 425–427.

Khaselev, O., Bansal, A. and Turner, J.A. (2001) High-efficiency integrated multijunction photovoltaic/electrolysis systems for hydrogen production. *International Journal of Hydrogen Energy*, 26, 127–132.

Kleiman-Shwarsctein, A., Hu, Y.-S., Forman, A.J., Stucky, G.D. and McFarland, E.W. (2008) Electrodeposition of α-Fe$_2$O$_3$ doped with Mo or Cr as photoanodes for photocatalytic water splitting. *The Journal of Physical Chemistry C*, 112, 15900–15907.

Krol, R., Liang, Y. and Schoonman, J. (2008) Solar hydrogen production with nanostructured metal oxides. *Journal of Materials Chemistry*, 18, 2311–2320

Krol, R. and Schoonman, J. (2008) *Sustainable Energy Technologies, in Photo-Electrochemical Production of Hydreogen*. R. Krol and M. Grätzel, Eds., Springer Netherlands.

Krol, R. and Grätzel, M. (2012) *Photoelectrochemical Hydrogen Production. Electronic Materials: Science & Technology*. Springer, USA.

Kruse, B., Grinna, S. and Buch, C. (2002) *Hydrogen-Status and Possibilities*. The Bellona Foundation, Oslo.

Leduc, J. and Ahmed, S.M. (1988) Photoelectrochemical and impedance characteristics of specular hematite. 2. Deep bulk traps in specular hematite at small a.c. frequencies. *The Journal of Physical Chemistry*, 92, 6661–6665.

Le Formal, F., Tetreault, N., Cornuz. M., Moehl, T., Grätzel, M. and Sivula, K. (2011) Passivating surface states on water splitting hematite photoanodes with alumina overlayers. *Chemical Science*, 2, 737–743.

Liang, Y., Enache, C.S. and Krol, R. (2008) Photoelectrochemical characterization of sprayed α-Fe$_2$O$_3$ Thin Films: Influence of Si doping and SnO$_2$ interfacial layer. *International Journal of Photoenergy*, 2008, 7.

Long, M. and Cai, W. (2008) Photoelectrochemical properties of BiVO$_4$ film electrode in alkaline solution. *Chinese Journal of Catalysis*, 29, 881–883.

Mendes, A., Lopes, T., Andrade, L. and Dias, P. Células Fotoeletroquímicas e Processo da sua construção, patent under protection.

Luo, H., Mueller, A.H., McCleskey, T.M., Burrell, A. K., Bauer, E. and Jia, Q.X. (2008) Structural and photoelectrochemical properties of BiVO$_4$ thin films. *The Journal of Physical Chemistry C*, 112, 6099–6102.

Macdonald, D.D. (2006) Reflections on the history of electrochemical impedance spectroscopy. *Electrochimica Acta*, 51, 1376–1388.

Mavroides, J.G., Kafalas, J.A. and Kolesar, D.F. (1976) Photoelectrolysis of water in cells with SrTiO3 anodes. *Applied Physics Letters*, 28, 241–243.

Meda, L., Tozzola, G., Tacca, A., Marra, G., Caramori, S., Cristino, V. and Alberto Bignozzi, C. (2010) Photo-electrochemical properties of nanostructured WO_3 prepared with different organic dispersing agents. *Solar Energy Materials and Solar Cells*, 94, 788–796.

Memming, R. (2001) *Semiconductor Electrochemistry*. 1st ed. 2001: Wiley-VCH.

Miller, E.L., Paluselli, D., Marsen, B. and Rocheleau, R.E. (2005) Development of reactively sputtered metal oxide films for hydrogen-producing hybrid multijunction photoelectrodes. *Solar Energy Materials and Solar Cells*, 88, 131–144.

Minggu, L.J., Wan Daud, W.R. and Kassim, M.B. (2010) An overview of photocells and photoreactors for photoelectrochemical water splitting. *International Journal of Hydrogen Energy*, 35, 5233–5244.

Nakamura, R., Tanaka, T. and Nakato, Y. (2004) Mechanism for visible light responses in anodic photocurrents at N-doped TiO_2 film electrodes. *The Journal of Physical Chemistry B*, 108, 10617–10620.

Nathan, S.L. (2005) Basic research needs for solar energy utilization. *In: Report on the Basic Energy Sciences Workshop on Solar Energy Utilization*. R.M. Nault, Ed., C.I.o. Technology.

Nowotny, J., Bak, T., Nowotny, M.K. and Sheppard, L.R. (2007a) Titanium dioxide for solar-hydrogen I. Functional properties. *International Journal of Hydrogen Energy*, 32, 2609–2629.

Nowotny, J., Bak, T., Nowotny, M.K. and Sheppard, L.R. (2007b) Titanium dioxide for solar-hydrogen II. Defect chemistry. *International Journal of Hydrogen Energy*, 32, 2630–2643.

Nowotny, J., Bak, T., Nowotny, M.K. and Sheppard, L.R. (2007c) Titanium dioxide for solar-hydrogen III: Kinetic effects at elevated temperatures. *International Journal of Hydrogen Energy*, 32, 2644–2650.

Nowotny, J., Bak, T., Nowotny, M.K. and Sheppard, L.R. (2007d) Titanium dioxide for solar-hydrogen IV. Collective and local factors in photoreactivity. *International Journal of Hydrogen Energy*, 32, 2651–2659.

Nozik, A.J. (1976) p-n photoelectrolysis cells. *Applied Physics Letters*, 29, 150–153.

Nozik, A.J. (1978) Photoelectrochemistry: Applications to solar energy conversion. *Annual Review of Physical Chemistry*, 29, 189–222.

Nozik, A.J. and Memming, R. (1996) Physical chemistry of semiconductor–liquid interfaces. *The Journal of Physical Chemistry*, 100, 13061–13078.

Ohashi, K., McCann, J. and Bockris, J.O.M. (1977) Stable photoelectrochemical cells for the splitting of water. *Nature*, 266, 610–611.

O'Hayre, R., Cha, S.-W., Colella W. and Prinz, F.B. (2006) *Fuel Cells Fundamentals*. John Wiley & Sons, Inc., New York, USA.

Paracchino, A., Laporte, V., Sivula, K., Grätzel, M. and Thimsen, E. (2011) Highly active oxide photocathode for photoelectrochemical water reduction. *Nature Materials*, 10, 456–461.

Peterson, P. (2012) Resources for Electrochemistry. http://www.consultrsr.com/resources/ref/select.htm.

Pine Instrument Company (2012) http://www.pineinst.com/echem/viewproduct.asp?ID=47799.

Rahman, M.M., Krishna, K. M., Soga, T., Jimbo, T. and Umeno, M. (1999) Optical properties and X-ray photoelectron spectroscopic study of pure and Pb-doped TiO_2 thin films. *Journal of Physics and Chemistry of Solids*, 60, 201–210.

Reichman, J. (1980) The current-voltage characteristics of semiconductor-electrolyte junction photovoltaic cells. *Applied Physics Letters*, 36, 574–577.

Reiss, H. (1978) Photocharacteristics for electrolyte-semiconductor junctions, *Journal of The Electrochemical Society*, 125, 937–949.

Santato, C., Ulmann, M. and Augustynski, J. (2001) Photoelectrochemical Properties of Nanostructured Tungsten Trioxide Films. *The Journal of Physical Chemistry B*, 105, 936–940.

Satsangi, V.R., Dass, S. and Shrivastav, R. (2010) Nanostructured α-Fe$_2$O$_3$ in PEC generation of hydrogen. *In: On Solar Hydrogen & Nanotechnology.* John Wiley & Sons, Ltd.

Sayama, K., Nomura, A., Arai, T., Sugita, T., Abe, R., Yanagida, M., Oi, T., Iwasaki, Y., Abe, Y. and Sugihara, H., (2006) Photoelectrochemical decomposition of water into H$_2$ and O$_2$ on porous BiVO$_4$ thin-film electrodes under visible light and significant effect of Ag Ion treatment. *The Journal of Physical Chemistry B*, 110, 11352–11360.

Sinn, C., Meissner, D. and Memming, R. (1990) Charge transfer processes at WSe$_2$ electrodes with pH-controlled stability. *Journal of The Electrochemical Society*, 137, 168–172.

Sivula, K., Le Formal, F. and Grätzel, M. (2011) Solar water splitting: Progress using hematite (α-Fe$_2$O$_3$) photoelectrodes. *ChemSusChem*, 4, 432–449.

Solarska, R., Jurczakowski, R. and Augustynski, J. (2012) A highly stable, efficient visible-light driven water photoelectrolysis system using a nanocrystalline WO$_3$ photoanode and a methane sulfonic acid electrolyte. *Nanoscale*, 4.

The Energy Report – 100% renewable energy by 2050 (2011) by WWF, ECOFYS, and OMA.

Tilley, S.D., Cornuz, M., Sivula, K. and Grätzel, M. (2010) Light-induced water splitting with hematite: Improved nanostructure and Iridium oxide catalysis. *Angewandte Chemie International Edition*, 49, 6405–6408.

Trieb, F. (2005) Concentrating solar power for the Mediterranean region. *In: Final Report*, DLR, Editor. German Aerospace Center (DLR).

Vinke, I.C., Diepgrond, J., Boukamp, B.A., de Vries, K.J. and Burggraaf, A.J. (1992) Bulk and electrochemical properties of BiVO4. *Solid State Ionics*, 57, 83–89.

Wilke, K. and Breuer, H.D. (1999) The influence of transition metal doping on the physical and photocatalytic properties of titania. *Journal of Photochemistry and Photobiology A: Chemistry*, 121, 49–53.

Wilson, R.H. (1977) A model for the current-voltage curve of photoexcited semiconductor electrodes. *Journal of Applied Physics*, 48, 4292–4297.

Wrighton, M.S., Morse, D.L., Ellis, A.B., Ginley, D.S. and Abrahamson, H.B. (1976) Photo-assisted electrolysis of water by ultraviolet irradiation of an antimony doped stannic oxide electrode. *Journal of the American Chemical Society*, 98, 44–48.

Würfel, P. (2005) *Physics of Solar Cells: From Principles to New Concepts.* Weinheim: Wiley-VCH Verlag GmbH.

Zerta, M., Schmidt, P.R., Stiller, C. and Landinger, H. (2008) Alternative world energy outlook (AWEO) and the role of hydrogen in a changing energy landscape. *International Journal of Hydrogen Energy*, 33, 3021–3025.

Chapter 11

Photobiohydrogen production and high-performance photobioreactor

Qiang Liao[1,2], Cheng-Long Guo[1,2], Rong Chen[1,2], Xun Zhu[1,2] & Yong-Zhong Wang[1,2]
[1]*Key Laboratory of Low-grade Energy Utilization Technologies and Systems, Chongqing University, Chongqing, China*
[2]*Institute of Engineering Thermophysics, Chongqing University, Chongqing, China*

11.1 INTRODUCTION

The carbon-based fossil fuels, including coal, oil and natural gas, are currently the world's primary energy sources, which have fueled global economic development over the past century. However, fossil fuels are finite resources and the fast depletion of them has been mainly responsible for the energy crisis and for causing global warming. In the face of such a situation, renewable energy resources inevitably become the only solution to resolve existing problems for the sustainable development. Among the renewable energy alternatives being explored, hydrogen is one of the ideal energy carriers because it has numerous advantages of excellent combustion performance, cleanness, high efficiency, and a three times higher caloric value than petroleum (Akkerman et al., 2002; Das and Veziroglu, 2001). In spite of its attractiveness, the complete replacement of fossil fuels by hydrogen is still far away from widespread industrial application as a result of the bottleneck of large-scale hydrogen production. Present hydrogen production methods that can be scaled up, including steam reforming of natural gas, thermal cracking of light oil, coal gasification and electrolysis of water, etc., still rely on fossil fuels or huge electricity consumption, which are clearly unsustainable. Besides, these methods have other intrinsic drawbacks: i) complex process and high cost of equipment investment, ii) environmental pollution (Asada and Miyake, 1999). Therefore, the development of a safe, economical, and sustainable way to produce hydrogen is the key to the realization of hydrogen energy. Biohydrogen production is such a technology that can overcome the above barriers and offer the significant advantages in that it is cost-effective, pollution-free and environmentally compatible (Levin et al., 2004). Typically, biohydrogen production technologies can be classified into two types: dark hydrogen fermentation and photobiohydrogen production. As compared to dark hydrogen fermentation, photobiohydrogen production can convert solar energy into hydrogen using water and simple organic compounds as the hydrogen sources, and enable carbon dioxide reduction and waste treatment. Over the past decades, therefore, extensive efforts have been devoted to the improvement in the performance of the photobiohydrogen production. This chapter will summarize recent advances in the photobiohydrogen production technology.

The organization of this chapter is as follows. First of all, the general descriptions of photobiohydrogen production as well as the critical issues related to this hydrogen production technology are given in Section 11.2. Then, the comprehensive introduction of researches for improving the hydrogen-producing microorganisms by genetic and metabolic engineering are provided in Section 11.3. In the following, Section 11.4 presents the high-performance photobioreactor designs and operating parameters optimization. Particular attention will be paid to the cell immobilization technique for the improvement in the biomass concentration. Finally, challenges and future directions on the photobiohydrogen production technology will be addressed in Section 11.5.

11.2 GENERAL DESCRIPTION OF PHOTOBIOHYDROGEN PRODUCTION

In general, two types of photosynthetic microorganisms can be used to produce hydrogen. One is photoautotrophic microorganisms mainly including green algae and cyanobacteria, which decompose water into hydrogen and oxygen using light as the energy source, so-called water photolysis, and carbon dioxide as the carbon source for microorganism growth. Another one is photoheterotrophic microorganisms (photosynthetic bacteria) that separate hydrogen atom from the organic compounds to produce hydrogen. Their respective photobiohydrogen production processes are described below.

11.2.1 Photoautotrophic hydrogen production

11.2.1.1 Green algae

Green algae with both the oxygenic photosynthesis and hydrogen metabolism are the only known eukaryotes that can produce hydrogen through water photolysis (Happe et al., 2002). Since Gaffron (1939) firstly found that *Scenedesmus obliguus* with oxygen-free cultivation could utilize hydrogen as a reductant and Gaffron and Rubin (1942) then revealed that unicellular algae could liberate hydrogen slowly in the dark if the surrounding air was replaced by nitrogen, after more than 70 years' development it has been found that over 1000 microorganisms among about 30 green algae genera are able to produce hydrogen by water photolysis (Kruse et al., 2005), including *Chlamydomonas reinhardti* (Happe and Kaminski, 2002), *Scenedesmus obliquus* (Wunschiers et al., 2001), *Platymonas subcordiformis* (Guan et al., 2004) and others, as summarized in Table 11.2.1. Photobiohydrogen production by green algae can provide many features of higher light efficiency than agricultural plants, short growth, automatic collection of light energy and cleanness (Das and Veziroglu, 2008). Besides, supplied carbon dioxide is for the microorganism growth so that it is carbon-free hydrogen production and enables carbon dioxide capture (Levin et al., 2004).

As illustrated in Figure 11.2.1, green algae contain two photosynthetic systems, photosystem I (PSI) and photosystem II (PSII) which are inserted in the thylakoid membrane with an enclosed inner space called lumen and operate in series (Akkerman et al., 2002; Allakhverdiev et al., 2010; Miyake et al., 1999; Srirangan et al., 2011). The reaction centre of PSII shows the strongest adsorption at 680 nm and thus is called P680, while the reaction center of PSI, P700, shows the strongest absorption at 700 nm. In the photobiohydrogen production process, water is transported from the bulk into green algae through the cell membrane and light is transmitted through the medium solution

Table 11.2.1 Characteristics of different photobiohydrogen production microorganisms.

Classifications	Microorganisms	Strains	Characteristics
Photoautotrophic microorganisms	Green algae	*Chlamydomonas reinhardti* *Scenedesmus obliquus* *Platymonas subcordiformis*	Carbon-free hydrogen production Short growth cycle High theoretical energy conversion efficiency
	Cyanobacteria	*Anabaena cylindrica* *Nostoc muscorum* *Plectonema boryanum* *Oscillotoria limnetica* *Aphanothece halophytica*	Simple nutritional requirements Inexpensive cultivation Separating hydrogen and oxygen evolution
Photoheterotrophic microorganisms	Photosynthetic bacteria	*Rhodospirillum rubrum* *Rhodobacter sphaeroides* *Rhodopseudomonas palustris*	Use of a wide range of the solar spectrum Lack of oxygen-evolving activity Ability to consume organic substrates derived from wastes

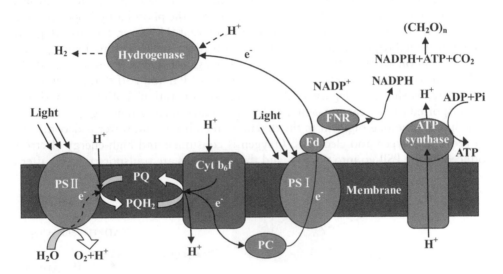

Figure 11.2.1 The photosynthetic system of green algae (Akkerman et al., 2002).

and absorbed by the antenna pigment; it is then transferred to the P680 reaction centre. Subsequently, the strong oxidant generated after the excitation of P680 reaction centre induces the splitting of water into oxygen, protons and electrons. Oxygen goes into the mitochondria through the chloroplast membrane and is consumed by the respiration of mitochondrial, and in the meantime supplied carbon dioxide is fixed. Protons are left in the lumen while electrons are used to reduce the reaction center and transported to plastoquinone (PQ) after renewed excitations. Then, protons are picked up from stroma producing fully reduced plastoquinone (PQH_2) that is diffused to the cytochrome complex (Cyt b_6f), via which electrons are transferred to the water-soluble electron carrier plastocyanin (PC). This complex also provides additional pumping force for protons across lipid bilayer. In addition, the proton gradient across the thylakoid membrane drives adenosine triphosphate (ATP) production via the action of ATP-synthase. Via the lumen, PC diffuses to P700 and releases the electrons which reduce the reaction center and are transported to the electron carrier ferredoxin (Fd) after renewed light-induced excitations. At last, nicotinamide adenine dinucleotide phosphate (NADP$^+$) is reduced to NADPH via the action of ferredoxin-NADP reductase (FNR), and the protons in stroma and the electrons transported are combined to generate hydrogen biocatalyzed by hydrogenase. Hydrogen produced then moves out of the phototbioreactor.

11.2.1.2 Cyanobacteria

Unlike green algae, cyanobacteria are simple prokaryotes possessing photosynthetic pigment. Many cyanobacteria are currently found to be capable of evolving hydrogen from water, including *Anabaena cylindrica* (Weissman and Benemann, 1977), *Nostoc muscorum* (Shah et al., 2003), *Plectonema boryanum* (Kashyap et al., 1996), *Oscillotoria limnetica, Aphanothece halophytica* (Belkin and Padan, 1978) and so on, which are summarized in Table 11.2.1. Due to the presence of different enzymes, i.e. hydrogenase and nitrogenase, cyanobacteria can further be divided into the non-nitrogen fixing or nitrogen fixing cyanobacteria. Consequently, the photobiohydrogen production pathways of cyanobacteria are noticeably different. However, the advantages of simple nutritional requirements, inexpensive cultivation and separating hydrogen and oxygen evolution are still offered by both types of cyanobacteria (Pinto et al., 2002).

Non-nitrogen fixing cyanobacteria also consist of PSI and PSII that are similar to green algae, as shown in Figure 11.2.2. During the operation, both water and light are transported to the reaction center of PSII in a reactor. After that, light is absorbed by the light-harvesting pigment in the thylakoid membrane and water is decomposed into protons, oxygen and electrons. Oxygen is consumed and high-energy electrons produced by the PSII go into the PQ, and then electrons are transported to Fd after a

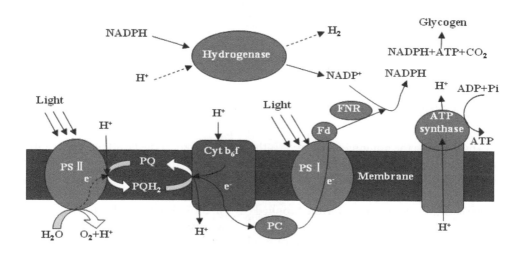

Figure 11.2.2 The photosynthetic system of non-nitrogen fixing cyanobacteria.

series of electron transfer via PSI and Cyt b_6f in the thylakoid membrane and finally reduce $NADP^+$ to NADPH. The NADPH generated by the catabolism of endogenous glycogen is oxidized to yield the electrons required for photobiohydrogen production using hydrogenase as the biocatalyst (Allakhverdiev et al., 2010; Appel and Schulz, 1998). Regarding the nitrogen fixing cyanobacteria, the photobiohydrogen production enzyme is nitrogenase instead of hydrogenase, which can fix nitrogen to ammonia accompanied by obligatory reduction of protons to hydrogen (Asada and Miyake, 1999; Benemann, 1997). Eventually, generated hydrogen is transported out of the photobioreactor and collected.

11.2.2 Photoheterotrophic hydrogen production

Photosynthetic bacteria are prokaryotes having only one photosynthesis system, which can utilize simple organic compounds as electron donor to produce hydrogen under anaerobic circumstances. Gest and Kamen (1949) found the hydrogen evolution and photosynthetic nitrogen fixation by illuminating *Rhodospirilum rubrum*. At present, the mainly used photosynthetic bacteria for the photobiohydrogen production by the photo-fermentation are: *Rhodospirillum rubrum* (Najafpour et al., 2004), *Rhodobacter sphaeroides* (Koku et al., 2002), and *Rhodopseudomonas palustris* (Chen et al., 2007; Suwansaard et al., 2010). As summarized in Table 11.2.1, hydrogen production by photosynthetic bacteria offers many unique features. Firstly, photosynthetic bacteria can use a wide range of the solar spectrum, making it more practical for photobiohydrogen production. Secondly, photobiohydrogen production by photosynthetic bacteria does not generate oxygen, avoiding the oxygen inhibition to enzymes. Thirdly, they are able to consume many organic substrates derived from wastes, such as acetic acid, lactic acid, butyric acid, glucose and so on. Hence, photosynthetic bacteria have been regarded as the most promising microorganisms for both the photobiohydrogen production and wastewater treatment (Basak and Das, 2007; Chen et al., 2011; Shi and Yu, 2005).

In the process of photobiohydrogen production by photosynthetic bacteria through photo-fermentation, substrates from the bulk solution as the electron donor and light as energy source are simultaneously transported into bacteria. Electrons are then liberated from the electron donor and transported through the photosynthetic apparatus driven by captured light energy. Protons are transferred through the membrane and a proton gradient developed is used by the ATP-synthase enzyme to generate ATP and reverse electron flow to produce high energy electrons. As shown in Figure 11.2.3, the bacteriochlorophyll excitation is stimulated by a photon in the reaction centre. The generated energy leads to the release of an electron that reduces the membrane quinone (Q) pool. The Q makes protons to be released to the periplasmic space and then to reduce the cytochrome bc_1 complex (Cyt bc_1) that enables the reduction of cytochrome c_2 (C_2). In turn, C_2 reduces the oxidized primary electron donors in the reaction centre, forming and closing a cycle. The protons accumulated in the periplasm form an electrochemical gradient which is used not only by the ATP-synthase to generate ATP but also to transport the electrons further to the electron acceptor Fd. When nitrogen is present, it can be reduced by nitrogenase to ammonium using the electrons derived from the Fd. However, under nitrogen-free circumstances, these electrons can be used to reduce protons into hydrogen with the help of nitrogenase (Akkerman et al.,

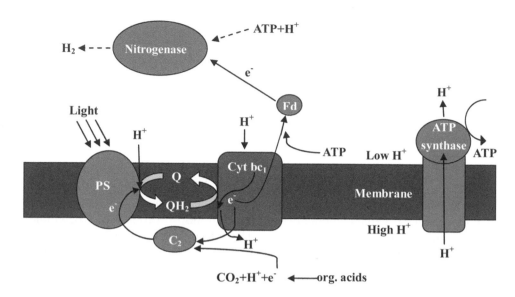

Figure 11.2.3 The photosynthetic system of photosynthetic bacteria (Akkerman et al., 2002).

2002; Keskin et al., 2011). The produced hydrogen ultimately transports out of the photobioreactor.

11.2.3 Critical issues in photobiohydrogen production

Although photobiohydrogen production is encouraging, there are still some major challenges that should be resolved. Firstly, the characteristics of microorganisms clearly play important roles in the photobiohydrogen production processes, which directly affect the rate of the photobiochemical reactions. Thus breeding microorganisms with high photobiohydrogen production performance is one of the critical issues in photobiohydrogen production. In addition to the microorganisms themselves, the photobiohydrogen production performances are also inherently related to the design of the photobioreactor and operating conditions. Basically, the photobioreactor design and operating conditions affect the mass transport that is coupled with the photobiochemical reaction during the photobiohydrogen production process. Too low reactant concentrations or too high hydrogen concentration can cause a low reaction rate and thus poor photobiohydrogen production performance. In addition to mass transport, inefficient light transport and non-uniform light distribution can also decrease the light utilization and the photobiohydrogen production rate. Therefore, efficient reactants transport and products removal along with high light transport and uniform distribution are keys to maintaining the physiological activity and metabolic stability of microorganisms and achieve continuous, stable and good photobiohydrogen production performance, which highly depends on the photobioreactor design and operating conditions. Besides, other operating conditions, such as temperature, pH value and so on, also significantly affect the photobiohydrogen production performance. As a result, the improvement in photobioreactor design together with the optimization of operating conditions is another critical issue in photobiohydrogen production. To improve photobiohydrogen production performance, genetic modification and metabolic engineering as well as the photobioreactor design and operating condition optimization have become the focus in existing researches.

11.3 GENETIC AND METABOLIC ENGINEERING

The currently-employed microorganisms have many problems, such as oxygen toxicity to enzymes, lower photosynthetic efficiency, insufficient supply of electrons and protons and presence of uptake hydrogenase, etc. Therefore various strategies, including genetic modification of microorganism and metabolic engineering, have been adopted for achieving a high photobiohydrogen production performance (Das et al., 2008; Mathews and Wang, 2009).

As for green algae and cyanobacteria, the photosynthetic rate is several times higher than the respiration rate under normal conditions (Melis and Melnicki, 2006). Excessive oxygen produced by the photosynthetic process lowers the photobiohydrogen production performance due to the extreme sensitivity of enzymes to oxygen (Manish and Banerjee, 2008; Melis, 2002). Melis (2007) attenuated the photosynthesis/respiration (P/R) capacity ratio in green algae by DNA insertional mutagenesis for the isolation and characterization of P/R aberrant mutants to reduce the oxygen generation rate, thereby stabilizing its metabolism and hydrogen-evolution. By using an independent approach that can lower photosynthesis and/or enhance cellular respiration, anaerobic conditions can thus be established when the P/R ratio drops below one, which increased the photobiohydrogen production performance. In addition, it has also been shown that engineering oxygen-tolerant hydrogenase genes of, *hydS* and *hydL* from *Thiocapsa roseopersicina* into sensitive organisms also helps to reduce the oxygen sensitivity. An expression vector pEX-Tran used for *Synechococcus sp.* PCC7942 transformation is readily available and suitable for other cyanobacterial systems as well (Xu et al., 2005). *Chlamydomonas reinhardtii* mutants obtained by random and oriented mutagenesis have also succeeded in alleviating the extreme oxygen sensitivity of the green algal reversible hydrogenase and the oxygen tolerance was enhanced by approximately 10-fold (Ghirardi et al., 2000).

The rate of electron transfer from PSII to PSI is about 10 times lower than the capture rate of photon by the antenna pigments. Photons captured by the antenna systems are not being fully used and accumulated, causing a low photosynthetic efficiency. Hence, the improvement in the photosynthetic efficiency by the enhancement of electron transfer has been studied by bioengineers (Hallenbeck and Benemann, 2002). Truncating the chlorophyll antenna size of PSII is an effective method to boost the photosynthetic efficiency of green algae because large antenna complexes can dissipate excessive photons as fluorescence or heat at saturating light conditions (Polle et al., 2003). Several genes have also been found to be able to confer a truncated antenna size in green algae by random mutagenesis (Polle et al., 2002). The tla1 strain of chlorophyll deficient with a functional chlorophyll antenna size of PSI and PSII being about 50% and 65% required a higher light intensity for the saturation of photosynthesis and thus showed larger solar conversion efficiencies and a higher photosynthetic productivity than the wild type under mass culture conditions (Polle et al., 2003). For the photosynthetic bacteria, a stable mutant of the photosynthetic bacterium *Rhodobacter sphaeroides* with an altered light-harvesting system by UV irradiation was found to be able to decrease the core antennal content by 2.7-fold and increase peripheral antennal content by 1.6-fold as compared to the wild-type strain. The resulting photobiohydrogen production rate in this mutant under the light of 800- and 850-nm corresponding to the absorption maxima of peripheral antennal was 1.5 times higher than the wild-type strain (Vasilyeva et al., 1999). In another novel mutant of MTP4 created from the wild-type strain *Rhodobacter sphaeroides* RV by UV irradiation, the contents of chlorophylls and carotenoids of the chromatophores were reduced to 41 and 49%, respectively. With this mutant, the photobiohydrogen production rate was increased 50% higher than the wild-type strain RV (Kondo et al., 2002).

According to the discussion in preceding section, it can be understood that the insufficient supply of electrons and protons can also seriously affect the performance of microorganisms for photobiohydrogen production. Lee and Greenbaum (2003) proposed a genetic insertion of a polypeptide protein channel with a hydrogenase promoter to reduce the proton gradient across the thylakoid membrane by avoiding the ATP-synthase channel to produce ATP. Kruse et al. (2005) isolated a strain Stm6 with modified respiratory metabolism in *C. reinhardtii* to increase the availability of electrons and protons for hydrogen evolution under anaerobic conditions. The results showed that the strain could accumulate a large amount of starch in the cells and provide low dissolved oxygen concentration, both of which contributed to the 5–13 times increase of the photobiohydrogen production rate (Kruse et al., 2005). Since excess reducing equivalents generated by organic acid oxidation in *Rhodobacter capsulatus* can be consumed to reduce protons into hydrogen by nitrogenase, the elimination of the cyt cbb_3 oxidase that serves as a redox signal to the RegB/RegA regulatory system can also enhance nitrogenase expression, significantly increasing the photobiohydrogen production rate (Ozturk et al., 2006).

Besides, produced hydrogen can be consumed by the uptake hydrogenase (NiFe-hydrogenase) in the process of the photobiohydrogen production. Therefore, the hydrogen consumption needs to be prevented by inactivating uptake [NiFe]-hydrogenase. Masukawa et al. (2002) constructed a hydrogenase mutants, hupL(-) mutant (deficient in the uptake hydrogenase) from *Anabaena sp.* PCC 7120, by which 4–7 times higher photobiohydrogen production rate than that of the wild strain under optimal conditions was achieved, indicating that the removal of the hupL gene is an effective way for the improvement of the photobiohydrogen production performance in a nitrogenase-based system. Other researchers have also obtained a hypF-deficient mutant to dramatically increase the hydrogen evolution capacity of *Thiocapsa roseopersicina* BBS under nitrogen-fixing conditions because of the inactivation of hydrogen uptake activity (Fodor et al. 2001). A suicide vector containing a gentamicin cassette in the hupSL genes into *Rhodobacter sphaeroides* O.U.001 has been introduced and the results showed that the uptake hydrogenase genes were destroyed by site directed mutagenesis. The wild type and the mutant cells showed similar growth patterns but the total volume of hydrogen gas evolved by the mutant was 20% higher than that of the wild type strain (Kars et al., 2008). In addition, ammonium ion can inhibit the synthesis of nitrogenase and reduce the energy level for photosynthetic bacteria, thereby affecting the photobiohydrogen production performance. In this case, the *glnB* and *glnK* play key roles in repressing the nitrogenase expression in the presence of ammonium ion.

Table 11.3.1 Genetic modification and metabolic engineering strategies to improve photobiohydrogen production.

Problems	Strategies	Microorganisms	References
Oxygen toxicity to enzymes	Attenuating the photosynthesis/ respiration capacity ratio by DNA insertional mutagenesis	*Chlamydomonas reinhardtii*	Melis (2007)
	Reducing the oxygen sensitivity of enzymes by engineering oxygen-tolerant hydrogenase genes	*Synechococcus sp.*	Xu et al. (2005)
	Enhancing oxygen tolerance of hydrogenase by random and directed mutagenesis	*Chlamydomonas reinhardtii*	Ghirardi et al. (2000)
Lower photosynthetic efficiency	Truncating the chlorophyll antenna size of PSII to dissipate excessive photons	*Chlamydomonas reinhardtii*	Polle et al. (2003)
		Rhodobacter sphaeroides	Vasilyeva et al. (1999) Kondo et al. (2002)
Insufficient supply of electrons and protons	Reducing the proton gradient across the thylakoid membrane	*Chlamydomonas reinhardtii*	Lee and Greenbaum (2003)
	Increasing the availability of electrons and protons for hydrogen evolution using a strain Stm6 with modified respiratory metabolism	*Chlamydomonas reinhardtii*	Kruse et al. (2005)
	Increasing nitrogenase expression	*Rhodobacter capsulatus*	Ozturk et al. (2006)
Presence of uptake hydrogenase	Removing the hupL gene	*Anabaena sp.*	Masukawa et al. (2002)
	Increasing the hydrogen evolution capacity by a hypF-deficient mutant	*Thiocapsa roseopersicina*	Fodor et al. (2001)
	Destroying the uptake hydrogenase genes by site directed mutagenesis	*Rhodobacter sphaeroides*	Kars et al. (2008)
Lower ammonia tolerance	Interrupting the genes coding for two PII-like proteins, *GlnB* and *GlnK*	*Rhodobacter sphaeroides*	Kim et al. (2008)

Hence, interruption of the genes coding for two PII-like proteins from the chromosome of *Rhodobacter sphaeroides* can make the *glnB-glnK* mutant exhibit less ammonium ion-mediated repression for nitrogenase compared with its parental strain, resulting in more hydrogen accumulation by the mutant under the conditions (Kim et al., 2008).

In summary, the incorporation of the genetic and metabolic engineering into the cultivation of hydrogen-producing microorganisms has significantly improved the performance of microorganisms in different aspects (Table 11.3.1). The oxygen toxicity to enzymes for green algae and cyanobacteria can be weakened by alleviating the extreme oxygen sensitivity to enzymes. Main measures include attenuating the photosynthesis/respiration (P/R) capacity ratio, engineering oxygen-tolerant hydrogenase genes and achieving the appropriate mutants. To enhance the transport of electrons, truncating the chlorophyll antenna size is an effective approach to boost the photobiohydrogen production rate for both photoautotrophic and photosynthetic bacteria. The sufficient supply of electrons and protons can also be achieved by a genetic insertion of a polypeptide protein channel. Additionally, to reduce the consumption of hydrogen by the uptake hydrogenase, construction of a hydrogenase mutants or the removal of the hupL gene has to be conducted. Specially, for photosynthetic bacteria, the *glnB-glnK* mutant can effectively resolve the problem of ammonium ion.

11.4 HIGH-PERFORMANCE PHOTOBIOREACTOR

The performance of a photobioreactor will directly determine the feasibility of large-scale photobiohydrogen production. Therefore, how to develop a high-performance photobioreactor is critically important and has received much attention from all over the world. Various strategies for achieving a high-performance photobioreactor, including modification of the photobioreactor configurations, optimization of the operating parameters, and application of the cell immobilization, have been carried out, which are discussed below.

11.4.1 Modification of photobioreactor configurations

As mentioned earlier, the mass transport and light transmission simultaneously occur in the photobioreactor and are inherently coupled with the photobiochemical reaction. To ensure the physiological activity and metabolic stability of microorganisms and thus achieve continuous, stable and photobiohydrogen production performance, a high-performance photobioreactor that enables efficient reactants transport and products removal and light transmission with good uniformity, is required. Various enhancing strategies have been proposed towards different types of photobioreactors shown in Figure 11.4.1, including column photobioreactors, flat panel photobioreactors and tubular photobioreactors, which are summarized in Table 11.4.1.

11.4.1.1 Column photobioreactor

A column photobioreactor (Figure 11.4.1a) usually consists of a vertically-oriented column made by transparent material with an external artificial light source. An inlet and an outlet are set at the bottom and top of the column photobioreactor for supplying the fresh medium and collecting the produced hydrogen, respectively. For green algae and cyanobacteria, the bottom of the column photobioreactor also retains an inlet for the gases such as argon and carbon dioxide to maintain anaerobic conditions and provide carbon source. Such a design possesses several advantages of simple structure, excellent liquid content and heat and mass transfer characteristics, high durability of the bioactivity and packing material, low costs (Dasgupta et al., 2010; Miron et al., 2002). For these reasons, extensive researches on the column photobioreactor have been reported. However, the drawbacks of the column photobioreactor are also obvious, that is, small illumination surface area and low surface-to-volume ratio for scaling up (Ugwa et al., 2008). Various measures to improve the column photobioreactor configurations have been proposed by changing the external light sources to the internal light sources and increasing the surface-to-volume ratio. An annular triple jacketed photobioreactor consisting of three coaxial glass cylinders with two closed chambers was developed by Liu et al. (2006) for this purpose. In this design, the chamber formed between the outer and middle cylinders was used for cultivation and fluorescent lamps were inserted in the inner chamber as an internal light source, which increased the surface-to-volume ratio, leading to a light conversion efficiency of approx. 1% and a photobiohydrogen

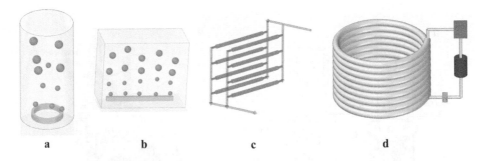

a   b   c   d

Figure 11.4.1 Different types of photobioreactors: (a) column photobioreactor; (b) flat panel photo-bioreactor; (c) horizontal tubular photobioreactor; (d) helical tubular photobioreactor (Dasgupta et al., 2010).

Table 11.4.1 Various enhancing strategies towards various photobioreactors

Types	Advantages	Disadvantages	Modification	References
Column photobioreactor	Excellent liquid content	Small illumination surface area	Annular triple jacketed photobioreactor	Liu et al., (2006)
	Good heat and mass transfer	Low surface-to-volume ratio upon scale-up	Column photobioreactor with internal light source	Chen and Chang (2006)
	High durability of the biocatalyst and packing material		Torus shaped photobioreactor	Fouchard et al., (2008)
	Low material and operating costs			
Flat panel photobioreactor	Large illumination surface area	Many compartments and support materials required upon scale-up	Flat plate photobioreactor with some baffles	Zhang et al., (2001)
	Suitability for outdoor cultures		Alveolar panel photobioreactor	Tredici et al., (1991)
	Good light path		V-shaped flat panel photobioreactor	Iqbal et al., (1993)
	Ease to clean up	Wall growth High power consumption for mixing	Flat panel photobioreactor with rocking motion	Gilbert et al., (2011)
Tubular photobioreactor	Large illumination surface area	Poor mass transfer	Near horizontal tubular photobioreactor	Tredici et al., (1998)
	Suitability for outdoor cultures	Accumulation of gas	Conical helical photobioreactor	Gebicki et al., (2009) Morita et al., (2000)
		Wall growth	An α-shape tubular photobioreactor	Lee et al., (1995)

production rate more than 20 mL/L/h. Another column photobioreactor with internal illumination (Figure 11.4.2a) was developed to enhance the illuminating condition, in which a glass-made vessel was closed and several optical fibers protected in glass tubes were then inserted into the liquid medium to provide illumination (Chen et al., 2010). In addition, Fouchard et al. (2008) designed an innovative torus-shaped photobioreactor (Figure 11.4.2b) using a marine impeller to circulate the culture. The results showed that the combination of the loop configuration caused by the torus geometry with the

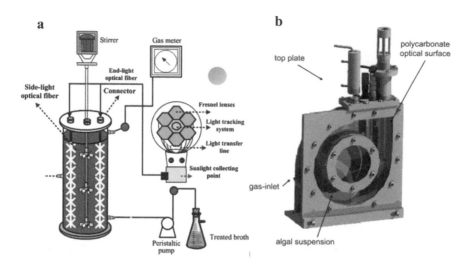

Figure 11.4.2 Column photobioreactor configurations: (a) column photobioreactor with the internal illumination (Chen et al., 2010), and; (b) torus-shaped photobioreactor (Fouchard et al., 2008).

impeller ensured a good mixing without dead zone, improving the photobiohydrogen production performance.

11.4.1.2 Flat panel photobioreactor

A flat panel photobioreactor is an enclosed rectangular transparent chamber with a shallow depth and relatively large length and width as depicted in Figure 11.4.1b, which can be placed vertically, leanly, or horizontally and illuminated from one side by the incident light. In large-scale photobiohydrogen production applications, several photobioreactors are usually arranged in-parallel to increase the system volume. This design offers many merits such as large illumination surface area, good light path and ease to clean up, all of which make it suitable for outdoor cultures (Ugwa et al., 2008). A good example of this type of photobioreactor is the one used for photobiohydrogen production with *Rhodopseudomonas sp*. HCC 2037 by Hoekema et al. (2002). In this design as shown in Figure 11.4.3, the flat panel photobioreactor consists of a stainless-steel frame and three polycarbonate panels, forming two compartments. One compartment contains the bacterial culture with the desirable temperature maintained by the circulated water in another compartment via a water bath. The results showed that light energy can be efficiently utilized using this flat-panel photobioreactor. Although promising, the flat panel photobioreactor still have some limitations of complex structure, support materials required upon scale-up, wall growth, high power consumption of mixing, etc. (Akkerman et al., 2002; Ugwa et al., 2008). Therefore, it is necessary to modify the configuration of the flat panel photobioreactor for improving the photobiohydrogen production performance. Zhang et al. (2001) designed a flat plate photobioreactor with the addition of baffles to improve the agitation for better mixing. Tredici et al. (1991) devised an alveolar panel photobioreactor which

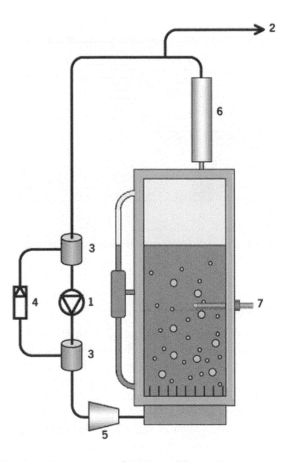

Figure 11.4.3 Flat panel photobioreactor configurations: (1) membrane gas pump; (2) gas bag for collection of produced gas; (3) pressure vessel; (4) pressure valve; (5) mass flow controller; (6) condenser; and; (7) pH/redox electrode (Hoekema et al., 2002).

was well suited to the outdoor mass cultivation. The advantages of the high surface-to-volume ratio, flexible orientation with respect to the sun's rays, effective mixing, oxygen removal and good control of environmental and nutritional conditions were provided by this design, allowing the operation at high cell concentrations and achieving high biomass productivity. Besides, a V-shaped flat panel photobioreactor that offered the above-mentioned advantages was also developed to improve the performance of the photobioreactor (Iqbal et al., 1993). Gilbert et al. (2011) developed a flat panel photobioreactor with rocking motion (Figure 11.4.4) to overcome the limitation caused by the poor level of agitation for improving light penetration. The flat panel photobioreactor with rocking motion consisted of a teflon frame sandwiched between two acrylic sheets with neoprene gaskets. The rocking motion was then achieved with the help of a motor and an eccentric motion arrangement. The results proved that the flat panel photobioreactor with rocking motion had a high potential for the large-scale photobiohydrogen production.

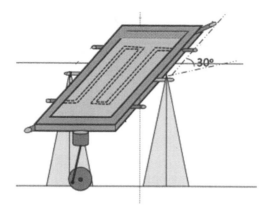

Figure 11.4.4 Flat panel photobioreactor with rocking motion (Gilbert et al., 2011).

11.4.1.3 Tubular photobioreactor

A tubular photobioreactor is usually constructed with either a glass or plastic tube, in which the culture liquid is pumped by mechanical or airlift pumps. In terms of the configuration, tubular photobioreactors can be divided into two major categories: horizontal and helical tubular photobioreactor, as shown in Figures 11.4.1c and 11.4.1d. The most significant feature of tubular photobioreactors is large illumination surface area so that it is particularly suitable for outdoor mass cultures of photosynthetic microorganisms (Posten, 2009; Ugwa et al., 2008). However, tubular photobioreactors also face some problems of poor mass transfer, gas accumulation and wall growth (Dutta et al., 2008). Hence a lot of modifications to the tubular photobioreactor configuration have been proposed to improve the performance of photobioreactors. Tredici and Zittelli (1998) designed a near horizontal tubular photobioreactor which was laid on a wooden framework with an angle of 5 degrees horizontally. The elevation reduced the gas holdup and improved the oxygen removal, resulting in a higher volumetric productivity and photosynthetic efficiency with *Arthrospira platensis*. Gebicki et al. (2009) also devised a near horizontal tubular photobioreactor with an inclination of below 10 degrees (Figure 11.4.5a) and found that a photobiohydrogen production rate of 3.3 mL/L/h with *R. capsulatus* was achieved. In addition to horizontal photobioreactors, investigations on the helical photobioreactor have also been widely performed. Morita et al. (2000) used PVC tube to be coiled in a conical framework, forming a conical helical photobioreactor with a cone angle of 60 degrees (Figure 11.4.5b) and examined the effect of the cone angle. It was shown that this photobioreactor had the maximal photosynthetic efficiency of 6.84% among all cone angle tested because 60 degrees gave the highest light harvesting efficiency with the same basal area. Lee et al. (1995) designed an α-shape tubular photobioreactor (Figure 11.4.5c), in which the algal culture was lifted by air to a receiver tank, and then flowed down through paralleled tubes with a horizontal angle of 25 degrees to reach another set of air riser tubes. Again the culture was lifted to another receiver tank and then flowed down through

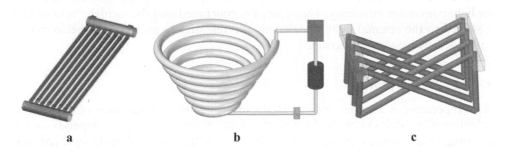

Figure 11.4.5 Tubular photobioreactor configurations: (a) near horizontal tubular photobioreactor (Gebicki et al., 2009); (b) conical helical photobioreactor (Morita et al., 2000), and; (c) α-shape tubular photobioreactor (Lee et al., 1995).

paralleled tubes connected to the base of the first set of riser tubes. Such design enables an increase in the surface-to-volume ratio and equable photosynthetically-available radiance, leading to a high biomass density.

11.4.2 Optimization of the operating parameters

To achieve a high performance photobioreactor, not only the improvement in photo-bioreactor design is required but also the operation parameters need to be optimized. Typically, the factors affecting the photobiohydrogen production performance of the photobioreactor include the illumination conditions, temperature, pH value and so on (Show et al., 2011).

11.4.2.1 Illumination conditions

The photobiohydrogen production by microorganisms is driven by light that functions as an energy source so that the illumination conditions, both light wavelength and light intensity, can affect the photobiohydrogen production performance (Carlozzi et al., 2010; Liu et al., 2010; Uyar et al., 2007). As the stimulation of the photobiohydrogen production reaction occurs at different light wavelengths, depending on the used microorganisms and operating conditions, the appropriate wavelength will be distinguished. Chen et al. (2006b) reported that the indigenous purple non-sulfur bacteria *Rhodopseudomonas palustris* WP3-5 yielded excellent performance by absorbing lights at the wavelengths of 522 and 860 nm. Tian et al. (2010) studied the effect of the light wavelength on the performance of photobiohydrogen production by *Rhodopseudomonas palustris* CQK 01 using monochromatic LED lamps. The results showed that the best photobiohydrogen production performance was obtained at 590 nm due to the existence of absorption maxima of bacteriochlorophyll α. Furthermore, Tian et al. (2010) found that photobiohydrogen production was saturated at around 5000 lx and high light intensity resulted in a striking decrease of the conversion efficiency from light energy to hydrogen. This team also demonstrated that the light conversion efficiency decreased monotonically with increasing the illumination intensity because not all light can be absorbed at high illumination intensity (Liao et al.,

2010). The energy supply exceeded the metabolic capability of nitrogenase, leading to low light conversion efficiency. Therefore, it is concluded that the incident light should be shifted to the optimal light wavelength and intensity to enhance the performance of photobiohydrogen production.

11.4.2.2 Temperature

The temperature is another critical factor affecting the performance of the photobiohydrogen production as the bioactivity of microorganisms highly depends on the operating temperature. Carlozzi and Lambardi (2009) reported that the optimal temperature to culture *R. palustris* was between 30°C and 35°C. Wang et al. (2010) studied the photobiohydrogen production performance of *Rhodopseudomonas palustris* CQK 01 by varying temperatures from 15 to 40°C and found that the photobiohydrogen production rate, substrate utilization efficiency and hydrogen bioconversion yield increased as the temperature increased from 15 to 30°C. With the temperature further increasing to 40°C, the trends reversed, indicating the optimal temperature for the biohydrogne production by *Rhodopseudomonas palustris* CQK 01 is 30°C. To attain an excellent photobiohydrogen production rate, therefore, the optimal temperature control by using the circulated water via temperature controlled water in existing studies was proposed (Hoekema et al., 2002). However, the resulting high energy consumption limits the application of this technique for industrialization. Wang et al. (2010) then utilized heaters arranged at the bottom of the photobioreactor and T-type thermocouples distributed at the photobioreactor to control the temperature of the influent solution. This method can effectively avoid heat dissipation and achieve energy-savings. In summary, for enhancing the performance of photobioreactors, the temperature should be controlled to maximize the bioactivity in terms of used microorganisms and but with low energy consumption.

11.4.2.3 pH value

It is well known that the pH value of the influent medium can affect the metabolic pathways and thus the performance of the photobiohydrogen production. The study with respect to the effect of pH value on the photobiohydrogen production by *Rhodobacter sphaeroides* O.U. 001 revealed that the optimum pH value leading to the maximum photobiohydrogen production rate was found to be around 7 (Sasikala et al., 1995). Studies by Tian et al. also (2010) showed that the optimum photobiohydrogen production rate by *Rhodopseudomonas palustris* CQK-01 occurred at a pH of 7.0. Hence, it can be found that the optimal pH value for photosynthetic bacteria is usually about 7. For phototrophic bacteria, the optimal pH value is slightly changed with respect to microorganisms. Works by Li et al. (2008) concluded the optimal photobiohydrogen production rate was achieved at the pH of 7.0–7.5 with the acetate and butyrate as the substrate. However, Fang et al. (2005) reported the optimal pH values were 8.0 and 9.0 with respect to acetate and butyrate in batch experiments, respectively. The performance of photobiohydrogen production significantly dropped when the pH was not in the optimum range because the activity of enzyme was inhibited. In some researches, to obtain the optimal pH, HCl/NaOH was added into the solution to stimulate the microorganisms for achieving high activity of enzyme (Skjanes et al., 2008). Therefore, a high performance photobioreactor requires to be operated at the optimal pH value that changes with the microorganism and substrate.

11.4.2.4 Substrate

Previous studies show that the performance of the photobiohydrogen production also depends on the type of substrate as the composition of the organic compounds is different, affecting the biodegradation reaction (Chen et al., 2008; Kapdan and Kargi, 2006). Barbosa et al. (2001) chose four substrates, i.e. lactate, malate, acetate and butyrate, for the photobiohydrogen production using the photosynthetic bacteria as the producers and the results showed that the highest photobiohydrogen production rate and light efficiency were achieved with acetate as the substrate. In addition to the substrate type, the concentration of the substrate also affects the photobiohydrogen production performance, whose optimal value depends on the photobioreactor design. Zhu et al. (2012) found that the photobiohydrogen production performance of a column photobioreactor with transparent packed materials increased with increasing glucose concentration from 0.02 to 0.1 M at a flow rate of 500 ml/h. For a flat panel photobioreactor, Liao et al. (2010) studied the photobiohydrogen production performance at a flow rate of 70 ml/h and found that the photobiohydrogen production rate increased with an increase in the glucose concentration from 0.02 to 0.06 mM as a result of enhanced mass transport, and then decreased gradually as the glucose concentration was further increased. The reason is that too high glucose concentration results in the substrate inhibition, lowering the photobiohydrogen production rate. However, Zhang et al. (2010) reported that the optimal substrate concentration of a groove-type flat panel photobioreactor was obtained at glucose concentration of 10 g/l due to the discrepancy of organic load carrying capacity for different photobioreactors. From these studies, it is revealed that the photobiohydrogen production performance is affected not only by the substrate type but also by the substrate concentration. Too low or too high concentration can cause the mass transfer limitation or the substrate inhibition, thereby lowering the performance of photobioreactor. It is essential to control the substrate conditions according to the photobioreactor design in real applications.

11.4.2.5 Nutrients

In addition to the organic compounds, nutrients, such as nitrogen, phosphate and other inorganic trace minerals, are also necessary for maintaining the cell cultivation and hydrogen production in the photobiohydrogen production process (Chen et al., 2010; Eroglu and Melis, 2011; Hakobyan, 2012). The appropriate nutrients can enhance the performance of the photobiohydrogen production through changing the metabolism of microorganisms. Tao et al. (2008) used L-glutamate, $(NH_4)_2SO_4$ or ethanolamine as nitrogen source for the photobiohydrogen production by a photosynthetic non-sulfur strain named ZX-5 and found that 7-mM L-glutamate was the best nitrogen source for photobiohydrogen production. Besides the selection of nitrogen source, the carbon to nitrogen ratio in the medium is also of importance in the photobiohydrogen production. Eroglu et al. (1999) used malic acid and glutamic acid as the carbon and nitrogen sources, respectively, and the results showed that a carbon to nitrogen ratio of 15:2 exhibited the highest photobiohydrogen production rate. Burrows et al. (2008) also optimized the concentrations of nutrients including nitrogen, phosphate and so on, for the photobiohydrogen production by *Synechocystis sp*. PCC6803 and obtained a 150-fold increase in the photobiohydrogen production rate. The effect of the concentrations of molybdenum and iron on the photobiohydrogen production by *Rhodobacter sphaeroides* O.U.001 was also investigated (Kars et al., 2006). The results showed that 16.5 μM sodium molybdate and 0.1 mM ferric citrate in nutrient solution yielded the highest total hydrogen accumulation. In summary, an appropriate nutrient composition is one of the keys to obtain a high photobiohydrogen production performance and the respective optimal nutrients concentrations depends on the used microorganism and other operating conditions.

11.4.2.6 Operational modes

Existing studies indicate that the distributions of biomass, substrate and product concentrations in photobioreactors are also affected by the operational modes. Hence, the effects of the operational modes, including batch, continuous and fed-batch modes, on the mass transfer and biomass distribution have been investigated and enhancement strategies have also been proposed (Argun and Kargi, 2011). Shi and Yu (2006) utilized continuous operational mode for the photobiohydrogen production with *Rhodopseudomonas capsulate*, the maximum photobiohydrogen production rate of 37.8 ml/g dry weight (dwt)/h, light conversion efficiency of 3.69% and substrate conversion efficiency of 45% were achieved at a mixture of acetate 1.8 g/l, propionate of 0.2 g/l and butyrate of 1.0 g/l. Ren et al. (2009) studied the photobiohydrogen production by *Rhodopseudomonas faecalis* strain RLD-53 in fed-batch operational mode using acetate as the sole carbon and the results demonstrated the fed-batch operational mode obtained a high efficiency. Chen et al. (2006a) compared three different operational modes with the same initial substrate concentration and culture volume to determine which operational mode was preferable in the photobiohydrogen production by indigenous purple nonsulfur bacterium *Rhodopseudomonas palustris* WP3-5. It was suggested that the fed-batch operational mode was a good mode due to the high cell concentration and adjustable substrate loading under similar culture conditions. Conclusions drew from previous studies are different may be due to the design of the photobioreactor. Recently, an enhancement method with the ultra-sonication pretreatment was proposed to improve the batch operational mode by increasing the mass transport and permeability of the cell membrane so that the performance of the photobiohydrogen production was greatly improved (Zhu et al., 2011). In summary, to improve the photobiohydrogen production rate, the enhancement strategies need to be incorporated into these operational modes to increase the mass transfer rate and biomass as well as their uniform distributions.

11.4.2.7 Anaerobic condition

In a photobiohydrogen production system, maintaining good anaerobic condition is critically important to prevent the toxic effect from the oxygen evolution, allowing for the activation of the enzymes involved in hydrogen metabolism (Melis, 2002). For instance, continuous sparging of inert gas into the photobiohydrogen production system can maintain a low partial pressure of oxygen (Hallenbeck and Benemann, 2002). However, the hydrogen content obtained is very low in such a system, which increases the difficulties in the separation and purification of hydrogen. The addition of 3-(3, 4-dichlorophenyl)-1, 1-dimethylurea (DCMU) that is a potent inhibitor of the direct PSII-dependent photobiohydrogen production pathway can lead to a complete anaerobiosis (Fouchard et al., 2005). Besides, the adsorbent and reductant of oxygen can also make the photobiohydrogen production system to be operated under anaerobic conditions by consuming the oxygen in the solution (Hansel and Lindblad, 1998). Unfortunately, these techniques do not possess practical value because of the high processing cost. In 2000, Melis et al. (2000) and his colleagues developed a novel two-stage hydrogen production method for sustaining photobiohydrogen production using the green alga of *Chlamydomonas reinhardtii* as the photobiohydrogen production microorganism. In this method, photosynthetic oxygen evolution and carbon accumulation (stage 1) and concomitant hydrogen production (stage 2) were temporally separated to circumvent the severe oxygen sensitivity of the reversible hydrogenase by the sulfur deprivation in culture, which can reversibly inactivate PSII and oxygen evolution to ensure a transition from stage 1 to stage 2 (Melis et al. 2000). Although this technique is effective, the system becomes complex. As a result, more cost-effective and simple methods are needed for industrialization.

11.4.2.8 Hydrogen partial pressure

As the photobiohydrogen production proceeds, hydrogen produced in the photobioreactor is accumulated, causing an increase of the hydrogen partial pressure and thereby lowering the photobiohydrogen production performance. Thus, reducing the hydrogen partial pressure by efficiently removing hydrogen produced from the system to facilitate the photobiochemical reaction is of importance to improve the performance of the photobioreactor. Currently, many strategies of removing or separating the produced hydrogen have been developed to weaken the negative effect of the hydrogen accumulation. Liao et al. (2012) found that feeding the sparging gas at 10 ml/min led to the increase of the photobiohydrogen production rate, hydrogen yield and light conversion efficiency by 2.2, 3.1 and 2.2 times than the case without gas sparging, respectively. But this method increases the cost of the hydrogen separation. Furthermore, they reported an ultrasonic treatment method that enabled the hydrogen concentration in the solution to be decreased from $300\,\mu mol/L$ to $50\,\mu mol/L$ due to the disturbance from the acoustic streaming (Wang et al., 2012). However, this technique can result in an increase in the operation cost. As a consequence, it is essential to develop a simple technique that can lower the hydrogen partial pressure with low operation cost.

In summary, to obtain a high photobiohydrogen production performance, the operating conditions need to be optimized. The optimal combination of the illumination conditions, temperature, pH value, nutrients, substrate concentration as well as operational modes in terms of both the photobioreactor design and the used microorganism can maximize the photobiohydrogen production rate. Besides, careful selection of substrate type along with the maintenance of good anaerobic condition and low hydrogen partial pressure also benefits for the photobiohydrogen production.

11.4.3 Application of cell immobilization

In an existing photobioreactor, microorganisms are usually suspended in culture solution. Although such operation can provide the advantage of good mass transfer between microorganisms and substrates, it is difficult to ensure long-term operation and high photobiohydrogen production performance due to low biomass caused by wash-out. Therefore, ever-increasing attention has recently been turned to the photobiohydrogen production by immobilized microbial cells for increasing the biomass concentration. As such, not only the volumetric productivity and stability and light utilization efficiency are increased, but also the ability to recover and reuse the cell mass is also enhanced (Das and Veziroglu, 2001). Typically, cells can be immobilized by cell entrapment and biofilm to dramatically increase the biomass concentration in the photobioreactor.

11.4.3.1 Cell entrapment

The features provided by cell entrapment include the lower cost and easy operation. As the cells are entrapped, an anaerobic environment is automatically created. Moreover, the entrapment materials allow the system to be stably operated at relatively high flow rate without suffering from cell wash-out. However, it should be pointed out that the cell entrapment requires strict entrapment materials, high mechanical strength, sufficient light supply and low mass transfer resistance, which are currently unsatisfied with the state-of-the-art technique. Therefore, many researchers have attempted to solve these problems for improving the photobiohydrogen production by the cell entrapment technique.

A variety of support materials such as agar, agarose, alginates, pectin and carrageenan have been tested (von Felten et al., 1985). The results showed that agar was the best immobilizing agent according to the photobiohydrogen production rate and stability, which were also superior to the suspension culture. The photobiohydrogen production performance with the purple nonsulphur bacterium *Rhodopseudomonas palustris* DSM 131 immobilized by agar, agarose, carageenan and sodium alginate were also explored (Fissler et al., 1995). It was shown that the cells immobilized in agar, agarose and carageenan yielded low photobiohydrogen production performance as compared to suspended cells due partly to cell damage at the temperature of 45–50°C during the immobilization process in these matrices. However, with sodium alginate beads produced at room temperature, a higher photobiohydrogen production performance than suspended cell was achieved. In addition, this work also indicated that the reduction of gel bead diameter can increase the surface-to-volume ratio, enable the easy access of the substrate to the cells and an increase in the light supply per immobilized cell, both of which improved the hydrogen yield.

As mentioned earlier, the mechanical strength and light supply of the entrapment technique are poor. Planchard et al. (1989) entrapped cells in a planar agar matrix bounded by a microporous membrane filter. The addition of a microporous membrane significantly enhanced the mechanical strength. However, although the mechanical strength is enhanced by this microporous membrane, the light supply is still inefficient. In order to improve the light supply, therefore, a gel layer/microporous membrane structure with an inner optical fiber illumination device was developed (Mignot et al., 1989), which can guide light from an external source to the entrapped cells, suppressing thermic effect and ensuring a homogeneous illumination and a large active surface. Another method to utilize optical fibers was to directly immobilize cells onto the surface of optical fibers using alginate gel (Yamada et al., 1996). In this design, the optical fibers

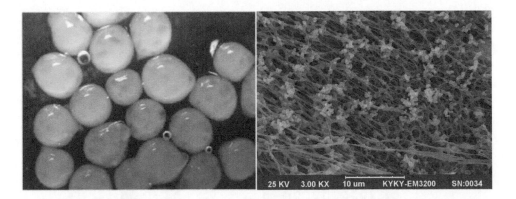

Figure 11.4.6 Immobilized solid support matrices and their scanning electron microscope images (Tian et al., 2009).

can emit light laterally from their surface, enhancing the efficient distribution of light to the cells and thereby improving the photobiohydrogen production performance. In addition, the mechanical stability and resistance to oxygen were also obtained by coating the immobilized cells with a transparent layer of gellan gum.

To reduce the mass transfer resistance by the cell entrapment, polyvinyl alcohol-boric acid gel granule was used to immobilize the indigenous photosynthetic bacteria *Rhodopseudomonas palustris* CQK 01 (Tian et al., 2009). As shown in Figure 11.4.6, this type of gel granule has numerous pores which are mutually connected and cells are then firmly attached to the surface of the solid support matrix. These stable net-work matrices created a biocompatible environment and in the meantime high porosity provided the sufficient space for the bacterial growth, facilitating the transfer of substrate, water, and products. This team also adopted a photobioreactor packed with sodium alginate/polyvinyl alcohol-124/carrageenan granules containing *Rhodopseudomonas palustris* CQK 01 for photobiohydrogen production in a continuous operational mode (Wang et al., 2010). With this design, the mass transfer resistance was reduced and the maximal photobiohydrogen production rate of 2.61 mmol/L/h was achieved. As mentioned in the above section, the photobiohydrogen production by photosynthetic bacteria can simultaneously realize the energy production and wastewater treatment. Zhu et al. (1999) used the wastewater of tofu factory as carbon source and anoxygenic phototrophic bacterium *Rhodobacter sphaeroides* immobilized in agar gels. The bacterium could be protected by the immobilization from the inhibitory effect of ammonium ion. The results showed that the maximum photobiohydrogen production rate obtained was $2100\,ml/h/m^2$ along with a yield of hydrogen of $0.24\,ml/mg$ the total organic carbon removal ratio reached 41% in 85 h.

11.4.3.2 Biofilm

Another immobilization technique is the biofilm technique. Compared with cell entrapment, the biofilm technique can not only increase the biomass concentration but also lower the mass transfer resistance as a result of thin biofilm thickness, causing high photobiohydrogen production performance, sufficient light supply and weakened product

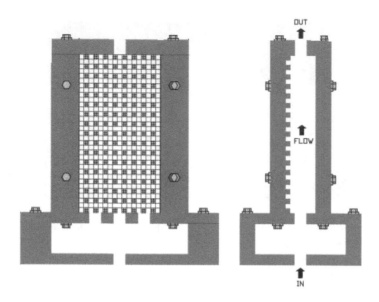

Figure 11.4.7 The groove-type photobioreactor (Zhang et al., 2010).

inhibition. In addition, long term retention of biomass promotes the application of the biofilm technique in industrialization. Therefore, the immobilized-cell system with the biofilm formation is regarded as a more promising technique for continuous photo-biohydrogen production (Hallenbeck and Benemann, 2002). Typically, the biofilm can be formed onto the inner surface of the photobioreactor surface or packed materials surface, which have been widely explored.

Tsygankov et al. (1993) found that the planar glass surface modified by silane cou-pling reagents could improve the hydrophobicity of the surface, which greatly increased the bacterial attachment. In addition, Zhang et al. (2010) developed a groove-type photobioreactor (Figure 11.4.7), in which the grooves were successfully incised on a transparent panel for the cell immobilization. Their results showed that the perfor-mance of the photobiohydrogen production of the groove-type photobioreactor was about 75% higher than that of the flat panel photobioreactor due to the enriched immobilized biomass, the higher specific surface area of the photobioreactor and enhanced convective mass transfer for both substrate and metabolic products of the photo-fermentation. Liao's team in Chongqing University has also carried out extensive works on the biofilm technique for the photobiohydrogen production. They used a thin glass slide as carrier to immobilize photosynthetic bacteria in a flat panel photobioreac-tor and manufactured a transparent microchannel photobioreactor (Figure 11.4.8) to visualize the colony formation of photosynthetic bacteria, *Rhodopseudomonas palus-tris* CQK 01, as well as the biogas bubble behavior within the microstructure (Liao et al., 2010; Qu et al., 2011; Wang et al., 2011). The results showed that the biomass concentration, light penetration and mass transport were significantly affected by the biofilm microstructure.

It should be mentioned that the cell immobilization on the planar surface has a problem of low specific surface area. Immobilizing cells on the solid particles or fibers

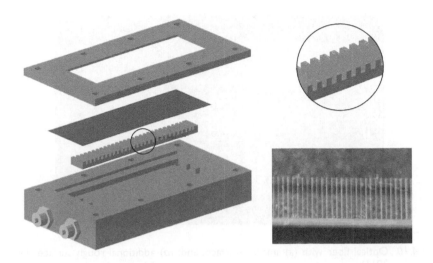

Figure 11.4.8 The transparent microchannel photobioreactor (Qu et al., 2011).

Figure 11.4.9 Scanning electron microscope images of the surface of activated carbon fiber: (a) before immobilization, and; (b) after immobilization (Ren et al., 2012).

can resolve this problem to some extent. Ren et al. (2012) developed an anaerobic fluidized-bed type photobioreactor with activated carbon fiber as support material. The activated carbon fibers have a high specific surface area, excellent biocompatibility and good adsorption capacity. When they fluidize in the reactor, cells can adhere onto the surface to form a biofilm (Figure 11.4.9). Such design could dramatically improve the light energy utilization and photobiohydrogen production performance. Xie et al. (2012) also developed surface modified activated carbon fibers as a solid carrier to improve the photobiohydrogen production. The oxidation of activated carbon fibers by HNO$_3$ increased the concentration of oxygen functional groups on the external surfaces and surface roughness. Consequently, the amount of immobilized cells was improved due to the increased surface roughness and abundant surface functional groups. Researchers in this group also used a novel bio-carrier to immobilize photo-fermentative bacteria for photobiohydrogen production (Xie et al., 2011). The light shading effect could be prevented and better fluidization of solid carriers could

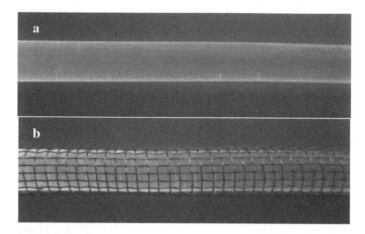

Figure 11.4.10 Optical fiber with (a) smooth-surface, and; (b) additional rough surface (Guo et al., 2011).

be obtained by controlling the proper size of bio-carrier. Moreover, three types of carriers, including activated carbon, silica gel, and expanded clay were also used to be additives to the medium (Chen and Chang, 2006). It was indicated that the addition of clay and silica gel provided extra surface area for attached cell growth to form biofilm, resulting in better hydrogen yield, and light conversion efficiency than those obtained from the carrier-free culture. However, the activated carbon-supplemented culture did not show a significant improvement in the performance of the photobiohydrogen production due to high cell density in the activated carbon-supplemented culture and the tiny fragments flaking off the activated carbon carriers, which both resulted in the blocking of light penetration from the external light sources and poor light conversion efficiency in the activated carbon-supplemented culture. In order to enhance the light conversion efficiency, therefore, some researchers utilized the materials with high transparence as a carrier for the photobiohydrogen production. Tekucheva et al. (2011) developed a novel photobioreactor using an inexpensive glass-fiber matrix to accelerate the immobilization process and to enhance light penetration. Furthermore, in order to enhance both the light penetration and the surface-to-volume ratio of the carrier for the high biomass concentration in the photobioreactor, Tian et al. (2010) designed a biofilm photobioreactor with packed glass beads on which *Rhodopseudomonas palustris* CQK 01 was immobilized on the surface of the transparent packed materials to form a biofilm. In addition, they also developed a biofilm photobioreactor with the external light source replaced by an internal light source of optical fibers (Figure 11.4.10a). Particularly, an additional rough surface (Figure 11.4.10b) was created to enhance the biofilm formation (Guo et al., 2011). The results showed that the biofilm photobioreactor with the added rough-surface optical fibers exhibited higher performance of photobiohydrogen production than did the biofilm photobioreactor with the smooth-surface optical fiber and the optical fiber suspension photobioreactor because it could furnish extra surface attachment sites, enhancing the biofilm formation and enriching biomass in photobioreactor. In the wake of this photobioreactor design,

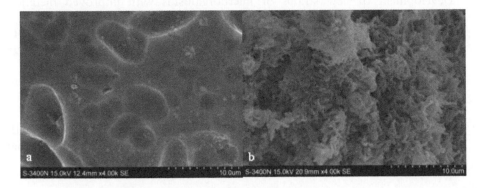

Figure 11.4.11   Scanning electron microscopy images of the surface of spherical glass bead: (a) before biofilm, and; (b) after biofilm (Zhu et al., 2012).

an unsaturated flow bioreactor consisted of a reaction bed packed with spherical glass beads and an internal optical fiber was also developed by this team to promote the rates of substrate and products transfer, to achieve efficient gas-liquid separation and to enhance the biofilm formation (Figure 11.4.11) through utilizing the three-phase interfaces (Zhu et al., 2012). Unfortunately, the surface of glass bead carriers in these studies is smooth, leading to difficulty in the bacterial adhesion and biofilm formation. Therefore, surface treatment is needed to improve this design of the photobioreactor.

In summary, the cell immobilization technique can greatly increase the biomass concentration in the photobioreactor, which is beneficial for the improvement in the photobiohydrogen production performance. However, the problems of low light penetration and mass transfer inside of the entrapment and biofilm still exist. To industrialize the cell immobilization technique for the large-scale photobiohydrogen production, the cell immobilization with high surface-to-volume ratio, mass transfer rate, light penetration and good mechanical strength and biocompatibility is needed.

11.5   CHALLENGES AND FUTURE DIRECTIONS

In the process of photobiohydrogen production, the coupling of the photosynthesis and mass transport and light transmission occurs. The photobiohydrogen production performance highly depends on the microorganisms used and their bioactivity as well as the optimizations of the photobioreactor design and operational conditions. There has been a substantial improvement in the hydrogen yield, photobiohydrogen production rate and light efficiency. This chapter gives an overview of past studies on the high-performance microorganisms, photobioreactor design and effect of the operational conditions. The past experimental data show that although the improvement has been achieved, the photobiohydrogen production performance still has to considerably surpass the present achievement. To attain a high and stable performance, therefore, future research into photobiohydrogen production should be directed to addressing the following critical issues for the widespread commercialization.

(1) Development of new strains of microorganisms with improved metabolic capability and high tolerance to oxygen and ammonium ion.
(2) Development of a high-performance photobioreactor with high surface-to-volume ratio, efficient mass transfer and products removal as well as high light transmission and uniform distribution.
(3) Optimization of the operating conditions to ensure a high photobiohydrogen production performance in terms of the employed microorganisms and photobioreactor design.
(4) Improvement in the biomass concentration in photobioreactor to maximize the hydrogen yields and production rates by effective cell immobilization technology.

REFERENCES

Akkerman, I., Janssen, M., Rocha, J. and Wijffels, R.H. (2002) Photobiological hydrogen production: photochemical efficiency and bioreactor design. *International Journal of Hydrogen Energy*, 27, 1195–1208.

Allakhverdiev, S.I., Thavasi, V. Kreslavski, V.D., Zharmukhamedov, S.K., Klimov, V.V., Ramakrishna, S., Los, D.A., Mimuro, M., Nishihara, H. and Carpentier, R. (2010) Photosynthetic hydrogen production. *Journal of Photochemistry and Photobiology C: Photochemistry Reviews*, 11, 101–113.

Appel, J. and Schulz, R. (1998) Hydrogen metabolism in organisms with oxygenic photosynthesis: hydrogenases as important regulatory devices for a proper redox poising? *Journal of Photochemistry and Photobiology B: Biology*, 47, 1–11.

Argun, H. and Kargi, F. (2011) Bio-hydrogen production by different operational modes of dark and photo-fermentation: An overview. *International Journal of Hydrogen Energy*, 36, 7443–7459.

Asada, Y. and Miyake, J. (1999) Photobiological hydrogen production. *Journal of Bioscience and Bioengineering*, 88, 1–6.

Barbosa, M.J., Rocha, J.M.S., Tramper, J. and Wijffels, R.H. (2001) Acetate as a carbon source for hydrogen production by photosynthetic bacteria. *Journal of Biotechnology*, 85, 25–33.

Basak, N. and Das, D. (2007) The prospect of purple non-sulfur (PNS) photosynthetic bacteria for hydrogen production: The present state of the art. *World Journal of Microbiology & Biotechnology*, 23, 31–42.

Belkin, S. and Padan, E. (1978) Hydrogen metabolism in the facultative anoxygenic cyanbacterial blue-green algae, *Oscillatoria limnetica* and *Aphanothece halophytica*. *Archives of Microbiology*, 116, 109–111.

Benemann, J.R. (1997) Feasibility analysis of photobiological hydrogen production. *International Journal of Hydrogen Energy*, 22, 979–987.

Burrows, E.H., Chaplen, F.W.R. and Ely, R.L. (2008) Optimization of media nutrient composition for increased photofermentative hydrogen production by *Synechocystis sp.* PCC 6803. *International Journal of Hydrogen Energy*, 33, 6092–6099.

Carlozzi, P., Buccioni, A., Minieri, S., Pushparaj, B., Piccardi, R., Ena, A. and Pintucci, C. (2010) Production of bio-fuels (hydrogen and lipids) through a photofermentation process. *Bioresource Technology*, 101, 3115–3120.

Carlozzi, P. and Lambardi, M. (2009) Fed-batch operation for bio-H_2 production by *Rhodopseudomonas palustris* (strain 42OL). *Renewable Energy*, 34, 2577–2584.

Chen, C.Y. and Chang, J.S. (2006) Enhancing phototropic hydrogen production by solid-carrier assisted fermentation and internal optical-fiber illumination. *Process Biochemistry*, 41, 2041–2049.

Chen, C.Y., Lee, C.M. and Chang, J.S. (2006a) Feasibility study on bioreactor strategies for enhanced photohydrogen production from *Rhodopseudomonas palustris* WP3-5 using optical-fiber-assisted illumination systems. *International Journal of Hydrogen Energy*, 31, 2345–2355.

Chen, C.Y., Lee, C.M. and Chang, J.S. (2006b) Hydrogen production by indigenous photosynthetic bacterium *Rhodopseudomonas palustris* WP3-5 using optical fiber-illuminating photobioreactors. *Biochemical Engineering Journal*, 32, 33–42.

Chen, C.Y., Lu, W.B., Wu, J.F. and Chang, J.S. (2007) Enhancing phototrophic hydrogenproduction of *Rhodopseudomonas palustris* via statistical experimental design. *International Journal of Hydrogen Energy*, 32, 940–949.

Chen, C.Y., Lu, W.B., Liu, C.H. and Chang, J.S. (2008) Improved phototrophic H_2 production with *Rhodopseudomonas palustris* WP3-5 using acetate and butyrate as dual carbon substrates. *Bioresource Technology*, 99, 3609–3616.

Chen, C.Y., Yeh, K.L., Lo, Y.C. Wang, H.M. and Chang, J.S. (2010) Engineering strategies for the enhanced photo-H_2 production using effluents of dark fermentation processes as substrate. *International Journal of Hydrogen Energy*, 35, 13356–13364.

Chen, C.Y., Liu, C.H., Lo, Y.C. and Chang J.S. (2011) Perspectives on cultivation strategies and photobioreactor designs for photo-fermentative hydrogen production. *Bioresource Technology*, 102, 8484–8492.

Das, D. and Veziroglu, T.N. (2001) Hydrogen production by biological processes: a survey of literature. *International Journal of Hydrogen Energy*, 26, 13–28.

Das, D. and Veziroglu, T.N. (2008) Advances in biological hydrogen production processes. *International Journal of Hydrogen Energy*, 33, 6046–6057.

Das, D., Khanna, N. and Veziroglu, T.N. (2008) Recent developments in biological hydrogen production processes. *Chemical Industry & Chemical Engineering Quarterly*, 14, 57–67.

Dasgupta, C.N., Gilbert, J.J., Lindblad, P., Heidorn, T., Borgvang, S.A., Skjanes, K. and Das, D. (2010) Recent trends on the development of photobiological processes and photobioreactors for the improvement of hydrogen production. *International Journal of Hydrogen Energy*, 35, 10218–10238.

Dutta, D., De, D., Surabhi, C. and Bhattacharya, S.K. (2005) Hydrogen production by cyanobacteria. *Microbial Cell Factories*, 4, 36.

Eroglu, E. and Melis, A. (2011) Photobiological hydrogen production: Recent advances and state of the art. *Bioresource Technology*, 102, 8403–8413.

Eroglu, I., Aslan, K., Gunduz, U., Yucel, M., and Turker, L. (1999) Substrate consumption rates for hydrogen production by *Rhodobacter sphaeroides* in a column photobioreactor. *Journal of Biotechnology*, 70, 103–113.

Fang, H.H.P., Liu, H. and Zhang T. (2005) Phototrophic hydrogen production from acetate and butyrate in wastewater. *International Journal of Hydrogen Energy*, 30, 785–793.

Fissler, J., Kohring, G.W. and Giffhorn, F. (1995) Enhanced hydrogen production from aromatic acids by immobilized cells of *Rhodopseudomonas palustris*. *Applied Microbiology and Biotechnology*, 44, 43–46.

Fouchard, S., Hemschemeier, A., Caruana, A., Pruvost, K., Legrand, J., Happe, T., Peltier, G. and Cournac, L. (2005) Autotrophic and mixotrophic hydrogen photoproduction in sulfur-deprived *Chlamydomonas* cells. *Applied and Environmental Microbiology*, 71, 6199–6205.

Fouchard, S., Pruvost, J. and Legrand, J. (2008) Investigation of H_2 production by microalgae in a fully-controlled photobioreactor. *International Journal of Hydrogen Energy*, 33, 3302–3310.

Gaffron, H. (1939) Reduction of carbon dioxide with molecular hydrogen in green algae. *Nature*, 143, 204–205.

Gaffron, H. and Rubin, J. (1942) Fermentative and photochemical production of hydrogen in algae. *Journal of General Physiology*, 26, 219–240.

Gebicki, J., Modigell, M., Schumacher, M., Burgb, J.V.D. and Roebroeck, E. (2009) Development of photobioreactors for anoxygenic production of hydrogen by purple bacteria. *Chemical Engineering Transactions*, 18, 363–366.

Gest, H. and Kamen, M.D. (1949) Photoproduction of molecular hydrogen by *Rhodospirillum rubrum*. *Science*, 109, 558–559.

Ghirardi, M.L., Zhang, L.P., Lee, J.W., Flynn, T., Seibert, M., Greenbaum, E. and Melis, A. (2000) Microalgae: a green source of renewable H_2. *Trends in Biotechnology*, 18, 506–510.

Gilbert, J.J., Subhabrata, R. and Das, D. (2011) Hydrogen production using *Rhodobacter sphaeroides* (O.U. 001) in a flat panel rocking photobioreactor. *International Journal of Hydrogen Energy*, 36, 3434–3441.

Guan, Y.F., Zhang, W., Deng, M.C., Jin, M.F. and Yu, X.J. (2004) Significant enhancement of photobiological H_2 evolution by carbonylcyanide m-chlorophenylhydrazone in the marine green alga *Platymonas subcordiformis*. *Biotechnology Letters*, 26, 1031–1035.

Guo, C.L., Zhu, X., Liao, Q., Wang, Y.Z., Chen, R. and Lee, D.J. (2011) Enhancement of photo-hydrogen production in a biofilm photobioreactor using optical fiber with additional rough surface. *Bioresource Technology*, 102, 8507–8513.

Hakobyan, L., Gabrielyan, L. and Trchounian, A. (2012) Ni (II) and Mg (II) ions as factors enhancing biohydrogen production by *Rhodobacter sphaeroides* from mineral springs. *International Journal of Hydrogen Energy*, 37, 7482–7486.

Hallenbeck, P.C. and Benemann, J.R. (2002) Biological hydrogen production: fundamentals and limiting processes. *International Journal of Hydrogen Energy*, 27, 1185–1193.

Hansel, A. and Lindblad, P. (1998) Towards optimization of cyanobacteria as biotechnologically relevant producers of molecular hydrogen, a clean and renewable energy source. *Applied Microbiology and Biotechnology*, 50, 153–160.

Happe, T. and Kaminski, A. (2002) Differential regulation of the [Fe]-hydrogenase during anaerobic adaptation in the green alga *Chlamydomonas reinhardtii*. *European Journal of Biochemistry*, 269, 1022–1032.

Happe, T., Hemschemeier, A., Winkler, M. and Kaminski, A. (2002) Hydrogenase in green algae: do they save the algae's life and solve our energy problems? *Trends in Plant Science*, 7, 246–250.

Hoekema, S., Bijmans, M., Janssen, M., Tramper, J. and Wijffels, R.H. (2002) A pneumatically agitated flat-panel photobioreactor with gas re-circulation: anaerobic photoheterotrophic cultivation of a purple non-sulfur bacterium. *International Journal of Hydrogen Energy*, 27, 1331–1338.

Iqbal, M., Grey, D., Stepan-Sarkissian, F. and Fowler, M.W. (1993) A flat-sided photobioreactor for continuous culturing microalgae. *Aquacultural Engineering*, 12, 183–190.

Kapdan, I.K. and Kargi, F. (2006) Bio-hydrogen production from waste materials. *Enzyme and Microbial Technology*, 38, 569–582.

Kars, G., Gunduz, U., Yucel, M., Turker, L. and Eroglu, I. (2006) Hydrogen production and transcriptional analysis of *nif*D, *nif*K and *hup*S genes in *Rhodobacter sphaeroides* O.U.001 grown in media with different concentrations of molybdenum and iron. *International Journal of Hydrogen Energy*, 31, 1536–1544.

Kashyap, A.K., Pandey, K.D. and Sarkar, S. (1996) Enhanced hydrogen photoproduction by non-heterocystous cyanobacterium *Plectonemaboryanum*. *International Journal of Hydrogen Energy*, 21, 107–109.

Keskin, T., Abo-Hashesh, M. and Hallenbeck, P.C. (2011) Photofermentative hydrogen production from wastes. *Bioresource Technology*, 102, 8557–8568.

Koku, H., Eroglu, I., Gunduz, U., Yucel, M. and Turker, L. (2002) Aspects of the metabolism of hydrogen production by *Rhodobacter sphaeroides*. *International Journal of Hydrogen Energy*, 27, 1315–1329.

Kondo, T., Arakawa, M., Hirai, T., Wakayama, T., Hara, M. and Miyake, J. (2002) Enhancement of hydrogen production by a photosynthetic bacterium mutant with reduced pigment. *Journal of Bioscience and Bioengineering*, 93, 145–150.

Kruse, O., Rupprecht, J., Mussgnug, J.H., Dismukesc, G.C. and Hankamer, B. (2005) Photosynthesis: a blueprint for solar energy capture and biohydrogen production technologies. *Photochemical & Photobiological Sciences*, 4, 957–970.

Lee, Y.K., Ding, S.Y., Low, C.S. and Chang, Y.C. (1995) Design and performance of an α-type tubular photobioreactor for mass cultivation of microalgae. *Journal of Applied Phycology*, 7, 47–51.

Levin, D.B., Pitt, L. and Love, M. (2004) Biohydrogen production: prospects and limitations to practical application. *International Journal of Hydrogen Energy*, 29, 173–185.

Li, R.Y., Zhang, T. and Fang, H.H.P. (2008) Characteristics of a phototrophic sludge producing hydrogen from acetate and butyrate. *International Journal of Hydrogen Energy*, 33, 2147–2155.

Liao, Q., Qu, X.F., Chen, R., Wang, Y.Z., Zhu, X. and Lee, D.J. (2012) Improvement of hydrogen production with *Rhodopseudomonas palustris* CQK-01 by Ar gas sparging. *International Journal of Hydrogen Energy*, 37, 15443–15449

Liao, Q., Wang, Y.J., Wang, Y.Z., Zhu, X., Tian, X. and Li, J. (2010) Formation and hydrogen production of photosynthetic bacterial biofilm under various illumination conditions. *Bioresource Technology*, 101, 5315–5324.

Liu, B.F., Ren, N.Q., Xie, G.J., Ding, J., Guo, W.Q. and Xing, D.F. (2010) Enhanced biohydrogen production by the combination of dark- and photo-fermentation in batch culture. *Bioresource Technology*, 101, 5325–5329.

Liu, J.G., Bukatin, V.E. and Tsygankov, A.A. (2006) Light energy conversion into H_2 by *Anabaena variabilis* mutant PK84 dense cultures exposed to nitrogen limitations. *International Journal of Hydrogen Energy*, 31, 1591–1596.

Manish, S. and Banerjee, R. (2008) comparison of biohydrogen production processes. *International Journal of Hydrogen Energy*, 33, 279–286.

Masukawa, H., Mochimaru, M. and Sakurai, H. (2002) Disruption of the uptake hydrogenase gene, but not the bidirectional hydrogenase gene, leads to enhanced photobiological hydrogen production by the nitrogen-fixing cyanobacterium *Anabaena sp.* PCC 7120. *Applied Microbiology and Biotechnology*, 58, 618–624.

Mathews, J. and Wang, G. (2009) Metabolic pathway engineering for enhanced biohydrogen production. *International Journal of Hydrogen Energy*, 34, 7404–7416.

Melis, A. (2002) Green alga hydrogen production: progress, challenges and prospects. *International Journal of Hydrogen Energy*, 27 1217–1228.

Melis, A. (2007) Photosynthetic H_2 metabolism in *Chlamydomonas reinhardtii* (unicellular green algae). *Planta*, 226, 1075–1086.

Melis, A. and Melnicki, M.R. (2006) Integrated biological hydrogen production. *International Journal of Hydrogen Energy*, 31, 1563–1573.

Melis, A., Zhang, L.P., Mare, F., Ghirardi, M.L. and Seibert, M. (2000) Sustained photobiologieal hydrogen gas production upon reversible inaetivation of oxygen evolution in the green alga *Chlamydomonas reinhardtii. Plant Physiology*, 122, 127–135.

Mignot, L., Junter, G.A. and Labbe, M. (1989) A new type of immobilized-cell photobioreactor with internal illumination by optical fibres. *Biotechnology Techniques*, 3, 299–304.

Miron, S.A., Garcia, M.C.C., Camacho, F.G., Grima, E.M. and Chisti, Y. (2002) Growth and characterization of microalgal biomass produced in bubble column and airlift photobioreactors: studies in fed-batch culture. *Enzyme and Microbial Technology*, 31, 1015–1023.

Miyake, J., Miyake, M. and Asada, Y. (1999) Biotechnological hydrogen production: research for efficient light energy conversion. *Journal of Biotechnology*, 70, 89–101.

Morita, M., Watanable, Y. and Saiki, H. (2000) Investigation of photobioreactor design for enhancing the photosynthetic productivity of microalgae. *Biotechnology and Bioengineering*, 69, 693–698.

Najafpour, G., Younesi, H. and Mohamed, A.R. (2004) Effect of organic substrate on hydrogen-production from synthesis gas using *Rhodospirillum rubrum*, in batch culture. *Biochemical Engineering Journal*, 21, 123–130.

Ozturk, Y., Yucel, M., Daldal, F., Mandaci, S., Gunduz, U., Turker, L. and Eroglu, I. (2006) Hydrogen production by using *Rhodobacter capsulatus* mutants with genetically modified electron transfer chains. *International Journal of Hydrogen Energy*, 31, 1545–1552.

Pinto, F.A.L., Troshina, O. and Lindblad, P. (2002) A brief look at three decades of research on cyanobacterial hydrogen evolution. *International Journal of Hydrogen Energy*, 27, 1209–1215.

Planchard, A., Mignot, L., Jouenne, T. and Junter, G.A. (1989) Photoproduction of molecular hydrogen by *Rhodospirillum rubrum* immobilized in composite agar layer/microporous membrane structures. *Applied Microbiology and Biotechnology*, 31, 49–54.

Polle, J.E.W., Kanakagiri, S., Jin, E., Masuda, T. and Melis, A. (2002) Truncated chlorophyll antenna size of the photsystems – a practical method to improve microalgal productivity and hydrogen production in mass culture. *International Journal of Hydrogen Energy*, 27, 1257–1264.

Polle, J.E.W., Kanakagiri, S. and Melis, A. (2003) tla1, a DNA insertional transformant of the green alga *Chlamydomonas reinhardtii* with a truncated light-harvesting chlorophyll antenna size. *Planta*, 217, 49–59.

Posten C. (2009) Design principles of photo-bioreactors for cultivation of microalgae. *Engineering in Life Sciences*, 9, 165–177.

Qu, X.F., Wang, Y.Z., Zhu, X., Liao, Q., Li, J., Ding, Y.D. and Lee, D.J. (2011) Bubble behavior and photo-hydrogen production performance of photosynthetic bacteria in microchannel photobioreactor. *International Journal of Hydrogen Energy*, 36, 14111–14119.

Ren, H.Y., Liu, B.F., Ding, J., Xie, G.J., Zhao L., Xing, D.F., Guo, W.Q. and Ren, N.Q. (2012) Continuous photo-hydrogen production in anaerobic fluidized bed photo-reactor with activated carbon fiber as carrier. *RSC Advances*, 2, 5531–5535.

Ren, N.Q., Liu, B.F., Zheng, G.X., Xing, D.F., Zhao, X., Guo, W.Q. and Ding, J. (2009) Strategy for enhancing photo-hydrogen production yield by repeated fed-batch cultures. *International Journal of Hydrogen Energy*, 34, 7579–7584.

Sasikala, C.H., Ramana, C.H.V. and Rao, P.R. (1995) Regulation of simultaneous hydrogen photoproduction during growth by pH and glutamate in *Rhodobacter sphaeroides* O.U. 001. *International Journal of Hydrogen Energy*, 20, 123–126.

Shah, V., Garg, N. and Madamwar, D. (2003) Ultrastructure of the cyanobacterium *Nostoc muscorum* and exploitation of the culture for hydrogen production. *Folia Microbiologica*, 48, 65–70.

Shi, X.Y. and Yu, H.Q. (2005) Response surface analysis on the effect of cell concentration and light intensity on hydrogen production by *Rhodopseudomonas capsulata*. *Process Biochemistry*, 40, 2475–2481.

Shi, X.Y. and Yu, H.Q. (2006) Continuous production of hydrogen from mixed volatile fatty acids with *Rhodopseudomonas capsulata*. *International Journal of Hydrogen Energy*, 31, 1641–1647.

Show, K.Y., Lee, D.J. and Chang, J.S. (2011) Bioreactor and process design for biohydrogen production. *Bioresource Technology*, 102 8524–8533.

Skjanes, K., Knutsen, G., Kallqvist, T. and Lindblad, P. (2008) H_2 production from marine and freshwater species of green algae during sulfur starvation and considerations for bioreactor design. *International Journal of Hydrogen Energy*, 33, 511–521.

Srirangan, K., Pyne, M.E. and Chou, C.P. (2011) Photofermentative hydrogen production from wastes. *Bioresource Technology*, 102, 8557–8568.

Suwansaard, M., Choorit, W., Zeilstra-Ryalls, J.H. and Prasertsan, P. (2010) Phototropic H_2 production by a newly isolated strain of *Rhodopseudomonas palustris*. *Biotechnology letters*, 32, 1667–1671.

Tao, Y.Z., He, Y.L., Wu, Y.Q., Liu, F.H., Li, X.F., Zong, W.M. and Zhou, Z.H. (2008) Characteristics of a new photosynthetic bacterial strain for hydrogen production and its application in wastewater treatment. *International Journal of Hydrogen Energy*, 33, 963–973.

Tekucheva, D.N., Laurinavichene, T.V., Seibert, M. and Tsygankov, A.A. (2011) Immobilized purple bacteria for light-driven H_2 production from starch and potato fermentation effluents. *Biotechnology Progress*, 27, 1248–1256.

Tian, X., Liao, Q., Liu, W., Wang, Y.Z., Zhu, X., Li, J. and Wang, H. (2009) Photo-hydrogen production rate of a PVA-boric acid gel granule containing immobilized photosynthetic bacteria cells. *International Journal of Hydrogen Energy*, 34, 4708–4717.

Tian, X., Liao, Q., Zhu, X., Zhu, X., Wang, Y.Z., Zhang, P., Li, J. and Wang, H. (2010) Characteristics of a biofilm photobioreactor as applied to photo-hydrogen production. *Bioresource Technology*, 101, 977–983.

Tredici, M.R., Carlozzi, P., Zittelli, G.C. and Materassi, R. (1991) A vertical alveolar panel (VAP) for outdoor mass cultivation of microalgae and cyanobacteria. *Bioresoure Technology*, 38, 153–159.

Tredici, M.R., Zittelli, G.C. and Benemann, J.R. (1998) A tubular integral gas exchange photobioreactor for biological hydrogen production. In: *Biohydrogen* Ed. O.R. Zaborsky. Plenum Press, London, UK.

Tsygankov, A.A., Hall, D.O., Liu, J. and Rao, K.K. (1998) An automated helical photobioreactor incorporating cyanobacteria for continuous hydrogen production. In: *Biohydrogen* ed. O.R. Zaborsky. Plenum Press, London, UK.

Tsygankov, A.A., Hirata, Y. Asada, Y. and Miyake, J. (1993) Immobilization of the purple non-sulfur bacterium *Rhodobacter sphaeroides* on glass surfaces. *Biotechnology Techniques*, 7, 283–286.

Ugwa, C.U., Aoyagi, H. and Uchiyama, H. (2008) Photobioreactors for mass cultivation of algae. *Bioresource Technology*, 99, 4021–4028.

Uyar, B., Eroglu, I., Yucel, M., Gndz, U. and Turker, L. (2007) Effect of light intensity, wavelength and illumination protocol on hydrogen production in photobioreactors. *International Journal of Hydrogen Energy*, 32, 4670–4677.

Vasilyeva, L., Miyake, M., Khatipov, E., Wakayama, T., Sekine, M., Hara, M., Nakada, E., Asada, Y. and Miyake, J. (1999) Enhanced hydrogen production by a mutant of *Rhodobacter sphaeroides* having an altered light-harvesting system. *Journal of Bioscience and Bioengineering*, 87, 619–624.

von Felten, P., Zurrer, H. and Bachofen, R. (1985) Production of molecular hydrogen with immobilized cells of *Rhodospirillum rubrum*. *Applied Microbiology and Biotechnology*, 23, 15–20.

Wang, Y.J., Liao, Q., Wang, Y.Z., Zhu, X. and Li, J. (2011) Effects of flow rate and substrate concentration on the formation and H_2 production of photosynthetic bacterial biofilm. *Bioresource Technology*, 102, 6902–6908.

Wang, Y.Z., Liao, Q., Zhu, X., Tian, X. and Zhang, C. (2010) Characteristics of hydrogen production and substrate consumption of *Rhodopseudomonas palustris* CQK 01 in an immobilized-cell photobioreactor. *Bioresource Technology*, 101, 4034–4041.

Wang, Y.Z., Xie, X.W., Zhu, X., Liao, Q., Chen, R., Zhao, X. and Lee, D.J. (2012) Hydrogen production by *Rhodopseudomonas palustris* CQK 01 in a continuous photobioreactor with ultrasonic treatment. *International Journal of Hydrogen Energy*, 37, 15450–15457.

Weissman, J.C. and Benemann, J.R. (1977) Hydrogen production by nitrogen-starved cultures of *Anabaena cylindrica*. *Applied and Environmental Microbiology*, 33, 123–131.

Wunschiers, R., Senger, H. and Schulz, R. (2001) Electron pathways involved in H$_2$-metabolism in the green alga *Scenedesmus obliquus*. *Biochimica et Biophysica Acta-Bioenergetics*, 1503, 271–278.

Xie, G.J., Liu, B.F., Ding, J., Ren, H.Y., Xing, D.F. and Ren, N.Q. (2011) Hydrogen production by photo-fermentative bacteria immobilized on fluidized bio-carrier. *International Journal of Hydrogen Energy*, 36, 13991–13996.

Xie, G.J., Liu, B.F., Xing, D.F., Nan, J., Ding, J., Ren, H.Y., Guo, W.Q. and Ren, N.Q. (2012) Photo-hydrogen production by *Rhodopseudomonas faecalis* RLD-53 immobilized on the surface of modified activated carbon fibers. *RSC Advances*, 2, 2225–2228.

Xu, Q., Yooseph, S., Smith, H.O. and Venter, C.J. (2005) Development of a novel recombinant cyanobacterial system for hydrogen production from water. In: *Genomics: GTL program projects*. J. Craig Venter Institute, Rockville, MD, 63.

Yamada, A., Takano, H., Burgess, J.G. and Matsunaga, T. (1996) Enhanced hydrogen production by a marine photosynthetic bacterium, *Rhodobacter marinus*, immobilized onto light-diffusing optical fibers. *Journal of Marine Biotechnology*, 4, 23–27.

Zhang, C., Zhu, X., Liao, Q., Wang, Y.Z., Li, J., Ding, Y.D. and Wang, H. (2010) Performance of a groove-type photobioreactor for hydrogen production by immobilized photosynthetic bacteria. *International Journal of Hydrogen Energy*, 35, 5284–5292.

Zhang, K., Miyachi, S. and Kurano, N. (2001) Evaluation of a vertical flat-plate photobiore-actor for outdoor biomass production and carbon dioxide bio-fixation: effects of reactor dimensions, irradiation and cell concentration on the biomass productivity and irradiation utilization efficiency. *Applied Microbiology and Biotechnology*, 55, 428–433.

Zhu, H., Suzuki, T., Tsygankov, A.A., Asada, Y. and Miyake, J. (1999) Hydrogen production from tofu wastewater by *Rhodobacter sphaeroides* immobilized agar gels. *International Journal of Hydrogen Energy*, 24, 305–310.

Zhu, X., Guo, C.L., Wang, Y.Z., Liao, Q., Chen, R. and Lee, D.J. (2012) A feasibility study on unsaturated flow bioreactor using optical fiber illumination for photo-hydrogen production. *International Journal of Hydrogen Energy*, 37, 15666–15671.

Zhu, X., Xie, X.W., Liao, Q., Wang, Y.Z. and Lee, D.J. (2011) Enhanced hydrogen production by *Rhodopseudomonas palustris* CQK 01 with ultra-sonication pretreatment in batch culture. *Bioresource Technology*, 102, 8696–8699.

Chapter 12

Decontamination of water by combined solar advanced oxidation processes and biotreatment

Sixto Malato, Isabel Oller, Pilar Fernández-Ibáñez &
Manuel Ignacio Maldonado
Plataforma Solar de Almería (CIEMAT), Carretera Senés, Tabernas (Almería), Spain

12.1 INTRODUCTION

One of the major threats to water quality is chemical pollution from heavy metals, solvents, dyes, pesticides, etc. Such chemicals enter the aquatic medium in several different ways, either by direct dumping, such as in industrial effluents, or from wastewater treatment plants (WWTP) that are unable to eliminate them. Another indirect source is plant health products, such as biocides and fertilizers, in agriculture. Discharge resulting from lax enforcement of legislation, illegal usage and inappropriate application of substances may also be considerable. In general, very water-soluble substances are transported and distributed more easily in the water cycle.

The main methods for destroying such toxic compounds in natural water are biodegradation and photodegradation. Photodegradation may be by direct or indirect photolysis. In indirect photolysis, a photosensitizer (such as nitrate or humic acids) absorbs the light and transfers the energy to the pollutants, which otherwise would not react, since they do not absorb light in the wavelength interval of solar photons on the Earth's surface (i.e. >300 nm). Biological degradation of a chemical refers to its elimination by the metabolic activity of living organisms, usually microorganisms, particularly the bacteria and fungi living in natural water and soil. In this context, conventional biological processes do not always provide satisfactory results, especially for industrial wastewater treatment, since many of the organic substances produced by the chemical industry are toxic or resistant to biological treatment (Muñoz and Guieysee, 2006; Lapertot and Pulgarin, 2006). Conventional methods of water decontamination that can address many of these problems are often chemically, energetically and operationally intensive, and when used in large systems require a considerable infusion of capital, engineering expertise and infrastructure. This practically precludes their use in much of the world. Furthermore, intensive chemical treatments (such as those involving ammonia, chlorine compounds, hydrochloric acid, sodium hydroxide, permanganate, alum and ferric salts, coagulation and filtration aids, anti-scalants, corrosion control chemicals, and ion exchange resins and regenerants) and residuals resulting from treatment (sludge, brines, toxic waste) can add to the problems of contamination and salting of freshwater sources. Air stripping and adsorption, which merely transfer toxic materials from one medium to another, are not long-term solutions. Incineration converts toxics into carbon dioxide, water and inorganic acids, but negative public perception has very often prevented its implementation. In view of all of the above, a feasible option for such biologically persistent wastewater is the use of the advanced technologies based on chemical oxidation called advanced oxidation processes (AOPs), which are widely recognized as highly efficient treatments for recalcitrant wastewater (Pera-Titus et al., 2004). These methods rely on the formation of highly reactive chemical species which degrade even the most recalcitrant molecules into biodegradable compounds. Although reacting systems vary (Comninellis et al., 2008), all of them are characterized by the production of hydroxyl radicals (OH), which are able to oxidize and mineralize almost any organic molecule, yielding CO_2 and inorganic ions. They are also non-selective, which is a useful attribute for wastewater treatment and solution of pollution problems. The versatility of the AOPs is also enhanced by the fact that hydroxyl radicals may be produced in different ways, facilitating compliance with the specific treatment requirements. Methods based on UV, H_2O_2/UV, O_3/UV and other combinations use photolysis of H_2O_2 and ozone to produce the hydroxyl radicals, but generation of UV radiation by lamps and ozone production are expensive. So future applications of these processes could be improved through the use of catalysis and solar energy. Therefore, research is focusing more and more on those AOPs which can be driven by solar irradiation. Of special interest is photo-Fenton, which is based on addition of H_2O_2 to dissolved iron salts and irradiation with UV-VIS light, because sunlight can be used for it (Pignatello et al., 2006).

12.2 SOLAR PHOTO-FENTON

Fenton and Fenton-like processes are probably among the advanced oxidation processes most applied in the treatment of industrial wastewater (Suty et al., 2004). The first proposals for wastewater treatment applications were reported in the 1960s. Yet it was not until the early 1990s that the first studies on the application of the photo-Fenton process for the treatment of wastewater were published by the groups of Pignatello, Lipcznska-Kochany, Kiwi, Pulgarín and Bauer (Pignatello et al., 2006). Much of the literature on photo-Fenton includes the possibility of driving the process with solar radiation because it seems to be the most suitable of all AOPs for being driven by sunlight, because soluble iron-hydroxyl and especially iron-organic acid complexes even absorb part of the visible light spectrum (Figure 12.2.1), not only ultraviolet radiation (Malato et al., 2009).

Hydrogen peroxide is decomposed to water and oxygen in the presence of iron ions in the Fenton reaction in aqueous solutions, Equation 12.2.1, as first reported by H.J.H. Fenton (Fenton, 1894). Mixtures of ferrous iron and hydrogen peroxide are called Fenton reagents. Equations 12.2.1–12.2.3 show the basic reactions in the absence of other interfering ions and organic substances. Regeneration of ferrous iron from ferric iron by Equations 12.2.2 and 12.2.3 is the rate limiting step in the catalytic iron cycle, if iron is added in small amounts.

$$Fe^{2+} + H_2O_2 \rightarrow Fe^{3+} + OH^- + OH^\bullet \tag{12.2.1}$$

$$Fe^{3+} + HO_2^\bullet \rightarrow Fe^{2+} + O_2 + H^+ \tag{12.2.2}$$

$$Fe^{3+} + O_2^{\bullet-} \rightarrow Fe^{2+} + O_2 \tag{12.2.3}$$

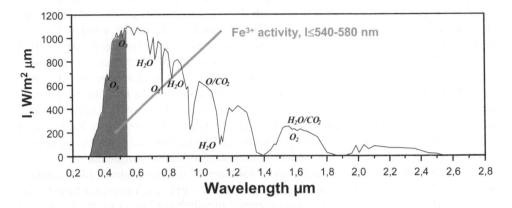

Figure 12.2.1 Normal solar irradiance (I) on the Earth's surface (ASTM E891-87, air mass 1.5), main light absorbing gases and light absorption of Fe^{3+} species.

If organic substances (quenchers, scavengers, or in the case of wastewater treatment, pollutants) are present in the system $Fe^{2+}/Fe^{3+}/H_2O_2$, they react in many ways with the hydroxyl radicals generated. The organic radicals generated continue reacting, prolonging the chain reaction and thereby contributing to reducing the consumption of oxidants in wastewater treatment by Fenton and photo-Fenton. In aromatic pollutants, the ring system is usually hydroxylated before it is broken up during oxidation, typically into intermediate degradation products containing quinone and hydroquinone structures. In any case, sooner or later, ring opening reactions further mineralize the molecule. One important drawback of the Fenton method, especially for total mineralisation of organic pollutants, is that carboxylic intermediates cannot be further degraded. Carboxylic and dicarboxylic acids (L: Mono- and Dicarboxylic acids) are known to form stable iron complexes, which inhibit the reaction with peroxide (Kavitha and Palanivelu, 2004). Hence, the catalytic iron cycle reaches a standstill before total mineralisation is accomplished (Equation 12.2.4).

$$Fe^{3+} + nL \rightarrow [FeL_n]^{x+} \overset{H_2O_2, \text{ dark}}{\longrightarrow} \text{ no further reaction} \qquad (12.2.4)$$

The primary step in the solar photoreduction of dissolved ferric iron is a ligand-to-metal charge-transfer reaction in which intermediate complexes dissociate as shown in the reaction in Equation 12.2.5. The ligand may be any Lewis base able to form a complex with ferric iron (OH^-, H_2O, HO_2^-, Cl^-, R-COO$^-$, R-OH, R-NH$_2$ etc.). Depending on the reacting ligand, the product may be a hydroxyl radical such as in Equation 12.2.6 or other radical derived from the ligand. The direct oxidation of an organic ligand is possible as well, as shown for carboxylic acids in Equation 12.2.7.

$$[Fe^{3+}L] + h\nu \longrightarrow [Fe^{3+}L]^* \longrightarrow Fe^{2+} + L^\bullet \qquad (12.2.5)$$

$$[Fe(OH)]^{2+} + h\nu \longrightarrow Fe^{2+} + OH^\bullet \qquad (12.2.6)$$

$$[Fe(OOC-R)]^{2+} + h\nu \rightarrow Fe^{2+} + CO_2 + R^\bullet \qquad (12.2.7)$$

The ferric iron complex has different light absorption properties depending on the ligand, so Equation 12.2.5 takes place with different quantum yields and also at different wavelengths. Consequently, pH plays a crucial role in the efficiency of the photo-Fenton reaction, because it strongly influences which complexes are formed. Thus, pH 2.8 has frequently been postulated as optimum for photo-Fenton treatment, because there is no precipitation yet and the predominant iron species in solution is $[Fe(OH)]^{2+}$, the most photoactive ferric iron-water complex. In fact, as shown in its general form in Equation 12.2.5, ferric iron can form complexes with many substances and undergo photoreduction. Carboxylic acids are of special importance because they are frequent oxidation intermediate products, and ferric iron-carboxylate complexes may have much higher quantum yields than ferric iron-water complexes.

Fe^{3+} complexes present in mildly acidic solutions absorb an appreciable amount of light in the UV and into the visible region, and may complex with certain target compounds or their by-products. These complexes typically have higher molar absorption coefficients in the near-UV and visible regions than aquo complexes. Polychromatic quantum efficiencies from 0.05 to 0.95 are common in the UV/visible range (Pignatello et al., 2006), making the photo-Fenton process suitable for being driven by sunlight.

12.2.1 Solar photo-Fenton hardware

Much solar detoxification system component equipment (Blanco and Malato, 2003) is identical to what is used for other types of water treatment, and construction materials are available on the market. Most piping may be made of polyethylene or polypropylene, but not metal or composite materials that could be degraded by the oxidizing conditions of the process. Neither may reactive materials that would interfere with the photocatalytic process be used. All materials used must be inert to degradation by solar UV light so they last the minimum required system lifetime. Photocatalytic reactors must transmit UV-Vis light efficiently because of the process requirements. The best reflecting/concentrating material is aluminium, because, while aluminium coated mirrors are low-cost and highly reflective in the UV-Vis band of the terrestrial solar spectrum, the reflectivity (reflected radiation/incident radiation) from 300 to 400 nm of traditional silver-coated mirrors is very low. Aluminium, which is the only metal surface that is highly reflective throughout the ultraviolet spectrum, has a reflectivity range from 92.3% at 280 nm to 92.5% at 385 nm. Comparable values for silver are 25.2% and 92.8%, respectively. Aluminium also reflects perfectly in the visible range. The photocatalytic reactor must be transparent to UV-Vis radiation. Visible range transmissivity of different materials is usually high, and it is in the UV range where restrictions appear. The choice of materials that are both transmissive to UV light and resistant to its destructive effects is limited. Common materials that meet these requirements are fluoropolymers, acrylic polymers and several types of glass. Quartz has excellent UV transmission as well as good temperature and chemical resistance, but its high cost makes it completely unfeasible for photocatalytic applications. Fluoropolymers are a good choice of plastics for photoreactors due to their good UV transmittance, excellent ultraviolet stability and chemical inertness. However, in order to achieve minimum pressure resistance, the wall thickness of the fluoropolymer tube has to be increased, which in turn lowers its UV transmittance. Other low cost polymeric materials are significantly more susceptible attack by hydroxyl radicals. Standard glass is not satisfactory because it absorbs part of the UV radiation that reaches it, due to its iron content. Low-iron borosilicate glass, which has good transmissive properties in the solar range with a cut-off at about 285 nm (Blanco et al., 2000), would seem to be the most adequate. Therefore, although both fluoropolymers and glass are valid photoreactor materials, if a large field with a considerable number of photoreactors is being designed, there will be a high system pressure drop. So in such cases, fluoropolymer tubes are not the best choice of material, and borosilicate glass is a better solution.

The original solar photoreactor designs (Dillert et al., 1999) for photochemical applications were based on line-focus parabolic-trough concentrators (PTCs). In part, this was a logical extension of the historical emphasis on trough units for solar thermal applications. Furthermore, PTC technology was relatively mature and existing hardware could be easily modified for photochemical processes. The main disadvantages are that these collectors (i) use only direct radiation (ii) are expensive (iii) have low efficiencies as they concentrate sunlight therefore increasing temperature and promoting iron precipitation, and (iv) a high iron concentration is needed to absorb concentrated sunlight. On the other hand, one-sun (non-concentrating) collectors have no moving parts or solar tracking devices. They do not concentrate radiation, so efficiency is not reduced by factors associated with concentration and solar tracking. As there is no concentrating system (with its inherent reflectivity), the efficiency is higher than for PTCs, and they are able to utilize the diffuse as well as the direct portion of the solar radiation. An extensive effort in the design of small non-tracking collectors has resulted in the testing of several different non-concentrating solar reactors (Blanco et al., 2007). Although one-sun collector designs possess important advantages, the design of a robust one-sun photoreactor is no simple matter, due to the need for weather-resistant and chemically inert ultraviolet-transmitting reactors. In addition, non-concentrating systems require significantly more photoreactor area than concentrating photoreactors and, as a consequence, full-scale systems must be designed to withstand the operating pressures for fluid circulation.

Design of a solar collector for a photo-Fenton reactor is subject to some major optimization constraints: (1) collection of maximum solar UV-Vis radiation, (2) working temperatures below 50°C, (3) efficiency at low iron concentrations, (4) its construction must be economical, and finally (5) the system pressure drop must be low. Tubular photoreactors therefore have a decisive advantage in the inherent structural efficiency of tubing for flowing water. Tubing is also available in a large variety of materials and sizes and is a natural choice for a pressurized fluid system. A particular type of low concentration collector called the Compound Parabolic Concentrator (CPC) is used in thermal applications. This combination of parabolic concentrators and static flat systems is also an attractive option for solar photochemical applications (Ajona and Vidal, 2000). CPCs are static collectors with an ideal reflective surface according to non-imaging optics that can be designed for any given reactor shape. The entire circumference of the receiver is illuminated, rather than just the front, as in conventional flat plates. The ideal optics of these concentrating devices thus combines both the advantages of the PTC and static systems (Colina-Márquez et al., 2010). The concentration factor (R_C) of a two dimensional CPC collector is given by Equation 12.2.8.

$$R_{C,CPC} = \frac{1}{\sin \theta_a} = \frac{A}{2\pi\, r} \tag{12.2.8}$$

The normal values for the semi-angle of acceptance (θ_a) for photochemical applications is 90 degrees, whereby $R_C = 1$ (non-concentrating solar system). If the CPC is designed for an acceptance angle of $+90°$ to $-90°$ (Figure 12.2.2), all incident solar direct and diffuse radiation can be collected. The light reflected by the CPC is distributed all around the tubular receiver so that almost the entire circumference of the receiver tube is illuminated. CPCs have the advantages of both the PTC and non-concentrating collector technologies and none of the disadvantages, so they seem to be the best option for solar photocatalysis. They can make highly efficient use of both direct and diffuse solar radiation, without the need for solar tracking.

An important factor in photoreactor design is its diameter. As mentioned above, the Fenton reagent consists of an aqueous solution of hydrogen peroxide and ferrous

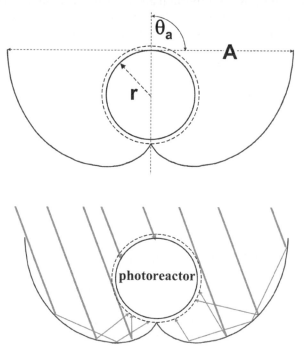

Figure 12.2.2 Schematic drawing and photo of CPC with a semi-angle of acceptance of 90°.

ions producing hydroxyl radicals. When UV/visible radiation is added, it is called photo-Fenton, which is a catalytic process. Fe^{3+} (and related species and organic complexes) absorbs solar photons as a function of its absorptivity. This must be taken into consideration when the optimal photoreactor load is calculated as a function of light-path length. After long experimentation with different photoreactors under sunlight at the Plataforma Solar de Almería (PSA) in Spain, the optimal concentration proposed is 0.2–0.5 mM of iron depending on photoreactor diameter (Malato et al., 2009).

Under the "SOLARDETOX" project (Solar Detoxification Technology for the Treatment of Industrial Non-Biodegradable Persistent Chlorinated Water Contaminants, Brite Euram III Program, 1997–2000, BRPR-CT97-0424), a European consortium coordinated by the PSA was formed for the development and marketing of solar detoxification of recalcitrant water contaminants. The main goal of the project was to develop a commercial non-concentrating solar detoxification system using the compound parabolic collector technology (CPC). A full-size demonstration plant for field demonstration was constructed at Hidrocen facilities (Madrid, Spain). The same collectors have also been used to treat paper mill effluents in Brazil and Germany, and paper mill effluents, surfactants, and textile dyes in Spain (Malato et al., 2007). In 2004, a new CPC plant was installed in the context of a project for the collection and recycling of plastic pesticide bottles using advanced oxidation process (AOP) driven by solar energy funded by the European LIFE-ENVIRONMENT programme. This plant, which is now in routine operation, has a total collector surface of 150 m^2 and photo-reactor volume of 1.06 m^3. More recently, in a new step forward, solar photo-Fenton and aerobic biological processes have been combined, in a 100 m^2 CPC solar photo-Fenton reactor and an aerobic biological treatment plant based on a 1 m^3 immobilized-biomass activated sludge reactor (Figure 12.2.3). The overall efficiency

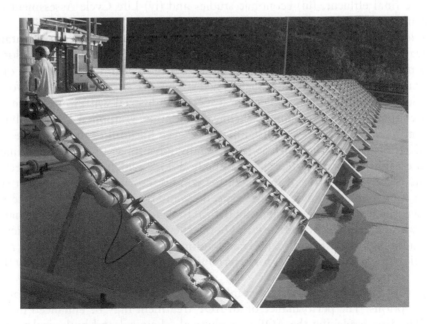

Figure 12.2.3 Solar photo-Fenton plant with 100 m^2 of CPC.

in the combined system was about 95% mineralization. 50% of the initial TOC was degraded in the photo-Fenton pre-treatment, while 45% was removed in the aerobic biological treatment (Blanco et al., 2009). The following sections of this chapter focus mainly on how to enhance the solar photo-Fenton efficiency by integration with biotreatment.

12.3 STRATEGY FOR COMBINING SOLAR ADVANCED OXIDATION PROCESSES AND BIOTREATMENT

Chemical oxidation for complete mineralization is usually expensive, because the oxidation intermediates formed during treatment tend to be more and more resistant to their complete chemical degradation, and furthermore, they all consume energy (radiation, ozone, etc.) and chemical reagents (catalysts and oxidizers) which increase with treatment time (Oller et al., 2011). One attractive potential alternative is to apply these chemical oxidation processes in a pre-treatment to convert the initially persistent organic compounds into more biodegradable intermediates, which would then be treated in a biological oxidation process at a considerably lower cost. Therefore, the main role of the chemical pre-treatment is partial oxidation of the biologically persistent part and produce biodegradable reaction intermediates. The percentage of mineralization should be minimized during pre-treatment to avoid unnecessary expenditure of chemicals and energy, thereby lowering the operating cost. The choice depends on the quality standards to be met and the most effective treatment at the lowest reasonable cost. Therefore, the main factors in the decision on which wastewater treatment technologies to be applied are: (i) the quality of the original wastewater, (ii) the quality of the final effluent, (iii) economic studies and (iv) Life Cycle Assessment of the treatment technology.

As information on AOPs efficiency in eliminating certain specific pollutants in wastewater compared to conventional options is necessary, bench-scale and pilot plant studies must be done to develop the technologies and generate information on new industrial wastewater treatments. Such scaled studies are even more decisive for combining decontamination technologies for a specific industrial wastewater, as described in this chapter. When preliminary chemical oxidation is applied in a combination treatment line, its effect may be insignificant or even harmful to the properties of the original effluent, even though it is conceptually advantageous. This underlines the need to establish a step-by-step research methodology which takes these effects into account, because the effect of the operating conditions on the pre-treatment stream (contact time, oxidant and/or catalyst type, dose and toxicity, temperature, etc.) must be known. Such studies must employ analytical tools to infer the reaction mechanisms, pathway and kinetics, evaluate the effect of the chemical pre-treatment on toxicity and biodegradability, the effect of cations and anions in the wastewater matrix, and the application of various techniques for determining biodegradability and toxicity (Rizzo, 2011).

Appropriate techniques must be combined to provide technically and economically feasible options. The performance of an AOP treatment may be enhanced in several ways. The first is placing the AOP in a physical, chemical and biological treatment sequence. Such an approach often involves at least one AOP step and one biological treatment step (Mandal et al., 2010). Whether the AOP or the biological process comes first in the treatment line, the overall purpose of reducing costs is nearly the same as minimizing AOP treatment and maximizing the biological stage. The individual biological and chemical oxidation efficiencies must be calculated to find the optimal operating conditions for the combined process. This involves detailed knowledge of both biological and chemical processes. Therefore, several analytical parameters must be monitored during each step of the treatment line. The usual chemical parameters measured are total organic carbon (and/or chemical oxygen demand), the concentration of specific pollutants in the target wastewater, and heteroatoms from contaminants completely degraded during the AOP treatment released (Cl, N, P,...) as inorganic species (Cl^-, NO_3^-, PO_4^{3-},...) into the medium. Toxicity analyses (with organisms like *Vibrio fischeri, Daphnia magna,* activated sludge, etc.) and biodegradability tests (using activated sludge) are very important to ensure that AOP effluent conditions are suitable for treatment by conventional biodegradation. The following sections highlight the main parameters necessary for proper evaluation of an AOP to determine the best way of combining it with a biotreatment.

12.3.1 Average oxidation state

One of the most widely used parameters in wastewater biodegradability assessment is the Average Oxidation State (AOS), which can be calculated with Equation 12.3.1 (Scott and Ollis, 1995), where TOC (total organic carbon) and COD (chemical oxygen demand) are expressed in moles of C/L and moles of O_2/L, respectively:

$$AOS = 4 \times \left(\frac{TOC - COD}{TOC} \right) \tag{12.3.1}$$

The AOS is from +4 (for the most oxidized state of carbon, CO_2) to −4 (for the most reduced state of carbon, CH_4). The AOS, which varies with treatment time and indicates the oxidation state of the organic compounds in the wastewater, can be used to determine how the AOP is modifying them. Figure 12.3.1 shows an example of how the AOS commonly evolves during AOP wastewater treatment. Although the AOS rises rapidly at first, this increase later slows down, suggesting that the chemical nature of the reaction intermediates generated did not vary significantly after certain stage. Furthermore, when contaminants are oxidised before mineralization, it usually means that biodegradability is increasing, as in the transformation of chlorophenol into phenol, and phenol into oxalic acid, for example. This is because when the AOS stabilizes, oxidation is producing mineralization (the last step of the process), and no further substantial change in wastewater biodegradability is expected.

So if high biodegradability is to be achieved, it must be before or right at the moment the AOS stabilizes. From that point on, the photocatalytic treatment only mineralizes organic carbon, even though the chemical nature of the organic compounds does not change substantially. It may therefore be concluded that the AOS provides indirect information about wastewater biodegradability. The example in Figure 12.3.1 shows the TOC, COD and AOS during solar photo-Fenton treatment of an industrial waste water. As observed, the AOS rises to a maximum of around 2 and remains there

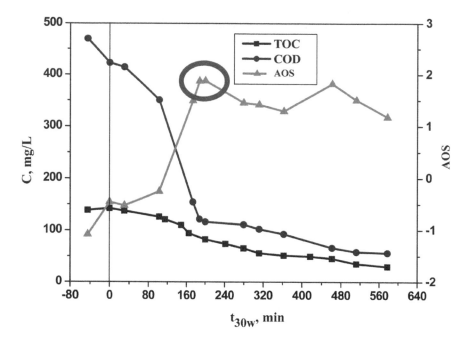

Figure 12.3.1 Evolution of TOC, COD and AOS during the solar photo-Fenton treatment of an industrial wastewater.

almost until the end of the experiment (circled in blue). This means that from this point on, the chemical nature (oxidation state) of the mixture is not going to change significantly, because oxidation is causing mineralization. Parallel behavior of the COD and TOC also makes this very clear. Before this point (treatment time less than 200 minutes), the AOS rose sharply (sharp drop in COD along with slow fall in TOC). It can therefore be assumed that the biodegradability of the effluent should change from 0 to 200 minutes, but not after, and therefore, any bioassay for determining biodegradability enhancement should be applied while AOS is still changing and only until it stabilizes but not afterwards. As bioassays are usually difficult, expensive and slow, reliable and rapid chemical analyses, such as COD and TOC are useful aids in deciding the best time for their application.

12.3.2 Activated sludge respirometry

Respirometry assays measure the Oxygen Uptake Rate (OUR) in live biomass and are an indicator of the microbiological activity present in it. The oxygen demand determined in respirometric assays has recently been found to be an excellent control parameter, as it represents a direct measure of the proper activity and viability of microorganisms in aerobic activated sludge. Furthermore, as this test directly assesses the primary function of an activated sludge process, it can be used for efficient measurement of any acute toxicity in industrial wastewater that could affect the activated sludge in a Municipal Wastewater Treatment Plant (MWWTP) (Gutiérrez et al., 2002). Aerobic biomass activity at different stages of AOP treatment of industrial wastewater

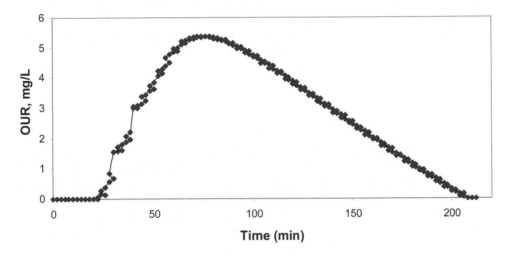

Figure 12.3.2 Measurement of readily biodegradable COD in a respirometric assay.

could be affected by the amount of biodegradable or hardly biodegradable substances present in the medium and the presence of any compounds which are toxic or inhibit cell activity.

The active organisms in activated sludge biomass require molecular oxygen to oxidize the organic load in the wastewater and provide organic carbon to synthesize the compounds necessary for their continuous growth. The two processes are concurrent and together determine the treatment's wastewater organic load elimination rate. The toxicity analysis consists of loading the respirometer with the desired amount of activated sludge from the WWTP and continuously aerating it until air saturation. Then, the sample is added and oxygen consumption measured. The percentage of inhibition can be calculated from the oxygen measurement according to the equipment protocol.

Apart from toxicity assays, activated sludge respirometry analyses are also employed for assessing the biodegradability of industrial wastewater partially treated by an AOP. It is used as a short-term biodegradability assay, to evaluate the oxygen uptake rate during the time the sample and the activated sludge are in contact. Biodegradation parameters such as the maximum oxygen uptake rate and dissolved oxygen consumption found in the respirometric tests are realistic analyses for evaluating the efficiency of partial wastewater oxidation (textile wastewater, landfill leachate wastewater, phenol wastewater, etc.) by an AOP (Goi et al., 2009). In this assay, the oxygen uptake rate from a mixture of a certain amount of the pretreated wastewater and activated sludge (in the endogenous phase and with inhibited autotrophic bacterial activity) is measured during a contact period of around 20 minutes. In the end, the readily biodegradable fraction of the COD (COD_{rb}) is found (as a function of the total oxygen consumption and the biomass growth rate). The COD_{rb}/COD ratio shows the sample's biodegradability. Over 0.1 means that it is biodegradable and below 0.05 is not. The different wastewater COD fractions (biodegradable, non-biodegradable, non-soluble, etc.) can also be determined by respirometric assay (Lagarde et al., 2003). Figure 12.3.2 is a typical oxygen uptake rate graphic. As observed, the area under the curve shows the readily biodegradable fraction of COD in a wastewater sample containing non-biodegradable contaminants partially treated by solar photo-Fenton. In this particular case, COD was 197 mg/L and the COD_{rb} found in the respirometric analysis was 37.4 mg/L. Thus COD_{rb}/COD is 0.29, so the sample is considered biodegradable and completely biocompatible for discharge into a conventional municipal WWTP.

12.3.3 Zahn-Wellens test

Study of the inherent biodegradability of a chemical compound enables its potential for biodegradation under optimal aerobic conditions, such as in a conventional WWTP, to be determined. In these assays, the chemicals are exposed to microorganisms, sometimes previously adapted to the substance for a long period of time to increase compound degradation. The Zahn-Wellens method (Z-W) is standardized by a European Union protocol (Directive 88/302/EEC), and is recommended much more for biodegradability testing than BOD for several reasons. First, the procedure is similar to a real activated sludge biological reactor, and the biomass can even adapt to the compounds in the sample, since the Z-W test lasts 28 days, and second, biodegradation efficiency can also be evaluated by TOC and HPLC-UV to ensure the reliability of the results. The main drawbacks of this biological assay are that it is not applicable to volatile or semi-volatile compounds, or to compounds with water solubility under 50 mg of carbon per litre (application range 50 to 400 mg/L organic carbon).

The Z-W procedure consists of placing activated sludge (preferably from the WWTP which is going to receive the industrial wastewater) and nutrients in contact with the target compound as the only carbon source in the medium. This mixture is kept under proper aeration (and agitation) and in the dark or under diffuse light for 28 days (ambient temperature around 20–25°C). The percentage of biodegradability can be determined at any given time by monitoring total organic carbon and using Equation 12.3.2:

$$D_t = \left[1 - \left(\frac{TOC_t - TOC_b}{TOC_a - TOC_{ba}}\right)\right] \times 100 \qquad (12.3.2)$$

where D_t is the percentage of biodegradability after time t, C_a is the TOC (mg/L) in the sample measured three hours after the beginning of the experiment (to take the effect of adsorption of the compound on the biomass into account), C_t is the TOC measured at time t (usually measured daily), C_b is the blank TOC (containing only the same amount of activated sludge as the samples, distilled water and nutrients in order to evaluate the TOC produced only by the biomass metabolism) measured at time t and C_{ba} is the blank TOC measured three hours after the beginning of the experiment. The biodegradability threshold is 70%. The biodegradability of a reference compound such as diethylene glycol should be evaluated as a control to ensure the method is working properly and the correct activity of the activated sludge. 70% elimination of the TOC in this substance in less than 14 days demonstrates proper activity of the activated sludge.

As a practical example of the use of biodegradability analyses, Figure 12.3.3 shows photo-Fenton degradation of an industrial wastewater. As photo-Fenton is an

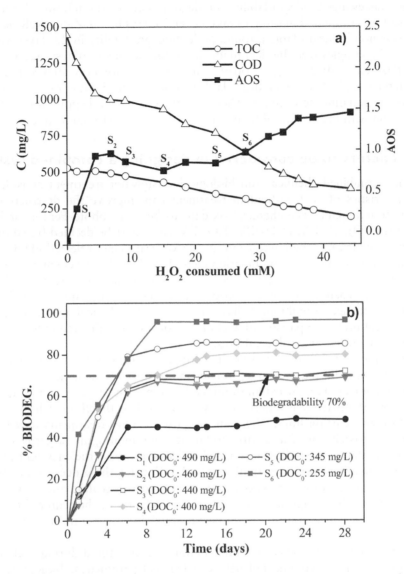

Figure 12.3.3 Photo-Fenton degradation of an industrial wastewater: (a) AOS during treatment, and; (b) Z-W biodegradability analyses of selected samples.

oxidation process, more oxidized organic intermediates are formed at the beginning (Note how fast COD drops until H_2O_2 is 10 mM) without any substantial mineralization (very low, as measured by TOC, until H_2O_2 is 10 mM). After a certain amount of H_2O_2 has been consumed, COD and TOC behave similarly, stabilizing the AOS. After 25 mM of H_2O_2, another increase in AOS, though not as sharp, and another steady state after 35 mM of H_2O_2 were observed, indicating that the very different intermediates formed go through different oxidation-mineralization steps. Formation of more oxidized intermediates is usually an indirect demonstration of improved wastewater biodegradability. Biodegradability of the mixture during phototreatment was evaluated in samples taken over 28 days by the Zahn-Wellens (Z-W) test. S1 was non-biodegradable (around 50% biodegradability in 28 days), whereas the rest of the samples were biodegradable according to the Z–W test. S2, S3 and S4 were biodegradable after a long period of 8 days or more. However, after S5 (345 mg/L TOC), all the samples became biodegradable in a short time (5 days or less). S6 had the highest biodegradation percentages. Therefore, the most suitable point for combining photo-Fenton with the biological treatment is somewhere between S5 and S6.

12.3.4 Factors to be considered in designing a combined system

Design of a combined chemical and biological wastewater treatment consider how the characteristics of each individual treatment can improve the destruction of a persistent contaminant. The chemical oxidant to be used (photo-Fenton or Fenton reagent, O_3/H_2O_2, O_3/UV, H_2O_2/UV, TiO_2/UV, etc.) must be decided based on tests to determine which has the highest rate in the key parameter selected (TOC, COD, biodegradability, toxicity or a combination thereof) with the lowest chemical consumption. The rest of the characteristics to be taken into consideration are widely known: the ability of the chemical oxidation process, its potential for forming toxic intermediates or not, change in pollutant behaviour, choice of biological agent, comparison of different cultures, comparison of acclimated and non-acclimated cultures, use of monospecific cultures, anaerobic cultures, etc.

Measurement of the combined process efficiency depends on the purpose of the treatment, but usually requires independent optimization of each chemical and biological step. For example, the extent of mineralization of the organic compounds may be a measure of efficiency if highly pure water is used or the effluent has a specific dissolved organic carbon limit. The main purpose of other treatments may be total elimination of toxicity or of a specific pollutant. Determining the target is an essential step in combination studies since it helps define process efficiency and provides a basis for comparing operating conditions and optimizing the process. Nevertheless, if the influent concentration is expected to change, correct scaling and design of the photoreactor may be complicated, because the correlation of the required treatment time and substrate concentration cannot be estimated directly, but must be determined experimentally.

Therefore, several analytical parameters must be monitored during each step of the treatment train, as commented before. Chemical parameters, biological assays for toxicity (with various organisms like *Vibrio fischeri*, *Daphnia magna*, but recommended with activated sludge, etc.) and biodegradability (always using activated sludge) are very important to ensure the optimal conditions for complete effluent treatment. In the biological system itself, and in addition to daily control analyses such as total suspended and volatile solids, total organic carbon or chemical oxygen demand, pH and dissolved oxygen in the system, etc., it is also essential to measure anions and cations in the biological medium, since nutrients are vital to the microorganisms that make up the activated sludge populations and monitoring of the nitrogen species provides much information on nitrification and denitrification that take place during biotreatment. This whole series of analytical parameters is necessary for engineering the design of the combined strategy. For further understanding of the underlying processes, additional analytical methods may be necessary for the identification of unknown intermediate degradation products (chromatography coupled with mass spectrometry). Considerable effort and sophisticated analytical equipment may be necessary to explain (Gómez-Ramos et al., 2011) why, for example, acute toxicity rises during treatment by pinpointing a single intermediate product much more toxic than the original pollutant, or when the purpose is to degrade certain contaminants to below a limit (usually $\mu g/L$) in complicated water containing other organics, and therefore COD, TOC or HPLC/UV cannot be used.

12.4   COMBINING SOLAR ADVANCED OXIDATION PROCESSES AND BIOTREATMENT: CASE STUDIES

12.4.1   Case study A: An unsuccessful AOP/biological process

In the first example the targets are two biorecalcitrant substances used as synthesis intermediates in the pharmaceutical industry, 2-(2, 4-dichlorophenyl)-2-(1H-imidazol-1-dylmethyl)-1,3-dioxolan-4-ylmethanol (CAS 84682-23-5) (DIM) and 2-(2,4-dichlorophenyl)-2-(1H-1,2,4-triazol-1-ylmethyl)-1,3-dioxolan-4-ylmethanol (CAS 67914-85-6) (DTIM). These two non-biodegradable compounds mixed in a distilled water matrix were degraded by solar Photo-Fenton. Each contaminant was dissolved at a concentration of 200 mg/L (COD of 700 mg/L) because they are usually found at this concentration in industrial wastewater. Figure 12.4.1 shows wastewater degradation by solar photo-Fenton, toxicity results and Zahn-Wellens biodegradability analyses.

It may be observed that the target compounds were susceptible to complete degradation and mineralization by photo-Fenton. Both substances were completely eliminated after 25 minutes of illumination (DIM and DTIM = 0) with 27.5 mM of hydrogen peroxide. The AOS was also calculated from TOC and COD results, but did not increase until around 80% of TOC had been mineralized. Toxicity analyses showed that inhibition remained the same until practically the end of the treatment. Furthermore, biodegradability monitored by Zahn-Wellens showed that DIM and DTIM biodegradability were only slightly enhanced when photo-Fenton pretreatment was extended until TOC was below 98 mg/L (at this point, biodegradability after 11 days was 60%). From these results, it can be concluded that the best treatment option for wastewater containing DIM and DTIM is to apply solar photo-Fenton (or other AOP) only until almost complete mineralization. In this case a combined AOP/biological process strategy is not feasible.

12.4.2   Case study B: A successful AOP/biological process

Industrial wastewater with a low organic load (TOC under 500 mg/L) can usually be treated in a combined AOP/biological process. This is the case in the following example, the successful treatment of saline industrial wastewater containing around 600 mg/L of a non-biodegradable compound (R-methylphenylglycine, MPG, also from the pharmaceutical industry) and 400 to 600 mg/L dissolved organic carbon (TOC). Tests performed in solar photo-Fenton pilot plants [Gernjak et al., 2006] were used to design a large-scale hybrid solar photocatalytic-biological plant with a 4 m^3 daily treatment

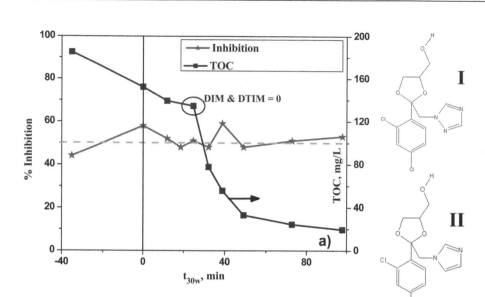

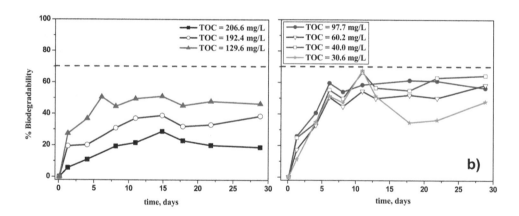

Figure 12.4.1 Degradation of wastewater containing two biorecalcitrant compounds by solar photo-Fenton: (a) TOC elimination and toxicity, and; (b) Biodegradability of several samples from intermediate stages of the treatment DIM (I) and DTIM (II) Structures are also shown.

capacity. This demonstration plant was erected in the grounds of a pharmaceutical company in the south of Spain (see Figure 12.2.3). Using the same protocol described in Section 12.4.1, and after performing photo-Fenton experiments, biodegradability was monitored by Zahn-Wellens tests with the results presented in Figure 12.4.2. Biodegradability was enhanced and the threshold was reached at TOC < 150 mg/L (70% biodegradability in 7 days). A sample with an initial TOC of around 180 mg/L was biodegradable, but after a longer period of around 15 days. This means that with

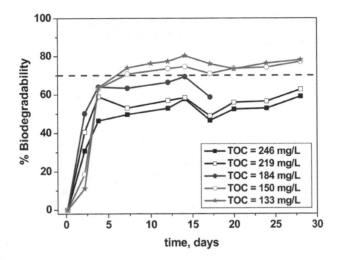

Figure 12.4.2 Biodegradability of several samples from intermediate stages of the photo-Fenton treatment of biorecalcitrant wastewater containing 550 mg/L of TOC. Samples at TOC > 250 mg/L are not shown as biodegradability below TOC = 254 mg/L.

a period of adaptation, higher organic loads could be fed to the bioreactor. This is discussed later, under full-sized plant results. From these results, it can be concluded that the best treatment option for this wastewater, containing around 550 mg/L of TOC was solar photo-Fenton until around 60% mineralisation followed by biotreatment.

The full-sized solar photo-Fenton reactor consists of a 3000-L buffer or recirculation tank and 100 m² solar collector field made up of three rows of compound parabolic collectors (CPCs) specially developed for photo-Fenton applications. The total system volume is 4000 L (1260 L of illuminated volume). The aerobic biological treatment part of the plant consists of three modules: a 5000 L neutralization tank, a 2000 L conditioner tank, and a 1000 L fixed-bed biofilm reactor colonized by activated sludge from the wastewater treatment plant installed at the pharmaceutical company itself. This biological system was operated directly in continuous mode. It was operated in batch mode only during the start-up phase (inoculation, bacteria fixation, and growing, etc.) and for the first experiments, such as the one shown in Figure 12.4.3. A detailed description can be found elsewhere (Oller et al., 2007). From the point of view of operation, the industrial saline wastewater partially oxidized by photo-Fenton is discharged into the neutralization tank where water is roughly neutralized with concentrated NaOH and iron is settled and removed when necessary. Then the photo pre-treated effluent is transferred to the conditioner tank, where the pH is automatically adjusted to between 6.8 and 7.5. Then the effluent is pumped to the bioreactor.

According to previous laboratory and pilot plant-scale biological tests performed (Figure 12.4.2), biodegradability enhancement of industrial wastewater by photo-Fenton is accomplished when TOC is reduced to approximately 40%. At that moment partially treated wastewater is transferred to the aerobic biological reactor. No mineral medium is usually added to the photo-Fenton pretreated effluent as the seawater

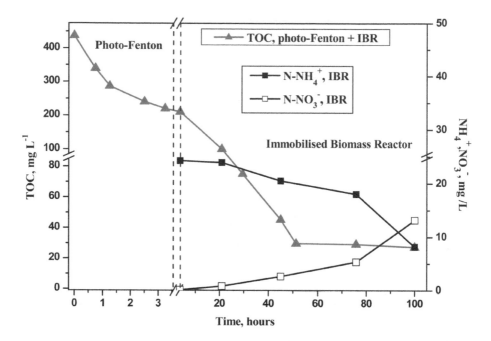

Figure 12.4.3 Degradation of wastewater containing biorecalcitrant compounds in a combined solar photo-Fenton and biotreatment.

matrix and the ammonium generated by the degradation of wastewater (approximately 4 mM), fulfill the C and N, P requirements (8–10 mg/L depending on wastewater composition). Figure 12.4.3 shows TOC and nitrogen concentrations in the photo-Fenton and IBR. It should be observed that the water was transferred from photo-Fenton to biotreatment at around 200 mg/L and that it fell to around 30 mg/L very quickly (48 hours) compared to the Z-W test, demonstrating that an adapted bioreactor produces more efficient results than the Z-W test carried out with fresh biomass. Nitrification was also complete. In this case the combined AOP/biological process strategy can be applied.

ACKNOWLEDGEMENTS

The authors wish to thank the Spanish Ministry of Science and Innovation for funding under the EDARSOL Project (Reference: CTQ2009-13459-C05-01).

REFERENCES

Ajona, J. A. and Vidal, A. (2000) The use of CPC collectors for detoxification of contaminated water: design, construction and preliminary results. *Solar Energy*, 68, 109–120.
Blanco, J., Malato, S., Fernández, P., Vidal, A., Morales, A., Trincado, P., de Oliveira, J.C., Minero, C., Musci, M., Casalle, C., Brunotte, M., Tratzky, S., Dischinger, N., Funken, K.-H.,

Sattler, C., Vincent, M., Collares-Pereira, M., Mendes, J.F. and Rangel, C.M. (2000) Compound parabolic concentrator technology development to commercial solar detoxification applications. *Solar Energy*, 67, 317–330.

Blanco, J. and Malato, S. (2003) *Solar Detoxification*, UNESCO Publishing, France.

Blanco-Galvez, J., Fernández-Ibáñez, P. and Malato-Rodríguez S. (2007) Solar photocatalytic detoxification and disinfection of water: Recent overview. *Journal of Solar Energy Engineering*, 129, 4–15.

Blanco J., Malato S., Fernández-Ibañez P., Alarcón D., Gernjak W. and Maldonado M.I. (2009) Review of feasible solar energy applications to water processes. *Renewable and Sustainable Energy Reviews*, 13, 1437–1445.

Colina-Márquez J., Machuca-Martínez F. and Li Puma G. (2010) Radiation adsorption and optimisation of solar photocatalytic reactors for environmental applications. *Environmental Science and Technology*, 44, 5112–5120.

Comninellis, C., Kapalka, A., Malato, S., Parsons, S.A., Poulios, I. and Mantzavinos, D. (2008) Advanced oxidation processes for water treatment: advances and trends for R&D. *Journal of Chemical Technology & Biotechnology*, 83, 769–776.

Dillert R., Cassano A. E., Goslich R. and Bahnemann D. (1999) Large scale studies in solar catalytic wastewater treatment. *Catalysis Today*, 54, 267–282.

Fenton, H.J.H. (1894) Oxidation of tartaric acid in presence of iron. *Journal of the Chemical Society*, 65, 899–910.

Gernjak W., Fuerhacker M., Fernández-Ibáñez P., Blanco J. and Malato S. (2006) Solar photo-Fenton treatment—Process parameters and process control. *Applied Catalysis B: Environmental*, 64, 121–130.

Goi D., Di Girogio G., Cimarosti I., Lesa B., Rossi G. and Dolcetti G. (2009) Treatment of landfill leachate by H2O2 promoted wet air oxidation: COD-AOX reduction, biodegradability enhancement and comparison with a fenton-type oxidation. *Chemical and Biochemical Engineering*, 2, 343–349.

Gómez-Ramos, M.D.M., Pérez-Parada, A., García-Reyes, J.F., Fernández-Alba, A.R. and Agüera, A. (2011) Use of an accurate-mass database for the systematic identification of transformation products of organic contaminants in wastewater effluents. *Journal of Chromatography A*, 1218, 8002–8012.

Gutiérrez, M., Etxebarria, J. and De Las Fuentes, L. (2002) Evaluation of wastewater toxicity: Comparative study between Microtox® and activated sludge oxygen uptake inhibition. *Water Research*, 36, 919–924.

Kavitha, V. and Palanivelu, K. (2004) The role of ferrous ion in Fenton and photo-Fenton processes for the degradation of phenol. *Chemosphere*, 55, 1235–1243.

Lagarde, F., Tusseau-Vuillemin, M-H., Lessard, P., Hèduit, A., Dutrop, F. and Mouchel, J.-M. (2003) Variability estimation of urban wastewater biodegradable fractions by respirometry. *Water Research*, 39, 4768–4778.

Lapertot, M. and Pulgarin, C. (2006) Biodegradability assessment of several priority hazardous substances: Choice, application and relevance regarding toxicity and bacterial activity. *Chemosphere*, 65, 682–690.

Malato S., Blanco J., Alarcón D.C., Maldonado M.I., Fernández-Ibáñez P. and Gernjak W. (2007) Photocatalytic decontamination and disinfection of water with solar collectors. *Catalysis Today*, 122, 137–149.

Malato, S., Fernández-Ibañez, P., Maldonado, M.I., Blanco, J. and Gernjak, W. (2009) Decontamination and disinfection of water by solar photocatalysis: Recent overview and trends. *Catalysis Today*, 147, 1–59.

Mandal, T., Maity, S., Dasgupta, D. and Datta, S. (2010) Advanced oxidation process and biotreatment: Their roles in combined industrial wastewater treatment. *Desalination* 250, 87–94.

Muñoz, R. and Guieysee, B. (2006) Algal-bacterial processes for the treatment of hazardous contaminants: A review. *Water Research*, 40, 2799–2815.

Oller I., Malato S., Sánchez-Pérez J.A., Maldonado M.I., Gernjak W., Pérez-Estrada L.A., Muñoz J.A., Ramos C. and Pulgarín C. (2007) Pre-industrial-scale combined solar photo-fenton and immobilised biomass activated-sludge bio-treatment. *Industrial & Engineering Chemistry Research*, 46, 7467–7475.

Oller I., Malato S. and Sánchez-Pérez J.A. (2011) Combination of advanced oxidation processes and biological treatments for wastewater decontamination-A review. *Science of the Total Environment*, 409, 4141–4166.

Pera-Titus, M., García-Molina, V., Baños, M.A., Giménez, J. and Esplugas, S. (2004) Degradation of chlorophenols by means of advanced oxidation processes: A general review. *Applied Catalysis B: Environmental*, 47, 219–256.

Pignatello, J.J., Oliveros, E. and MacKay, A. (2006) Advanced oxidation processes for organic contaminant destruction based on the fenton reaction and related chemistry. *Critical Reviews in Environmental Science and Technology*, 36, 1–84.

Rizzo, L. (2011) Bioassays as a tool for evaluating advanced oxidation processes in water and wastewater treatment. *Water Research*, 45, 4311–4340.

Scott J.P. and Ollis D.F. (1995) Integration of chemical and biological oxidation processes for water treatment: review and recommendations. *Environmental Progress*, 142, 88–103.

Suty, H., De Traversay, C. and Cost, M. (2004) Applications of advanced oxidation processes: present and future. *Water Science and Technology*, 49, 227–233.

Chapter 13

Solar driven advanced oxidation processes for water decontamination and disinfection

Erick R. Bandala[1] *& Brian W. Raichle*[2]

[1]*Energy and Environmental Research Group., Universidad de las Américas, Puebla. Sta., Catarina Mártir, Cholula 72820 Puebla, Mexico*
[2]*Department of Technology and Environmental Design, Appalachian State University, Katherine, Harper Hall, Boone, NC, USA*

13.1 INTRODUCTION

The industrial revolution of the late 18th century brought about new paradigms, including unprecedented and sustained population growth and a shift from a manual labor economy towards machine-based manufacturing. As a result, generations of industrial and domestic waste started to accumulate, resulting in new problems related to waste management and site contamination. As industrialization increases and populations rise, the amount of waste inevitably will surpass the natural capacity of ecosystems to self-purifiy. The continued production of waste during the last few decades has exceeded this capacity and has caused disorder, instability, harm, or discomfort to these ecosystems.

Efforts to mitigate the negative effects of waste from anthropogenic activities fall into two main categories. One category strives to decrease or eliminate waste generation through design and implementation of cleaner industrial processes. The second involves site restoration using novel state-of-art technologies able to remove waste with less impact on the surroundings. Many technological approaches for improving water, air and soil quality have been developed over the last few decades. With an increasing emphasis placed on sustainability, technological solutions are evaluated not only by their cost-effectiveness but also by their ability to withdraw pollutants from the environment without generating by-products and, preferably, by their use of renewable sources of energy. Among the different technological approaches developed, Advanced Oxidation Processes (AOPs) have recently emerged as very interesting alternatives for water treatment.

AOPs were initially defined as processes involving the generation of highly reactive oxidizing species able to attack and degrade organic substances (Bolton, 2001). Nowadays, AOPs are considered physical-chemical processes with high thermodynamic viability and the ability to produce deep changes in the chemical structure of contaminants as a result of the participation of free radicals in Redox reactions (Domenech et al., 2004). Free radicals, mainly hydroxyl radicals (HO), are of particular interest for environmental restoration because of their high oxidation capability. However, other studies have suggested that, besides hydroxyl radicals, AOPs can also generate other oxidizing species (Anipsitakis and Dionysiou, 2004). Generated radicals are able to oxidize organic pollutants mainly by hydrogen abstraction or by electrophylic addition to double bonds that generates organic free radicals ($R^\bullet$). These free radicals can react with oxygen molecules forming peroxy-radicals and initiate oxidative degradation chain reactions that may lead to the complete mineralization of the organics. AOP-generated free radicals involved in the degradation process may be produced by photochemical and non-photochemical procedures as has been widely reported previously (Quiroz et al., 2011).

In particular, photochemical AOPs have generated great interest in the last decade since these procedures have led to the use of renewable sources of energy to promote the chemical procedures involved. Solar radiation has been identified as a potential source for driving photochemical AOPs with interesting potential for real applications, specifically for water detoxification and disinfection (Orozco et al., 2008; Bandala et al., 2008a,b). The most studied technological approaches to water disinfection using solar radiation are homogeneous and heterogeneous photocatalysis. Both processes have been widely tested at the laboratory, bench and pilot-plant scale for water detoxification and disinfection with interesting results that will be discussed in later sections of this chapter.

13.2 SOLAR RADIATION COLLECTION FOR AOPs APPLICATIONS

Solar driven AOPs possess interesting advantages when compared with artificial light promoted AOPs reactions. Solar radiation availability, reduced cost, and simplicity are the most commonly cited. Nevertheless, use of solar radiation also has some challenges that must be faced being solar radiation collection probably the most significant. Despite the availability of free solar radiation almost everywhere around the world, its use requires an efficient and cost effective optical system that focuses and uniformly distributes the radiation of a surface. This optical system is known as a solar radiation collector.

The first reported attempts to collect solar radiation for the promotion of AOPs were in the early 90's at Sandia National Laboratories (USA). These early efforts used parabolic trough collectors, usually employed for solar thermal applications. The initial project objectives were not achieved at that time since the high concentration optical system required an expensive sun tracking system for optimal operation.

After initial use of high concentration solar systems for driving AOPs, interest shifted toward the use of low or non-concentrating, i.e. non imaging systems, since it was identified that non-tracking systems may be able to promote AOPs without the disadvantages of high concentrating, i.e. tracking, collectors. Since the first use of the non-tracking, low concentration solar collectors for AOP applications, a wide variety of different solar collection geometries have been tested. Some of the important geometries are shown in cross section in Figure 13.2.1. The main differences, advantages and disadvantages of these different geometries have been discussed in the past and the discussion continues today. Some of the main findings have been summarized by Bandala et al. (2004). These authors reported that reactors based on non-imaging collectors have attracted interest (Blanco et al., 1994) since these reactors share some of the advantages of tracking parabolic troughs and non-concentrating reactors (Malato et al., 1997), which has been confirmed by several studies comparing

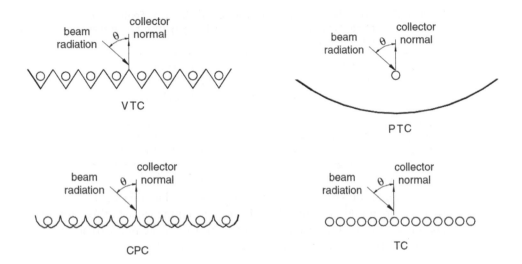

Figure 13.2.1 Schematic representation of different concentrating and non-concentrating solar collection geometries usually reported for photocatalytic applications.

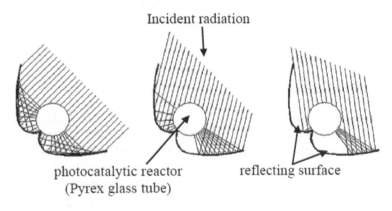

Figure 13.2.2 Ray tracing approach showing the capability of CPC geometry to concentrate diffuse solar radiation.

non-concentrating reactors with parabolic troughs (Curco et al., 1996; Malato et al., 1997; Gimenez et al., 1999).

Probably, the most studied non-imaging reactor for AOP applications is the compound parabolic collector (CPC). The most frequently quoted advantages of non-imaging CPC reactors are their ability to use global solar radiation (radiation coming from all directions in the sky, as shown in Figure 13.2.2), simplicity of operation, and ability to operate in the turbulent flow regime (which improves mass transfer). However, other non-imaging reactors offer these advantages in principle, to different degrees. There are many possibilities for the geometry of the reflectors that illuminate the tubes. Reflector geometries other than the CPC that have received attention in the

Table 13.3.1 Main chemical reactions involved in the dark Fenton and photo-Fenton processes; k values in every case represents rate constant.

(1) $Fe^{2+} + O_2 \rightarrow O_2^{\bullet-} + Fe^{3+}$	$k = 1.15\,M^{-1}\,s^{-1}$
(2) $H_2O_2 \leftrightarrow HO_2^- + H^+$	$k = 1.26 \times 10^{-2}\,M^{-1}\,s^{-1}$
(3) $Fe_{2+} + H_2O_2 \leftarrow Fe^{3+} + HO^{\bullet} + HO^-$	$k = 55\,M^{-1}\,s^{-1}$
(4) $Fe(OH)^{2+} + hv \rightarrow Fe^{2+} + HO^{\bullet}$	$k = 2 \times 10^{-3}\,M^{-1}\,s^{-1}$
(5) $H_2O_2 + hv \rightarrow 2HO^{\bullet}$	$\phi_{HO.} = 0, 98, 254\,nm$
(6) $HO^{\bullet} + H_2O_2 \rightarrow HO_2^{\bullet} + H_2O$	$k = 2.7 \times 10^7\,M^{-1}\,s^{-1}$
(7) $Fe_{2+} + HO^{\bullet} \rightarrow Fe_{3+} + HO^-$	$k = 2.7 \times 10^7\,M^{-1}\,s^{-1}$

past include V-trough and L-shaped collectors, as shown in Figure 13.2.1. The performance of other non-imaging geometries have been found quite close to those found for CPCs, due mainly to a more uniform photon distribution in the photorreactor rather than radiation intensity reaching the receiver (Brito et al., 2012).

13.3 SOLAR HOMOGENOUS PHOTOCATALYSIS

The Fenton process is one of the most widely used AOPs for water and wastewater treatment. Fenton's reaction uses hydrogen peroxide and ferrous salt to generate hydroxyl radicals (HO), chemical species possessing inherent properties that enable them to mineralize dissolved organic pollutants into CO_2, water, and mineral acids (Bandala et al., 2007). When this process is driven by ultraviolet (UV) radiation, visible light, or both it is known as the photo-Fenton process. The photo-Fenton process possesses several advantages over the dark Fenton reaction, mainly an increased reaction rate and the possibility of using a cheap, clean, and widely distributed energy source: solar radiation. Table 13.3.1 depicts the main chemical reactions involved in both dark Fenton and photo-Fenton processes along with reaction rate constants for each reaction.

In agreement with the information in Table 13.3.1, it is worthy to note Equations 13.3.3 and 13.3.4 since they are the main processes involved in the solar photo-Fenton reaction. Equation 13.3.3 shows the actual decomposition of hydrogen peroxide catalyzed by ferrous salt. Hydrogen peroxide decomposition catalyzed by Fe^{2+} generates one hydroxyl ion and one hydroxyl radical. Hydroxyl radicals are the most important specie generated during the reaction since it may react with organic matter to oxidize it. In the case of dark Fenton reactions, ferrous salt may be considered the limiting reagent anytime the reaction described in Equation 13.3.3 stops once all the available ferrous iron has been oxidized to ferric iron independent of the amount of hydrogen peroxide added to the reaction. The advantage of including radiative energy in the process is shown in Equation 13.3.4. In the case of photo-assisted Fenton reactions, radiation of a certain wavelength (usually UV and part of the visible radiation) may generate the so-called photo-reduction of Fe^{3+} to Fe^{2+} depicted in Equation 13.3.4 plus one mole of hydroxyl radicals. Photo-reduced iron can participate again in the decomposition of hydrogen peroxide as in Equation 13.3.3, and the process may keep occurring if enough hydrogen peroxide is provided to the reaction mixture. As stated earlier, the useful wavelength range for carrying out the catalytic

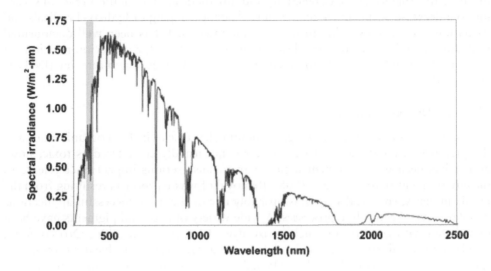

Figure 13.3.1 The solar spectrum.

process (300–500 nm) may be provided by solar radiation since it has been widely reported that the sun is an important light source for driving photo-Fenton process in the wavelength above 300 nm as shown in Figure 13.3.1.

13.3.1 Degradation of organic pollutants by solar driven photo-Fenton processes

Several different types of organic pollutants have been tested for the application of solar driven photo-Fenton processes including dyes and textile wastewater effluents, surfactants and algal toxins, among many others.

The following are some of the most widely reported organic pollutants tested for solar driven photo assisted Fenton and Fenton-like processes.

13.3.1.1 Pesticide degradation

Solar driven photo assisted Fenton (SDPAF) and Fenton-like processes have been tested for pesticide degradation. Recently, Quiroz et al., (2011) published a complete review on the application of advanced oxidation processes for pesticide removal in aqueous media, including interesting tabulated data on references related with Fenton and Fenton-like processes. Specifically, triazinic pesticides have been successfully removed from water by using solar driven Fenton-like processes (Bandala et al., 2007; Perez et al., 2006) as well as dimethylurea pesticides (Farre et al., 2007; Perez et al., 2006); organophosphorous (Farre et al., 2007; Hincapié et al., 2005); chloracetic acid (Bandala et al., 2007a); and phenol and phenolic derivatives (Farre et al., 2007). An interesting effect that should be taken into account when applying homogeneous photocatalysis for pesticide degradation is the presence of salt counterions. Inorganic anions (Cl^-, SO_4^{2-}, HPO_4^{2-}) present in the water or added as reagents have a significant effect on the reaction rates of Fenton processes such as complexation with Fe^{2+}

or Fe^{3+}, affecting iron species reactivity and distribution; precipitation reactions leading to a decrease of the active dissolved Fe^{3+}; or scavenging of hydroxyl radicals and oxidation reactions involving these inorganic radicals. It has been well documented that, for example, chloride ions show an inhibitory effect for oxidation reactions of phenols (Tang and Huang, 1996), dichlorvos (Lu et al., 1997), and atrazine (De Laat et al., 2004).

13.3.1.2 Dye degradation

Textile wastewater is considered highly polluting because of its high organic loads and the presence of color. Colored wastewater is not usually considered as toxic, however it has been well documented that it may cause serious impact once release to the environment (Orozco et al., 2008). Removal of waste products resulting from the textile industry, specifically dyes, is probably one of the most successful applications of solar driven Fenton-like processes. A wide variety of dyes and pigments have been remediated with good results such as azo-dyes (Chacón et al., 2006; Orozco et al., 2008) and benzidine-based dyes (Bandala et al., 2008c), as well as dye mixtures in real textile wastewater (Bandala et al., 2008c). As with pesticide degradation, several variables affecting dye degradation have being identified related with iron salt counterions, reaction pH, and reagent concentration, among many others (Orozco et al., 2008).

13.3.1.3 Surfactants degradation

Surfactants (surface active agents) are molecules which include in their chemical structure a hydrophilic head and a hydrophobic tail. This structure allows surfactants to increase the aqueous solubility of hydrophobic compounds by solubilization. Surfactants are frequently used in soap and as active ingredients in detergent formulations like shampoos and dishwashing liquids. They play an important role in the paper, food, polymers, cosmetics, food, pharmaceuticals, and oil recovery industries (Bandala et al., 2008a). Besides their environmentally undesirable characteristics, surfactants produce aesthetic effects after being released into the natural water courses, inhibit gas transference between the water and the atmosphere, and consume dissolved oxygen. Removal of up to 99% of surfactants from wastewater has been also successfully carried out using SDPAF and Fenton-like processes in short reaction time (Bandala et al., 2008b; Lin et al., 1999; Amat et al., 2004). This process has been applied to the treatment of real wastewater from the surfactant enhanced soil washing (SESW) process commonly applied to the restoration of oil polluted sites (Bandala et al., 2008a). Results obtained in this case were also very interesting since the SDPAF and Fenton-like processes tested were capable not only to completely remove the surfactant but also eliminate 69% of the TPHs and other pollutants in the influent, measured as chemical oxygen demand (COD). The resulting effluent has a high potential to be further treated with conventional, i.e. biological, wastewater treatments.

13.3.2 Microorganisms inactivation by solar driven photo-Fenton processes

Despite several different examples of the application of SDPAF for organic pollutant remediation, relatively little work have been done on its application for inactivation of microorganism in water. The removal of pathogenic species from water is important due to their ability to generate immediate adverse health effects on the population forced to consume non-safe water. Just to have an idea of the problem's magnitude, it has been estimated that in Africa, Latin America and the Caribbean alone, nearly one billion people no not have access to safe water supplies (WHO/UNICEF, 2000). As a result of this situation, waterborne diseases result in the death of 1.5 million children every year (Montgomery and Elimelech, 2007). Besides health concerns, lack of access to safe drinking water is also associated with poverty and limits sustainable development (Bandala et al., 2011a).

13.3.2.1 Bacteria

Solar water disinfection (SODIS) has been identified in the past decade as a simple, environmentally friendly, and low cost point-of-use treatment technology for drinking water purification capable to use the bacteriostatic effect of the UV-A part of the solar spectrum (320–400 nm) and dissolved oxygen in water to inactivate pathogenic species through the production of reactive forms of oxygen (ROS). According to previous systematic studies, the best bacteria inactivation effect is reached on sunny days when heat and UV radiation combine synergistically (EWAG, 2002; Schmid et al., 2008).

Application of advanced oxidation processes to water disinfection using solar radiation, coined as enhanced photocatalytic solar disinfection (ENPHOSODIS) by Bandala et al. (2011a), has improved SODIS disinfection efficiency, solved some disadvantages identified previously with SODIS, and allows the efficient inactivation of not only common waterborne bacteria but other highly resistant microorganisms (Guisar et al., 2007). In this regard, SDPAF processes have been tested for the inactivation of bacterial strains commonly infecting water such as *Escherichia coli* and *Pseudomonas aeruginosa* with very satisfactory results (Bandala et al., 2011a,c).

13.3.2.2 Helminth egg

Species with the ability to resist adverse conditions may survive conventional disinfection processes even after long periods of treatment. One example of these kinds of undesirable species is helminth eggs. The WHO has estimated that about 1 billion people in developing countries are infected by Ascaris and that 25–33% of this population is affected by helminthiasis alone. These diseases are importantly related to poor physical growth and development, as well as retardation of intellectual and cognitive development in children less than 15 years of age (Bandala et al., 2012).

Inactivation of helminth eggs in water is not an easy task mainly because their basic structure makes them resistant to external agents. Conventional sanitary engineering processes have been applied to the removal of helminth eggs from wastewater. For these cases, 80–100% helminth egg removal within 20–35 h of treatment has been achieved. Presence of these microorganisms, however, is not exclusive to wastewater. Several studies have reported the presence of helminth eggs in surface and even ground water and it is well documented that they possess high resistance to disinfection, resisting conventional drinking water treatment and emerging live from domestic taps (Bandala et al., 2012).

13.3.2.3 Spores as indicators

Helminth eggs are not the only microorganisms with the capability to resist conventional water disinfection processes. Other well-known pathogenic species with such high resistance are *Cryptosporidium parvum* oocyst and *Giardia lamblia* cysts, among many others. Resistance of *C. parvum* oocysts to chlorine disinfection has been extensively documented. Chlorine concentrations of 4 mg/L, the highest residual allowed by U.S. regulations, require more than 15 hours of contact time to inactivate 99% of *C. parvum* oocysts at pH 6.0 and 20°C. The chlorine dosage necessary for treatment increases dramatically with increasing pH and decreasing temperature (Guisar et al., 2007).

Having in mind all these considerations, it is clear that using *E. coli* as pathogenic indicator of drinking water disinfection is, by far, not enough. In the search for a more reliable indicator, *Bacillus* spores have emerged as an interesting alternative. It is known that spores of Bacillus spp., are highly resistant to inactivation by different physical stress conditions such as toxic chemicals or biocidal agents, desiccation, extreme pressure and temperature, as well as exposure to high doses of UV or ionizing radiation. Spores of *Bacillus subtilis* are commonly used test organisms for inactivation studies due to their high degree of resistance to various sporicidal treatments, reproducible inactivation response, safety (*B. subtilis* is not pathogenic to humans) and resistance to UV radiation, and have become a useful conservative index for water disinfection anytime it is considered that, once *B. subtilis* spores has been inactivated, anything else with less resistance will surely be inactivated (Bandala et al., 2011b).

SDPAF processes have been demonstrated as highly efficient for *Bacillus subtilis* spore inactivation in water as a surrogate microorganism for *C. parvum* oocysts. Using relatively low Fenton reagent concentrations and low solar radiation intensity (about 1 sun), Guisar et al., (2007) obtained up to 96% spore viability reduction in only 1.5 hours of exposure to solar radiation. Bandala et al. (2011a) used heminth eggs to demonstrate that the use of highly resistant microorganisms as a conservative index for water disinfection is desirable and that an adequate solar radiation dose is required to ensure the final required water quality. In their work, these authors used *Ascaris suum* eggs, very similar to *A. lumbricoides,* the actual specie infecting humans, to assess the amount of radiation necessary to inactivate > 5-log (99.999% removal) of helminth eggs. They found that approximately $140 \, kJ \, L^{-1}$ was required to achieve this task. When they tested the same experimental conditions (Fe^{2+} and H_2O_2 concentration) for the inactivation of *E. coli* and *P. aeruginosa*, they found that less than $10 \, kJ \, L^{-1}$ were required to reach up to > 6-log inactivation (99.9999% removal) of both bacteria. Finally, they found no significant increase in the inactivation dose required when up to $5 \, mg \, L^{-1}$ natural organic matter (NOM) was added to the bacterial suspension.

In relation to *Bacillus subtilis* spore inactivation, several different Fenton reagent concentrations were tested in combination with solar UV-A radiation ($\lambda_{max} = 365$ nm) for spore inactivation. The best spore inactivation conditions were found to be $[Fe^{2+}] = 2.5$ mM and $[H_2O_2] = 100$ mM and radiation. Under these conditions, over 9-log inactivation was reached after only 20 min of reaction. The effect of ionic strength and natural organic matter (NOM) on spore inactivation kinetics was also tested. In both cases, an important decrease of the inactivation rate, fitted to the delayed Chick-Watson inactivation kinetics, was observed (Bandala et al., 2011b).

13.3.2.4 Sequential processes

In other recent work (Bandala et al., 2012), the same research group reported the use of sequential coupled disinfection processes, in this case SDPAF processes followed by chlorine. In their work they suggested that the high efficiency of ozone–chlorine sequential disinfection previously reported could be generalized to different reactive oxygen species (i.e. hydroxyl radicals) which could synergistically enhance the oxidative properties of chlorine, thus improving the inactivation process. If true, it means that other methods to produce hydroxyl radicals might produce a similar synergistic effect in sequential processes. They assessed the photo-assisted Fenton process alone under different H_2O_2 and Fe^{2+} concentrations to test its ability to inactivate *Ascaris suum* eggs. The effect of free chlorine alone was also tested. Using the best reaction conditions, free chlorine only treatment achieved 83% egg inactivation after 120 min of reaction time, while the sequential photo-assisted Fenton process plus chlorine treatment achieved over 99% of egg inactivation after $120 \, kJ \, L^{-1}$ (about one hour of solar radiation). No effect on helminth eggs inactivation was observed with free chlorine alone after $550 \, mg \, min \, L^{-1}$, whereas egg inactivation in the range of 25–30% was obtained for sequential processes (ENPHOSODIS then chlorine) using only $150 \, mg \, min \, L^{-1}$.

13.4 SOLAR HETEROGENOUS PHOTOCATALYSIS

Heterogeneous photocatalytic degradation is usually related with the use of a stable, solid semiconductor capable of stimulating, under the effect of irradiation, a redox reaction at the solid/solution interface, while remaining unchanged after many turnovers of the redox system. When the semiconductor is in contact with a liquid electrolyte solution containing a redox couple, charge transfer occurs across the interface to balance the potentials of the two phases. An electric field is formed at the surface of the semiconductor, bending its energetic bands from the bulk of the semiconductor toward the interface. During the absorption of a photon with appropriate energy (photo excitation) by the semiconductor, band bending provides the conditions for carrier separation. The two generated charge carriers should react at the semiconductor/electrolyte interface with the species in solution and, under steady state conditions; the amount of charge transferred to the electrolyte must be equal and opposite for the two types of carriers. When the charge carriers, usually denominated electron/hole pairs, are generated in the semiconductor the electron moves away from the surface to the bulk of the semiconductor as the hole migrates towards the surface.

Metal oxides are common semiconductor materials suitable for photocatalytic purposes. Table 13.4.1 lists some selected semiconductor materials, which have been used for photocatalytic reactions in the past, along with their band gap energy and the wavelength range required to activate the catalysts. Among all these possibilities, TiO_2 is a widely analyzed, low-cost, nontoxic, stable, highly photoreactive, and chemically and biologically inert photocatalyst. Solar driven heterogeneous photocatalysis (SDHP) using titanium dioxide is the AOP most widely used as an alternative to conventional water decontamination and disinfection as well as air remediation technologies (Castillo et al., 2011).

Table 13.4.1 Optical properties for some photocatalytic semiconducting materials.

Semiconductor	Band gap (eV)	Band gap equivalent wavelength (nm)
BaTiO$_3$	3.3	375
CdO	2.1	590
CdS	2.5	497
CdSe	1.7	730
Fe$_2$O$_3$	2.2	565
GaS	1.4	887
GaP	2.3	540
SnO$_2$	3.9	318
SrTiO$_3$	3.4	365
TiO$_2$	3.0	390
WO$_3$	2.8	443
ZnO	3.2	390
ZnS	3.7	336

TiO$_2$ has proven to be highly effective in the nonselective decomposition of organic molecules and inactivation of microorganisms due to high decomposition and mineralization rates when used in combination with specific radiation sources (see Table 13.4.1). Furthermore, water treatment with SDHP neither requires the addition of other chemicals reactants nor generates hazardous waste by-products.

The basic reactions occurring within the photocatalyst particles after absorption of the proper wavelength radiation are as follows:

$$TiO_2 \xrightarrow{h\nu} e^- + h^+ + TiO_2 \tag{13.4.1}$$

$$e^- + h^+ + TiO_2 \rightarrow TiO_2 + h\nu' \tag{13.4.2}$$

$$(TiO^{IV} - O_{2-} - Ti^{IV}) - OH + h_{BV}^+ \rightarrow (TiO^{IV} - O_{2-} - Ti^{IV}) - OH + H^+$$

$$O_{2(ads)} + e_{BC}^- \rightarrow O_{2(ads)}^- \tag{13.4.3}$$

As shown in Equations 13.4.1 to 13.4.4, several oxidizing species may be generated as a result of the interaction between the photocatalyst and the radiation, although evidence supports the idea that the hydroxyl radical ($^{\cdot}OH$) is the main oxidizing specie responsible for the photo-oxidation of most organic compounds studied. Once the absorption of one photon with the required energy (ultraviolet radiation, $\lambda < 390$ nm) occurs, the first step is the generation of electron/hole pairs, which are separated between the conduction and valence bands (Equation 13.4.1). Recombination of the generated charge carriers (Equation 13.4.2) may occur. However, if the dissolvent is redox active (i.e. water) it may acts as a donor and acceptor of electrons avoiding recombination and improving the photonic efficiency, producing that on a hydrated and hydroxylated TiO$_2$ surface, the holes trap $^{\cdot}OH$ radicals linked to the surface and avoid recombination reactions (Equation 13.4.3). It should be emphasized that even trapped electrons and holes can rapidly recombine on the surface of a particle (Equation 13.4.2). The recombination process can be partially avoided through the capture of the electron by pre-adsorbed molecular oxygen, forming a superoxide radical (Equation 13.4.4).

13.4.1   Degradation of organic pollutants by solar driven heterogeneous photocatalysis

Degradation of organic pollutants using photocatalytic processes is probably the most investigated AOP over the last three decades. It has been widely tested for the degradation of mono aromatics (i.e. benzene, dimethoxybenzenes, halobenzenes, nitrobenzene, chlorophenols, nitrophenols, benzamide, and aniline) and, consequently, these pollutants appear as model compounds in a wide variety of scientific papers. In addition to these, several other types of molecules have been investigated as substrates for photocatalytic degradation. Some of the most frequently reported are water-miscible solvents (i.e. ethanol, alkoxyethanol), haloaliphatics (i.e. trichloroethylene, tetrachloromethane), pesticides and surfactants, among many others.

Despite the wide variety of reports on the application of heterogeneous photocatalytic processes for water and air treatment, relatively few reports are available dealing with the application of solar radiation to drive these processes. Solar driven heterogeneous photocatalytic processes (SDHPC) have been, however, applied to the treatment of organic contaminants. Some of the most important applications are described below.

13.4.1.1   Pesticide degradation

Strongly colored compounds can be removed from surface or wastewater by conventional water treatment processes (i.e. adsorption), however these phase change processes have to be discouraged anytime they result in solid matrix waste. It is highly desirable to degrade dyes by oxidation to avoid the risk of contamination. Solar driven heterogeneous photocatalytic processes have been extensively applied for pesticide degradation. Quiroz et al. (2011) have recently published an interesting review about the different applications of SDHPC for the removal of these compounds. To mention some examples, chlorinated insecticides (Bandala et al., 2002; Hincapié et al., 2005); triazine derivatives (Parra et al., 2004); carbamate derivatives (Arancibia et al., 2002); haloacetic acid derivatives (Terashima et al., 2006) and organophosphorus (Pichat et al., 2007), among many others, have been successfully removed from water (Bandala and Torres, 2008d).

13.4.1.2   Dye degradation

As mentioned early for homogeneous photocatalytic processes, SDHPC degradation of dyes and pigments has been extensively tested. The most recent published works dealing with SDHPC dye degradation focus mainly on azo dyes (Sajjad et al., 2010; Chung and Chen, 2009; Malato et al., 2009). However, a few reports dealing with SDHPC removal of other dye types have been published.

13.4.1.3 Surfactant degradation

Use of SDHPC for surfactant removal in aqueous media has been also widely reported for many different surfactant structures. Among the most recently reported are non-ionic (Du et al., 2008; Eng et al., 2010) and cationic (Han et al., 2009; Naldoni et al., 2009; Natoli et al., 2012) surfactants; relatively few reports are available dealing with anionic surfactants.

13.4.2 Microorganisms inactivation by solar driven heterogeneous photocatalysis

TiO_2 semiconductor photocatalyst is widely reported, as suspended powder or thin film, to inactivate different organisms such as viruses, vegetative cells, and spores of organisms with a high resistance to desiccation and radiation (i.e. *Escherichia coli, Lactobacillus acidophilus, Saccharomyces cerevisiae, Bacillus atrophaeus, Aspergillus niger,* and *Kocuria rhizophila*) with very interesting results (Castillo et al., 2011; Muranyi et al., 2009). Some other specific strains have been tested for SDHPC disinfection, for example Fernandez-Ibañez et al. (2008) tested the inactivation of *Fusarium solani* and *Fusarium sp* spores, a pathogen infecting food crops using slurry TiO_2 and a solar CPC photorreactor. They found that these pathogens are susceptible to solar photocatalytic disinfection with TiO_2 in distilled water not only at the laboratory scale but also at the pilot plant scale (Fernandez-Ibañez et al., 2008; Polo-Lopez et al., 2010). The effect of some water quality parameters on disinfection processes has been estimated at pilot plant scale and some authors have proposed that natural organic matter (NOM) and hardness may inhibit the SDHPC process due to the hydroxyl radical scavenging effect of NOM (Bandala et al., 2011b), the formation of calcium carbonate film adhering to the internal glass wall of the photoreactor which is in contact with the liquid being treated, and to the presence of calcium carbonate precipitates on catalyst surface (Acevedo et al., 2012). In a very recent work, Byrne et al. (2011) reviewed the available literature of SDHPC enhancement for solar disinfection of water, including an analysis of parameters affecting the process and a comparison with the widely known solar disinfection (SODIS) technology.

13.5 CHALLENGES AND PERSPECTIVES

Solar driven water disinfection technologies hold great promise as low cost, effective replacements for conventional water treatments, potentially bringing clean drinking water to a large number of people. However, before widespread adoption is realized, several significant technological hurdles must be overcome.

13.5.1 Photorreactor design

UV at ground level includes both direct beam and diffuse radiation at almost similar portions. During cloudy days, this proportion in solar UVA spectrum may change to 60% diffuse and 40% direct. Since direct beam radiation may represent as little as 40% of the total radiation available, non-imaging solar collecting systems capable to use global radiation arriving from any direction have a clear advantage, as stated earlier, compared to expensive imaging/concentrating optic-based systems which cannot harvest diffuse radiation.

The major advantage with non-imaging collectors is that the collection factor remains constant for all values of sun zenith angle within the acceptance angle limit. Therefore, many different non-imaging geometries have been used in the design of larger-scale solar disinfection systems, including up to small community scale (Bandala and Estrada, 2007). A solar collecting system for use in, for example, rural, isolated communities in developing countries should have many specific attributes: high illuminated volume/total volume ratio; low flow operation to maximize residence time; serve as a UVA dosimetric indicator (considering that both, photocatalytic detox and disinfection are dose dependent); high UVA reflectivity; high (90%) UVA transmission in the receiver; robust to harsh environmental conditions; minimal lifecycle cost; low environmental impact; low maintenance requirements and ease of access to replacement parts; and minimal external power requirement. In addition, the photoreactor design must also provide electron acceptors, typically dissolved oxygen, since the concentration of dissolved oxygen will be rapidly consumed in the initial stages of the reaction as water temperature increases in static batch systems.

As said earlier, several technological approaches have been reported that strive to offer the characteristics described previously. However, more research is required to demonstrate satisfactory full scale performance before widespread deployment of this technology for point-of-use water or industrial wastewater treatment.

13.5.2 Suspended vs. immobilized photocatalyst

One of the main disadvantages for heterogeneous photocatalytic processes in water treatment is the generation of catalyst slurries. In agreement with previous studies of pilot plant scale systems, suspended TiO_2 is more effective than immobilized TiO_2, probably due to limitations imposed by mass transfer processes on the latter's reaction rate (Bandala and Torres, 2008). Immobilized TiO_2, however, present some important advantages when compared with suspensions, such as reduced material loss, cost reductions, and the possibility to escape further recovery steps after the water treatment process. Many materials have been tested for the immobilization of TiO_2 as well as a wide variety of immobilization methodologies (Gelover et al., 2006). However, the controversy remains over the performance of immobilized TiO_2 photocatalyst in comparison with suspended TiO_2, since some authors have found no advantages at all for the use of immobilized systems (Sordo et al., 2010) whereas others claims important improvements when using immobilized applications (Gelover et al., 2006; Acevedo et al., 2012). It is probable that this controversy is related to the lack of a proper comparison methodology rather than a real difference between the tested materials. Several chemical techniques for TiO_2 film deposition on solid surfaces have been described in the past: chemical vapor deposition (CVD), serigraphy, galvanoplasty, anodization, electrophoresis, electroless deposition, spray pyrolysis, controlled precipitation or chemical deposition, sol-gel chemical deposition, and magnetron sputtering. In the same way, a wide variety of solid matrix materials for TiO_2 immobilization are reported. It is reasonable to speculate that the variation in results is at least in part due to the range of experimental conditions. Development of a standardized testing protocol must be considered another interesting research challenge to the field deployment of solar driven photocatalytic processes.

13.5.3 Visible light active photocatalyst materials

As a result of its large band gap (3.2 eV), TiO_2 semiconductors exhibits photocatalytic activity only within UV radiation wavelengths (≤ 400 nm). This specific characteristic limits the photosensitivity to the UV part of the solar spectrum and is an important technological limitation. Since sunlight consists of only about 5% UV radiation, efficiency enhancements are needed to enhance the viability of SDHPC processes. A recent emerging field of research is the development of photocatalytic materials excitable by visible solar radiation, which account for 45% of the solar spectrum. Several modifications to TiO_2 have produced visible radiation active materials with improved photosensitivity and quantum yield. Several ways to get this so-called daylight photocatalysis have been reported recently including dye sensitization, coupling TiO_2 with other semiconductors possessing favorable band gaps and potentials, surface deposition of metal clusters, and doping the crystal lattice with metals (Fe, Co, Ag) and/or nonmetal foreign atoms (N, C, F, S).

According to the literature, one of the more promising approaches to achieve visible light active TiO_2 is by doping with nonmetal elements including N and S. After initial reports of visible-light photo-active nitrogen-doped TiO_2, many groups have demonstrated that anion-doped TiO_2 has extended optical absorbance into the visible region (Asahi et al., 2001). However, the number of publications concerning the photocatalytic activity of these materials for SDPC processes is still limited. In many cases, the UV activity of undoped TiO_2 has been reported much greater than the visible light activity of the doped material (Reginfo-Herrera and Pulgarin, 2010). Therefore, photocatalysts developed for SDHOC applications should be tested under simulated solar irradiation or, preferably, under real sun conditions. However, in agreement with these authors, N-doped TiO_2 materials did not exhibit enhanced photocatalytic degradation of phenol or the photocatalytic inactivation of *E. coli* under simulated solar light, as compared to Degussa P25. They suggest that although N, or N-S co-doped TiO_2 may show visible light response, the localized states responsible for visible light absorption are not important in the photocatalytic activity. Other studies, however, report that although solar visible radiation displays a lower activity than solar UV radiation it is possible to observe interesting photocatalytic activity for N-doped TiO_2 under these conditions and the overall effect of using complete (UV+visible) radiation is higher than that observed for regular TiO_2 using only UV solar radiation (Castillo et al., 2011). More research is required to determine if visible light active materials can deliver an increased efficiency of photocatalytic processes under solar radiation.

13.6 CONCLUSIONS

Solar driven AOPs were demonstrated as cost-effective emerging methodologies for water decontamination and disinfection with very interesting advantages compared with conventional water and wastewater treatment processes such as higher efficiency, lower cost, easy operation and maintenance but mainly the opportunity of using a wide available, cheap and interesting alternative source of energy: the Sun.

Several different applications for both, homogeneous and heterogeneous photo-catalytic processes, solar driven AOPs have been reported in the past. Some of them are included in this work as a representative example for applications at laboratory, bench and full scale of these technologies in the removal of organic pollutants or pathogenic microorganisms.

An interesting approach analyzed is related with the use of such technologies coupled sequentially with conventional water and wastewater treatment procedures or even with other advanced oxidation technologies. As shown, preliminary results suggest that application of sequential processes, AOPs + conventional, is useful for the improvement of the performance of conventional water treatment processes, decreasing costs and generating confidence on the application of non-conventional technologies.

The analysis carried out, including main highlights in novel solar collection optical approaches, use of slurry or immobilized photocatalysts, application of doped materials with radiation absorbance shifted to higher wavelength, have shown that most research is necessary in order to generate the proper application of such technologies in the improvement of the environment.

REFERENCES

Acevedo, A., Carpio, E.A. and Rodriguez, J. (2012) Disinfection of natural water by solar photocatalysis using immobilized TiO_2 devices: Efficiency in eliminating indicator bacteria and operating life of the system. *Journal of Solar Energy Engineering*, 134, 1–10.

Amat, A., Arques, A., Miranda, M.A. and Segui, S. (2004) Photo-Fenton reaction for the abatement of commercial surfactants in a solar pilot plant. *Solar Energy*, 77, 559–566.

Asahi, R., Morikawa, T., Ohwaki, T., Aoki, K. and Taga, Y. (2001) Visible-light photocatalysis in nitrogen-doped titanium oxides. *Science*, 293, 269–271.

Arancibia, C., Bandala, E.R. and Estrada, C.A. (2002) Radiation absorption and rate constants for carbaryl photocatalytic degradation in a solar collector. *Catalysis Today*, 76, 149–159.

Bandala, E.R., Gelover, S., Leal, M.T., Arancibia, C., Jiménez, A. and Estrada, C.A. (2002) Solar photocatalytic degradation of Aldrín. *Catalysis Today*, 76, 189–199.

Bandala, E.R., Arancibia, C.A., Orozco, S.L. and Estrada, C.A. (2004) Solar photoreactors comparison based on oxalic acid photocatalytic degradation. *Solar Energy*, 77, 503–512.

Bandala, E.R., Pelaez, M.A., García, J.A., Dionysiou, D.D., Gelover, S. and Macías, D. (2007a) Degradation of 2,4-dichlorophenoxyacetic acid (2,4-D) using cobalt-peroximonosulfate in Fenton-like process. *Journal of Photochemistry and Photobiology A: Chemistry*, 186, 357–363.

Bandala, E.R., Domínguez, Z., Rivas, F. and Gelover, S. (2007b) Degradation of atrazine using solar driven Fenton-like advanced oxidation technologies. *Journal of Environmental Science and Health B*, 42, 21–26.

Bandala, E.R., Velasco, Y. and Torres, L.G. (2008a) Decontamination of soil washing wastewater using solar driven advanced oxidation processes. *Journal of Hazardous Materials*, 160, 402–407.

Bandala, E.R., Peláez, M.A., Salgado, M.J. and Torres, L.G. (2008b) Decontamination of sodium dodecyl sulfonate using solar driven Fenton like Advanced Oxidation Processes. *Journal of Hazardous Materials*, 151, 578–584.

Bandala, E.R., Peláez, M.A., García-López, A.J., Salgado, M.J. and Moeller, G. (2008c) Photocatalytic decolourization of synthetic and real textile wastewater containing benzidine-based azo dyes. *Chemical Engineering and Processing*, 47, 169–176.

Bandala, E.R. and Torres, L.G. (2008) Pesticide removal from water using advanced oxidation technologies: Challenges and Perspectives. In: *Pesticide research trends*, Ed. A.B. Tennefy. Nova Publishers Press. New York, USA.

Bandala, E.R., Brito, L. and Pelaez, M.A. (2009) Degradation of domoic acid toxin by UV promoted Fenton-like process. *Desalination*, 245, 135–145.

Bandala, E.R., Gonzalez, L., De la Hoz, F., Pelaez, M., Dionysiou, D.D., Dunlop, P.S.M., Byrne, J.A. and Sanchez, J.L. (2011a) Application of azo dyes as dosimetric indicators for enhanced photocatalytic solar disinfection (ENPOSODIS). *Journal of Photochemistry and Photobiology A: Chemistry*, 218, 185–191.

Bandala, E.R., Perez, R., Velez-Lee, A.E., Sanchez-Salas, J.L., Quiroz, M.A. and Mendez-Rojas, M.A. (2011b) *Bacillus subtilis* spore inactivation in water using photo-assisted Fenton reactions. *Sustainable Environmental Research*, 21, 285–290.

Bandala, E.R., Castillo-Ledezma, J.H., González, L. and Sánchez-Salas, J.L. (2011c) Solar driven advanced oxidation processes for inactivation of pathogenic microorganisms in water. *Recent Research Developments in Photochemistry and Photobiology*, 8, 1–16.

Bandala, E.R., Gonzalez, L., Sanchez-Salas, J.L. and Castillo-Ledezma, J.H. (2012) Inactivation of Ascaris eggs in water using sequential solar driven photo-Fenton and free chlorine. *Journal of Water and Health*, 10, 20–30.

Blanco, J., Malato, S., Bahnemann, D., Bockelmann, D., Weichgrebe, D. and Goslich, R. (1994) Effective industrial waste water treatment by solar photocatalysis; application to fine chemicals spanish company. In: *Proceedings of the 7th International Symposium on Solar Thermal Concentrating Technologies*, September 26–30, 1994, Moscow, Russia.

Brito, J.A., Bandala, E.R. and Raichle, B. (2012) Modeling optimization of fixed arbitrary concentrators. In: *Proceedings of the Word Renewable Energy Forum*. May 13–17, 2012, Denver, Colorado.

Byrne, J.A., Fernandez-Ibañez, P., Dunlop, P.S.M., Alrousan, D.M.A. and Hamilton, J.W.J. (2011) Photocatalytic enhancement for solar disinfection of water: A review. *International Journal of Photoenergy*, 2011, 12.

Chacón, J.M., Leal, M.T., Bandala, E.R. and Sánchez, M. (2006) Solar photocatalytic degradation of azo-dyes by photo-Fenton process. *Dyes and Pigments*, 69, 144–150.

Chung, Y. and Chen, C. (2009) Degradation of azo dye reactive violet 5 by TiO_2 photocatalysis. *Environmental Chemistry Letters*, 7, 347–352.

Curco, D., Malato, S., Blanco, J., Gimenez, J. and Marco, P. (1996) Photocatalytic degradation of phenol: comparison between pilot-plant-scale and laboratory results. *Solar Energy*, 56, 387–400.

De Laat, J., Lee, T.G. and Legube, B. (2004) A comparative study of the effects of chloride, sulfate and nitrate ions on the rates of decomposition of H_2O_2 and organic compounds by Fe(II)/H_2O_2 and Fe(III)/H_2O_2. *Chemosphere*, 55, 715–723.

Du, Z., Feng, C., Li, Q., Zhao, Y. and Tai, X. (2008) Photodegradation of NPE-10 surfactant by Au-doped nano-TiO_2. *Colloids and Surfaces A: Physicochemical and Engineering Aspects* 315, 254–258.

EAWAG (Swiss Federal Institute of Environmental Science and Technology) and SANDEC (Department Water and Sanitation in Developing Countries), *Solar Water Disinfection, A Guide for the Application of SODIS*. Swiss Centre for Development Cooperation in Technology, 2002, p. 88.

Eng, Y.Y., Sharma, V.K. and Ray, A.K. (2010) Photocatalytic degradation of non-ionic surfactants, Brij 35 in aqueous TiO_2 suspensions. *Chemosphere*, 79, 205–209.

Farre, M.J., Franch, M.I., Ayllon, J.A., Peral, J. and Domenech, X. (2007) Biodegradability of treated aqueous solutions of biorecalcitrant pesticides by means of photocatalytic ozonation. *Desailinization*, 211, 22–33.

Fernandez-Ibañez, P., Sichel, C., Polo-López, M.I., de Cara-García, M. and Tello, J.C. (2008) Photocatalytic disinfection of natural well water contaminated by *Fusarium solani* using TiO_2 slurry in solar CPC photo-reactors. *Catalysis Today*, 144, 62–68.

Gelover, S., Gomez, L.A., Reyes, K. and Leal, M.T. (2006) A practical demonstration of water disinfection using TiO2 films and solar light. *Water Research*, 40, 3274–3280.

Gimenez, J., Curco, D. and Queral, M.A. (1999) Photocatalytic treatment of phenol and 2,4-dichlorophenol in a solar plant in the way to scaling-up. *Catalysis Today*, 54, 229–243.

Guisar, R., Herrera, M.I., Bandala, E.R., García, J.L. and Corona, B. (2007) Inactivation of waterborne pathogens using solar photocatalysis. *Journal of Advanced Oxidation Technologies*, 10, 1–4.

Han, C., Li, Z. and Shen, J. (2009) Photocatalytic degradation of dodecyl-benzenesulfonate over TiO_2-Cu_2O under visible radiation. *Journal of Hazardous Materials*, 168, 215–219.

Hincapie, M., Maldonado, M.I., Olleri, I., Gernjak, W., Sánchez, J.A., Ballesteros, M. and Malato, S. (2005) Solar photocatalytic degradation and detoxification of UE priority substances. *Catalysis Today*, 101, 203–210.

Lin, S.H., Lin, C.M. and Leu, H.G. (1999) Operating characteristics and kinetic studies of surfactant wastewater treatment by Fenton oxidation. *Water Research*, 33, 1735–1741.

Lu, M.C., Chen, J.N. and Chang, C.P. (1997) Effect of inorganic ions on the oxidation of dichlorvos insecticide with Fenton's reaction. *Chemosphere*, 35, 2285–2293.

Malato, S., Blanco, J., Richter, C., Curco, D. and Gimenez, J. (1997) Low concentrating CPC collectors for photocatalytic water detoxification: comparison with a medium concentrating solar collector. *Water Science and Technology*, 35, 157–164.

Malato, S., Fernandez, P., Maldonado, M.I., Blanco, J. and Gernjak, W. (2009) Decontamination and disinfection of water by solar photocatalysis: Recent overview and trends. *Catalysis Today*, 147, 1–59.

Montgomery, M.A. and Elimelech, M. (2007) Water and sanitation in developing countries: including health in the equation. *Environmental Science and Technology*, 41, 17–24.

Muranyi, P., Scrami, C. and Wunderlich, J. (2009) Antimicrobial efficiency of titanium dioxide-coated surfaces. *Journal of Applied Microbiology*, 108, 1966–1973.

Naldoni, A., Bianchi, C., Ardizzone, S., Cappellettia, G., Ciceri, L., Schibuola, A., Pirola, C. and Pappinia, M. (2009) Photocatalysis for the degradation of ionic surfactants in water: The Case of DPC. *MRS Proceedings*, 1171, 1171-S07-19–26.

Natoli, A., Cabeza, A., De La Torre, A.G., Aranda, M.A.G. and Santacruz, I. (2012) Colloidal processing of macroporous TiO_2 materials for photocatalytic water treatment. *Journal of the American Ceramic Society*, 95, 502–508.

Orozco, S.L., Bandala, E.R., Arancibia-Bulnes, C.A., Serrano, B. and Suárez-Parra, R. (2008) Effect of iron SALT on the color removal of water containing the azo-dye reactive blue 69 using photo assisted Fe(II)/H_2O_2 and Fe(III)/H_2O_2 systems. *Journal of Photochemistry and Photobiology A: Chemistry*, 198, 144–149.

Parra, S., Stanea, S.E., Guasaquillo, I. and Thampi, K.R. (2004) Photocatalytic degradation of atrazine using suspended and supported TiO_2. *Applied Catalysis B: Environmental*, 51, 107–116.

Perez, M.H., Penuela, G., Maldonado, M.I., Malato, O., Fernández, P., Oller, I., Gernjak, W. and Malato, S. (2006) Degradation of pesticides in water using solar advanced oxidation processes. *Applied Catalysis B: Environmental*, 64, 272–281.

Pichat, P., Vanner, S., Disaud, J. and Rubio, J.P. (2007) Field solar photocatalytic purification of pesticide-containing rinse waters from tractors cisterns used for grapevine treatment. *Solar Energy*, 77, 533–542.

Polo-López, M. I., Fernández-Ibáñez, P., García-Fernández, I., Oller, I., Salgado-Tránsito, I. and Sichel, C. (2010) Resistance of *Fusarium sp* spores to solar TiO$_2$ photocatalysis: influence of spore type and water (scaling-up results). *Chemical Technology and Biotechnology*, 85, 1038–1048.

Quiroz, M.A., Bandala, E.R. and Martínez-Huitle, C. (2011) Advanced oxidation processes (AOPs) for removal of pesticides from aqueous media. *In: Pesticides- Formulations, effects, fate*, Ed. M. Stoycheva. InTech Press.

Reginfo-Herrera, J.A. and Pulgarin, C. (2010) Photocatalytic activity of N, S co-doped and N-dope commercial anatase TiO$_2$ powders towards phenol oxidation and E. coli inactivation under simulated solar light irradiation. *Solar Energy*, 84, 37–43.

Sajjad, A.K.L., Shamalia, S., Tian, B., Chen, F. and Zhang, J. (2010) Comparative studies of operational parameters of degradation of azo dyes in visible light by highly efficient WOx/TiO$_2$ photocatalyst. *Journal of Hazardous Materials*, 177, 781–791.

Schmid, P., Kohler, M., Meierhofer, R., Luzi, S. and Wegelin, M. (2008) Does the reuse of PET bottles during solar water disinfection pose a health risk due to the migration of plasticizers and other chemicals into the water? *Water Research*, 42, 5054–5060.

Shi, N., Li, X., Fan, T., Zhou, H., Ding, J., Zhang, D. and Zhu H. (2010) Biogenic N-I-co-doped TiO$_2$ photocatalyst derived from kelp for efficient dye degradation. *Energy and Environmental Science*, 4, 172–180.

Sordo, C., Van Grieken, R., Marugan, J. and Fernandez-Ibañez, P. (2010) Solar photocatalytic disinfection with immobilized TiO$_2$ at pilot-plant scale. *Water Science and Technology*, 61, 507–512.

Tang, W.Z. and Huang, C.P. (1996) 2,4-dichlorophenol oxidation kinetics by Fenton's reagent. *Environmental Technology*, 17, 1371–1378.

Terashima, Y., Ozaki, H., Giri, R.R., Tano, T., Nakatsuji, S., Takami, R. and Taniquechi, S. (2006) Photocatalytic oxidation of low concentration 2,4-dichlorophenoxyacetic acid solutions with new TiO$_2$ fibber catalyst in a continuous flow reactor. *Water Science and Technology*, 54, 55–63.

WHO/UNICEF. (2000) *Global water supply and sanitation assessment report*, New York/Geneva.

Chapter 14

Solar energy conversion with thermal cycles

Giampaolo Manzolini & Paolo Silva
Dipartimentio di Energia, Politecnico di Milano, Milano, Italy

14.1 INTRODUCTION

The increasing concentrations of gases such as carbon dioxide (CO_2) and methane (CH_4) – producing the so-called "greenhouse effect" – in the atmosphere are regarded by many members of the scientific community as a consequence of human activities.

The use of energy represents the largest source of emissions, accounting for over 90% of anthropogenic greenhouse gases, while electricity production produces about 35% of the total CO_2 emissions (Key World Energy, 2012).

Renewable energies (i.e. hydropower, biomass, solar, wind) for electricity production are seen as one of the best options for reducing the impact of human activities on the environment. Electricity production from renewable energy is by definition CO_2 neutral[1], with no resulting impacts on global CO_2 emissions and concentration.

Among renewable energies, solar energy could play a fundamental role in satisfying energy demand in countries with high solar radiation. In particular, solar thermal power could easily cover the commercial demand for bulk electricity in the range of tens to hundreds of MW. Focusing on power production, photovoltaic and thermal systems are the available technologies. Photovoltaics consist of the direct conversion of solar radiation into electricity by means of photovoltaic effect. In solar thermal power plants solar radiation is first converted into thermal energy through a concentrator, then into electricity through a thermodynamic cycle, as in fossil fuel-based plants. These kinds of plants are usually called Concentrated Solar Power plants (CSP).

Whilst photovoltaics seems to be a promising and suitable technology for distributed generation (i.e. small-size plants in the range of 1–100 kW), thermodynamic plants might be an attractive solution for centralized large-scale electricity production in the range of tens to hundreds of MW, with predictable low costs and relatively low land demand. The most significant advantage of solar thermal plants over photovoltaics is the adoption of thermal energy storage (TES) which can decouple the electricity production from the energy source. Moreover, TES increases the operating hours of the plant, with economic benefits.

---

[1] While solar, wind and hydro systems do not emit CO_2 during the power production process, biomass combustion processes, the emissions from which were previously subtracted from the atmosphere, do emit CO_2. The overall balance depends on the non-renewable resources used.

Depending on the type of concentrator, the CSP can be based on linear receivers (i.e. parabolic trough or Fresnel collectors), or point focus receivers, such as solar towers or parabolic dish systems. Linear concentrators, mainly parabolic trough collectors, are currently the most proven solar thermal electric technology and are becoming the reference technology for commercial applications. A summary of the CSP plants operating globally is given in the Appendix.

Today, several companies are active in the field of solar thermal technologies. These include Schott (Schott Solar), Acciona (Acciona Energy), Abengoa (Abengoa Solar), Areva (Areva CSP), Siemens (Siemens Concentrated Solar Power) and Archimede Solar Energy (Archimedes Solar Energy). This chapter introduces the principles of solar thermal energy and discusses the advancements, as well as the potentiality, in power production. The last section makes a comparison between the actual costs of CSP versus competitive technologies.

14.2 SOLAR CONCENTRATION CONCEPT IN THERMAL SYSTEMS

The sun has been worshipped as a life-giver to our planet since ancient times. Every hour the amount of energy that the Earth intercepts is about 500 million TWh (values at the top of the atmosphere) which corresponds to the total energy consumption of the entire world over a thousand years. However, this energy is diluted in space: the total integrated spectral irradiance has been made to conform to the value of the solar constant accepted by the space community, i.e. $1366.1\ W/m^2$.

The atmosphere along with atmospheric gases such as ozone (O_3), oxygen (O_2), water (H_2O) and carbon dioxide (CO_2) strongly absorb light for some wavelengths. Other causes of energy reduction occur once the irradiance has passed through the atmosphere include scattering by aerosols and dust particles. By the time light reaches the earth, total energy density is about $1000\ W/m^2$ for Air Mass 1.5.

As anticipated in the introductory paragraph, of the two technologies which convert solar radiation into electricity, this chapter will focus on Concentrated Solar Power plants. The power production process in CSP is based on two steps: the conversion of solar energy into heat and then into power via a conventional thermodynamic cycle.

The overall conversion efficiency can be seen as the product of each step:

$$\eta_{ideal} = \eta_{th} \cdot \eta_{carn} \tag{14.2.1}$$

where η_{th} is the efficiency conversion of solar radiation in thermal power and η_{carn} stands for power cycle efficiency.

In order to fully understand the solar concentration concept and the advantages of concentrating solar radiation, the two terms will now be discussed.

Starting with power cycle efficiency (η_{carn}), the second law of thermodynamics states that the ideal conversion from thermal to mechanical power depends only on the temperature of the heat source. The real conversion cycle will never achieve the efficiency of the ideal cycle. However, the higher the efficiency of the thermodynamic limit, the higher can be the real cycle efficiency. Using Carnot's cycle, which can be representative of a Rankine cycle[2], the conversion efficiency (η_{carn}) is as follows:

$$\eta_{Carn} = 1 - \frac{T_{amb}}{T_{max}} \qquad (14.2.2)$$

where T_{amb} is the ambient temperature and T_{max} is the maximum temperature in the cycle.

From this formula, it can be deduced that the higher the maximum temperature of the cycle, the higher the ideal conversion efficiency from thermal power to electricity. For example, moving from a maximum temperature of 600 K to 1000 K, the Carnot efficiency increases from 50% to 70%.

Moving on to the second term of ideal conversion efficiency, η_{th} represents the ratio between the net heat absorbed by the collector, $\dot{Q}_{ABS}$, and the heat concentrated on the collector itself $\dot{Q}_{CONCENTRATED}$:

$$\eta_{th} = \frac{\dot{Q}_{ABS}}{\dot{Q}_{CONCENTRATED}} \qquad (14.2.3)$$

In a CSP plant, solar energy is concentrated on an absorber by a collecting structure; the entire system is designed to capture most of the solar energy and transfer it to a fluid inside the absorber. In order to produce useful heat for the power cycle, the absorber must be at a high temperature and consequently it will emit energy to the environment as infrared emissions[3]. Therefore the net heat absorbed by the collector ($\dot{Q}_{ABS}$) can be written as the difference between the energy received from the collecting system and the energy emitted to the environment, as expressed by the following equation:

$$\dot{Q}_{ABS} = \dot{Q}_{CONCENTRATED} - \dot{Q}_{EMITTED} = \alpha \cdot C \cdot G - \sigma \cdot \varepsilon \cdot (T_{abs}^4 - T_{amb}^4) \quad (14.2.4)$$

where α is the hemispherical absorptivity of the absorber, C is the geometrical concentration ratio of the collector, G is the direct normal irradiance [W/m^2], epsilon is the hemispherical emissivity of the absorber, sigma is the Stefan-Boltzmann constant [5.67e–08 W/m^{2}K^4], T_{abs} is the average temperature of the absorber and T_{amb} is the sky temperature or the temperature viewed by the absorber [K].

With regard to the non-dimensional coefficient, the thermal efficiency of a collector (η_{th}) can be expressed as the ratio between the net heat absorbed by the collector and the heat concentrated on the collector itself (all these formulae assume a unitary surface):

$$\eta_{th} = \frac{\dot{Q}_{ABS}}{\dot{Q}_{CONCENTRATED}} = \frac{\dot{Q}_{ABS}}{C \cdot G} - \frac{\dot{Q}_{EMITTED}}{C \cdot G} = \alpha - \frac{\sigma \cdot \varepsilon \cdot (T_{abs}^4 - T_{amb}^4)}{CG}$$

$$(14.2.5)$$

---

[2]See section 14.5 for a detailed discussion of a Rankine cycle.
[3]Any body at a temperature above 0 K emits radiation energy expressed by the Stefan-Boltzmann Law (Incropera & DeWitt, 2007).

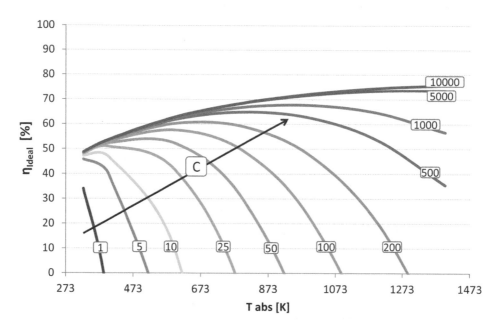

Figure 14.2.1 Ideal conversion efficiency from solar energy to mechanical work assuming different concentration ratios (C) and absorber temperatures.

From this equation it can be noted that the thermal efficiency and, consequently, the heat recovered from the collecting system increases with the concentration ratio; $\dot{Q}_{EMITTED}$ is constant since it depends only on absorber temperature, but, being divided by the concentration factor, its relative contribution decreases.

In order to better explain these concepts, the resulting ideal conversion efficiency (see Equation 14.2.1) as a function of maximum temperature and concentration ratio is summarized in Figure 14.2.1 (for this example α and ε are assumed equal to 0.94 and G to 800 W/m^2).

In addition to the thermodynamic advantages previously discussed and shown in Figure 14.2.1, the adoption of a concentration system substitutes expensive components working at high temperature, like the absorber, with cheaper mirrors at ambient temperature. Moreover, since the performance of the absorber is fundamental and the amount required is reduced, research activity can focus on its improvement, pushing its performances to higher values.

An example of the material absorptivity/emissivity impact on system efficiency is shown in Figure 14.2.2. For simplicity a constant absorptivity is assumed, while three different values of emissivity are considered (0.94, 0.5 and 0.1). Reducing the emissivity while keeping a high absorptivity reproduces the properties of an advanced material with high performance values, though probably at higher cost. However, a very high concentration ratio can significantly reduce the economic impact because of the overall limited influence on total plant costs. Moreover it can be noted that low emissivity is fundamental for a low-medium concentration ratio (in the range of 100) and high absorber temperature (>800 K).

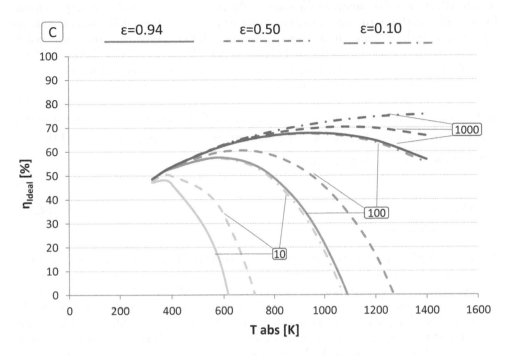

Figure 14.2.2 Ideal conversion efficiency from solar energy to mechanical work for different concentration ratios (C), absorber temperatures and at different epsilon.

As a final remark on the concentration concept, we should underline that concentrating technologies require tracking systems in order to maximize collected solar energy; however, tracking systems increase plant complexity as well as costs, with a potential negative impact on the overall availability of the plant. However, as will be clearly described in the following sections, the advantages of solar concentration overcome the few above-mentioned drawbacks.

14.3 CONCENTRATING SOLAR TECHNOLOGIES

The previous section described the basics of solar power concentration. This section will deal with existing solar collectors used with the two different types of concentration. Solar energy can be concentrated on a line, called linear focus collectors, or on a point, called point focus collectors. Another characteristic which distinguishes the two concentrations is that linear collectors are single-axis tracking while point focus collectors are two-axis tracking.

Solar tracking is fundamental in maximizing the capture of solar energy. Without it, solar radiation would be collected on the absorber only when the beam incidence angle is equal to the design conditions of the collector, i.e. for just a few minutes per day.

Another aspect of solar collectors is the system of concentration, which can be continuous or discrete. Continuous concentrators are mirrors in the form of a parabola which reflect the solar energy at the focus of the parabola; the parabolic shape is made by a rigid metallic structure. Discrete concentration is achieved by several mirrors which move independently in order to collect the solar energy at the same focus point.

Most of the existing solar thermal plants (2050 MW vs. 2120 MW data from 2012 ("NREL Database," n.d.)) are based on a linear focus technology, but the point focus is seen as the most attractive because of its potential for cost reduction.

Before discussing the technology in detail, we need to define solar multiple (SM) and introduce the general approach to CSP design. SM is the ratio between the power delivered by the solar field at design conditions ($\dot{Q}_{SF,design}$) and the nominal thermal input of the power cycle $\dot{Q}_{PB,design}$.

$$SM = \frac{\dot{Q}_{SF,design}}{\dot{Q}_{PB,design}} \qquad (14.3.1)$$

A solar multiple equal to one corresponds to a solar field aperture area which delivers at nominal conditions the design thermal energy input of the power cycle. The SM coefficient is one of the optimization parameters of a CSP plant. It is usually above one in order to: (i) mitigate solar transient and fluctuation in irradiance; and, if available, (ii) store part of the collected thermal energy in a Thermal Energy Storage. The advantage of a SM above one, even without thermal storage, is to increase plant operating hours. One drawback, during high radiation days, is that part of the solar field might be defocused in order to respect the thermal input to the power block, thus wasting potential solar radiation.

Existing plants are usually designed with SM higher than one and with a TES for the above-mentioned advantages.

Moving to the energy conversion process, all concentrating solar technologies convert solar energy into thermal energy supplied to a fluid (i.e. the fluid increases its temperature in the solar field). Thus, solar field efficiency can be introduced as the ratio between the thermal power absorbed by the fluid and the direct normal solar radiation available on the collector area. Solar field efficiency (η_{SF}) is defined as:

$$\eta_{SF} = \frac{\dot{Q}_{FLUID}}{\dot{Q}_{SUN}} = \frac{\dot{Q}_{FLUID}}{G \cdot A} \qquad (14.3.2)$$

where $\dot{Q}_{FLUID}$ [kW] is the thermal power transferred to the fluid and $\dot{Q}_{SUN}$ [kW] is the solar energy coming from the sun. $\dot{Q}_{SUN}$ can be expressed as a product of the direct beam radiation, G (W/m^2) and the collector surface area A [m^2]. Solar field efficiency can be further divided into two different efficiencies: optical and thermal. In order to better explain this concept, a schematic of a concentrating system with main conversion steps[4] is represented in Figure 14.3.1.

---

[4]A more rigorous representation would also consider other phenomena such as absorbance of the receiver.

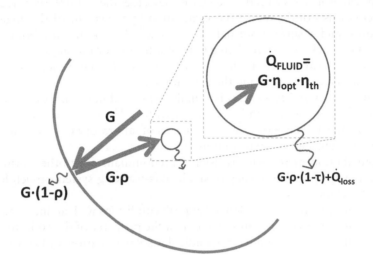

Figure 14.3.1 Schematic of conversions and losses in a generic concentrating system.

Optical efficiency accounts for the difference between solar radiation available at the reflector and the amount of radiation effectively transferred to the receiver, while thermal efficiency accounts for the thermal losses to the ambient mainly as radiative losses. The resulting solar field efficiency can be written as:

$$\eta_{SF} = \frac{\dot{Q}_{FLUID}}{G \cdot A} = \frac{\dot{Q}_{FLUID}}{\dot{Q}_{REC}} \times \frac{\dot{Q}_{REC}}{G \cdot A} = \eta_{opt} \times \eta_{th} \qquad (14.3.3)$$

where $\dot{Q}_{REC}$ is the heat concentrated on the receiver, η_{opt} is the optical efficiency and η_{th} is the thermal efficiency.

Solar field efficiency strongly depends on the operating conditions: ambient temperature, wind speed, sun position, solar radiation and material properties, etc. In particular, optical efficiency varies significantly with the position of the sun, while thermal efficiency is dependent on ambient temperature and solar radiation.

Optical efficiency is usually defined at the design conditions (nominal optical efficiency) and then corrected during different operating conditions to take account of the variation in material properties according to the incidence angle of radiation.

The overall optical efficiency depends on several contributions, which are usually split to outline their differences: nominal optical efficiency, geometrical losses and the $K(\theta)$ as expressed in equation 3.4.

$$\eta_{opt} = \eta_{opt-peak} \cdot \eta_{geom} \cdot K(\theta) \qquad (14.3.4)$$

where $\eta_{opt-peak}$ is the nominal optical efficiency, η_{geom} is the geometrical efficiency, which includes shadowing, tail-end losses and blocking, and $K(\theta)$ is the correction of the efficiency for the incidence angle. $K(\theta)$ takes into account the variation in material optical properties with the incidence angle, together with the cosine effect. The cosine effect consists of the reduction of radiation by the cosine of the angle between solar radiation and a surface normal. The cosine effect occurs for both single-axis and double-axis tracking systems being more important for the single-axis technology (about 60% on a yearly base) than in the double-axis (about 80%). This is because the former can reduce to zero only the azimuth angle.

Variation in material properties is usually expressed in the incidence angle modifier (IAM), which includes τ and α variation with θ. The best way to predict optical efficiency at different solar positions is by means of laboratory experiments, owing to the difficulty of splitting the contribution of τ and α.

The contribution of the $K(\theta)$ is usually predominant over other terms, making the average yearly optical efficiency of single axis-tracking systems much lower than double-axis tracking ones.

Nominal optical efficiency is defined at peak conditions with an incidence angle (θ) equal to zero. It can be seen as representative of the property of the collector materials (i.e. reflector, absorber, etc.) under perpendicular solar radiation and is defined as:

$$\eta_{opt-peak} = \rho \cdot \gamma \cdot \tau \cdot \alpha|_{\theta=0} \tag{14.3.5}$$

where ρ is the mirror reflectivity, γ the intercept factor, τ the vacuum glass transmissivity and α the receiver absorptivity.

After this short general introduction on concentration systems, the description will be divided between adopted tracking systems.

14.3.1   Linear focus

Linear focus concentration is based on parabolically curved mirrors or segmented mirrors, according to the Fresnel principle, which concentrates solar energy onto a receiver pipe. This configuration allows single-axis tracking. A collector field comprises many collectors (usually named trough) in parallel rows aligned with the tracking axis orientation.

The fluid is heated in the receiver and then sent to the power block to convert the thermal energy into electricity. The distribution of the fluid along the field is made through pipes called headers. The header that brings the fluid from the power block to the trough is named "cold header", since it is at lower temperature, while the header that collects the fluid is named "hot header" because of the higher temperature. The particularity of these pipes is in the variation in diameter along with the variation in fluid flow: the target is to keep an almost constant fluid velocity. The connection between the header and the power cycle is made by connecting pipes which transfer the fluid from the solar field to the power cycle.

An example of a solar field layout is given in Figure 14.3.2. In particular, the two typical configurations are shown: on the left side the "H" configuration is reported. On the same side of the power block there are two different sections of the solar field, with two cold and two hot headers. This configuration is adopted in Andasol project (Herrmann and Geyer, 2002), SEGS VIII and IX. In the "I" configuration (used in SEGS VI, (Patnode, 2006)), reported on the right side, there is just one section. In both the "H" and "I" configurations one loop is composed of two solar collector rows, connected by

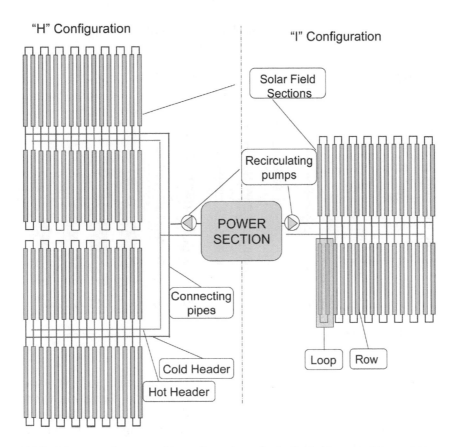

Figure 14.3.2 Schematic of two possible configurations of solar fields (the central line must be seen as symmetric axis).

a "U shape" pipe: the cold and hot headers can be on the same side leading to the so-called "central-feed configuration". This layout minimizes the piping and allows direct access to each collector row without buried pipes. One drawback is that this configuration is not balanced from a pressure drop point of view; therefore pressure balancing valves are required at each row, adding significant pressure drops. An alternative layout can be represented by the "direct return" and "reverse-return" configurations (Kreith and Goswami, 2007), where the cold and hot headers are on different sides of the rows. In particular, the inverse return configuration balances the pressure drop but requires longer connecting pipes.

To summarize, the typical layout configuration is the central feed since it reduces piping length and thermal losses, keeping a good access to the collectors.

The tracking orientation of the trough can be either in a N-S direction, an E-W direction or anywhere in between; the direction which guarantees the highest collection of solar energy is N-S. As an example, the electricity produced in a year from a parabolic trough-based solar plant located in the United States at a latitude of 34° is shown in Figure 14.3.3. The only difference between the two cases is in the axis orientation.

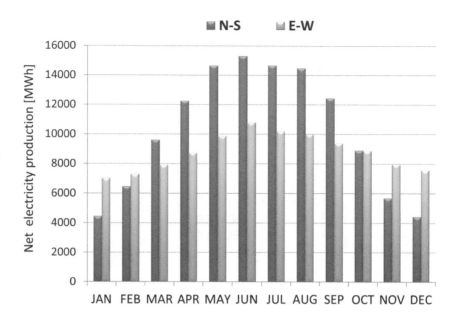

Figure 14.3.3 Monthly electricity output for N-S and E-W orientations.

North-south tracking produces 15% more electricity yearly than an E-W orientation: production is higher during the summer, while the situation changes in winter. The qualitative result can be extended at different latitudes, with changes from quantitative standpoint: at lower latitudes the difference between tracking orientation increases, while at higher latitudes it reduces.

Since in the two cases (N-S tracking and E-W tracking) the ambient conditions, power cycle configuration and performance, and the collecting system are the same, the only difference can be due to optical efficiency. As discussed before, optical efficiency depends on the solar incidence angle: N-S tracking follows the solar azimuth angle, hence reducing its impact, while solar altitude remains. Consequently, in winter at moderate latitude, the solar incidence angle can be significant also at midday, where the solar altitude is lower than 40° (during the winter solstice it is equal to 90° − latitude − ecliptic (23°27′)). E-W tracking does the opposite, following solar altitude, with advantages during the middle hours of the day in winter. This effect explains the variation in the electricity production of different tracking axes as a function of the season.

14.3.2 Parabolic trough

Commercial applications of solar plants based on parabolic trough (PT) technology began in the mid-1980s with the construction of Solar Energy Generating Systems (usually known by the acronym SEGS) by LUZ in California. Nine different SEGS plant were built, named SEGS I to SEGS IX, for a total installed capacity of 354 MW (Cohen and Kearney, 1999). Afterwards, this type of technology was forgotten until

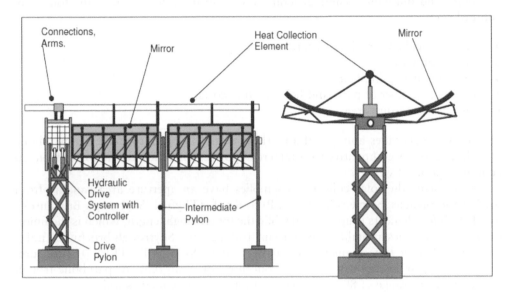

Figure 14.3.4 Schematic of a parabolic trough. Main components are outlined.

1998 when the Eurotrough project started. This project was cost-shared by a group of European companies with the aim of developing an innovative parabolic trough collector with high performances and reduced costs. Then, after 2000, the construction of solar thermal plants started again with Nevada Solar One and the installation of "Plataforma Solar de Almeria" in Spain. In fact, the "Plataforma Solar de Almeria" is not a power plant but a test facility where different solar technologies can be tested and compared.

Nowadays, several companies, such as Siemens (Siemens Concentrated Solar Power), Archimede Solar Energy (Archimede Solar Energy), Schott (Schott Solar), Abengoa (Abengoa Solar) to mention just a few, develop parabolic trough systems or manufacture components.

Between 2009 and 2011 total installed capacity increased significantly, reaching almost 2.1 GW. Most of this capacity was installed in only two nations: the United States and Spain. An example of recently built solar power plants is the Andasol project which has three plants of 50 MW (*The parabolic trough power plants Andasol 1 to 3*, 2008).

The basic component of the solar field is the Solar Collector Assembly (SCA), a schematic of which is shown in Figure 14.3.4.

Each SCA consists of trough-shaped mirrors, also named parabolic trough reflectors, supported by a structure and the tracking systems. The parabolic trough collectors have a characteristic cylindrical shape with a parabolic curvature. Solar radiation is reflected and concentrated on an absorber tube/pipe (also called the Heat Collection Element) which transfers the thermal power to a fluid flowing inside. The parabolic trough collectors are designed to achieve high performance at low cost with high reliability and durability. These general targets can be translated into the following sub-tasks:

- High optical and tracking accuracy;
- Low heat losses;
- Manufacturing simplicity;
- Reduced number of parts and field erection costs;
- Increased aperture area.

Hence, R&D activities aim at fulfilling these targets in order to make the parabolic trough technology competitive to commercial power generation technologies from an economic point of view.

At present, the solar collector assemblies have an aperture area ranging from 5.77 m for Siemens (Siemens Sunfield LP) to 6 m for Skyfuel (Skytrough Brochure)[5]. The length for elements is up to 12 m while the total trough length, which is the union of elements, reaches a total length of about 100–120 m ("Skytrough brochure," n.d.).

The absorber tube diameter is usually 70 mm (SCHOTT PTR® 70 Brochure; Archimede Solar Energy), making the concentration ratio (CR) of parabolic trough collectors in the range of 80. The concentration ratio is defined as follows[6]:

$$CR = \frac{aperture\ width}{adsorber\ diameter} = \frac{W}{D_{abs}} \tag{14.3.6}$$

After this short and general introduction, a detailed description of parabolic trough technology is now presented, in which the three main components – reflectors, heat collection element and structure – are discussed.

14.3.3 Reflectors

Reflectors must reflect and concentrate solar direct beam radiation onto the linear receiver located at the focus of the parabola, called the heat collection element (HCE).

Aside from pure performances whilst new and in clean conditions, the reflectors must maintain constant reflectivity over the years. Since solar plants are usually installed in places with high solar radiation, often in desert locations, the ambient is abrasive because of sand and dust transported by the wind. Receivers must be resistant to these conditions, making the endurance test really challenging. For this reason, protective coatings are deposited on the reflector surface to keep high reflectivity and reduce wear effects.

In addition, the reflectors must guarantee high reflectivity at all solar incidence angles: reflectivity τ is usually defined at zero incidence angle where it has the maximum values. Usually, it is known that reflectivity varies as a function of solar incidence angle. For higher incidence angles, mirror reflectivity lowers, affecting optical and overall plant efficiency. Variation in performance with the incidence angle is usually

---

[5]It should be noted that Skyfuel is testing a solar collector with an aperture area of about 8 m.
[6]There is discussion on how to define the concentration ratio. Some authors prefer to adopt the receiver diameter instead of the circumference as in this work.

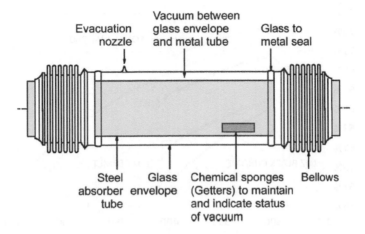

Figure 14.3.5 Heat collection element (Flabeg Solar; Price et al., 2002).

defined by the incidence angle modifier (IAM). However, the IAM not only takes into account reflectivity variation, but also glass and absorber tube property variations with the angle. Examples of IAM trends will be presented later, in the parabolic trough performance section.

Trough-shaped mirrors can be made by a glass layer with low iron concentration and a reflective silvered film. For example, glass-based technology is developed by Flabeg (Flabeg). Another option is multiple layers of polymer film with a layer of pure silver to provide for high specular reflectance (Skytrough Brochure). Other companies which are active in reflector development are Alanod (Alanod), 3M (3M). Alanod products, for example, are based on polished aluminium which has a slightly lower reflectivity but a higher resistance to wear and lower weight, with advantages for structure and plant reliability.

Typical reflectivity values are 94.4% for glass-based reflectors and slightly lower for polymer-based ones (93%) and aluminium (92%). The higher investment cost and weight of glass-based reflectors is nowadays balanced from an economic point of view by a higher efficiency compared to polymer and aluminium-based reflectors.

14.3.4 Heat collection element

The absorber tube, also called heat collection element (HCE), consists of a steel tube with a selective coating layer. The coating is added in order to increase the absorption properties within the solar spectrum while keeping emissivity low.

The development of new coatings has increased the performance of absorber tubes thanks to higher absorptivity and lower emissivity. An example of the improvement in this technology is given in Figure 14.3.6 which shows the efficiencies of collectors with different selective coatings. The coatings adopted for use in SEGS plants were LUZ Black Chrome and Luz Cermet with efficiency in the range of 60%. R&D activity has increased absorber performance, reaching an overall collector efficiency of more

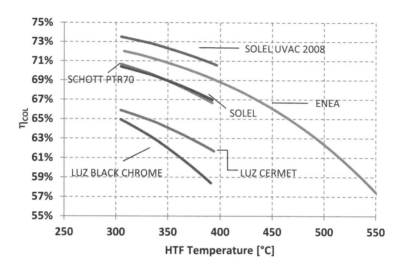

Figure 14.3.6 Commercial collectors efficiency as function of fluid temperature (DNI = 800W/m^2) (Manzolini et al., 2011a). In this case, Schott PTR 70 performances do not correspond to the latest version.

Table 14.3.1 Features of commercial absorbers (Price et al., 2002; Forristall, 2003).

	Luz Black Chrome	Luz Cermet	Solel UVAC	Solel UVAC 2008	Schott PTR 70*	Archimede Solar Energy
Absorption	0.94	0.92	0.955	0.97	0.955	0.945
Transmittance	0.935	0.935	0.965	0.97	0.965	0.97
Absorber's Length [m]	4.06	4.06	4.06	4.06	4.06	4.06
D_{in} absorber [mm]	64	64	64	64	64	64
D_{out} absorber [mm]	70	70	70	70	70	70
D_{in} glass envelope [mm]	109	109	109	109	119	109
D_{out} glass envelope [mm]	115	115	115	115	125	115

than 70% at temperatures below 400°C. Moreover, recent coatings can withstand tube temperatures of about 550°C, with significant thermodynamic advantages. The main properties of HCE are summarized in Table 14.3.1.

Because the absorber tube works at high temperature, a glass tube surrounding the absorber tube is also adopted. A schematic of a heat collection element with all the fundamental components is shown in Figure 14.3.4.

The glass tube, usually made by Pyrex®, makes a vacuum annulus between the glass and the tube which prevents coating oxidation and reduces heat loss; pressure within the annulus is typically about 1×10^{-2} Pa (Price et al., 2002). The glass tube has an antireflective coating on both surfaces to maximize solar transmittance while limiting reflective losses. Along with the antireflective coating, significant developments have been made in improving absorber performance.

Most of heat collection elements are applied to stand-alone concentrated solar power plants with synthetic oil as the heat transfer fluid. The synthetic oil decomposes at high temperature, producing hydrogen which permeates across the tubes and ends up in the vacuum annulus. The presence of hydrogen in the vacuum annulus is detrimental for the thermal performance of the collector. For this reason the vacuum is usually filled with getters which absorb hydrogen and other gases that eventually permeate during operation. The getters are barium-based and, as additional components required for the HCE, they affect the overall costs.

The adoption of vacuum glass is beneficial for HCE efficiency, but this requires a dedicated glass-to-metal sealing. In addition, the adoption of bellows is necessary to equalize the thermal expansion of metal and glass in order to guarantee the vacuum conditions in the annulus. A drawback of the bellows is the shadow thrown onto the absorber tubes, with consequent efficiency penalties. However, recently SCHOTT PTR 70® have reduced the impact of bellows, leading to an active area above 96.7% (SCHOTT PTR® 70 Brochure). Active length is defined as the active aperture area of the receiver on the total receiver area.

Commercial HCE efficiency as a function of heat transfer fluid (HTF) temperature is shown in Figure 14.3.6. Finally, in order to limit bending of the tube, HCE length is usually equal to 4 m. For this reason, several HCEs are placed in series to make a trough.

14.3.5 Structure

The structure, usually of metal, serves to hold the heat collection element and the reflectors at the correct position. The parabolic shape of the reflector is usually formed by the structure. The structure has to fulfil the following requirements: efficient use of material, ease of transportation onto the site, easy to assemble, and able to withstand atmospheric conditions for at least 30 years. Moreover, it must have a high torsional and bending stiffness under wind loads (the structure must be designed to work under wind conditions, typical of desert locations). For example, Andasol plant can continue operating in winds up to 13.6 m/s (about 50 km/h); above this wind speed, the collectors are put in a wind-protected position. It must be remembered that the aperture length of the parabola is in the range of 6 m and wind drag and lift can be really significant. For these reasons, research activity on structure development is carried out using computational fluid dynamics (CFD) analysis and wind tunnel testing to determine the required torsional and bending stiffness in order to achieve a desired interception factor at defined conditions. An example of CFD analysis on parabolic troughs performed at Politecnico di Milano is shown in Figure 14.3.7.

Research activity has led to different types of structure becoming available on the market, each based on a different concept. Flagsol's Eurotrough and Ultimate Trough are based on a torque box design and made from galvanized steel (Graf and Nava, 2011). The torque box guarantees savings on materials, uses thick and reliable mirrors, and reduces wind loads compared to other configurations. SENER proposes a torque tube plus stamped steel cantilever, with the mirrors supported by arms. The design of Acciona solar power is based on recycled aluminium or steel struts and geo

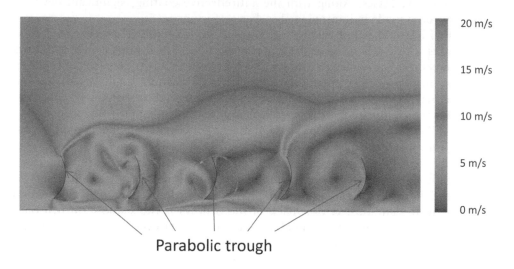

Parabolic trough

Figure 14.3.7 Wind velocity field around five different parabolic troughs.

hubs. The advantages of this configuration are higher rigidity via interlinking and no cutting or welding during construction. Lastly, the ENEA design is based on a torque tube with precise reflector supported by arms.

The structure has to be fixed to the ground with foundations which add significantly to the overall cost of solar fields. To reduce these associated costs, research activity is also being done on advanced foundations, an example of which is steel-reinforced, drilled pier foundations.

14.3.6 Parabolic trough performance

This section tries to give an overview of parabolic trough performances during a typical year. As mentioned above, the parabolic trough transfers solar radiation to a fluid flowing in the HCE: this conversion is affected by optical and thermal losses. Optical losses depend mainly on $K(\theta)$ contribution, and secondarily on end-losses and shadowing. In order to give an idea of the impact of $K(\theta)$ on overall performance, the optical efficiency ratio (OR) is introduced. This is calculated as the ratio between off-design optical efficiency and nominal optical efficiency. The hourly OR calculated for a parabolic trough-based solar plant located in the United States at a latitude of 34° with N-S axis tracking is shown in Figure 14.3.8.

It can be noted that during the winter, even at midday, optical efficiency is about half the nominal optical efficiency. This is due to the above-mentioned high incidence angle consequence of low solar altitude. For example, the cosine effect, which is only one part of the optical penalty, is equal to 0.6 (during winter, solar altitude at midday is about 40°–50°). On the contrary, during summer, optical efficiency at noon is close to nominal conditions because the incidence angle is about 10°.

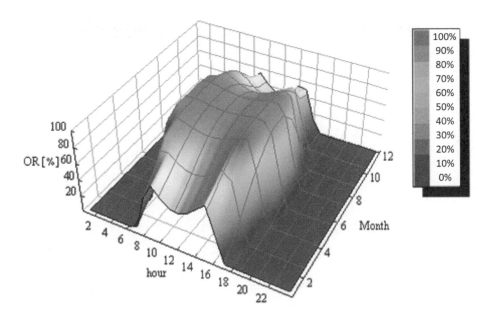

Figure 14.3.8 Annual maps of optical efficiency ratio (OR) for PT technology (Giostri et al., 2013).

For the United States location at latitude north 34°, the yearly optical efficiency is 53% (meaning that optical losses account for almost half of the solar radiation that is lost), while the nominal optical efficiency was about 75%.

In addition to optical losses, parabolic troughs are affected by thermal losses. As a general consideration, the solar field is operated such that inlet and outlet temperatures are constant; hence, absolute heat losses are constant. Focusing on thermal efficiency, which was defined in Equation 14.3.3 as the heat transferred to the fluid divided by the solar radiation impinging on the receiver, it can be seen also as:

$$\eta_{th} = \frac{\dot{Q}_{FLUID}}{\dot{Q}_{REC}} = \frac{\dot{Q}_{REC} - \dot{Q}_{HT\,LOS}}{\dot{Q}_{REC}} = 1 - \frac{\dot{Q}_{HT\,LOS}}{\dot{Q}_{REC}} = 1 - \frac{\dot{Q}_{HT\,LOS}}{G \cdot A \cdot \eta_{opt}} \qquad (14.3.7)$$

where $\dot{Q}_{HT\,LOS}$ is the heat losses [W], G is the solar normal radiation [W/m²], A is the collector area [m²] and η_{opt} is the optical efficiency as defined in Equation 14.3.5.

Keeping in mind that absolute heat losses are constant, depending only on absorber size and temperature, it can be noted that thermal efficiency drops when $G \cdot A \cdot \eta_{opt}$ is lower than nominal conditions. Hence, during spring, fall and mostly in winter, thermal efficiency is significantly lower than in summer. Thermal losses depend also on ambient temperature since they are proportional to the temperature difference between the receiver and the ambient (Equation 14.2.3). However, this is a second order effect, because the temperature variation in typical solar plant sites is about 30–40°C, while the average temperature difference between the receiver and the ambient is in the range of 300°C.

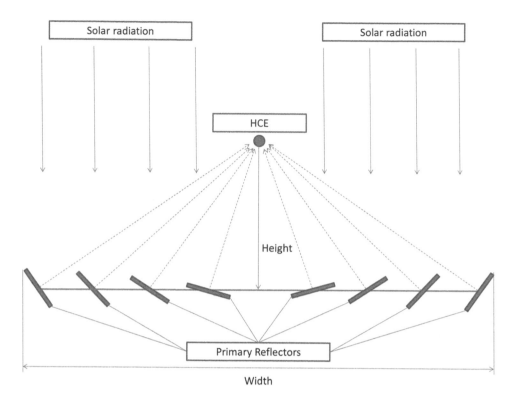

Figure 14.3.9 Linear Fresnel Reflector concept.

The yearly thermal efficiency of commercial collectors is in the range of 90% (10% of radiation concentrated on the receiver is lost) while, at nominal conditions, thermal efficiency is about 95%.

This section has briefly described parabolic trough technology and has given an idea of the expected efficiency of this kind of technology. Taking into account only the two main losses (optical and thermal) related to the solar field, yearly efficiency is below 50%. Piping losses and consumption by fluid recirculating pumps further penalize efficiency, even if they account for less than 6%.

14.3.7 Linear Fresnel

The second type of linear focus technology is the linear Fresnel reflector (LFR). This concept is based on several mirrors which can be moved independently and which collect radiation on the HCE: the parabolic shape is substituted by several ground-based mirrors which have different curvatures and which move independently. A schematic of the LFR technology is shown in Figure 14.3.9.

Compared to the parabolic trough, this is a rather new technology. Actually, there are only two commercial manufacturers for LFR: Novatech Biosolar (Novatech Biosol, 2012) and, more recently, Areva (Areva Solar, 2012). However, several companies (e.g. Skyfuel) and research centres (e.g. NREL and Fraunhofer) are investigating the application of LFR technology since it is considered promising, with potential economic advantages for the following reasons: (i) ground-based mirrors allow the adoption of lighter structures thanks also to reduced wind drag effect; (ii) land area is minimized due to reduced shading between collector rows; (iii) the receiver is fixed and tracking energy consumption is decreased; (iv) the absence of ball joints lowers pumping losses; and (v) ground-based mirrors are easier to clean.

Moreover, the concentration ratio (CR) of LFR can be higher than with parabolic troughs since it is not limited by parabola aperture width. The adoption of more mirrors would not change the wind drag effect, hence the same structure can be kept.

The CR for LFR is defined in Equation 14.3.8[7].

$$CR = \frac{aperture\ width}{adsorber\ diameter} = \frac{n \cdot W}{D_{abs}} \qquad (14.3.8)$$

where D_{abs} [m] is the absorber diameter, n is the number of primary mirrors and W [m] is the aperture of each mirror.

The aperture width in this case is the aperture width of each mirror times the number of mirrors. Commercial LFRs have a concentration ratio of about 160 (which is about twice that of parabolic trough systems).

LFRs can have a secondary reflector which collects solar radiation from the primary mirrors to the absorber tube. The concentration ratio definition for this case is given in Equation 14.3.8; the only difference is that the solar radiation can be subjected to multiple reflections. The HCE can be the same as for parabolic trough, while the glass tube can also be absent where there is a secondary receiver.

The first application of LFR was in direct steam generation plants (DSG, see Section 14.5 for a detailed discussion of this technology) with saturated steam. In order to increase the conversion efficiency of collected heat in the power section, the production of superheated steam in the solar field was recently investigated (Novatech biosol, 2012), as was the adoption of molten salts as HTF (Areva CSP).

At the time this chapter was written (end of 2012), there are just a few solar plants in operation based on LFR technology. The most important is Puerto Errado 2, built by Novatec Solar with a 30 MW electric power output (Puerto Errado 2). In this plant, saturated steam at 55 bar and 270°C is produced in the solar field. Other operating plants, though with significantly lower power output, are the Kimberlina solar thermal power plant (Kimberlina Solar Plant) and the Liddel plant by Areva, formerly Ausra, (Liddel Solar plant).

In terms of size, a commercial Fresnel linear collector has a width of about 16.6 m and a height of 7–8 m. In general, the width affects the concentration ratio, while the height affects optical efficiency. This part will be discussed in detail later.

The next section describes the main components of Fresnel collectors. The LFR structure is not treated in detail for the sake of brevity, and because Fresnel mirrors are ground-based with consequent less complexity than parabolic trough reflectors.

---

[7]As for parabolic troughs, there is a debate on how to define the concentration ratio. Some authors prefer to adopt the receiver diameter instead of the circumference, as in this work.

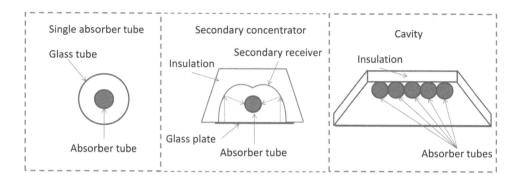

Figure 14.3.10 Schematic of heat collection element adopted in LFR.

14.3.8 Heat collection element

The heat collection element transfers concentrated solar energy to the fluid. Compared to trough technology, there are several heat collection element concepts available, all of which can be merged into three different categories:

– Single absorber tube;
– Secondary reflector/concentrator;
– Cavity receiver.

A schematic of each category is shown in Figure 14.3.10. The single absorber tube concept is similar to that of parabolic trough. Solar radiation is concentrated by the primary mirrors on the heat collection element, which has a glass tube to limit thermal losses. Compared to PT, the HCE is much simpler since it does not move with the structure; connections and differential thermal expansions are much simpler. In this case, the size of each mirror is similar to the heat collection element. Mirrors are usually flat or have a small curvature.

The second category is based on the adoption of a secondary reflector or concentrator which further concentrates solar radiation coming from the primary mirrors on the absorber tube. Considering that the mirror's width is about the same as the secondary reflector's (i.e. larger than HCE), this configuration can increase the concentration ratio of the single absorber tube configuration. Novatec (Novatec Biosol, 2012) and ENEA concepts are based on the secondary receiver (Grena & Tarquini, 2011).

Besides the higher concentration ratio, the advantages of the secondary reflector concepts are lower thermal losses: the upper part of the absorber tube does not see the sky temperature, but the temperature of the reflectors which is higher. Moreover, the secondary reflector can be insulated and a glass plate can be used to close the secondary reflector cavity, reducing convective heat loss. A disadvantage of this configuration is the additional reflection of solar radiation, with optical penalties.

Optimization of the secondary reflector shape has been rigorously investigated and optimized in the literature, since the optical analysis has much higher degrees of freedom than the single absorber tube. A picture showing the optical and thermal

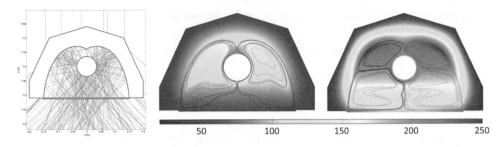

Figure 14.3.11 Optical assessment (left side), thermal analysis (center and right) for the Novatech secondary reflector configuration (Binotti et al., 2011).

assessment for Novatec's secondary receiver is shown in Figure 14.3.11. Similar assessments were performed by Barale et al. (2010), Veynandt et al. (2010) and Morin & Dersch (2009).

The third category involves a cavity receiver, which places this concept in between the single absorber and the secondary concentrator. The configuration is based on multiple absorber tubes being placed in a cavity, which provides thermal insulation, thus reducing heat loss.

In general, the single absorber tube has higher heat losses but is much simpler and cheaper than the secondary reflector and cavity concepts. Adoption of a cavity or secondary reflector system increases complexity, but also cost. Moreover, both cavity receiver and secondary reflector systems shadow part of the primary reflector, reducing optical efficiencies.

The absorber tubes in LFR must have the same physical properties as PT applications. High absorptivity and low emissivity are required in order to reduce as much as possible the heat losses. For this reason the same technology adopted in parabolic trough systems can be used in Fresnel (i.e. Schott PTR 70® (SCHOTT PTR® 70 Brochure).

14.3.9 Reflectors

Reflectors in LFR are also called primary to distinguish them from the secondary reflector. Their function is to reflect and concentrate solar direct beam radiation onto the HCE. Compared to PT, where just one parabolic reflector is assumed, LFR is based on several reflectors, which can be flat or have a very small curvature. The reflector width is generally close to the absorber diameter for the single tube configuration; the width of secondary reflectors is equal to the aperture width of the secondary receiver.

Reflectors must have a high reflectivity when new and operating in clean conditions, as well as over their entire lifetime. Since the reflectors are ground-based, i.e. below 1–2 m altitude, wind speed over the system is lower than in parabolic trough installations, thus reducing wear, especially by sand in desert locations.

Since the function of mirrors is the same in both LFR and PT applications, their reflectors share the same manufacturers. Therefore mirrors can be made from: (i) a glass layer with low iron concentration and a reflective silvered film (Flabeg); (ii) multiple layers of polymer film with a layer of pure silver to provide for high specular reflectance (Skytrough Brochure); and (iii) polished aluminium (Alanod).

14.3.10 Linear Fresnel performance

This section describes the performance of a solar field using LFR technology and compares it to PT systems. As before, performance will be described from both optical and thermal perspectives. With regard to thermal analysis, it is much more difficult to provide a general indication since LFR technology involves more than one receiver design. Moreover, the receivers are strongly affected by the concentration ratio as well as by the absorber tube operating temperature, which also varies with manufacturer. As a general consideration, Fresnel collectors usually have a higher concentration ratio than PT, hence the potential for thermal loss reduction (see Equation 14.2.3). As regards receiver configuration, the single absorber tube case is similar to PT technology and will have the highest thermal losses due to the radiative contribution. The secondary receiver and cavity configuration limit radiative losses thanks to the insulation layer. Since further general considerations are not possible, thermal efficiency of the Novatech collector will be presented. Experimental results (Novatech Biosol. Technical data – NOVA 1) show that heat losses for the Fresnel collector at design conditions are in the range of 15–30 W/m^2, with a thermal efficiency of 95% (whereas parabolic trough efficiency was in the range of 90%). However, these losses were calculated at lower temperature than the corresponding PT case.

Moving to optical efficiency, this is defined as the collected radiation divided by the DNI multiplied by the total mirror surface[8] (see Equation 14.3.9).

$$\eta_{SF} = \frac{\dot{Q}_{REC}}{G \cdot A_{total}} = \frac{\dot{Q}_{REC}}{G \cdot n \cdot A_{mir}} \tag{14.3.9}$$

where the total reflecting area (A_{total}) is equal to the number of mirrors (n) times the single mirror reflecting area (A_{mir}).

Assuming this efficiency (in the literature another definition of optical efficiency is proposed), LFR has a lower optical efficiency than PT even at design conditions or where the incidence angle is equal to 0°. A schematic explaining the above-mentioned concept is shown in Figure 14.3.12. This is because even with solar incidence angle equal to 0°, the primary mirror must be inclined to centre the solar radiation on the absorber tube (the angle is equal to θ which corresponds to the angle between the solar and the reflected radiation). Hence, optical efficiency is penalized because of the cosine effect on the mirror's area ($A \cdot \cos(\theta_i)$). This effect is more significant for mirrors further from the receiver.

This effect was not present in PT because the parabolic shape makes the mirrors parallel to the solar radiation at any given point. In order to increase the optical efficiency in LFR, the average incidence angle should be minimized; minimization can be performed in two different ways: (i) by increasing the height of the receiver;

---

[8]There is discussion on this definition, since in parabolic trough systems the aperture area is considered and not the reflective area as in Fresnel. A recent study showed that a different definition has limited impact on final results (Giostri et al., 2013).

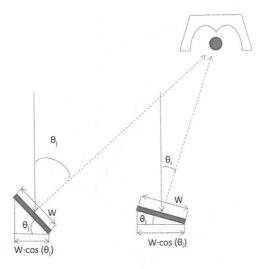

Figure 14.3.12 Characteristic angles of Fresnel collectors at 0° incidence angle.

or (ii) by reducing the concentration ratio. As a drawback, both solutions increase the cost of the solar field.

For this reason, when developing a Fresnel technology, it is not correct to focus on only one parameter, but the minimization of the cost of electricity must be the final target.

Moving to yearly optical efficiency, the IAM for the linear Fresnel concentrators will now be discussed. The starting point is that LFR requires two projections of the incidence angle: one on the longitudinal plane and the other on the transversal plane (see Figure 14.3.13): $\theta_\parallel$ is defined as the angle between the vertical axis and the beam vector projection on the longitudinal plane, and $\theta_\perp$ is defined as the angle between the vertical axis and the beam projection on the transversal plane. In addition, another characteristic angle, θ_i, can be defined as the angle between the sunray vector and its projection on the transversal plane. This angle corresponds to the above-described incidence angle of PT technology. Relations between angles are summarized in Equation 14.3.10 to 14.3.12:

$$\theta_\perp = \arctan(|\sin(\gamma)|\tan(\theta_z)) \tag{14.3.10}$$

$$\theta_\parallel = \arctan(\cos(\gamma)\tan(\theta_Z)) \tag{14.3.11}$$

$$\theta_i = \arcsin(\cos(\gamma)\sin(\theta_z)) \tag{14.3.12}$$

where γ is the azimuth angle and θ_Z is the zenith angle.

The incidence angle modifier which is a function of the zenith and azimuth angle includes the cosine effect, the primary mirrors mutually blocking and shading, the secondary reflector and support shading, variation in optical properties and modification of the interception factor.

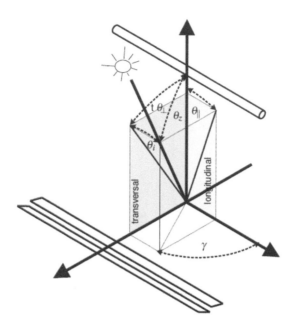

Figure 14.3.13 Angles definition of a linear Fresnel reflector with horizontal N-S orientation tracking axis (Mertins, 2009).

For simplicity, the overall IAM of the collector is calculated as the product of two different IAMs related to $\theta_{\parallel}$ and $\theta_{\perp}$ characteristic incidence angles, as follows:

$$IAM(\theta_z; \gamma) = IAM(\theta_{\perp}) \cdot IAM(\theta_i) \qquad (14.3.13)$$

This methodology was introduced by McIntire (1982) and Ronnelid et al. (1997) and recently confirmed by Mertins (2009). As an example, the IAM (θ_i) and IAM $(\theta_{\perp})$ of a commercial collector are shown in Figure 14.3.14. These data are taken from a commercial simulation tool (Thermoflex® database). From the result, it can be noted that IAM (θ_i), which is not the tracking axis, gives the larger contribution to the IAM, while IAM $(\theta_{\perp})$ exhibits an irregular trend for incidence angles between 0° and 45° because of the secondary reflector shading over primary mirrors, reducing effective mirror aperture area.

To summarize, the optical efficiencies of LFR are defined as:

$$\eta_{optical\_LFR} = \eta_{optical\_LFR}|_{0°} \, IAM(\theta_{\perp})IAM(\theta_i)\theta_{end\_loss} \qquad (14.3.14)$$

The LFR optical efficiency ratio shown in Figure 14.3.15 summarizes the optical efficiency of the collector for every month of the year. IAM (θ_i) presents a maximum during the day at 10 h and 16 h, while the presence of IAM $(\theta_{\perp})$ in LFR leads to a smoother shape, with lower efficiency, in particular for high incidence angles. Compared to a parabolic trough collector (see Figure 14.3.8), which is affected only by θ_i, linear Fresnel has a lower optical efficiency.

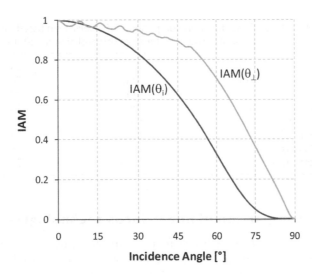

Figure 14.3.14 Longitudinal and transversal incidence angle modifier (IAM) trend as a function of incidence angle (Giostri et al. 2013).

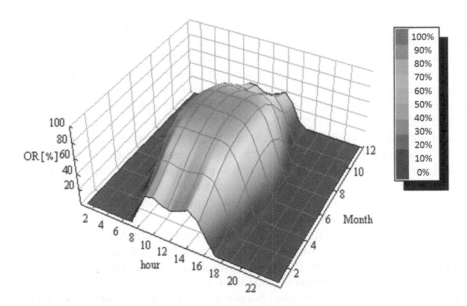

Figure 14.3.15 Annual maps of optical efficiency ratio (OR) for LFR technology (Giostri et al. 2013).

The yearly optical efficiency of the commercial collector with the performances taken from the Thermoflex® database is in the range of 38%; this is 15% points lower than PT. With regard to thermal losses, thermal efficiency is about 90% like PT.

To summarize, linear Fresnel collectors have good potentiality to reduce costs, but they are affected by lower efficiencies than PT. In particular, the concept of several flat mirrors introduces significant optical penalties compared to the parabola system.

Table 14.3.2 Costs for main components of a solar field based on parabolic trough technology (Manzolini et al., 2011a).

Support Structure	$€/m^2$	64
Reflecting mirror	$€/m^2$	54
HCE	$€/m$	200
Driver/controls	$€/m^2$	15.7
Foundations	$€/m^2$	19.2
Assembling	$€/m^2$	22.8
Contingencies	$€/m^2$	8
Solar field BOP	%	30

14.3.11 Cost comparison of linear focus technologies

The last section, dedicated to the linear focus technology, tries to give an idea about the costs of parabolic trough and linear Fresnel. Before going into details, it must be admitted that the two technologies have a different level of development: 30 years have passed since the first solar plant based on PT was built, while the older Fresnel plant is less than five years old. Obviously, this level of development affects the cost of the technology.

Besides this, the cost of a parabolic trough field is in the range of 220 to 300 $€/m^2$ (Giostri et al., 2013; Graf and Nava, 2011; Morin et al., 2012). The cost of the solar field can be split into the component parts shown in Table 14.3.2.

For every linear metre of trough, where the aperture width is about 6 m, the cost of the support structure and reflecting mirrors accounts for more than 50% of the overall cost of the solar field. The remainder is due to the HCE (15%) and to the civil works and drivers.

Current research activity focuses on increasing the aperture area of the parabola in order to reduce the ancillary costs of drive units, sensors, control systems and pylon foundations (Graf and Nava, 2011). In addition, cost reductions can be made in solar field assembly: fewer collectors and smaller labour costs. Finally, another topic is the improvement of mirrors in terms of performance and weight; lighter mirrors can reduce the cost of the structure and foundations.

Where LFR is concerned, it is much more difficult to find reliable cost information in the literature. Few studies discuss the maximum costs that the technology can support to be competitive with PT systems. Results show that Fresnel must be at least 50% cheaper than PT (Giostri et al., 2013; Morin et al., 2012). The target is feasible since the support structure would be cheaper, along with the foundations, drivers and controls. Moreover, the higher concentration ratio can reduce the HCE share of total costs.

14.3.12 Point focus

The second type of solar field is based on point-focusing systems. This technology is based on two-axis tracking systems which have a higher ratio of concentration than single-axis tracking (hundreds of sun vs. tens to sun) with potentially higher working temperatures. In particular, the yearly optical efficiency of two-axis tracking is more constant than single-axis collectors. On the other hand, the very same aspects penalize this configuration from a cost point of view and land occupation of the solar field.

The concept of point focus is to transfer the collected solar radiation to a fluid; the thermal power produced is then converted into electricity by a thermodynamic cycle.

Considering the very high concentration ratio, the thermal losses of central receiver systems are usually less important than optical losses when compared to linear focus systems. On the contrary, since heat fluxes are significant (500–$1000\,kW/m^2$), heat transfer phenomena on the receiver, together with transient conditions, become more important and must be carefully investigated.

Point focus technologies can be divided into two main categories featuring different power outputs: (i) the solar tower (also named power tower) concept for net power output above $10\,MW$; and (ii) the solar dish or dish Stirling system which has a power output of up to $200\,kW$.

Since all of the solar radiation is collected at the same point, there is no necessity for a piping system as in the trough configuration. In particular, in the solar dish configuration the conversion from heat to electricity is performed directly at the focus of the collecting systems, while with power towers, thermal power is transferred to a fluid and then converted to electricity at the bottom of the tower.

By the end of 2012, very few solar plants based on point focus technology were operating worldwide, producing a total installed capacity close to $40\,MW$, with another $17\,MW$ under construction in Spain and $110\,MW$ in Nevada, US (NREL). Over the next few years the construction of 12 new heavy-duty plants in the United States and Spain will yield a total installed capacity of more than $1.5\,GW$. The most important power tower plants are the PS-10 and PS-20 as well as Gemasolar plants in Siviglia (E), though there are several plants under construction, in particular in the United States. With regard to dish Stirling technology, there is just one plant running in the US, producing a total output of $1.5\,MW$. It must be noted that dish Stirling holds the record for solar-to-electricity efficiency, with a value of 31.25%.

From these numbers, it can be noted how this technology is still under development and far behind linear focus technology.

14.3.13   Central receiver systems

Central receiver systems, sometimes named power tower systems, are based on several sun-tracking mirrors (usually called heliostats) which reflect incidence radiation onto a receiver. The concentration ratio of this type of technology, defined as receiver area divided by mirrors area, is usually in the range of 500–1000 suns.

The CR in this particular case is defined as:

$$CR = \frac{total\ mirror\ aperture\ area}{receiver\ area} = \frac{n \cdot A_{mirr}}{A_{rec}} \qquad (14.3.15)$$

where n is the number of heliostats and A the area of each heliostat [m^2].

Since there are few operating plants based on central receiver technology, most of the example described will refer to the PS-10 plant in Spain.

14.3.14 Collector field

The collector field is based on a large number of heliostats with a tracking control system to continuously focus direct solar radiation onto the receiver aperture area. Heliostats are usually supported by a metallic structure with the tracking systems. They can be flat or also a parabolic shape with small curvature (Buck and Teufel, 2009).

The optical efficiency of the solar field, which depends on the heliostat field performances, is equal to the ratio of the net power intercepted by the receiver and the product of the direct insolation and the total mirror area. This parameter includes the cosine effect, mirror properties as reflectivity, shadowing, blocking, aberration and atmospheric attenuation, and receiver spillage.

Since optical efficiency is fundamental for achieving a high solar-to-electricity conversion, and the solar field complexity is significant (the solar field counts hundreds of heliostats), several modelling codes have been developed in order to determine plant performances and to find the optimal configuration (Belhomme et al., 2009; Delsol Modelling Tool; Noone et al., 2012; Pitz-Paal et al., 2011; Wei et al., 2010).

These models can predict optical efficiency as well as the thermal flux density with a good degree of accuracy. For example, in the following Figure 14.3.16 there is a comparison of experimental measurement and the simulation results for the PS-10 power plant at noon on 21 March. The simulation required detailed information about the heliostat number, positions and optical properties, as well as receiver characteristics.

All these simulation tools are based on optical and geometrical correlations which aim at maximizing the yearly optical efficiency. In the literature, there are several solar field configurations which belong to three different groups: surround fields (the heliostats surround the receiver), north fields (the heliostats are just on the north side of the receiver) and a configuration in between these two. In general, north fields have a higher efficiency for high latitude or high incidence angles, while surround fields are typical of locations close to the equator. Examples of two configurations are shown in Figure 14.3.17.

As anticipated, the optical efficiency of central tower systems is higher than single-axis tracking systems since the solar incidence angle on the heliostats is limited. An example of the incidence angle modifier for a central tower system is shown

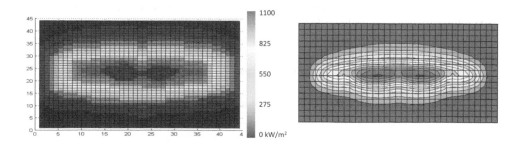

Figure 14.3.16 Measured flux on PS-10 tower (top) and simulation results with Delsol for the same plant on the 21st of March at midday (Colzi et al., 2010).

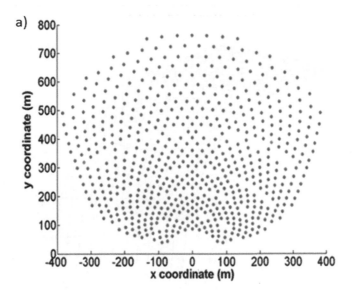

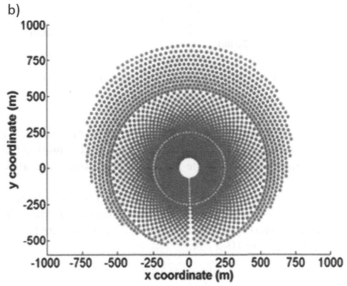

Figure 14.3.17 Representation of optimized field with surround field (a) and north field (b) configurations (Kreith and Goswami, 2007).

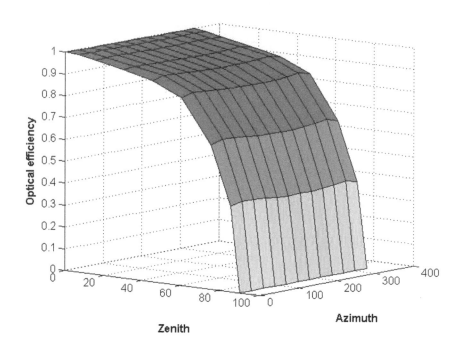

Figure 14.3.18 IAM for PS-10 type central receiver plant.

in Figure 14.3.18: this IAM is taken from the Thermoflex™ database. It can be noted that the resulting optical efficiency is affected by high zenith angles, while it is almost constant with the azimuth angle. Single-axis tracking systems are affected by both angles.

Yearly optical efficiency can be in the range of 61% for a location in France (Garcia et al., 2008), to 64% for Spain; for comparison, the Spanish case has an optical efficiency at a nominal operational rate of 77%. As expected, these efficiencies are higher than linear focus by 10–17%.

14.3.15 Central receiver

Solar radiation is concentrated from heliostats to the central receiver, where it is transferred to a fluid. There are different types of central receiver: external tubular, cavity tubular, billboard tubular and volumetric. The central receiver design depends on the fluid heated and the type of application. For example, volumetric receiver has been suggested for thermochemical applications at high temperature (Pitz-Paal et al., 2011), while tubular design is usually applied to water boilers and heat transfer fluid such as molten salts. This is because in tubular design the high heat transfer coefficient of the fluid (either water or molten salts) can restrain the temperature of the tube. Conversely, when a gas fluid is used, the tubular configuration cannot be adopted because of the

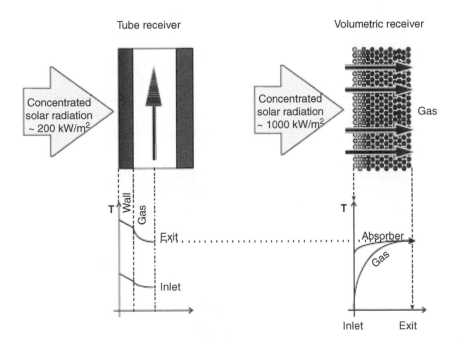

Figure 14.3.19 Tubular receiver (left side) vs. volumetric receiver (right side) (Kreith and Goswami, 2007).

poor heat transfer coefficient. For this reason, a volumetric receiver, which is characterized by a porous structure, increases the transfer surface, reducing the temperature on the receiver even with gas fluids. The two options are shown in Figure 14.3.19.

The height of the receiver is usually in the range of 100–120 m. For example, the centre of the receiver of the PS-10 tower is 100.5 m. In general, the taller the tower, the higher the optical efficiency of the solar field since the cosine effect, shading and blocking among heliostats are minimized. On the other hand, the cost of the tower increases. It should be said that the tower is usually made of concrete. In addition to these configurations, there is also a more exotic design where the tower acts as secondary receiver and solar radiation is transferred to the fluid on the ground (see Figure 14.3.20). This configuration penalizes the optical efficiency of the solar field but reduces the cost of the tower since it has to support the secondary mirror instead of the boiler, with advantages in terms of weight. An example of a beam-down concept is the Tokyo Tech CSP project in Abu Dhabi (Beam-Down Solar Concentration).

The maximum temperature on the receiver depends on the particular heat transfer fluid, the receiver technology and the power cycle adopted to convert the thermal power into electricity. For example, the PS-10 plant is based on a saturated steam boiler working at 250°C and 40 bar. Different fluids have been investigated in order to increase the temperature: molten salts have a maximum temperature of 565°C as in Gemasolar plant (ref), while air can go up to 800°C. Suggested operating conditions for tubular receivers are summarized in Table 14.3.3.

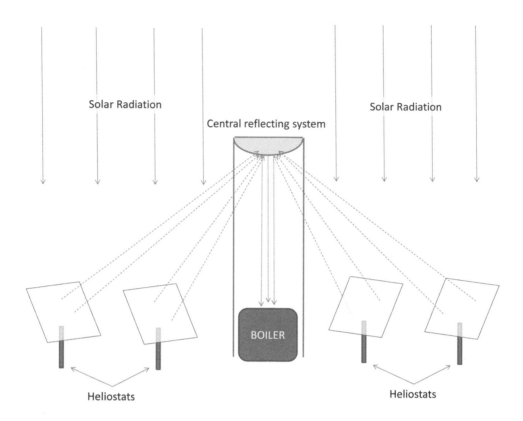

Figure 14.3.20 Beam-Down solar concentration concept.

Table 14.3.3 Main operating conditions for solar tower receivers for two different fluids.

	Water steam	Molten Salt Receiver
Outlet Temperature [°C]	250/525	566
Incident Flux [kW/m²]	350	550
Peak flux [kW/m²]	700	800
Maximum pressure [bar]	100–135	–
Thermal efficiency [%]	80–93	85–90

The application of central receiver technology to thermochemical processes requires even higher temperature: zinc oxide thermal reduction (which is required in water-splitting processes for hydrogen production) works at 1700°C (Schunk et al., 2009), while coal gasification for syngas production works at 1100°C (Z'Graggen et al., 2006).

It is difficult to provide figures for the thermal efficiency of central receivers since there are several different design configurations of the receiver and working

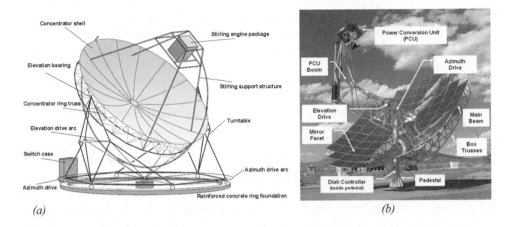

Figure 14.3.21 Principal components and tracking mechanisms adopted in a solar dish system: (a) turntable ring – Eurodish (Schlaich, 2001) and (b) pedestal based mechanism (Stirling Energy Systems, 2013).

temperatures (ranging from 300°C to 800°C) and the type of receiver. As a general consideration, cavity receivers will probably have a higher efficiency than external tubular receivers, but they will probably be penalized from an optical point of view.

To give an idea of a real application, the thermal efficiency of the PS-10 is 92% at nominal rated operation, while the yearly average reduces to 90.2%.

14.3.16 Solar dish

A solar dish system is composed of a reflector obtained by a solid of revolution surface, namely a paraboloid, and a receiver positioned at the focal point of the reflector. The solar collector is oriented towards the sun by a rotational movement along two orthogonal axes by a two-axis tracking system. A thermal engine is placed at the top of the receiver with working fluid heated by the concentrated radiation.

Parabolic dish systems are generally characterized by high efficiency, modularity and flexibility. Another advantage is that the conversion process does not consume water, as with most thermal-powered generating systems. Actually, the highest solar-to-electricity conversion efficiency among solar technologies (31.25%) has been achieved by a solar dish Stirling collector (Stirling Energy Systems) tested at Sandia National Laboratories in the United States.

However, these systems are still experiencing relatively high specific costs and some reliability issues, related to the receiver/engine block working at high temperature. It should be noted that, in relation to costs, there is the potential for economies of scale with mass production in the future, especially where large industrial production volumes are concerned.

The parabolic dish can be manufactured by discrete elements (facets) that approximate the geometry of a paraboloid or by a continuous reflector made with metal membranes or glass mirrors approaching the ideal geometry. Figure 14.3.21 shows the two typical structures of solar dish with tracking mechanisms based on a turntable or pedestal support, respectively. Figure 14.3.21a shows the components and working principle of a 10 kWe prototype, Eurodish, developed in Europe[9]. Figure 14.3.21b, shows a more diffused configuration based on pedestal support (Stirling Energy Systems, USA); in this case the solar concentrator dish structure supports an array of curved glass mirror facets.

With a solar dish collector, concentration can be as high as 3000, with potential high temperatures on the absorber and, consequently, on the thermal engine. The optical design of components and the accuracy of their manufacture determine solar radiation interception and limit the concentration factor. Cost optimization indicates 10–12 m as a maximum diameter for the reflector, therefore limiting the single solar dish to a net power output of about 25–30 kW with a solar radiation of 1000 W/m^2.

Nevertheless, solar dish technology can be applied even to multi-MW solar power plants thanks to its modularity: two examples consist of Maricopa Plant in Arizona and a Power Purchase Agreement (2010) for the project of a 664 MW plant in Calico (USA) involving about 26,000 dish Stirling "Sun Catcher" (25 kWe) made by Stirling Energy Systems. Recent market evolution seemed to convince the project's owner to convert the Calico project entirely to PV technology.

14.3.17 Receiver

The receiver of a dish Stirling has two functions: absorbing solar radiation reflected by the concentrator and transferring this energy to the working fluid of the thermal engine. Usually, the receivers used in parabolic dishes are of the cavity type to reduce radiative and convective losses (Mancini et al., 2003). In commercial solar dishes, two kinds of receiver have been used:

- *Tube receiver*: the absorber consists of several tubes which transfer the heat directly to the working fluid of the thermal cycle. The high temperatures of these absorbers (up to 800°C) make it difficult to use selective coatings due to the great overlap between absorbed and emitted radiation.
- *Reflux receiver* (heat pipes): these receivers use a primary loop with liquid metal (usually Sodium) that evaporates on the absorbing surface. The liquid metal then condenses, transferring the heat to the working fluid of the thermal cycle. This solution implies two heat transfer loops, but takes advantage of the high heat transfer coefficients (800 W/cm^2) and the metal condensation with a more uniform heating of the working fluid.

The absorbing surface of the receiver is generally positioned behind the concentrator focus in order to limit the intensity of the incident thermal solar flux to values of approximately 750 kW/m^2.

Recently, some models have been proposed in the power range of a few kW in order to obtain cost reductions through lighter structures and standardized low-maintenance engines. Infinia is a Stirling manufacturer that proposes a 3.2 kWe solar dish Stirling (Infinia Company), based on a Beta-free piston configuration (see Section 14.5 for details on the Stirling cycle) with helium as the working fluid. The diameter of the

---

[9]Project co-financed by the European Community (under Contract N. Jor3-CT98-0242).

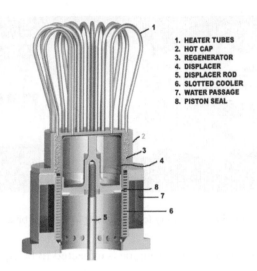

1. HEATER TUBES
2. HOT CAP
3. REGENERATOR
4. DISPLACER
5. DISPLACER ROD
6. SLOTTED COOLER
7. WATER PASSAGE
8. PISTON SEAL

Figure 14.3.22   Heat pipes heat exchanger with Sodium for a Stirling engine.

solar dish is 4.4 m, the concentration factor 800 and the receiver metal temperature about 650°C, resulting in an overall electrical efficiency of 21%.

14.3.18   Power system

The heat-to-electricity conversion in solar dish technology is usually based on a Stirling reciprocating engine, while a Brayton cycle gas turbine is rarely adopted. In Stirling engines for solar applications (Kongtragool and Wongwises, 2003), the working fluid is usually hydrogen or helium. The principle of Stirling engines consists of fluid compressed to about 20 MPa, heated to temperatures generally higher than 700°C owing to the high concentration ratio and then expanded to produce power. The cycle concludes with the cooling of the fluid (further details about the Stirling engine can be found in Section 14.5). Heat pipes with sodium as the intermediate fluid are usually adopted for these engines so as to achieve uniform and controlled temperatures of the fluid (Figure 14.3.22).

Brayton engines, on the other hand, use air as the working fluid, with typical maximum pressures of 0.25 MPa and turbine inlet temperatures of 850°C. In this kind of application conversion efficiencies are limited to 25–27%. In Brayton cycle systems the receiver is a volumetric absorber where solar concentrated radiation passes through a quartz window and is absorbed by a honeycomb-like matrix which provides a high exchange surface (Figure 14.3.23).

An example that represents an attempt to apply a gas turbine cycle to a solar dish system is the adaptation of the Garrett Turbine Engine Company's automotive gas turbine (Stine and Diver, 1994). The gas turbine is based on an open cycle with a centrifugal compressor and a radial turbine operating at 87,000 rpm. Because of the target operating temperatures of this engine, 1371°C (2500°F), a ceramic turbine and ceramic hot-section components are currently under development. At such operating

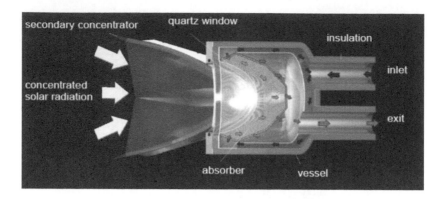

Figure 14.3.23 Receiver cavity for air heating in a gas turbine engine.

conditions the net power output of the engine is expected to be equal to 75 kW with a cycle efficiency of 47%.

14.4 HEAT TRANSFER FLUIDS AND STORAGE

This section focuses on the heat transfer fluids used in concentrated solar power plants and on the thermal energy storage systems. The HTFs collect solar energy in the solar field and transfer it to the power block where it is converted into electricity. Part of this energy can be stored in the TES, hence decoupling the solar radiation from the power production with advantages for the dispatchability of the electricity.

The HTFs from the solar field can be directly used in a turbine to produce power, in this case also becoming the working fluid of the thermodynamic cycle (direct configuration). Alternatively, the fluid can be used to transfer heat to the working fluid through heat exchangers (indirect configuration). In this second case the fluid circulating in the solar receiver can be properly referred to as heat transfer fluid.

Direct configuration, for example, occurs when water is evaporated and superheated inside the absorber tubes of linear collectors or within the receiver of solar tower systems. The resulting system is defined as a direct steam generation (DSG) plant. Because of the high pressure required, as well as related control issues, this technology has so far been applied in only a few commercial plants. DSG plants have occasionally been applied to central receiver towers or to Fresnel reflector systems producing saturated steam. Parabolic trough DSG plants have recently been studied in the DISS project at the "Plataforma Solar de Almeria" producing superheated steam at 400°C/100 bar.

Other direct configuration plants are based on compressed gas, heated inside the receiver and then expanded in a gas turbine (central receiver) or in a Stirling engine (solar dish).

Indirect configuration with HTF is the commonly used configuration in most commercial solar plants. Heat transfer fluids are employed in parabolic troughs, in linear Fresnel reflectors or in central receiver plants as well. In all these systems HTF circulates

Table 14.4.1 Characteristics of the nitrate salts and most used synthetic oil (Kearney et al., 2003).

Property	Solar Salt	Hitec	Hitec XL (Calcium Nitrate Salt)	LiNO$_3$ mixture	Therminol VP-1
Composition, %	–	–	–	–	biphenyl/diphenyl oxide
NaNO$_3$	60	7	7	–	–
KNO$_3$	40	53	45	–	–
NaNO$_2$	–	40	–	–	–
Ca(NO$_3$)$_2$	–	–	48	–	–
Freezing point, °C	220	142	120	120	13
Upper temperature, °C	600	535	500	550	400
Density @300°C, kg/m^3	1899	1640	1992	–	815
Viscosity @300°C, cp	3.26	3.16	6.37	–	0.2
Heat capacity@300°C, J/kgK	1495	1560	1447	–	2319

through the receiver tubes, increasing its temperature. The HTF is then used to generate high-pressure superheated steam in a steam generator coupled with a conventional reheated steam turbine to produce electricity.

Moving to thermal energy storage, in general, solar power plants can operate as stand-alone with TES in order to increase their operating hours, or alternatively they can be coupled with a back-up fossil-fuelled boiler or integrated in conventional power stations. Besides operating hours perspective, thermal energy storage is an inherent capability of indirect configuration plants with HTF, allowing increased electricity production and usually a decrease of its costs.

14.4.1 Heat transfer fluids

HTFs in a solar plant should satisfy many technological requirements: stability at high temperature, low vapour pressure at high temperature, low freezing point, high boiling point, low flammability, low corrosivity and relatively low cost. Firstly, the selection of HTF is crucial for increasing the operating temperature of a solar thermal plant, and hence the efficiency conversion from heat to power (see Section 14.2). Conventional HTFs in commercial solar power plants are synthetic oil and molten salts; they allow good cycle performance and assure trade-off between the above-mentioned issues. Table 14.4.1 lists the operating temperatures and the main characteristics of some HTFs used in commercial parabolic troughs and power towers.

In existing parabolic trough plants, the most common HTF is synthetic oil (Therminol VP-1), in spite of its characteristics of flammability and toxicity, and temperature limitations (up to 400°C). The use of molten salts would primarily increase the maximum temperature of the solar field to 500–600°C, thereby increasing the cycle efficiency of the power plant. Moreover, molten salts are non-flammable, environmentally friendly and cheaper than other HTFs. The main drawback of molten salts, which definitely limit their application, is the requirement for expensive anti-freezing systems because of salt's high solidification temperature (about 120–220°C depending on the fluid). Advanced HTFs as ionic liquids have been already studied and proposed (Moens et al., 2003), with the aim of reducing the freezing point below even room temperature.

Alternatively, liquid metals like sodium can be used in CSP plants as HTF. Sodium's low melting point (97.7°C) and high boiling point (873°C) allows a much larger range of operational temperatures and the use of advanced cycles such as combined Brayton/Rankine cycles in central receiver systems. Nevertheless, the use of sodium poses many technological issues due to its high flammability. Another option is the adoption of gases as HTFs, however their application involves high pressure drops and low heat transfer coefficients.

In the following the commonly used and most promising HTFs are described.

14.4.1.1 Synthetic oil

Synthetic oil is by far the most common solution adopted in solar plants. Commercially adopted synthetic oil is typically a diphenyl/biphenyl oxide. Dowtherm A and Solutia Therminol VP-1 are industrial products that have been used in SEGS plants in the United States or in Andasol plants in Spain. On the other hand, mineral oils are not used in CSP because of their temperature limitations to 300°C, in spite of their low cost (San Diego Regional Renewable Energy Study Group, 2005). The field of exploitable temperature with synthetic oil (see Table 14.4.1) varies between 13°C, where solidification takes place, and a maximum of 400°C, beyond which the phenomenon of thermal cracking occurs. The maximum working temperature in the solar field is therefore typically limited to about 390°C, while the operating pressure of oil is about 12–15 bar in order to keep it in a liquid state and avoid evaporation at normal working temperatures (Giostri et al., 2012). The oil is generally expensive (about 5 €/kg (Manzolini et al., 2011b)), flammable and highly toxic for life and the environment, so that spills and leakages should be avoided. Although collector design has advanced to excellent levels of performance and reliability, occasional spills of HTF may occur, primarily because of piping or equipment failure. Existing plants have reduced HTF spills to very low levels (Cohen and Kearney, 1999); good maintenance practices and the use of ball-joint assemblies rather than flexible hoses between trough collectors are the major contributors to this improvement. In case of any spill or release, the affected collector loop is immediately separated from the rest of the circuit and shut down. An appropriately equipped crew will repair the damage and remove any hazardous wastes, moving it to an on-site bioremediation facility which employs indigenous bacteria to digest the hydrocarbon contamination, and in two to three months restore the soil to a normal condition. Following these operations and maintenance (O&M) best practices, the average fluid losses for SEGS plants at Kramer Junction (USA) between 1996 and 2002 was 2–3% of the site inventory per year (Cohen and Kearney, 1999).

14.4.1.2 Molten salts

Molten nitrate salts are, typically, a mixture of $NaNO_3$ and KNO_3 of variable composition, even if the most commonly used is a mixture known as "solar salt" (respectively 60% $NaNO_3$ and 40% KNO_3). The use of molten salt HTF in a trough plant has several advantages, as has already been pointed out. Depending on the fluid, the solar field output temperature can be raised up to 550°C (see Table 14.4.1), thereby increasing the Rankine cycle efficiency with respect to synthetic oil plants. In addition to this, molten salts have excellent heat transfer properties, are non-flammable, non-toxic, environmentally friendly and cheaper than other HTFs. Furthermore, using molten salt in both the solar field and thermal energy storage system eliminates the need for expensive heat exchangers and also allows for a substantial reduction in the cost of the storage system. In fact, with respect to synthetic oil plants, the HTF temperature variation in the solar field can increase up to a factor of 2.5, reducing the physical size of the storage system for a given capacity. The first example of a parabolic trough plant using a molten salt mixture as HTF and thermal storage fluid is "Progetto Archimede" in Italy, based on ENEA technology collectors reaching 550°C (Manzolini et al., 2011a). Regarding other CSP technologies, molten salts have also been employed in central receiver plants with storage systems (Gil et al., 2010).

The main drawbacks of molten salts are corrosivity at high temperatures and, more importantly, high freezing points (120°C–220°C depending on salt composition). Corrosion issues can be easily solved by adopting stainless steel (AISI 316 or 321) for tubes, piping and storage systems, together with other devices such as a layer of nitrogen at atmospheric pressure in the upper part of storage tanks.

As far as high solidification is concerned, this should be avoided at all costs because it can disrupt circuits and cause mechanical failures in the reverse process of liquefaction. In fact, the specific volume of the mixture increases by about 5% when changing from solid to liquid state. Therefore freeze protection methods should be adopted along with specific operational and maintenance requirements.

Molten salt technology was first developed in the United States for central receiver systems, thanks to the operation of the 10 MWe "Solar Two" plant in Barstow, California. In that plant, before filling the boiler with salt each morning, the receiver was heated to approximately 290°C to reduce thermal stresses and to ensure that solidification of salts did not take place inside the tubes. This pre-heating was achieved by focusing a selected subset of the heliostat field onto the receiver so as to reach a uniform temperature distribution both vertically and circumferentially.

For parabolic trough plants the technical issues are more challenging. During the first start-up of the power plant a boiler or heater is necessary to obtain melting of the HTF; at the same time pre-heating of the tubes and piping in the whole solar field should prevent thermal stresses during plant filling. The solution is to employ extensive heat-tracing equipment on piping and collector receivers (heating is obtained through resistive Joule effect). The same operation can be used in instances of a restart of one loop following a failure or for routine loop maintenance that requires HTF removal.

On the other hand, during operation the solar field cannot be drained and solidification must therefore be avoided. Freeze protection during night-time is achieved by means of a low-flow circulation of hot salt in the solar field: a fossil-fuelled boiler can be used for heating the salt or, alternatively, molten salts can be taken from thermal storage tanks (from a cold tank in case of a two-tank storage system – see the following paragraph). In this way, critical thermal gradients during start-up are prevented. Assuming overnight heat losses of approximately 25 W/m^2, a storage capacity of 1 hour is suitable for freeze protection operation, according to an annual performance calculation (Kearney et al., 2003).

A direct comparison among the various types of molten salts suggests that solar salt has the highest operating temperature limit: it can be used for temperatures up to 600°C. Beyond this temperature solar salt degrades and nitrite formation occurs, creating solid precipitates. In addition, it is one of the lowest-cost nitrate salts (about 0.6 €/kg (Manzolini et al., 2011b)). Mechanical integrity studies for parabolic troughs (Hasuike et al., 2006; Yang et al., 2010) have shown that, with salt temperatures of 600°C and heat flux on the receiver up to 800 kW/m^2, a maximum surface temperature of 700°C is ensured for every geometry of the receiver tube, demonstrating the technical feasibility of high-temperature collectors. However, a major disadvantage of solar salt is its relatively high freezing point of 220°C. Hitec salt offers a lower freezing point (about 140°C) but at higher cost. Finally, HitecXL is a calcium nitrate salt mixture with a lower freezing point of about 120°C. The density, viscosity and heat capacity properties are comparable for all nitrate salts, as shown in Table 14.4.1.

Research into HTFs has led to experiments with a new category of fluids that exhibit very low freezing points: the so-called room temperature ionic liquids (RTIL) (Moens et al., 2003). They are essentially salt-like materials, usually in the classes of quaternary ammonium compounds, that are composed of organic cations combined with organic or inorganic anions, and which are liquid at or near ambient temperature.

14.4.2 Storage

Thermal energy storage systems allow efficient storing of solar energy as heat. Thermal energy storage can avoid the effects of variation of the solar source, which by its nature is highly variable, thereby making the system more flexible and meeting the needs of productive processes. In this way, in CSP plants, the thermal input to the power section can be more constant and in general electricity production can be independent from the collection of solar energy. Alternatively, it is possible to integrate it with fossil fuels or renewable fuels, such as oil, natural gas or biomass, obtaining a so called "hybrid" plant.

With regard to CSP plants, the sizing of thermal storage can be carried out according to different design philosophies:

- **Buffering.** Here, thermal storage is designed to cancel out the effect of clouds transiting over the power block. The quick variation in steam flow and quality could cause severe dynamic variations of the turbine's working conditions. In particular, load variations faster than a few MW/minute can be unacceptable for the engine, bringing about a degradation or affecting the lifetime of the turbine itself. The required storage capacity for "buffering" purposes is relatively small (typically delivering up to 1 hour of power at full load).
- **Displacement of the production period.** In this case, storage allows a decoupling of the electricity production from sunny periods, when energy demand or prices on the grid can be higher. The displacement of production generally involves the use of a medium-high storage capacity (typically delivering between 3 and 6 hours of power at full load), and does not necessarily require an increase in the surface of the solar field.
- **Extension of the production period.** This type of storage is aimed at extending the operating hours of the plant beyond the insolation period. This solution requires a proper sizing of the storage together with an increase in the solar field surface

(in principle the storage can be designed to deliver up 12 hours of power at full load).

When TES is employed to increase the number of operating hours of the plant, the optimization parameter "solar multiple" (SM) is introduced (see Equation 14.2.1). Without TES, the SM is close to 1 (usually between 1 and 1.25) and all the collected thermal power is immediately used. Values higher than 1 mean that the plant can store the excess thermal energy (SM higher than 2.5 normally allows continuous operation throughout the day). From an economic perspective, the advantage of increasing SM is to extend the working hours of the plant, which also operates at higher efficiency thanks to higher load fraction values. On the other hand, the construction cost of the plant increases proportionally to the capacity of the thermal storage system and SM. For these opposing tendencies there is an optimal size for both TES and SM that maximizes the revenues from sales of electricity or equivalently minimizes the levelized cost of energy (LCOE). An optimum has to be found on a case-by-case basis by means of economic analysis (for further details see Section 14.6).

14.4.2.1 Thermal energy storage materials

There are different TES technologies which are characterized by the type of storage medium and its integration into the power plant. The TES medium can be different from the solar field fluid, requiring a heat exchanger to transfer the stored heat to the power plant and a separated loop for the solar field.

Commonly used TES exploit "sensible heat" variations of a substance, which can be measured by the change in its internal energy/temperature. This type of TES consists of a storage medium, a container (usually a tank) and inlet/outlet devices. Tanks must both contain the storage material and prevent losses of thermal energy. The amount of energy input to TES by a sensible heat device is proportional to the difference between the final and the initial storage temperature, the mass of the storage medium and its heat capacity. The amount of stored heat can be expressed as:

$$Q = mc_p\Delta T = \rho V c_p \Delta T \tag{14.4.1}$$

where c_p [kJ/kg K] is the specific heat at constant pressure of the storage material, ΔT [°C] is the temperature variation in the storage, V [m^3] is the material volume and ρ [kg/m^3] is the density of the material. However, besides the density and the specific heat of the storage material, other properties are significant for sensible heat storage: namely, allowable operational temperatures; thermal conductivity and diffusivity; vapour pressure; compatibility among materials; thermal stability; heat transfer coefficients; and cost. Sensible heat storage media can be classified as solid or liquid materials:

- **Solid media** (mainly high-temperature concrete and castable ceramics) are usually used in packed beds, requiring a fluid to exchange heat with the solar field or the power block. When the fluid is a liquid, heat capacity of the liquid in the packed bed is not negligible, and the system is called dual-storage. Packed beds also favour thermal stratification, which can be exploited in a profitable way. Another advantage of the dual system is the potential use of inexpensive solids such as rock, sand or concrete as storage materials. Concrete, for example, is chosen because of its low cost, availability throughout the life of the plant and easy processing. Moreover, concrete is a material with high specific heat, good mechanical properties (especially when subjected to compression strain), thermal expansion with a coefficient close to one of steel, and high mechanical resistance to cyclic thermal loading. When concrete is heated, a number of reactions and transformations take place which influence its strength and other physical properties: resistance to a compression strain at 400°C is about 20% lower than its value at ambient temperature; the specific heat decreases in the range of temperature between 20°C and 120°C; and the thermal conductivity decreases between 20°C and 280°C (Gil et al., 2010). Resistance to thermal cycling depends on the thermal expansion coefficients of the materials used in the concrete. To minimize such problems, a basalt concrete is sometimes used. Steel needles and reinforcement are sometimes added to the concrete to prevent cracking. At the same time, by doing so, thermal conductivity is increased by about 15% at 100°C and 10% at 250°C. Another material that can be employed is rock, which is a more inexpensive TES material costwise, even if its physical and mechanical properties are not as good as concrete (see Table 14.4.2).

• **Liquid media:** liquids (mainly molten salts, mineral oils and synthetic oils) are more commonly used for sensible thermal storage than solids. One important design consideration of a liquid storage system is the need to maintain a separation between the colder fluid and the warmer fluid. There are mainly two ways of separating temperatures: two-tank systems and stratified thermal storage tanks (thermocline tanks). *Two-tank systems* separately store the hot and cold fluid in different tanks; as the hot tank is being filled, the cold is being emptied and vice versa. In this way mixing is avoided but a higher storage volume is required compared to the stratified thermal storage tanks. In fact, in a *thermocline tank* liquid medium maintains natural thermal stratification because of density differences between hot and cold fluid. Thermal stratification ensures the existence of separate volumes of liquid at different temperatures inside the tank and the temperature gradient occurs in a small portion of the tank height, the so-called thermocline. Significant volume and cost reductions are generally achieved with respect to the two-tank configuration. The requirements of this type of TES are that the hot and cold fluids have to be supplied in different parts of the tanks in order to limit fluids mixing: the fluid enters from the upper part of storage during charging, and the cold fluid has to be extracted from the bottom part during discharging. In any case a minimum mixing of hot and cold fluids takes place. For this reason, an accurate design of the geometry of inlet and outlet ducts is necessary to limit fluid velocity. As regards the shape of the tank, a slim storage container is desirable to improve thermal stratification (Dincer and Rosen, 2002). However, the optimum value of the ratio between the tank height and diameter cannot be determined whatever the techno-economic optimization of the power plant.

In addition to storage tanks based on sensible heat storage media, there are other TES solutions employing **latent heat storage media**. In fact, thermal energy can be stored almost isothermally in some substances as latent heat of phase change, as heat of fusion, exploiting solid-to-liquid transition, heat of vaporization, exploiting liquid-to-vapour transition, or even solid structure change (transition from amorphous

Table 14.4.2 Materials and fluids applicable for thermal storage (NREL, 2000).

Storage Medium	Temperature Cold [°C]	Hot [°C]	Average Density [kg/m³]	Average heat conductivity [W/mK]	Average heat capacity [kJ/kgK]	Volume specific capacity [kWh$_{th}$/m³]	Media cost per kg [$/kg]	Media cost per kWh$_{th}$ [$/kWh$_{th}$]
Solid Media								
Sand-rock-mineral oil	200	300	1.7	1.0	1.30	60	0.15	4.2
Reinforced concrete	200	400	2.2	1.5	0.85	100	0.05	1.0
NaCl	200	500	2.16	7.0	0.85	150	0.15	1.5
Cast iron	200	400	7.2	37.0	0.56	160	1.0	32.0
Cast steel	200	700	7.8	40.0	0.60	450	5.0	60.0
Silifica fire bricks	200	700	1.82	1.5	1.00	150	1.0	7.0
Magnesia fire bricks	200	1200	3	5.0	1.15	600	2.0	6.0
Liquid media								
Mineral oil	200	300	770	0.12	2.6	55	0.3	4.2
Synthetic oil	250	350	900	0.11	2.3	57	3.0	43.0
Silicone oil	300	400	900	0.10	2.1	52	5.0	80.0
Nitrite salts	250	450	1.825	0.57	1.5	152	1.0	12.0
Nitrate salts	265	565	1.87	0.52	1.6	250	0.7	5.2
Carbonate salts	450	850	2.1	2.0	1.8	430	2.4	11.0
Liquid Sodium	270	530	850	71.0	1.3	80	2.0	21.0
Phase change media								
NaNO$_3$	308		2.257	0.5	200	125	0.2	3.6
KNO$_3$	333		2.11	0.5	267	156	0.3	4.1
KOH	380		2.044	0.5	150	85	1.00	24.0
Salts-ceramics								
NaCO$_3$-BaCO$_3$/MgO	500–850		2.6	5.0	420	300	2.00	17.0
NaCl	802		2.16	5.0	520	280	0.15	1.2
Na$_2$CO$_3$	854		2.533	2.0	276	194	0.20	2.6
K$_2$CO$_3$	897		2.29	2.0	236	150	0.60	9.1

to crystalline structure). Nowadays, mainly the solid-to-liquid transition has been applied, and substances used under this technology are called phase change materials (PCM). Storage systems utilizing PCM can be reduced in size compared to single-phase sensible heating systems. However, heat transfer design and media selection are more difficult, and experience with low-temperature salts has shown that the performance of the materials degrades after a moderate number of freeze–melt cycles. Phase change materials allow large amounts of energy to be stored in relatively small volumes as a result of a higher energy density, theoretically resulting in cost reduction in comparison with the other storage concepts. For these reasons their application in CSP plants seems to be attractive, even if the potential reduction of costs and technical feasibility are still to be demonstrated.

A further possibility is to exploit the chemical heat of reaction of specific media, obtaining a chemical storage. This type of thermal storage exploits appropriate chemical reactions that are fully reversible. The heat produced is used to promote an endothermic chemical reaction and, if this reaction is completely reversible, the stored heat can be completely recovered by the reverse reaction. Usually, catalysts are required to control the reverse reaction. The main advantages of thermal storage with thermochemical reversible reactions are the high density of storage and the possibility of maintaining the energy stored at a temperature close to that of the environment. Major issues are the complexity and the costs of the system itself and of materials, the effective control of the kinetics of reactions with different operating conditions and aspects related to the properties of the components involved in the reactions (thermodynamic properties, toxicity, flammability, etc.). This type of storage will be the object of study and experimentation in the near future but at present it can be considered far from commercialization.

14.4.2.2 Integration schemes between storage and power plant

There are different ways in which TES systems can be integrated into a solar plant. These depend on the storage medium, heat transfer fluid in the solar field and the type of process coupled to the solar plant. For simplicity, in the following we refer to the case in which the process is a power plant producing electricity.

The integration of TES can use a direct or indirect system. In a direct system, the heat transfer fluid is also used as the storage medium, while in an indirect system a second medium is used for storing the heat.

A **direct storage system** is composed of a plant configuration with two tanks where the HTF produced by the solar field is directly stored in a hot tank. The cooled HTF is pumped to the other, cold tank where it remains until the solar field starts to operate. The adoption of the very same fluid in the solar field and in storage tanks implies that it must have, at the same time, the characteristics of a good HTF and good storage medium. The use of molten salts as an HTF and storage medium allows the solar field to be operated at temperatures higher than current heat transfer fluids such as synthetic oil. This configuration also allows a substantial reduction in the costs of TES systems, owing to the elimination of expensive heat exchangers, improving the performance of the plant (no temperature difference due to the heat exchange) and reducing the LCOE. Moreover it is worth noting that the maximum temperature of HTF can be increased with respect to synthetic oil thanks to the higher thermal stability of salts. Some complications, in the case of molten salts, are due to the high freezing point (from 120°C to 240°C depending on the type of salt used, see also Table 14.4.1) and this means that special care has to be taken to avoid the salt freezing in the solar field. Hence, freeze protection operations must be undertaken, increasing O&M costs. Figure 14.4.1 describes the layout of the Solar Tres plant, a solar tower plant which uses molten salts ($NaNO_3$ and KNO_3) as HTF (Gil et al., 2010).

In an **indirect storage system**, a second fluid is used for storing the thermal power produced by the solar field. Within this configuration, both the two-tank and single-tank systems (the thermocline system, as stated earlier) are the adoptable solutions. The main disadvantages of this configuration relate to the presence of heat exchangers, meaning additional cost, heat transfer irreversibility and the temperature limitation of one of the fluids, usually the one circulating in the solar field.

Figure 14.4.2 depicts a power plant based on the indirect *two-tank storage system*, in which the heat transfer fluid which circulates in the solar field is different from the storage medium. Here, the energy is stored by another medium in the TES (generally

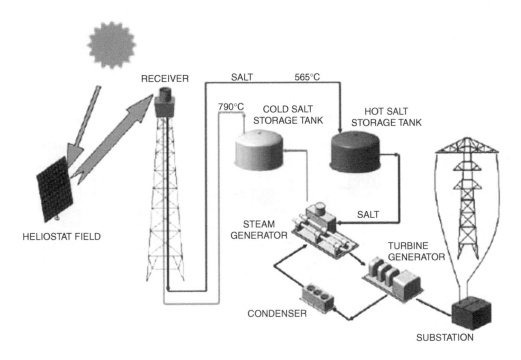

Figure 14.4.1 Example of power plant layout with the two tanks direct storage system (Gil et al., 2010).

molten salts), heated by the HTF circulating in the solar field and pumped through heat exchangers (generally synthetic oil). During a thermal storage charge cycle, a portion of the oil from the solar field is directed to the oil-to-salt heat exchanger: molten salts are taken from the cold storage tank and flow in a counter current arrangement through the heat exchanger. During the discharge cycle, the oil and salt flow paths are reversed in the oil-to-salt heat exchanger; heat is then transferred from the salt to the oil to provide the thermal energy for the steam generator.

Another active indirect storage system is the *single-tank system*, in which hot and cold fluids are stored in the same tank (see Figure 14.4.3). Equipping the system with a thermally stratified tank (thermocline) is a potential way of reducing the cost of the plant. The HTF from the solar field passes through a heat exchanger, heating the thermal storage fluid media. Usually, a filler material can be used to aid stratification and reduce the quantity of heat transfer fluid. Experimental studies performed up to now have found that this filler material, depending on the type, can also act as a thermal energy storage medium. Recent research projects conducted by Sandia National Laboratories have identified quartzite rock and silica sands as potential filler materials (Brosseau et al., 2005). Depending on the cost of the storage fluid, the thermocline can result in a substantially lower cost storage system (Gil et al., 2010). The advantages of the single-tank thermocline system are: reduced storage tank costs and lower cost of the filler materials (rocks and sand). A thermocline system has been estimated to be about 35% cheaper than the two-tank storage system (Brosseau et al., 2005).

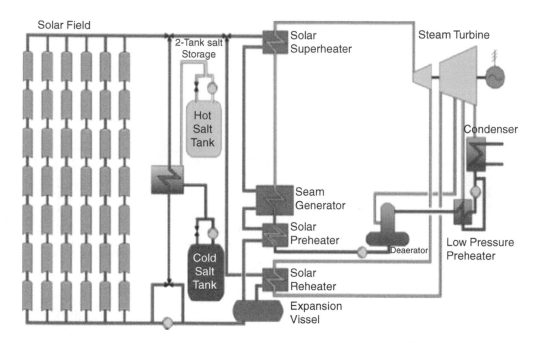

Figure 14.4.2 Example of power plant layout with the two tanks indirect storage system (Gil et al., 2010).

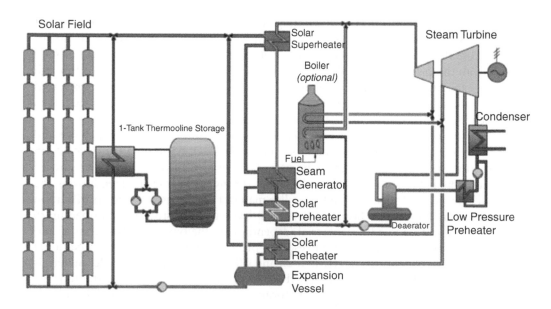

Figure 14.4.3 Example of power plant layout with the one tank indirect storage system (thermocline) (Gil et al., 2010).

However, the design of the storage system is more complex because of the aforementioned mixing issues.

Lastly, an indirect configuration can adopt a **solid as storage medium**. When using concrete, for example, the solar energy of the solar field is transferred through the HTF to the solid storage material system. The storage material comprises a tube heat exchanger which transfers the thermal energy from the HTF to the storage and vice versa. This heat exchanger has a significant impact on overall investment costs and, moreover, the design of geometry parameters such as tube diameter; the number of pipes is also a key issue, increasing engineering costs. While experimental tests have, however, shown long-term instability of the media (Gil et al., 2010), this technology is potentially advantageous for the very low cost of thermal energy storage media.

Another possibility is phase-change materials as storage media. The overall concept of this type of storage system is the same as in solid systems, but the storing material has a melting temperature within the range of the charging and discharging temperatures of the HTF (Bauer, Laing, & Tamme, 2011; Steinmann, Laing, & Tamme, 2010).

For the sake of completeness, there is also the possibility of obtaining an indirect storage system in which the **secondary fluid is steam** (Beckam and Gilli, 1984; Slocum et al., 2011). In principle the advantage of steam storage lies in its direct use in the power block Rankine cycle. The disadvantage is related to the high storage pressures, thus limiting its application to small-size storage tanks, exploiting the buffering of the system. Examples of steam storage are the central-tower commercial plants PS10 (10 MW$_{el}$) and PS20 (20 MW$_{el}$) built in Spain and which started operating in 2007 and 2009 respectively; both of them feature sliding pressure steam accumulators which are able to provide approximately 1 hour of operation at 50% load (Gil et al., 2010).

14.5   FROM HEAT TO POWER

This section discusses the part of the plant dedicated to the conversion into electricity of the thermal power collected in the solar field. Operating plants are based on two different technologies: Rankine cycle and Stirling cycle. In large-scale CSP, which can be either linear focus or power tower, technologies, the power cycle is based on the water Rankine cycle, which is used also in other solid fuel-based technologies such as coal plants and waste-to-energy plants. In solar dish, thermal conversion is based on the Stirling cycle, which, though it has a fairly high efficiency even at small power output, is difficult to scale up.

One of the main characteristics required to power cycles in solar plant is a high conversion efficiency even at partial load; solar-based plant are characterized by daily start-up and shut-down procedures, as well as by partial load operating conditions as a consequence of the variation in solar radiation during the day. In reality, the adoption of large storage systems (i.e. 7.7 hours as in Andasol) reduces the operating time at partial load, something which cannot be avoided.

An example of thermal input for a power cycle with different storage sizes and solar multiple, and in different seasons, is shown in Figure 14.5.1. At large storage size, power production can begin before sunrise because the thermal input stored during the previous day can be used.

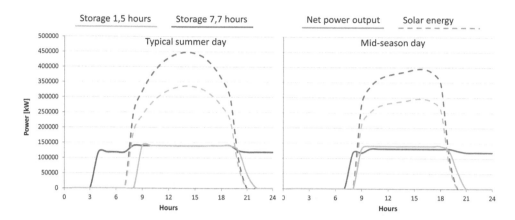

Figure 14.5.1   Power production (solid lines) vs. solar radiation (dashed lines) for two different solar field and storage sizes during a summer day (left side) and mid-season day (right side).

Solar dish, which does not have a storage system, is more affected by the variation in solar radiation during the day.

The efficiency of power section is defined as follows:

$$\eta_{PS} = \frac{Net\ Power\ Output}{Thermal\ input} = \frac{W_{net}}{\dot{m}_{HTF} \cdot (h_{HTF,in} - h_{HTF,out})} \tag{14.5.1}$$

where W_{net} is the net power output generated in the thermodynamic cycle [W], $\dot{m}_{HTF}$ is the HTF mass flow rate [kg/s] and h_{HTF} are the enthalpies of the heat transfer fluid at the inlet and outlet of the power section. W_{net} depends on the thermodynamic cycle adopted for thermal conversion.

Another classification of solar plants is between stand-alone plant and hybrid plant. In stand-alone plant the entire power output of the plant is related to the input of solar power. Another configuration consists of the integration of the solar field in a solid fuel-based power plant. Solid fuel can be either fossil (coal, natural gas) or renewable (e.g. biomass). Integrated solar combined cycle (ISCC) plants, which integrate the solar field in a combined cycle, are already commercially available, as in Indiantown (Florida, USA). Usually, hybrid power plants have a limited solar energy share of total electricity production. The plant mentioned above, for example, has a net power output from the solar plant of 75 MW out of 3.7 GW of the entire plant (Martin Next Generation Solar Snergy Center). Hybrid plants have a higher flexibility than stand-alone, since fossil fuel can be regulated to maintain a constant thermal input to the power section by balancing solar input fluctuation. Moreover, this configuration exploits an economic advantage as a consequence of the power section having a higher number of operating hours, hence reducing its cost per kWh produced.

Since the power section is fundamental for the success of CSP technology and, with actual technologies, has a significant impact on overall plant costs (about 40–45%), research activity has been focusing on other options with the aim of cost reduction and faster start-up, as well as higher conversion efficiency. One of the most investigated candidates is the Brayton cycle, which is cheaper than Rankine and it has faster start-up; however, this technology to be competitive requires temperatures in the solar field above 700°C and/or the adoption of fluid in the solar field in gas phase. These two requirements put Brayton cycle still far from the commercialization phase.

14.5.1 Rankine cycle

The power section (which converts thermal power into electricity) in all operating CSP plants with a net power output above 1 MW is based on water Rankine cycle. Water Rankine cycles are based on a closed loop where heat is supplied externally (i.e. by the solar field). The power output is obtained by the expansion of water in the gaseous phase in a steam turbine. Compression is performed in the water liquid phase, hence requiring less power.

The same cycle is applied in coal-based plant (which accounts for half of the world's power production (Key World Energy, 2012)) and nuclear plant. Even if the concept is the same, thermodynamic conditions and performance of the Rankine cycle in CSP plant are significantly different than other applications. For example, modern coal-based plant – usually called advanced supercritical or ultra supercritical pulverized coal plant – works with live steam pressure at turbine inlet of between 250 and 300 bar, and temperatures of 600–620°C (Manzolini et al., 2011). This is because the temperature of coal combustion is about 1500–2000°C.

The temperature in the solar field ranges between 400°C (linear focus technologies) and 800°C (point focus technologies), hence limiting the maximum temperature of the steam. It must be noted that the only operating solar tower plant works with saturated steam at a temperature of 250°C (ref. PS-10).

In general, the higher is the evaporation pressure and the steam temperature, the higher is the conversion efficiency of the Rankine cycle. For this reason, current research activity is focusing on the development of innovative HCE (see Section 14.3) and the adoption of innovative heat transfer fluids (see Chapter 14.4) which can withstand temperatures of up to 550–600°C for linear collectors and 1000°C for point focus concentrators.

Two different types of power cycle configurations can be adopted: the indirect cycle and the direct cycle. A schematic of the two configurations is shown in Figure 14.5.2.

The indirect cycle configuration is characterized by the adoption of different working fluids between the solar field and the power cycle. In detail, the heat collected in the solar field is transferred to a heat transfer fluid which is sent to the power section. In the power section, the HTF is cooled in heat exchangers by evaporating and super-heating steam. This configuration allows a degree of freedom since the mass flow rate and the pressure in the solar field differ from the power cycle; however, it does require the adoption of expensive heat exchangers.

The direct cycle configuration is based on direct steam generation inside the solar field. In this configuration the adoption of the HTF and all its associated components, in particular the boiler, can be avoided. The maximum temperature in the solar field coincides with the steam temperature at the turbine inlet, with advantages from an efficiency point of view. Moreover, lower circulating pump consumption occurs in the solar field. Obviously there also some drawbacks, namely: (i) higher fluid pressure inside the solar field (about 100 bar vs. about 35 bar of synthetic oil), with penalties

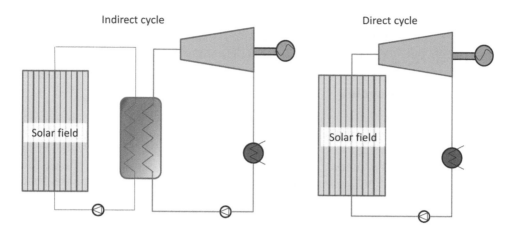

Figure 14.5.2 Schematic of the two configurations for the power section: indirect cycle (left side) and direct cycle (right side).

Table 14.5.1 Stream properties for two indirect layouts (Giostri et al., 2012).

Stream	Fluid type	Synthetic oil (Andasol type)			Molten Salts (Archimede type)		
		M [kg/s]	p [bar]	T [°C]	M [kg/s]	p [bar]	T [°C]
1	HTF	725.8	25.0*	308.0	355.9	15.0*	312.7
2	HTF	618.7	17.6	390.0*	296.6	3.7	550.0*
3	Steam	63.5	95*	370.0*	47.1	115.0*	540.0*
4	Steam	51.6	14.5*	370.0*	37.9	14.5*	540.0*
5	Steam	49.1	0.096	45.0*	30.8	0.096	45.0*
6	Steam	2.6	9.2	312.8	1.6	9.2	471.4
7	Water	49.1	8.7*	143.9	36.3	8.7*	143.9
8	Water	63.5	117.6	260.4*	47.1	141.2	278.9*
9	Water	63.5	100	299.5	47.1	120.0	313.4

(*) Assumptions

from piping and absorber thickness; (ii) potential hot spots on the selective coating, as a consequence of difficult superheater (SH, see below) steam temperature control; and (iii) the absence of reheating (RH, see below) because of the large volumetric flow, which would require a great number of parallel absorber tubes. The only operating plant based on direct cycle configuration is the linear Fresnel collector developed by Novatec solar; this has a maximum steam temperature of 550°C and no RH (Novatec Biosol, 2012).

A schematic of the power section for a conventional indirect cycle is given in Figure 14.5.3. This shows a state-of-the-art power section, where the steam turbine has both a super-heater and a re-heater and seven regenerative bleedings (including one for the deareator). This configuration is adopted in most operating plants, including those of Andasol (*The parabolic trough power plants Andasol 1 to 3*, 2008) and SEGS.

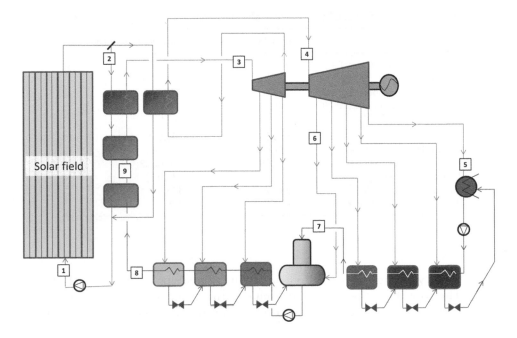

Figure 14.5.3 Power block scheme for an indirect cycle.

In general, a simpler layout can also be adopted, with fewer regenerative bleedings or without RH. However, the higher the number of regenerators, the higher the power cycle efficiency since all the heat collected in the solar field is used for steam evaporation and superheating, rather than economization. Reheating is fundamental to achieving high conversion efficiencies, as well as a high vapour fraction at the turbine outlet.

HTF thermal power is transferred to the power cycle in four different heat-exchangers: an evaporator, an economizer, and a superheater and a reheater. Different heat exchanger configurations can be adopted: the RH section can be in parallel with the SH only, or with the whole boiler, as in SEGS VI plant (Patnode, 2006). Optimal configuration depends on heat transfer and steam temperatures, and is usually selected to minimize HTF boiler outlet temperature and heat transfer irreversibilities inside the boiler.

The steam turbine is usually divided into a high pressure (HP) section and a low pressure (LP) section; between the two sections the steam is sent to the boiler for reheating. The maximum steam temperature that occurs at the turbine inlet, and after reheating, is in the range of 371°C when synthetic oil is used as the heat transfer fluid; this temperature occurs in all existing plant with indirect cycle configuration because of the maximum working temperature of the synthetic oil (400°C) – a conservative maximum temperature in the solar field of 391°C is usually taken. If molten salts are adopted as the heat transfer fluid, steam temperature can be increased up to 500–540°C. For example, the Archimede plant in Sicily (Italy) works with a steam temperature of 525°C at the turbine inlet. The drawbacks of this HTF have already been discussed in the previous section. Because of the moderate maximum temperature

Table 14.5.2 Comparison between air cooled and evaporative tower condenser.

	Air cooled condenser	Evaporative tower condenser
Ambient temperature (°C)		23
Humidity (%)		60
Air dry bulb temperature (°C)		23
Air wet bulb temperature (°C)		17.74
Heat rejected (MW)		86.2
Air flow (kg/s)	5880.5	1386.9
Air outlet temperature (°C)	37.41	32.74
Condensing pressure (bar)	0.096	0.0689
Water make-up (kg/s)	–	36.84
Fan consumption (kW)	842.2	346.6
Pump consumption (kW)	–	298.3

of the steam cycle, the evaporating pressure is in the range of 100 bar and RH around 25 bar. The pressures are set in order to limit moisture content in the last stage of of the steam turbine because water droplets during expansion can be detrimental for turbine blades.

The steam cycle, as a closed loop, requires a heat exchanger to discharge the heat: in Rankine cycle this heat exchanger is called a condenser because the steam condenses. Condensers can be based on evaporative cooling towers as well as dry coolers. Considering that solar thermal plants are usually placed in arid places with high solar radiation and low precipitation, dry cooling condensers are more typical. Air condensers are more expensive than evaporative cooling systems and are more sensitive to the ambient conditions: the air temperature at the condenser inlet is the dry bulb temperature, and the entire heat rejected is transferred as sensible heat to the air. Conversely, evaporative towers transfer the heat rejected from the condenser also as latent heat by evaporating water in the air, requiring lower amounts of air. In addition, the air temperature at evaporative tower inlet is the wet bulb, which is lower than dry bulb particularly in dry conditions.

Comparisons between evaporative tower and air cooled condenser conditions are summarized in Table 14.5.2. The evaporative tower has lower fan consumptions (346.6 kW vs. 842.2 kW) because of the lower air mass flow. This advantage is partially balanced by the consumption of the recirculating pump, which is not present in air condensers, and the required water make-up of 36.8 kg/s; the cost of water replacement can be significant because of solar field location. Both condensate extraction and feedwater compression are performed by dedicated pumps which have a motor driver.

The number of regenerative bleedings can be variable (between four and seven) with the pressure selected in order to optimize heat exchange in the regenerators. The target is to minimize the temperature difference between the hot stream and the cold stream in order to limit entropy production, hence maximizing efficiency.

When direct steam generation configuration is adopted, the power section has a much simpler layout since the entire heat exchanger section can be avoided. The layout of a DSG configuration and the thermodynamic properties of main streams are shown

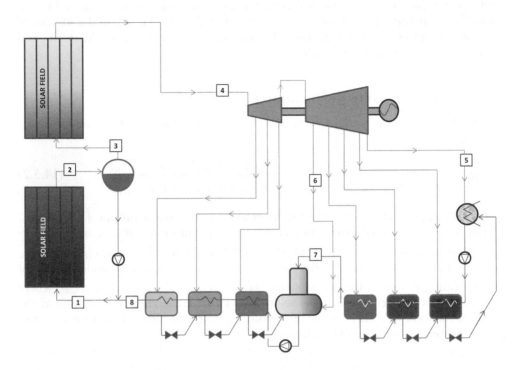

Figure 14.5.4 Power block scheme for a direct cycle.

Table 14.5.3 Stream properties for the direct layout (DSG) (Giostri et al., 2012).

Stream	Fluid type	M [kg/s]	p [bar]	T [°C]
1	Water	73.7	100*	279.3*
2	Steam	73.7	91.3	304.7
3	Steam	56.8	89.3	302.8
4	Steam	56.8	81.0	540.0*
5	Steam	36.7	0.096	45.0*
6	Steam	2.4	9.2	256.1
7	Water	44.1	8.7*	143.9
8	Water	56.8	100*	260.4*

(*) Assumptions

in Figure 14.5.5 and Table 14.5.3. The DSG plant considered has steam SH, like the more recent Novatec technology, but no RH.

The feedwater line is composed of seven regenerators, one being the deareator. The steam turbine, with no RH, has a lower vapour fraction at the outlet compared to the indirect cycle. Evaporation in the solar field is based on a reboiler concept: steam quality at the evaporation section outlet is kept below 1 (i.e. 0.77 (Forristall, 2003)) to guarantee good wettability of the collector, thus preventing the formation of hot spots. The separated liquid fraction is sent back at the inlet of the solar field by a recirculation pump and is mixed with feedwater at 260°C. The main drawbacks of the DSG configuration are the absence of available storage suitable for the steam, and temperature control in the SH section.

14.5.2 Rankine cycle performance

When a Rankine cycle is adopted for power conversion, the power section efficiency is defined as:

$$\eta_{PS} = \frac{W_{net}}{\dot{m}_{HTF} \cdot (h_{HTF,in} - h_{HTF,out})} = \frac{W_{gross,tur} - \sum W_{pumps} - W_{aux}}{\dot{m}_{HTF} \cdot (h_{HTF,in} - h_{HTF,out})} \quad (14.5.2)$$

where $W_{gross,tur}$ is the power output generated in the steam turbine [W], W_{pumps} accounts for pump (feedwater and condensate) consumptions, W_{aux} considers auxiliaries consumption as condenser fan, $\dot{m}_{HTF}$ is the HTF mass flow rate [kg/s] and h_{HTF} are the enthalpies of the heat transfer fluid at the inlet and outlet of the boiler. In case of direct steam generation, the denominator will be $\dot{m}_{steam} \cdot (h_{steam,TV} - h_{water,in})$, where m_{steam} is the steam generated in the solar field [kg/s], $h_{steam,TV}$ is the enthalpy at the outlet of the solar field (i.e. turbine inlet) and $h_{water,in}$ is the enthalpy of the water after the feedwater section (i.e. solar field inlet).

Thermodynamic conditions and the performance of different power cycles at design conditions are summarized in Table 14.5.3 and Table 14.5.4. The examples given reproduce existing plant beside indirect molten salts; this is because there is no stand-alone plant running with this HTF.

Molten salts can work at higher temperature, increasing the power cycle efficiency by 12% compared to conventional synthetic oil configurations. Direct cycle configuration with superheating can achieve almost the same efficiency owing to higher temperature at the turbine inlet, whilst it is penalized for the absence of RH. Lastly, saturated steam cycle in direct configuration is penalized from an efficiency point of view by about 3% points versus commercial synthetic oil.

A comparison of the yearly power block efficiency is more difficult since, besides cycle configuration and HTF, this depends also on condensing technology and on thermal storage size: the bigger the thermal storage, the closer power block efficiency is to the nominal working conditions at constant thermal input. Condensing technology strongly affects power block efficiency as, during the summer when solar radiation is higher, ambient temperature also increases, together with the condensing temperature. An evaporative tower is less penalized than an air-cooled condenser, since the wet bulb temperature is more constant than that of the dry bulb, in particular in sites with low relative humidity, such as deserts.

14.5.3 Stirling cycle

The Stirling engine is a closed cycle that has been designed for small solar power applications (from a few kW up to 100 kW). It has a potential for high cycle efficiency, the ideal Stirling cycle equalling the efficiency of the Carnot cycle. Efficiency is one

Table 14.5.4 Performances at on-design conditions for five selected technologies

	Indirect Andasol (A. Giostri et al., 2013)	Indirect Molten Salts (Andrea Giostri et al., 2012) Archimedes	Direct Superheating (Andrea Giostri et al., 2012) Supernova	Direct saturated (A. Giostri et al., 2013) Novatech	Direct saturated (Colzi et al., 2010)
HTF working temperature (°C)	297.3–391.0	300.0–550.0	–	–	–
T steam SH (°C)	371	525	540.0	270.0*	260.0*
Pressure solar field (bar)	25.0	15.0	100.0	66.0	45.0
Pressure at turbine inlet (bar)	95.0	115.0	81.0	55.0	40.0
Steam mass flow @turbine inlet (kg/s)	63.5	47.1	56.8	77.2	18.0
RH temperature (°C)	371	525	–	–	–
RH pressure (bar)	14.5	14.5	–	–	–
Condensing pressure (bar)	0.096	0.096	0.096	0.096	NA
N° of regenerators	6	7	7	3	4
Temperature at boiler inlet (°C)	234.8	278.9	–	–	–
Temperature at solar field inlet (°C)	–	–	260.4	195.9	NA
Net power output (MW)	50.0	50.0	50.0	50.0	11.0
Thermal input (MW)	144.5	128.9	130.2	156.7	35.8
Power block efficiency (%)	34.6	38.8	38.4	31.9	30.7

*Saturated steam.

of the main targets for solar power cycle design because of the reduced size of the collector area, and thus lower costs, for a given power output. Most proposed Stirling applications (Stine & Diver, 1994) are for small (10 to 100 kW) engines placed at the focus of a parabolic dish concentrator. In point of fact, for small power output, the net efficiency of Rankine or Brayton cycle-based engines is seriously degraded, favouring the high efficiency potential of the Stirling engine. On the other hand, component size and costs increase significantly for large-scale Stirling engines.

The ideal Stirling cycle results from two constant-temperature and two constant-volume processes with working fluid in the gas phase. Figure 14.5.5 shows the four processes in pressure-volume and temperature-entropy diagrams. Energy is exchanged (either produced or absorbed) by the cycle only during the constant-temperature processes; however, heat must be transferred during all four processes.

Because the processes at constant volume (2-3 and 4-1) involve an equal amount of heating and cooling of the working fluid, in the hypothesis of ideal gas behaviour, a regenerator may be used which transfers the heat internally to the cycle, between expanded and compressed gas. As has already been pointed out, with an ideal heat exchanger, heat would only be introduced and discharged from the cycle at constant temperatures, ideally obtaining the same efficiency as the Carnot cycle. However, the

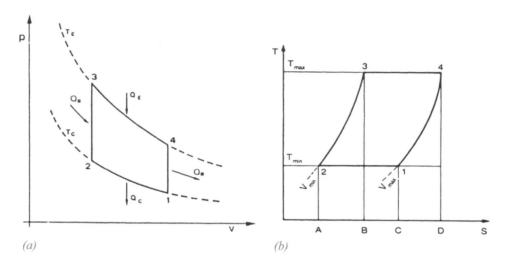

Figure 14.5.5 Ideal Stirling cycle representations in p-V (a) and T-s (b) diagrams.

processes occurring in the engines are not ideal, leading to lower efficiency than Carnot. The main penalties of a real engine that cause performance decay (Walker, 1980) are:

- **Heat exchange losses:** (i) non-isothermal processes (engines are adiabatic in nature; isothermal processes are unfeasible in practice); (ii) ΔT in heat exchangers due to finite surfaces;
- **Fluid dynamic losses:** (i) pressure losses (especially in the regenerator); (ii) losses due to the dead volumes (volumetric efficiency); (iii) leakages of fluid;
- **Losses due to kinematics:** isochores are not perfectly executed in normally adopted mechanisms;
- **Electrical and mechanical losses.**

As regards the mechanical device, a reciprocating piston-cylinder arrangement is normally used. The kinematic transmission is based on a rod-crank mechanism in which the pistons move according to a sinusoidal law. Another kinematic transmission involves a Wobble-Joke or a Swash-plate mechanism, with similar effects. As a result the four transformations that compose the cycle are partially overlapped in a real cycle, with consequent reduction of overall efficiency (Figure 14.5.6).

14.5.4 Stirling configurations

Depending on the mechanical scheme adopted, existing Stirling engines can be classified as one of three basic arrangements (Martini, 1983; Stine, 2007), as represented in Figure 14.5.7.

The **Alpha configuration** is characterized by the presence of two distinct cylinders with two corresponding pistons. Both pistons have the same pressure at any given time. The crank mechanism is generally kinematic, usually of a rod-crank type. In this way, a sinusoidal motion is induced on the pistons, with a phase shift close to 90°.

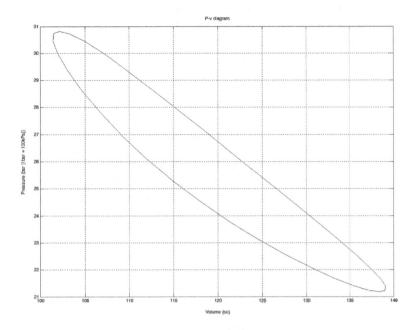

Figure 14.5.6 Representation of a real Stirling cycle in a p-V diagram.

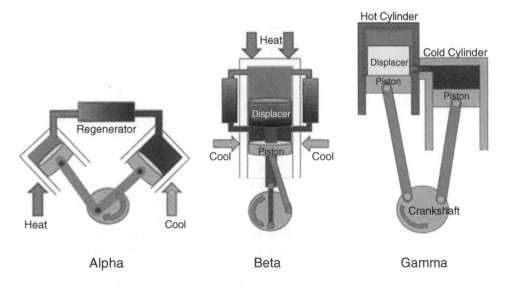

Figure 14.5.7 Main types of Stirling engine component arrangements.

The Alpha configuration characterizes most of the Stirling engines made in the last century: simplicity of construction and the possibility of using technologies derived from the automotive industry. In spite of this simple design, the Alpha configuration has several drawbacks that currently limit its spread. These are mainly related to:

(i) sealing problems on the hot cylinder; (ii) lubrication issues due to the need to lubricate both the pistons and the mechanism; and (iii) high specific volumes derived from the particular geometry of the engine and by the presence of two pistons. The latter is a factor that increases the specific costs of the system and precludes its use for applications requiring compact solutions. The Eurodish collector discussed in Section 14.3 exploits the Alpha scheme.

The **Beta configuration** is characterized by the presence of a single cylinder and two components moving inside the same cylinder. The first is the piston, responsible for the phases of compression and expansion, and the second is the displacer accomplishing the processes at constant volumes. The displacer is a mechanical device similar to the piston but, unlike the latter, it is not equipped with seals: its surfaces must support only the pressure difference related to the fluid dynamic losses in the regenerator. Piston and displacer may have a different rotating mechanism. In general, they have a sinusoidal motion, associated with a phase shift of 90° between piston and displacer. Recently, a novel configuration called "Beta free piston" has been developed by some manufacturers and is characterized by the absence of kinematic connection between piston and displacer (Thombare & Verma, 2008). The motion of piston and displacer is entrusted to resonance phenomena and regulated by the natural frequency of a linear generator, connected to the grid directly or through inverters. This technology is employed, for example, by Infinia in its 3.2 kWe solar dish collector described in Section 14.3. In general, the main advantages of the Beta configuration are: (i) its compactness due to the presence of a single cylinder for both the compression and expansion phase; (ii) high power density; (iii) reduced dead volumes; and (iv) the possibility of developing solutions intrinsically sealed with the adoption of a free-piston mechanism (i.e. it is possible to pressurize both the working fluid and the chamber containing the generator, avoiding leakages). Figure 14.5.8 shows a schematic view of a Beta free-piston Stirling generator. It can be noted that the whole engine is encapsulated to avoid leakages of fluid.

Another option is the **Gamma configuration**. This differs from the Beta configuration by having two separate cylinders which contain, respectively, the displacer and the power piston. The Alpha configuration also has two cylinders but both feature a power piston to implement the compression and expansion phases. Like the Beta configuration, the displacer has to overcome only the pressure losses generated during the displacement of the fluid from the hot side to the cold side of the engine, and vice versa. The power piston generates the phases of compression and expansion. The Gamma configuration is rarely used because of its modest efficiency, low specific power and high vibrations generated by the kinematics required to operate the system.

In addition to these three classic configurations there is one further scheme that is a variant of Alpha, namely the **double-effect Alpha configuration**. This configuration derives from the Alpha with three or four interconnected pistons: the lower part of each piston acts as volume of compression and is interconnected to the expansion chamber (upper volume) of the next piston. Figure 14.5.9 shows the working principle of an engine based on a double-effect Alpha configuration. This configuration retains the simple design feature of the Alpha engine, bringing about some major improvements, namely: (i) an increase in the specific power (W/cm^3) owing to the double action of the pistons, which allows the displacement of the motor to be halved: (ii) a reduction in leakages (the only seal towards the outside is the ring which allows sliding of the connecting rods of each piston); (iii) high continuity in the movements; and (iv) reduced

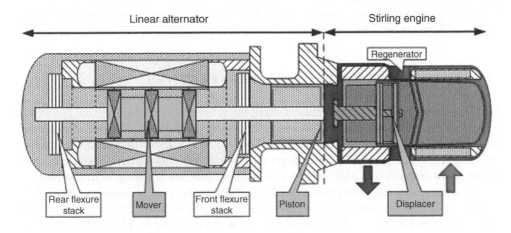

Figure 14.5.8 Schematic representation of a Beta free-piston Stirling generator.

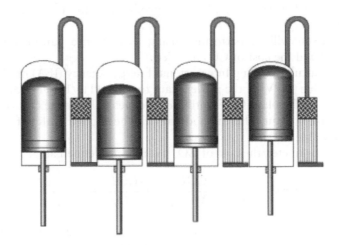

Figure 14.5.9 Double-effect Alpha configuration.

inertial effects due to the presence of three or four pistons, respectively phase-shifted by 120° or 90°, which improve mechanical efficiency since the dual action works the crank in one direction while compression is directly subtracted from expansion.

14.5.5 Stirling working fluids

The working fluid in a Stirling engine is employed in a closed cycle and has to satisfy the following requirements: (i) good heat transfer coefficients; (ii) low viscosity; (iii) high thermal stability; and possibly (iv) low cost (Urieli and Berchowitz, 1984).

Light molecule fluids like H_2 or He are considered suitable fluids for solar dish Stirling applications because they meet the majority of these requirements; however, a small caveat concerns leakages. In fact, due to the small size of their molecules, these fluids tend to escape through seals. The problem is more serious for H_2 because of its highly flammable nature. Two of the examples presented in Section 14.3 employ helium as a working fluid, namely Eurodish and Infinia engines, while Stirling Energy Systems "Sun catcher" uses hydrogen.

14.6 ECONOMICS AND FUTURE PERSPECTIVES

This last chapter tries to give an indication of the economics of solar thermal plants and compares it with commercial technologies for power production. The selected competitive technologies are natural gas combined cycles (NGCC) and advanced supercritical pulverized (ASC) coal plants: the adoption of fossil fuel-based plants occurs because of the more certain costs and, besides renewables, they are the only kind installed.

In addition to the above-mentioned reference cases, two technologies with CO_2 capture will also be considered. Fossil fuel-based plants with CO_2 capture are the competitive technologies of renewable for power production. This is because CO_2 is seen as one of the main issues of fossil-fuelled power plants: CO_2 concentration in the atmosphere amplifies global warming (some believe that it is also the main reason behind global warming) and power production accounts for 35% of world CO_2 emissions.

The comparison will be performed using as reference parameters the cost of electricity (COE) and the cost of CO_2 avoided.

The COE is calculated using International Energy Agency (IEA) models by setting the net present value (NPV) of the power plant to zero (PH3/14; PH4/33). This can be achieved by varying the plant COE until the revenues balance the cost over the whole life time of the power plant. This methodology can be applied both to fossil fuel-based plants and renewable ones.

The second parameter, the cost of CO_2 avoided, is defined as:

$$Cost\ of\ CO_2\ avoided = \frac{(COE)_{inn} - (COE)_{ref}}{(CO_2 kWh^{-1})_{ref} - (CO_2 kWh^{-1})_{inn}} \qquad (14.6.1)$$

where *ref* is the reference technology for power production and *inn* is the innovative plant which can be either renewable-based or fossil fuel-based with CO_2 capture. The cost of CO_2 avoided represents the additional cost of electricity consumed to avoid the emission of 1 kg of CO_2 into the atmosphere. Another interpretation of the cost of CO_2 avoided is the value of carbon tax that makes the COE for innovative plants equal to the reference plant.

The COE and cost of CO_2 avoided assumed for the reference cases are summarized in Table 14.6.1 (Franco et al., 2010; Gazzani et al., 2012a, 2012b; Manzolini et al., 2012). The calculated cost of electricity was determined assuming 7500 hrs/y, which is typical of base load plants such as NGCC and ASC. Considering recent renewable energy diffusion (in Europe at least), it is difficult to predict what the operating hours of a power plant are going to be, and whether power plants with CO_2 capture will be assimilated into green-energy sources or not.

Results show that the COE for conventional NGCC and ASC plants is similar; moreover, the cost of CO_2 avoided for the two reference cases is also pretty close and in the range of 50 €/t$_{CO2}$. To give an idea about the current price of CO_2 emissions

Table 14.6.1 Summary of reference cases performances and economics (Franco et al., 2010; Gazzani et al., 2012a, 2012b; Manzolini et al., 2012).

	NGCC	NGCC with CO_2 capture	ASC	ASC with CO_2 capture
Net electric efficiency [%]	58.34	49.9	45.25	33.55
CO_2 emissions [kg_{CO2}/MWh_{el}]	352	41	772	104
Specific investment costs [$€/kW_{net}$]	630	970	166	2556
COE [$€/MWh$]	54.1	69.1	54.8	85.5
Investment costs [$€/MWh$]	9.6	15.6	23.1	35.7
Fixed costs [$€/MWh$]	3.9	5.2	4.9	10.7
Consumables costs [$€/MWh$]	0.6	1.4	2.8	7.6
Fuel costs [$€/MWh$]	40.1	46.9	23.9	32.2
Cost of CO_2 avoided [$€/t_{CO2}$]	–	48.5	–	46

trading on the market, the average value in 2012 was between 6 and 9 $€/t_{CO2}$: today, CO_2 capture technologies are not economically sustainable.

The COE for fossil fuel-based plants can be split into four different cost centres: (i) investment costs (i.e. equipment and installation costs); (ii) fixed costs (i.e. labour and maintenance costs); (iii) consumables (i.e. water make-up and chemicals); and (iv) fuel costs. Coal-based plants are characterized by high investment costs and low fuel costs, which is the opposite for NGCC.

In solar plants, almost the total cost of electricity arises from investment costs since the fuel is "free" and labour and maintenance have a limited impact.

At current technology development rates, CSP plants have higher investment costs than fossil fuel-based plants. Moreover, the yearly operating hours of CSP are in the range of 2000-4000 hrs rather than 7500 hrs for fossil fuel-based plants. This is because the energy source is not always available (there is no solar radiation during the night) and it varies during the day even in clear sky conditions (solar radiation is much lower in the morning than at noon). Because of the limited amount of CSP installed worldwide, there is little information about the actual plant costs; even less information is available for solar tower and linear Fresnel technologies which are still under development. Another parameter affecting the cost of electricity is plant location: electricity production is almost proportional to the yearly available solar energy, which can vary from 2100 kWh/m² in Seville (Spain) to 2600 kWh/m² in the Mohave Desert in California (USA). Finally, plant design and storage optimal size depend on the solar energy available throughout the year. Examples of relative COE variation as a function of the site – Las Vegas (USA), direct normal irradiation (DNI) equal to 2600 kWh/m²; Seville (Spain) and Darwin (Australia), DNI equal to 2100 kWh/m² – storage size and solar multiple are shown in Figure 14.6.1 (Astolfi et al., 2011).

Observing the LCOE curves displayed in Figure 14.6.1, the best results are obtained by the Las Vegas site, which has high storage capacity: the adoption of large storage is necessary to limit defocusing during summer days. LCOE in Las Vegas is equal to 139 €/MWh, which is about 20% lower than the Seville plant where LCOE is equal to 168 €/MWh); this value is very close to the difference in terms of available

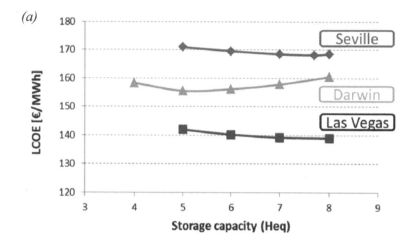

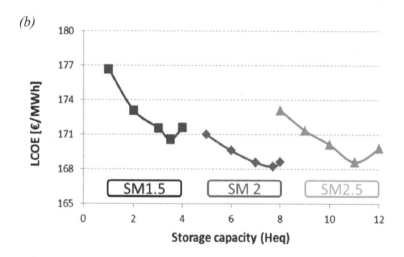

Figure 14.6.1 LCOE variation with storage size for three different solar plant location designed for SM=2 (a) and different SM and storage size in Seville site (b).

solar radiation. Seville is the less attractive site because of its lower solar radiation. Finally, Darwin falls in between the other two sites (LCOE equals 155 €/MWh). It requires a smaller storage (5 h_{eq} compared to 7.7 h_{eq} for Seville and 8 h_{eq} for Las Vegas) because of its lower latitude and a more homogeneous solar distribution through the year. A second study was carried out for three different solar multiple sites (1.5, 2, 2.5) in Seville – see the right side of Figure 14.6.1. LCOE curves have a minimum for all three sizes. The optimal heat storage capacity is a compromise between higher plant cost and lower defocusing. As expected, Andasol I storage size (7.7 equivalent hours) proved to be the best of the considered SM.

Table 14.6.2 Summary of three selected CSP plants. performances and economics for Las Vegas site: Morin (Morin et al., 2012) and Politecnico (Giostri et al., 2013)

		Nevada Solar One Type		Novatec type		Andasol Type	
Solar field Technology		PT		LFR		PT	
Plant configuration		Indirect Cycle		Direct cycle		Indirect Cycle	
Storage		No		No		Yes (7,7 hours)	
Nominal net power	MW	50		50		44.5	
Turbine gross power	MW	54.6		52.5		50.2	
Thermal storage	hours	–		–		8	
Total electricity production	MWh/y	97818		76136		206484	
Overall efficiency	%	16.05		10.2		15.6	
Total mirror surface	m^2	235,527		289,101		509,204	
Total land surface	m^2	704,252		593,205		1,522,577	
Equivalent hours	kWh/kW$_{nom}$	1,956		1,522		4,666	
Assumptions		Politecnico	Morin	Politecnico	Morin	Politecnico	Morin
Land specific cost	€/m^2	0	7	0	7	0	7
Solar field	€/m^2	220	275	125.73*	150.1*	220	275
Solar field overall costs	M€	51.8	64.7	36.3	43.4	112.2	150.5
Power block specific cost	€/kW$_{gross}$	667.5	882.64[a]	605	800.00	667.5	882.64[a]
Power block overall costs	M€	36.5	44.1	30.21	40.0	33.5	44.3
Storage costs	M€	–	–	–	–	42.8	42.8
Indirect cost & Contingencies	% plant cost	31	20	31	20	31	20
Total plant costs	M€	115.7	137.5	89.3	113.9	247.0	272.5
O&M	M€	2.5	3.6	2.1	3.1	4.1	4.6
Consumables	€/MWh	2.9	–	2.9	–	2.9	–
Reduction	%	–	–	42.85	45.42	–	–
COE	€/kWh	156.3	176.1	156.3	176.1	139.0	161.4

*Calculated to achieve the same COE of the reference parabolic trough case Nevada Solar One type.

A tentative economic comparison of existing linear CSP systems is summarized in Table 14.6.2 using Las Vegas as the reference site (2600 kWh/m^2). An economic assessment is made for three different cases: Nevada Solar One type plant (synthetic oil as HTF, no storage), Andasol type plant (synthetic oil as HTF, thermal storage) and linear Fresnel technology developed by Novatech (DSG, no storage). For linear Fresnel, with no information available about the collector costs, the costs are calculated based on the LCOE of the Nevada Solar One type plant (neither has any storage).

For all of these cases, two different sets of assumptions were made in order to determine their impacts on the overall results.

The calculated COE ranges between 139.0 €/MWh and 176.1 €/MWh; as expected, these figures are higher than reference cases based on fossil fuels both with and without capture. Focusing on capital costs, solar field has the larger contribution, in particular when TES is considered. The specific investment costs for the Nevada

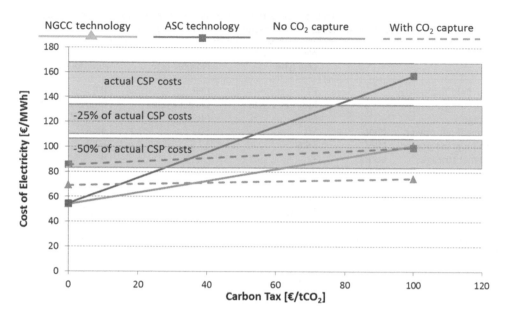

Figure 14.6.2 Cost of Electricity [€/MWh] for different solar field costs; reference cases with COE as function of the carbon tax is also shown.

Solar One case are about 2300 €/kW$_{net}$ higher than ASC and NGCC cases. This is a consequence of solar field costs as well as limited power output (50 MW): CSP suffers from a small-scale power section compared to conventional fossil-fuel plants, with drawbacks from an economic point of view. As a term of comparison, fossil fuel-based plants have a net power output between 750 and 850 MW, which is more than 10 times larger than the typical CSP size, with scale-up cost advantages. With regard to Fresnel technology, the equivalent cost should be about 45–50% lower than parabolic trough. This level of cost saving might be achieved as a consequence of improved receiver and tracking systems, cheaper mirrors and structure. Moreover, the investigated LFR has a concentration ratio twice that of PT, reducing absorber-specific costs.

The same analysis could also be carried out on the power tower system, however the overall results would show no change since the system has yet to show economic benefits to match PT, and it shares some negative aspects with the linear focus technology in having a small-scale power section.

Turning to CSP plants, assuming technological improvements of the solar field and thermal storage system, this could bring cost reductions of between 25% and 50% of current prices. No cost variation is assumed for the power section as this is a mature technology and the only significant breakthrough can come from scaling up. In addition, the cost of electricity is reported as a function of carbon tax price (€/t$_{CO2}$). The results of this analysis are summarized in Figure 14.6.2.

The COE bandwidth considers the resulting cost of electricity set against the two different sets of assumptions (Morin et al., 2012; A. Giostri et al., 2013). This analysis assumed the same power plant efficiency and electricity production shown in Table 14.6.1. Obviously, solar field improvement can also come from a performance standpoint, not only from an economic perspective. Figure 14.6.2 shows that assuming a cost reduction of 50% moves CSP technology closer to the reference fossil fuel-based plants.

As regards competitive renewable technologies, photovoltaics have recently become cheaper than CSP, thanks to a higher installed capacity globally (in the range of 100 GW, which is 50 times more than CSP) and economies-of-scale effects owing to larger production volumes by manufacturers. For example, PV costs reduced from about 3000 €/kW in 2010, when they were similar to CSP plants, to 1500 €/kW in 2012. Admittedly, however, the operating hours of PV systems are also shorter since they have no storage capability.

However, unlike photovoltaic and wind energy, concentrated solar power systems have a great advantage in terms of "dispatchability", namely the capability of decoupling electricity production from the availability of the source through thermal energy storage systems. Dispatchability will be the key factor for CSP plants in future scenarios in which renewable energy sources will gain in importance in electricity grids.

REFERENCES

3M website. http://solutions.3m.com/wps/portal/3M/en_US/Renewable/Energy/Applications/CSP/

Abengoa Solar. http://www.abengoasolar.com/web/en/nuestros_productos/plantas_solares/

Acciona Energy. http://www.acciona-energia.com/activity_areas/csp.aspx

Alanod. http://alanod.com/opencms/opencms/en/areas_of_application/solar_applications/

Archimede Solar Energy. http://www.archimedesolarenergy.com/

Areva CSP. http://www.areva.com/EN/operations-3640/concentrated-solar-power-technology.html

Areva Solar. http://www.areva.com/EN/operations-3422/concentrated-solar-power-technology.html

Astolfi, M., Binotti, M., Giostri, A., Manzolini, G., Marzo, A. De and Merlo, L. (2011) Indirect molten salts storage management and size optimization for different solar multiple and sites in a parabolic trough solar power plant. In:*Proceedings of Solar paces conference*. 20–23 September 2011 Granada, Spain.

Barale G., Heimsath, A., Nitz, T.A. (2010) Optical design for a linear Fresnel in Sicily. In: *Proceedings of Solar paces conference*. 21–24 September 2010 Perpignan, France.

Bauer, Laing & Tamme, (2011) Recent Progress in Alkali Nitrate/Nitrite Developments for Solar Thermal Power Applications. In: *Procedings of Molten salts Chemistry and Technology*. Trondheim, Norway, 5–9 June 2011.

Beam-Down Solar Concentration. http://www.gov-online.go.jp/pdf/hlj_ar/vol_0019e/28-29.pdf

Beckam, G. and Gilli, P. (1984) *Thermal Energy Storage*. Wien – New York: Springer – Verlag.

Belhomme, B., Pitz-Paal, R., Schwarzbözl, P. and Ulmer, S. (2009) A new fast ray tracing tool for high-precision simulation of heliostat fields. *Journal of Solar Energy Engineering*, 131, 031002.

Blake, D. and Prince, H. (2003) Assessment of a molten salt heat transfer fluid in a parabolic trough solar field. *Journal of Solar Energy Engineering*, 125, pp. 170–176.

Brosseau, D., Kelton, J., Ray, D., Edgar, M., Chisman, K. and Emms, B. (2005) Testing of thermocline filler materials and molten-salt heat transfer fluids for thermal energy storage systems in parabolic trough power plants. *Journal of Solar Energy Engineering*, 127, 109–116.

Buck, R. and Teufel, E. (2009) Comparison and optimization of heliostat canting methods. *Journal of Solar Energy Engineering*, 131, 011001.

Cohen G. and Kearney, D.K.G. (1999) Final Report on the operation and maintenance improvement program for concentrating solar power plants (Report Sandia 99–1290).

Colzi, F., Petrucci, S., Manzolini, G., Chacartegui, R., Silva, P., Campanari, S. and Sánchez, D. (2010) Modeling on/off-design performance of solar tower plants using saturated steam. In: *Proceedings of the ASME 2010 4th International Conference on Energy Sustainability*. 17–22 May 2010, Phoenix (Az), USA. ES2010-90399.

Delsol Modelling Tool. http://energy.sandia.gov/?page_id=6530

Dincer, I. and Rosen, M. (2002) *Thermal Energy Storage Systems and Application*. J. W. & Sons, Ed. Chichester.

Key World Energy (2012). http://www.iea.org/publications/freepublications/publication/kwes.pdf

Flabeg. http://www.flabeg.com/en/solar/mirrors-for-csp-cpv/cpv.html.

Flabeg Solar. http://www.flabeg.com/en/solar/solar-engineering/ultimate-trough.html.

Forristall, R. (2003) Heat transfer analysis and modeling of a parabolic trough solar receiver implemented in engineering equation solver heat transfer analysis and modeling of a parabolic trough solar receiver implemented in Engineering Equation Solver. http://large.stanford.edu/publications/coal/references/troughnet/solarfield/docs/34169.pdf

Franco, F., Anantharaman, R., Bolland, O., Booth, N., Dorst, E. Van and Ekstrom, C. (2010) Common framework and test cases for transparent and comparable techno-economic evaluations of CO_2 capture technologies – the work of the European Benchmarking Task Force. In: *Proceedings of the 10th International Conference on Greenhouse Gas control*, 19–23 September 2010, Amsterdam (NL).

Garcia, P., Ferriere, A., Flamant, G., Costerg, P., Soler, R. and Gagnepain, B. (2008) Solar field efficiency and electricity generation estimations for a hybrid solar gas turbine project in France. *Journal of Solar Energy Engineering*, 130, 014502.

Gazzani, M., Macchi, E. and Manzolini, G. (2012a) CO_2 capture in integrated gasification combined cycle with SEWGS – Part A: Thermodynamic performances. *Fuel*. doi:10.1016/j.fuel.2012.07.048.

Gazzani, M., Macchi, E. and Manzolini, G. (2012b) CO_2 capture in natural gas combined cycle with SEWGS. Part A: Thermodynamic performances. *International Journal of Greenhouse Gas Control*. doi:10.1016/j.ijggc.2012.06.010.

Gil, A., Medrano, M., Martorell, I., Lazaro, A., Dolado, P., Zalba, B. and Cabeza, L.F. (2010) State of the art on high temperature thermal energy storage for power generation part 1- concepts, materials and modellization. *Renewable and Sustainable Energy Reviews*, 14, 31–55.

Giostri, A., Binotti, M., Astolfi, M., Silva, P., Macchi, E. and Manzolini, G. (2012) Comparison of different solar plants based on parabolic trough technology. *Solar Energy*, 86, 1208–1221.

Giostri, A., Binotti, M., Silva, P., Macchi, E. and Manzolini, G. (2013) Comparison of two linear collectors in solar thermal plants: Parabolic trough versus fresnel. *Journal of Solar Energy Engineering*, 135, 011001. doi:10.1115/1.4006792.

Graf, D. and Nava, P. (2011) Ultimate trough – The new parabolic trough collector generation for large scale solar thermal power plants. In:*Proceedings of the ASME 2011 5th International Conference on Energy Sustainability*.7–10 August 2011 Washington DC, US. ES2011-54.

Grena, R. and Tarquini, P. (2011) Solar linear Fresnel collector using molten nitrates as heat transfer fluid. *Energy*, 36, 1048–1056.

Hasuike, H., Yoshizawa, Y., Suzuki, A. and Tamaura, Y. (2006) Study on design molten salt solar receivers for beam-down solar concentrator. *Solar Energy*, 80, 1255–1262.

Herrmann, U. and Geyer, M. (2002) The AndaSol Project. Workshop on thermal storage for trough systems. http://www.nrel.gov/csp/troughnet/wkshp_2003.html

IEA (2000). Leading options for the capture of CO_2 emissions at power stations. IEA, 419 Februray. PH3/14. 420.

IEA (2004). Improvement in power generation with post-combustion capture of CO_2. 421IEA, November. PH4/33.

Incropera, F. P. and DeWitt, D. P. (2007) Fundamentals of Heat and Mass Transfer. Water (Vol. 3, p. 997). John Wiley. http://www.osti.gov/energycitations/product.biblio.jsp?osti_id=6008324

Infinia Company. http://www.infiniacorp.com/

Kearney, D., Herrmann, U., Kelly, B., Mahoney, R., Cable, R., Blake, D., Price, H., Potrovitza, N., Pacheco, J. and Nava, P. (2003) Assessment of a molten salt heat transfer fluid in a parabolic trough solar field. *Journal of Solar Energy Engineering*, 125, 170–176.

Kreith, F. and Goswami, Y. G. (2007) *Handbook of Energy Efficiency and Renewable Energy*. Boca Raton, London, New York: CRC Press, Taylor and Francis Group.

Kimberlina Solar Plant. http://www.areva.com/mediatheque/liblocal/docs/pdf/activites/energ-renouvelables/Areva_Kimberlina_flyer.pdf

Kongtragool, B. and Wongwises, S. (2003) A review of solar-powered Stirling engines and low temperature differential Stirling engines. *Renewable and Sustainable Reviews*, 7, 131–154.

Kreith, F. and Goswami, Y. G. (2007) *Handbook of Energy Efficiency and Renewable Energy*. Boca Raton, London, New York: CRC Press, Taylor and Francis Group.

Liddel Solar Plant. http://www.ausra.com/pdfs/LiddellOverview.pdf

Mancini, T., Heller, P., Butler, B. and Al., E. (2003) Dish-stirling Systems: An overview of development and status. *Journal of Solar Energy Engineering*, 125, 135–151.

Manzolini, G., Giostri, A., Saccilotto, C., Silva, P. and Macchi, E. (2011a) Development of an innovative code for the design of thermodynamic solar power plants part B: Performance assessment of commercial and innovative technologies. *Renewable Energy*, 36, 2465–2473.

Manzolini, G., Giostri, A., Saccilotto, C., Silva, P. and Macchi, E. (2011b) Development of an innovative code for the design of thermodynamic solar power plants part B: Performance assessment of commercial and innovative technologies. *Renewable Energy*, 36, 2465–2473.

Manzolini, G., Macchi, E., Binotti, M. and Gazzani, M. (2011) Integration of SEWGS for carbon capture in natural gas combined cycle. Part B: Reference case comparison. *International Journal of Greenhouse Gas Control*, 5, 214–225.

Manzolini, G., Macchi, E., and Gazzani, M. (2012) CO_2 capture in integrated gasification combined cycle with SEWGS – Part B: Economic assessment. *Fuel*. doi:10.1016/j.fuel.2012.07.043.

Martin Next Generation Solar Energy Center. http://www.cleanenergyactionproject.com/Clean EnergyActionProject/CS.FPL_Martin_Next_Generation_Solar_Energy_Center___Hybrid_Renewable_Energy_Systems_Case_Studies.html

Martini, W. (1983) *Stirling Engine Design Manual* (p. NASA CR–135382).

McIntire, W. (1982) Factored approximations for biaxial incident angle modifiers. *Solar Energy*, 29, 315–322.

Mertins, M. (2009) Technische und wirtschaftliche Analyse von horizontalen Fresnel-Kollektoren. University of Karlsruhe, Karlsruhe, Germany.

Moens, L., Blake, D., Rudnicki, D. and Hale, M. (2003) Advanced thermal storage fluids for solar parabolic trough systems. *Journal of Solar Energy Engineering*, 125, 112–116.

Morin, D.E. (2009) Comparison of linear fresnel and parabolic trough collector systems – Influence of linear fresnel collector design variations on break even cost. In: *Proceedings of Solar Paces Conference*. 15–18 September 2009, Berlin (D).

Morin & Dersch, 2009

Morin, G., Dersch, J., Platzer, W., Eck, M. and Häberle, A. (2012) Comparison of linear fresnel and parabolic trough collector power plants. *Solar Energy*, 86, 1–12.

Noone, C. J., Torrilhon, M. and Mitsos, A. (2012) Heliostat field optimization: A new computationally efficient model and biomimetic layout. *Solar Energy*, 86, 792–803.

Novatech Biosol. Technical data – NOVA 1. 2009. http://www.novatecsolar.com/files/mne0911_pe1_broschure_english.pdf

Novatech Biosol (2012) http://www.novatecsolar.com/

NREL. (2000) Survey of thermal storage for parabolic trough power plants. Cologne, Germany. NREL/SR–550–27925. www.solarreserve.com/what-we-do/csp-projects/crescent-dunes

NREL Database. http://www.nrel.gov/csp/solarpaces/operational.cfm

Patnode, A. (2006) Simulation and performance evaluation of parabolic trough solar power plants. Thesis at Madison: University of Wisconsin.

Pitz-Paal, R., Botero, N.B. and Steinfeld, A. (2011) Heliostat field layout optimization for high-temperature solar thermochemical processing. *Solar Energy*, 85, 334–343.

Price, H., Lüpfert, E., Kearney, D., Zarza, E., Cohen, G., Gee, R. and Mahoney, R. (2002) Advances in parabolic trough solar power technology. *Journal of Solar Energy Engineering*, 124, 109.

Puerto Errado 2. http://www.novatecsolar.com/56-1-PE-2.html

Ronnelid, M., Perers, B. and Karlsson, B. (1997) On the factorisation of incidence angle modifiers for CPC collectors. *Solar Energy*, 59, 281–286.

San Diego Regional Renewable Energy Study Group (2005) Potential for renewable energy in the San Diego Region. San Diego, Ca. http://www.renewablesg.org/docs/Web/Renewable_Study_AUG2005_v4.pdf

SCHOTT PTR® 70 Brochure. http://www.google.it/url?sa=t&rct=j&q=&esrc=s&source=web&cd=1&cad=rja&ved=0CCAQFjAA&url=http%3A%2F%2Fwww.schottsolar.com%2Fglobal%2Fproducts%2Fconcentrated-solar-power%2Fschott-ptr-70-receiver%2F&ei=oA58UMnGG8vhtQbdp4CwCg&usg=AFQjCNHeHVBtSjxaQkc2ig5sAH570Ecwwg

Schott Solar. http://www.schott.com/solar/english/index.html

Schunk, L.O., Haeberling, P., Wepf, S., Meier, A. and Steinfeld, A. (2009) A solar receiver-reactor for the thermal dissociation of zinc oxide. *Journal of Solar Energy Engineering*, 130, 021009-1–021009-6.

SENER. Case Study: GEMASOLAR Central Tower Plant [Online] 2010. www.sener/es.

Siemens Concentrated Solar Power. http://www.energy.siemens.com/hq/en/power-generation/renewables/solar-power/concentrated-solar-power/?stc=wwecc120556

Siemens Sunfield LP. http://www.energy.siemens.com/hq/pool/hq/power-generation/renewables/solar-power/concentrated-solar-power/SiemensSunFieldLP.pdf

Skytrough Brochure. http://www.skyfuel.com/downloads/brochure/SkyTroughBrochure.pdf

Slocum A.H., D.S. Codd, J. Buongiorno, C. Forsberg, T. McKrell, J. Nave, C.N. Papanicolas, A. Ghobeity, C.J. Noone, S. Passerini, F. and Rojas, A. M. (2011) Concentrated solar power on demand. *Solar Energy*, 85, 1519–1529.

Solar Dish Tech. http://www.solarpaces.org/CSP_Technology/docs/solar_dish.pdf

Steinmann, WD., Laing, D. and Tamme, R. (2010) Latent heat storage systems for solar thermal power plants and process heat applications. *Journal of Solar Energy Engineering*, 132, 0210031–0210035.

Stine, W. (2007) *Stirling Engines*. Taylor & Francis Group. LLC.

Stine, W. and Diver, R. (1994) A compendium of solar dish/stirling technology. http://www.dtic.mil/cgi-bin/GetTRDoc?Location=U2&doc=GetTRDoc.pdf&AD=ADA353041

The Parabolic Trough Power Plants Andasol 1 to 3. (2008) http://www.google.it/url?sa=t&rct=j&q=&esrc=s&source=web&cd=1&ved=0CCAQFjAA&url=http://www.solarmillennium.de/includes/force_download.php?client=1&path=upload/Download/Technologie/eng/Andasol1-3engl.pdf&ei=3NV7UOKoCo_SsgbWnoGgCw&usg=AFQjCNGxgRI4m6V1blFjxMdW3RiIht3dVw&cad=rja

Thombare, D. and Verma, S. (2008) Technological development in the stirling cycle engines. *Renewable and Sustainable Energy Reviews*, 12, 1–38.

Urieli, I. and Berchowitz, D. (1984) Stirling Cycle Engine Analysis (p. 256). Bristol: Adam Hilger.

Veynandt, De la T. and Bezian, R.K. (2010) Design optimization of a solar power plant based on linear fresnel refelctor. In: *Proceedings of Solar Paces Conference*. 21–24 September 2010 Perpignan, France.

Walker, G. (1980) *Stirling Engines*. Clarendon Press, Oxford 1980.

Wei, X., Lu, Z., Yu, W. and Wang, Z. (2010) A new code for the design and analysis of the heliostat field layout for power tower system. *Solar Energy*, 84, 685–690.

Yang, M., Yang, X. and Ding, J. (2010) Heat transfer enhancement and performance of the molten salt receiver of a solar power tower. *Applied Energy*, 87, 2808–2811.

Z'Graggen, A., Haueter, P., Trommer, D., Romero, M., de Jesus, J. C. and Steinfeld, A. (2006) Hydrogen production by steam-gasification of petroleum coke using concentrated solar power – II reactor design, testing, and modeling. *International Journal of Hydrogen Energy*, 31, 797–811.

Appendix 14.1 Operating Concentrated Solar Power on a worldwide scale.

Database existing plants	Location	Start	Net Turbine capacity	Electricity production (MWh/yr)	Technology	HTF	Storage (h)	Type
Andasol -1	Spain	2008	49.9	158000	PT	Diphenyl/ Biphenyl oxide	7.5	2 tank indirect
Andasol -2	Spain	2009	49.9	158000	PT	Diphenyl/ Biphenyl oxide	7.5	2 tank indirect
Andasol -3	Spain	2010	49.9	175000	PT	Diphenyl/ Biphenyl oxide	7.5	2 tank indirect
Archimede	Italy	2010	4.72	9200	PT	Molten salts	8	2 tank direct
Aste 1A	Spain	2012	50	170000	PT	Thermal oil	8	2 tank indirect
Aste 1B	Spain	2012	50	170000	PT	Thermal oil	8	2 tank indirect
Astexol II	Spain	2012	50	170000	PT	Thermal oil	8	2 tank indirect
Augustin Fresnel I	France	2012	0.25	–	LFR	Thermal oil	8	2 tank indirect
Colorado Integrated Solar Project	USA	2010	2	49	PT			
Extresol-1	Spain	2010	50	158000	PT	Diphenyl/ Biphenyl oxide	7.5	2 tank indirect
Extresol-2	Spain	2010	49.9	158000	PT	Diphenyl/ Biphenyl oxide	8.5	2 tank indirect
Gemasolar Thermosolar plant	Spain	2011	19.9	110000	ST	Molten salts	15	2 tank direct
Helioenergy 1	Spain	2011	50	95000	PT	Thermal oil	none	
Helioenergy 2	Spain	2012	50	95000	PT	Thermal oil	none	
Helios I (Helios I)	Spain	2012	50	97000	PT	Thermal oil	none	

(Continued)

Appendix 14.1 (Continued)

Database existing plants	Location	Start	Net Turbine capacity	Electricity production (MWh/yr)	Technology	HTF	Storage (h)	Type
Holaniku at Keahole Point	USA, Haweii	2009	2	4030	PT	Xceltherm-600	2	non fornito
Ibersol Ciudad Real (Puertollano)	Spain	2009	50	103000	PT	Diphenyl/ Biphenyl oxide	NA	
ISCC Hassi R'mel (ISCC Hassi R'mel)	Algeria	2011	25	NA	PT	Thermal oil	none	
ISCC Kuraymat (ISCC Kuraymat)	Egypt	2011	20	3400	PT	Therminol VP-1	none	
ISCC Morocco (ISCC Morocco)	Morocco	2010	20	NA	PT	Thermal oil	none	
Jülich Solar Tower	Germany	2008	2.5	NA	ST	Air	1.5	Ceramic heat sink
Kimberlina Solar Thermal Power Plant	USA, California	2008	5	NA	LFR	Water	none	
La Dehesa	Spain	2011	49.9	175000	PT	Diphenyl/ Biphenyl oxide	7.5	2-tank indirect
La Florida	Spain	2010	50	175000	PT	Diphenyl/ Biphenyl oxide	7.5	2-tank indirect
La Risca (Alvarado I)	Spain	2009	50	105200	PT	Diphenyl/ Biphenyl oxide	none	
Lebrija I (LE-1)	Spain	2011	50	120000	PT	Therminol VP1	none	
Majadas I	Spain	2010	50	104500	PT	Diphenyl/ Biphenyl oxide	NA	
Manchasol-1 (MS-1)	Spain	2011	49.9	158000	PT	Diphenyl/ Biphenyl oxide	7.5	2-tank indirect
Manchasol-2 (MS-2)	Spain	2011	50	158000	PT	Diphenyl/ Biphenyl oxide	7.5	2-tank indirect
Maricopa Solar Project (Maricopa)	USA, Arizona	2010	1.5	NA	Dish/ Engine	non fornito	none	
Martin Next Generation Solar Energy Center (MNGSEC)	USA, Florida	2010	75	155000	PT	Thermal oil	none	
Morón	Spain	2012	50	100000	PT	Thermal oil	none	
Nevada Solar One (NSO)	USA, Nevada	2007	72	134000	PT	Dowtherm A	0.5	0.5 hours full-load storage
Palma del Río I	Spain	2011	50	114500	PT	Diphenyl/ Biphenyl oxide	none	
Palma del Río II	Spain	2010	50	114500	PT	Diphenyl/ Biphenyl oxide	NA	
Planta Solar 10 (PS10)	Spain	2011	11	23400	ST	Water	1	other
Planta Solar 20 (PS20)	Spain	2009	20	48000	ST	Water	1	other

(Continued)

Appendix 14.1   (Continued)

Database existing plants	Location	Start	Net Turbine capacity	Electricity production (MWh/yr)	Technology	HTF	Storage (h)	Type
Puerto Errado 1 Thermosolar Power Plant (PE1)	Spain	2009	1,4 gross	2000	LFR	Water	NA	Single-tank thermocline
Puerto Errado 2 Thermosolar Power Plant (PE2)	Spain	2012	30	49000	LFR	Water	0.5	Single-tank thermocline
Saguaro Power Plant	USA, Arizona	2006	1	2000	PT	Xceltherm 600 (solar field); n-pentane (ORC working fluid)	NA	
Sierra SunTower (Sierra)	USA, California	2009	5	NA	ST	Water	none	
Solaben 3	Spain	2012	50	100000	PT	Thermal oil	none	
Solacor 1	Spain	2012	50	100000	PT	Thermal oil	none	
Solacor 2	Spain	2012	50	100000	PT	Thermal oil	none	
Solar Electric Generating Station I (SEGS I)	USA, California	1984	13.8	NA	PT	Therminol	3	2-tank direct
Solar Electric Generating Station II (SEGS II)	USA, California	1985	30	NA	PT	Therminol	NA	
Solar Electric Generating Station III (SEGS III)	USA, California	1985	30	NA	PT	Therminol	NA	
Solar Electric Generating Station IV (SEGS IV)	USA, California	1989	30	NA	PT	Therminol	NA	
Solar Electric Generating Station V (SEGS V)	USA, California	1989	30	NA	PT	Therminol	NA	
Solar Electric Generating Station VI (SEGS VI)	USA, California	1989	30	NA	PT	Therminol	NA	
Solar Electric Generating Station VII (SEGS VII)	USA, California	1989	30	NA	PT	Therminol	NA	
Solar Electric Generating Station VIII (SEGS VIII)	USA, California	1989	80	NA	PT	Therminol	NA	

(Continued)

Appendix 14.1   (Continued)

Database existing plants	Location	Start	Net Turbine capacity	Electricity production (MWh/yr)	Technology	HTF	Storage (h)	Type
Solar Electric Generating Station IX (SEGS IX)	USA, California	1990	80	NA	PT	Therminol	NA	
Solnova 1	Spain	2009	50	113520	PT	Thermal oil	none	
Solnova 3	Spain	2009	50	113520	PT	Thermal oil	none	
Solnova 4	Spain	2009	50	113520	PT	Thermal oil	none	
Termesol 50 (Valle 2)	Spain	2011	49.9	175000	PT	Diphenyl/ Biphenyl oxide	7.5	2-tank indirect
Thai Solar Energy 1 (TSE1)	Thailand	2012	5	8000	PT	Water/Steam	none	

http://www.nrel.gov/csp/solarpaces/operational.cfm
PT = Parabolic Trough, LFR = Linear Fresnel, ST = Solar tower, DS = Dish Stirling

Chapter 15

Solar hybrid air-conditioning design for buildings in hot and humid climates

Kwong-Fai Fong

Division of Building Science and Technology, College of Science and Engineering,
City University of Hong Kong, Hong Kong, China

15.1 INTRODUCTION

In hot and humid cities, space conditioning and refrigeration generally consume close to half of energy use in office and residential buildings. In conventional air-conditioning provision, electricity-driven equipment is commonly applied. The projection of energy supply finds that fossil fuels may still play a very significant role in 2035 if there is no abrupt change of energy use (IEA, 2010). The Intergovernmental Panel on Climate Change (IPCC) has recently issued a special report about the role of renewable energy sources in climate change mitigation (IPCC, 2011). It indicates renewable energy, including solar energy, has a large potential to mitigate greenhouse gas emissions. But how can renewable energy be widely used in the building sector?

In 1977, the International Energy Agency (IEA) set up the Solar Heating and Cooling Programme to promote the technology development and standardization of solar heating and cooling since (IEA, 2012). In recent decades, solar air-conditioning has been broadly advocated (Altener, 2002; Eicker, 2009), aiming mainly at small and medium applications in buildings. Although solar air-conditioning has been promoted, most of the demonstration projects are found in temperate and cold climate regions where the emphasis is placed on heating rather than cooling. The demand for cooling systems is not a priority and the holistic system design for both cooling and dehumidification is seldom touched.

In fact, the market for solar-thermal collectors is growing rapidly around the world. Mainland China has the largest total capacity of evacuated tube and flat-plate collectors in operation, accounting for almost 60% of the world market (IEA, 2011). The capital cost of solar-thermal collectors has been reduced by about 20% for each doubling of installed capacity (IEA, 2007). Therefore, the application potential of solar-thermal energy becomes economically and technically viable from such blooming production. With continuous population and economic growth, the strategic design of solar air-conditioning should secure the increasing energy demand and attain low carbon urbanization in the long run.

General design guidelines and demonstration projects have been established to promote wider use of solar air-conditioning (Delorme et al., 2004; Wang et al., 2009). In addition, basic design calculations and discussions have been provided for various common solar refrigeration and air-conditioning systems (Eicker, 2003; Henning, 2004). In recent years, solar energy systems have been designed in building systems through solar heating and cooling, as well as different integrated methods (Wang and Zhai,

2010). Solar-thermal desiccant cooling systems with appropriate auxiliary heating can be operated during both daytime and night-time (Enteria et al., 2009). New system design and material for solar desiccant cooling was proposed to enhance its energy performance (Ge et al., 2010). The design and operation of a solar air-conditioning system using parabolic solar collectors and double-effect absorption chiller was evaluated for the application in a conditioned space (Qu et al., 2010). By combining an ejector cooling system and an inverter-type heat pump, a prototype of a hybrid solar cooling and heating system was designed and constructed (Huang et al., 2010). The energy and economic feasibility of a solar air-conditioning system for buildings in temperate and Mediterranean climates has also been evaluated (Calise, 2010).

In this chapter, Section 15.2 discusses the possible design approaches and features of solar air-conditioning. Section 15.3 presents the cooling and energy performances of the various solar air-conditioning systems, including the principal systems and different hybrid designs. Section 15.4 demonstrates the application potential of solar hybrid air-conditioning (SHAC) in the various hot and humid cities in Southeast Asia. In conclusion, Section 15.5 looks at future development.

15.2 DESIGN APPROACHES OF SOLAR AIR-CONDITIONING

In the context of solar air-conditioning, various types of solar collectors, refrigeration and air-conditioning cycles are interlinked or hybridized so that the indoor design conditions of the buildings can be fulfilled throughout the year. Solar air-conditioning can be developed in the following four approaches according to system complexity, as consolidated in Figure 15.2.1.

In any one of the design approaches, electricity is still required to drive the parasitic equipment – such as pumps, fans and cooling towers – of the various solar air-conditioning systems. In the study of solar air-conditioning systems, it is necessary to take a holistic view of the energy consumption of all the equipment involved.

15.2.1 The solar-electric approach

Photovoltaic (PV) panels are used to generate electricity, which can be in turn be used to drive conventional vapour-compression refrigeration. The compression chiller driven by direct current can also be considered in order to prevent losses from current conversion. Both roof-mounted and building-integrated strategies can be applied for the installation of PV panels. In this approach, an auxiliary electricity supply, generally from the power grid, is required in case of electricity deficit to the solar-electric air-conditioning system. Although it is relatively straightforward to apply the conventional monocrystalline and polycrystalline PV panels for driving the compression chiller, their environmental impacts from production and disposal have been the subject of study in recent years (Gottessfeld and Cherry, 2011; EC, 2011), apart from the new thin film PV which is technologically mature enough to supersede the crystalline types.

15.2.2 The solar-thermal approach

Solar-thermal collectors, like flat-plate collectors, evacuated tubes or parabolic concentrators, are applied to generate heat for the thermally driven refrigeration or

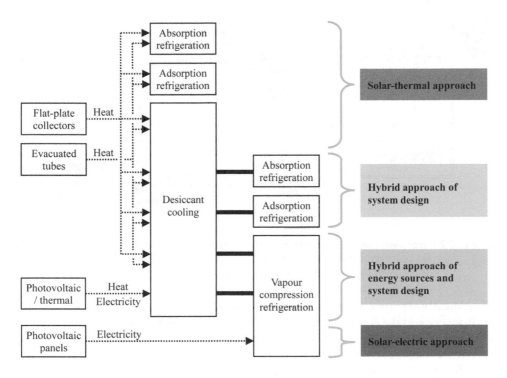

Figure 15.2.1 Feasible design approaches of solar air-conditioning.

air-conditioning cycle. Similar to the solar-electric approach, both roof-mounted and the building-integrated strategies can be considered for the collector installation. In this approach, auxiliary heating, typically using fuel gas, is needed in case of thermal deficit in driving the solar-thermal air-conditioning system. The basic system designs include solar absorption refrigeration, solar adsorption refrigeration and desiccant cooling, each of which is described below.

15.2.2.1   Solar absorption refrigeration

Absorption chillers have been developed now for almost a century with various working pairs of absorbent and refrigerant, as well as different system configurations, including single-effect, double-effect and even triple-effect. For solar energy application, single-effect absorption refrigeration is the most popular due to its relatively low driving temperature. Figure 15.2.2 illustrates the solar absorption refrigeration system. Solar-thermal gain is firstly collected in the hot water storage tank by the hot water pump. To drive the absorption chiller, the regenerative water pump feeds in the hot water from the storage tank. If the driving temperature is not enough, an auxiliary heater can be used. Chilled water from the chiller is delivered to the air handling unit so that the conditioned supply of air can be provided to the building zone accordingly. As the required outdoor air flow rate is less than that of the supply air, return air from the building zone is drawn back to the air handling unit for the sake of air balancing.

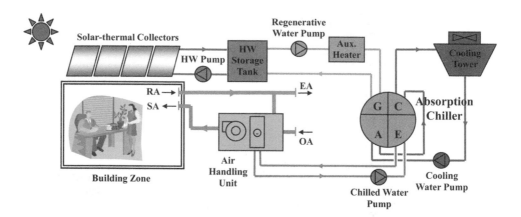

Figure 15.2.2 Solar absorption refrigeration system for space conditioning. (Abbreviation: A: absorber; C: condenser; E: evaporator; EA: exhaust air; G: generator; HW: hot water; OA: outdoor air; RA: room air; and SA: supply air).

When the absorption chiller is operating, the cooling tower removes heat from both the absorber and the condenser in series. The common working pairs are lithium bromide/water and water/ammonia. For a typical single-effect LiBr-H_2O absorption chiller the driving temperature is generally between 70°C and 90°C, which can be primarily achieved by solar-thermal gain, together with the assistance of auxiliary heating.

15.2.2.2 Solar adsorption refrigeration

Adsorption chillers have a relatively low driving temperature by using an appropriate working pair of adsorbent and refrigerant. The schematic diagram of a solar adsorption refrigeration system is presented in Figure 15.2.3. Generally it is similar to that of solar absorption refrigeration, except that an adsorption chiller is used. The economical adsorption pair is silica gel and water, where silica gel is the adsorbent and water the refrigerant. Other effective adsorption pairs include zeolite/water and activated carbon/ammonia. Compared to the absorption cycle, its driving temperature can be down to about 60°C. Typically, there are two chambers containing adsorbent in the adsorption chiller. While one chamber is used for adsorption, the other is used for desorption. Their roles are interchanged according to the period of adsorption/desorption process (usually 6 minutes for the silica gel/water pair). A pseudo-continuous operation is therefore formed in the refrigeration cycle. Cooling water and regenerative water are fed into the adsorption chamber and the desorption chamber respectively, and these two water circuits are alternatively changed according to the role of the chamber.

15.2.2.3 Solar desiccant cooling

A solar desiccant cooling system can directly provide conditioned air to the building zone, as shown in Figure 15.2.4. The core part of this system is the desiccant component, and both solid and liquid sorbents, such as silica gel and lithium chloride respectively, can be applied. Although desiccant cooling using liquid sorbent has the

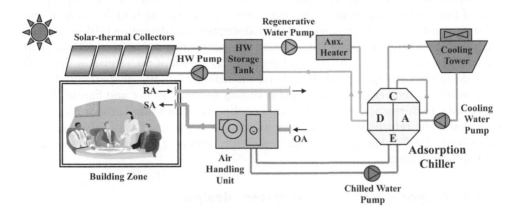

Figure 15.2.3 Solar adsorption refrigeration system for space conditioning (New abbreviation: A: adsorber; and D: desorber).

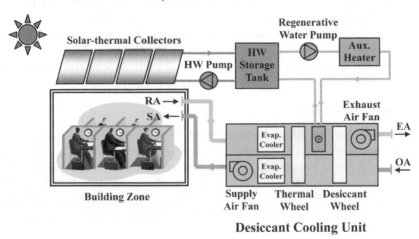

Figure 15.2.4 Solar desiccant cooling system for space conditioning.

merit of thermal storage, there are health and safety concerns to the building occupants. Since the supply air would be in direct contact with the slightly corrosive liquid sorbent, there is a risk of carry-over to the conditioned space. Desiccant cooling using solid sorbent, however, is stable in the processes of adsorption and desorption, so it is more suitable for direct application in the supply air stream. In this regard, a solid desiccant wheel is the major component of desiccant cooling. The other components include a thermal wheel, direct evaporative coolers, solar collectors, a hot water storage tank and an auxiliary heater. The thermal wheel is used to remove the sensible heat of the process air after passing the desiccant wheel. The evaporative cooler at the supply air stream is used to cool down the process air to become supply air, which still has a sufficiently low humidity ratio for handling the space latent load. The evaporative cooler at the exhaust air stream is used to cool down the room air for better sensible heat recovery at the thermal wheel.

Solar desiccant cooling systems can enhance indoor air quality owing to the provision of full outdoor air. However, primary energy consumption is higher than in other solar air-conditioning systems, such as solar absorption or adsorption refrigeration systems and conventional vapour compression refrigeration systems, in which the return air scheme can be used at the air handling unit. Adoption of the return air design for solar desiccant cooling depends heavily on the climatic conditions, particularly solar irradiation, air temperature and humidity during summer. In hot and humid climates the cooling performance of solar desiccant cooling systems using outdoor air is better than that using return air, since a larger amount of conditioned air is involved (Fong and Chow, 2007). Auxiliary heating is therefore involved to achieve a satisfactory cooling performance.

15.2.3 A hybrid approach to system design

In general, the building cooling load can be divided into zone cooling load (mainly sensible load) and ventilation load (mainly latent load). If the refrigeration cycle is used to handle the former, and the hygroscopic nature of desiccant cooling used to tackle the latter, the cooling load can be effectively handled in such a load-sharing approach. Due to the separate handling of the cooling load, individual controls for zone temperature and zone relatively humidity become more practical. The hybrid approach of system design can be the basic SHAC or SHAC enhanced by high temperature cooling. The details of these two alternatives are described as follows.

15.2.3.1 Principal SHAC

Figure 15.2.5 presents the schematic diagram of the principal SHAC system fully driven by solar-thermal energy. In the configuration of the desiccant cooling unit the two evaporative coolers can be omitted and a cooling coil adopted at the supply air stream instead. In this case the absorption/adsorption chiller generates chilled water to the cooling coil located at the downstream of the thermal wheel. As both the heat-driven chiller and desiccant cooling are involved, there would be two sets of regenerative

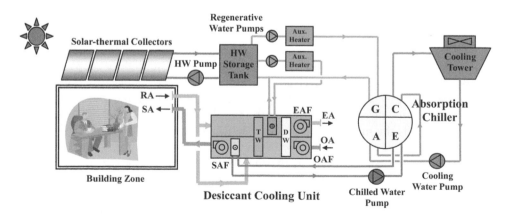

Figure 15.2.5 SHAC system using absorption chiller (New abbreviation: DW: desiccant wheel; EAF: exhaust air fan; OAF: outdoor air fan; SAF: supply air fan; and TW: thermal wheel).

water pumps and auxiliary heaters, one for the heat-driven chiller and the other for the desiccant cooling cycle.

15.2.3.2 SHAC enhanced by high temperature cooling

In this approach the energy performance of SHAC is further facilitated by using the appropriate strategies of high temperature cooling, such as radiant ceiling cooling or a specific indoor ventilation method. Since this allows a higher chilled water supply temperature from the heat-driven refrigeration system, a better solar fraction of the solar energy system and a higher coefficient of performance of the refrigeration cycle is achieved.

15.2.4 A hybrid approach to energy sources and system design

15.2.4.1 SHAC with dual solar energy sources

In this design, photovoltaic/thermal (PV/T) panels are utilized, and cogeneration of electricity and heat can happen. The solar electric gain can be used to drive the compression chiller for the zone cooling load, while the solar-thermal gain can regenerate the desiccant cooling for the ventilation load (Fong et al., 2010b), as depicted in Figure 15.2.6. A power regulator is used to allow the power to come from either the PV panels or the regional grid. In this case, both auxiliary electricity supply and auxiliary heating are involved.

15.2.4.2 SHAC system with separate thermal and electrical energy sources

An alternative of this hybrid approach is to make use of the solar-thermal gain for fully regenerating the desiccant cooling, using electricity from the grid for the conventional compression chiller. Figure 15.2.7 illustrates the SHAC system energized by the separate thermal and electrical energy sources.

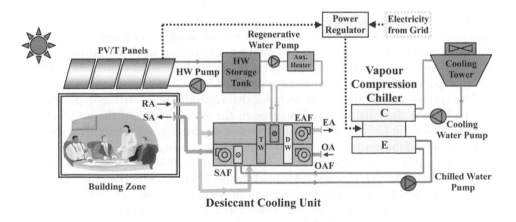

Figure 15.2.6 SHAC system with both electricity and heat generation.

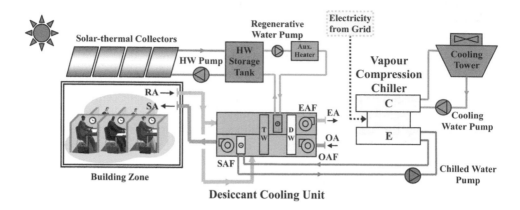

Figure 15.2.7 SHAC system with separate thermal and electrical energy sources.

15.3 PERFORMANCE EVALUATION OF VARIOUS SOLAR AIR-CONDITIONING SYSTEMS

With the knowledge of different system configurations of solar air-conditioning approaches, their cooling and energy performances for buildings in hot and humid climates are of great interest. The operating energy would be determined for the various solar air-conditioning systems against conventional types, since energy saving is the primary concern of any newly proposed air-conditioning (AC) system. As a start to investigating solar air-conditioning systems in hot and humid climates, the study area is subtropical Hong Kong (22.32°N, 114.17°E), using its weather data of the typical meteorological year (Chan et al., 2006). In this section, we look at dynamic simulation conducted for a typical office building zone (except Section 15.3.5 which considers premises with high latent load), in which the ratio of the installed collector area and the conditioned space is 1:2. A typical office has an area of 196 m^2 and an occupant density of 8 m^2/person. Daily occupancy is 10 hours between 8:00 a.m. to 6:00 p.m. The outdoor air amount is based on 0.01 m^3/s/person. The lighting heat gain is 17 W/m^2 (with 70% radiative) and the heat gain of office equipment is 230 W/person. The fenestration to wall ratio is 0.5. Based on indoor design conditions of 25.5°C and 60% in relative humidity, the estimated design zone cooling load and ventilation load are 20 kW and 9 kW respectively. The total net area of the solar collectors is 100 m^2 and the capacity of the hot water storage tank is 5 m^3.

The system simulation model includes the appropriate control components, so as to realize the dynamic interaction between the AC system and the building zone under the changing loading and climatic conditions throughout a year. The solar air-conditioning and the conventional AC systems are built using the validated component models of the plant simulation program TRNSYS (SEL, 2006) and its associated component library TESS (TESS, 2004). Meanwhile the dynamic component models of the absorption chiller, the adsorption chiller and the desiccant wheel are specifically constructed according to those validated by Kim and Infante Ferreira (2008), Cho and Kim (1992) and Zhang et al. (2003) respectively. Dynamic simulation is carried out to determine the annual total energy consumption of system operation, as well as other kinds of performance indicators, such as solar fraction (SF) and coefficient of performance (COP). In order to have more accurate evaluation of responses and operation of the solar air-conditioning system arising from changing boundary conditions, the simulation time step is set at 6 minutes, so that a total of 87,600 simulations are run for a year-round study. As electrical energy and thermal energy are involved in the various solar air-conditioning systems, they would be converted to primary energy consumption (E_p) for comparison purposes. The conversion of electrical energy to primary energy takes into account the local fuel mix.

15.3.1   Principal solar-thermal air-conditioning systems

The basic design of solar air-conditioning is in accordance with the solar-thermal approach, including the solar absorption refrigeration system, the solar adsorption refrigeration system and the solar desiccant cooling system, as described in Section 15.2.2. From the previous study (Fong et al., 2010a), the cooling and energy performances of these solar-thermal air-conditioning systems for office building application are consolidated in Table 15.3.1. Their results are also contrasted with those of the conventional AC systems using air-cooled vapour compression chillers (ACVCC) and water-cooled vapour compression chillers (WCVCC). Although the common solar-thermal collectors include the flat-plate collectors, evacuated tubes and parabolic concentrators, the last type is the least effective, as found in the aforementioned study. In principle, parabolic concentrators can harness thermal energy at higher temperature; however, in a fixed space for collector accommodation and the changing incidence of solar irradiation, they have the lowest overall solar-thermal gain in a year. As such, only the flat-plate collectors and the evacuated tubes are involved in the study.

Compared to the conventional air-conditioning system in the return air scheme, the solar absorption refrigeration system with evacuated tubes has the best energy performance, with 35.1% and 33.6% less year-round primary energy consumption than the ACVCC and the WCVCC respectively. The solar adsorption refrigeration system, which has about 20% more energy consumption, does not have better energy performance than the conventional system here. Nevertheless, its application potential still exists in the hybrid design of solar air-conditioning systems which are discussed in Section 15.3.3. For the same kind of solar air-conditioning system, evacuated tubes can have better energy performance than the flat-plate collectors in the hot and humid climate from a year-round perspective.

Solar desiccant cooling has relatively high year-round primary energy consumption, since it has to handle the extra ventilation load due to the inherent nature of full outdoor air design. Accordingly, the total cooling capacity of desiccant cooling system is much higher than that of the other systems. In addition, the parasitic energy consumption is comparatively high, particularly the supply air fan and exhaust air fan. Compared to conventional AC systems in the outdoor air scheme, solar desiccant cooling systems with evacuated tubes can have 4.9% and 1.5% less primary energy consumption than ACVCC and WCVCC respectively. The feature of solar desiccant cooling in effect more than satisfies the required cooling load. It is able to supply the outdoor air amount far above the minimum requirement of the functional area,

Table 15.3.1 Summary of cooling and energy performances of principal solar-thermal air-conditioning systems.

Type of system	Type of solar collectors	Year-round averaged SF	Year-round averaged COP	Year-round total E_p per AC area (kWh/m^2)	Energy saving vs. corresponding ACVCC/ WCVCC
Solar absorption refrigeration (RA)	Flat-plate collectors	0.497	0.769	371	4.4%/2.2%
	Evacuated tubes	0.818	0.763	252	35.1%/33.6%
Solar adsorption refrigeration (RA)	Flat-plate collectors	0.313	0.435	657	−69.0%/−72.9%
	Evacuated tubes	0.577	0.437	478	−23.0%/−25.9%
Solar desiccant cooling (OA)	Flat-plate collectors	0.336	1.066	762	−11.0%/−14.8%
	Evacuated tubes	0.552	1.059	653	4.9%/1.5%
ACVCC (RA)	NA	NA	2.859	389	NA
WCVCC (RA)	NA	NA	3.195	380	NA
ACVCC (OA)	NA	NA	2.802	687	NA
WCVCC (OA)	NA	NA	3.177	664	NA

Remarks:
1. RA refers to the return air scheme, in which a conventional air handling unit is applied with return air and outdoor air mixed together to form the supply air for the building zone.
2. OA refers to the outdoor air scheme, in which the air handling unit takes the full outdoor air to form the supply air for the building zone.
3. NA means not applicable.

and this merit can guarantee good indoor air quality and ventilation effectiveness. Therefore the application potential of solar desiccant cooling remains valid.

15.3.2 SHAC with load sharing

The SHAC system (i.e. the principal SHAC) with load sharing has been introduced in Section 15.2.3.1. In the study by Fong et al. (2010b) the absorption chiller is designed to handle the zone cooling load of the office building, while desiccant cooling tackles the ventilation load. Solar-thermal gain is used to drive both the related refrigeration cycle and desiccant unit. The performance results also cover the year-round averaged zone temperature (T_z) and zone relative humidity (RH_z), the coefficient of performance of chiller (COP_{ch}) and the coefficient of performance of desiccant cooling (COP_{dc}), as summarized in Table 15.3.2. As described in Section 15.3.1, evacuated tubes are more effective than flat-plate collectors in hot and humid climates, so only the former type is involved here.

Compared to the year-round primary energy consumption of conventional ACVCCs and WCVCCs shown in Table 15.3.1, the SHAC-Ab system with load sharing clearly gives substantial savings of 36.8% and 35.3% respectively. Although the primary energy consumption of the SHAC-Ab system is only 2.4% lower than that of solar absorption refrigeration shown in Table 15.3.1, the SHAC-Ab system has tight control of the design zone temperature and relative humidity. While the SHAC-Ad

Table 15.3.2 Summary of cooling and energy performances of SHAC with load sharing.

Type of system	Type of solar collectors	Year-round averaged $T_z(°C)/$ RH_z (%)	Year-round averaged SF	Year-round averaged COP_{ch} COP_{dc}	Year-round total E_p per AC area (kWh/m^2)	Energy saving vs. ACVCC/ WCVCC in Table 1
SHAC with absorption chiller (SHAC-Ab)	Evacuated tubes	24.9/58.3	0.804	0.779/0.904	246	36.8%/35.3%
SHAC with adsorption chiller (SHAC-Ad)	Evacuated tubes	24.9/58.4	0.590	0.460/0.919	445	−14.5%/−17.2%

system does not have primary energy-saving potential, it does have slightly better energy performance than solar adsorption refrigeration, as shown in Table 15.3.1.

15.3.3   SHAc with radiant cooling

The SHAC with radiant cooling is one of the hybrid systems enhanced by high temperature cooling, as discussed in Section 15.2.3.2. Although solar energy is able to power up heat-driven refrigeration, its contribution is quite limited because of the conventionally low chilled water supply temperature at around 6°C. If this temperature could be raised, it would enhance the solar fraction and the solar refrigeration would rely less on auxiliary heating. In the SHAC with radiant cooling, 15°C to 18°C of chilled water can be supplied from the absorption/adsorption chiller to the indoor radiant ceilings and the cooling coil of the desiccant cooling unit. The radiant ceilings would handle the zone sensible load, while the desiccant cooling tackles the zone latent load and the ventilation load. Radiant ceilings apply both the means of radiation and convection for cooling purposes, and their types include chilled panels, passive chilled beams and active chilled beams. The schematic diagram of the SHAC with the chilled panels and the passive chilled beams is shown in Figure 15.3.1, while that for the active chilled beams appears in Figure 15.3.2. The latter SHAC demands additional fans to maintain the effective induction of room air, so as to facilitate indoor convective heat transfer. As the cooling coil is used at the supply air stream, the two evaporative coolers of the original desiccant cooling unit become obsolete. In previous studies (Fong et al., 2010c; 2011a), the three kinds of radiant ceilings are associated with the SHAC-Ab/SHAC-Ad using evacuated tubes, the results for which are presented in Table 15.3.3.

As can be seen from Table 15.3.3, whether for the SHAC-Ab or the SHAC-Ad system, both passive and active chilled beams provide satisfactory zone temperatures and relative humidity. However, the chilled panels cannot be adopted, since its average T_z is up to 29.0°C, thus depriving thermal comfort for both types of SHAC systems. SHAC systems with passive or active chilled beams have definite energy saving potential compared to conventional AC systems. The yearly E_p of the SHAC-Ab system using passive

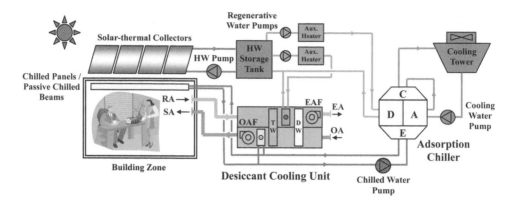

Figure 15.3.1 SHAC using adsorption chiller for radiant cooling with chilled panels or passive chilled beams.

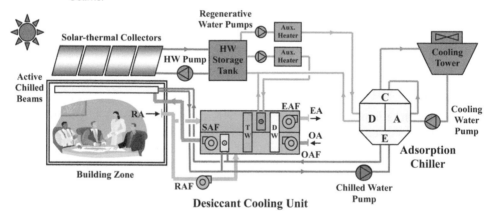

Figure 15.3.2 SHAC using adsorption chiller for radiant cooling with active chilled beams (New abbreviation: RAF: return air fan).

Table 15.3.3 Summary of cooling and energy performances of SHAC systems with radiant cooling.

Type of system	Type of chilled collectors	Year-round averaged $T_z(^\circ C)$/ RH_z (%)	Year-round averaged SF	Year-round averaged COP_{ch}	Year-round total E_p per AC area (kWh/m^2)	Energy saving vs. ACVCC/ WCVCC in Table 1
SHAC-Ab	Chilled panels	29.0/42.8	0.782	0.840	209	46.2%/45.0%
	Passive chilled beams	25.4/52.2	0.882	0.830	137	64.7%/63.9%
	Active chilled beams	25.1/52.2	0.828	0.823	191	50.8%/49.6%
SHAC-Ad	Chilled panels	29.0/43	0.614	0.564	365	6.1%/3.9%
	Passive chilled beams	25.4/53	0.754	0.558	228	41.2%/39.9%
	Active chilled beams	25.0/53	0.689	0.548	311	20.0%/18.1%

chilled beams are 64.7% and 63.9% less than that of ACVCC and WCVCC respectively, while those of the SHAC-Ad system are 41.2% and 39.9% less respectively. On the other hand, the yearly E_p of the SHAC-Ab system using active chilled beams are 50.8% and 49.6% less than the two conventional systems, while those of the SHAC-Ad system are 20.0% and 18.1% less. It has been found that passive chilled beams have higher SF and lower yearly E_p compared to active ones. This is because the better COP_{ch} of absorption/adsorption chillers results in less frequent heat demand for generation/desorption from the solar-thermal gain. On the other hand, active chilled beams require substantial energy demand from the additional supply air and return air fans. In this hybrid design it is clear that the adsorption chiller also has an essential role in overall energy merit. By adopting an appropriate high temperature cooling approach, the feature of low driving temperature of adsorption chillers can be effective.

From this study it can be seen that using passive chilled beams is the best choice for SHAC systems working in hot and humid regions. Passive chilled beams also have other merits, such as silent operation and free from drafts, making this option more attractive. Of course, prevention of condensation is of primary importance during humid weather. The involvement of desiccant cooling in SHAC can control indoor humidity effectively. The problem of infiltration of the building envelope can be avoided by use of good-quality building materials and workmanship, together with positive air pressurization in the building zone.

15.3.4 SHAC coordinated with new indoor ventilation strategies

In the conventional design of indoor air distribution of supply air, mixing ventilation (MV) is used in order to have homogenous air conditions within the entire building zone. The supply air temperature of MV is generally around 15°C. If the supply air temperature is raised without sacrificing thermal comfort, this can help to enhance overall energy performance. Therefore displacement ventilation (DV), which allows a supply air temperature of 19°C for office use (Lin et al., 2005), has been promoted. This provides a useful strategy for high temperature cooling for solar air-conditioning. In this sense, the supply air flow rate of DV can be maintained or even reduced compared to that of MV, but a higher return/exhaust air temperature results. This can also reduce ventilation load, and hence the cooling capacity of the entire air-conditioning system. As such, SHAC combined with appropriate indoor ventilation strategies gives another alternative, as described in Section 15.2.3.2. In this hybrid design, either the absorption or adsorption chiller can be paired with the desiccant cooling unit for the return air scheme. As the latent cooling load of the building zone can be handled by the desiccant cycle, the problem of insufficient latent capacity due to the higher supply air temperature of DV can be solved.

Figure 15.3.3 depicts the configuration of SHAC coordinated with DV. In DV, air is supplied to the building zone at floor level and exhausted at ceiling level. A temperature gradient, and subsequently a humidity gradient, is developed along the zone height, maintaining thermal comfort mainly within the occupied zone, regardless of the level above it. A common hot water storage tank is used to provide the driving heat for both the desiccant cycle and the heat-driven chiller, but separate auxiliary heaters can be added. The results of cooling and energy performances of SHAC

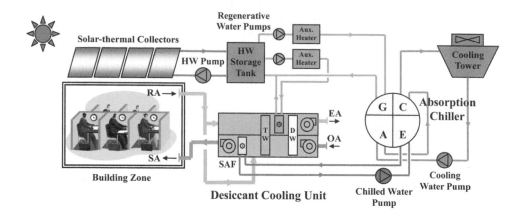

Figure 15.3.3 SHAC using absorption chiller for displacement ventilation.

Table 15.3.4 Summary of cooling and energy performances of SHAC and conventional systems for DV and MV.

Type of system	Type of ventilation collectors	Year-round averaged $T_z(°C)/$ RH_z (%)	Year-round averaged SF	Year-round averaged COP_{ch}/COP_{dc}	Year-round total E_p per AC area (kWh/m^2)	Energy saving vs. WCVCC using MV
SHAC-Ab	DV	25.1/59.1	0.905	0.809/0.845	192	49.5%
	MV	24.9/58.3	0.804	0.779/0.904	246	35.3%
SHAC-Ad	DV	25.0/59.3	0.741	0.475/0.880	310	18.3%
	MV	24.9/58.4	0.590	0.460/0.919	445	−17.2%
WCVCC	DV	24.4/70.3	NA	3.556/NA	273	NA
WCVCC	MV	24.8/58.9	NA	3.195/NA	380	NA

Remarks: NA means not applicable.

and conventional AC systems, as shown by Fong et al. (2011b), are consolidated in Table 15.3.4.

Table 15.3.4 summarizes the year-round performances of the different systems. All of them can maintain satisfactory indoor conditions of averaged T_z and RH_z, except the WCVCC for DV, which has a RH_z of 70.3% due to the relatively high supply air humidity ratio. When compared to the conventional WCVCC for MV, the SHAC systems are technically feasible, with a primary energy saving of 49.5% for SHAC-Ab and 18.3% for SHAC-Ad. In the same ventilation strategy the SHAC-Ab has a primary energy saving of 29.7% against the conventional AC system for DV, and 35.3% against that for MV. Even the SHAC-Ab for MV has an energy saving of 9.9% against the conventional system for DV. This really demonstrates the effectiveness of the hybrid design of solar air-conditioning systems using absorption chillers.

When compared to MV counterparts, the adoption of DV can reduce total primary energy consumption from 246 kW/m² to 192 kW/m² (a 21.9% drop) in SHAC-Ab; and from $445\,kW/m^2$ to $310\,kW/m^2$ (a 30.3% drop) in SHAC-Ad. A higher percentage reduction in primary energy consumption can be achieved by the hybrid system with the adsorption chiller; this is because there is a more significant percentage rise of SF in the adsorption chiller, and hence the decrease in the driving energy is more substantial.

Recently, a novel indoor ventilation strategy, called stratum ventilation (SV), has been advocated (Lin et al., 2009). Stratum ventilation uses an even higher supply temperature than displacement ventilation, and the energy saving potential of SHAC can be further advanced. This will undoubtedly enhance solar energy deployment in air-conditioning for buildings.

15.3.5 SHAC for premises with high latent load

In conventional AC design, a cooling coil is used to conduct both the cooling and dehumidification processes for the supply air. This is suitable for typical offices and residential units, where the zone sensible load is more significant in the total cooling load. However, some commercial premises have a high zone latent load with which conventional AC provision cannot cope effectively. Such commercial premises are common and include restaurants, indoor food markets and entrance lobbies. The latent heat gain of these building zones comes from humid fresh air, the building's occupants and indoor services such as hot food, spas and water ponds. Additional latent heat gain can be caused by excessive infiltration and frequent opening to the outdoors.

Because of the high latent load for this kind of building zone, substantial sub-cooling followed by reheating of the supply air is needed in conventional AC design, causing high energy requirements. In addition, it is common for these AC systems to suffer over-cooling problems in these premises. As AC equipment is designed to suit the latent load, so its sensible cooling capacity naturally becomes over-provided. If there is no reheat provision, and no simultaneous temperature and humidity control, this can cause thermal comfort problems. As a result of this over-cooling potential, it is common for people to wear jackets or additional clothing, thus defeating the primary objective of air-conditioning: i.e. thermal comfort. In addition, there is risk of condensation at the supply air grilles because of low supply air temperature. To alleviate the high energy demand and thermal discomfort problem associated with conventional AC design, SHAC with desiccant cooling can fit the purpose; the appropriate system for premises with high latent load is illustrated in Figure 15.2.7 of Section 15.2.4.2. Solar-thermal energy is primarily used for desiccant cooling, and electricity from the power grid for the VCC, so this solar hybrid design is represented by SHAC-VCC.

In a previous study carried out by Fong et al. (2011c), dynamic simulation was used for energy evaluation of a Chinese restaurant in which the ratio of the installed collector area and the conditioned space was again 1:2. The restaurant used in the stuady – typical of premises with a high latent load in subtropical Hong Kong – had an area of $196\,m^2$ and an occupant density of $1\,m^2$/person. The daily occupancy schedule covered 17 hours between 6:00 a.m. and 11:00 p.m. The design outdoor air amount was determined at $0.01\,m^3$/s/person. The lighting heat gain was $20\,W/m^2$ (with 70% radiative). The other sensible and latent heat gains were $1.23\,kW$ (with 50% radiative) and $1.77\,kW$ respectively. The total net area of the evacuated tubes was $100\,m^2$ and the size of the hot water storage tank was $5\,m^3$. Based on the design indoor conditions of 22°C and 60%RH, the estimated design zone sensible and latent loads came out at 19 kW and 13 kW respectively, reflecting the relatively high latent component. The performance results of the SHAC-VCC and conventional systems for the restaurant are presented in Table 15.3.5.

For the design RH of 60% in Table 15.3.1, the total primary energy consumption of the SHAC is lower than that of the conventional system by 49.5%. The huge difference is due to the substantial sub-cooling and reheating of the supply air, as well as the large supply air flow rate. It is also apparent that the SHAC-VCC can offer T_z without overcooling and close to the design value of 22°C. In the SHAC-VCC, the zone thermostat and the zone humidistat can control the supply air cooling coil (mainly for handling the sensible cooling load) and the heating coil (mainly for the latent load) separately, so the zone temperature and humidity can be maintained more steadily throughout different loading and climatic conditions in the course of a year. However, this is not the case in the conventional AC system. Although zone thermostat and zone humidistat are provided, they are used to control the one, and only one, cooling coil, which handles both the sensible and latent load simultaneously. As such, the supply air temperature, and hence the zone temperature, tends to be sub-cooled very low in order to fulfil the design humidity. Thus, the independent humidity and temperature controls are not as effective as that of the SHAC. Table 15.3.5 also shows that the annually averaged COP_{ch} of the SHAC is better than that of the conventional system by 5%. The solar fraction of the SHAC is about 0.3, indicating that auxiliary heating has a role in supporting the year-round operation.

In order to prove better performance by the SHAC system, a scenario of a higher indoor RH of 70% is involved, since this may be favourable to the conventional AC system. To evaluate this, the minimum dewpoint temperature is decreased to 10°C and the supply air temperature is lowered to about 12°C. Since the reheat demand is cut, the capacity of the chiller is therefore reduced and the supply air flow rate drops accordingly. The lower part of Table 15.3.5 presents the performance results of this scenario. It can be seen that the total primary energy consumption of the conventional system is still higher, but much closer to that of the SHAC. Now the SHAC can have 22.7% less E_p at 70%RH, instead of 49.5% less at 60%RH.

Table 15.3.5 Summary of cooling and energy performances of SHAC and conventional systems for Chinese restaurant at 60%RH and 70%RH.

Type of system	Design relative humidity	Year-round averaged $T_z(°C)/$ RH_z (%)	Year-round averaged SF	Year-round averaged COP_{ch}/COP_{dc}	Year-round total E_p per AC area (kWh/m²)	Energy saving vs. WCVCC
SHAC-VCC	60%	22.2/56.2	0.295	3.39/0.63	1,699	49.5%
WCVCC	60%	21.4/63.0	NA	3.23/NA	3,362	NA
SHAC-VCC	70%	22.4/59.8	0.441	3.41/0.80	1,348	22.7%
WCVCC	70%	20.2/72.4	NA	3.23/NA	1,743	NA

Remarks: NA means not applicable.

15.4 APPLICATION POTENTIAL OF SHAC IN VARIOUS HOT AND HUMID CITIES IN SOUTHEAST ASIA

With the assurance of the energy and cooling merits of the SHAC system in subtropical Hong Kong, it is interesting to consider its application potential in the different hot and humid cities in Southeast Asia and South China, where solar irradiation is abundant. Air-conditioning is essential in maintaining the economic and commercial activities of these urban areas. In this study, six additional cities are included: Bangkok, Guangzhou, Kuala Lumpur, Manila, Singapore and Taipei – see Figure 15.4.1. Particularly in the cities of the Southeast Asia, there is rapid economic growth and increasing energy demand, and with it comes rising fossil fuel consumption which is increasing environmental pressures. Resource availability varies greatly from place to place in this region, which offers large potential for developing renewable energy initiatives. Many Southeast Asian countries have already adopted medium- and long-term targets for renewable energy. For instance, Indonesia, Singapore and Thailand have recently announced reduction targets for carbon dioxide emissions in support of the Copenhagen Accord (Ölz and Beerepoot, 2010).

Year-round dynamic simulations of both SHAC and conventional AC systems are being carried out for buildings in these cities, including the same office building zone mentioned in Section 15.3. In these simulations, only the absorption chiller is involved in the SHAC since it is more effective than the adsorption chiller in earlier hybrid designs. The two types of air-conditioning system are specifically designed according to the climatic conditions of these cities, for each of which weather data for a typical meteorological year (DOE, 2011a; DOE, 2011b) are being used. As all the cities are located in the northern hemisphere, the installation of solar collectors faces south, with the tilt angle the same as the latitude of the respective cities in order to harness a maximum of solar irradiation throughout a year. Table 15.4.1 summarizes the results of cooling and energy performances of the SHAC and the conventional AC systems.

Figure 15.4.1 Hot and humid cities of Southeast Asia and South China in this study.

Table 15.4.1 Summary of cooling and energy performances of SHAC and conventional systems in the cities in Southeast Asia and South China.

City (latitude, longitude)	Annual global horizontal irradiation (kWh/m²)	Type of system	Year-round averaged SF	Year-round averaged COP_{ch}/ COP_{dc}	Year-round total E_p per AC area (kWh/m²)	Energy saving vs. VCC
Bangkok (13.92°N, 100.60°E)	1756.2	WCVCC	NA	3.117	548	NA
		SHAC-Ab	0.857	0.773/0.968	273	50.3%
Guangzhou (23.17°N, 113.33°E)	1073.4	WCVCC	NA	3.175	366	NA
		SHAC-Ab	0.714	0.782/0.958	271	25.8%
Hong Kong (22.32°N, 114.17°E)	1268.8	WCVCC	NA	3.195	380	NA
		SHAC-Ab	0.804	0.779/0.904	246	35.3%
Kuala Lumpur (3.12°N, 101.55°E)	1466.4	WCVCC	NA	3.092	536	NA
		SHAC-Ab	0.777	0.777/0.915	313	41.7%
Manila (14.52°N, 121.00°E)	1537.9	WCVCC	NA	3.090	550	NA
		SHAC-Ab	0.753	0.774/0.993	330	39.9%
Singapore (1.37°N, 103.98°E)	1587.0	WCVCC	NA	3.067	555	NA
		SHAC-Ab	0.809	0.772/0.977	302	45.7%
Taipei (25.07°N, 121.55°E)	1388.1	WCVCC	NA	3.171	395	NA
		SHAC-Ab	0.862	0.778/0.859	222	43.9%

Remarks: NA means not applicable.

From the table it is clear that there is substantial primary energy saving of the SHAC against the conventional system, ranging from 25.8% to 50.3%. Particularly in tropical cities, like Bangkok, Kuala Lumpur and Singapore, the energy saving is well above 40%. This indicates the direct contribution that annual global horizontal irradiation to the SHAC can make, leading to relatively high SF in these places. As the SHAC system has independent temperature and humidity controls, it is suitable for those premises even with high space sensible and latent loads. Along with the various renewable energy incentives and policy implementation advocated in the related cities, solar air-conditioning systems should become technologically attractive and economically competitive.

15.5 CONCLUSION AND FUTURE DEVELOPMENT

The cooling and energy performances of solar hybrid air-conditioning systems are assured in hot-humid climates. With its hybrid design and independent temperature and humidity controls, the SHAC system can be designed for sharing zone load and ventilation load; for radiant cooling with passive and active chilled beams; for new indoor ventilation strategies (both DV and SV); and for premises with high latent load. In all these SHAC systems, solar-thermal gain can be the primary energy source. Due to the intermittent nature of solar irradiation, it is inevitable that SHAC will require auxiliary heating, which may demand fossil fuel consumption. Despite this, the primary energy saving of the SHAC is still guaranteed compared to the conventional compression refrigeration system. For buildings with high sensible and latent loads, this merit becomes more prominent. As a result, solar air-conditioning is an effective means to reduce the carbon footprint of buildings.

In fact, SHAC systems can be further facilitated by building integrated installations of solar-thermal collectors, incorporating technological advancements of sorption refrigeration cycles and component integration in hybrid design. There are now absorption and adsorption chillers with a cooling capacity of below 10 kW, which can compete with conventional AC systems in small- and medium-scale applications. For solid desiccant wheels, new sorbent materials, such as inorganic zeolitic adsorbent, are appearing, which would be more effective in regeneration by low temperature heat source. More research work is being carried out on liquid desiccant cooling with the inherent feature of thermal storage. Although adsorption chillers do not currently perform as well as absorption chillers, continual improvement is being made in the working pairs, the method of desorption and the configuration of multiple chambers. This would help in raising the COP of the adsorption refrigeration while keeping the advantage of low driving temperature. In point of fact, the advancement of sorption refrigeration and air-conditioning can be beneficial beyond solar air-conditioning: in combined cooling, heating and power, heat-driven refrigeration cycles are also essential.

There are still a number of barriers to wider application of solar air-conditioning systems, among them spatial requirement and economic consideration. It will take time for building practitioners to feel comfortable with this new air-conditioning technology. More demonstration projects and operational experience of solar air-conditioning would certainly help in convincing policy makers, building developers, engineering practitioners and market players. Were the life cycle assessments of solar air-conditioning better known, their environmental merit would also become known more widely. In the trend towards sustainable urbanization, solar cooling can play an essential role in realizing zero carbon deployment in regions with hot and humid climates. With appropriate system integration of cooling and heating, together with a strategy of hybrid use of renewable energy sources, more and more opportunities for solar-thermal technologies for building use should emerge in the near future.

ACKNOWLEDGEMENT

The work described in this paper was fully supported by a grant from the City University of Hong Kong (Strategic Research Grant, Project No. 7002765).

Nomenclature

A	Absorber of absorption chiller or adsorber of adsorption chiller
AC	Air-conditioning
ACVCC	Air-cooled vapour compression chiller
C	Condenser
COP	Coefficient of performance
COP_{ch}	Coefficient of performance of chiller
COP_{dc}	Coefficient of performance of desiccant cooling
D	Desorber
DV	Displacement ventilation
DW	Desiccant wheel

E	Evaporator
EA	Exhaust air
EAF	Exhaust air fan
EC	Evaporative cooler
E_p	Primary energy consumption (kWh)
G	Generator
HW	Hot water
MV	Mixing ventilation
NA	Not applicable
OA	Outdoor air or outdoor air scheme
OAF	Outdoor air fan
PV	Photovoltaic
PV/T	Photovoltaic/thermal
RA	Room air or return air scheme
RAF	Return air fan
RH	Relative humidity (%)
RH_z	Zone relative humidity (%)
SA	Supply air
SAF	Supply air fan
SF	Solar fraction
SHAC	Solar hybrid air-conditioning
SHAC-Ab	Solar hybrid air-conditioning with absorption chiller
SHAC-Ad	Solar hybrid air-conditioning with adsorption chiller
SHAC-VCC	Solar hybrid air-conditioning with vapour compression chiller
SV	Stratum ventilation
TW	Thermal wheel
T_z	Zone temperature (°C)
VCC	Vapour compression chiller
WCVCC	Water-cooled vapour compression chiller

REFERENCES

Altener (2002) *Promoting Solar Air Conditioning – Technical Overview of Active Techniques*, ALTENER Project Number 4.1030/Z/02-121/2002.

Calise, F. (2010) Thermoeconomic analysis and optimization of high efficiency solar heating and cooling system for different Italian school buildings and climates. *Energy and Buildings*, 42, 992–1003.

Chan, A.L.S., Chow, T.T., Fong, S.K.F. and Lin, J.Z. (2006) Generation of a typical meteorological year for Hong Kong. *Energy Conversion and Management*, 47, 87–96.

Cho, S.H. and Kim, J.N.K (1992) Modeling of a silica gel/water adsorption-cooling system, *Energy*, 17, 829–839.

Delorme, M., Six, R., Mugnier, D., Quinette, J.-Y., Richler, N., Heunemann, F., Wiemken, E., Henning, H.-M., Tsoutsos, T., Korma, E., Dall'O, G., Fragnito, P., Piterà, L., Oliveira, P., Barroso, J., Ramón-López, J., Torre-Enciso, S. (2004) *Solar air conditioning*, the European Commission and Rhônalpénergie-Environment.

DOE (2011a) Weather Data, All Regions: Asia WMO Region 2, EnergyPlus Energy Simulation Software, Energy Efficiency and Renewable Energy, U. S. Department

of Energy, website: http://apps1.eere.energy.gov/buildings/energyplus/cfm/weather_data2.
cfm/region=2_asia_wmo_region_2.

DOE (2011b) Weather Data, All Regions: Southwest Pacific WMO Region 5, EnergyPlus
Energy Simulation Software, Energy Efficiency and Renewable Energy, U. S. Department
of Energy, website: http://apps1.eere.energy.gov/buildings/energyplus/cfm/weather_data2.
cfm/region=5_southwest_pacific_wmo_region_5.

EC (2011) *Study on Photovoltaic Panels Supplementing the Impact Assessment for a Recast of
the WEEE Directive – Final Report*, European Commission DG ENV, 14 April 2011.

Eicker, U. (2003) *Solar Technologies for Buildings*. Chichester: Wiley.

Eicker, U. (2009) *Low Energy Cooling for Sustainable Buildings*. Chichester: Wiley.

Enteria, N., Yoshino, H., Mochida, A., Takaki, R., Satake, A., Yoshie, R., Teruaki, M. and
Baba, S. (2009) Construction and initial operation of the combined solar-thermal and electric
desiccant cooling system. *Solar Energy*, 83, 1300–1311.

Fong K.F. and Chow T.T. (2007) Application potential of solar-assisted desiccant cooling system
in sub-tropical Hong Kong. *In: Proceedings of Clima 2007 WellBeing Indoors,* June 10–14,
2007 Helsinki, Finland.

Fong, K.F., Chow, T.T., Lee, C.K., Lin, Z. and Chan, L.S. (2010a) Comparative study of different
solar cooling systems for buildings in subtropical city. *Solar Energy*, 84, 227–244.

Fong, K.F., Chow, T.T., Lee, C.K., Lin, Z. and Chan, L.S. (2010b) Advancement of solar
desiccant cooling system for building use in subtropical Hong Kong. *Energy and Buildings*,
42, 2386–2399.

Fong, K.F., Lee, C.K., Chow, T.T., Lin, Z. and Chan, L.S. (2010c) Solar hybrid air-
conditioning system for high temperature cooling in subtropical city. *Renewable Energy*, 35,
2439–2451.

Fong, K.F., Chow, T.T., Lee, C.K., and Lin, Z. and Chan, L.S. (2011a) Solar hybrid cooling system for high-tech offices in subtropical climate – Radiant cooling by absorption
refrigeration and desiccant dehumidification. *Energy Conversion and Management*, 52,
2883–2894.

Fong, K.F., Lee, C.K., Lin, Z., Chow, T.T. and Chan, L.S. (2011b) Application potential
of solar air-conditioning systems for displacement ventilation. *Energy and Buildings*, 43,
2068–2076.

Fong, K.F., Lee, C.K., Chow, T.T. and Fong, A.M.L. (2011c) Investigation on solar hybrid
desiccant cooling system for commercial premises with high latent cooling load in subtropical
Hong Kong. *Applied Thermal Engineering*, 31, 3393–3401.

Ge, T.S., Li, Y., Dai, Y.J. and Wang, R.Z. (2010) Performance investigation on a novel
two-stage solar driven rotary desiccant cooling system using composite desiccant materials.
Solar Energy, 84, 157–159.

Gottessfeld, P. and Cherry, C.R. (2011) Lead emissions from solar photovoltaic energy systems
in China and India. *Energy Policy*, 39, 4939–4946.

Henning, H.-M., (2004) *Solar-Assisted Air-Conditioning in Buildings, A Handbook for
Planners*. Springer-Verlag: Vienna and New York.

Huang, B.J., Wu, J.H., Hsu, H.Y. and Wang, J.H. (2010) Development of hybrid solar-assisted
cooling/heating system. *Energy Conversion and Management*, 51, 1643–1650.

IEA (2007) *Renewables for Heating and Cooling, Untapped Potential*. International Energy
Agency, IEA Publications, November 2007.

IEA (2010) *World Energy Outlook 2010*, International Energy Agency, website:
http://www.iea.org/weo/index.asp.

IEA (2011) *Solar Heat Worldwide, Markets and Contribution to the Energy Supply 2009*, Solar
Heating & Cooling Programme, International Energy Agency, 2011 edition.

IEA (2012) *Solar Heating and Cooling Programme, International Energy Agency*, website:
http://www.iea-shc.org.

IPCC (2011) *IPCC Special Report on Renewable Energy Sources and Climate Change Mitigation, Summary for Policymakers*. Intergovernmental Panel on Climate Change, June 2011.

Kim, D.S. and Infante Ferreira, C.A. (2008) Analytical modeling of steady state single-effect absorption cycles, *International Journal of Refrigeration*, 31, 1012–1020.

Lin, Z., Chow, T.T., Tsang, C.F., Chan, L.S. and Fong, K.F. (2005) Effect of air supply temperature on the performance of displacement ventilation (part I) – thermal comfort. *Indoor and Built Environment*, 14, 103–115.

Lin, Z., Chow, T.T., Tsang, C.F., Fong, K.F. and Chan, L.S. (2009) Stratum ventilation – a potential solution to elevated indoor temperatures. *Building and Environment*, 44, 2256–2269.

Ölz, S. and Beerepoot, M. (2010) *Deploying Renewables in Southeast Asia – Trends and Potentials*, International Energy Agency.

Qu, M., Yin, H. and Archer, D.H. (2010) A solar-thermal cooling and heating system for a building: Experimental and model based performance analysis and design. *Solar Energy*, 84, 166–182.

SEL (2006) *TRNSYS 16, a TRaNsient SYstem Simulation program*. The Solar Energy Laboratory, University of Wisconsin-Madison, WI, USA.

TESS (2004) *TESS Component Libraries – Version 2.0*. The Thermal Energy System Specialists.

Wang, R.Z., Ge, T.S., Chen, C.J., Ma, Q. and Xiong, Z.Q. (2009) Solar sorption cooling systems for residential applications: Options and guidelines. *International Journal of Refrigeration*, 32, 638–660.

Wang, R.Z. and Zhai, X.Q. (2010) Development of solar-thermal technologies in China. *Energy*, 35, 4407–4416.

Zhang, X.J., Dai, Y.J. and Wang, R.Z. (2003) A simulation study of heat and mass transfer in a honeycombed rotary desiccant dehumidifier, *Applied Thermal Engineering*, 23, 989–1003.

Chapter 16

Solar-desiccant air-conditioning systems

Napoleon Enteria
Enteria Grün Energietechnik, Davao, Philippines

16.1 INTRODUCTION

16.1.1 Energy and environment

One of the alarming situations regarding the current energy supply and demand scenario is the peaking of supply due to rapid utilization (Greene et al., 2006). The scenario also causes instabilities and unpredictability for the long-term energy supply situation owing to the domination of energy suppliers from specific regions and groupings. Hence, the main source of global conventional energy sources comes from the Middle East and Russia. A decade ago, the Middle East accounted for 33% of the world's conventional energy supply (IEA, 2004). In addition, the Russian Federation is becoming the global supplier, particularly of natural gas (Gelb, 2006). Due to the above situation, energy and economic politics have collided, examples being the 1973 oil crisis, the Gulf War and the Russia-Ukraine crisis (Gelb, 2006).

Since the start of the industrial revolution, large amounts of greenhouse gases (GHG) have been deposited in the atmosphere (Brown et al., 2009), the increase in global pollutants coming from human activities of urbanization, industrialization and so on (Lin et al., 2008). Greenhouse gases are the primary cause of global warming (IPPC). The ozone layer depleting substance such as CFCs, halons, and other ozone-depleting chemicals are the cause of ozone layer thinning (Calm and Didion, 1998). All of these have consequences for the increase in global temperature, which has serious effects on climate patterns – flooding, cyclones and other weather disturbances (Barrios et al., 2006). The above situation has serious consequences for global sustainability, as concluded by numerous studies (Bard and Frank, 2006).

The building sector is one of the primary energy consumers, accounting for 50% inclusive of commercial and industrial buildings (Zimmermann et al., 2005). Energy sector output is used to support electrical appliances, thermal comfort and other requirements such as lighting. As the global population grows, urbanization spreads and standards of living increase, it is expected that energy consumption by the building sector will increase (IEA, 2004; York, 2007; Solecki and Leichenko, 2006). Studies show that population size and age structure have effects on energy consumption (York, 2007). Power sector energy consumption will increase by 119% between 2002 and 2030 (IEA, 2004), while energy consumption in the residential sector, including agriculture, was 56.7% in 2006 compared to 44.2% in 1973 (IEA, 2008).

The provision of optimum human living conditions results in large amounts of energy consumption. The energy required to maintain human thermal comfort approaches 50% of total building energy consumption. In most cases, the energy needed is high-grade electric power (IEA, 2008). As most power plants consume large amounts of fossil fuel in generating power, so provision for optimal human living standards has contributed to high carbon-based energy consumption and greenhouse gas emissions (IEA, 2004). Alternative methods are therefore needed when talking about reducing conventional energy consumption and cutting greenhouse gases emissions (Coiante and Barra, 1996). However, it is imperative not to sacrifice healthy indoor and thermal comfort conditions for the sake of energy consumption reduction (Costa and Costa, 2006; Day et al., 2009).

16.1.2 The building environment

At present, maintaining clean indoor thermal comfort conditions is done using refrigerant-based air-conditioning systems (AC). The operation of refrigerant-based or vapour compression systems is by means of altering the pressure of the working fluid to change the boiling point and thus release or absorb latent heat. However, the operation of changing the pressure can only be done by mechanical means, for which pumps or compressors are used; thus, the so-called mechanical vapour compression system (ASHRAE, 1989). In addition, at present, most of the working substances are made from halocarbon compounds such as CFC and HCFC which affect the ozone layer through emissions of greenhouse gas (GHG) (Calm and Didion, 1998). These materials have a long-term effect on the general environment. Hence, the systems maintaining indoor thermal comfort conditions are gradually harming the natural environmental, causing ozone-layer depletion and greenhouse gases emissions.

The provision of indoor thermal comfort conditions for buildings, either through heating or cooling, is done by heat pump systems. These devices are widely called mechanical vapour compression systems. Several studies have been conducted to improve system performance through efficiency and reduce environmental damage. However, these systems still consume large amounts of energy in the form of high-grade electricity. The main energy source of mechanical vapour compression systems comes from electricity power grids. These air-conditioning systems play a major role in the energy consumption of buildings, most particularly in hot and humid climates. In the Middle East, more than 70% of building energy consumption is to support cooling (El-Dessouky et al., 2004). In Europe, 10% of building sector energy consumption is likewise to support cooling demand (Kolokotroni and Aronis, 1999). In Hong Kong, 45% of commercial building energy consumption is also for cooling (Zain et al., 2007). In Japan, 3% of building sector energy consumption is for cooling applications (Murakami et al., 2009). It is expected that in tropical countries which are hot and humid, energy demand for cooling and dehumidification will be very high (Wong and Li, 2007).

Commercial, office and industrial buildings commonly use centralized air-conditioning systems with heat pumps or refrigerant chillers. However, split-type air-conditioning systems provide an alternative. The application of centralized air-conditioning systems will introduce fresh air inside the buildings. However, it will also increase energy consumption as treatment of outdoor air latent and sensible energy

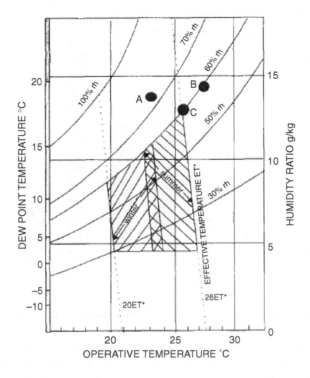

Figure 16.1.1 Singapore buildings indoor temperature and humidity (Sekhar, 1995).

is very high. Hence, recirculation methods are commonly applied to AC systems. Refrigerant-based chillers are commonly used in buildings, as evidenced by the many cooling towers that are installed on rooftops. The chillers are used to cool the water used in the cooling of air in the air cooling unit (ACU). Chillers are common in buildings to cool water distributed in buildings in which cool air is produced through a fan coil unit (FCU). In addition, centralized air-conditioning is commonly applied in large building spaces through a network of air ducts. As open spaces such as commercial establishments have a large volumetric supply of air, the treatment of outdoor air sensible and latent load has an impact on air-conditioning system performance. In addition, as in Singapore, buildings operate at lower temperature and higher relative humidity (approximately 23°C and 70% – see Figure 16.1.1). This is due to the non-reheating of cold air to reduce the energy consumption of the air handling unit (AHU).

Residential buildings, both public and privately owned, use window and split types of air-conditioning system. However, split-type air conditioning systems are the most common due to the flexibility, unlike window types, of locating the compressor/condenser unit and the evaporator/expander unit. Residential buildings use air-conditioning systems during both day and night. However, natural ventilation and air fans can be used simultaneously during daytime, when air-conditioning systems are not commonly used. As shown in Figure 16.1.2, in the case of Singapore,

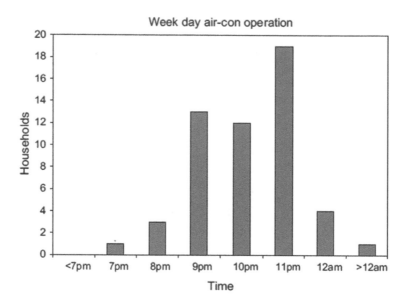

Figure 16.1.2 Singapore pattern of the residential air-conditioning operation (Chua and Chou, 2010).

residential buildings commonly operate air-conditioning systems at night, during the hours of sleep. As most residential buildings operate air-conditioning systems at this time, in terms of electric energy consumption, this is the off-peak period. However, in terms of indoor air quality, problems arise due to the recirculation of indoor air, resulting in a poor quality indoor environment.

16.2 THE BASIC CONCEPT

16.2.1 Thermodynamic processes

Thermally operated air-conditioning systems run by means of applying heat energy for the production of cooling effect (Grossman, 2002). However, as technologies are varied in operation principles and heat requirements, some thermally operated air-conditioning technologies have limited applications (Henning, 2007; Fan et al., 2007). The main advantage of these systems is the direct application of thermal energy for system operation. Hence, low-grade thermal energy can be used to operate the system. In addition, several thermal energy sources can be utilized for the system operation, such as waste heat (Henning et al., 2007). The concept of a thermally operated air-conditioning system is the utilization of a higher thermal energy source to drive the air-conditioning system and provide cooling effect. Figure 16.2.1 shows the general thermodynamic principle of the thermally operated cooling system for air-conditioning applications. The concept is based on four temperatures – the Carnot heat engine (Abrahamsson and Jernqvist, 1993; Hellman, 2002). However, the system still utilizes electric energy for the operation of fans, pumps and the control system. Combined thermal energy and electric energy can be generated from solar energy through a

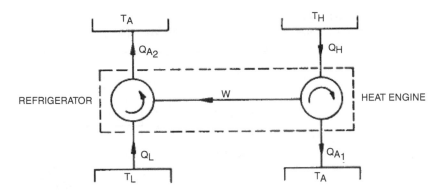

Figure 16.2.1 Thermodynamic principle of the thermally-operated air-conditioning system (Grossman and Johannsen, 1981).

thermo-electric collector (Thermal/Photovoltaic System) (Charalambous et al., 2007). This concept had been described by Mittelman et al. (2007) and Kribus et al. (2006).

From a thermodynamic point of view, the system performance is dependent on the thermal energy source temperature and on the cooling effect temperature (Boehm, 1986). The system relies on ambient temperature conditions (Grossman and Johannsen, 1981). Hence, system thermodynamic performance can be obtained based on the system's operational temperature conditions (Abrahamsson and Jernqvist, 1993; Hellman, 2002). Thermally activated/operated systems have wide potential for application, not only for buildings but also for other systems which produce thermal energy (such as waste heat). Transportation and industrial sectors are sectors with potential for thermally operated air-conditioning systems. Hence, Mazzei et al. (2005) discuss the provision of thermal comfort of desiccant-based thermally activated air-conditioning systems, which can save up to 50% conventional energy (Henning et al., 2001). Thus, the system has potential for further development and application through utilization of thermal energy resources and application of desiccant materials. Figure 16.2.2 shows the operational concept of the desiccant-based air-conditioning system. The processed air from the desiccant dehumidifier becomes hot due to the release of heat through condensation and sorption. Heat recovery devices are used to recover this energy for application again in the desiccant dehumidifier in conjunction with other sources of thermal energy (renewable energy, non-conventional energy and conventional energy). The condition of the air after the heat recovery becomes warm and dry. In hot and humid climates, the air condition is still above the thermal comfort temperature, so that an evaporative cooling process is applied by either direct addition of air moisture (direct evaporative cooling) or indirect addition of air moisture in a secondary air stream (indirect evaporative cooling). Furthermore, application of other air cooling techniques, both conventional and non-conventional, can also be used as an additional air cooler prior to introduction of the air to the indoor environment. Conventional air coolers such as sorption chillers and vapour compression systems are used with an increase in their performance. Non-conventional coolers, such as ground source heat pumps and water source heat pumps, can also be used. These auxiliary

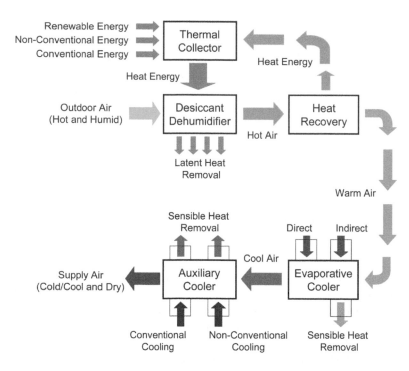

Figure 16.2.2 General concept of the thermally activated desiccant cooling technologies (Enteria and Mizutani, 2011).

coolers are applied when the required temperature of the air after the evaporative cooler is still insufficient to support indoor thermal comfort conditions.

16.2.2 Advantages of the open systems

The chemical contents of the air, such as indoor volatile organic compounds (VOCs), are readily absorbed by desiccants (Wolfrum et al., 2008), thus solving the problem of the recirculation of indoor air pollutants in the recirculation method of air-conditioning systems. In addition, desiccant-based air-conditioning systems have controlled indoor air quality, as reported by Zhang et al. (2007). Based on studies conducted by Fang et al. (2008), desiccant-coated wheels have removed VOCs from air, in particular the most common VOCs, toluene and n-hexane (Wolfrum et al. 2008). Therefore, desiccant-based air-conditioning systems have not only solved health problems related to moisture in buildings (Ahman et al. 2000), but have also improved the quality of indoor air. This applies also to the management of the indoor environment for office buildings, as reported by Shaw et al. (2005) of the National Research Council of Canada. Most importantly, desiccant-based air-conditioning systems can be applied to solve the problem of energy, comfort, environment and indoor air for those of sensitive dispositions, such as people in nursing care (Theodosiou and Ordoumpozanis, 2008).

The open-cycle desiccant-based air-conditioning system utilizes desiccants which are mostly salt-based. Development of other aqueous absorbent desiccants is progressing well. As sorption processes are natural and occur on the surface of the desiccant, the desiccant has another advantage: it can treat the biological and chemical contents of air owing to direct contact of the air with the desiccant. Airborne micro-organisms can be treated by desiccants (Goswami et al., 1997). Wang et al. (2011) show that desiccants can make airborne fungi inactive. Thus, serious problems in air quality can also be resolved by the open-cycle desiccant-based air-conditioning system. Nevertheless, it was shown by Goswami et al. (1997) that titanium dioxide (TiO_2) desiccant material can be used to control air micro-organisms through a photocatalytic process.

16.2.3 Desiccant materials

The adsorption process is a surface phenomenon occurring at the interface of two phases in which cohesive forces, including Van der Waals forces and hydrogen bonding, act between the molecules of all substances irrespective of their state of aggregation (Srivastava and Eames, 1998). This process is called physisorption. Absorption is a chemical process caused by valency forces called chemisorption (Low, 1960). The process of attracting moisture from the air is done either by adsorption or by absorption: the adsorption process is a physical process in which the property of the desiccant material remains the same; while in the absorption process, upon attracting moisture, the physical characteristic of the material changes. The desiccant materials can be either solid or liquid: the solid desiccant and hydrophilic desiccants are silica gel, activated alumina, and zeolites, while calcium chloride is an absorbent desiccant. Commercial hydrophobic solid desiccants are activated carbons, metal oxides, specially developed porous metal hydrides and composite adsorbents (Srivastava and Eames, 1998).

Some desiccant materials combine absorbent and adsorbent desiccants to form composites which enhance their physical properties and sorption capacity (Tokarev et al., 2002). The basic mechanism in the sorption of moisture between air moisture and the desiccant material is the difference in the water vapour pressure on the surface of the desiccant and of the material. The uptake of moisture from the air to the desiccant occurs when vapour pressure in the air is high; the removal of vapour from the desiccant material is done when the vapour pressure in the air is lower than on the desiccant material. When the vapour pressure is the same both in the air and on the desiccant material, an equilibrium is reached and the sorption process stops. The only means to make the sorption process proceed is to use outside forces such as increasing the air pressure, decreasing the temperature or by artificial electromotive force (Low, 1960). The same procedure, but in reverse, is applied for the removal of moisture from the desiccant material.

The most common absorbents are lithium bromide, lithium chloride, calcium chloride and triethelene glycol. Other possible candidates as absorbents are salt-based solutions or related materials which attract water molecules. Examples of alternative absorbents are potassium chloride and sodium chloride. Other candidates are a mixture of the commonly used absorbents mentioned above. Table 16.2.1 summarizes the common absorbents and their properties and compares them for their thermo-chemical, environmental, human toxicity and cost properties. Lithium chloride (LiCl) has a low vapour pressure at a given temperature but the material cost is high (Mei and Dai, 2008); Figure 16.2.3 shows LiCl in the psychometric chart. According to

Table 16.2.1 Comparison of the properties of six liquid desiccants (absorbents) at 25°C, to allow a fair comparison, a concentration giving an equilibrium relative humidity of ERH (Equilibrium Relative Humidity) = 50% has been chosen in each case, with the exception of sodium chloride where ERH = 75%, this being the minimum achievable (Davies and Knowles, 2006).

Property	Unit	Aqueous solution					
		$CaCl_2$	LiBr	LiCl	MgCl	ZnCl	NaCl
Cocentration (mass solute/ mass solution)		0.36	0.39	0.26	0.31	0.52	0.26
Hygroscopicity (equilibrium RH)	%	50	50	50	50	50	75
Cost	US$/m³	560	7300	4600	450	1400	180
Abundance in seawater[a]	m³/m³	2.3×10^{-3}	4.0×10^{-6}	3.0×10^{-6}	1.3×10^{-2}	1.0×10^{-9}	9.0×10^{-2}
Density	kg/m³	1.35	1.38	1.4	1.29	1.58	1.2
Viscosity	mPa-s	4.6	1.8	2.5	6	4.7	1.8
Specific heat capacity	kJ/kg-°C	2.6	2.6	3	2.1	2.3	3.4
Thermal conductivity	W/m-°C	0.56	0.48	0.56	0.52	0.46	0.58
Diffusivity of water in the solution	10^{-9}m²/s	0.54	1.17	0.9	0.91	0.8	1.86
Differential heat of dilution[b]	kJ/kg	80	no data	65	65[d]	no data	no data
Water absorption capacity	kg/m³	85	84	91	76	120	n.a.
Human toxicity[c]	L	0.14	0.23	0.10	0.49	0.03	0.66
Ecotoxicity (Daphnia magna)	ml/L	4.9(2)	no data	0.06 (2)	4.3 (1)	0.001 (6)	20 (5)

[a] Volume of absorbent that could theoretically be extracted from unit volume of seawater, assuming 100% recovery speed.
[b] Mass of water that, on absorption in the absorbent, will cause a 10% relative increase in equilibrium relative humidity.
[c] Estimated lethal dose in humans scaled from LD50 values for rats.
[d] At 50°C.

Ameel et al. (1995), the absorber utilizing LiCl solution at 35°C requires about five times the area of an absorber utilizing lithium bromide to achieve the same sorption performance. Also, lithium chloride solutions are not practical at an absorber temperature of 45°C or higher due to solubility limitations (Ameel et al., 1995). At the same absorbent mass flow rate, the dehumidification performance of lithium chloride is better. Where absorbent volumetric flow rate is the same, the dehumidification performance of lithium chloride is almost the same as lithium bromide. Using structured packing for dehumidification and regeneration studies of three absorbents (calcium chloride, lithium chloride and a mixture of 50% calcium chloride and 50% lithium chloride). Al-Farayedhi et al. (2002) show that lithium chloride has a higher rate of liquid-phase mass transfer coefficient than the other two absorbents, owing to its molecular weight.

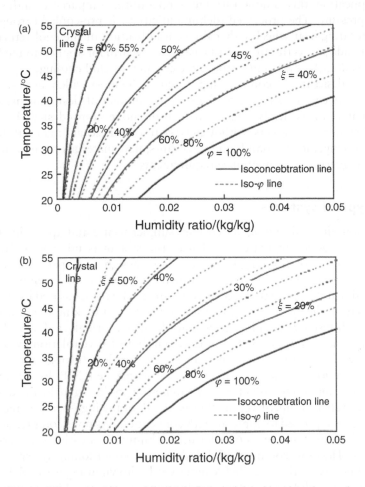

Figure 16.2.3 Status of the commonly used liquid desiccants (absorbents) in the psychrometric chart: a) Lithium Bromide (LiBr); b) Lithium Chloride (LiCl) (Liu et al., 2010).

16.3 SOLID-BASED SYSTEM

16.3.1 Basic concept

Solid desiccant air-conditioning systems are primarily based on the application of solid-based desiccant materials in controlling air moisture content. The sorption mechanism in the solid material is either through absorption or adsorption. Cooling by means of heat recovery, evaporative cooling or other means is applied to the system.

The design of the system is based on the fixed-bed type in alternative operation of moisture sorption and desorption. When the encapsulated phase change materials (EPCM) in the desiccant bed are applied, this absorbs the heat of sorption released during the dehumidification process and lowers the air temperature. However, its humidity is higher compared to the pure desiccant (Rady et al., 2009). For building cooling application, this is done through temperature, compared to industrial applications of pressure. The processed (dehumidified) air is pre-cooled through, in most cases, the rotating heat wheel, either through utilization of the cool return air or by means of outside air. As the air in most cases is still warm for application indoors, final air cooling is done by means of evaporative cooling, or chill cooling.

The solid desiccant air-conditioning system is the most widely used desiccant air-conditioning system. This is due to the simple handling of desiccant materials. The desiccant material is typically impregnated to the honeycomb designed wheels or to the cross-flow heat exchangers. Although typical solid desiccant materials have higher regeneration temperatures than liquid desiccants, new materials have been developed with lower regeneration temperature requirements.

16.3.2 Typical systems

Solid desiccant air-conditioning systems are simpler to use and apply due to the easy handling of the desiccant material. Hence, the system is not complicated, unlike liquid desiccant air-conditioning systems in both design and operation. Farooq and Ruthven (1991) investigated the desiccant bed for solar air-conditioning application. They showed that the optimal choice of desiccant can be compensated by the appropriate adjustment of the cyclic time. In addition, the cost of making the desiccant wheel and moisture diffusivity should be given consideration. Jurinak et al. (1984) presented the open-cycle desiccant air-conditioning system both for ventilation cycle and recirculation cycle. They showed that unbalancing the air flow through the dehumidifier improved the desiccant system coefficient of performance (COP) by 10–15% for the ventilation cycle and up to 50% for the recirculation cycle. To make the desiccant air-conditioning system competitive, the thermal COP of the high-performance desiccant systems must be improved to compete with conventional vapour compression systems, such as very high heat and mass transfer unit dehumidifiers with large thermal capacitance matrix. The most common solid desiccant air-conditioning system is composed of the two-wheel type, the so-called Munter Cycle shown in Figure 16.3.1. This is the basic design of the solid desiccant air-conditioning system, in which application of the air cooling in both the supply air and in the return air is implemented. Hence, several modifications to this cycle, implementing different operating strategies, are presented by Henning et al., 2007, Jain et al., 1995 and Henning et al., 1995.

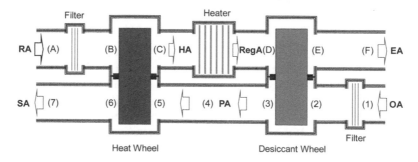

Figure 16.3.1 The double wheels type solid desiccant air-conditioning system (Enteria et al., 2010a).

Application of the desiccant wheel as the air dehumidifier has factors to be considered. Kang and Maclaine-cross (1989) show that the performance of the desiccant-based air-conditioning system relies much on the desiccant material's moisture sorption capacity. Kodama et al. (2001) show that there is an optimal speed at which a high sorption rate occurs in the rotating desiccant wheel. Optimal speed increases with increasing regeneration air flow rate, decreasing desiccant wheel depth, and decreasing bulk density of the rotor. Optimal wheel speed decreases with higher humidity and lower regeneration temperature. They also show that the sorption rate is relative to the relative humidity of the air. Zhang and Niu (2002) show that the sorption performance of the desiccant wheel depends on the wheel rotational speed and the number of transfer units; they suggest that the desiccant wheel should have 2.5 transfer units. Subramanyan et al. (2004a) show that increasing the air flow rate reduces the specific cooling load (difference in the enthalpy of the outdoor air and of the processed air). The cooling load increases due to the amount of air mass flow rate. In addition, Subramanyan et al. (2004b) show that increasing the air flow rate increases the supply air moisture content. Harse et al. (2005) show that for higher humidity air the optimal speed of the wheel is greater than for the air with lower humidity content. They show that at higher regeneration temperature, the performance of the desiccant wheel improves. The depth of the wheel affects the dehumidification rate and as the depth increases the dehumidification rate also increases, resulting in a lowering of the optimum wheel speed. Gao et al. (2005) show that the thickness of the desiccant material affects sorption capacity. At higher desiccant material thickness in the channel, higher sorption rate is attained owing to more time to reach the steady state. In addition, a lower desiccant rotor speed makes for optimum wheel speed. Their study shows that channel shapes affect rotor sorption capacity. Hence, for the same cross-sectional area, sinusoidal channel is the best performer due to its lower hydraulic diameter, resulting in higher air velocity and heat transfer coefficient. Furthermore, the study shows that increasing the outdoor air relative humidity increases the processed air temperature. However, the humidity content of both the processed and the exit air increases as relative humidity of the outdoor air increases. Enteria et al. (2010a) presented parameters affecting the performance of the desiccant wheel and performance evaluation for the desiccant wheel dehumidification capability. La et al. (2010) reviewed the development of the rotary desiccant wheel-based system.

16.3.3 Modified systems

In most designs, operation of the solid desiccant air-conditioning system is through a dehumidification-humidification process. In this process, air dehumidification is done at very low humidity content to achieve evaporative cooling. For this, the required regeneration temperature is increased. Accordingly, application of constant dehumidification will help to prevent the deep dehumidification in regions with hot and humid climates. Enteria et al. (2010b, 2010c) looked at the constant humidity air cooling cycle of the desiccant, as presented in Figure 16.3.2. Ando et al. (2005) show the double stage dehumidification process, in which two desiccant wheels are employed (Figure 16.3.3). The main purpose of this process is to reduce the air moisture content in the case of humid air with lower regeneration temperature requirements. Ge et al. (2009) investigated the two-stage desiccant air-conditioning system. They show

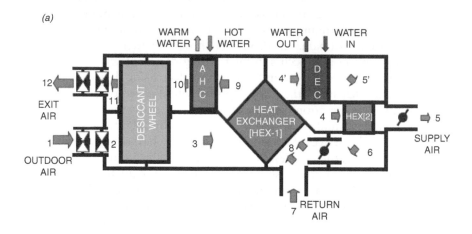

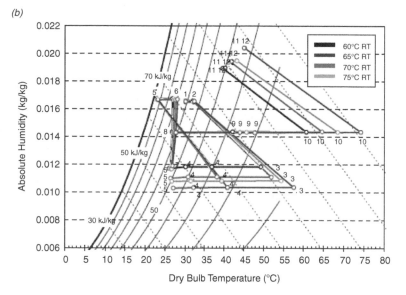

Figure 16.3.2 Constant humidity ratio supply air solid desiccant air-conditioning system: a) system diagram; b) psychrometric chart (Enteria et al., 2010b; Enteria et al., 2010c).

it has lower regeneration temperature requirements with higher COP. Kodama et al. (2005) show the multi-pass desiccant wheel presented in Figure 16.3.4. This shows that a 50°C regeneration temperature is enough for the desiccant wheel. Furthermore, Kodama et al. (2003) presented several designs of the desiccant air-conditioning system for humid climates, conditions in which the 4-wheel cycle (two desiccant wheels and two heat wheels) can be used. However, the 3-wheel cycle (1 desiccant wheel, 1 heat wheel and 1 total heat exchanger) proved better than the 4-wheel cycle.

The fixed-bed solid desiccant air-conditioning system is another type of modified system, on which several design studies have been conducted. The advantage of the fixed-bed system is the sorption process, which can be done in an isothermal way.

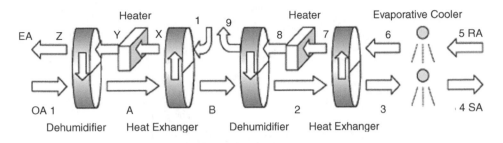

Figure 16.3.3 Double stage dehumidification solid desiccant air-conditioning system (Ando et al., 2005).

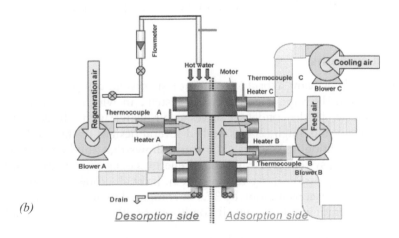

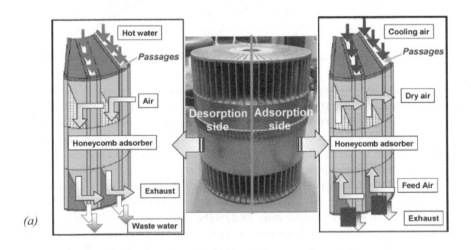

Figure 16.3.4 Principle and diagram of the multi-pass desiccant wheel: a) desiccant wheel design; b) installation diagram (Kodama et al., 2005).

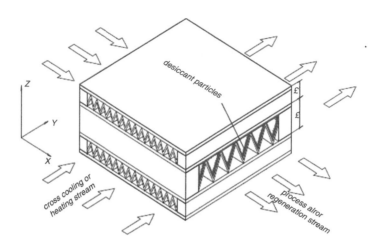

Figure 16.3.5 Desiccant coated cross-cooled compact dehumidifier (Yuan et al., 2008).

This means that the temperature of the air after passing the desiccant material is not increased. Hence, the sorption process is increased due to the removal of sorption heat. Yuan et al. (2008) proposed a cross-cooled compact solid desiccant dehumidifier. The aim of this design is to make the dehumidification process cooler due to the heat exchange between the dehumidified air and the secondary air. The performance of the design was shown to be better without secondary cooling. Majumdar and Worek (1989) investigated the open-cycle cooled-bed desiccant air-conditioning system using a cooled-bed dehumidifier. Here, system performance is more sensitive to regeneration and indoor temperature and to the outdoor humidity ratio than to the indoor humidity ratio and outdoor temperature. The study showed that the cooled-bed system can be regenerated at a low temperature of 50°C. Figure 16.3.5 shows the cross-flow fixed-bed type air-dehumidifier with cooler. Henning et al. (2007) show the application of fixed-bed through a cyclic operation called evaporative cooled sportive heat exchanger (ECOS).

16.3.4 Hybrid systems

The typical solid-based desiccant hybrid air-conditioning system comprises the rotating desiccant wheel and the vapour compression system, in which the evaporator serves as the air cooler while the condenser serves as air heater, with a back-up thermal energy source. Dhar and Singh (2001) studied several designs of the solid-based hybrid desiccant air-conditioning systems. The study shows that solid-based hybrid desiccant air-conditioning system gives substantial energy saving as compared to the conventional vapour compression refrigeration system. Jia et al. (2006) show that this system can reduce electricity consumption by 37.5% compared to the ordinary vapour compression system operating in the same air conditions of 30°C and 55% humidity (Figure 16.3.6).

Sheridan and Mitchell (1985) investigated the use of solid desiccant hybrid air-conditioning systems in Australian climatic conditions. With high sensible load air,

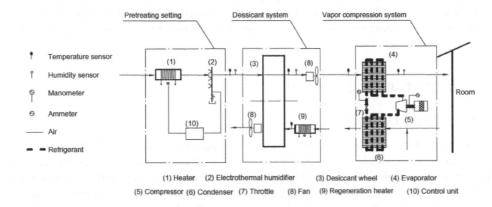

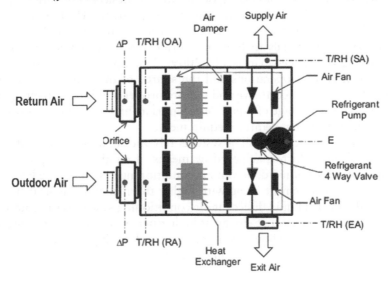

Figure 16.3.6 Schematic diagram of the hybrid solid desiccant air-conditioning system using desiccant wheel (Jia et al., 2006).

Figure 16.3.7 Schematic diagram of the hybrid solid desiccant air-conditioning system using desiccant-coated heat exchanger (Enteria et al., 2011).

the hybrid cycle saved 24% to 40% electric energy compared to the conventional cycle. Combining the hybrid system with an indirect evaporative cooler can bring significant savings in energy consumption. The hybrid system saves more energy in hot and dry climates than in hot and humid climates where more energy is consumed than for the conventional cycle. The system can save energy when the air to be treated has a higher sensible load fraction than latent load fraction. However, solar energy can be combined to meet the higher latent load of the air. Enteria et al. (2011) presented the combined desiccant-refrigerant air-conditioning system using the cyclic processes of air cooling + dehumidification; this can be used as an air heater + humidifier during the winter (see Figure 16.3.7). The result shows that the system can support a higher

COP of above 5. It also shows that the coefficient of performance increases as the air humidity increases. This means that the system is suitable in humid climates where air-conditioning is important. The system is also more compact than other solid desiccant air-conditioning systems.

16.4 LIQUID-BASED SYSTEM

16.4.1 Basic concept

The design of the liquid desiccant air-conditioning system uses the falling film type in the membrane, with air passing to its surface (Ren et al., 2007). Some designs apply the spray type to increase the surface area of air-desiccant contact. The design of the air dehumidifier uses an isothermal process which passes cool air/water at the back of the falling desiccant film (Yin et al., 2008). Since the regeneration of the desiccant material is by means of heat, many designs of liquid desiccant regenerators are done with solar energy. The cooling of air after the desiccant material is performed in the same way as in solid-desiccant air-conditioning systems. The liquid-based system utilizes liquid desiccant materials in removing air moisture content. The most widely used liquid desiccant materials are lithium chloride, lithium bromide, calcium chloride and glycol-based substances (Yin et al., 2009). The application of these materials depends on cost, type of operations and the source of thermal energy. In addition, some liquid desiccants are corrosive and require proper handling in their application. However, the main advantage of liquid desiccant is its high moisture removal capacity with lower regeneration temperature requirement.

Liquid desiccant air-conditioning systems rely on the liquid desiccant to control air moisture content, which is achieved by means of an absorption process. One of the main advantages of liquid desiccant air-conditioning systems is the lower regeneration temperature requirement along with higher thermal and chemical storage. The advantage of the hybrid liquid desiccant air-conditioning system is the complete operation of the system using electric energy at higher performance. This means, for small applications, that the hybrid liquid desiccant air-conditioning system will prevail over the pure liquid desiccant air-conditioning system.

16.4.2 Typical systems

Dehumidifiers or regenerators based on spray, wetted wall (falling film) and packed bed tower are the typical arrangements (Jain and Bansal, 2007). The wetted wall system uses a falling film absorbent in the plate while the air is in contact with the absorbent. This concept is practical for applications which do not need complex air dehumidifiers, such as for low thermal capacity buildings. Mesquita et al. (2006) developed the numerical models for simultaneous heat and mass transfer in parallel-plate dehumidifiers. Here, two polypropylene twin-wall plates form the channel, inside which cooling water flows in a cross-flow configuration with respect to the absorbent and air streams. Water mass flow rate is maintained high enough to keep the plates' walls essentially isothermal, with water temperature gains throughout the plate of less than 0.4°C. The constant thickness and simplified model under-predict the dehumidification, especially for low absorbent flow rates. The numerical model can be adapted

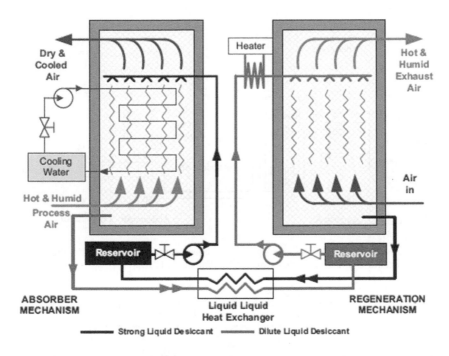

Figure 16.4.1 Schematic diagram of the open-cycle liquid desiccant air-conditioning system either with packed bed tower, spray tower or wetted wall column (Jain and Bansal, 2007).

for non-isothermal conditions, with the introduction of cooling water flow equations. Figure 16.4.1 shows the schematic design of the absorbent dehumidifier/regenerator.

16.4.3   Modified systems

Thermo-chemical storage is an option when the available source of thermal energy is not in phase with cooling/dehumidification demand. With this scenario, conventional energy usage can be reduced. Kessling et al. (1998) shows that the absorbency of hygroscopic salts such as LiCl and CalCl₂ is up to more than three times greater in energy storage than other adsorbents such as zeolite and silica gel. The schematic diagram in Figure 16.4.2 shows that the high storage capacity is based on the concentration between the strong and diluted salt solutions. Furthermore, Miller (1983) reported that energy storage via absorbents was competitive with phase-change materials, rock-bed storage and water systems.

16.4.4   Hybrid systems

Hybrid installation of open-cycle liquid desiccant air-conditioning systems is used to increase system performance. A two-stage open-cycle liquid desiccant air-conditioning system with CaCl2 and LiCl has a COP and exergy efficiency of 0.73% and 23.0% compared to the basic open-cycle liquid desiccant air-conditioning system (Xiong et al., 2010). Figure 16.4.3 shows a multi-stage open-cycle liquid desiccant air-conditioning

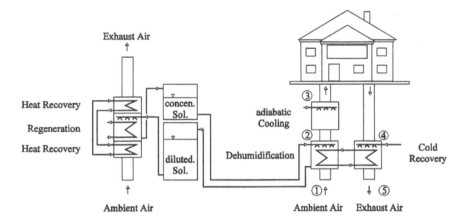

Figure 16.4.2 Schematic diagram of the open-cycle liquid desiccant air-conditioning system with thermo-chemical storage capacity (Kessling et al., 1998).

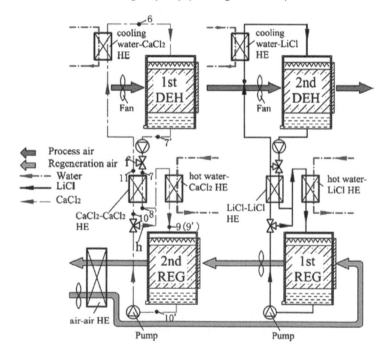

Figure 16.4.3 Schematic diagram of the two-stage open-cycle liquid desiccant air-conditioning system (Xiong et al., 2010).

system. Dai et al. (2001) studied the multi-cycle open-cycle liquid desiccant air-conditioning system, in particular the refrigerant, liquid desiccant and cooling water cycles. Lazzarin and Castellotti (2007) looked at the combination of an open-cycle liquid desiccant air-conditioning system and vapour compression system. The system

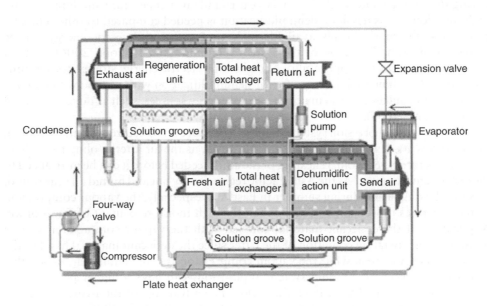

Figure 16.4.4 Schematic diagram of the hybrid open-cycle liquid desiccant air-conditioning system with vapour compression system (Zhu et al., 2010).

produced 20% to 30% more cooling than the vapour compression system alone. Xiong et al. (2010) also presented a novel open-cycle liquid desiccant air-conditioning system with CaCl₂. Compared to the basic system, the thermal coefficient of performance and exergy efficiency of the system increased from 0.24% to 0.73% and from 6.8% to 23.0% respectively. Khalid Ahmed et al. (1997), in a simulation model of the hybrid open-cycle liquid desiccant air-conditioning system, showed that the system provides an excellent alternative to conventional vapour absorption systems, particularly in hot and humid climates. The COP obtained was about 50% higher than that of a conventional vapour absorption system. A maximum COP of 1.25 was obtained during the study and the unit was found to be best suited for hot and humid areas. Figure 16.4.4 shows the hybrid open-cycle liquid desiccant air-conditioning system with vapour compression system.

16.5 SYSTEM APPLICATION

16.5.1 Countries

Thermally operated desiccant air-conditioning technologies are a promising alternative to the vapour compression system in handling air sensible and latent energy contents. This is due to the possibility of operating the system by energy sources other than fossil fuel-based electricity – solar energy, waste heat, etc. Research into the solar desiccant ventilation and air-conditioning system is very important owing to the fact that the amount of air thermal energy content is almost in phase with the amount of solar radiation (Tabor, 1962). In hot and humid climates, such as in East Asia during the summer and South East Asia year round, air temperature and humidity are high. In addition, as day-long dehumidification is needed compared to other climatic conditions, the more cheaply available night-time (off-peak) electric energy can be stored for daytime operation of the system (Hammou and Lacroix, 2006). Enteria et al. (2010b) show the applicability of night-time electric energy storage for daytime utilization. Combined solar energy for air dehumidification with ground water source for air cooling makes the system utilize natural energy sources, such as done in London (Ampofo et al., 2006).

The vapour compression system operates to remove air moisture content by cooling the air below its dew-point temperature. However, as the air after cooling to its dew-point temperate is very cold, reheating of the air is needed before it can be introduced to the indoor environment. As the Asia-Pacific region is very hot and humid all year round, in South East Asia and during summer in East Asia especially, the vapour compression system operates thoroughly to reduce the very high moisture content of outdoor air. Application of the desiccant material coupled with the vapour compression system minimizes the operating condition of the latter since the desiccant material handles the air latent energy content while the vapour compression system handles the air sensible load (hybrid desiccant). Liquid system applications can have a higher performance (44.5%) in green building (Ma et al., 2006). The advantage of the hybrid desiccant air-conditioning system is its operation in part loading (Jia et al., 2006).

Table 16.5.1 shows the development and application of desiccant-based air-conditioning systems, which are expanding globally. However, in the hot and humid climate of the Asia-Pacific Region, South America and Africa the system is still not fully utilized. Therefore, investigations of the system for applications in these regions should expand the potential of the system for more wide-scale use. The system has the potential to be a leading air-conditioning technology for energy-efficient, healthy buildings in hot and humid climates (Sekhar, 2007). As such, it should contribute significantly to reducing conventional energy consumption and GHG emissions by the building sector while also providing human thermal comfort conditions.

16.5.2 Temperate regions

Solid-based desiccant air-conditioning systems have actually been applied in many different climatic conditions. In addition, feasibility studies through numerical studies on the applicability of the system have also been carried out. White et al. (2009) conducted numerical investigation of the solar-powered solid-based desiccant air-conditioning system in different Australian climatic conditions. The investigation centred on direct application of solar energy for the regeneration of the desiccant wheel. The study showed that the system is applicable in the warm temperate climate of Melbourne and Sydney, but not in the tropical climate of Darwin due to the hot and humid outdoor air. The research also showed that solar energy can support building comfort condition by means of a high ventilation rate.

Bourdoukan et al. (2008) conducted numerical investigation of the solar-powered desiccant air-conditioning system using an evacuated tube collector. The researchers showed that collector areas vary with different location due to the required cooling load. For higher outdoor air humidity content, a higher solar collector area was needed in each of three geographical locations: La Rochelle (France), 13.4 g/kg; Bolzano

Table 16.5.1 Global development and application of the desiccant-based air-conditioning technologies (Enteria and Mizutani, 2011).

Continent	Country	Solid Desiccant System	Liquid Desiccant System	Hybrid Desiccant System
Africa	Egypt	o		
	Kenya	o		
Asia	China	o	o	o
	India	o	o	o
	Iran	o		
	Iraq	o		
	Israel	o	o	o
	Japan	o	o	o
	Kuwait		o	
	Lebanon	o		
	Pakistan	o		
	Quatar		o	
	Saudi Arabia	o	o	
	Singapore	o		
	South Korea	o		
	Thailand	o	o	
	Turkey	o		
Europe	France	o		
	Germany	o	o	o
	Italy	o	o	o
	Poland	o		
	Spain	o		
	Sweden	o	o	
	Switzerland		o	
	United Kingdom	o		
North America	Canada	o	o	
	Mexico	o		
	USA	o	o	o
Oceania	Australia	o	o	o
	New Zealand		o	
South America	Cuba	o		
	Brazil	o		

*Other countries may have research, development and application but no published literatures.

(Italy), 11.4 g/kg; and Berlin (Germany), 9.5 g/kg. Furthermore, the study showed that evacuated tube is better than the flat-plate collector owing to the small size of the required back-up thermal energy. Sand and Fischer (2005) investigated the application of the solid desiccant air-conditioning system with a package of HVAC equipment. This showed that the active desiccant module delivers the required air condition with lower cost.

Henning et al. (2001) considered the application of solar desiccant air-conditioning in Europe. They showed the possibility for a simple design of desiccant air-conditioning that is totally dependent on solar energy. However, the condition of the indoor air sometimes exceeded the required level. The system is best suited to a temperate climate. Further solar desiccant with chiller is feasible in terms of economic and energy point of view in the warm-humid climate in which energy saving of 50% is possible. Mavroudaki et al. (2002) presented a numerical investigation with regard to the application of single-stage desiccant air-conditioning system in European cities. The system is applicable in some parts of southern Europe as long as latent load is not high; this is due to the high regeneration temperature requirement for high relative humidity air. The system is feasible in most of central Europe. Atlantic and inland regions of southern Europe appear to be much more suitable to this technology than Mediterranean costal regions.

Smith et al. (1994) investigated the application of solar-powered solid desiccant air-conditioning system in residential buildings in the United States through transient system simulation (TRNSYS) simulation. The study focused on the Pittsburg (Massachusetts), Macon (Georgia) and Albuquere (New Mexico). It showed that building cooling demand was met, and that solar energy is suited to the operation of desiccant AC in the southwest of the US, with 72.7% of energy from solar. In the southeast of the county, 18.0% of desiccant air-conditioning was provided by solar energy. Casa and Schmitz (2005) investigated the application of borehole heat exchanger in desiccant air-conditioning with a gas engine (Figure 16.5.1). The system, installed in a demonstration building, saved 70% of energy for desiccant with a borehole heat exchanger. In the case of desiccant with chiller, it can save up to 30%. Cler (1992), investigating the possibility of applying desiccant dehumidification in military facilities, showed that it is recommended when additional cooling capacity is needed in existing HVAC systems. Also, for higher quantities of outdoor air make-up, a desiccant-based system is ideal for this type of application. In new construction, desiccant dehumidification equipment should be considered. This would reduce the size of chiller and electric energy demand. In addition, when designing new desiccant air-conditioning systems, desiccant regeneration from vapour compression, solar energy, cogeneration and others should be considered in the early phase of design.

Halliday et al. (2003) looked at the feasibility of applying solar desiccant air-conditioning systems in the UK. This study showed that the solid desiccant with solar power is feasible for application in buildings as long as the system is applied in a proper manner. Henning et al. (2007) investigated the application of desiccant air-conditioning system in a tri-generation system (power + heating and cooling). This used a vapour compression chiller and silica gel desiccant with the electricity to drive the chiller coming from combined cooling, heating and power CCHP, while the regeneration of the desiccant wheel was powered by waste heat from CCHP. An electric saving of more than 30% was made compared to the conventional air handling system.

Enteria et al. (2012) conducted a numerical investigation of the solar-powered desiccant air-conditioning system in East Asia (Northeast Asia and Southeast Asia). The system ventilation rate increased from the temperate Northeast Asia to tropical Southeast Asia. The solar desiccanr air-conditioning system is applicable under East Asian climatic conditions as long as the proper specifications are applied, such as the size of the flat-plate collector, inclination of the collector plate, thermal storage tank volume and the required air flow rates going to the building. In addition, an alternative desiccant air-conditioning system in which air cooling can be done independently can reduce the air flow rate requirement, as a pure desiccant air-conditioning system cannot support a lower supply air temperature.

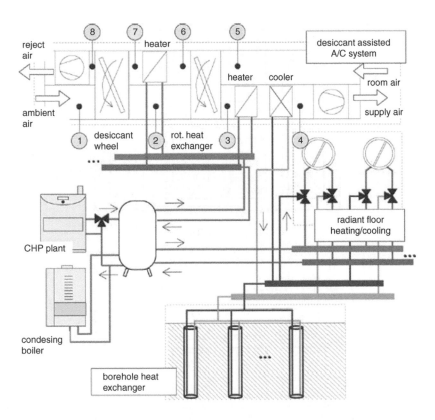

Figure 16.5.1 Desiccant assisted HVAC system with borehole heat exchanger (Casa and Schmitz, 2005).

Liu et al. (2004) investigated an office building in Beijing (China) with a total floor area of 20,000 m² over 10 stories and an installed open-cycle absorbent air-conditioning system. During the summer, the system is used to control the air latent load while the absorption chiller and compression chiller are used to control the air sensible load. With this combined usage of open-cycle absorbent air-conditioning and chillers, the chilled water temperature is increased from 15°C to 18°C. In this situation, the COP of the chiller is increased due to the increased evaporative temperature. The schematic design of this system is shown in Figure 16.5.2. It has an efficiency of over 80% compared to the conventional HVAC system. The thermal storage tank and the absorbent storage tank provide long operating hours, due to the thermo-chemical storage capability of the absorbent.

16.5.3   Sub-temperate regions

Fong et al. (2010) conducted investigations with regard to the application of solar desiccant air-conditioning system in the subtropical climate of Hong Kong. Although a typical desiccant air-conditioning system is not energy efficient, it can supply the required fresh air to a building, resulting in good indoor air quality and ventilation

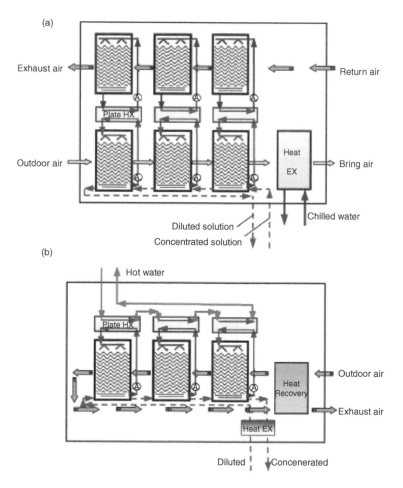

Figure 16.5.2 Schematic of open-cycle liquid desiccant air-conditioning system with multiple air-absorbent contact exchanger (a) air dehumidifier and (b) absorbent regenerator (Liu et al., 2004).

effectiveness. Hao et al. (2007) investigated the application of desiccant dehumidification with chilled ceiling and displacement ventilation. This showed it to be feasible in hot and humid climates owing to its capability of responding consistently to cooling demand. In addition, the system reduces building energy consumption by 8.2% compared to conventional air-conditioning. Niu et al. (2002) investigated application of the desiccant dehumidification by desiccant wheel with chilled-ceiling (Figure 16.5.3). The aim of the installation is for the desiccant to reduce air moisture content and thus avoid the condensation of moisture in the ceiling panel, and at the same time cool the air by means of the chilled ceiling. The results showed the combined system can save up to 44% of primary energy consumption in which 70% of the operating hour of the desiccant dehumidification can be provided with less than 80°C regeneration temperature.

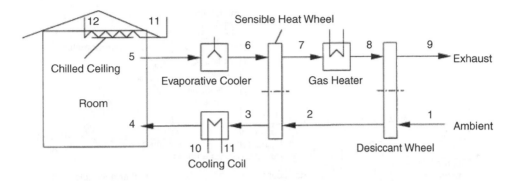

Figure 16.5.3 Chilled-ceiling with solid desiccant air-conditioning system (Niu et al., 2002).

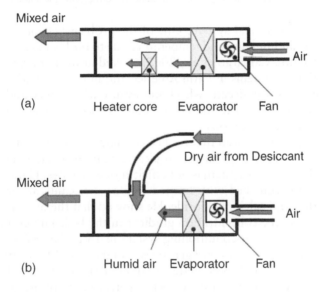

Figure 16.5.4 Automobile air conditioning system: a) conventional and b) with desiccant (Nagaya et al., 2006).

16.5.4 Hot and humid regions

Khalid et al. (2008) conducted numerical investigations with regard to the application of solar desiccant air-conditioning in Pakistan. The system had a higher COP in Lahore's climatic conditions, without auxiliary cooling. Hirunlabh et al. (2007) considered the applicability of the solid fixed-bed desiccant air-conditioning in Thailand, showing that it can save 24% of electric energy. Furthermore, it is practical for application in large buildings as a centralized air-conditioning system. Nagaya et al. (2006) investigated the application of the solid desiccant wheel in an automobile air-conditioning system (Figure 16.5.4). This showed that the system is energy efficient

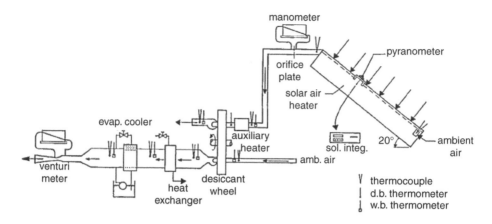

Figure 16.5.5 Solar-desiccant air-conditioning system (Joudi and Madhi, 1987).

compared to the conventional system. One of the problems encountered is the difficulty of controlling air humidity and temperature due to the heat exchange and coolant
flow to the evaporator. Camargo et al. (2005) looked at the application of solid desiccant air-conditioning systems in Latin America and in tropical and equatorial cities.
This system comprises a desiccant wheel and evaporative cooler and was shown to
be applicable as an alternative to the vapour compression system since it can provide
human thermal comfort conditions.

Dupont et al. (1994) showed that a silica compact-bed desiccant air-conditioning
system powered by solar energy in the tropical climate of Guadeloupe can produce
cooling power. However, the system is not efficient due to losses. Hamed (2003) investigated the packed porous bed with burned clay as desiccant carrier, and desiccant
impregnated with liquid calcium chloride. He showed that the mass transfer rate had
a significant effect on the concentration gradient in the bed. Jain et al. (1995) investigated the solid desiccant air-conditioning system in 16 Indian cities, showing that a
cycle with a wet surface heat exchanger gives a higher COP than other cycles. The
Dunkle cycle has been found to be better in all climatic conditions. Heat exchanger
effectiveness of above 0.8 is desirable for better performance of the cycles not using a
wet surface heat exchanger. The effect on COP of the evaporative cooler is insignificant
but it can control the room sensible load factor. Joudi and Madhi (1987) investigated
the applicability of the solar desiccant air-conditioning system in Basrah, Iraq (Figure
16.5.5). For the local weather conditions, they showed that a regeneration temperature
of 70°C can be provided using solar energy in clear skies. Kabeel (2007) investigated
the application of the calcium chloride desiccant wheel constructed from iron wire and
a cloth layer. The system uses sole solar energy for the regeneration of the desiccant.
Tested in the climate of Egypt, the system had a high performance after the solar noon,
with the wheel's effectiveness dependent on solar radiation and air flow rate.

The tri-generation plant installed in Politecnico di Torino (Italy) was developed to
support the air-conditioning system shown in Figure 16.5.6 (Badami and Portoraro,
2009). The layout of the system, presented in Figure 16.5.7, shows the internal combustion co-generator, open-cycle absorbent air-conditioning system, cooling tower, two

Figure 16.5.6 Building heated and air-conditioned by the tri-generation plant in Italy (Badami and Portoraro, 2009).

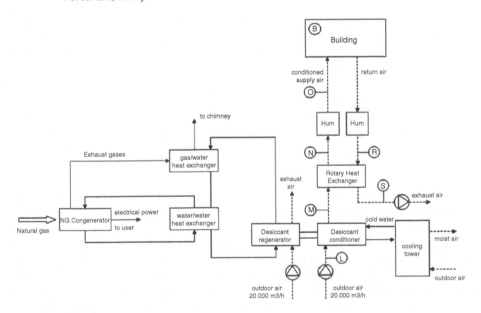

Figure 16.5.7 Layout of the tri-generation plant in Italy (Badami and Portoraro, 2009).

heat exchangers and the connecting pipes. Based on the economic analysis, the system payback time was assessed as 6.8–7.7 years. The system is an interesting alternative to common AC technologies and further investigation and analysis will be done on the system's actual operation.

Figure 16.5.8 Actual view of the open-cycle liquid desiccant air-conditioning system showing the absorber/dehumidifier (1), desorber/regenerator (2); air ducts (3), fan (4), rotary air/air heat exchanger (5), control cabinet (6), solar collector field (7) and hot water storage tank (8) (Gommed and Grossman, 2007).

An open-cycle liquid desiccant air-conditioning system was installed at the Energy Engineering Centre at the Technion, in the Mediterranean city of Haifa, as depicted in Figure 16.5.8 with schematic diagram presented in Figure 16.5.9 (Gommed and Grossman, 2007). The system uses solar energy for regeneration, with thermal and chemical storage tanks (hot water and absorbent). The analysis of the system revealed a thermal COP of approximately 0.8 with a parasitic loss of around 10%. Based on the analysis, parasitic losses could be minimized with an improvement of the overall COP.

A liquid desiccant system was installed at the Energy Park of the Asian Institute of Technology (AIT) in Pathumthani, Thailand (Katejanekarn et al., 2009). Figure 16.5.10 shows the actual installation of the solar-regenerated absorbent ventilation preconditioning system, while Figure 16.5.11 shows the schematic of the installed system. Results from the study showed that the solar open-cycle absorbent air-conditioning system can work in tropical climates like that of Thailand. The humidity of outdoor air can be reduced by 11%, while the temperature of the supply air is almost the same as the outdoor air. This means that an auxiliary air-cooling system is still needed to control the air temperature. Zhao et al. (2012) reviewed the open-cycle absorbent air-conditioning system in a $21,960\,m^2$ building in Shenzhen, China. The system can provide the required indoor thermal comfort and air quality based on temperature, humidity and CO_2 concentration. With a COP of 4.0, the system is much better than an ordinary AC system.

In Singapore, modern applications of liquid-based and solid-based desiccants are used in AC systems in commercial buildings. In one example where $540\,m^2$ solar hot water collector panels are used to support a $3000\,m^2$ factory hall using liquid-desiccant air-conditioning, the annual energy saving amounted to S\$ 50,000 (Singapore Liquid Desiccant).

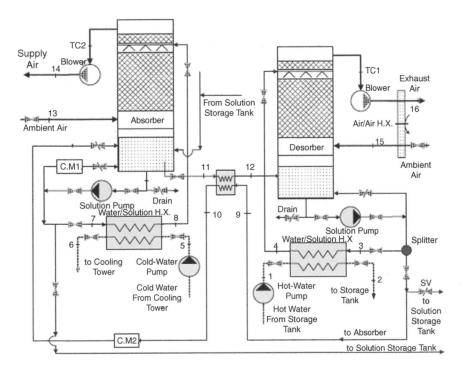

Figure 16.5.9 Schematic description of the open-cycle liquid desiccant air-conditioning system: the solid thick lines indicate air flow; the solid thin lines indicate solution flow; the dotted thin lines indicate water flow (Gommed and Grossman, 2007).

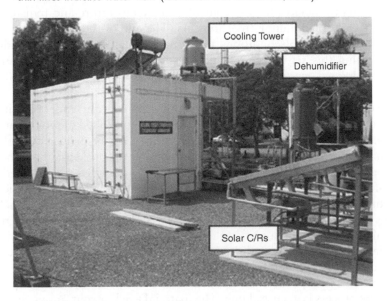

Figure 16.5.10 The solar-regenerated open-cycle liquid desiccant air-conditioning system installed at Asian Institute of Technology (AIT) (Katejanekarn et al., 2009).

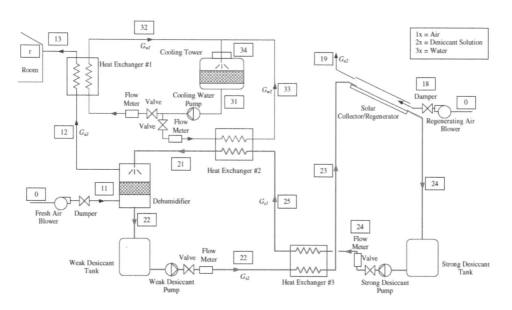

Figure 16.5.11 Schematic of the solar-regenerated open-cycle liquid desiccant air-conditioning system installed at Asian Institute of Technology (AIT) (Katejanekarn et al., 2009).

16.6 FUTURE AND PERSPECTIVES

Thermally operated desiccant-based air-conditioning technologies are being conceptualized as alternative devices for the control of air temperature and moisture content in artificially controlled indoor environments. The aim of these technologies is leave the general environmental conditions unaffected while controlling the indoor environment. Within this set-up there is potential for the utilization of clean energies converted to thermal energy, as well as the use of clean and environmentally friendly substances.

The technology has potential for industrial, commercial and residential sector applications so as to provide comfortable artificially controlled indoor environmental conditions. To achieve this potential the technology must be properly developed and redesigned for practical application. Several studies have considered how the technology may be modified, for example in size reduction, and practically applied. Internal scrutiny of is important to evaluate the behaviour of the technology. Aside for the internal study of the system, the external factors affecting the processes, such as thermal energy sources and how it will be properly coupled to the technology to make it economical and easily maintained, are important when applied in real situations.

The desiccant-based air-conditioning technology operated thermally utilizes the capability of desiccant material in controlling air moisture content. As the process is a physical or chemical phenomenon, the uptake of moisture does not require energy and, thus, the process is a natural one. The material needs a continuous re-sorption of moisture, either by pressure or temperature variation, to remove the uptake moisture in the material. The process is already utilized in chemical processes and industrial controls. The most practical application is the desiccant material used in the control of moisture in sensitive products such as foods. The uptake of moisture is natural as can be observed. Since the process is exothermic, removal of moisture is done mostly using thermal energy, while some is removed through reduction of pressure.

For the control of large spaces, such as industrial areas, assembly of complicated and sensitive electronic, electrical, chemical and mechanical systems, and application of the desiccant-based air-conditioning system is done using thermal energy. It should be noted that application of the technology in the commercial sector is undertaken in the same way as for industry, but with some variations.

For residential buildings the technology is still at the infant stage, but as the development of larger, ideal systems reaches the middle stages of development, smaller-scale systems will surely follow on in due course, along with the incorporation of other processes for optimization.

Several combined systems have been developed, such as Combined Heat and Power (CHP) and Combined Cooling, Heating and Power (CCHP) along with the Distributed Energy System (DES), the purpose being cost-effective implementation and optimization of the technology. In the combined systems the electricity generated is used for different applications, while thermal energy is used for the operation of sorption air-conditioning. This means that the systems provide heating, cooling and power production, and thus have higher overall system efficiency. In other systems the technology is coupled with other devices such as waste-thermal energy from from industrial, commercial or natural processes. This is dependent on the availability of thermal energy sources, and thus design is based upon it.

Some are coupled with thermal energy produced from the utilization of clean energy sources such as solar energy, biogas or biomass, or other forms. In addition, utilization of cheaper night-time electricity is used through thermal storage done in sensible or latent storage systems. Storage of unutilized thermal energy from different thermal energy sources, particularly given its variable availability, is indeed an approach that should be optimized from both economic and practical standpoints. Other designs are based on the combined operation and utilization of available clean energy, waste-thermal energy sources or conversion of night-time electricity for thermal storage. This is dependent on the needs and considerations within the design process.

REFERENCES

Abrahamsson, K. and Jernqvist, A. (1993) Carnot comparison of multi-temperature level absorption heat cycles. *International Journal of Refrigeration*, 16, 240–246.

Ahman, M., Lundin, A., Musabasic, V. and Soderman, E. (2000) Improved health after intervention in a school with moisture problems. *Indoor Air*, 10, 57–62.

Al-Farayedhi, A.A., Gandhidasan, P. and Al-Mutairi, M,A. (2002) Evaluation of heat and mass transfer coefficients in a gauze-type structured packing air dehumidifier operating with liquid desiccant. *International Journal of Refrigeration*, 25, 330–339.

Ameel, T.A., Gee, K.G. and Wood, B.D. (1995) Performance predictions of alternative, low cost absorbents for open-cycle absorption solar cooling. *Solar Energy*, 54, 65–73.

Ampofo. F., Maidment, G. and Missenden, J. (2006) Review of groundwater cooling systems in London. *Applied Thermal Engineering*, 26, 2055–2062.

Ando, K., Kodama, A., Hirose, T., Goto, M. and Okano, H. (2005) Experimental study on a process design for adsorption desiccant cooling driven with a low-temperature heat. *Adsorption*, 11, 631–636.

ASHRAE Fundamentals (1989). American Society of Heating, Refrigerating and Air-Conditioning Engineers, Atlanta, Georgia

Badami, M. and Portoraro, A. (2009) Performance analysis of an innovative small-scale trigeneration plant with liquid desiccant cooling system, *Energy and Buildings*, 41, 1195–1204.

Bard, E. and Frank, M. (2006) Climate change and solar variability: What's new under the sun? *EPSL Frontiers*, 248, 480–493.

Barrios, S., Bertinelli, L. and Strobl, E. (2006) Climatic change and rural-urban migration: The case of Sub-Saharan Africa. *Journal of Urban Economics*, 60, 357–371.

Boehm, R.F. (1986) Maximum performance of solar heat engines. *Applied Energy*, 23, 196–281.

Bourdoukan, P., Wurtz, E., Joubert, P. and Sperandio, M. (2008) Potential of solar heat pipe vacuum collectors in the desiccant cooling process: Modeling and experimental results. *Solar Energy*, 82, 1209–1219.

Brown, F.A., Salas, N.S., Jimenez, M.A.B. and Rosales, M.A. (2009) Possible future scenarios for atmospheric concentration of greenhouse gases: A simplified thermodynamics approach. *Renewable Energy*, 34, 2344–2352.

Calm, J.M and Didion, D.A. (1998) Trade-offs in refrigerant selections: past, present, and future. *International Journal of Refrigeration*, 21, 308–321.

Camargo, J.R., Godoy, E. and Ebinuma, C.D. (2005) An evaporative and desiccant cooling system for air conditioning in humid climates. *Journal of the Brazil Society of Mechanical Science and Engineering*, XXVII, 243–247.

Casa, W. and Schmitz, G. (2005) Experiences with a gas driven, desiccant assisted air conditioning system with geothermal energy for an office building. *Energy and Buildings*, 37, 493–501.

Charalambous, P.G., Maidment, G.G., Kalogirou, S.A. and Yiakoumetti, K. (2007) Photovoltaic thermal (PV/T) collectors: a review. *Applied Thermal Engineering*, 27, 275–286.

Chua, K.J. and Chou, S.K. (2010) Energy performance of residential buildings in Singapore. *Energy*, 35, 667–678.

Cler, G.L. (1992) *Desiccant-Based Dehumidification for Army Facilities*. US Army Construction Engineering Research Laboratories (USACERL), Champaign, IL, USA/TR FE-93/10.

Coiante, D. and Barra, L. (1996) Renewable energy capability to save carbon emissions. *Solar Energy*, 57, 485–491.

Costa, M.F.B. and Costa, M.A.F. (2006) Indoor air quality and human health. *Journal of Integrated Management of Occupational Health and the Environment*, 1, 1–10.

Dai, Y.J., Wang, R.Z., Zhang, H.F. and Yu, J.D. (2001) Use of liquid desiccant cooling to improve the performance of vapour compression air conditioning. *Applied Thermal Engineering*, 21, 1185–1202.

Day, A.R., Ogumka, P., Jones, P.G. and Dunsdon, A. (2009) The use of the planning system to encourage low carbon energy technologies in buildings. *Renewable Energy*, 34, 2016–2021.

Davies, P.A. and Knowles, P.R. (2006) Seawater bitterns as a source of liquid desiccant for use in solar-cooled greenhouses. *Desalination*, 196, 266–279.

Dhar, P.L. and Singh, S.K. (2001) Studies on solid desiccant based hybrid air-conditioning systems. *Applied Thermal Engineering*, 21, 119–134.

Dupont, M., Celestine, B. and Beghin, B. (1994) Desiccant solar air conditioning in tropical climates: Field testing in Guadeloupe. *Solar Energy*, 52, 519–524.

El-Dessouky, H., Ettouney, H. and Al-Zeefari, A. (2004) Performance analysis of two-stage evaporative coolers. *Chemical Engineering Journal*, 102, 255–266.

Enteria, N. and Mizutani, K. (2011) The role of the thermally activated desiccant cooling technologies in the issue of energy and environment. *Renewable and Sustainable Energy Reviews*, 15, 2095–2122.

Enteria, N., Yoshino, H., Satake, A., Mochida, A., Takaki, R., Yoshie, R., Mitamura, T. and Baba, S. (2010a) Experimental heat and mass transfer of the separated and coupled rotating desiccant wheel and heat wheel. *Experimental Thermal and Fluid Science*, 3, 603–615.

Enteria, N., Yoshino, H., Satake, A., Mochida, A., Takaki, R., Yonekura, H., Yoshie, R., Mitamura, T. and Baba, S. (2010b) Initial operation and performance evaluation of the developed solar thermal and electric desiccant cooling system. *Experimental Heat Transfer*, 24, 59–87.

Enteria, N., Yoshino, H., Satake, A., Mochida, A., Takaki, R., Yoshie, R. and Baba, S. (2010c) Development and construction of the novel solar thermal desiccant cooling system incorporating hot water production. *Applied Energy*, 87, 478–486.

Enteria, N., Mizutani, K., Monma, Y., Akisaka, T. and Okazaki, N. (2011) Experimental evaluation of the new solid desiccant heat pump system in Asia-Pacific climatic conditions. *Applied Thermal Engineering*, 31, 243–257.

Enteria, N., Yoshino, H., Mochida, A., Satake, A., Yoshie, R., Takaki, R., Yonekura, H., Mitamura, T. and Tanaka, Y. (2012) Performance of solid-desiccant cooling system with Silica-Gel (SiO_2) and Titanium Dioxide (TiO_2) desiccant wheel applied in East Asian climates. *Solar Energy*, 86, 1261–1279.

Fan, Y., Luo, L. and Souyri, B. (2007) Review of solar sorption refrigeration technologies: development and applications. *Renewable and Sustainable Energy Reviews*, 11, 1758–1775.

Fang, L., Zhang, A. and Wisthaler, A. (2008) Desiccant wheels as gas-phase absorption (GPA) air cleaners: Evaluation by PTR-MS and sensory assessment.*Indoor Air*, 18, 375–385.

Farooq, S. and Ruthven, D.M. (1991) Numerical simulation of a desiccant bed for solar air conditioning applications. *Journal of Solar Energy Engineering*, 113, 80–88.

Fong, K.F., Chow ,T.T., Lee, C.K., Lin, Z. and Chan, L.S. (2010) Comparative study of different solar cooling systems for buildings in subtropical city. *Solar Energy*, 84, 227–244.

Gao, Z., Mei, V. and Tomlinson, J. (2005) Theoretical analysis of dehumidification process in a desiccant wheel. *Heat Mass Transfer*, 41, 1033–1042.

Ge, T.S., Li, Y., Wang, R.Z. and Dai, Y.J. (2009) Experimental study on a two-stage rotary desiccant cooling system. *International Journal of Refrigeration*, 32, 498–508.

Gelb, B.A. (2006) *Russian Oil and Gas Challenges*. Congressional Research Service, USA.

Gommed, K. and Grossman, G. (2007) Experimental investigation of a liquid desiccant system for solar cooling and dehumidification.*Solar Energy*, 81, 131–138.

Goswami, D., Trivedi, D. and Blocks, S. (1997) Photocatalytic disinfection of indoor air. *Journal of Solar Energy Engineering*, 119, 92–96.

Greene, D.L., Hopson, J.L. and Li, J. (2006) Have we run out of oil yet? Oil peaking analysis from an optimist's perspectives. *Energy Policy*, 34, 515–531.

Grossman, G. (2002) Solar-powered systems for cooling, dehumidification and air-conditioning. *Solar Energy*, 72, 53–62.

Grossman, G. and Johannsen, A. (1981) Solar cooling and air conditioning. *Progress in Energy and Combustion Science*, 7, 185–228.

Halliday, S.P., Beggs, C.B. and Sleigh, P.A. (2003) The use of solar desiccant cooling in the UK: a feasibility study. *Applied Thermal Engineering*, 22, 1327–1338.

Hamed, A.M. (2003) Desorption characteristics of desiccant bed for solar dehumidification/humidification air conditioning systems. *Renewable Energy*, 28, 2099–2111.

Hammou, Z. and Lacroix, M. (2006) A new PCM storage system for managing simultaneous solar and electric energy. *Energy and Buildings*, 38, 258–265.

Hao, X., Zhang, G., Chen, Y., Zou, S. and Moschandreas, J. (2007) A combined system of chilled ceiling, displacement ventilation and desiccant dehumidification. *Building and Environment*, 42, 3298–3308.

Harse, Y., Utikar, R., Ranade, V. and Pahwa, D. (2005) Modeling of rotary desiccant wheel. *Chemical Engineering Technology*, 28, 1473–1479.

Hellman, H.M. (2002) Carnot-COP for sorption heat pumps working between four temperature levels. *International Journal of Refrigeration*, 25, 66–74.

Henning, H.M. (2007) Solar assisted air conditioning of building – an overview. *Applied Thermal Engineering*, 27, 1734–1749.

Henning, H.M., Erpenbeck, T., Hindenburg, C. and Santamaria, I.S. (2001) The potential of solar energy use in desiccant cooling cycles. *International Journal of Refrigeration*, 24, 220–229.

Henning, H.M., Pagano, T., Mola, S. and Wiemken, E. (2007) Micro tri-generation system for indoor air conditioning in the Mediterranean climate. *Applied Thermal Engineering*, 27, 2188–2194.

Henning, S., Dhar, P.L. and Kaushik, S.C. (1995) Evaluation of solid-desiccant-based evaporative cooling cycles for typical hot and humid climates. *International Journal of Refrigeration*, 18, 287–296.

Hirunlabh, J., Charoenwat, R., Khedari, J. and Teekasap, S. (2007) Feasibility study of desiccant air-conditioning system in Thailand. *Building and Environment*, 42, 572–577.

IEA. (2004) *World Energy Outlook*. International Energy Agency, Paris, France.

IEA. (2008) *Key World Energy Statistics*. International Energy Agency, Paris, France.

IPCC. *IPCC/TEAP Special Report. Special Report on Safeguarding the Ozone Layer and the Global Climate System*. Intergovernmental Panel on Climate Change. http://www.ipcc.ch/pdf/special-reports/sroc/sroc_full.pdf

Jain, S., Dhar, P.L. and Kaushik, S.C. (1995) Evaluation of solid-desiccant-based evaporative cooling cycles for typical hot and humid climates. *International Journal of Refrigeration*, 18, 287–296.

Jain, S. and Bansal, P.K. (2007) Performance analysis of liquid dehumidification systems. *International Journal of Refrigeration*, 30, 861–872.

Jia, C.X., Dai, Y.J., Wu, J.Y. and Wang, R.Z. (2006) Analysis on a hybrid desiccant air-conditioning system. *Applied Thermal Engineering*, 26, 2393–2400.

Joudi, K.A. and Madhi, S.M. (1987) An experimental investigation into a solar assisted desiccant-evaporative air-conditioning system. *Solar Energy*, 39, 97–107.

Jurinak, J.J., Mitchell, J.W. and Beckman, W.A. (1984) Open-cycle desiccant air conditioning as an alternative to vapor compression cooling in residential applications. *Journal of Solar Energy Engineering*, 106, 252–260.

Kabeel, A.E. (2007) Solar powered air conditioning system using rotary honeycomb desiccant wheel. *Renewable Energy*, 32, 1842–1857.

Kang, T. and Maclaine-Cross, I. (1989) High performance solid desiccant, open cooling cycles. *Journal of Solar Energy Engineering*, 111, 176–183.

Katejanekarn, T., Chirarattananon, S., and Kumar, S. (2009) An experimental study of a solar-regenerated liquid desiccant ventilation pre-conditioning system. *Solar Energy*, 83, 920–933.

Kessling, W., Laevemann, E. and Kapfhammer, C. (1998) Energy storage for desiccant cooling systems components development. *Solar Energy*, 64, 209–221.

Khalid, A., Mahmood, M., Asif, M. and Muneer, T. (2008) Solar assisted, pre-cool hybrid desiccant cooling system for Pakistan. *Renewable Energy*, 34, 151–157.

Khalid Ahmed, C.S., Gandhidasan, P. and Al-Farayedhi, A.A. (1997) Simulation of a hybrid liquid desiccant based air-conditioning system, *Applied Thermal Engineering*, 18, 125–134.

Kodama, A., Andou, K., Ohkura, M., Goto, M. and Hirose, T. (2003) Process configurations and their performance estimations of an adsorptive desiccant cooling cycle for use in a damp climate. *Journal of Chemical Engineering of Japan*, 36, 819–826.

Kodama, A., Hirayama, T., Goto, M., Hirose, T. and Critoph, R. (2001) The use of psychrometric charts for the optimization of a thermal swing desiccant wheel. *Applied Thermal Engineering*, 21, 1657–1674.

Kodama, A., Watanabe, N., Hirose, T., Goto, M. and Okano, H. (2005) Performance of a multipass honeycomb adsorber regenerated by a direct hot water heating. *Adsorption*, 11, 603–608.

Kolokotroni, M. and Aronis, A. (1999) Cooling-energy reduction in air-conditioned offices by using night ventilation. *Applied Energy*, 63, 241–253.

Kribus, A., Kaftori, D., Mittelman, G., Hirshfeld, A., Flitsanov, Y. and Dayan, A. (2006) A miniature concentrating photovoltaic and thermal system. *Energy Conversion and Management*, 47, 3582–3590.

La, D., Dai, Y.J., Li, Y., Wang, R.Z. and Ge, T.Z. (2010) Technical development of rotary desiccant dehumidification and air conditioning: a review. *Renewable and Sustainable Energy Review*, 14, 130–147.

Lin, M., Oki, T., Holloway, T., Streets, D.G., Bengtsson, M. and Kanae, S. (2008) Long-term transport of acidifying in East Asia-Part I: Model evaluation and sensitivity studies. *Atmospheric Environment*, 42, 5939–5955.

Liu, X.H., Geng, K.C., Lin, B.R. and Jiang, Y. (2004) Combined cogeneration and liquid-desiccant system applied in a demonstration building, *Energy and Buildings*, 36, 945–953.

Liu, X.H., Yi, X.Q. and Jiang, Y. (2011) Mass transfer comparison of two commonly used liquid desiccants: LiBr and LiCl aqueous solutions. *Energy Conversion and Management*, 52, 180–190.

Low, M. (1960) Kinetics of chemisorption of gases on solids. *Chemical Reviews Journal*, 60, 267–312.

Ma, Q., Wang, R.Z., Dai, Y.J. and Zhai, X.Q. (2006) Performance analysis on a hybrid air-conditioning system of a green building. *Energy and Buildings*, 38, 447–453.

Majumdar, P. and Worek, W.M. (1989) Performance of an open-cycle desiccant cooling system using advanced desiccant matrices. *Heat Recovery System and CHP*, 9, 299–311.

Mavroudaki, P., Beggs, C.B., Sleigh, P.A. and Halliday, S.P. (2002) The potential for solar powered single-stage desiccant cooling in southern Europe. *Applied Thermal Engineering*, 22, 1129–1140.

Mei, L. and Dai, Y.J. (2008) A technical review on use of liquid-desiccant dehumidification for air-conditioning application. *Renewable and Sustainable Energy Review*, 12, 662–89.

Mesquita, L.C.S., Harrison, S.J. and Thomey, D. (2006) Modeling of heat and mass transfer in parallel plate liquid-desiccant dehumidifiers. *Solar Energy*, 80, 1475–1482.

Miller, W.M. (1983) Energy storage via desiccants for food/agricultural applications. *Energy in Agriculture*, 2, 341–354.

Mittelman, G., Kribus, A. and Dayan, A. (2007) Solar cooling with concentrating photovoltaic/thermal (CPVT) systems. *Energy Conversion and Management*, 48, 2481–2490.

Murakami, S., Levine, M.D., Yoshino, H., Inoue, T., Ikaga, T., Shimoda, Y., Miura, S., Sera, T., Nishio, M., Sakamoto, Y. and Fujisaki, W. (2009) Overview of energy consumption and GHG mitigation technologies in the building sector of Japan. *Energy Efficiency*, 2, 179–194.

Nagaya, K., Senbongi, T., Li, Y., Zheng, J. and Murakami, I. (2006) High energy efficiency desiccant assisted automobile air-conditioner and its temperature and humidity control system. *Applied Thermal Engineering*, 26, 1545–1551.

Niu, J.L., Zhang, L.Z. and Zuo, H.G. (2002) Energy savings potential of chilled-ceiling combined with desiccant cooling in hot and humid climates. *Energy and Buildings*, 34, 487–495.

Rady, M.A., Huzayyin, A.S., Arquis, E., Monneyron, P., Lebot, C. and Palomo, E. (2009) Study of heat and mass transfer in a dehumidifying desiccant bed with micro-encapsulated phase change materials. *Renewable Energy*, 34, 718–126.

Ren, C.Q., Tu, M. and Wang, H.H. (2007) An analytical model for heat and mass transfer processes in internally cooled or heated liquid desiccant-air contact units. *International Journal of Heat and Mass Transfer*, 50, 3545–3555.

Sand, J.R. and Fischer, J.C. (2005) Active desiccant integration with packaged rooftop HVAC equipment. *Applied Thermal Engineering*, 25, 3138–3148.

Sekhar, S.C. (1995) Higher space temperatures and better thermal comfort – A tropical analysis. *Energy and Buildings*, 23, 63–70.

Sekhar, S.C. (2007) A review of ventilation and air-conditioning technologies for energy-efficient healthy buildings in the tropics. *ASHRAE Transactions*, 113, 426–434.

Shaw, C., Won, D. and Reardon, J. (2005) Managing volatile organic compounds and indoor air quality in office buildings – An engineering approach, *National Research Council of Canada*, Institute of Research in Construction.

Sheridan, J.C. and Mitchell, J.W. (1985) A hybrid desiccant cooling system. *Solar Energy*, 34, 187–193.

Singapore Liquid Desiccant. http://www.sgc.org.sg/fileadmin/ahk_singapur/Energies/L-DCS_-_Michael_Hinterbrandner.pdf

Srivastava, N.C. and Eames, I.W. (1998) A review of adsorbents and adsorbates in solid-vapour adsorption heat pump systems. *Applied Thermal Engineering*, 18, 707–714.

Smith, R.R., Hwang, C.C. and Dougall, R.S. (1994) Modeling of a solar-assisted desiccant air conditioner for a residential building. *Energy*, 19, 679–691.

Solecki, W.D. and Leichenko, R.M. (2006) Urbanization and the metropolitan environment: Lessons from New York and Shanghai. *Environment*, 48, 8–23.

Subramanyan, N., Maiya, M. and Murthy, S. (2004a) Parametric studies on a desiccant assisted air-conditioner. *Applied Thermal Engineering*, 24, 2679–2688.

Subramanyan, N., Maiya, M. and Murthy, S. (2004b) Application of desiccant wheel to control humidity in air-conditioning systems. *Applied Thermal Engineering*, 24, 2679–2688.

Tabor, H. (1962) Use of solar energy for cooling purposes. *Solar Energy*, 6, 136–142.

Theodosiou, T. and Ordoumpozanis, K. (2008) Energy, comfort and indoor air quality in nursery and elementary school buildings in the cold climatic zone of Greece. *Energy and Buildings*, 40, 2207–2214.

Tokarev, M., Gordeeva, L., Romannikov, V., Glaznev, I. and Aristov, Y. (2002) New composite sorbent $CaCl_2$ in mesopores for sorption cooling/heating. *International Journal of Thermal Sciences*, 41, 470–474.

Wang, Y.F., Chung, T.W. and Jian, W.M. (2011) Airborne fungi inactivation using an absorption dehumidification system. *Indoor and Built Environment*, 20, 333–339.

White, S.D., Kohlenbach, P. and Bongs, C. (2009) Indoor temperature variations resulting from solar desiccant cooling in a building without thermal backup. *International Journal of Refrigeration*, 32, 695–704.

Wolfrum, E., Peterson, D. and Kozubal, E. (2008) The volatile organic compound (VOC) removal performance of desiccant-based dehumidification systems: testing at sub-PPM VOC concentrations. *HVAC&R Research*, 14, 129–140.

Wong, N. and Li, S. (2007) A study of the effectiveness of passive climate control in naturally ventilated residential buildings in Singapore. *Building and Environment*, 42, 1395–1405.

Xiong, Z.Q., Dai, Y.J. and Wang, R.Z. (2010) Development of a novel two-stage liquid desiccant dehumidification system assisted by $CaCl_2$ solution using exergy analysis method. *Applied Energy*, 87, 1495–1504.

Yin, Y., Zhang, X., Wang, G. and Luo, L. (2008) Experimental study on a new internally cooled/heated dehumidifier/regenerator of liquid desiccant system. *International Journal of Refrigeration*, 31, 857–866.

Yin, Y.G., Zhang, X.S., Peng, D.G. and Li, X.W. (2009) Model validation and case study on internally cooled/heated dehumidifier/regenerator of liquid desiccant systems. *International Journal of Thermal Sciences*, 48, 1664–1671.

York, R. (2007) Demographic trends and energy consumption in European Union Nations, 1960–2025. *Social Science Research*, 36, 855–872.

Yuan, W., Zheng, Y., Liu, X. and Yuan, X. (2008) Study of a new modified cross-cooled compact solid desiccant dehumidifier. *Applied Thermal Engineering*, 28, 2257–2266.

Zain, Z.M., Taib, M.N. and Baki, S.M.S. (2007) Hot and humid climate: Prospect for thermal comfort in residential building. *Desalination*, 209, 261–268.

Zhang, L. and Niu, J. (2002) Performance comparisons of desiccant wheels for air dehumidification and enthalpy recovery. *Applied Thermal Engineering*, 22, 1347–1367.

Zhao, K., Liu, X.H., Zhang, T. and Jiang, Y. (2012) Performance of temperature and humidity independent control air-conditioning system in an office building. *Energy and Buildings*, 43, 1895–1903.

Zhu, W.F., Li, Z.J., Liu, S., Liu, S.Q. and Jiang, Y. (2010) In-situ performance of independent humidity control air-conditioning system driven by heat pumps. *Energy and Buildings*, 42, 1747–1752.

Zimmermann, M., Althaus, H.J. and Hass, A. (2005) Benchmarks for sustainable construction a contribution to develop a standard. *Energy and Buildings*, 37, 1147–1157.

Chapter 17

Building integrated concentrating solar systems

Daniel Chemisana[1] *& Tapas K. Mallick*[2]

[1]*Polytechnic School, University of Lleida, 25001, Spain*
[2]*College of Engineering, Mathematics and Physical Sciences, University of Exeter, TR10 9EZ, United Kingdom*

17.1 INTRODUCTION TO BUILDING INTEGRATION OF SOLAR ENERGY SYSTEMS

Building integration of solar systems can refer to the roof or the façade of a building. Building Integrated (BI) solutions are of great interest since they have several advantages, such as aesthetically pleasing roof or facade integration, on-site energy generation, higher electrical and/or thermal conversion efficiencies, and better use of space. The term "building integration" is generally classified in two ways: (i) Building Integrated Solar Energy (BISE); and (ii) Building Applied Solar Energy (BASE). These are defined as:

- BISE – when a solar energy system such as photovoltaics or solar thermal system is directly integrated within the building as a replacement of building fenestrations; for example, in a building integrated photovoltaics (BIPV) system where PV is the replacement for an existing building component or where no added building components are required to integrate such PV systems within the new buildings.
- BASE – when a solar energy system such as PV and solar thermal collectors are installed within the existing building or new building without any replacement of building fenestrations; a good example of a building applied photovoltaics (BAPV) system is PV integrated into a roof or a solar thermal collector installed into a roof without any replacement of building materials.

In the literature and in the commercial sector there are several BI systems; however, their architectural quality can be further improved. In this way, the use of solar technologies can be increased. Towards this direction, solar concentrating systems could offer several advantages in comparison with non-concentrating ones. Nevertheless, at present the use of concentrating technologies is limited, while most of the existing installations have devices of considerable size (for example, solar power towers, parabolic-trough concentrators, parabolic-dish concentrators, large Fresnel concentrators with 2-axis tracking). With regard to Concentrating Photovoltaics (CPV), more than 30 companies are developing these systems; many are start-ups, while there is a tendency for rapidly increasing production (Kurtz, 2009).

The use of concentrating systems requires the development of reliable systems from the producers. On the other hand, utilization of a solar concentrator usually means the necessity of tracking. An important issue is the tracker to be simple in order to reduce the complexity and the cost of the system. When comparing such a system with a flat-panel PV device built for the same application, the additional cost of the tracker and its maintenance must be compensated by the advantages that are provided by the use of a concentrating technology. On the other hand, flat-panel systems can be used to replace structural elements of a building and this (in most cases) is not possible by means of concentrating technologies (Swanson, 2000).

In terms of the concentrators (reflectives or refractives), their integrability depends on its concentration factor, C (defined as the ratio between the aperture area of the primary concentrator and the active cell area). Concentrating systems with $C > 2.5X$ generally use tracking, whereas systems with $C < 2.5X$ can be static. However, in the long term static concentrators with higher ratios which make use of luminescence and photonic crystals may appear (Luque-Heredia et al., 2007). Low concentrating ratio systems ($C < 10X$) are of great interest as they are mostly of linear geometry and thus one tracking axis is sufficient for efficient operation (Tripanagnostopoulos, 2008). The combination of improved sheet metal capability with the high capacity of the PV industry can lead to a large deployment of low concentration PVs (Kurtz, 2008). CPV is a feasible method to reduce the high initial cost of PV solar energy since concentrating solar radiation onto solar cells means that the area of semiconductor devices is diminished. Considering that a higher concentration factor has higher cost reduction, within the concentration range where single-axis tracking may be used, the most desirable concentration factor is that which approaches the upper limit of single-axis concentrators.

On the other hand, solar thermal systems can be considered as an alternative solution for instances where the priority is the production of heat. Several promising technologies are included in this category, including solar collectors with vacuum tubes, reflectors combined with simple concentrating thermal (CT) systems. Moreover for solar thermal, solutions with low cost and low complexity should be preferred.

For all applications of BI systems, certain requirements must be fulfilled. These requirements are associated with factors such as the design of the building, the conformity to the context of the building, an architecturally pleasing design, etc.

The first part of this chapter describes the characteristics of BI systems without concentration, and later focuses on concentrating systems specifically. We begin by considering general concepts relating to building integration of solar thermal and photovoltaic systems. In the second part, all the terms and characteristics are adapted and applied to the concentring systems, thermal and photovoltaic.

17.1.1 Solar thermal systems and building integration requirements

Building Integration of Solar Thermal (BIST) systems involve incorporating a solar heat generator while preserving and considering the other functions of the building envelope. Some of these functions are, for example, to protect the building interior space from weather conditions, to prevent noise, control daylight, regulate air renovation, and to achieve good insulation conditions leading to energy efficiency.

The characteristics of solar thermal collectors present difficulties when being integrated into buildings, due to their size, materials, rigidity, colour and auxiliary installations. Accordingly, thermal collectors are mostly added to fulfil only a technical function (BASE) in a manner such that their visual impact is kept low and poor integration is minimized. This constrains the placement to the roof top (flat or inclined). Within the roof integration, the tilted configuration maximizes annual efficiency; however, overheating problems may occur in the warmest seasons. In these cases proper system dimensioning must be performed taking into account both the heat demand and overheating possibilities.

BIST systems on the façade, at present, are practically non-existent because the main use of thermal collectors to date has been to provide heat, with no or minimum involvement in the building environment and envelope. In this way, installation of the solar panel in the most visible part of the building requires more attention than simply to cover a technical function. At this point a balance becomes necessary, sometimes pitting architecture and engineering against each other – in other words, a conflict between efficiency and aesthetics.

To state how to increase the architectural quality of BIST, a clear definition of a successful integration must be established. In this sense, Munari Probst and Roecker (2007) conducted a survey of how architects and engineers perceive integration quality. The authors inferred a set of guidelines synthesizing the criteria highlighted in the survey:

1   The use of the solar energy system as a construction element (facade cladding, roof covering, etc.) facilitates the work of integration design. Certainly, the "logic" of the building design is easier to follow when the architect has to balance fewer elements which fulfil more functions.
2   The position as well as the dimensions of the collector field should be evaluated by considering the building as a whole (important issues are: energy production goals, formal integration needs, etc). Another option is the use of dummy elements (non-active elements with a similar appearance).
3   The choice of colours and materials for the system should match with colours and materials characterizing building and context. The initial choice of technology is fundamental because it imposes the material of the external-visible-system layer (glass, metal, plastic, etc). In the frame of the chosen technology, material treatments (surface colour, texture) offered by the various available products can be considered.
4   Module size and shape should be chosen by taking into consideration building and facade/roof composition grids (or vice versa). The proposed jointing types should also be considered while choosing the product (different jointing types differently underline the modular grid of the system in relation to the building).

The criteria and guidelines imply that the designer should have knowledge of the existing technologies and available products in order to make the right choice (of products and technologies) to realize a successful building integration. Nevertheless, the above-mentioned integration requirements are difficult to achieve because the currently available collectors on the market have been developed with insufficient awareness of building integration aspects (Munari Probst et al., 2005). In order to properly design collectors oriented to their integration in buildings, a methodology has been defined by Munari Probst and Roecker (2007). Figure 17.1.1 shows the different steps involved in designing the collector: integration, efficiency, users' preferences and feasibility – all need to be taken into account to cover all points of view.

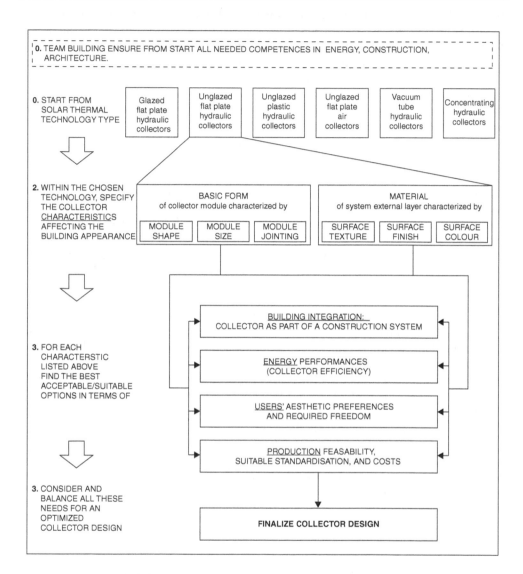

Figure 17.1.1 Design methodology (Munari Probst and Roecker, 2007).

Towards BIST systems, the Solar Heating and Cooling program (SHC) of the International Energy Agency (IEA) created a Task called the Successful Architectural Implementation of Solar Thermal Systems Project Database. The project database was initiated by Task 39 with support from Task 41 (Solar Energy and Architecture) and Task 37 (Advanced Housing Renovation with Solar and Conservation) and is now online. Some representative cases of BIST systems are included in Figure 17.1.2.

From the four examples in the figure, some characteristics are apparent regarding the grade of integration, and pose two questions: (i) has the collector participating in the building envelope any additional function other than to produce heat?; and (ii) how does the collector fit in with the building's aesthetics?

(a)

(b)

(d)

(c)

Figure 17.1.2 Four different configurations of BIST systems. a) Multifamily house and commercial building with roof integrated solar collectors in Graz, Austria (Picture source: D. Chemisana);
b) Roof integrated solar collectors ina a single family house in Pietarsaari, Finland (Picture source: A.G.. Hestnes, 1999); c) Transparent façade collectors in Ljubljana, Slovenia
(Picture source: C. Maurer et al., 2012), and; d) Student residence with solar collectors
integrated in the façade in Dornbirn, Austria (Picture source: Munari Probst and Roecker,
2007).

In order to answer the first question, it is necessary to see how multifunctional the solar collector is. In Figures 17.1.2a and 17.1.2c the collector is acting as a shading element, whilst in the other two cases (Figures 17.1.2b and 17.1.2d) it works as a double skin roof or façade respectively. As can be seen, the solar system provides additional positive functions to the building envelope. It is impossible from the pictures to see the effects of the insulation or the structure on the building, both of which aspects could also be satisfied by the solar technologies depending on the configuration used. In this sense the cost-effectiveness ratio becomes better when compared to a scenario in which a heat only benefit is gained.

The second question concerns the formal integration of the collector, where the module's characteristics are essential. In the first configuration (Figure 17.1.2a) the solar collectors are totally standard, placed on a specific structure to hold the systems at the required angle for both illumination control and electrical production. In Figures 17.1.2b and c the collectors also have quite standard characteristics. By contrast, in the another case the module's active area is divided into strips to allow partial lightening (Figure 17.1.2c). IEA SHCP Task 41, concerning the influence and characteristics of the collectors in the appearance of the building, defines a set of key criteria for all types of solar collectors: module shape and size, jointing, visible materials, surface texture, colours, field size and position. When the collector's response to these aspects is positive, the solar system's flexibility ensures a proper building integration from a formal point of view.

Next, Figure 17.1.3 synthesizes the two questions posed above. As can be seen, inside the house schematic a second column entitled "Constructive" is defined. This refers to the constructive properties of the solar collector: insulating properties, waterproofing, resistance to impacts and wind/snow loads, etc. This point has not been considered in the previous explanations due to as a building element, it is clear that must fulfil all constraints as a constructive element in order to be architecturally incorporated.

A summary of the requirements for building integration of solar thermal systems are presented in Table 17.1.1.

17.1.2 Solar photovoltaic systems and building integration requirements

In Table 17.1.2, several requirements for the building integration of non-concentrating PVs are presented along with their advantages. For grid-connected PV systems, some of their advantages are: they do not require additional land; the cost of the PV wall or roof can be offset against the cost of the building element it replaces; power is generated on site and thus replaces electricity that would otherwise be purchased at commercial rates; and by connecting to the grid, the high cost of storage is avoided and security of supply is ensured.

The way people deal with PVs in architecture differs from country to country depending on factors such as the influence of the government on house building. However, in all cases of PV system integration into buildings there are some important issues which should be taken into consideration (Table 17.1.2).

At this point it should be mentioned that the integration of PV systems in architecture can be divided into five categories, based on the increasing extent of architectural

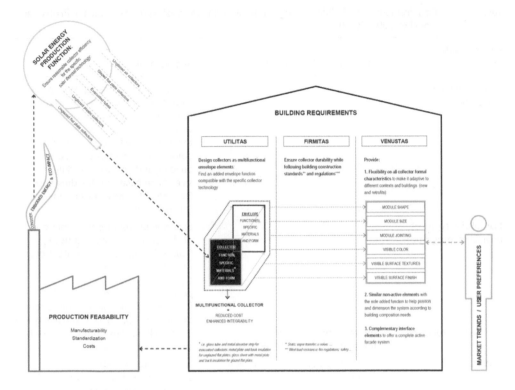

Figure 17.1.3 Development methodology for new systems building integration (Munari Probst and Roecker, 2012).

integration (Reijenga, 2003): (ii) Applied invisibly; (ii) Added to the design; (iii) Adding to the architectural image; (iv) Determining architectural image; and (v) Leading to new architectural concepts.

Several available urban-scale BIPV products and projects, along with details about their applications and characteristics, can be found at: http://www.pvdatabase.org/products_viewall.php, from IEA PVPS Task 10. Some representative examples of BIPV are included in the following (Figure 17.1.4).

From the pictures shown in Figure 17.1.4, the different grades of building integration can be distinguished, according to the criteria defined by IEA PVPS task 7 (see Table 17.1.2) and by Reijenga (2003). In the case of the first photograph (Figure 17.1.4a), the PV modules are placed on the flat roof in such a manner that they cannot be seen from the street. This example reflects the lowest grade of building integration, where the PV modules are applied to the building without replacement of any building material and without any extra functionality other than to produce electricity (BAPV). However, in Figure 17.1.4b the photovoltaic modules can be observed to be part of the façade and to participate directly in the architectural design. The modules are working as a building element at the same time than as an energy source. The PVs are manufactured using transparent encapsulation in order to allow partial illumination in the interior space. In Figure 17.1.4c, the inclined roof PV system can be seen

Table 17.1.1 Requirements for building integration of solar thermal systems (Munari Probst and Roecker 2007; 2012).

Type of building integration	Description of the system	Requirements
Facade and roof integrations		use of the solar energy system as a construction element; evaluation of position and dimensions of the collector field by considering the building as a whole; choice of colours and materials based on building and context; module size/shape based on building and facade/roof composition grids (or vice versa)
For all the types of envelope integration		ensuring/preserving the functions of the envelope: protection (e.g. from rain), insulation, comfort etc. new designs based on: type of building; energy performance; users' required freedom and aesthetic preferences; production feasibility/standardisation
Functional Integration	multifunctional collectors integration into building envelope	envelop functions compatible with solar heat collection
Constructive Integration	integrating solar collectors in a façade	to take into consideration the constructive characteristics of the specific technology to be integrated along with the specificities of the constructive system hosting the collectors
Formal Integration		flexibility in terms of all collector characteristics affecting building appearance: module shape/size/ jointing, collector colour, visible surfaces textures etc.

to work with the transparency of an atrium and to provide architectural continuity as a roof. The PV panels are performing wholly and perfectly a building function, defining the image of the house and leading to a new architectural concept. The "Schott Ibérica" building (Figure 17.1.4d) in Barcelona combines a semi-transparent photovoltaic façade in which the coloured windows are transparent and the cells are opaque, resulting in a singular shading system. The façade constitutes an important element of the building's aesthetics, both for its interior space and outwardly.

As described previously, PV modules can be in the form of: (i) roofing materials; (ii) wall materials; and (iii) photovoltaic flexible modules applicable for construction materials (Shinjo, 1994; Toyokawa and Uehara, 1997). PV building roof integration includes exchangeable PV shingles, prefabricated PV roof panels and insulated PV roof panels (Shinjo, 1994). PV glass curtain walls and PV metal curtain walls are used for integration of PV modules with wall materials (Shinjo 1994; Toyokawa and Uehara,

Table 17.1.2 Requirements for building integration of non-concentrating photovoltaics (Reijenga, 2000a; 2000b; 2003; Reijenga and Kaan, 2011).

Type of building integration	Description of the system	Requirements	Advantages (along with the production of electricity)
Into the roof	the system is part of the external skin	the system should be part of an impermeable layer in the construction	
	the system is above the impermeable layer	the impermeable layer has to be pierced in order to mount the system on the roof	
	if the PVs are transparent → serve as water and sun barriers and also transmit daylight		sun protection (avoid overheating in summer) in glass-covered areas, such as sunrooms and atriums
Into the façade	glass or frameless PVs for sunshades, louvers and canopies	orientation is important: facade systems might be suitable in certain countries, especially at a northern (above 50°N) or a southern (below 50°S) latitude	shade, protection from rain
PV systems as part of a passive cooling strategy (specific case)			PVs replace building elements; PVs are very well ventilated at the back; a separate mounting construction is not necessary; the air-conditioning system is eliminated
All the types of building integration		technical aspects of PV, cables and inverters a structure strong enough to withstand wind or snow loads aesthetic quality → criteria formulated by the IEA PVPS Task 7 workgroup for the evaluation of the aesthetic quality: natural integration, architecturally pleasing designs, good composition of colours and materials, dimensions that fit the gridula, harmony, composition, PV systems that match the context of the building, well-engineered design, use of innovative design ventilation at the back of the modules (not important for thin film a-Si) shadow is not allowed on the modules; the mounting and removing of the modules should be easy; the modules should stay clean (or can be cleaned); the electrical connections should be easy; the wiring should be sun-proof and weather-proof	

Figure 17.1.4 Four different configurations of BIPV systems. a) Flat roof installation in Lleida, Spain (Picture source: D. Chemisana); b) Solar façade with a detail of its view from inside the building in Lleida, Spain. (Picture source: D. Chemisana); c) Glass ceiling with transparent BIPV modules (Picture source: Petter Jelle et al, 2012); d) Solar transparent façade "Schott Ibérica" in Barcelona, Spain (Picture source: Munari Probst and Roecker, 2012).

1997). Most roof-mounted systems are retrofitted and hence are not fully integrated into the roof structure but are mounted onto existing roofs (Watt et al., 1999). Fully integrated BIPV roofing systems must perform the function of a standard roof and provide water tightness, drainage and insulation. These characteristics could also be extended to BIST systems.

In terms of the aesthetics of the modules, frameless modules look very similar to window glass and the individual module is hard to recognize. Its smooth surface has a high aesthetic value. On the other hand, framed modules give a totally different effect. Firstly, the frames can be heavy and thus determine the total impression of the surface. Moreover, the highly visible frames divide the surface into modules and every individual module is very recognizable. As a solution, smaller frames in the same colour as the cells can be used and are less visible. Also, the soldering between the cells is a small detail but is important for the image of very visible PV systems. Older techniques had very visible and not very smooth soldering. However, new techniques have hidden the soldering better, e.g. by moving it towards the back. In addition, modules vary significantly in size, while the glazing is available as single and double (insulating) glass. In general, thin-film modules allow greater freedom to select size and colour than c-Si modules (Reijenga, 2003).

Regarding the temperature of the PV panels, module efficiency and thus the amount of electricity produced, decreases as the temperature increases for mono and polycrystalline silicon cells (amorphous silicon cells present little temperature dependence). In many non-BIPV applications, modules are mounted on free-standing frames with ambient air on both sides (for cooling of both sides). In contrast, some BIPV applications install the modules in close contact to building material, such as roofs or wall insulation, and the lack of circulating air increases the module's temperature. Hence, a good design criterion for mono or poly silicon applications is to allow as much cooling as possible by providing air flow behind the module and minimizing the effect of insulation. This is not an important issue for amorphous silicon modules (Fanney et al., 2001; Reijenga, 2003).

In terms of cooling of PVs and thus increasing their efficiency, hybrid PVT (Photovoltaic Thermal) panels have been developed. These systems combine a PV panel with a thermal collector and in this way they produce electricity and heat simultaneously. In parallel, they offer other advantages, such as generation of higher electricity output than a standard PV panel, maximization of the available roof space, etc. PVTs are classified according to the kind of heat removal fluid: PVT/water or PVT/air and according to the type of fluid circulation: passive or active. In the literature there are several studies about these types of systems (Tripanagnostopoulos et al., 2002; Tonui and Tripanagnostopoulos, 2007; Ibrahim, 2011).

Finally, it should be mentioned that high-efficiency operation requires substantial changes to the traditional inverter technologies. For example, the use of micro-inverters can reduce the overall cost of BIPV systems (Ericsson and Rogers, 2009).

From this brief introduction to building integration of solar energy, it can be noted that research concerning BIPV started earlier than BIST. In fact, as an indicator, all the IEA Tasks focused on BIPV have now been concluded, while new Tasks regarding BIST are ongoing. Globally, the characteristics of how to integrate a photovoltaic or a solar thermal system reflect the same principles. The main difference is centred on the materials and the auxiliary systems needed, which with regard to building integration in the case of PV were more beneficial at the outset. At present, research conducted into new materials and configurations for solar thermal systems seeks to overcome difficulties for their building integration and make strong advances in this field.

In the next section, the ideas described above are discussed in depth for building integration of concentrating systems.

17.2 BUILDING INTEGRATED CONCENTRATING SYSTEMS

In the field of Building Integrated Concentrating Solar systems (BICS), solar thermal as well as Photovoltaics (PVs) are included. First, some previous concepts regarding solar concentration are introduced ahead of the presentation of BICPV systems. Some of the configurations presented (e.g. Fresnel concentrators) can also refer to a Building Integrated Concentrating Solar Thermal (BICST) system, since the receiver CPV (or CPVT) can be replaced by a Concentrating Thermal (CT) unit. Later, some representative BICST technologies are included. Lastly, building integration requirements for concentrating systems are summarized.

17.2.1 Physics of concentrating solar system

17.2.1.1 Why solar concentration?

The general concept of the PV solar concentrator is to reduce the amount of expensive solar cell by low-cost optical material. The sunlight either focused to a point or to a line is reflected by or refracted through an optical element to increase the solar flux at the solar cell, thus the electrical power of the system. The solar flux at the solar cell can be increased by light trapping using the total internal reflection using polymer material which has properties like glass. A PV solar concentrator increases insolation intensity at the PV surface, reducing the area of photovoltaic material required per unit of power output. A cost reduction can be achieved for the overall photovoltaic/concentrator system when the concentrator cost is lower than the displaced PV material cost. In the case of thermal receivers, the use of solar concentration enables the attainment of higher working fluid temperatures. This leads to the more effective use of thermal concentrating systems for some applications, such as solar cooling with double-effect absorption chillers or concentrating solar power with higher-efficiency cycles. Optical concentrators can be reflective, refractive, diffractive or a combination of these.

17.2.1.2 The concentration ratio of a CPV system

The concentration ratio determines the increase in relative radiation at the surface of the exit aperture/absorber. The concentration ratio can be defined in several ways, as described below.

17.2.1.2.1 Geometric concentration ratio

The geometric concentration ratio is defined as the ratio of the area of aperture to the area of the receiver (Duffie and Beckmann, 1991), i.e. $C = A_a/A_r$. This ratio has an upper limit that depends on whether the concentrator is three-dimensional, such as a paraboloid, or two-dimensional, such as a compound parabolic concentrator. In terms of the half acceptance angle, the concentration ratio is defined as (Rabl, 1976b):

$$\left.\begin{array}{ll} C = \dfrac{1}{\sin\theta_s} & \text{for a two-dimensional system} \\[2ex] C = \left(\dfrac{1}{\sin\theta_s}\right)^2 & \text{for a three-dimensional system} \end{array}\right\} \qquad (17.2.1)$$

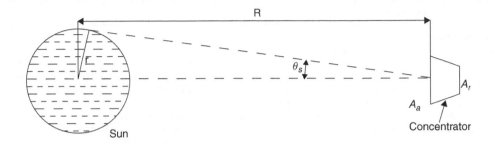

Figure 17.2.1 The half-angle subtended by the sun at a distance R from a concentrator with aperture area A_a and receiver area A_r.

17.2.1.2.2   Optical concentration ratio

The optical concentration ratio for an actual system is the proportion of incident rays within the collecting angle that emerge from the exit aperture. This yields an optical concentration ratio defined as (Winston, 1980):

$$C_{op} = \frac{G_r}{G_a} \tag{17.2.2}$$

In other words, it can be defined as the average irradiance (G) over the receiver area, divided by the insolation incident on the collector aperture.

17.2.1.2.3   Limits to concentration

An attainable concentration limit follows from physical optics (Rabl, 1976b). The disk of the Sun subtends at the surface of the Earth an angle of $2\theta_s$, as shown in Figure 17.2.1. Concentration is achieved by making a small image of the Sun with a given diameter optical device. Rays forming the smallest image make a cone with the largest semi-angle φ. When the semi-angle of the image-forming cone is $\varphi = \pi/2$, the maximum theoretical limit for the concentration ratio is achieved (Winston and Welford, 1979); i.e. the area of the input aperture of the device divided by the area of the Sun's image. In three dimensions the maximum concentration ratio is $C_{max} = 1/\sin^2 \theta_s$, for a value of $\theta_s = 0.27°$, C_{max} is 45,031.

17.2.2   Types of concentrators

Imaging types of solar concentrator largely depend on image formation of the Sun to the receiver at the exit area. Alternatively, in some cases the Sun's image does not form at the exit area, hence the term non-imaging solar concentrator. In Table 17.2.1, an indication of the different types of solar concentrator based on geometrical concentration ratios and their application types is given. This clearly shows that higher-concentration ratio systems are primarily used for large-scale power generation, which requires precise single- or two-axis tracking mechanisms. Low-concentration systems, such as compound parabolic concentrators (CPC) and quasi-stationary devices, are used for building integration and in domestic heating/cooling.

Table 17.2.1 Classification of solar concentrators based on the concentration ratios and their applications.

Concentration ratio (X)	Traking requirements, type of system	Application
1–2	Stationary, CPC	Heating, cooling, building integration
2–10	Quasi-stationary, CPC and Parabolic trough	Power generation, heating and cooling
10–100	1-axis tracking, Parabolic trough	Power generation
100–10000	2-axis tracking, Parabolic dish, Fresnel lens	Power generation, CPV
10000–100000	2-axis, solar tower, solar furnace	Power generation, materials assessment, laser

17.2.2.1 Non-imaging optics

Non-imaging optics provide effective and efficient collection, concentration, transport and distribution of energy in applications where image forming is unnecessary (Welford and Winston, 1982). In imaging optics an image is formed at the exit aperture or on a screen, whereas for non-imaging optics no image of the object is formed (Winston 1980). In an "ideal" non-imaging concentrator the first concentrator aperture is radiated uniformly from a Lambertian source. The absorber then receives a uniform flux (Leutz, 1999b). The Sun approximates to a Lambertian source, although its brightness is not uniform and its wavelength-dependent brightness changes significantly across its disc. Practical non-imaging concentrators are designed with one or two pairs of acceptance half-angles that accept light (for example, diffuse insolation) incident at angles other than the almost paraxial rays of the Sun. Concentrated solar fluxes are thus non-uniform (Winston and Hinterberger, 1995) and flux densities at the absorber in a non-imaging solar concentrator are influenced by solar disc size and solar spectral irradiance (i.e. colour dispersion) (Leutz et al., 2000) and by the proportion of diffuse insolation, particularly at low concentrations (Rabl, 1985).

For non-imaging optical systems, the edge-ray principle (Winston, 1974) states that extreme rays entering a concentrating system through an entrance aperture must be extreme rays when leaving this system through another aperture (i.e. receiver or absorber) for maximal optical concentration.

Non-imaging systems can be made either by using a refracting lens or by using reflective mirrors (Boes and Luque, 1992). Fresnel lenses may offer flexibility in non-imaging optical design. For photovoltaics, uniformity of solar flux maintains electrical efficiencies by minimizing electrical energy losses (Leutz, 1999b). Non-imaging Fresnel lenses allow uniformity of flux at the photovoltaic material to be achieved as manufacturing errors at the back and front faces of Fresnel lenses are partially self-correcting. In contrast, an angular error in the plane of a mirror leads to twice this error in the reflected beam.

17.2.2.2 Non-imaging optics: examples

a) Compound Parabolic Concentrator (CPC)

Developed originally for the detection of Cherenkov radiation in particle physics experiments (Hinterberger and Winston, 1966), a CPC for solar energy applications consists

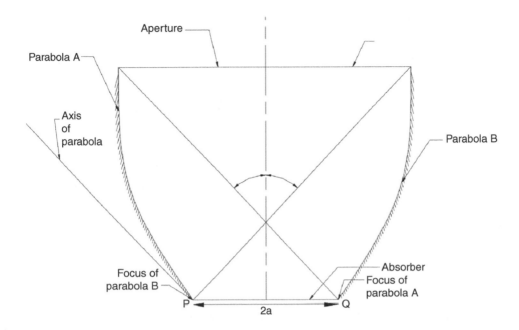

Figure 17.2.2 Schematic diagram of a compound parabolic concentrator.

of two different parabolic reflectors that can reflect both direct and a fraction of the diffuse incident radiation at the entrance aperture onto the absorber, in addition to the direct solar radiation absorbed directly by the absorber. The axis of the parabola makes an angle θ_a or $-\theta_a$ with the collector mid-plane and its focus at P (or Q), as shown in Figure 17.2.2 (Rabl, 1976b). The slope of the end point of the parabola is parallel to the collector mid-plane. A CPC reflector shape can be designed in different ways according to the absorber shape. A basic form for a flat one-sided absorber is shown in Figure 17.2.2.

The equation of a CPC with a flat absorber

For the coordinates in Figure 17.2.3, by rotation of the axis and translation of the origin, in terms of the diameter (2a) and the acceptance angle (θ_{max}), the equation for a meridian section CPC reflector is (Welford, 1978) is:

$$(r\cos\theta_{max} + y\sin\theta_{max})^2 + 2a(1 + \sin\theta_{max})^2 r - 2a\cos\theta_{max}(2 + \sin\theta_{max})z$$
$$-a^2(1 + \sin\theta_{max})(3 + \sin\theta_{max}) = 0 \tag{17.2.3}$$

In polar coordinates, the complete parametric equation becomes (Welford, 1978)

$$r = \frac{2f\sin(\theta - \theta_{max})}{1 - \cos\theta} - a'; \quad z = \frac{2f\cos(\theta - \theta_{max})}{1 - \cos\theta}$$

where $$\tag{17.2.4}$$

$$f = a'(1 + \sin\theta_{max})$$

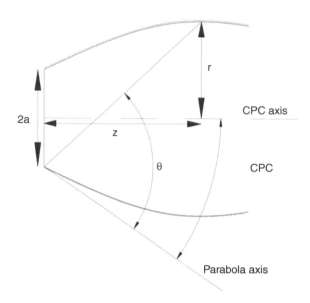

Figure 17.2.3 The angle θ used in the parametric equations of the CPC.

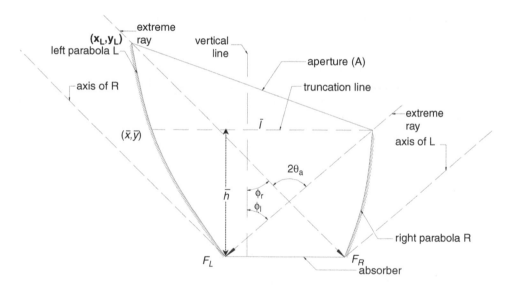

Figure 17.2.4 Asymmetric CPC with half acceptance angle $2\theta_a = \phi_l + \phi_r$.

b) Asymmetric compound parabolic concentrator (ACPC)

The foci and end points of the two parabolas of an ACPC make different angles with the absorber surface, as shown in Figure 17.2.4. A is the aperture of the concentrator, R is the right parabola, L is the left parabola, F_R is the focus of R and F_L is the focus of L. For the ACPC, effective concentration ratio varies with the angle of incidence (Rabl, 1976b).

Rabl (1976b) and Smith (1976) state that this type of concentrator has a maximum concentration ratio of $[\sin(\theta_{max}/2)]^{-1}$, where $\theta_{max} = \phi_l + \phi_r$. However, (Mills and Giutronich, 1978) have shown that the maximum concentration ratio for a parabolic asymmetric concentrator is

$$C_{PA_{max}} = \left[\frac{1 + \sin \phi_r}{\tan(\theta_{max}/2)} - \cos \phi_r\right][\cos(\phi_r - \omega)^{-1}] \qquad (17.2.5)$$

and the minimum concentration ratio is

$$C_{PA_{min}} = \left[\frac{1 + \sin \phi_l}{\tan(\theta_{max}/2)} - \cos \phi_l\right] \qquad (17.2.6)$$

where

$$\cos \omega = \frac{\sin \theta_{max} - \cos \phi_r + \cos \phi_l}{2\sqrt{(1 - \cos \phi_l \cos \phi_r)}} \qquad (17.2.7)$$

Truncation of the reflectors of an ACPC reduces the size and cost of a system but results in a loss of concentration. The degree of truncation for a given ACPC can be determined in terms of the coordinates of a full ACPC. As Figure 17.2.4 illustrates, the left half of the ACPC is terminated at the point $(\bar{x}, \bar{y})$, instead of the end point (x_L, y_L) of the full ACPC. The right half of the ACPC, is of course, truncated in an analogous manner. Truncation does not change the absorber area. The width $(\bar{l})$, height $(\bar{h})$, and the position coordinates $(\bar{x})$ of the truncated ACPC are (Rabl, 1976b)

$$\bar{l} = 2\bar{x}\cos\theta - \frac{\bar{x}^2}{s(1 + \sin\theta)}\sin\theta + s(\sin\theta - \cos^2\theta) \qquad (17.2.8)$$

$$\bar{h} = \bar{x}\sin\theta + \frac{\bar{x}^2\cos\theta}{2s(1 + \sin\theta)} - \frac{s}{2}\cos\theta(1 + \sin\theta) \qquad (17.2.9)$$

and

$$\bar{x} = s\left[\frac{(1 + \sin\theta)}{\cos\theta}\right]\left[-\sin\theta + \left(1 + \frac{\bar{h}}{h}\cot^2\theta\right)^{\frac{1}{2}}\right] \qquad (17.2.10)$$

17.2.3 Building integrated concentrating photovoltaics

The following is an analysis of the suitability for architectural integration of the principle types of existing concentrators, categorized by concentration factor.

17.2.3.1 High concentration systems (C >100X)

High concentration systems require two-axis tracking with high precision (tolerances below 0.2°). The integrability of such a system will be highly compromised by the fact that it is mobile and by its size and dimensions which, even when minimized, are considerable. Incorporation is best achieved on the roof of the building (particularly

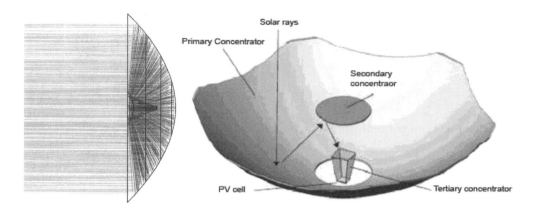

Figure 17.2.5 Schematic of the Cassegrain concentrator. In the left, 2D ray tracing view and in the right, 3D model of the system (Gordon and Feuermann, 2005). (Picture source: Chemisana, 2011).

for flat roofs) where the system is invisible from the exterior. This group is currently dominated by point focus Fresnel systems.

There are a number of companies producing high concentration systems, some of whom are mentioned below.

The Spanish company Sol3g, now absorbed by Abengoa Solar, produced a modular system with a row of 10 Fresnel lenses per module (solar aperture $1200 \times 120\,\text{mm}^2$) which may be custom-designed according to space and consumption requirements (Chellini et al., 2007). The array of modules is positioned on a high precision tracker fabricated by Feina Ltd. Green and Gold Energy offers a system called SunCube^{TM} which consists of a device of approximately $1\,\text{m}^2$ ($1064 \times 1064\,\text{mm}^2$) aperture formed of 9 Fresnel lenses divided into 3 rows. Each system is coupled to a small two-axis tracker (Green and Goldenergy, 2011). Emcore commercializes a module formed by 8 Fresnel lenses in two lines of 4 (Emcore Soliant 1000). The dimensions of each module are $733 \times 378\,\text{mm}^2$ (Emcore, 2012). Using similar technologies to those previously mentioned and employing Fresnel lenses, Whitfield Solar designed the WS-Si24 system. This achieves a concentration factor of 70X and is therefore a medium concentration system (Anstey et al., 2007). However, the optical and tracking technology used imply characteristics of integration that place it within this section.

Using point concentration reflectors, Menova developed PowerSpar in two confingurations: the RFP 20 and the RFP 40 which consist of four RFP 20 units (Menova, 2008), and SVV Technology Innovations a Ring Array Concentrator (RAC), which emulates a point focus lens using reflectors (Vasylyev and Vasylyev, 2002; 2003).

In addition to the above systems, it is worth mentioning the concentrator based on Cassegrain Optics (Figure 17.2.5) which has been commercialized by SolFocus (Gordon and Feuermann, 2005; Winston and Gordon, 2005; Mcdonald and Barnes, 2008) and the Light-guide Solar Optics (LSO) system presented by Morgan Solar Inc. (Morgansolar, 2012), which is based on optical light guides (Figure 17.2.6). As with the systems described in the previous paragraph, these can be installed on flat roofs. Both have a minimal receiver size, encapsulating the PV cells within the concentrator

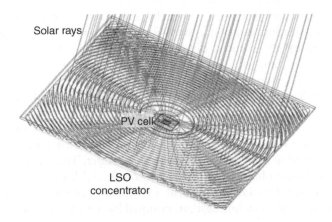

Figure 17.2.6 Light-guide Solar Optics (LSO) system presented by Morgan Solar Inc. (Morgansolar, 2012). (Picture source: Chemisana, 2011).

itself. In 2011, Morgan Solar Inc. developed the Sun Simba, which is based on their standard, mass-produced LSO. Sun Simba is a fully engineered, modular concentrating solar panel, optimally designed to perform in high heat, high wind-load and extreme moisture conditions, leading to lower maintenance costs. Experimental performance results of the Sun Simba were presented by Myrskog et al. (2012).

Given that the practical degree of integration of high concentration systems is limited by the need to incorporate them onto a high precision two-axis tracker, their means of integration are analogous between each case. It is therefore considered that the architectural issues are already well explained.

Finally, in the field of high concentration systems it is worth mentioning the BICPV for box-window curtain wall assemblies, a day-lighting system within 'double-skin' and with a reduction in unwanted solar gain (the DOE Solar Energy Technologies Multi-Year Program Plan). The goals of this façade were: maximization of solar energy utilization, reduction of the overall energy consumption profile of the building (by means of the synergistic combination of power generation (using PV cells) and high-quality heat capture with a simultaneous reduction in building cooling and lighting loads). The system consists of multiple concentrator modules which are situated within a glass façade or glass atrium roof of a building and are mounted on a highly accurate, inexpensive tracking mechanism (Dyson et al., 2007).

17.2.3.2   Medium concentration systems (10X < C < 100X)

Medium concentration systems can generally be divided into two groups: parabolic troughs and those using Fresnel optics in the form of lenses or mirrors. Concentrators which achieve the higher end of this concentration range (60–85X) and which, due to their size, make integration in buildings impossible are the Concentrating Solar Power (CSP) devices. In these type of systems, when decreasing the concentration ratio, building integration is facilitated.

An important problem associated with linear CPVs is overheating produced by the high density of light flux received by the cells, the majority of which is transformed into heat. The high concentration systems mentioned above use a passive cooling system, facilitated by the reduced dimensions of the cell which allow the use of a fin-based heat sink. Contrarily, in the case of linear concentration systems, passive cooling is complicated due to the larger surface area of the solar cells. This results in less cost-effective dissipators than in the case of insulated cells (Edenburn, 1980). For solar receivers which receive linearly concentrated light, the most adapted means of cooling is by active dissipation using liquids such as heat conducting fluids (Florschuetz, 1975). A new group of solar generators has appeared which take advantage of the evacuated heat stored in the thermal fluid as a bi-product. These are known as hybrid or co-generation Photovoltaic Thermal Concentrators (CPVT).

Medium concentration systems present a wide range of possible building integration configurations. The principle designs, grouped by their integration characteristics, are described below.

i) Parabolic trough concentrators

The installation of parabolic trough concentrators in buildings is similar to that of high concentration systems; they are generally placed on flat roofs and are ideally hidden from view. Solar tracking is achieved by rotation of the entire concentrator/receiver ensemble about a single axis. The majority of devices which use parabolic concentrators are thermal generators (Weiss and Rommel, 2008). Exponents of parabolic CPV systems are: the Combined Heat and Power Solar System (CHAPS) developed at the Australian National University, with a concentration factor of 37X, which employs a photovoltaic/thermal (PVT) module (Coventry, 2005) and the Euclides system designed at the Polytechnic University of Madrid, with a concentration factor of 38.2X, in which the cells are passively refrigerated using fins (Luque et al., 1997; Antón and Sala, 2007). In 2009, Niedermever patented a new concentrating system for PVT generation (Niedermeyer, 2008).

ii) Linear Fresnel reflectors

With parabolic troughs, daily solar tracking is achieved by moving the entire concentrator/receiver ensemble. However, within this range of concentrations good versatility is offered by systems which work using Fresnel reflection, some of which are worthy of note (some of the systems described below are included owing to their importance as concentrating technologies, despite being thermal collectors):

1 Concentrators with 2-axis trackers in which tracking is achieved by movement of the entire system, such as the BiFres system developed at the University of Lleida (equipped with a PVT receiver), whose integration in buildings would be restricted to flat (horizontal) roofs (Rosell et al., 2005) (Figure 17.2.7).
2 Static concentrators in which solar tracking is achieved by movement of the receiver. This option offers greater scope for integration in buildings as it can easily be installed on either flat or inclined roofs. Installation on façades, however, presents certain problems: the mirrors prevent light from passing into the building, and the mobile receiver, which protrudes outward from the building, creates strain

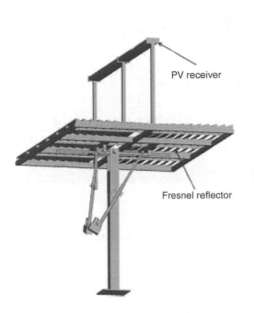 

Figure 17.2.7 Two-axis Bifres Fresnel reflector (Rosell et al., 2005). The PV receiver is water cooled, getting benefit of the thermal energy (PVT module). (Picture source: Chemisana, 2011).

on the building structure and presents an unaesthetic appearance. The main exponent of this technology is the CCStaR system (equipped with a thermal receiver) developed at the University of the Balearic Islands. It should be mentioned that in the most recently presented CCStaR prototypes the Fresnel reflectors are replaced by parabolic reflectors (Pujol et al., 2006).

3   Concentrators in which the tracking is achieved by the movement of the individual mirrors. The possibilities for integrating such systems are similar to those for the previous case of a stationary concentrator. The most important design within this group is the Compact Linear Fresnel Reflector (CLFR) presented in 1997 by Mills and Morrison (Mills and Morrison, 1997) and commercialized by Ausra. The CLFR system is used for direct steam generation. Similar systems to the CLFR have been developed, these being the solar collector Solarmundo presented by (Häberle et al., 2002) and commercialized by Power Group GMBH, and the Mirroxx Fresnel collector commercialized by Mirroxx GMBH, a spin-off of PSE-AG (Berger et al., 2009). Using the same concentration principle, the company Helio-Dynamics have presented a collector, HD211, for integration in buildings with a receiver which can be thermal or PVT (currently the HD211, renamed HD10, only incorporates a thermal receiver) (HelioDynamics, 2004). In 2009, the University of Lleida constructed another such system with a PVT receiver, in collaboration with *NUFRI* Corporation and Trigen Solar S.L. More recently, research was conducted at the Australian National University on a PVT receptor incorporated into the concentrating system developed by Chromasun (Vivar et al., 2012); more details concerning this device are described below in section 17.2.4.

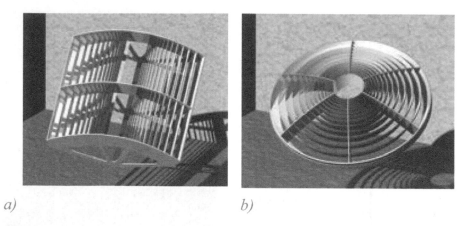

a)   *b)*

Figure 17.2.8 a) Ring Array Concentrator (RAC), and; b) Slat Array Concentrator (SAC) (Vasylyev, 2002; 2005). (Picture source: Chemisana, 2011).

A new concept in the field of Fresnel reflection systems is the so called Non-imaging Reflective Lens (NIRL) concentrator, of which there are two types: the axially symmetric Ring Array Concentrator (RAC) and the linearly symmetric Slat Array Concentrator (SAC) (Vasylyev, 2005). These operate by using mirrors to direct and concentrate light onto a receiver behind the optical element, thus emulating a lens (Figure 17.2.8). The high concentration RAC requires two-axis tracking, whereas the medium concentration SAC can be employed with either one- or two-axis tracking (Vasylyev, 2004). This type of concentrator combines the high optical efficiency achievable by mirrors with the flexibility of design which is characteristic of lenses. The principle drawback of these systems is that solar tracking is achieved by movement of the whole system, incurring the aforementioned restrictions with regard to architectural integration.

The University of Lleida is currently developing concentration technology which uses reflection, in a similar way to the systems developed by Chemisana and Rosell (2009), but with a design which prioritizes architectural integrability. The system consists of a linear Fresnel reflector which focuses radiation in a manner analogous to a lens. The receiver remains static and solar tracking is achieved by a simple and effective way by rotation of the individual mirrors. Thus, overall movement is minimized, facilitating incorporation into buildings and offering different possibilities for suiting the varied requirements of specific installations (Figure 17.2.9). High and medium concentration reflective systems are summarized in Table 17.2.2.

iii) Linear Fresnel lenses

Firstly, before commenting on the different properties and characteristics of Fresnel lenses when applied to BICPV, two systems must be mentioned. Although of low architectural integrability, as per the systems described previously, these are the first references of this kind of linear concentrator.

These products are both formed by arched Fresnel lenses situated on a solar tracker. The first, designed by Entech Solar (USA) (O'Neill et al., 1990), uses a two-axis tracker and a PV or PVT receiver. The second, designed by SEA Corp. (later Photovoltaics

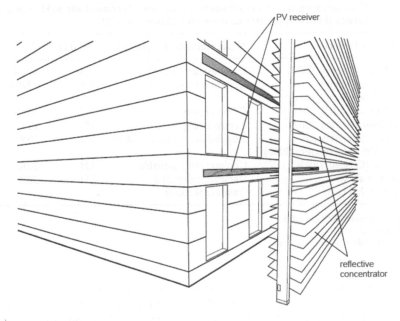

a)

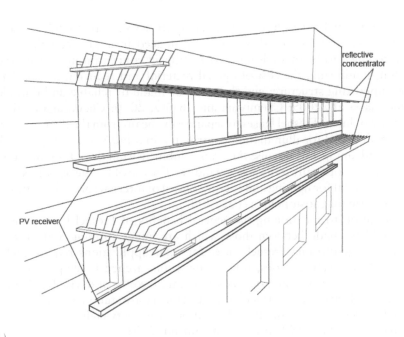

b)

Figure 17.2.9 Building integrated system presented by Chemisana and Rosell (Chemisana and Rosell, 2009), a) Curtain wall architectural design, and; b) Parasol architectural design. (Picture source: Chemisana, 2011) .

Table 17.2.2 Concentrating systems which use aspheric/Puntual Fresnel Lenses (PFL) or Linear Fresnel Lenses (LFL) as a primary concentrator (Chemisana, 2011)

Company or reference	Actual status of the system	PFL/LFL	C[1]	Cell type
Abengoa Solar (Chellini, 2007)	Commercially available	PFL	476X	3J[2]
Green and Gold Energy (Green & Goldenergy, 2011)	Commercially available	PFL	1370X	3J
Emcore (Emcore, 2012)	Commercially available	PFL	500X	3J
Whitfield Solar (Anstey, 2007)	Commercially available	PFL	70X	c-Si[3]
Photovoltaics International (Kaminar, 1991; Bottenberg, 2000)	Stopped production in 2000	LFL	10X	c-Si
Entech Solar (O'Neill, 1990)	Commercially available	LFL	20X	c-Si
Chemisana et al. (Chemisana, 2011a,b)	Demonstration and test installations	LFL	30X	c-Si

[1] C: Geometric concentration ratio.
[2] 3J: triple-junction solar cell.
[3] c-Si: monocristaline silicon solar cell.

Internacional) (Kaminar et al., 1991; Bottenberg et al., 2000) uses a one-axis tracker and a PVT receiver. Recently, Entech Solar announced two new systems: TermaVolt™ II (PVT) and SolarVolt™ II (PV). Both systems are based on the same technology but use different receivers. Entech has resized the initial prototypes designed in the 1980s into these two smaller, low-cost devices applicable for both ground and roof-mount applications.

Linear Fresnel lenses have a number of attractive features when used for solar concentration applications: they may be produced in large sizes; their aspect ratio can be designed to be small, leading to a compact concentrating system; they may be very thin, minimizing the cost of optical material and reducing the mechanical load on the supporting structure; and they may be made of reliable and durable material (Chemisana et al., 2009; Chemisana and Ibañez, 2010; Chemisana et al., 2011a,b). The ability of linear Fresnel lenses to separate the beam from the diffuse solar radiation makes them useful for illumination control in a building's interior space. The Fresnel lenses are advantageous because they can combine within them both the concentrating element and the optically transparent window. The use of Fresnel lenses as a transparent covering material for lighting and energy control of internal spaces has attracted special attention (Tripanagnostopoulos et al., 2007).

In addition to mentioning the general benefits of Fresnel lenses, some comparison should be made between those which are image forming and those which are anidolic. Image-forming Fresnel lenses for solar applications require high precision tracking. Non-imaging lenses, often convex and arched in shape and designed for medium concentration using one-axis tracking, have been devised as highly competitive solar collectors. If the tracking requirements are minimized, the cost reduction achieved by reduction of the PV cells' surface area outweighs the cost of the optical elements (Leutz et al., 1999a; Leutz and Suzuki, 2001).

The concept of using a fixed concentrator with a tracking absorber has been mentioned in the past (Kritchman et al., 1979; 1981a; 1981b). It is based on a stationary wide angle optical concentrator that, whatever the location of the sun, transmits the

Figure 17.2.10 Linear Fresnel lens concentrating system developed at the University of Lleida. Right: detail of the concentrated spot on the PVT receiver.

input radiation onto a small moving focal area which, in turn, is tracked by the PV receiver. Following this approach, the University of Lleida has developed a prototype (Figure 17.2.10) based on a stationary Fresnel lens which focuses solar radiation onto a PVT receiver which tracks the moving focal area (Chemisana and Ibañez, 2010). The advantages of this type of CPV make it architecturally versatile, allowing integration onto flat or inclined roofs or as lightweight façades, windows etc. Thus, their characteristics correspond perfectly to the requirements of well-integrated systems described by the IEA PVPS Task 7 workgroup (Luque-Heredia et al., 2007; Tripanagnostopoulos, 2008).

Refractive systems under high- and medium-concentration ratios are summarized in Table 17.2.3.

17.2.3.3 Low concentration systems (C < 10X)

Within this group fall an extremely large number of systems and variations based on very distinct technologies.

From an intuitive point of view, the simplest system is the V-trough reflector which directs light onto the receiver using flat mirrors (Tabor, 1958; Hollands, 1971; Fraidenraich, 1998a; Rabl, 1976a; Fraidenraich and Almeida, 1991). The V-trough can achieve at most 3X concentration. To ensure uniform illumination of the PV cells, planar reflectors require solar tracking (Freilich and Gordon, 1991; Gordon et al., 1991; Klotz et al., 1995; Fraidenraich, 1998b; Klotz, 2000; Poulec and Libra, 2000; Dobon et al., 2001). If solar tracking is not continuous, the V-trough behaves as an anidolic (non-imaging) optical system. Use of such devices, as with all low-concentration systems, is beneficial as commercial cells may be used and as cell heating is reduced (King et al., 2000; Klotz et al., 2001). However, despite the low light flux,

Table 17.2.3 Concentrating systems which use Puntual Reflectors (PR), Parabolic Trough Reflectors (PTR), Linear Fresnel Reflectors (LFR) as a primary concentrator. (Chemisana, 2011)

Company or reference	Actual status of the system	PR/PTR/ LFR	C^1	Cell Type
Menova Energy (Menova, 2008)	Commercially available	PR	1450	3J[2]
SVV Technology Innovations (RAC) (Vasylyev S. and Vasylyev V. 2002; 2003)	Commercially available (only concentrator provided)	PR	2500X	3J
Solfocus (Gordon and Feuermann, 2005; Winston and Gordon, 2005; Mcdonald and Barnes, 2008)	Commercially available	PR	500X	3J
Aronstis Solar (Bernardo, 2008)	Commercially available	PTC	10X	c-Si[3]
Euclides system (Luque, 1997; Anton and Sala, 2007)	Demonstration and test installations	PTC	37X	c-Si
CHAPS system (Coventry, 2005)	Demonstration and test installations	PTC	38.2X	c-SI
BiFres system (Rosell, 2005)	Demonstration and test installations	LFR	11X	c-Si
HelioDynamics (Vasylyev, 2004)	Commercially available (currently only thermal module)	LFR	10X	p-Si[4]
Trigen Solar	Demonstration and test installations	LFR	20X	c-Si
SVV Technology Innovations (RAC) (Vasylyev and Vasylyev, 2002; 2003)	Commercially available (only concentrator provided)	LFR	40X	c-Si
Chemisana and Rosell (Chemisana and Rosell, 2009)	Demonstration and test installations	LFR	18X	c-Si

[1] C: Geometric concentration ratio.
[2] 3J: triple-junction solar cell.
[3] c-Si: crystalline silicon solar cell.
[4] p-Si: polycrystalline silicon solar cell.

cells may overheat to temperatures above 80°C. Operation is considerably improved through use of a thermal dissipator (Solanki et al., 2008). By taking advantage of the extracted heat, such a system can be converted into a PVT generator.

In this range of concentrations, as in medium concentration devices, there are parabolic trough systems. An example is the PVT concentrating system (Bernardo et al., 2011) commercialized by companies Arontis Solar and Absolicon Solar, which concentrates with a ratio of 7.6X onto a Photovoltaic/Thermal (PVT) receiver. Figure 17.2.11 shows an experimental installation and a schematic where the module triangular cross section, which is actively refrigerated using a fluid, is illustrated.

Compound Parabolic Concentrators (CPC) form a category of reflectors largely used for static systems. When used to illuminate PV cells, high losses are suffered due to the non-uniform illumination pattern produced on the cell surface. V-trough systems are less prone to producing detrimental hot-spots than are CPC systems (Swanson, 2003). Many works can be mentioned within this category (Almonacid et al., 1987; Goetzberger, 1988; Zanesco and Lorenzo, 2002; Mohedano et al., 1998; Uematsu et al., 1998; Garg and Adhikari, 1999; Brogren et al., 2000; Brogren and Karlsson, 2002; Helgesson, 2004; Nilsson, 2005; Mallick et al., 2004). With the objective of

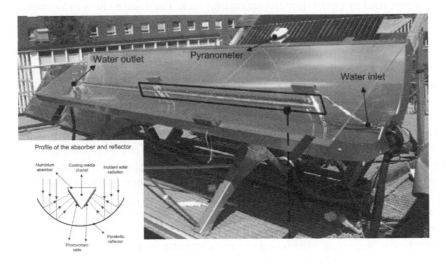

Figure 17.2.11 Parabolic trough PVT concentrator. Experimental installation in Sweden and schematic of the PVT module (Bernardo et al., 2011).

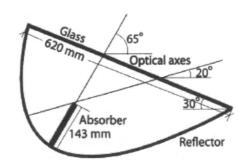

Figure 17.2.12 Section of the stand-alone MaReCo for Stockholm conditions (aperture tilt 30°; optical axes 20° and 65° defined from the horizon) (Adsten et al., 2005).

improving the system, many authors have opted for incorporation of bi-facial cells (Bowden et al., 1993; Mayregger et al., 1995; Ortabasi, 1997; Hernandez et al., 2000; Libra and Poulek, 2004; Weber et al., 2006; Parada et al., 1991). These double the amount of radiation or concentration that can be realized at the PV receiver. However, their use is not possible in high concentration systems as they have no un-illuminated surface onto which the essential heat dissipator may be attached. Designs of static concentration systems (with a typical acceptance angle of 30°) are normally intended for use with bifacial cells. Their concentration factor may be increased by use of a dielectric (Winston et al., 2005).

In terms of the asymmetric compound parabolic concentrators, Figure 17.2.12 illustrates an asymmetric CPC known as "Maximum Reflector Collector" (MaReCo). This system was characterized experimentally for high-latitude bi-facial cell BIPV applications (Adsten, 2002) and different MaReCo configurations were developed (stand-alone, roof integrated, wall integration etc). The figure shows the cross-section of roof integrated MaReCo designed for Stockholm conditions (Adsten et al., 2005). The highest optical efficiency which was reported for a bi-facial based MaReCo was 56%. Other examples of optical efficiencies of similar systems are: 91% for dielectric-filled BIPV covers (Zacharopoulos et al., 2000) and 85% for an air-filled asymmetric CPC BIPV system (Mallick et al., 2002a).

Some other examples of static concentrators which use dielectrics are presented by Edmonds et al. (1987). The cells are positioned in a V-trough concentrator filled with oil or water (the dielectric) which also serves as a cooling function. The design presented by Shaw and Wenham (2000) uses an anidolic lens to reach a concentration factor of 2X and optical efficiency of 94%. The flat static concentrator described in (Uematsu et al., 2001a; 2001b; 2001c; 2003) has been used to analyse various possible configurations, including use of monofacial cells (1.5X) or bifacial cells (2X) and different types of illumination of the rear face. However, Uematsu et al. did not take into account thermal effects in the PV cells.

Two linear dielectric non-imaging concentrating designs (symmetric and asymmetric) for PV integrated building façades were analysed using 3D ray-tracing analysis (Zacharopoulos et al., 2000). A "slim line" design was reported to achieve a concentration ratio of 4X (Wenham et al., 1995). Thermal analysis indicated that performance loss through additional heating of the PV cells was more than offset by the gains achieved through concentration. The efficiency of the module was reported to be 15% greater than that of the flat-plate module. Static concentrators offer a compromise between high concentration systems that require tracking and one-sun flat-plate modules (Wenham et al., 1995).

Some additional studies in the field of dielectric non-imaging concentrating covers for PV integrated building facades are those of Zacharopoulos (2001) and Korech et al. (2007), where the total internal reflection (within the dielectric material) is used to provide optimal optical efficiency.

An image of the second generation of the Photovoltaic Facades of Reduced Costs Incorporating Devices with Optically Concentrating Elements (PRIDE) dielectric-filled system, based on the first studies conducted by (Zacharopoulos et al., 2000) is shown in Figure 17.2.13. This system was studied and did show excellent power output compared to a similar non-concentrating system (they were characterized indoors by using both a flash and continuous solar simulator). Nevertheless, durability and instability (of the dielectric material) occurred under long-term outdoor characterization when the concentrator was made by means of casting technology. With regard to large-scale manufacturing, durability and reduction of the weight and cost of the concentrator, second generation PRIDE designs use 6 mm wide solar cells at the absorber of dielectric concentrators. PV concentrator modules achieved a power ratio of 2.01 when compared to a similar non-concentrating system. The solar to electrical conversion efficiency for the PV panel was 10.2% when characterized outdoors. It should be mentioned that in large-scale manufacturing, a module cost reduction of over 40% is potentially achievable by using this concentrator technology (Mallick and Eames, 2007).

Systems which concentrate radiation using elements opaque to visible light (CPC, V-trough) cannot be installed on areas of a building through which light is supposed to enter without reducing natural lighting in the interior. To reach a certain concentration

Figure 17.2.13   Second Generation dielectric PRIDE Photovoltaic Concentrator Module (Mallick and Eames, 2007).

factor, the reflecting surface area used by these systems is elevated compared to the surface area of PV cells. Given the detrimental effect this has on illumination of interior spaces, low-concentration static concentrators are preferable for architectural integration. These form the vast majority of CPC systems. They may be designed to be installed at any inclination or position which receives solar radiation (flat roofs, inclined roofs, façades, etc.). What's more, although the reflector area is high with respect to the PV cell area, the volume of the entire system is relatively reduced, the geometry approximately tending to a parallelepiped.

The aforementioned linear Fresnel systems with one-axis tracking (Kaminar et al., 1991; Bottenberg et al., 2000) can also be included within the low-concentration group. Leutz et al. (1999a) designed a convex-shaped non-imaging stationary Fresnel lens (1.5–2X) intended for warming up evacuated tubes, but able to be used with PV including secondary optics.

Other low-concentration systems which are currently less used but are the subject of some study are: fluorescent/luminescent concentrators, quantum dot concentrators and holographic concentrators.

The idea of using fluorescent concentrators (Figure 17.2.14) to concentrate both direct and diffuse radiation without tracking systems first appeared in the late 1970s (Weber and Lambe, 1976; Goetzberger and Greubel, 1977). In a fluorescent concentrator, a matrix of dye molecules absorbs radiation and emits light with a longer wavelength. Most of the emitted light is internally totally reflected and therefore

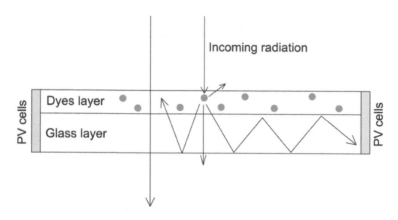

Figure 17.2.14 Fluorescent technology. The main part of reemitted photons are trapped in the layers and guided by total internal reflection to the PV cells placed on the edges. Photon loss occurs because of nontrapped emission or absorption by other dyes. Radiation frequencies noncaptured by the dyes are transmitted through the layers and can be captured and collected by a second fluorescent device or other system which take benefit of them.

trapped and guided to the edges of the concentrator, where solar cells convert it into electricity. This concept was investigated intensively in the early 1980s (Wittwer et al., 1981; Seybold and Wagenblast, 1989). After 20 years of progress in the development of solar cells, fluorescent dyes and new concepts, several groups (Luque et al., 2005; Van Roosmalen, 2004; Goldschmidt et al., 2006; Richards and Shalav, 2005; Rau et al., 2005; Goldschmidt et al., 2007; Danos et al., 2006; Debije et al., 2007; Slooff et al., 2007) are currently reinvestigating the potential of fluorescent concentrators. In the quantum dot concentrator, the luminescent dye is replaced by quantum dots. Quantum dots are crystalline semiconductors which degrade less than organic dyes. Quantum dots can be tuned to the absorption threshold by the choice of dot diameter. Red shift between absorption and luminescence is primarily determined by the variance of dot sizes, which in turn can be optimized by choice of growth conditions. Reabsorption can therefore be minimized and high efficiencies and high concentration ratios achieved (Barnham et al., 2000). Of the systems which use this kind of technology, the organic dye-based Organic Solar Concentrator (OSC), designed in the Massachusetts Institute of Technology (MIT) and commercialized by Covalent Solar, has had the most impact, particularly within the field of building integration. The primary advantages of this system are that it does not require tracking and that its geometry is completely planar. It is formed of a stack, the principle layers being the OSC and the PV cells. This system is aesthetically superior to conventional PV systems; the colour is tuneable, better views through, transparent metal oxide contacts are not required and they may be formed with flexible plastics. Owing to their versatility, their position in buildings varies from atriums and roofs to windows. The concentration for which the system works with best efficiency, when combined with a variety of PV cell types, is 3 suns (Currie et al., 2008).

The idea of holographic solar concentrators was first proposed in the early 1980s (Horner and Ludman, 1981; Ludman, 1982a,b; Bloss et al., 1982). Holographic elements have a number of advantages over conventional optical elements: they are lightweight, easy to reproduce and one holographic element can be used to perform several different functions. For example, there is a demonstration project which utilizes light-directing holograms for both daylighting and PV power generation (Müller, 1994). Holograms can be fabricated, which concentrate the spectrally disperse solar radiation (Ludman et al., 1992).

There has been a surge of interest in this kind of technology thanks to the recent appearance of the Holographic Planar Concentrator™ (HPC) designed by the American company Prism Solar Technologies, Inc. This is the key technology in Prism Solar products. The HPC acts as an extremely low-cost concentrator (3 suns) without mechanical tracking or cooling systems. The bi-facial HPC configuration uses 72% less silicon than a standard module, leading to a more cost-effectiveness product. Furthermore, this new type of concentrator can be installed on rooftops or even incorporated into windows and glass doors (Castro et al., 2010; PrismSolar, 2012).

Finally, it should be mentioned that for concentrating systems an important factor is their economic viability. In this direction there are some companies which aim at the development of cheap devices. An example is the "Cool Earth Solar" company with its innovative design of an inflated, balloon-shaped concentrator. Each 8-foot-diameter concentrator is made of plastic film with a transparent upper hemisphere and a reflective lower hemisphere. When the concentrator is inflated with air, it forms a shape which focuses or concentrates sunlight onto a PV cell placed at the focal point (Coolearthsolar, 2012). Another example is the "Pacific Solar Tech" company with its MicroPV TM Concentrator Photovoltaic Modules and a "silicone-short environment" (Pacificsolartech, 2006).

17.2.4 Building integrated solar thermal (concentrating)

As mentioned previously, the category of BICST may include configurations similar to the aforementioned BICPV, provided that there is a CT (instead of a PV) receiver. Of the examples of BICST systems which follow, some of the technologies have the potential for BI applications, while other devices have low-level potential.

1) Parabolic troughs (on a small-scale)

The POWER ROOF™ is a new concept in solar energy because it is a high-temperature solar collector which at the same time also serves as an insulating system. The system integrated building approach is an important characteristic of this technology, along with energy cost savings. POWER ROOF™ is an attractive solution for large industrial and commercial energy users and is designed for new as well as for existing facilities. A basic advantage is the fact that its temperature range can fulfil an important array of issues. Collection temperatures above 398°C can be achieved, thus the system can provide energy in the form of steam for uses such as space heating and domestic hot water (as well as for industrial applications: desalination, absorption cooling, water purification, etc). In a commercial-building level, Solargenix Power Roof was installed in 2002 on a 930 m² office building in Raleigh, North Carolina. This system utilizes a *fixed* parabolic reflector and tracking receiver and provides 50 tons (176 kW) of cooling as well as heat. It produces 170–175°C water which powers a double-effect absorption chiller. Where needed, the thermal energy can be converted into electricity (Gee et al., 2003).

The companies Absolicon Solar and Arontis Solar offer two parabolic trough systems (T10 and MT10) very similar to the one described in 17.2.3.3, but with thermal absorbers. The T10 system is prepared for domestic solar heating and the MT10 for steam and process heat industrial applications (Absolicon Solar Concentrator AB, 2012).

2) Solar collectors with Micro-Concentrators (MCT)

A representative example of this type of technology is the system commercialized by Chromasun Solar company. Chromasun and its partners in 2011 activated a 60-collector SCT system at Santa Clara University (SCU) to heat water (the largest rooftop concentrating solar thermal installation in California). The installation is shown in Figure 17.2.15a. From Figure 17.2.15b it can be seen the Fresnel reflector shape which focuses irradiance onto the absorber tube under a concentration ratio of 20–30X. From outside, the MCT has a flat-panel format. Like flat-panel collectors, MCT is easy to install, weighs about the same and is easy to look after. MCTs can produce much higher temperatures and efficiencies than a flat-plate collector and has a much smaller package. The MCT has been designed purposely for rooftop integration (Sultana et al., 2012).

In relation to "micro" systems, it should be mentioned that in the field of CSP there is a tendency for development of "compact" devices (Micro-scaled Concentrated Solar Power (MicroCSP)) in order to facilitate their building integration. An example is a Micro CSP system for air-conditioning from the company Sopogy® (Sopogy, 2012).

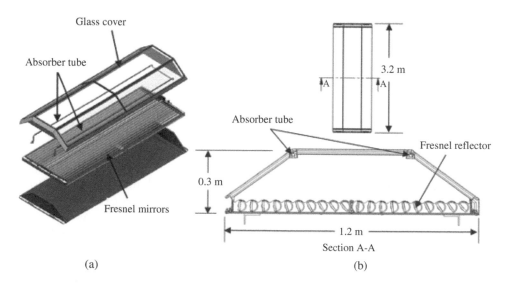

(a)   (b)

Figure 17.2.15 (a) Exploded view of a solar micro-concentrator system and (b) Cross-section of the MCT collector (Sultana et al., 2012).

3) Integrated collector storage solar water heaters (ICSSWH) with asymmetric CPC reflector

ICSSWHs are simple, low-cost solar devices. However, their disadvantage is the significant ambient heat losses, especially during night-time and non-collection periods (Tripanagnostopoulos and Yianoulis, 1992). Thereby, several studies have been carried out focusing on the improvement of thermal performance of these systems, especially during night-time operation. Horizontal water storage tanks are less effective in terms of water temperature stratification than vertically orientated ones (Smyth et al., 2003); however, they can achieve Concentration Ratios (CR) >1 when combined with CPC reflector troughs. In this way, water storage thermal losses are reduced, due to smaller absorber than aperture surface area (Tripanagnostopoulos and Souliotis, 2004). An example of a recently developed, novel ICSSWH system can be found in the study by Souliotis et al. (2011), who investigated an ICSSWH experimentally and theoretically. The goal of the study was the achievement of low thermal losses during night-time. The unit was based on a heat-retaining ICS vessel design consisting of two concentric cylinders mounted horizontally inside a stationary truncated asymmetric CPC reflector trough. The annulus between the cylinders was partially evacuated and contained a small amount of water, which changed phase at low temperature, producing a vapour and thus creating a thermal diode transfer mechanism from the outer absorbing surface to the inner storage vessel surface (Figure 17.2.16).

The optical study, in conjunction with the experimental study of the ICS system, revealed that even if the absorbed solar radiation is distributed non-uniformly on the outer vessel throughout the year, thermal stratification within the inner vessel during daily operation is not significant. The thermal diode mechanism in the annulus plays an important role for the effective operation the device, with the pressure of the water vapour in annulus being an important parameter. Comparison of experimental results between the ICS system and a commercial Flat-Plate Thermosiphonic Unit (FPTU) device did show that the studied ICS system has an effective thermal performance and

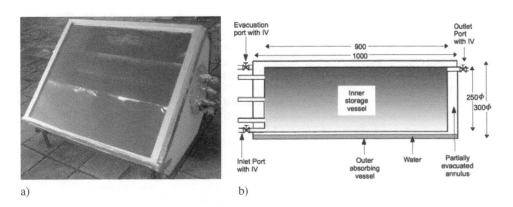

Figure 17.2.16 a) ICS experimental model mounted at the test field, and; b) Cross-sectional front view of the vessel (Souliotis et al., 2011).

thermal losses close to the FPTU, making it a promising solar system for domestic water-heating applications.

Another possibility is the combination of stationary CPC collectors with multiple absorber segments and several configurations such as tubular and flat-fin type absorbers with pipe (Tripanagnostopoulos and Yianoulis, 1996). These types of solar devices are appropriate for applications in the temperature range of 100–200°C.

At this point it should be mentioned that among the systems presented above, only system 1, parabolic troughs (on a small-scale), has the potential to be actually integrated into buildings (maximum level of BI). The other systems show a low level of BI potential since they are merely placed or installed on the roof without replacing elements of the roof.

17.2.5 Concentrating systems and building integration requirements

Building integration is directly related with the concentration ratio C of a system. Thereby, the evaluation of the BI potential of a device is based on its concentration. In Table 17.2.4 the requirements for building integration of several concentrating systems are given.

Table 17.2.4 Requirements for building integration of several concentrating solar systems (Chemisana, 2011).

Description of the system	Type of building integration	Tracking requirements	Restrictions/
High concentration: C > 100x (mainly point focus Fresnel systems)	Flat roofs mainly	Two-axis tracking with high precision (tracking, cleaning, etc.).	Whole system movement. Control and management High temperatures management (case, surrounding air, etc.) Possible failures require immediate actions due to the very high fluxes effects
Medium concentration: 10x < C < 100x (parabolic troughs; Fresnel lenses or mirrors)	Flat or inclined roofs Façades	Two-axis tracking in some of the systems Single-axis tracking for the whole system Single-axis tracking for the receiver or for the concentrator	Whole system movement (if applicable) Maintenance, but much less restrictive than if high concentration Possible effects derived from the high fluxes. These are closer to the restrictions of high concentration systems when C approaches 100x
Low concentration: C < 10x (V-trough, CPC, Fluorescent systems, holographic systems, etc.)	All	Do not require tracking → this facilitates their BI at any location of the building	The same than conventional solar panels

17.3 CONCLUSIONS

Concentrating solar energy systems can reduce system cost if the cost of the reflector or the lens system (and tracking system where applicable) is less than the cost of the replaced receiver material (i.e. PV). A concentrating panel separates the functions of light collection and conversion into electricity that are integrated in a flat photovoltaic panel (Luque, 1986) or converted into thermal energy in flat-plate thermal modules. In solar thermal collection the benefit of using concentrating technologies is mostly related to the achievement of higher fluid temperatures. In relation to higher temperatures, the materials involved must be adequate for temperature ranges and present proper thermal tolerance. In PV applications, an advantage of concentrating panels is the tendency for solar cell efficiency to increase when the cell is under high irradiance. However concentrators for photovoltaic applications present some drawbacks when compared to flat PV panels, as follows:

- Complexity of the tracking mechanism that may for remote applications (either geographically or in terms of façade accessibility) make them less attractive (Luque, 1989);
- Inability at high concentration ratios to collect diffuse light, which limits their use in locations where high diffuse radiation prevails;
- Large size of the basic concentrator module (due to the need to spread the structure and tracking costs by producing a larger amount of electricity) as compared to flat-panel is too large for small applications where photovoltaic electricity is cost-competitive today;
- High cost of the more sophisticated cells and the more complex structures used previously with concentrators jeopardizes the advantages obtained in cell area reduction;
- Increased solar cell temperatures decrease PV efficiency and differential heating (hot spot) increases mismatch errors;
- Lack of radiation uniformity leading to mismatch errors.

These drawbacks can be avoided by using low or low-medium concentration ratio systems, as for instance the cylindrical Fresnel lens with CPC or the asymmetric compound parabolic concentrator described above. In cloudy and overcast sky conditions, low concentration systems accept a significant component of the diffuse solar radiation because of their high acceptance-half angle (Mills and Giutronich, 1978).

In summary, detailed technological advancements have been discussed in this chapter, including description of low, medium and high concentrating solar energy systems, their applications and the current issues for building integration.

REFERENCES

Absolicon Solar Concentrator AB (2012). www.absolicon.com.
Adsten M. (2002) *Solar thermal collectors at high latitudes: Design and performance of non-tracking concentrators*. PhD Thesis, Uppsala University, Sweden.
Adsten, M., Helgesson, A. and Karlsson, B. (2005) Evaluation of CPC-collector designs for stand-alone, roof- or wall installation. *Solar Energy*, 79, 638–647.

Almonacid, P.G., Luque, A., Aguilar, J.D., Almonacid, L. and Lara M. (1987) Analysis of a phtotovoltaic static concentrator prototype. *Solar and Wind Technology*, 4, 145–149.

Anstey, B.D., Bentley, R.W., Bonner, T., Hughes, C., McNicholl, R., Muñoz, H., Norton, M., Oliva, J., Teruel, J., Vicente, M., Weatherby, C. and Wheeler, S. (2007) Progress with the Whitfield Solar PV Concentrator. In: *Proceedings of the 4th International Conference on Solar Concentrators for the Generation of Electricity or Hydrogen (ISCS-4)*. 12–16 March, El Escorial (Spain).

Antón, I. and Sala, G. (2007) *The EUCLIDES Concentrator*. Chapter 13 in Concentrator Photovoltaics. Springer-Verlag Berlin Heidelberg, Germany.

Barnham, K.W.J., Marques, J.L., Hassard, J. and O'Brien, P. (2000) Quantum-dot concentrator and thermodynamic model for the global redshift. *Applied Physics Letters*, 76, 1197–1199.

Berger, M., Häberle, A., Louw, J., Schwind, T. and Zahler, C. (2009) Mirroxx fresnel process heat 2009. Collectors for industrial applications and solar cooling. In: *Proceedings of SolarPACES 2009*. 20–23 June, Sevilla (Spain).

Bernardo, L.R., Perers, B., Håkansson, H. and Karlsson, B. (2011) Performance evaluation of low concentrating photovoltaic/thermal systems: A case study from Sweden. *Solar Energy*, 85, 1499–1510.

Bloss, W.H., Griesinger, M. and Reinhardt, E.R. (1982) Dispersive concentrating systems based on transmission phase holograms for solar applications. *Applied Optics*, 21, 3739–3742.

Boes, E.C. and Luque, A. (1992) *Photovoltaic concentrator technology*, Chapter 8 in Renewable Energy, Sources for Fuel and Electricity. Island Press, Washington. D.C., USA.

Bottenberg W. R., Kaminar, N., Alexander, T., Carrie, P., Chen, K., Gilbert, D., Hobden, P., Kalaita, A. and Zimmerman, J. (2000) Manufacturing technology improvements for the PVI SUNFOCUS™ concentrator. In: *Proceedings of 16th European Photovoltaic Solar Energy Conference*. 1–5 May, Glasgow (United Kingdom).

Bowden, S., Wenham, S.R., Coffey, P., Dickinson, M. and Green, M.A. (1993) High efficiency photovoltaic roof tile with static concentrator. In: *Proceedings of 23rd IEEE Photovoltaic Specialist Conference*. 10–14 May, Louisville (USA).

Brogren, M. and Karlsson, B. (2002) Low-concentrating water-cooled PV-thermal hybrid systems for high latitudes. In: *29th IEEE Photovoltaic Specialists Conference*. 19–24 May, New Orleans (USA).

Brogren, M., Nostell, P. and Karlsson, B. (2000) Optical efficiency of a PV-thermal hybrid CPC module. In: *Eurosun 2000*. 19–22 June, Copenhagen (Denmark).

Castro, J.M., Zhang, D., Myer, B. and Kostuk, R.K. (2010). Energy collection efficiency of holographic planar solar concentrators. *Applied Optics*, 49, 858–870.

Chellini, S., Vallribera, J. and Pardell, R. (2007) Technical highlights of a solar simulator for PV concentration modules. In: *Proceedings of the 4th International Conference on Solar Concentrators for the Generation of Electricity or Hydrogen (ISCS-4)*. 12–16 March, El Escorial (Spain).

Chemisana, D. (2011) Building integrated concentrating photovoltaics: A review. *Renewable Sustainable Energy Reviews*, 15, 603–611.

Chemisana D. and Ibáñez, M. (2010) Linear Fresnel concentrators for building integrated applications. *Energy Conversion and Management*, 51, 1476–1480.

Chemisana, D., Ibáñez, M. and Barrau, J. (2009) Comparison of Fresnel concentrators for building integrated photovoltaics. *Energy Conversion and Management*, 50, 1079–1084.

Chemisana, D. and Rosell, J.I. (2009) Design and optical performance of a nonimaging fresnel reflective concentrator for building integration applications. *Energy Conversion and Management*, 52, 3241–3248.

Chemisana, D., Ibáñez, M. and Rosell, J.I. (2011a) Characterization of a photovoltaic-thermal module for Fresnel linear concentrator. *Energy Conversion and Management*, 52, 3234–3240.

Chemisana, D., Vossier, A., Pujol, L., Perona, A. and Dollet, A. (2011b) Characterization of Fresnel lens optical performances using an opal diffuser. *Energy Conversion and Management*, 52, 658–663.

Coolearthsolar (2012) Product specifications. www. coolearthsolar.com.

Coventry, J.S. (2005) Performance of a concentrating photovoltaic/thermal collector. *Solar Energy*, 78, 211–222.

Currie, M.J, Mapel, J.K., Heidel, T.D., Goffri, S. and Baldo, M.A. (2008) High-efficiency organic solar concentrators for photovoltaics. *Science*, 321, 226–228.

Danos, L., Kittidachachan, P., Meyer, T.J.J., Greef, R. and Markvart, T. (2006) Characterisation of fluorescent collectors based on solid, liquid and langmuir blodget (LB) films. In: *Proceedings of the 21st European Photovoltaic Solar Energy Conference*. 4–8 September, Dresden (Germany).

Debije, M.G., Broer, D.J. and Bastiaansen, C.W.M. (2007) Effect of dye alignment on the output of a luminescent solar concentrator, In: *Proceedings of the 22nd European Photovoltaic Solar Energy Conference*. 3–7 September Milan (Italy).

Dobon, F., Monedero, J., Valera, P., Acosta, L., Marichal, G.N., Osuna, R. and Fernandez, V. (2001) Very low concentration system (VLC). In: *17th European Photovoltaic Solar Energy Conference*. 22–26 October, Munich (Germany).

Duffie, J. A. and Beckman, W. A. (1991) *Solar Engineering of Thermal Processes*. John Wiley & Sons, New York, USA.

Dyson, A., Jensen, M. and Stark, P. (2007) Integrated concentrating (IC) solar façade system. *DOE Solar Energy Technologies Program Review Meeting*, April 17–19, Colorado, USA.

Edenburn, M.W. (1980) Active and passive cooling for concentrating photovoltaic arrays. In: *Conference Record, 14th IEEE PVSC*. 7–10 January, San Diego (USA).

Edmonds, I.R., Cowling, I.R. and Chan, H.M. (1987) The design and performance of liquid filled stationary concentrators for use with photovoltaic cells. *Solar Energy*, 39, 113–122.

Emcore (2012) Emcore Soliant 1000 specifications. www.emcore.com/wp-content/themes/emcore/pdf/EMCORE_Soliant_1000.pdf

Ericsson, R.W. and Rogers A.P. (2009) A microinverter for building-integrated photovoltaics. *IEEE*, 911–917.

Fanney, A.H., Dougherty, B.P. and Davis, M.V. (2001) Measured performance of building integrated photovoltaic panels. *ASME Journal of Solar Engineering*, 123, 187–192.

Florschuetz, L.W. (1975) On heat rejection from terrestrial solar cell arrays with sunlight Concentration. In: *Conference Record, 11th IEEE PVSC*. 6–8 May, Scottsdale (USA).

Fraidenraich, N. (1998a) Applications design procedure of V-trough cavities for photovoltaic systems. *Progress in Photovoltaics: Research and Applications*, 6, 43–54.

Fraidenraich, N. (1998b) Design procedure of V-trough cavities for photovoltaic systems. *Progress in Photovoltaics: Research and Applications*, 6, 43–54.

Fraidenraich, N. and Almeida, G.J. (1991) Optical properties of V-trough concentrators. *Solar Energy*, 47, 147–155.

Freilich, J. and Gordon, J.M. (1991) Case study of a central-station grid-intertie photovoltaic system with V-trough concentration. *Solar Energy*, 46, 267–273.

Garg, H.P. and Adhikari, R.S. (1999) Performance analysis of a hybrid photovoltaic/thermal (PV/T) collector with integrated CPC troughs. *International Journal of Energy Research*, 23,1295–1304.

Gee, R., Cohen, G., Greenwood, K. (2003) Operation and Preliminary Performance of the Duke Solar Power Roof™: A Roof-Integrated Solar Cooling and Heating System. In: *Proc. ASME 2003 International Solar Energy Conference (ISEC2003)*. 15–18 March, Hawaii (USA).

Goetzberger, A. (1988) Static concentration systems with enhanced light concentration. In: *Proc. 20th IEEE Photovoltaic Specialists Conference*. 26–30 September, Las Vegas (USA).

Goetzberger, A. and Greubel, W. (1977) Solar energy conversion with fluorescent collectors. *Applied physics*, 14, 123–139.

Goldschmidt, J.C., Glunz, S.W., Gombert, A. and Willeke, G. (2006) Advanced fluorescent concentrators, In: *Proceedings of the 21st European Photovoltaic Solar Energy Conference*. 4–8 September, Dresden (Germany).

Goldschmidt, J.C., Peters, M., Löper, P., Schultz, O., Dimroth, F., Glunz, S.W., Gombert, A. and Willeke, G. (2007) Advanced fluorescent concentrator system design. In: *Proceedings of the 22nd European Photovoltaic Solar Energy Conference*. 3–7 September, Milan (Italy).

Gordon, J.M. and Feuermann, D. (2005) Optical performance at the thermodynamic limit with tailored imaging designs. *Applied Optics*, 44, 2327–2331.

Gordon, J.M., Kreider, J.F. and Reeves, P. (1991) Tracking and stationary flat plate solar collectors: Yearly collectible energy correlations for photovoltaic applications. *Solar Energy*, 47, 245–252.

Green & Gold Energy (2011) SunCube™ specifications. http://www.greenandgoldenergy. com.au/Documents/GGESunCubeSpecSheet20110310a.pdf

Häberle, A., Zahler, C., Lerchenmüller, H., Mertins, M., Wittwer, C., Trieb, F. And Dersch, J. (2002) The Solarmundo line focussing Fresnel collector. Optical and thermal performance and cost calculations. In: *proceedings of SolarPACES 2002*. 4–6 September, Zurich (Switzerland).

Helgesson, A., Krohn, P., Karlsson, B., Svensson, L. and Broms, G. (2004) Evaluation of MARECO-hybrid placed in HAMMARBY SJÖSTAD (SWEDEN). In: *19th European Photovoltaic Solar Energy Conference*. 7–11 June, Paris (France).

HelioDynamics (2004) HD211 product sheet. www.heliodynamics.com/HD10specsheet2.pdf

Hernandez M., Mohedano R., Munoz F., Sanz A., Benitez P. and Miñano J.C. (2000) New static concentrator for bifacial photovoltaic solar cells. In: *16th European Photovoltaic Solar Energy Conference*. 1–5 May, Glasgow (United Kingdom).

Hestnes, A.G. (1999) Building Integration Of Solar Energy Systems, *Solar Energy*, 67, 181–187

Hinterberger, H. and Winston, R. (1966) Efficient light coupler for threshold Èerenkov counters. *Review of Scientific Instruments*, 37, 1094–1095.

Hollands, K.G.T. (1971) Concentrator for thin-film solar cells. *Solar Energy*, 13, 149–163.

Horner, J.L. and Ludman, J.E. (1981) Single holographic element wavelength demultiplexer. *Applied Optics*, 20, 1845–1847.

Ibrahim A., Yusof Othman M., Hafidz Ruslan M., Mat S. and Sopian K. (2011) Recent advances in flat plate photovoltaic/thermal (PV/T) solar collectors, Renewable and Sustainable Energy Reviews 15, 352–365.

Kaminar, N., McEntee, J., Stark, P. and Curchod, D. (1991) SEA 10X concentrator development progress. In: *Proc. 22nd IEEE Photovoltaic Specialists Conference*. 7–11 October, Las Vegas (USA).

King, D.L., Quintana, M.A., Kratochvil, J.A., Ellibee, D.E. and Hansen, B.R. (2000) Photovoltaic module performance and durability following long-term field exposure. *Progress in Photovoltaics: Research and Applications*, 8, 241–256.

Klotz, F.H. (2000) European Photovoltaic V-trough concentrator system with gravitational tracking (ARCHIMEDES). In: *16th European Photovoltaic Solar Energy Conference*. 1–5 May, Glasgow (United Kingdom).

Klotz, F.H., Mohring, H.D., Gruel, C., Alonso, M., Sherborne, J., Bruton, T. and Tzanetakis, P. (2001) Field test results of the Archimedes photovoltaic V-trough concentrator systems. In: *Proceedings of the 17th European Photovoltaic Solar Energy Conference and Exhibition*. 22–26 October, Munich (Germany).

Klotz, F.H., Noviello, G. and Sarno, A. (1995) PV V-trough systems with passive tracking: Technical potential for mediterranean climate. In: *13th European Photovoltaic Solar Energy Conference*. 23–27 October, Nice (France).

Korech, O., Gordon, J.M., Katz, E.A., Feuermann, D. and Eisenberg, N. (2007) Dielectric microconcentrators for efficiency enhancement in concentrator solar cells, *Optics Letters*, 32, 2789–2791.

Kritchman, E.M., Friesem, A.A. and Yekutieli, G. (1979) Efficient Fresnel lens for solar concentration. *Solar Energy*, 22, 119–123.

Kritchman, E.M., Friesem, A.A. and Yekutieli, G. (1981a) A fixed Fresnel lens with tracking collector. *Solar Energy*, 27, 7–13.

Kritchman, E.M., Friesem, A.A. and Yekutieli, G. (1981b) Convex Fresnel lens with large grooves. *Solar Energy*, 27, 129–37.

Kurtz, S. (2008) Opportunities and challenges for development of a mature concentrating photovoltaic power industry. National Renewable Energy Laboratory. Technical report NREL/TP-520-43208; September.

Kurtz, S. (2009) Opportunities for Development of a Mature Concentrating Photovoltaic Power Industry. In: CS MANTECH Conference. 18–20 May, Tampa (USA).

Leutz, R. and Suzuki, A. (2001) *Nonimaging Fresnel Lenses. Design and Performance of Solar Concentrators*. Ed. Springer.

Leutz, R., Suzuki, A., Akisawa, A. and Kashiwagi, T. (1999a) Design of a nonimaging Fresnel lens for solar concentrators. *Solar Energy*, 65, 379–387.

Leutz, R., Suzuki, A. and Kashiwagi, T. (1999b) Nonimaging Fresnel lens concentrators for photovoltaic applications. In: *Proceedings ISES Solar World Congress*, 4–9 July, Jerusalem (Israel).

Leutz, R., Suzuki, A., Akisawa, T. and Kashiwagi, T. (2000) Flux densities in optimum nonimaging Fresnel lens solar concentrators fors space. In: *Proceedings of the 28th IEEE Photovoltaic Specialist Conference*, 15–22 September, Anchorage (USA).

Libra, M. and Poulek, V. (2004) Bifacial PV modules in solar trackers and concentrators. In: *Proceedings 19th European Photovoltaic Solar Energy Conference*. 7–11 June, Paris (France).

Ludman, J.E. (1982a) Approximate bandwidth and diffraction efficiency in thick holograms. *American Journal of Physics*, 50, 244–246.

Ludman, J.E. (1982b) Holographic solar concentrator. *Applied Optics*, 21, 3057–3058.

Ludman, J.E., Sampson, J.L., Bradbury, R.A., Martin, J.G., Riccobono, J., Sliker, G. and Rallis, E. (1992) Photovoltaic systems based on spectrally selective holographic concentrators, In: *Proceedings of SPIE, Conference:* Practical Holography VI. 11–13 February, San Jose (USA).

Luque, A. (1986) Analysis of high efficiency back point contact silicon solar cells. *Solid-State Electronics*, 31, 65–79.

Luque, A. (1989) *Solar Cells and Optics for Photovoltaic Concentration*. Adam Hilger, Bristol, UK.

Luque, A., Martí, A., Bett, A., Andreev, V.M., Jaussaud, C., Van Roosmalen, J.A.M., Alonso, J., Räuber, A., Strobl, G., Stolz, W., Algora, C., Bitnar, B., Gombert, A., Stanley, C., Wahnon, P., Conesa, J.C., Van Sark, W.G.J.H.M., Meijerink, A., Van Klink, G.P.M., Barnham, K., Danz, R., Meyer, T., Luque-Heredia, I., Kenny, R., Christofides, C., Sala, G. and Benítez, P. (2005) Fullspectrum: a new PV wave making more efficient use of the solar spectrum. *Solar Energy Materials and Solar Cells*, 87, 467–479.

Luque, A., Sala, G., Arboiro, J.C., Bruton, T., Cunningham, D. and Mason, N. (1997) Some results of the EUCLIDES photovoltaic concentrator prototype. *Progress in Photovoltaics: Research and Applications*, 5, 195–212.

Luque-Heredia, I., Moreno, J.M., Magalhaes, P.H., Cervantes, R., Quemere, G. and Laurent, O. (2007) *Inspira's CPV Sun Tracking*, Chapter 11 in Concentrator Photovoltaics. Springer-Verlag Berlin Heidelberg, Germany.

Mallick, T.K. and Eames, P.C. (2007) Design and fabrication of low concentrating second generation pride concentrator. *Solar Energy materials and Solar Cells*, 91, 697–698.

Mallick, T.K., Eames, P.C. and Norton, B. (2002a) Asymmetric compound parabolic photovoltaic concentrators for building integration in the UK: an optical analysis. In: *World Renewable Energy Congress*, 29 June–5th July, 2002, Köln, Germany.

Mallick, T.K., Eames, P.C., Hyde, T.J. and Norton, B. (2004) The design and experimental characterisation of an asymmetric compound parabolic photovoltaic concentrator for building facade integration in the UK. *Solar Energy*, 77, 319–327.

Mallick, T.K., Eames, P.C. and Norton, B. (2002b) The application of computational fluid dynamics to predict the thermo-fluid behaviour of a parabolic asymmetric photovoltaic concentrator. In: *World Renewable Energy Congress*, 29 June–5th July, 2002, Köln, Germany.

Maurer, C., Pflug, T., Di Lauro, P., Hafner, J., Knez, F., Jordan, S., Hermann, M. and Kuhn, T.E. (2012) Solar Heating and Cooling with Transparent Façade Collectors in a Demonstration Building, *Energy Procedia*, 30, 1035–1041.

Mayregger, B., Auer, R., Niemann, M., Aberle, A.G. and Hezel, R. (1995) Performance of a low-cost static concentrator with bifacial solar cells. In: *13th European Photovoltaic Solar Energy Conference*, 23–27 October, Nice (France).

Mcdonald, M. and Barnes, C. (2008) Spectral optimization of CPV for integrated energy output. In: *Proceedings of SPIE*, Conference: Optical Modeling and Measurements for Solar Energy Systems II. 13–14 August, San Diego (USA).

Menova (2008) Power-Spar RFP 20 specifications. www.powerspar.com/images/RFP%2020%20Specs_R9-Apr-22-08.pdf

Mills, D.R. and Giutronich, J.E., (1978) Asymmetrical non-imaging cylindrical solar concentrators. *Solar Energy*, 20, 45–55.

Mills, D.R. and Morrison, G.L. (1997) Modelling of compact linear Fresnel reflector powerplant technology: Performance and cost estimates. In: *Proceedings of the International Solar Energy Society Conference*. 27–30 April, Washington (USA).

Mohedano, R., Benitez, P. and Miñano, J.C. (1998) Cost Reduction of building integrated PV's via static concentration systems. In: *2nd World Conference and Exhibition on Photovoltaic Solar Energy Conversion*. 6–10 July, Vienna (Austria).

Morgansolar (2012) Sun Simba specifications. www.morgansolar.com.

Müller, H.F.O. (1994) Application of holographic optical the bandwidth of holographic solar concentrators. In elements in buildings for various purposes like daylighting, solar shading and photovoltaic power generation. *Renewable Energy*, 5, 935–941.

Munari Probst, M.C. and Roecker, C. (2007) Towards an improved architectural quality of building integrated solar thermal systems (BIST), *Solar Energy*, 81, 1104–1116.

Munari Probst, M.C. and Roecker, Ch. (2012) Criteria for Architectural Integration of Active Solar Systems IEA Task 41, Subtask A, *Energy Procedia*, 30, 1195–1204.

Munari Probst, M.C., Roecker, C. and Schueler, A., (2005) Architectural integration of solar thermal collectors: results of a European survey. In: *the Proceedings of ISES Solar World Congress 2005*, 8–12 August, Orlando (USA).

Myrskog, S., Dufour, P., Guo, Y., Drew, K. and Morgan, J.P. (2012) Experimental results of Morgan Solar Inc.'s HCPV Sun Simba. In: *Proceedings of 8th International Conference on Concentrating Photovoltaic Systems. 16–18 April, Toledo (Spain)*.

Niedermeyer, W. (2008) *Parabolic Trough Solar Collector for Fluid Heating and Photovoltaic Cells*. U.S. Patent 7343913.

Nilsson, J. (2005) *Optical Design and Characterization of Solar Concentrators for Photovoltaics*. PhD Thesis, Lund University (Sweden).

O'Neill, M.J., Walters, R.R., Perry, J.L., McDanal, A.J. Jackson, M.C. and Hess, W.J. (1990) Fabrication, installation and initial operation of the 2000 sq. m. linear fresnel lens photovoltaic concentrator system at 3M/Austin (Texas). In: *Proc. 21th IEEE Photovoltaic Specialists Conference. 21–25 May, Kissimee (USA)*.

Ortabasi, U. (1997) Performance of a 2X cusp concentrator PV module using bifacial solar cells. In: *Proc. 26th IEEE Photovoltaic Specialist Conference*. 29 September–3 October, Anaheim (USA).

Pacificsolartech, 2006. Product specifications. www. pacificsolartech.com

Parada, J., Miñano, J.C. and Silva, JL. (1991) Construction and measurement of a prototype of P.V. module with static concentrator. In: *Proceedings of the 10th EC Photovoltaic Solar Energy Conference*. 8–12 April, Lisbon (Portugal).

Petter Jelle, B., Breivik, C. and Drolsum Røkenes, H. (2012) Building integrated photovoltaic products: A state-of-the-art review and future research opportunities. *Solar Energy Materials and Solar Cells*, 100, 69–96.

Poulek, V. and Libra, M. (2000) TRAXLETM the new line of trackers and tracking concentrators for terrestrial and space applications. In: *16th European Photovoltaic Solar Energy Conference*. 1–5 May, Glasgow (United Kingdom).

PrismSolar (2012) Product specifications. www.prismsolar.com/homepage.html

Pujol, R., Marínez, V., Moià, A. and Schweiger, H. (2006) Analysis of stationary Fresnel like linear concentrator with tracking absorber. In: *13th SolarPaces Symposium*. 20–23 June, Sevilla (Spain).

Rabl, A. (1976a) Comparison of solar concentrators. *Solar Energy*, 18, 93–111.

Rabl, A. (1976b) Solar concentrators with maximum concentration for cylindrical absorbers. *Applied Optics*, 15, 1871–1873.

Rabl, A. (1985) Otical and thermal properties of Compound Parabolic Concentrators. *Solar energy*, 18, 497–511.

Rau, U., Einsele, F. and Glaeser, G.C. (2005) Efficiency limits of photovoltaic fluorescent collectors. *Applied Physics Letters*, 87, 171101-1–171101-3.

Reijenga, T. (2000a) Photovoltaic building integration concepts – What do architects need? In: *Proceedings of IEA PVPS Task7 Workshop Lausanne Featuring A Review of PV Products, IEA PVPS Task7*, 11–12 February, Halcrow Gilbert, Swindon.

Reijenga, T. (2000b) Photovoltaics in the built environment. In: *Proceedings of 2nd World Solar Electric Buildings Conference, ESAA, ANZSES*, 8–10 March, Sydney, Australia.

Reijenga, T.H. (2003) *Handbook of Photovoltaic Science and Engineering*. Edited by A. Luque and S. Hegedus, John Wiley & Sons, NY.

Reijenga, T.H. and Kaan, H.F. (2011) *Handbook of Photovoltaic Science and Engineering*. Edited by A. Luque and S. Hegedus, John Wiley & Sons, NY.

Richards, B.S. and Shalav, A. (2005) The role of polymers in the luminescence conversion of sunlight for enhanced solar cell performance. *Synthetic Metals*, 154, 61–64.

Rosell, J.I., Vallverdu, X., Lechon, M.A. and Ibanez, M. (2005) Design and simulation of a low concentrating photovoltaic/thermal system. *Energy Conversion and Management*, 46, 3034–3046.

Seybold, G. and Wagenblast, G. (1989) New perylene and violanthrone dyestuffs for fluorescent collectors. *Dyes and Pigments*, 11, 303–317.

Shaw, N.C. and Wenham, S.R. (2000) Design of a novel static concentrator lens utilising total internal reflection surfaces. In: *16th European Photovoltaic Solar Energy Conference*. 1–5 May, Glasgow (United Kingdom).

Shinjo, F. (1994) R&D of photovoltaic modules integrated with construction material. In: *Conference Record of the IEEE Photovoltaic Specialist Conference, vol. 1. December*, Waikoloa, HI, USA, 778–780.

Slooff, L.H., Budel, T., Burgers, A., Bakker, N., Büchtemann, A., Danz, R., Meyer, T. and Meyer, A (2007) The luminescent concentrator: Stability issues. In: *Proceedings of the 22nd European Photovoltaic Solar Energy Conference*. 3–7 September Milan (Italy).

Smith, R.H. (1976) *Conference on Heliotechnique and Development*. Development Analysis Associates Inc., Cambridge, Massachusetts, USA, 251.

Smyth M., Eames P.C. and Norton B. (2003), Heat retaining integrated collector/storage solar water heaters, Solar Energy 75, 27–34.

Solanki, C.S., Sangani, C.S., Gunashekar, D. and Anthony, G. (2008) Enhanced heat dissipation of V-trough PV modules for better performance. *Solar Energy Materials and Solar Cells*, 92, 1634–1638.

Sopogy (2012) Product specifications. www.sopogy.com

Souliotis, M., Quinlan, P., Smyth, M., Tripanagnostopoulos, Y., Zacharopoulos, A., Ramirez, M. and Yianoulis, P. (2011) Heat retaining integrated collector storage solar water heater with asymmetric CPC reflector, *Solar Energy*, 85, 2474–2487.

Sultana, T., Morrison, G.L, Rosengarten, G (2012) Thermal performance of a novel rooftop solar micro-concentrating collector, *Solar Energy*, 86, 1992–2000.

Swanson, R.M. (2000) The promise of concentrators. *Progress in Photovoltaics: Research and Applications*, 8, 93–111.

Swanson, R.M. (2003) Photovoltaic concentration. In: *Handbook of photovoltaic science and engineering*. Eds. Luque, A. and Hegedus, S., Wiley.

Tabor, H. (1958) Stationary mirror systems for solar collectors. *Solar Energy*, 2, 27–33.

Tonui, J. K. and Tripanagnostopoulos, Y. (2007) Air-cooled PV/T solar collectors with low cost performance improvements. *Solar Energy*, 81, 498–511.

Toyokawa, S. and Uehara, S. (1997) Overall evaluation for R&D of PV modules integrated with construction materials. In: *Conference Record of the IEEE Photovoltaic Specialist Conference*, October, Anaheim, CA, USA, pp. 1333–1336.

Tripanagnostopoulos, Y. (2008) Building integrated concentrating PV and PV/T systems. In: *proceedings of the Eurosun 2008*. 7–10 October, Lisbon (Portugal).

Tripanagnostopoulos, Y. and Souliotis, M. (2004) ICS solar systems with horizontal cylindrical storage tank and reflector of CPC or involute geometry. *Renewable Energy*, 29, 13–38.

Tripanagnostopoulos, Y. and Yianoulis, P. (1992) Integrated collector-storage systems with suppressed thermal losses. *Solar Energy*, 48, 31–43.

Tripanagnostopoulos, Y. and Yianoulis, P. (1996) CPC solar collectors with multichannell absorber. *Solar Energy*, 58, 49–61.

Tripanagnostopoulos, Y., Nousia, Th., Souliotis, M. and Yianoulis, P. (2002) Hybridphotovoltaic/thermalsolarsystems. *Solar Energy*, 72, 217–234.

Tripanagnostopoulos, Y., Siabekou, Ch. and Tonui, J.K. (2007) The Fresnel lens concept for solar control of buildings. *Solar Energy*, 81, 661–675.

Uematsu, T., Tsutsui, K., Yazawa, Y., Warabisako, T., Araki, I., Eguchi, Y. and Joge, T. (2003) Development of bifacial PV cells for new applications of flat-plate modules. *Solar Energy Materials and Solar Cells*, 75, 557–566.

Uematsu, T., Warabisako, T., Yazawa, Y. and Muramatsu, S. (1998) Static micro-concentrator photovoltaic module with an acorn shape reflector. In: *2nd World Conference and Exhibition on Photovoltaic Solar Energy Conversion*. 6–10 July, Vienna, Austria.

Uematsu, T., Yazawa, Y., Joge, T. and Kokunai, S. (2001b) Fabrication and characterisation of a flat-plate static concentrator photovoltaic module. *Solar Energy Materials and Solar Cells*, 67, 425–434.

Uematsu, T., Yazawa, Y., Miyamura, Y., Muramatsu, S., Ohtsuka, H., Tsutsui, K. and Warabisako, T.. (2001c) Static concentrator photovoltaic module with prism array. *Solar Energy Materials and Solar Cells*, 67, 415–423.

Uematsu, T., Yazawa, Y., Tsutsui, K. Miyamura, Y., Ohtsuka, H., Warabisako, T. and Joge, T. (2001a) Design and characterisation of flat-plate static-concentrator photovoltaic modules. *Solar Energy Materials and Solar Cells*, 67, 441–448.

Van Roosmalen, J.A.M. (2004) Molecular-based concepts in PV towards full spectrum utilization. *Semiconductors*, 38, 970-975.

Vasylyev, S. (2004) Performance measurements of a slat-array photovoltaic concentrator. In: *Proceedings of the American Solar Energy Society,* 9–14 July, Portland (Oregon).

Vasylyev, S. (2005) Nonimaging Reflective Lens Concentrator. In: *Proceedings of the International Conference on Solar Concentrators for the Generation of Electricity or Hydrogen.* 1–5 May, Scottsdale (Arizona).

Vasylyev, S. and Vasylyev, V. (2003) *Non-Imaging System for Radiant Energy Flux Transformation.* U.S. Patent # 6620995.

Vasylyev. V. and Vasylyev, S. (2002) Expected optical performances of novel type multi-element high-heat solar concentrators. In: *Proceedings of the American Solar Energy Society Conference.* 15–20 June, Reno (Nevada).

Vivar, M., Everett, V., Fuentes, M., Blakers, A., Tanner, A., Le Lievre, P. and Greaves M. (2012) Initial field performance of a hybrid CPV-T microconcentrator system. *Progress in Photovoltaics: Research and Applications* (doi: 10.1002/pip.2229).

Watt, M., Kaye, J., Travers, D. and MacGill, I. (1999) *Opportunities for the Use of Building Integrated Photovoltaic in NSW.* Report to the University of New South Wales, Photovoltaics Special Research Centre, Sydney, Australia.

Weber, K.J., Everett, V., Deenapanray, P.N.K., Franklin, E. and Blakers, A.W. (2006) Modeling of static concentrator modules incorporating lambertian or v-groove rear reflectors. *Solar Energy Materials and Solar Cells,* 90, 1741–1749.

Weber, W.H. and Lambe, J. (1976) Luminescent greenhouse collector for solar radiation. *Applied Optics,* 15, 2299–2300.

Weiss, W. and Rommel, M. (2008) Process heat collectors. State of the art within Task 33/IV. *IEA SHC-Task 33 and SolarPACES-Task IV: Solar Heat for Industrial Processes.*

Welford, W.T. (1978) *The Optics of Nonimaging Concentrators: Light and Solar Energy.* Academic Press, New York, USA.

Welford, W.T. and Winston, R. (1982) Nonconventional optical systems and the brightness theorem. *Applied Optics,* 21, 1531–1533.

Wenham, S.R., Bowden, S., Dickinson, M., Largent, R., Jordan, D., Honsberg, C.B. and Green, M. (1995) Prototype Photovoltaic Roof Tiles. In: *Proceedings of 13th European Photovoltaic Solar Energy Conference.* 23–27 October, Nice (France).

Winston, R. (1974) Principles of solar collectors of a novel design. *Solar Energy,* 16, 89–95.

Winston, R. (1980) Light collection within the framework of geometrical optics. *Journal of the Optical Society of America,* 60, 245–247.

Winston, R. and Gordon, J.M. (2005) Planar concentrators near the etendue limit. *Optics Letters,* 30, 2617–2619.

Winston, R. and Hinterberger, H. (1995) Principles of cylindrical concentration for solar energy. *Solar Energy,* 17, 255–258.

Winston, R. and Welford, W.T. (1979) Geometrical vector flux and some new nonimaging concentrators. *Journal of the Optical Society of America A,* 69, 532–536.

Winston, R., Miñano, J.C. and Benítez, P. (2005) *Nonimagin Optics.* Ed. Elsevier Academic Press.

Wittwer, V., Heidler, K., Zastrow, A. and Goetzberger, A. (1981) Theory of fluorescent planar concentrators and experimental results. *Journal of Luminescence,* 24/25, 873–876.

Zacharopoulos, A. (2001) *Optical Design Modelling and Experimental Characterisation of Line-Axis Concentrators for Solar Photovoltaic and Thermal Applications.* PhD Thesis, University of Ulster, UK.

Zacharopoulos, A., Eames, P.C., McLarnon, D. and Norton, B. (2000) Linear dielectric non-imaging concentrating covers for PV integrated building facades. *Solar Energy,* 68, 439–452.

Zanesco, I. and Lorenzo, E. (2002) Optimisation of an asymmetric static concentrator: the PEC-44D. *Progress in Photovoltaics: Research and Applications,* 10, 361–376.

Chapter 18

Solar energy use in buildings

Ursula Eicker
Center for Sustainable Energy Technology, Stuttgart University of Applied Sciences, Stuttgart, Germany

18.1 INTRODUCTION

Buildings account today for about 40% of final energy consumption worldwide, and they are responsible for about one-third of overall CO_2 emissions (36% in Europe, 39% in the USA, about 20% in China). Figure 18.1.1 describes the distribution of end energy consumption in the European Union (EU). Especially in urban structures, building energy consumption is typically twice as high as transport, e.g. approximately factor 2.2 in London. The energy consumption of buildings consists of heating, cooling, ventilation, lighting and other electrical appliances. Solar energy contributes to the supply of heating energy by passive solar gains, to the reduction of electrical lighting consumption by "daylighting" (i.e. illumination of indoor spaces) and contributes to heating, cooling and electricity supply by active solar components such as solar thermal collectors or photovoltaic modules.

Meanwhile, most countries have significantly increased their energy standards: average heat transfer coefficients for new buildings are today about 0.3 and $0.4\,\mathrm{W\,m^{-2}\,K^{-1}}$ in Europe. In 2009, the EU tightened the Energy Performance of Buildings Directive (EPDB) and now demands nearly zero energy standards for new buildings until 2020. Here, building energy demand is balanced with local renewable energy production, mainly solar and biomass resources, resulting in a net zero energy demand. For existing public buildings, the zero energy standard will come into effect in 2018.

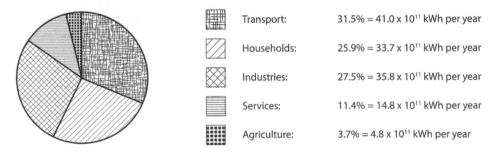

	Transport:	31.5% = 41.0 x 10¹¹ kWh per year
	Households:	25.9% = 33.7 x 10¹¹ kWh per year
	Industries:	27.5% = 35.8 x 10¹¹ kWh per year
	Services:	11.4% = 14.8 x 10¹¹ kWh per year
	Agriculture:	3.7% = 4.8 x 10¹¹ kWh per year

Figure 18.1.1 Distribution of end energy consumption within the European Union with a total value of 1.3×10^{13} kWh per year.

With high heat insulation standards and the heat recovery ventilation concept of passive houses, a low limit of heat consumption has meanwhile been achieved, which is around 10 times lower than today's average consumption values. A crucial factor for the low consumption of passive buildings was the development of new glazing and window technologies, which have enabled windows to become passive solar elements, at the same time causing only low transmission heat losses. In buildings with low heating requirements, other energy consumption in the form of electricity for lighting, power and air conditioning, as well as warm water in residential buildings, is becoming more and more dominant. In this area, renewable sources of energy, especially solar energy, can make an important contribution to the supply of electricity and heat.

In warm climatic regions and in non-residential buildings with high internal loads, cooling is the dominant energy requirement. Here, solar protection is the main issue to reduce energy demand and high glazing fractions, which might be useful in cold climates but which now cause problems for summer comfort.

The majority of the world's new buildings are constructed in Asia. The Asian building sector today accounts for about 25% of final energy consumption and this is expected to rise to 32% by 2030. A World Bank study showed that China and India could cut their current energy consumption by 25% with cost-effective retrofitting of lighting, air conditioning, boilers and heat recovery. The Chinese Ministry of Construction states that 95% of all buildings are highly energy consuming and that energy consumption is currently two to three times that of developed countries in achieving the same comfort.

18.2 PASSIVE SOLAR GAINS IN COLD AND MODERATE CLIMATIC REGIONS

Passive solar energy use contributes significantly to the heating energy demand of every building. The main energy supply is via short-wave solar irradiance transmitted by glazing, which provides daylight and is converted into heat by absorption on wall surfaces. This form of energy transfer is described as *passive* and takes place solely by solar irradiance absorption, thermal conduction, long-wave radiative exchange and free convection; i.e. transfer is not line-bound and requires no auxiliary mechanical energy for moving a heat carrier.

Solar irradiance is absorbed without transport losses directly by the building shell or by means of internal storage. Besides windows and the associated internal storage, the possibilities of passive use also include transparent thermal insulation on a heat-conducting external wall. Despite having a lower solar efficiency compared to windows, transparent thermal insulation in connection with a massive building component enables a temporal phase shift between irradiance and utilization of heat. This characteristic reduces overheating problems that can occur with large glazing.

Unheated winter gardens rank among the classical forms of passive solar use. As elements placed in front of the building shell, they contribute to the insulation of the building, but at the same time they reduce the amount of solar irradiance available to the main building. Thus, the amount of both daylight and direct heat entry by solar irradiance into the adjoining heated rooms are clearly reduced. Furthermore, indirect heating of glazed conservatories by adjoining rooms often leads to an increase in the

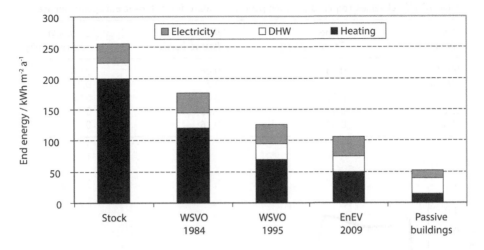

Figure 18.2.1 Energy demand for the existing building stock and subsequent legal requirements in residential buildings per square metre of heated floor space Germany.

heating requirement of buildings. Only with very energy-conscious use of an unheated conservatory can energy gains for the building actually be achieved.

How efficiently solar gains can be used by a building depends on the ratio of gains to heat losses. The higher the heat losses, the better solar gains can be used. In well-insulated buildings with low losses, however, solar heat gains often cannot be used, as they might overheat the buildings, especially when the thermal building mass is low.

Most countries now regulate heat losses. In China, which has severely cold regions (between 5500 and 8000 heating degree days), exterior walls for three-storey buildings are supposed to have heat transfer coefficients (U-values) below $0.33\,\mathrm{W\,m^{-2}\,K^{-1}}$, and for high rise buildings below $0.48\,\mathrm{W\,m^{-2}\,K^{-1}}$, whereas in temperate regions $0.5\,\mathrm{W\,m^{-2}\,K^{-1}}$ are sufficient. In Japan, today's wall U values are between 0.39 and $1.76\,\mathrm{W\,m^{-2}\,K^{-1}}$ depending on climatic condition, and in Korea between 0.47 and $0.76\,\mathrm{W\,m^{-2}\,K^{-1}}$. Figure 18.2.1 shows the development of energy requirements in Germany.

In Europe, with its wide geographical extent of nearly 35° geographical latitude difference (36° in Greece, 70° in northern Scandinavia), a wide range of climatic boundary conditions is covered. In Helsinki (60.3° northern latitude), average exterior air temperatures reach $-6°$ C in January, while southern cities such as Athens at 40° latitude still have averages of $+10°$ C. Consequently, building standards vary widely: whereas average heat transfer coefficients (U-values) for detached houses are $1\,\mathrm{W\,m^{-2}\,K^{-1}}$ in Italy, they are only $0.4\,\mathrm{W\,m^{-2}\,K^{-1}}$ in Finland. The heating energy demand determined is comparable in both cases at about $50\,\mathrm{kWh\,m^{-2}\,a^{-1}}$. The necessary U value to achieve the passive house standard with less than $15\,\mathrm{kWh\,m^{-2}\,a^{-1}}$ heating energy demand for several climate zones is shown in Table 18.2.1. In addition to heating energy requirements, residential buildings require energy for domestic hot water, electricity consumption and cooling in hot climates.

Table 18.2.1 U values required to reach passive standard for different European climates

U values /W m^{-2} K^{-1}	Rome Passive building	Helsinki Passive building	Stockholm Passive building
Wall	0.13	0.08	0.08
Window	1.4	0.7	0.7
Roof	0.13	0.08	0.08
Ground	0.23	0.08	0.1
Mean	0.33	0.16	0.17

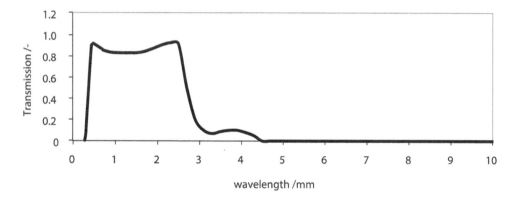

Figure 18.2.2 Wavelength-dependent transmittance of a single glazing.

18.2.1 Passive solar gains by glazing

Glazing is characterized by high transmittances for short-wave solar irradiance up to wave lengths of 2.5 μm and an impermeability to long-wave radiant heat emitted from building components with a maximum intensity at wavelengths of around 10 μm, as demonstrated by Figure 18.2.2. This transmission characteristic results in a greenhouse effect and the conversion of solar irradiance by absorption in structural elements into heat, the long-wave radiation proportion of which is not transmitted through the glazing.

18.3 TOTAL ENERGY TRANSMITTANCE OF GLAZING

Aside from the direct transmission of short-wave solar radiation with a transmittance τ, part of the irradiance is absorbed in the panes, as illustrated by Figure 18.3.1. Heating the panes causes a heat flow towards the room, which contributes to the total energy transmission factor (g-value). The absorption coefficient of single glazing can be up to 50% for special sun-protection glazing. The secondary heat emission degree q_i is defined as the relation of the heat flow on the room side $\dot{Q}_i$ per square metre of window area A_w to the impacting solar radiation G. The value of q_i is calculated by

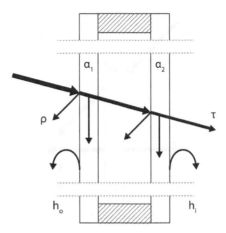

Figure 18.3.1 Transmission τ, reflection ρ, absorption α and heat transfer coefficients at the inside (h_i) and outside (h_o) of a double-glazing.

solving heat balance equations. The degrees of transmission, absorption and reflection must be known for each pane of a multi-pane system. The total energy transmission factor then results from the relation of the total heat flow $\dot{Q}_{total}/A_w$ into the room to the irradiance:

$$g = \frac{\dot{Q}_{total}/A_w}{G} = \frac{\tau G + \dot{Q}_i/A_w}{G} = \tau + q_i \qquad (18.3.1)$$

The transmittance τ is calculated based on EN 410 by integration of the wavelength-dependent transmission over the solar spectrum. With perpendicular incidence, transmittance is approximately 90% for an uncoated simple float glass and about 80% for a two-pane system. Transmittance falls and the total energy transmission factor of the two-pane system rarely exceeds 65% due to the metallic coating in thermally insulating glazing commonly used today on the room-side panes. The absorption factor α of the short-wave solar radiation is similarly calculated for each pane by integration over the spectrum, taking into account inter-reflections in multi-pane systems.

The solutions of the heat balance equations for single, double and triple glazing are indicated in EN 410. Using the example of single glazing, a heat balance is shown in Figure 18.3.2 and can be described as follows: the intensity αG absorbed by the pane of surface A_w is divided into a heat flow inward $\dot{Q}_i/A_w$ and outward $\dot{Q}_o/A_w$.

$$\alpha G = \frac{\dot{Q}_i}{A_w} + \frac{\dot{Q}_o}{A_w} \qquad (18.3.2)$$

These heat flows can be calculated with the help of the heat transfer coefficients h_i (standard value $7.7 \text{ W m}^{-2} \text{ K}^{-1}$) and h_o (standard value $25 \text{ W m}^{-2} \text{ K}^{-1}$), and the temperature

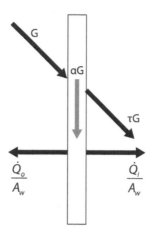

Figure 18.3.2 Irradiance G, absorption αG and secondary heat flows of single glazing outward and inward.

difference between pane surface T_s and room air T_i or outside air T_o. Temperature differences between the outside and inside surface are neglected.

$$\frac{\dot{Q}_i}{A_w} = h_i(T_s - T_i) \quad \frac{\dot{Q}_o}{A_w} = h_o(T_s - T_o) \tag{18.3.3}$$

From the heat flow balance, first determine the pane surface temperature T_s:

$$\alpha G = (h_i + h_o)T_s - h_iT_i - h_oT_o$$

$$\Rightarrow T_s = \frac{\alpha G + h_iT_i + h_oT_o}{h_i + h_o} \tag{18.3.4}$$

With this pane temperature T_s, the heat flow inward can be calculated:

$$\frac{\dot{Q}_i}{A_w} = h_i(T_s - T_i) = \frac{h_i}{h_i + h_o}(\alpha G - h_o(T_i - T_o))$$

$$= \underbrace{\frac{h_i}{h_i + h_o}\alpha G}_{\text{secondary heatflow}} - \underbrace{\frac{1}{\frac{1}{h_i} + \frac{1}{h_o}}(T_i - T_o)}_{\text{Transmission losses}} \tag{18.3.5}$$

Transmission heat losses of the glazing are calculated separately via the U-value. Thus, for the definition of the secondary heat emission degree q_i, the ambient temperature can be set equal to the outside temperature The result for q_i is:

$$q_i = \frac{\dot{Q}_i/A_w}{G} = \alpha\frac{h_i}{h_i + h_o} \quad for\ T_i = T_o \tag{18.3.6}$$

For double-glazing, the characteristic values are calculated similarly, although an additional heat balance for the outside pane must now be created. The absorption coefficient for the outside pane is defined as α_1 and for the internal pane as α_2. The secondary heat emission degree also depends on the layer thicknesses s_1, s_2 and heat conductivities λ_1, λ_2 of the two panes and on the thermal resistance R_{air} of the standing air layer between the panes:

$$q_i = \frac{\left(\dfrac{\alpha_1 + \alpha_2}{h_o} + \dfrac{\alpha_2}{\dfrac{s_1}{\lambda_1} + R_{air} + \dfrac{s_2}{\lambda_2}}\right)}{\dfrac{1}{h_i} + \dfrac{1}{h_o} + \dfrac{s_1}{\lambda_1} + R_{air} + \dfrac{s_2}{\lambda_2}} \qquad (18.3.7)$$

The solar radiation let through (transmitted by) the glazing into the room $\dot{Q}_{trans}$ results directly from the product of the g-value and the solar irradiance:

$$\dot{Q}_{trans} = g\, G \qquad (18.3.8)$$

The calorific loss through the window is deducted from the transmitted power, which is characterized by the heat transfer coefficient of the glazing U_g or of the entire window including the frame U_w. Double-glazing coated panes filled with heavy noble gases achieve a minimum U_g value of $1.0\,\mathrm{W\,m^{-2}\,K^{-1}}$ and triple glazing at best reaches a U_g value of $0.4\,\mathrm{W\,m^{-2}\,K^{-1}}$. Even at a glazing with a U_g value of $1.3\,\mathrm{W\,m^{-2}\,K^{-1}}$, a wooden or plastic frame increases the window's U_w value slightly. For passive house concepts, specially insulated expanded polystyrene frameworks must be used to avoid worsening the low glazing values of triple glazing by the frame proportion.

The passive solar gain $\dot{Q}_u$ usable in a room results from the balance of losses and gains. The losses are calculated from the U_w value of the window of surface A_w and from the temperature difference between the room air T_i and the outside air T_o:

$$\frac{\dot{Q}_u}{A_w} = U_w(T_i - T_o) - g\, G \qquad (18.3.9)$$

From the available energy balance, an effective U-value U_{eff} can be defined, which is often used for monthly or annual balance calculations with mean temperature differences and irradiances.

$$U_{eff} = \frac{\dot{Q}_u}{A_w(T_i - T_o)} = U_w - g\frac{G}{T_i - T_o} \qquad (18.3.10)$$

Balanced over a heating season, for example about $400\,\mathrm{kWh\,m^{-2}\,a^{-1}}$ of solar irradiance is available on a south-facing façade in Germany. The so-called heating degree day number is obtained by multiplying the mean temperature difference between the inside and outside, about $17°\mathrm{C}$, by the number of days in the heating season. There are about 3500 Kelvin-days per year on average in Germany. Thus, the maximum usable

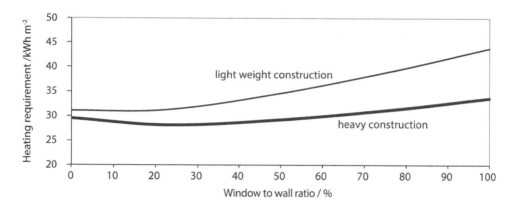

Figure 18.3.3 Influence of the window area proportion on the heating requirement.

energy per square metre of glazing surface for two-pane low-e coated glazing with a
U_w value of $1\,W\,m^{-2}\,K^{-1}$ and $g = 0.65$ is:

$$\frac{Q_u}{A_w} = 0.65 \times 400 \times 10^3 \frac{Wh}{m^2 a} - 1.0 \frac{W}{m^2 K} \times 3500 \frac{K\,days}{a} \times 24 \frac{h}{day}$$
$$= 260 \frac{kWh}{m^2} - 84 \frac{kWh}{m^2} = 176 \frac{kWh}{m^2} \qquad (18.3.11)$$

The amount of heat effectively usable in the room depends greatly on the storage
capability of the structural elements inside. High passive solar heat gains can easily
lead to overheating of the interior and thus do not contribute to covering the heating
requirement.

According to the EN 832 monthly balance procedure for calculating the heating
requirement, the efficiency of the solar irradiance transmitted by windows is a function
of the relation of the monthly gains to the transmission and ventilation heat losses.
For lowenergy buildings with heat-storing heavy building construction and an annual
heating requirement of between about 30 and $70\,kWh\,m^{-2}\,a^{-1}$, the result is a flat
minimum of the heating requirement for a window area proportion on the south-
facing façade of approximately 25%. In administrative buildings with mostly higher
internal loads, the window area proportion should be even lower to avoid overheating
in summer; this is described in Figure 18.3.3. The minimum heating requirement is
obtained for 0–20% of the window area proportion for a light building with a small
storage capacity and between 20 to 30% for a building with high internal mass.

18.4 NEW GLAZING SYSTEMS

For flexible control of total energy transmission, glazing systems are being developed that modify their transparency degree temperature-dependently (thermo-tropic)
or electrically controlled (electro-chromic).

Thermotropic layers are polymer mixtures or hydrogels which are inserted in a homogeneous mixture between two windowpanes and reduce the light permeability by up to 75% with rising temperature.

For example, electrochrome thin films made from tungsten oxide can be evaporated on windowpanes with conductive oxide coatings. The two thin film electrodes are connected by a polymer ion conductor. When an external electrical field is applied and causes cat-ions (e.g. Li+) from the counter-electrode to accumulate onto the tungsten oxide, the transmittance falls dependent on wavelength to 10–20%. A tight seal at the edges is very important for long-term stability. The first commercially available glass with a maximum surface of $0.9\,m \times 2.0\,m$ achieve a reduction of the total energy transmission factor from 44% in the bright status to 15% in the dark status (U-value $= 1.6\,W\,m^{-2}\,K^{-1}$). In systems with a lower U-value of $1.1\,W\,m^{-2}\,K^{-1}$, the g-value falls from 36% to 12%.

18.5 TRANSPARENT THERMAL INSULATION (TTI)

Since the early 1980s, several thousand square metres of transparent thermally insulated façade systems have been installed throughout Germany. Compared to conventional thermal insulation in buildings, transparently insulated external walls can well utilize incoming solar radiation. The energy-saving potential for the application of such solar systems is high. For example, if a fifth of all existing building façades in Germany were equipped with transparent insulating systems, approximately 15% of the heat needed for room heating could be supplied. The technology is particularly interesting for the renovation of old buildings with heavy, very heat-conducting walls. Transparent insulation can be attached directly onto an external wall and a transparent plaster is used to protect the material from weather. The potentially high costs of glass composite structures can be significantly abated by foregoing complex frame constructions. Transparent thermal insulation can also be applied to provide daylighting of building interiors. The elements of the thermal insulation can scatter and direct light and provide for the even illumination of a room. Combined with very good heat-insulating properties, transparent thermal insulation has the potential for large-scale application on external building walls.

18.6 OPERATIONAL PRINCIPLE OF TRANSPARENT THERMAL INSULATION

If short-wave solar radiation hits an external wall, the radiation is absorbed and converted into heat. Although the external surface warms up, most of the heat produced is transferred to the outside air. Only a small portion of the heat produced reaches the building interior. If a transparent insulating layer is attached in front of the wall (in the simplest case, as a windowpane), heat emission to the outside is made more difficult.

In addition to the transmission coefficient for solar radiation and the heat resistance of the transparent thermal insulation, the main parameters influencing the extent of the useful heat gain are the absorption coefficient, heat conductivity and storage capability of the adjoining wall. The wall itself produces a time delay in the heat flow

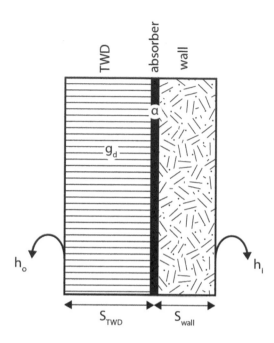

Figure 18.6.1 System structure of a transparently insulated wall.

such that the maximum values of the solar-induced heat flow reach the inside when the direct solar gains through the windows have already decreased and outside temperatures are falling. In addition, the thermal characteristics of the entire building play a role; however, above all it is the heat-storing capability of the interior structural elements that aids in the avoidance of overheating.

For example, an external wall even with 10 cm external insulation under German climatic conditions will still exhibit annual heat losses of over 30 kWh per square metre. However, with the application of transparent thermal insulation, this same wall acts as a solar collector during the heating season and produces around 50–100 kWh m^{-2} of useful heat for the building.

The heat transfer coefficient of an external wall insulated with transparent insulation results from the total of the thermal resistances of the existing external wall and the transparent insulation, with layer thickness s_{TTI} and heat conductivity λ_{TTI} of the transparent insulating material, layer thickness s_{wall} and heat conductivity λ_{wall} of the external wall and also the heat transfer coefficients inside h_i and outside h_o.

$$U_{eff} = U_{wall-TTI} - \eta_0 \frac{G}{T_i - T_o} \qquad (18.6.1)$$

The wall absorber (see Figure 18.6.1) is characterized by the absorption coefficient α and the transparent insulation by the diffuse total energy transmission factor g_d. The U-values of 10 cm TTI are typically about 0.8 W m^{-2} K^{-1}, which are lower than the best double-glazed heat-protection glass, but still twice as high as the heat transfer coefficients of 10 cm of conventional insulating material with U-values of $0.4\,\mathrm{W\,m^{-2}\,K^{-1}}$.

Just as for the energy balance of windows, an effective U-value can also be defined for transparent thermal insulation as the difference between losses and solar gains with efficiency η_0.

$$U_{eff} = U_{wall+TTI} - \eta_0 \frac{G}{T_i - T_o} \qquad (18.6.2)$$

Solar efficiencies η up to 50% with a simultaneous low heat transfer coefficient lead to low or even negative effective U-values. Given favourable wall orientation, negative U-values can be achieved, which means the wall produces heat gains for the building. Weekly averaged measurements of a 10 cm transparently insulated building in the city of Freiburg, Germany exhibited effective U-values of between 0 and $-3.5\,\mathrm{W\,m^{-2}\,K^{-1}}$.

The solar efficiency η corresponds to the total energy transmission factor of a glazing and consists of the g_d-value of the transparent insulating material, the absorption factor of the absorber α and the proportion of the heat flow inward to the total heat flow. The heat flow from the absorber inward is calculated from the temperature node of the absorber at temperature T_a to the room air temperature T_i via the heat transfer coefficients of the wall.

$$U_{wall} = \left(\frac{s_{wall}}{\lambda_{wall}} + \frac{1}{h_i} \right)^{-1}$$

$$\frac{\dot{Q}_i}{A} = U_{wall}(T_a - T_i) \qquad (18.6.3)$$

The heat flow from the absorber outward $\dot{Q}_o$ is calculated via the heat transfer coefficient U_{TTI} of the TTI material.

$$\frac{\dot{Q}_o}{A} = U_{TTI}(T_a - T_o) \qquad (18.6.4)$$

The solar efficiency is then proportional to the ratio of the interior heat flux to the total heat flux as the sum of interior and exterior flux.

$$\eta_0 = \alpha g_d \frac{U_{wall}(T_a - T_i)}{U_{wall}(T_a - T_i) + U_{TTI}(T_a - T_o)} \qquad (18.6.5)$$

Assuming identical temperatures inside and outside, a constant solar efficiency can be defined (as shown in Figure 18.6.2) that is very suitable for material comparisons and estimates of the energy yield.

$$\eta_0 = \alpha g_d \frac{U_{wall}}{U_{wall} + U_{TTI}} \qquad (18.6.6)$$

With 5 cm transparent capillary tubes, the U_{TTI} value is $1.3\,\mathrm{W\,m^{-2}\,K^{-1}}$ at a g-value of 0.67; with 10 cm it is $0.8\,\mathrm{W\,m^{-2}\,K^{-1}}$ and $g = 0.64$. The aerogel material shows a U_{TTI} value of $0.8\,\mathrm{W\,m^{-2}\,K^{-1}}$ with a very small layer thickness of 2.4 cm at a g-value of 0.5.

For optimal use of solar heat, one must ensure that transparently insulated rooms do not overheat. Conventional window areas usually bring sufficiently high solar gains

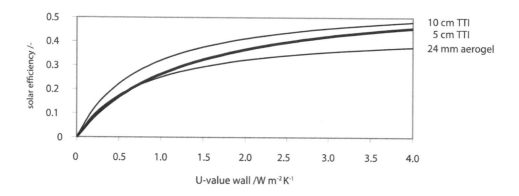

Figure 18.6.2 Solar efficiency as a function of the heat transfer coefficient of the opaque wall.

into the room during the day; additional gains from the TTI wall must be stored and then used in the evening hours. The temporal shift between heat production at the absorber and the maximum heat flow into the room rises with external wall thickness and also depends on the density, thermal capacity and heat conductivity of the wall. Phase shifts of 6 to 8 hours are achieved with a 24 cm brick or lime-sandstone wall while a little more than 5 hours are possible with concrete walls. With sufficiently thick wall constructions, the result is typically 100% efficiencies of the heat produced by the TTI wall within the core months of the heating season; in the transition months efficiency is around 30%.

The high absorber temperatures of the external wall, which can reach peak values of between 70° and 80°C, are effectively dampened at wall thicknesses over 20 cm and are rarely higher than 30°C at the interior surface, even in the summer. The absorber temperatures are lower with heavy, heat-conducting components than with light. The heat that develops can penetrate quickly into the heavy external wall and be led into the interior. Any thermal tensions are thus correspondingly low.

The thermal deformations of an external wall on an experimental house in Stuttgart were measured on a long-term basis. Compression stresses and slight swelling of the wall due to the high temperature difference did not pose a problem. Fine cracks of about 1 to 2 mm in the plaster resulted from accelerated drying of the new building's brickwork dampness around the absorber; however, these cracks did not influence the load-carrying capacity of the wall. It is worth planning for defined joints at the edges of the TTI surfaces for new buildings. The use of shading systems such as blinds or shutters prevents heating of the external wall during the transition period and in the summer months, although the use of these systems is complex in terms of construction and maintenance. In order to avoid exposure to the sun during the transition or summer period, constructional shadings such as balconies or roof projections must be planned very carefully. Foregoing shading mechanisms is possible if the transmittance of the transparent insulation is strongly angle-dependent, such that with a high summer sun position having angles of incidence over 60° at the south-facing façade, less than 20% of the irradiance reaches the absorber wall. The transmittance of a transparent insulation system falls, for example, from approximately 50% with perpendicular incidence to 15% with a sun elevation angle of 60°.

Simulations for highly insulated low-energy buildings have shown that even when the south-facing façade is largely covered with a TTI heat insulation system, the excess heat in summer can be expelled by night ventilation. The number of hours with room temperatures over 26°C was below 300 hours per year under German conditions and only 50 hours higher than a conventionally insulated building.

On the other hand, at angles of incidence of 60°, transparent insulation with a glass covering can still have a very high total energy transmission factor of 35%. In this case, sun protection is unavoidable when the façade is largely covered. If nocturnal ventilation is not possible, e.g. in office buildings, shading of the TTI surfaces in the summer should likewise be provided.

The orientation of the transparently insulated façade is crucial both for energy gain in the winter and for protection against overheating in the summer. Façade orientations between south-east and south-west are suitable. Twice as much energy, some $400 \, \text{kWh} \, \text{m}^{-2}$, falls on a south-facing façade in the winter heating season than on an east or west-facing one. In addition, the low winter sun position facilitates good light transmission by the TTI material. In summer, on the other hand, only the south-facing façade offers a certain natural sun protection with low transmittance.

18.7 MATERIALS USED AND CONSTRUCTION

TTI capillary or honeycomb structures are assembled from thin-walled plastic tubes and welded by a hot wire section or manufactured into strips of any width from extruder nozzles with an almost square cell cross-section. The typical cell diameter is 3 mm.

Two polymer types are in use today: polymethyl methacrylate (PMMA) and polycarbonates (PC). PMMA is characterized by high transmittance and by good UV stability. Due to the brittleness of the material and its poor fire retardance (class B3), PMMA is bound between windowpanes, which requires a complex mullion-transom construction costing between 400 and $750 \, \text{€} \, \text{m}^{-2}$. The cost of the 10–12 cm transparent insulation is typically only around $50 \, \text{€} \, \text{m}^{-2}$. It is the glazing and attachment, at approximately $250 \, \text{€} \, \text{m}^{-2}$, plus shading items such as blinds at around $150 \, \text{€} \, \text{m}^{-2}$, which drive up costs.

Polycarbonates are mechanically more stable and can be processed without the use of a glass covering; however, they are not very UV-resistant. PC fire retardance is better (class B1) than PMMA and the material is temperature-resistant to about 125°C. PC materials can be used in heatinsulating compound systems. The covering plaster is an acryl adhesive mixed with 2.5–3 mm diameter glass balls, which is applied in the factory directly onto the capillary material. Additional UV absorbers can likewise be brought into the cover plaster. Such heat-insulating compound systems can be manufactured with substantially reduced costs of around $150 \, \text{€} \, \text{m}^2$, since there are no complex glazing and shading systems. The weight of capillary materials is around $30 \, \text{kg} \, \text{m}^{-3}$.

Capillaries made of glass are manufactured like the polymer structures, but are complicated to produce due to high processing temperatures and associated engineering problems. Glass capillaries are much more temperature- and UV-resistant but also mechanically not very stable. The recycling ability of glass, which is also possible with PMMA, is advantageous. On the other hand, polycarbonates are recyclable only with high energy expenditure and with quality losses.

For glazing systems with smaller thicknesses of 2–3 cm, aerogels are suitable. Aerogels are highly porous, open-pored solids made of silica gel that consist of more than 90% air and 10% silicate and exhibit a very low heat conductivity ($\lambda = 0.02\,\mathrm{W\,m^{-1}\,K^{-1}}$). Aerogels can easily be poured into the cavity of double-glazing, are not inflammable, and are easy to dispose of and recycle. Significant disadvantages include a light transmission only about half that of capillary materials and a sensitivity to water. Aerogel material absorbs water that penetrates into the edge network of double-glazing and the sensitive structure is broken down by capillary forces.

Transparent heat-insulating systems are mainly used in two types of construction: mullion-transom or element construction with framed TTI panel elements. To avoid dirtying of the TTI materials, the external covers usually consist of highly transparent, iron-poor single glass panes. Element constructions are characterized by a higher degree of prefabrication, which has a potential for cost-reduction. From the outside, it is often impossible to distinguish between element and a mullion-transom construction installation.

Shading mechanisms such as blinds or shutters are preferably inserted between the outside windowpane and the TTI material. Lamella type systems can also be used in front of the façade and are highly reliable when movements between open and closed positions are minimal or when maintained in a lowered position.

Heat-insulating compound system construction with frameless direct installation. The transparently plastered capillary structures are supplied with a fabric for attaching the plaster to the conventional insulation and fastened to the external wall with a black adhesive that serves as an absorber.

18.8 HEAT STORAGE BY INTERIOR BUILDING ELEMENTS

The heat storage capacity of the interior components is decisive for the degree of useful energy produced by both passive solar use via windows and transparent heat-insulating systems. Heating demand can only be reduced if solar gains do not lead to overheating of the interior. The heat storage capacity of interior components can be roughly estimated from the storage mass, the thermal capacity and the possible rise in temperature of the storage mass. Thus, for example, a solid concrete wall with a thickness d of 30 cm, a heat capacity c of $1\,\mathrm{kJ\,kg^{-1}\,K^{-1}}$ and a gross density ρ of $2100\,\mathrm{kg\,m^{-3}}$ can store, with a rise in temperature of 5°C, 0.875 kWh of heat per square metre of surface.

$$\frac{Q}{A} = \rho d c \Delta T = 2100\,\frac{\mathrm{kg}}{\mathrm{m^3}} \times 0.3\,\mathrm{m} \times 1.0\,\frac{\mathrm{kJ}}{\mathrm{kgK}} \times 5\,\mathrm{K} = 3150\,\frac{\mathrm{kJ}}{\mathrm{m^2}} = 0.875\,\frac{\mathrm{kWh}}{\mathrm{m^2}}$$

$$(18.8.1)$$

This approach assumes that the component is completely warmed or cooled to the temperature levels forming the basis of the calculation. It also assumes very high heat

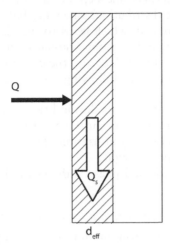

Figure 18.8.1 Amount of heat flowing into a component and the effective storage mass thickness d_{eff}.

transfer coefficients and high heat conductivities, which in practice is not the case. To what extent the storage capacity can be used depends, apart from the materials values, primarily on the duration of a rise in temperature.

Figure 18.8.1 illustrates the heat flow in a building element. If the amount of heat Q per surface A that flows in a given period into the wall is determined by dynamic calculation methods or by measurement, then an effective thickness d_{eff} wherein full storage capability is utilized can be calculated for the wall.

$$d_{eff} = \frac{Q}{Ac\rho\Delta T} \tag{18.8.2}$$

During a three-hour rise in temperature, a concrete wall (with $A = 1\,\text{m}^2$ surface) can, largely irrespective of its thickness, take up approximately 33 Wh for each Kelvin of temperature rise (with heat take-up on both sides). This corresponds to an effective thickness of approximately 5 cm. With a six-hour rise in temperature this value is approximately 9 cm.

If the heating up of a room is to be calculated, then rough estimates of the amount of heat Q flowing into the component can be calculated by using the heat transfer coefficient h_i and the temperature difference between the component surface $T_{s,1}$ (at the beginning of a time step Δt) and the room air T_i.

$$Q = h_i A\Delta t(T_i - T_{s,1}) \tag{18.8.3}$$

After the time step, the new temperature of the component $T_{s,2}$ results from the stored amount of heat $Q_s = Q$:

$$T_{s,2} = T_{s,1} + \frac{Q_s}{Ac\rho d_{eff}} \tag{18.8.4}$$

This process is repeated with each time step. For a more exact calculation of the temporally variable temperature distribution, an energy balance for a volume element of the storage mass must be created, which leads to the classical thermal heat conduction equation. For simplicity, only onedimensional temperature distributions will be derived; i.e. from the air over the surface into the component depth.

For passive solar energy use, the following boundary conditions play a role:

- the storage capacity of components during brief variations in temperature in the room caused by solar irradiance or air temperature modification,
- the potential for night cooling by utilization of the periodic modification of the air temperature between day and night,
- the temperature amplitude and phase shift on the inner side of a transparently insulated wall.

In all cases, heat is only absorbed or dissipated via the surface of the component. There are no heat sources in the component interior, so very simple energy balances for each volume element results. From an entering heat flow $\dot{Q}_{in}$, part of the energy leads to the rise in temperature in the volume element (heat storage $\dot{Q}_{st}$) and the remainder is passed on by thermal conduction into the next element $\dot{Q}_{out}$

$$\dot{Q}_{in} = \dot{Q}_{st} + \dot{Q}_{out} \qquad (18.8.5)$$

Based on Fourier's law of thermal conduction, the heat flow $\dot{Q}_{in}$ entering through the surface A is proportional to the temperature gradient at the point x_0.

$$\dot{Q}_{in} = -\lambda A \frac{dT}{dx}\bigg|_{x_0} \qquad (18.8.6)$$

The exiting heat flow $\dot{Q}_{out}$ at the point $x + dx$ is, at constant heat conductivity λ, only different from $\dot{Q}_{in}$ if the temperature gradient has changed in the volume element; e.g. has become flatter due to heat storage in the element.

$$\dot{Q}_{out} = -\lambda A \frac{dT}{dx}\bigg|_{x_0+dx} \qquad (18.8.7)$$

A Taylor series expansion of the temperature gradient at the point $x + dx$ leads to the following simplification, if all higher-order members are ignored:

$$\dot{Q}_{out} = -\lambda A \left(\frac{dT}{dx}\bigg|_{x_0} + \frac{d}{dx}\left(\frac{dT}{dx}\bigg|_{x_0}\right) dx + \cdots \right)$$
$$\approx -\lambda A \left(\frac{dT}{dx}\bigg|_{x_0} + \frac{d^2T}{dx^2}\bigg|_{x_0} dx \right) \qquad (18.8.8)$$

The amount of heat $\dot{Q}_{st}$ stored in the volume element $dV = A dx$ is given by:

$$\dot{Q}_{st} = \rho dV c \frac{dT}{dt} \qquad (18.8.9)$$

Thus, the energy balance equation leads to:

$$-\lambda A \frac{dT}{dx}\bigg|_{x_o} = \rho dV c \frac{dT}{dt} - \lambda A \frac{dT}{dx}\bigg|_{x_o} - \lambda A \frac{d^2T}{dx^2}\bigg|_{x_o} dx$$

$$\frac{\lambda}{\rho c} \frac{d^2T}{dx^2} = \frac{dT}{dt}$$

(18.8.10)

where $a = \frac{\lambda}{\rho c} [\frac{m^2}{s}]$ is termed the thermal diffusivity, which lies between $10^{-7}\,\mathrm{m^2\,s^{-1}}$ for wood and $10^{-4}\,\mathrm{m^2\,s^{-1}}$ for metals.

18.9 COMPONENT TEMPERATURES FOR SUDDEN TEMPERATURE INCREASES

The differential equation can be solved by a product approach for periodic boundary conditions or with temperature-equalizing processes, wherein one function is dependent only on time and the other only on place.

The heat flow is proportional to the so-called heat penetration coefficient $b = \sqrt{\lambda \rho c}$ and falls with $1/\sqrt{t}$ as shown in Figure 18.9.1.

The integration of the heat flow over time results in the total amount of heat penetrating into the component when there is a surface temperature jump.

$$\frac{Q}{A} = \int \frac{dQ}{A} = \int_0^{t_0} \left(-\sqrt{\frac{\lambda \rho c}{\pi}} T_c \frac{1}{\sqrt{t}} \right) dt = -\frac{2}{\sqrt{\pi}} \underbrace{\sqrt{\lambda \rho c}}_{b} \sqrt{t_0} T_c$$

(18.9.1)

Figure 18.9.2 shows that the amount of energy stored within a component is, like the heat flow, directly proportional to the heat penetration coefficient b and to the temperature jump ΔT at the surface, but it also rises with the root of time.

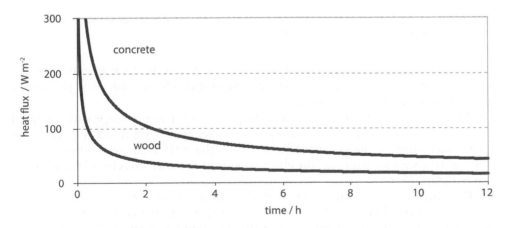

Figure 18.9.1 Heat flux as a function of time for a concrete and a wooden floor with a 10 K temperature jump on the surface.

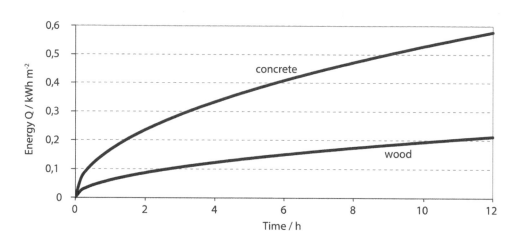

Figure 18.9.2 Amount of energy led into the component with a rise in temperature at the surface of 10 K.

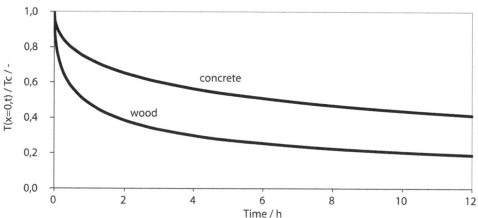

Figure 18.9.3 Relation of the surface temperature to the colder initial body temperature T_o during an air temperature jump of $+10$ K.

Usually, however, it is not the surface temperature but the air temperature which is known. Between the air and the surface temperature change there is a phase shift and a dampening of the amplitude.

With air temperature modifications, the potential for heat storage with a limited duration of the temperature jump is clearly smaller than with direct impact of the temperature jump on the surface. After 12 hours the surface has only taken up 60% of the air temperature jump; i.e. the effective storage capacity sinks by 40%, as can be seen in Figure 18.9.3.

The surface temperature changes during an air temperature jump essentially depend on the relation of the heat transfer coefficient h_i to the heat conductivity λ of the component. Component surfaces of materials with low heat conductivity clearly assume the air temperature faster (i.e. $T(x=0,t)/T_c$ becomes zero) than components

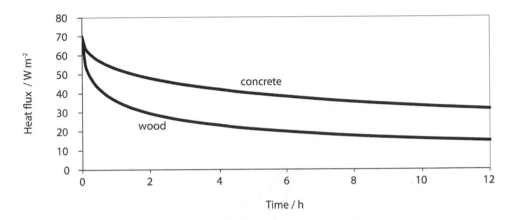

Figure 18.9.4 Heat flux into wood and concrete at an air temperature jump of 10 K.

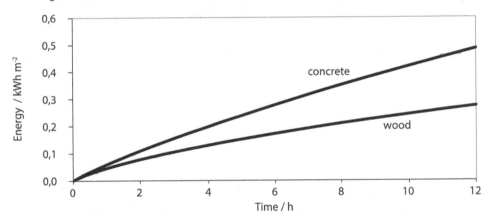

Figure 18.9.10 Amount of heat stored in the materials on an air temperature jump of 10 K.

that conduct heat well. Deep in the component, on the other hand, a temperature jump of air continues only very slowly with poorly conducting materials (see Figure 18.9.4). Since the temperature difference of the surface $T(x = 0)$ and air T_o (here zero) is larger at the concrete surface than at the wood surface, the larger heat flows and stored amounts of energy occur there.

To calculate the amount of heat Q (as shown in Figure 18.9.10) for each unit area A that has flowed into the component until the point in time t_1, the heat flux density $\dot{Q}/A$ must be integrated over the time. A solution to this problem is probably very complex. It is simpler to calculate the heat flow density in smaller time intervals and total afterwards.

Apart from temperature jumps, periodic changes in temperature caused by the external climate are of particular importance for practical applications. The outside temperature T_o and the irradiance converted to a fictitious sol-air temperature can be approximated as periodic functions of time t with a period t_0 of 24 hours and an amplitude T_{om}.

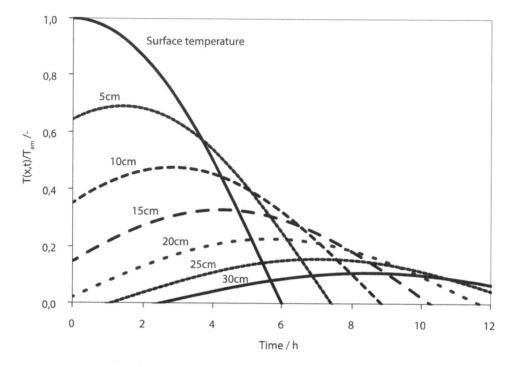

Figure 18.9.11 Variation in temperature T(x, t) normalised to the amplitude T_{sm} around the average value as a function of time and component depth x in 5 cm steps for a concrete building component.

The use of a simple energy balance model is recommended if the irradiance on a component surface is also to be considered in addition to fluctuations of air temperature. The energy balance model provides for the conversion of short-wave solar irradiance into a so-called sol-air temperature, which allows for the usage of the aforementioned analytic solutions to the thermal conduction equations.

The heat flow supplied to a component surface consists of the absorbed irradiance αG plus the heat flow transferred by the air (temperature T_o) to the surface (T_s) with the heat transfer coefficient h. This supplied heat flow is combined in a simple model, ignoring temperature-dependent modifications of the heat transfer coefficient h, into a purely temperature-dependent heat flow described by the sol-air temperature T_{So}.

$$\alpha G + h\,(T_o - T_s) = h(T_{So} - T_s)$$
$$T_{So} = T_o + \frac{\alpha G}{h} \tag{18.9.2}$$

A periodic change in temperature at a component surface occurs for example in transparently insulated components with absorption of solar radiation on the wall surface.

The periodic change in temperature (as seen in Figure 18.9.11) within the component can be calculated with exponentially dampened amplitude and period duration t_0.

The temperature field $T(x, t)$ is related for illustration purposes to the amplitude of the surface temperature fluctuation T_{sm}. Even at just 5 cm component depth, the surface amplitude of a concrete wall is reduced by 30%. The phase shift between the maximum surface temperature and the temperature at a depth of 5 cm is 1.5 h. The heat flow into a component with a periodic temperature boundary condition is again calculated using Fourier's Law, from the temperature gradient at the surface. The heat flow $\dot{Q}/A$ is proportional to the surface temperature amplitude T_{sm}.

18.10 SOLAR GAINS, SHADING STRATEGIES AND AIR CONDITIONING OF BUILDINGS

Existing office and administrative buildings in cool or moderate climates have approximately the same consumption of heat as residential buildings and most have a higher electricity consumption. In warm climates, air conditioning needs dominate consumption. Both heat and electricity consumption depend strongly on the building's use. In terms of the specific costs, electricity almost always dominates.

More than half of the running costs are accounted for by energy and technical services. A large part of the energy costs is due to ventilation and air conditioning. Heat consumption in administrative buildings can be reduced without difficulty, by improved thermal insulation, adoption of low-energy standards, which can reduce consumption to a few kWh per square metre and year in a passive building. Related to average consumption in the stock, a reduction to 5–10% is possible. Electricity consumption dominates total energy consumption where the building shell is energy-optimized. The measured electricity consumption for German office buildings was between 30 and 130 kWh m^{-2} a^{-1}.

The detailed measurements shown in Figure 18.10.1 over several years in the first passive office building in Germany (Weilheim/Teck), completed in 2000, illustrate ways of energy optimization: the passive house standard is realizable at low additional costs; hot water consumption is insignificant in office buildings; and the electrical energy consumption for building services (ventilation, lighting, pumps) can be limited to low target values (<15 kWh m^{-2} a^{-1}). The main consumer of electric energy is office equipment, which is responsible for more than 40% of the total energy consumption, with a rising trend during the three measurement years even though energy-saving equipment was used.

It is also possible to achieve the passive house standard through building rehabilitation. Detailed measurements at an office building in Tübingen, Germany (see Figure 18.10.2) show that a very low thermal heat consumption of less than 25 kWh m^{-2} a^{-1} can be achieved, although not all building elements, such as the ground floor, can be well insulated due to the low ceiling heights.

Non-residential buildings often have high internal loads due to a high amount of electrical equipment. About 50% of internal loads are caused by office equipment such as computers, printers, photocopiers, etc., which leads to an area-related load of about 10–15 W m^{-2}. Modern office lighting has a typical connected load of 10–20 W m^{-2} at a luminance level of 300 to 500 lx. The heat given off by people, around 5 W m^{-2} in an enclosed office or 7 W m^{-2} in open-plan offices, is also not negligible. Typical mid-range internal loads are around 30 W m^{-2}, resulting in a daily

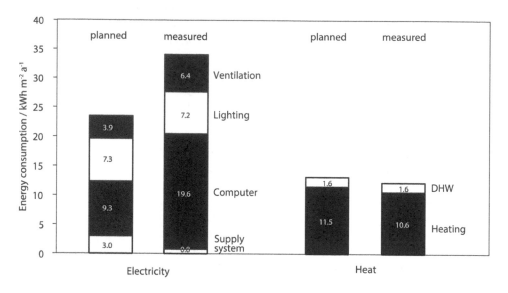

Figure 18.10.1 Measured consumption of electricity, heat and water heating in an office building with a passive house standard in Weilheim/Teck, Germany (Eicker, 2006).

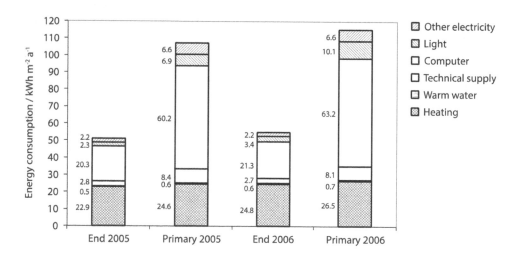

Figure 18.10.2 Measured energy consumption in the renovated offices of the engineering firm ebök in Tübingen, Germany.

cooling energy of $200\,\mathrm{Wh\,m^{-2}\,day^{-1}}$, in the high range between $40\text{–}50\,\mathrm{W\,m^{-2}}$ and $300\,\mathrm{Wh\,m^{-2}\,d^{-1}}$.

The detailed measurements in a passive standard office building in Weilheim, Germany, described above, show $30\text{–}35\,\mathrm{W\,m^{-2}}$ internal loads, in a south-facing office with 2 people and a computer workstation. In a north-facing office with 2 computer workstations, the loads were about $50\,\mathrm{W\,m^{-2}}$. The resulting daily intern loads in

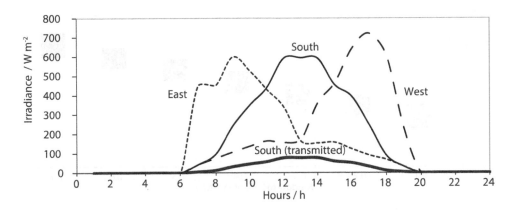

Figure 18.10.3 Diurnal variation of irradiance on different façade orientations and transmitted irradiance by a sun-protected south facade on a day in August (Stuttgart).

Table 18.10.1 Energy reduction coefficients of internal and external sun protection

Sun shading system	Colour	Energy reduction coefficient [-]
External sun shades	Bright	0.13–0.2
External sun shades	Dark	0.2–0.3
Internal sun shades	Bright	0.45–0.55
Reflection glazing	–	0.2–0.55

the south-facing office were between 200 and 300 Wh m^{-2} compared to 400 and 500 Wh m^{-2} in the north-facing office.

External loads caused by solar gains depend greatly on the surface area of the glazing as well as the sun-protection concept. Also, as demonstrated by Figure 18.10.3, the façade orientation plays a significant role. On a south-facing façade, a maximum irradiation of about 600 W m^{-2} occurs on a sunny summer day. The best external sun protection reduces this irradiation by 80%. Together with the total energy transmission factor (g-value) of low-E coated double-glazing of typically 0.65, the transmitted external loads are about 78 W per square metre of glazing surface. In the case of a 3 m^2 glazing surface of an office room, the result is a load of 234 W, which creates an external load of 20 W m^{-2} for a room surface area of 12 m^2. This situation is illustrated for south-, east- and west-facing façades in the summer.

The shading coefficients of sun-protection devices depend particularly on their arrangement: external sun protection can reduce the energy transmission of solar radiation by 80%, whereas with sun protection on the inside a reduction of at most 60% is possible. Table 18.10.1 describes this relation.

External loads depend on the relation of window surface to floor space, as well as the chosen shading system. For an area ratio between 0.1 and 0.7, the typical external loads are between 8 and 60 W m^{-2}. Together with the internal loads there are 25–90 W m^{-2} total cooling loads, as can been seen in Figure 18.10.4.

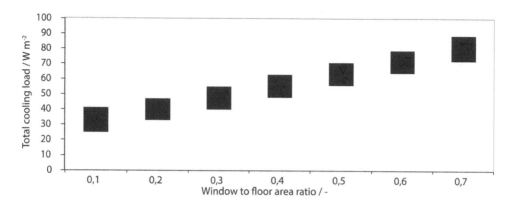

Figure 18.10.4 Cooling load as a function of the window to floor area ratio to net area.

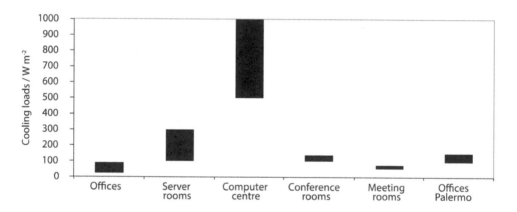

Figure 18.10.5 Typical cooling loads for buildings in Germany and offices in Palermo, Italy.

In the case of very high energy-intensive use, such as computer centres or server rooms, the cooling loads could increase by up to $1000\,\mathrm{W\,m^{-2}}$, as demonstrated by Figure 18.10.5.

The sum of external and internal loads leads to an average cooling load in administrative buildings of about $50\,\mathrm{W\,m^{-2}}$. The cooling load is dominated in many cases by the external loads (see Figure 18.10.6).

Air-conditioning in non-residential buildings is often more important than heating even in moderate climates. Air-conditioning in buildings and refrigeration is responsible for about 15% of total energy consumption worldwide, and in hot and humid climates for 30%. Cooling energy is often required in commercial buildings, the highest consumption worldwide being in the USA with up to $150\,\mathrm{kWh\,m^{-2}\,a^{-1}}$, and values quoted between $20\,\mathrm{kWh\,m^{-2}\,a^{-1}}$ for Sweden, 40–$50\,\mathrm{kWh\,m^{-2}\,a^{-1}}$ for China and $61\,\mathrm{kWh\,m^{-2}\,a^{-1}}$ for Canada.

Our own overview of the cooling energy requirement of different building projects shows a typical cooling energy consumption for administrative buildings of between 20 and $60\,\mathrm{kWh\,m^{-2}\,a^{-1}}$ in Europe, as shown in Figure 18.10.7.

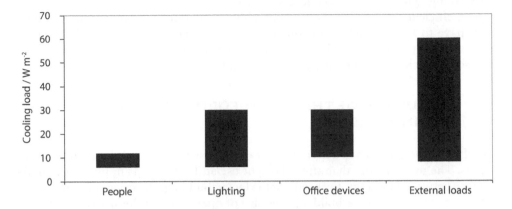

Figure 18.10.6 Cooling load distribution in offices buildings.

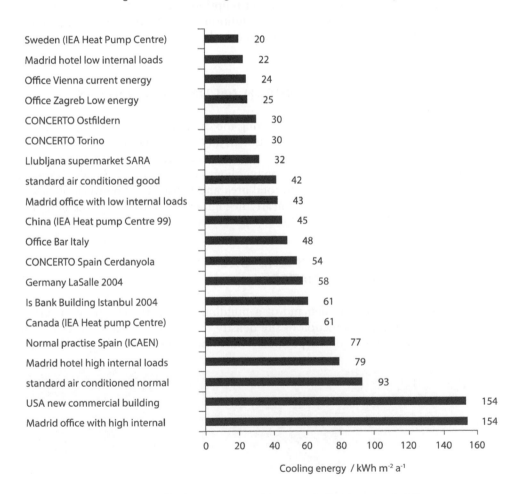

Figure 18.10.7 Cooling energy requirements of administration buildings.

Cooling loads should be reduced as much as possible by a good external shading system, efficient electrical appliances and night ventilation if possible. The remaining loads have to be removed by air-conditioning equipment. Solar photovoltaic energy can power electrical compression chillers and solar thermal collectors can be combined with sorption chillers to provide cooling of buildings.

18.11 INFLUENCE OF THE URBAN FORM ON SOLAR ENERGY USE IN BUILDINGS

The effects caused by urban design on solar energy use vary according to climatic conditions. The main factors that affect daylight use and solar gains in buildings are the distance between buildings, the height of the facing building, the orientation of and the reflectance from the facing buildings, the size of openings and the size of the shading device. The daylighting performance is especially significant for office buildings which are characterized by high lighting energy consumption and where the productivity of the employees is highly affected by lighting conditions. In residential building, electricity consumption for lighting is much less influenced by daylight performance due to a higher evening use profile.

18.12 RESIDENTIAL BUILDINGS IN AN URBAN CONTEXT

The urban shading effect was simulated using the dynamic building simulation software EnergyPlus (see Figure 18.12.1) along with all obstructing buildings around the simulated one. The model includes the reflectance from the obstruction surfaces and an albedo factor is defined for obstruction buildings and the ground. Site densities are defined as the ratio of built-up area to total area and these varied between 30% and 60%, as below 30% obstructions are nearly negligible and above 60% the buildings become unrealistically close.

18.13 SITE DENSITY EFFECT AND URBAN SHADING IN MODERATE CLIMATES

First the heating and cooling demand of a building type in the centre of an urban structures is analysed as a function of site density (building 9 in Figure 18.13.1). In a second step the influence of the position of each building within the urban setting

Figure 18.12.1 EnergyPlus models of multi family buildings for shading simulations with site densities varying between 60 and 30%.

is shown. Table 18.13.1 describes the geometric properties of the different building types.

For single-family houses with a building standard corresponding to the construction years between 1995 and 2001, the heating demand increases by 17% (to $84\,\text{kWh}\,\text{m}^{-2}\,\text{a}^{-1}$) for a 60% site density compared to the unshaded situation (with $72\,\text{kWh}\,\text{m}^{-2}\,\text{a}^{-1}$). Using an increased albedo of 0.7 for the surroundings instead of 0.2, the heating demand slightly reduces to $79\,\text{kWh}\,\text{m}^{-2}\,\text{a}^{-1}$ for the 60% site coverage, as shown in Figure 18.13.2.

Similar results are obtained for the multi-family houses as illustrated in Figure 18.13.3. Without any shading effect the heating consumption of a multi-family house is $43.7\,\text{kWh}\,\text{m}^{-2}\,\text{a}^{-1}$ and increases by 20% for 60% site coverage. The rather low cooling demand of $11.4\,\text{kWh}\,\text{m}^{-2}\,\text{a}^{-1}$ reduces with site density to $6.4\,\text{kWh}\,\text{m}^{-2}\,\text{a}^{-1}$, i.e. by 44%.

The heating consumption of the high-rise blocks with today's high insulation standards is $53\,\text{kWh}\,\text{m}^{-2}\,\text{a}^{-1}$ and cooling demand is $10\,\text{kWh}\,\text{m}^{-2}\,\text{a}^{-1}$. The heating demand is 25.5% higher for 60% site coverage. For high-rise blocks constructed between 1981 and 1985, the heating demand of $83\,\text{kWh}\,\text{m}^{-2}\,\text{a}^{-1}$ rises by 16.8% for 60% site coverage.

The heating consumption of terraced houses is highly influenced by the surface area proportion exposed to ambient conditions. The heating consumption of corner houses

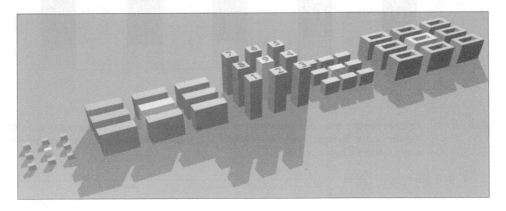

Figure 18.13.1 EnergyPlus models of building types and the urban quarter constituted of 9 generic building blocks from each building type. The distance between the buildings varies according to site densities. The building position numbers (1–9) are analogous for each cluster.

Table 18.13.1 Geometric properties and U values of building types

	Single family house	Multifamily house	Old highrise block	Row houses	Old apartment (courtyard)
Geometry	$10.5^x\ 10.5^x\ 3.5\,\text{m}$	$20^x\ 14^x\ 10.8\,\text{m}$	$24.4^x\ 24.4^x\ 30\,\text{m}$	$7^x\ 10^x\ 7\,\text{m}$	$10^x\ 10^x\ 13\,\text{m}$
U values/$\text{W}\,\text{m}^{-2}\,\text{K}^{-1}$					
Wall	0.5	0.3	0.88	0.5	1.45
Roof	0.3	0.3	0.97	0.3	1.3
Floor	0.22	0.22	0.85	0.22	2.6
Window	1.6	1.6	2.57	1.6	2.9

is 17% higher than the other buildings in the middle of the terrace, as shown in Figure 18.13.4. The heating demand increases with site density by 11% for the corner houses and 14% for the middle houses. The reflection effect with density barely influences the heating demand of the terraced houses.

When considering the position of one building type with a given building compactness (such as multi-family, single-family, courtyard, etc.) within the urban structure, the influence of shading becomes evident.

The building most affected from mutual shading is the one located in the middle of the urban structure that was considered above (building 9). Compared to the unshaded

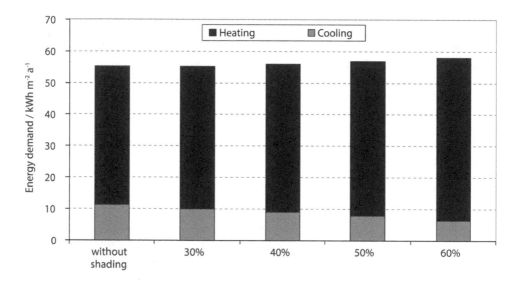

Figure 18.13.2 Average heating and cooling demand of multi-family houses with increasing urban site density under Stuttgart weather conditions. Similar tendencies are observed for the single family or high rise blocks.

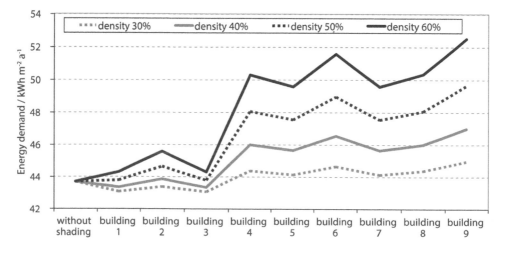

Figure 18.13.3 Heating demand of the multifamily house with different site coverage for the moderate climatic conditions of Stuttgart/Germany. Building number refers to the position in the cluster, as described in Figure 18.13.1.

situation, its heating energy demand increases by 3% for 30% site density, up to 20% for 60% site density.

For lower-height single-family houses, the shading effect is less. Building 9 in the centre of the site has a 13% higher demand for 60% site coverage than the unshaded buildings.

18.14 CLIMATE EFFECT

To understand the impact of weather together with site density effects, apartment blocks are simulated for Stuttgart/Germany, Ankara/Turkey and Hong Kong/China weather data.

In all climates the cooling demand decreases with increasing site density and the heating demand increases. In heating-dominated climates such as Germany, the overall energy demand for heating and cooling thus slightly increases by 6% for the highest site density. In Ankara/Turkey, with comparable heating demand but double the cooling demand than in Stuttgart/Germany, there is an optimum for the total energy demand at 30% site density. In Hong Kong, with its cooling-dominated climate, the energy demand drops by 20% compared to the unshaded situation (see Figure 18.14.1).

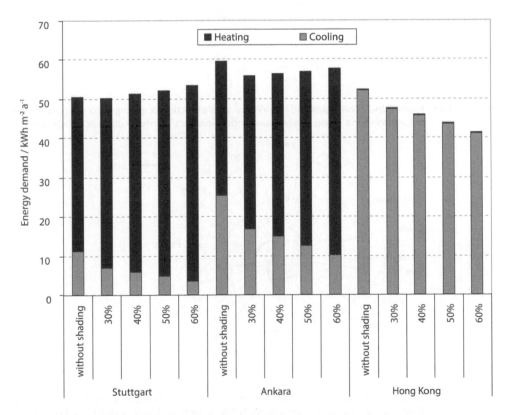

Figure 18.14.1 Comparison of heating and cooling demand of apartment blocks with U values according to the new German legislation (EnEV 2009) for different settlement densities in Stuttgart, Ankara and Hong Kong climate.

18.15 SOLAR GAINS AND GLAZING

Direct solar gains are significantly influenced by the glazing ratio. The simplest and most effective way for more solar gains is to have bigger glazing areas, which on the other hand also cause heat loss depending on the glazing type. For all building types the heating and cooling demand was calculated with various glazing ratios for different orientations. At first, the glazing ratio was only changed on one façade, with fixed other glazing ratios. As a result, the heating demand increases strongest for the northern façade, but hardly so for a southern façade. As can be seen in Figure 18.15.1, a single-family house with no windows and a north façade has 26% less heating demand

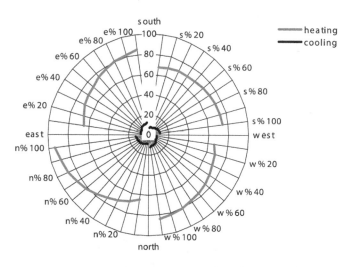

Figure 18.15.1 Heating and cooling demand (in kWh/m² a) as a function of glazing ratio of a given façade orientation for a single family house under Stuttgart/Germany climatic conditions.

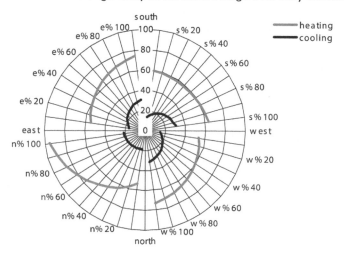

Figure 18.15.2 Heating and cooling demand (in kWh m⁻² a) as a function of glazing ratio of a given façade orientation for a multi family apartment block under Stuttgart climatic conditions.

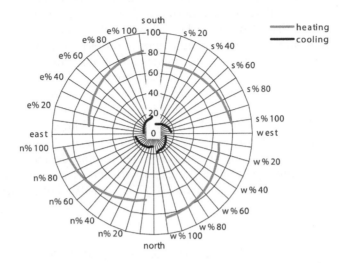

Figure 18.15.3 Dynamic simulation results of glazing ratio effect for highrise blocks under Stuttgart climatic conditions.

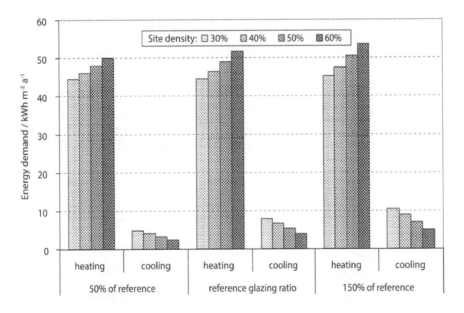

Figure 18.15.4 Influence of solar gains and glazing ratio on heating and cooling demand for multifamily houses with different site densities.

than the single-family house with a totally glazed north façade (65 kWh m^{-2} a^{-1} with no north façade glazing compared to 93 kWh m^{-2} a^{-1} for the fully glazed northern façade), whereas on the southern façade the demand reduces by only 7% with reducing window area.

In multi-family houses (see Figure 18.15.2) the heating demand decreases from 67 to 38 kWh m^{-2} a^{-1} with decreasing northern glazing fraction; i.e. by 42%, and by 24% for high-rise buildings (from 62 to 47 kWh m^{-2} a^{-1}), as shown in Figure 18.15.3.

In order to understand the importance of glazing ratios in relation to urban density, three types of glazing ratios were evaluated in four different site densities and the results are represented in Figure 18.15.4. As a reference scenario, typical glazing ratios from the German building typology were used for the different building types and their construction year (30% for the single-family building, 30% for the multi-family building, 32% for the high-rise apartment blocks, 16% for terraced housing and 17% for the old apartment house). When the glazing ratio is increased on all façades, the heating demand increases with increasing site density. The cooling demand nearly doubles with high glazing fractions and again decreases with increasing site density. Higher site densities increase the heating demand by 12% to 18%, but reduce cooling by up to 50%.

18.16   OFFICE BUILDINGS IN AN URBAN CONTEXT

The heating, cooling and lighting energy demands of office buildings were analysed as a function of urban density (see Figure 18.16.1). As the characteristic parameter, the height of the building H related to the distance between buildings W was used. The office building type investigated is a cellular plan office with a building envelope designed to a low-energy standard following the German Energy Saving Ordinance 2009.

Figure 18.16.2 shows total annual energy demand of a south-oriented sample office space under four different aspect ratio (H/W) scenarios in southern German climatic conditions. Figure 18.16.2 shows that the total energy demand increases by 20% when the H/W ratio changes from 0.5 to 1 due to an increase in heating energy demand. The total energy demand then slightly decreases by 6% with aspect ratios above 1 as the cooling energy demand decreases.

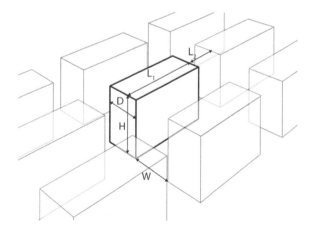

Frontal length (L$_1$)
Depth (D)
Height (H)
Distance between units (L$_2$)
Street width (W)

Spacing distance (L$_1$/L$_2$)
Aspect ratio (H/W)
Building depth to frontal length (D/L$_2$)

Figure 18.16.1   The form structure labels H, D and L$_1$ refer to the height, depth and frontal length of each unit, L$_2$ refers to the spacing between the units and W to the width of the street.

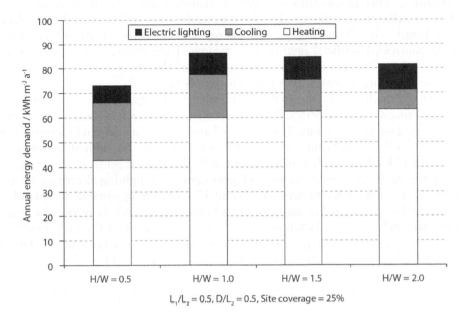

Figure 18.16.2 Annual energy demand of an office building room a function of aspect ratios (H/W) under German climate conditions.

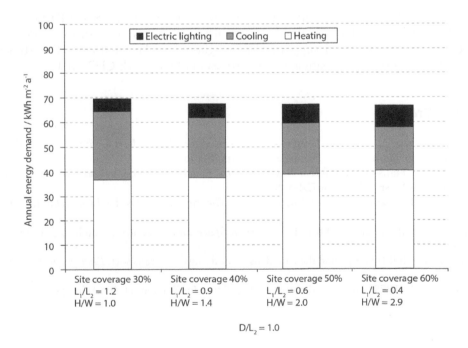

Figure 18.16.3 Heating, cooling and lighting demand of high-rise office blocks for different densities in Stuttgart climate.

High-rise office blocks with different site densities were also analysed for southern German climatic conditions. The 10-storey reference building has 24.4 m depth, 24.4 m length and 30 m height. The energy consumption of the office room's daylight-controlled artificial lighting was evaluated for different site coverage (see Figure 18.16.3). The required illumination level of the office room is 500 lux and the artificial lighting system was designed to supply this level.

When the shading effect due to the surrounding buildings is taken into account, the electric lighting demand increases. The shading effect results in less cooling requirement due to a decrease in solar gains. The simulated annual heating demand increases from $36\,kWh\,m^{-2}\,a^{-1}$ to $40\,kWhm^{-2}\,a^{-1}$. At 60% site coverage, the annual cooling loads decrease to $17\,kWh\,m^{-2}\,a^{-1}$, which is about 36% less than at 30% site coverage.

For non-residential buildings with a higher demand in lighting electricity, the simulation results showed that between 4.5% and 35% of lighting electricity demand can be saved by the site design. For high-rise office buildings the daylight responsive electric lighting demand rises from $5\,kWh/m^2$ for 30% site coverage to $8\,kWh/m^2$ for 60% site coverage. If no daylight responsive strategy is used, the electric lighting demand would be $36\,kWh/m^2$, which is 5 times as much in the worst case. For cooling-dominated climate conditions, the site design should focus on reducing the cooling load, but the daylighting illuminance level and the electric lighting loads should also be considered.

ACKNOWLEDGEMENTS

The graphical design, text editing and corrections were done by Silvio Barta who is thanked for his valuable contributions. The urban analysis simulations were done by PhD student Aysegül Tereci, supported by daylighting simulations done by PhD student Dilay Kesten, both participating in the European PhD network CITYNET.

REFERENCES

EN 12464-1:2002, *Light and lighting—lighting of work places—Part 1: indoor work places.* Brussels, Belgium: Comite' Europe'en de Normalisation.

Buschendorf, Hans-Georg (Hrsg.): *Lexikon Licht- und Beleuchtungstechnik.* Verlag Technik, Berlin 1989, ISBN 3-341-00724-5, S. 64.

Hentschel, H.-J. (2002) *Licht und Beleuchtung: Theorie und Praxis der Lichttechnik*, Hüthik Verlag.

Martin, L. and March, L. (Eds) (1972) *Urban Space and Structures*, Cambridge University Press, UK.

Steemers, K. (2003) Energy and the city: density, buildings and transport. *Energy and Buildings*, 35 3–14.

Steadman, J.P (1979) Energy and patterns of land use. In: Watson, D. (Ed.) Energy.

Chapter 19

The contribution of bioclimatic architecture in the improvement of outdoor urban spaces

Konstantina Vasilakopoulou[1,2], *Dionysia Kolokotsa*[2] &
Mattheos Santamouris[1]

[1]*Group Building Environmental Studies, Physics Department, University of Athens, Athens, Greece*
[2]*Environmental Engineering Department, Technical University of Crete, GR 73100, Crete, Greece*

19.1 INTRODUCTION

Cities are increasingly expanding their boundaries and populations and, as stated, "from the climatological point of view, human history is defined as the history of urbanization" (Nations HABITAT, 2011; *The State of the World Cities*, 2001). Increased industrialization and urbanization in recent years have dramatically affected the number of urban buildings, with major effects on the environmental quality of the urban environment.

Urbanization leads to a very high increase in energy use. A recent analysis showed that a 1% increase in per capita gross national product (GNP) leads to an almost equal (1.03%) increase in energy consumption. However, as reported, an increase in the urban population of 1% increases energy consumption by 2.2%; i.e. the rate of change in energy use is twice the rate of change in urbanization (Santamouris, 2012; *The State of the World Cities*, 2001).

Overconsumption of resources, mainly energy (which is associated with increased air pollution from motor vehicles, increases the ambient temperature because of the positive heat balance in cities, heat island (see below), noise pollution and solid waste management) is one of the more important environmental problems in urban areas of developed countries. In parallel, poverty, increasing unemployment, environmental degradation, lack of urban services, overburdening of existing infrastructure and lack of access to land, finance and adequate shelter are among the most important environmental, social and economic problems in cities of less developed countries.

Achieving sustainability in the urban environment is a compromise of different parameters and involves appropriate actions for urban management, policy integration towards a holistic planning approach, ecosystem thinking and strong cooperation and partnership between the different actors.

An increase in energy efficiency and in the use of renewable resources to supply cities, improvement of the urban thermal microclimate and adoption of sustainable consumption policies seem to be the main tools to reduce energy consumption in cities of the developed world.

Air temperatures in densely built urban areas are higher than the temperatures of the surrounding rural countryside. The phenomenon known as "heat island" arises from many factors, the more important of which are summarized by Oke et al. (1991). Urban heat island studies refer usually to the "urban heat island intensity", which is the maximum temperature difference between the city and the surrounding area. Data compiled by various sources show that heat island intensity can be as high as 15°C. Extensive studies on the heat island intensity in Athens, involving more than 30 urban stations, show that urban stations present higher temperatures, of between 5°C and 15°C, compared to reference suburban stations.

Heat island, the most documented phenomenon of climate change, has a very important impact on the energy consumption of buildings (Akbari et al., 1992). Increased urban temperatures exacerbate the cooling load of buildings, increase peak electricity demand for cooling and decrease the efficiency of air conditioners (Santamouris et al., 2001). In parallel, high urban temperatures considerably decrease the cooling potential of natural and night ventilation techniques and increase pollution levels. Recent research has provided data on the amplitude and the characteristics of heat island phenomenon in many European and US cities.

The impact of heat island on the cooling energy consumption of buildings is quite important. Studies have shown that cooling energy consumption may be doubled due to increased ambient temperatures in the affected areas (Santamouris et al., 2001; Hassid et al., 2000). At the same time, the environmental quality in the overheated zones is worsening as pollution is increasing (Stathopoulou et al., 2008), and the ecological footprint of the city is growing seriously (Santamouris et al., 2007a). As reported, for US cities with populations larger than 100,000, the peak electricity load will increase 1.5% to 2% for every 1°F increase in temperature. Taking into account that urban temperatures during summer afternoons in the US have increased by 2–4°F over the last 40 years, it can be assumed that 3–8% of the current urban electricity demand is used to compensate for the heat island effect alone.

The heat island phenomenon may occur during the day or the night. Its intensity is mainly determined by the thermal balance of the urban region and can result in up to 10°C of temperature difference.

To counterbalance the impact of the heat island phenomenon, efficient mitigation techniques have been proposed and applied (Santamouris et al., 2007b). This involves: the use of advanced cooling materials for the urban environment that are able to reflect solar radiation and amortize and dissipate heat (Doulos et al., 2004; Zinzi, 2010); strategic landscaping of cities, including appropriate selection and placing of green areas, use of vegetative roofs (Niachou et al., 2001; Santamouris et al., 2007c; Sfakianaki et al., 2009); solar control systems; and the use of heat sinks such as the ground, water and ambient air, to dissipate excess heat.

Urban regeneration may be a very powerful tool to meet the objectives of sustainable development through the rehabilitation of existing cities and building stock, the recycling of previously developed land and the retention of greenfield sites. In particular, refurbishment of existing urban environments should be seen as an excellent opportunity to implement sustainability notions and as a start to adopting these principles as a guide within which other considerations may trade off.

Although mitigation techniques have been extensively tested in various small-scale applications, existing data on their potential to mitigate heat islands when used in pavements and other urban structures are very limited. Some applications of cooling materials combined with other mitigation techniques to improve the environmental quality of open spaces are already reported for Tirana (Albania) and Athens (Greece) (Fintikakis et al., 2011; Gaitani et al., 2011).

This chapter aims to present the existing knowledge on heat island mitigation techniques and to provide information on existing applications. The main mitigation techniques are presented in a simple and comprehensive way, while some of the well-known and evaluated applications of different mitigation technologies are described.

19.2 MITIGATION STRATEGIES

The gathering of a great proportion of the world's population in cities, the need for cheap and readily available housing, the extended infrastructure and the alienation from traditional ways of living and working have caused the expansion of the manmade landscape. The consequences of urbanization in the natural landscape are important: the coverage of soil and permeable surfaces with non-permeable materials that absorb heat, and the decrease of forests and planted areas dramatically increase surface and air temperatures. These two parameters cause thermal discomfort to the inhabitants of cities, simultaneously leading to increased electricity consumption of urban buildings.

Heat island is one of the most documented climatic phenomena of the contemporary urban environment, mainly caused by urbanization, while global warming is another aggravating factor and an inevitable reality. Heat island effect is present in urban and suburban areas, causing city centres to have considerably higher temperatures than the suburbs. The phenomenon is present both during day and night, during all seasons of the year and concerns cities in every part of the world. The intensity of the phenomenon, meaning the temperature difference between the centre and the periphery of cities, depends on various factors, including local weather, geomorphology, urban geometry, anthropogenic heat, quality of the materials used for groundcover and buildings, the quantity and type of green spaces, etc.

Planners and scientists can do little to harness weather, geomorphology and even urban geometry. However, there are advances in new technologies that can contribute to the improvement of the thermal properties of the built environment and to reduce the intensity and the impacts of heat islands. The aim of these mitigation techniques is to reduce the energy consumption of urban buildings and the concentration of pollutants, so as to reduce temperature levels in the city centres and to establish comfort conditions for city dwellers, users, etc. These goals are usually achieved by ameliorating the thermal and optical properties of urban materials, in particular solar reflectance, thermal emissivity and heat capacity, by blocking the incident solar radiation and by increasing the cooling load of open spaces. These parameters largely determine the urban air and surface temperatures.

The main mitigation techniques for reducing urban heat islands include:

an increase in planted areas;
use of materials of high albedo values, and cooling materials;
shading of the surfaces covered with high heat capacity – low thermal emissivity materials;
use of thermal sinks.

19.2.1 Planted areas

Planted urban surfaces, except from being aesthetically appealing, help to decrease urban temperatures and reduce the energy consumption of buildings (Georgi et al., 2010). Green spaces that help cool the environment can be applied in two forms: (i) planted public areas, such as parks, gardens, and trees on the sides of roads or on squares; and (ii) planted roofs (green roofs).

19.2.1.1 Green public spaces

Trees and vegetation contribute to the reduction of air and surface temperatures through evapotranspiration and shading of adjacent urban surfaces. Evapotranspiration is "the combined loss of water to the atmosphere by evaporation and transpiration" (Santamouris et al., 2001, Shashua-Bar et al., 2011). Evaporation is the process through which a liquid is converted into a gas and transpiration is the process through which the water inside the body of a plant is released as water vapour in the atmosphere. External temperatures can be reduced by up to 6°C by using groundcover or lawn instead of paving, thus improving both comfort and energy efficiency of buildings, so that heating and cooling requirements are reduced (Sustainability Victoria, Landscape Design).

Shade provided by trees contributes to temperature reduction, as it blocks the solar radiation reaching paved surfaces and building envelopes. Shaded surfaces may be 5°–20°C cooler than unshaded ones (EPA1, 2007). Cooler surfaces also lessen heat island effect by reducing heat transfer to the surrounding air.

Cooler walls decrease the quantity of heat transmitted to buildings, thus lowering cooling costs. The savings associated with shading from trees can be up to US$200/tree, depending on the climate of the specific area (Akbari, 2002).

Joint studies by the Lawrence Berkeley National Laboratory (LBNL) and the Sacramento Municipal Utility District (SMUD) placed varying numbers of trees around houses to shade windows and then measured the buildings' energy use. The cooling energy savings ranged between 7% and 47% and were greatest when trees were planted to the west and southwest of buildings (EPA2).

Apart from cooling buildings during the warm months, deciduous trees allow solar radiation to reach building façades through their branches during winter. At the same time, when properly situated, trees can create wind barriers, protecting buildings and reducing heat losses during the cold period.

The impact of trees largely depends on their number and positioning. A small number of trees has a limited effect, compared to urban parks and gardens, which extend their positive effects to the surrounding built environment. A study of two urban parks in Singapore showed that the average temperatures obtained in both parks were lower and the range of the standard deviation obtained from the parks was smaller compared to those of the built environment. Also, a maximum 1.3 K average temperature difference was observed around parks. It was also shown that energy consumption for cooling can be reduced by almost 10% for buildings situated close to urban parks (Chen and Wong, 2006).

However, planting trees comes with problems that need to be taken into account, so as to select appropriate species and position trees so that they can have maximum effect. Such problems include interfering with aboveground and underground utilities, requiring maintenance and water, blocking part of the solar access during winter, trapping pollutants in the urban canopy layer and taking a significant period to yield the desired results (McPherson, 1994).

19.2.1.2 Green roofs and façades

Usually green spaces are found in open public areas. However, the technique of green roofs, public or private, has gained ground during the last few decades, as they reduce the energy consumption of the building while improving the microclimate of the wider urban space in which the building is situated.

The most important advantage, or at least the most popular reason for installing green roofs, is their contribution to a building's insulation, which usually results in energy savings both for heating and for cooling. The properties that increase the energy efficiency are: the shading that the plants provide to the roof; the temperature reduction from evapotranspiration; and insulation from the plants and the growing medium. Niachou et al. (2001) found that the contribution of green roofs significantly increases the insulation properties of a structure. The effect of green roofing on the energy savings of buildings, both for cooling and for heating, is greater in the case of a non-insulated building, less significant in moderately insulated buildings and smaller in well-insulated ones.

Apart from their impact on building energy consumption, planted roofs contribute to the mitigation of heat islands. Their first advantage is that they increase the total of a city's water-permeable surface area, helping water to be retained in the soil and allowing larger quantities to be available for evapotranspiration. At the same time, planted roofs present much higher albedo values than dark urban roof surfaces, thus reflecting off a greater proportion of the incident solar radiation and not transforming it into heat (Getter and Rowe, 2006). According to the United States Environmental Protection Agency (USEPA), green roofs can be cooler than the ambient air, whereas conventional roof surfaces can exceed ambient air temperatures by up to 50°C (EPA3).

Wong et al. (2007) studied the effect of an intensive rooftop garden system. It was found that surface temperature may be reduced up to 3.1 K and the ambient temperature at 1 m may reduce by up to 1.5 K.

Other advantages of green spaces, including planted roofs, are improved air quality and lower greenhouse gas emissions, resulting from the reduced energy consumption and from the removal of air pollutants, enhanced stormwater management and water quality, and the improved aesthetic experience of the users of the area.

19.2.2 Cool materials

The physical properties of the materials used for urban structures largely determine the thermal conditions in the city and the energy consumption of buildings. The albedo of a material or surface is its solar reflectivity, meaning the ability of the material to reflect off the incident solar radiation. The ability of the material to emit long-wave radiation is called emissivity (infrared emittance); high emissivity materials release energy that has been absorbed as short-wave radiation. Albedo and emissivity are the key factors which determine the temperature of an urban surface, thus having huge impact on ambient air temperature.

Materials with high albedo values are usually light-coloured, natural or artificial materials, widely used in vernacular architecture. However, in recent decades, "cool" artificial materials have been developed. The main types of cool materials are membranes, paints, anticorrosive and waterproofing coatings, cement products, pre-painted steel and mortar. The characteristic of these products is that their albedo is much higher than a conventional material of the same colour.

Cool materials are usually associated with "cool roofs". Products for cool roofing are made of highly reflective and emissive materials that can remain approximately 28–33°C cooler than conventional materials during peak summer weather (EPA4). The large-scale use of cool materials in urban areas might lead to indirect energy savings, due to the reduced air temperature brought about by increased solar reflectance. The indirect benefits that arise from this ambient cooling of a city or neighbourhood will in turn decrease the need for air-conditioning. Computer simulation results, calibrated with actual measurements, have shown that an increase in roof albedo from 0.2 to 0.78 reduced the cooling energy consumption of a house in Sacramento (USA) by 78% (Bretz et al., 1989). Multi-year observations show a temperature reduction of 0.3 K/decade because of the massive construction of high-albedo greenhouses in the Almeria area of Spain (Campra et al., 2008).

Cool roofs typically yield measured summertime daily air-conditioning savings and peak demand reductions of 10–30% (Kolokotsa, 2008).

Although highly reflective materials have been extensively tested in cool roof applications, existing data on their potential to mitigate heat islands when used in pavements and other urban structures are very limited. Conventional paving materials, such as asphalt and concrete tiles, have solar reflectances of 5–40%. That means that they absorb a great amount of the incident energy (EPA5). Ground-covering materials of high solar reflectance, emittance and permeability, when applied on a part of a pavement that is not shaded, can help reduce surface and ambient temperatures. In Flisvos Park, in the municipality of Paleo Faliro, Greece, 4500 m^2 of cool paving materials were applied. It was found that, under specific climatic conditions, the used materials may reduce the peak daily ambient temperature of a typical summer day by up to 1.9 K, while surface temperatures were reduced by up to 12°C (Santamouris et al., 2012).

Cool roofs and cool pavement technologies were considered in combination in research by Millstein and Menon (2011). It was assumed that roofs and pavements represent 25% and 35% of the urban area respectively, while the considered albedo increase for roofs and pavements was +0.25 and +0.15 respectively. It was calculated that cool roofs and cool pavements contribute to decreasing afternoon summertime temperatures by 0.11–0.53 K. For some of the urban locations studied, no statistically significant temperature reductions were found.

Rosenfeld et al. (1995) assumed an increase in the average albedo of 0.13, and in particular from 0.13 to 0.26 for an area of 100,000 km^2 in Los Angeles. It was calculated that the peak impact of the albedo change occurs in the early afternoon, the potential cooling exceeding 3 K at 3 p.m. Simulations carried out under different boundary conditions indicated that the expected peak summertime temperature reductions are between 2 and 4 K.

19.2.3 Shadings

Shading of urban surfaces can be provided by shading devices and structures, such as tents, pergolas, canopies, etc., or by trees. Shaded urban surfaces receive smaller amounts of solar radiation, thus having lower temperatures. This results in lower ambient temperatures in open spaces and in reduced rate of heat convection to the building interior, when shading is applied on a building façade. Papadakis et al. (2001) reported that the radiation incident on building façades reduces by 70–85% when trees are planted close to the building. At the same time, temperatures in the area around the trees were significantly lower than those in non-shaded areas.

19.2.4 Thermal sinks

Ground is characterized by its ability to store incident solar radiation as heat, conserving a constant temperature throughout the year. In this way, the underground temperature is lower than the aboveground air temperature during summer and higher than the air temperature during winter. The fact that the ground has a much lower temperature than that of the environment aboveground is enabling it to be used as a natural heat sink, during summer, by dissipating heat from aboveground constructions to the ground. Taking advantage of the ground temperature is achieved through ground-to-air heat exchangers, meaning tubes that are buried in the earth. The air enters the tubes, circulates underground and exits the tube having a lower temperature. The designation of this system varies, being referred to as both "earth tubes" or "ground-coupled air heat exchangers" (CRES, 2011).

Fintikakis et al. (2011), through simulations and field measurements, evaluated the microclimatic modifications that were applied in the historic centre of Tirana (Albania). The mitigation techniques that were applied were cool materials, green spaces, water, solar control and earth-to-air heat exchangers. The researchers found that the maximum temperature drop due to all the applied techniques was 3°C, while the maximum contribution of the earth-to-air heat exchangers was found to be close to 0.7°C.

19.2.5 Combination and interplay of mitigation strategies

Each of the above-mentioned mitigation techniques have positive results in reducing the heat island effect when applied in the urban environment. However, the combination of some or all of the strategies, depending on the specific site and weather patterns, will have multiple benefits.

One of the few studies in which different mitigation techniques were tested independently and in combination was carried out by Rosenzweig et al. (2006). The main aim of the study, which concerns the city of New York, was to test various mitigation techniques, like urban forestry, green roofs and light surfaces, and to understand how they could affect New York City's surface and near-surface air temperatures. The researchers studied the implications of the interventions both on the city scale and on six smaller case study areas, including on appropriate open spaces where the techniques could be applied.

Results of this study show that the mitigation strategies tested can reduce surface and near-surface air temperatures, but there is substantial variability in the magnitude of their effect across scenarios, case study areas and heat wave days. The study showed that vegetation cools surfaces more effectively than increases in albedo, and that the most effective mitigation strategy per unit area redeveloped is kerbside planting.

However, it was found that for New York City the greatest absolute temperature reductions are possible with the application of light surfaces; 64% of New York City's surface area could be redeveloped from dark, impervious surfaces to lighter high-albedo surfaces, along with the combined application of light roofs and urban forestry (Rosenzweig et al., 2006).

19.3 EXPERIMENTAL ANALYSIS OF OUTDOOR SPACES

As mentioned in the previous sections, the environmental quality of urban open space has become a fundamental subject for town planners, architects and climatologists. Recent research trends show the creation of a link between the complicated results of urban climatology on the one hand and town planners' more design-oriented outcomes on the other.

By carefully studying the state of the art in the specific sector, the assessment of the microclimate in outdoor spaces, as well as the implementation of various measures for the improvement of outdoor comfort, follows the steps indicated in Figure 19.3.1. Following the procedure depicted in Figure 19.3.1, the various case studies are analysed.

Therefore, the various experimental tests found in the bibliography can be classified into three categories, namely:

1 Assessment of outdoor comfort conditions in various regions and climates;
2 Proposal for improvement of outdoor comfort conditions using bioclimatic approaches and pre-implementation assessment;
3 Implementation of a series of outdoor comfort improvements and post-implementation assessment.

Sub-categorization may include climatic conditions, technologies applied, etc.

19.3.1 Assessment of outdoor comfort conditions

The assessment of outdoor comfort conditions through a series of measurements and questionnaires has been carried out by various researchers. The methodology followed

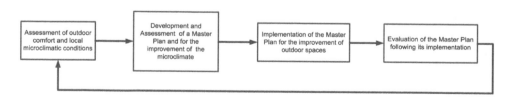

Figure 19.3.1 The outdoor spaces improvement procedure.

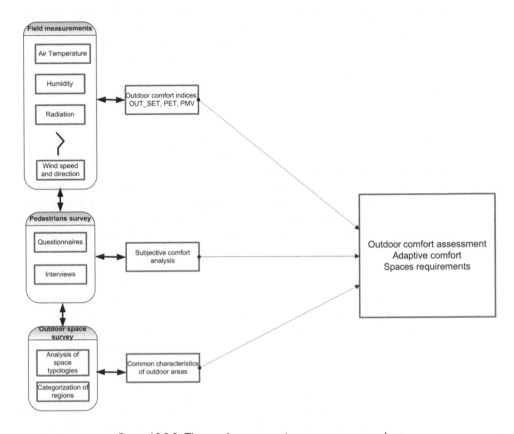

Figure 19.3.2 The outdoor spaces improvement procedure.

is depicted in Figure 19.3.2. In general the overall procedure is split into three major parallel activities: (i) field measurements; (ii) pedestrians and users' subjective comfort analysis; and (iii) outdoor space typology and characteristics. Specific indicators are then extracted and the outdoor comfort is assessed.

Indicatively, the following studies can be found:

- An urban park in Cairo (Egypt) close to the city centre was monitored by Mahmoud (2011). The author performed a series of field measurements coupled with the distribution of survey questionnaires to park visitors in order to assess their thermal comfort in a subjective and objective way. The study divided the park into six zones according to their activities and characteristics; i.e. peak zone, spine zone, entrance plaza zone, fountain zone, lake zone, canopy zone, pavement, seating and cascade zones. A total of 300 questionnaires from various zones of the park were collected measuring the subjective thermal sensation votes (TSV) as well as the calculated physiologically equivalent temperature (PET). During the hot season, maximum dissatisfaction was found in the spine, peak, seating and fountain spaces. The most preferred zones were: seating, pavement, canopy, fountain and cascade zones in terms of thermal comfort sensation.

- Bourbia and Boucherbia (2010) studied the thermal comfort conditions in the street canyons of Constantine (Algeria), where the impact of the streets geometry was analysed. Both air and surface temperature were measured in various streets in order to assess the interrelation between the urban streets geometry and thermal comfort. The physical parameters for the representation of the urban geometry were the height to width (H/W) ratio and the sky view factor. The authors showed that the larger the sky view factor, the higher the air and surface temperature measured and the higher the H/W ratio, the lower the air and surface temperature. Therefore, open spaces with no vegetation are exposed to the sun for most of the day, leading to discomfort problems. As a result, the geometry of urban canyons plays a critical role in urban heat island mitigation.

- In a study at the campus of University Putra (Malaysia), Makaremi et al. (2012) assessed the outdoor comfort conditions in shaded outdoor areas under hot and humid tropical climatic conditions. The research in this case was based also on subjective and objective measurements, including wind speed and direction, and temperature and humidity in the open campus areas, from which the PET index was extracted. In addition, a survey questionnaire was circulated among students. Analysis of the outdoor comfort survey indicated that although the PET values of the areas under study were higher than the comfortable range, a significant number of students declared that the comfort conditions were acceptable. There were thus some differences between measured outdoor comfort and people's perceptions, showing the ability of people to adjust to less comfortable environments.

- The outdoor comfort conditions in Dhaka (Bangladesh) were studied by (Ahmed, 2003) The methodology used was based on a survey questionnaire and measurement of environmental parameters, i.e. air temperature, globe temperature (mean radiant temperature) and relative humidity. The urban spaces were categorized in terms of their position and sheltering. The study showed that comfort boundaries varied significantly with: (i) presence of air flow; (ii) a person's activity; (iii) variations of globe and radiant temperatures; and (iv) shading. Shading was considered to be the most important aspect between noon and 3 p.m. in the specific climatic conditions. From the study, the author recommended that "drinking water, pedestrian access and street furniture-shading should be considered as a public amenity in the Tropics".

- Seven European cities – Athens (Greece), Thessaloniki (Greece), Milan (Italy), Fribourg (Switzerland), Kassel (Germany), Cambridge (UK) and Sheffield (UK) – were studied for their outdoor comfort conditions in the framework of the RUROS (Rediscovering the Urban Realm and Open Spaces) project (Nikolopoulou and Lykoudis, 2006). The results of the microclimatic and human monitoring were based on field surveys in which meteorological parameters were measured, assisted by the subjective results acquired from nearly 10,000 interviews. The comfort temperature range varied significantly among the various European regions; for example from around 23°C for Athens to 13°C for Fribourg. The study proved the strong relationship of outdoor comfort with air temperature, solar radiation and wind speed patterns, although comfort adaptation influenced the responses from pedestrians.

Analyses of the various case studies for outdoor comfort are tabulated in Table 19.3.1.

Table 19.3.1 The outdoor spaces improvement procedure.

Region	Climatic conditions	Survey methodology	Comfort indices	Findings	Reference
Crete	Mediterranean	Measurements of temperature, relative humidity and wind speed. Focus on shaded areas by vegetation.	DI, ET	The addition of 8 trees in an area of $100\,m^2$ can reduce the temperature for up to 3.1°C	Georgi et al., 2010
Malaysia	Tropic (hot and humid)	Measurement of ambient air temperature (Ta), Relative humidity (RH), wind speed (v) and mean radiant temperature (Tmrt) Questionnaires. Focus on shading areas.	PET	The acceptable conditions are higher than the ones accepted by the (PET < 34°C). The locations with shading obtained from plants and surrounding buildings had a longer thermal acceptable period.	Makaremi et al., 2012
Wellington New Zealand	Temperate	Ambient temperature, MRT in exposed and MRT in shaded areas	Bedford scale (Bedford, 1961)	People actively adapt to microclimatic conditions in outdoor environments. Maximum wind gust is the most important parameter in predicting comfort adaptivity, followed by, the mean wind speed.	Walton et al., 2007
Szeged, Hungary	Central European	Measurement of air temperature relative humidity, wind speed global radiation. Visitors positioning via GIS.	PET	The visitors mapping facilitated the detection of preferred areas versus the outdoor environmental conditions.	Kántor et al., 2010
Dhaka, Bangladesh	Tropic (hot and humid)	Measurement of radiant temperature, humidity, wind speed and direction, air temperature and lux.	–	At relative humidity of 70%, the comfort is between 28.5 and 32 8C and the radiant temperature between 28.71 and 32.57 8C, The airflow extends the comfort temperature but not significantly. Behavioural sinks, should be provided.	Ahmed, 2003

(Continued)

Table 19.3.1 Continued

Region	Climatic conditions	Survey methodology	Comfort indices	Findings	Reference
Algeria	Hot and arid	Measurement of surface temperature, sky view factor and air temperature	SVF	The geometry of urban canyons plays a decisive role in urban heat island mitigation. Moreover, SVF is suitable to be incorporated into urban design.	Bourdia et al., 2010
Negev, Israel	Hot and arid	Measurement of dry and wet bulb temperature and wind speed in two adjacent, semi-enclosed courtyard spaces.	ITS	The highest contribution in mid-day thermal stress is provided by a combination of shade trees and grass. Intermediate-level moderations of thermal stress were made by single landscape elements (grass, trees or mesh.	Shashua-Bar, et al., 2011
The Hague Netherlands Eindhoven Grote Markt, Netherlands	Northern European	Interviews with visitors. Air temperature, globe temperature, short-wave radiation, wind speed and wind direction measurements.	Balance comfort experience	Uncomfortable squares were considered: (i) the very wide ones; (ii) the very open and (iii) the ones comprising by cool materials during the heating season.	Lenzholzer et al., 2010
Curitiba, Brasil	Subtropical, wet and temperate	Air temperature, wind solar angles and height. Questionnaires survey	UTCI	UTCI allows for adequate comparisons of climatic conditions and can serve as a suitable planning tool for urban thermal comfort in sub-tropical regions.	Bröde et al., 2011

19.3.2 Assessment of bioclimatic technologies

Advanced computerized fluid dynamic tools have been used to simulate the micro-climatic conditions in the area during the monitoring period. This has permitted the validation of the accuracy of the used simulation code. Simulations have been performed to evaluate the microclimatic conditions in the considered area during the peak temperature summer period. Based on the results of the audit and the performed

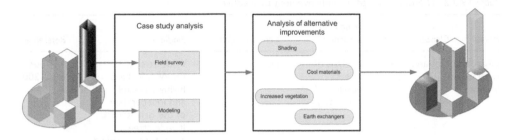

Figure 19.3.3 The assessment of bioclimatic technologies for outdoor spaces.

simulations, architectural and engineering measures have been proposed to improve climatic conditions in the area. Additional simulations have been carried out to evaluate the impact and the potential of the proposed measures on the climatic conditions in the area. Based on the results of the simulations, the overall design has been optimized (see Figure 19.3.3).

Advanced computerized fluid dynamic tools can be used to simulate the microclimatic conditions of an urban area. This permits validation of the accuracy of the simulation code used. Simulations can be performed to evaluate microclimatic conditions during peak temperature periods in summer. Based on the results obtained from the audit and the simulations, architectural and engineering measures can be proposed for improvement of climatic conditions in urban areas. Additional simulations can be carried out to evaluate the impact and the potential of proposed measures on the climatic conditions. Based on the results of simulations, the overall design of urban areas can be optimized (see Figure 19.3.3).

Some examples are included below and are tabulated in Table 19.3.2.

- Ng et al. (2011, 2012) studied the thermal effects of greening in improving the urban microclimate in the city of Hong Kong. It was emphasized that the cooling potential of the different greening strategies is related to building morphology. As a general rule of thumb, a reduction of outdoor temperature equal to 1 K is possible when tree coverage is larger than one-third of the total land area for areas where the building coverage ratio is almost 44%, which is the average value in Hong Kong. The greening effect on the thermal environment is considerable when greening of building façades rather than on rooftops is done.
- A rehabilitation of Albania's capital city of Tirana's historic centre was performed by Fintikakis, et al. (2011). The area under study has hot and dry summer climatic conditions and an analysis of the outdoor environment showed that there are up to 3°C local temperature differences due to different local thermal balance. Moreover, ambient temperatures in well-shaded areas may be 1.4–2.0°C lower that in neighbouring zones of similar configuration. The area was modelled using Computational Fluid Dynamics (CFD) modelling and a series of technologies were proposed for the improvement of the outdoor conditions, including:
  - Extended use of shading and solar control to reduce the surface temperature of the materials and the corresponding heat convection;

Table 19.3.2 The outdoor spaces improvement procedure.

Region	Climatic conditions	Mitigation measures	Comfort indices	Findings	Reference
Ghardaia, Algerian Sahara	Hot and arid	Analysis of streets design and orientation	PET	Wide streets (H/W = 0.5) are highly stressful and independent of the orientation. If wide streets are planned, shading strategies should be implemented directly at street level.	Ali-Toudert et al., 2006
Madrid, Spain	Mediterranean	Evaporative towers	Air temperature, humidity and wind speed	Reduction of air temperature in the pedestrian level by 3.5°C	Soutullo et al., 2011
Dubai, UAE	Hot and arid	Ventilation		25.3%, 26.6%, and 27.7% comfort on yearly basis when wind speed is 1.5, 3, and 6 m/s respectively	Al-Sallal et al., 2012
Athens, Greece	Mediterranean	Vegetation, Cool materials, shading	Air temperature	Reduction of air temperature almost 2°C.	Gaitani et al., 2011
Athens, Greece	Mediterranean	Cool materials	Air temperature	Reduction of air temperature almost 2°C.	Santamouris et al., 2012
Tirana, Albania	Mediterranean	Vegetation, Cool materials, Earth to air heat exchangers	Air temperature	Reduction of air temperature almost 2°C.	Fintikakis et al., 2011
Milan, Italy	Mediterranean	Field measurements, analysis of vegetation. Focus on the impact of vegetation	COMFA, Thermal budget (W/m^2)	The vegetation should be considered as a real tool for the control of microclimatic conditions in external spaces.	Picot, 2004
Hong Kong	Subtropical	Ventilation	PET, TSV, wind speed	Ensuring wind environment with wind speed of 0.9–1.30 m/s in the city through better planning. Reduction of solar radiation gains through shading and vegetation.	Ng, et al., 2012
Crete, Greece	Mediterranean	Spay fans to increase evapotranspiration	DI, ET	The addition of 8 trees in an area of 100 m^2 can reduce the temperature for up to 3.1°C	Georgi et al., 2010

- o   Use of large-size trees to enhance shading and evapotranspiration in the considered area;
- o   Use of light-coloured materials to decrease the absorption of solar radiation;
- o   Reduction of anthropogenic heat from traffic;
- o   Maximum possible use of environmental heat sinks and in particular of the ground to offer low-temperature fresh air.

The reduction of the ambient temperature achieved was up to 3°C, while the surface temperature reduction was 6–8°C.

- Improvement of microclimatic conditions of a square in the centre of Athens (Greece) was studied by Gaitani et al. (2011). The area's environmental parameters were initially monitored, and surface and air temperature, wind speed and direction, humidity and pollution levels were measured. Based on CFD model analysis the following conclusions were made: (i) lack of spatial homogeneity of the air temperature; (ii) increase in surface temperatures due to highly absorbing materials; (iii) high wind speeds in specific areas of the region under study; (iv) high concentration of particulate matter. Through a series of interventions, such as increased greenery, installation of cool materials, increased shading and use of earth-to-air heat exchangers, the area's comfort conditions were significantly improved.

- An assessment of wind flow in the urban area of Dubai (United Arab Emirates) and its contribution to passive cooling was performed by Al-Sallal et al. (2012). The overall assessment was based on CFD analysis of laminar and turbulent flow in the specific region. The study revealed the importance of street design for the bioclimatic urban design. In the case study, the wind speed during the laminar analysis showed that it was within comfort levels in the pedestrian area. Turbulent wind flow was channelled by the street canyons and increased in various areas, contributing to an improvement in thermal comfort sensation. Analysis showed that passive cooling through natural ventilation can be effective in providing comfort for 25%, 26% and 27% of the year with wind speeds of 1.5, 3 and 6 m/s respectively. Moreover, wind speed contributed to comfort when the aspect ratio of the canyons was 1.75.

- The contribution of vegetation in urban structures was evaluated by Picot (2004) for Piazza della Scienza in Milan (Italy). A set of field measurements, including air and radiant temperature, wind velocity and relative humidity, were taken, followed by an evaluation of thermal comfort coupled with a vegetation growth hypothesis. A series of simulations were performed, taking into consideration different vegetation growth scenarios without changing the site configuration. The screening potential of tree foliage on the reduction of solar radiation was also considered. The analysis showed that the presence of vegetation can reduce the heat budget of a human body by 50–150/m^2 depending on the type of vegetation and foliage.

- The influence of evaporative towers in the outdoor comfort levels for Madrid (Spain) was studied by Soutullo et al. (2011). An experimental campaign was carried out to analyse the thermal influence of the specific towers by installing a series of wind, temperature and humidity sensors in different positions and at different heights. At the pedestrian level a temperature reduction of almost 3.5°C and a cooling efficiency of about 32% was revealed.

- Almost 4500 m² of cool pavements were used to rehabilitate a major urban park in the greater Athens area. CFD were used to simulate the specific climatic conditions in the area before and after the installation of the new pavements. After validation against two sets of the collected experimental data, comparative calculations were performed with and without the cool pavements under the same climatic boundary conditions. It was found that the extensive application of reflective pavements, under specific climatic conditions, may reduce the peak daily ambient temperature during a typical summer day by up to 1.9 K, while surface temperatures were reduced by up to 12°C. The overall analysis showed that the use of cool pavements presents an efficient mitigation strategy to reduce the intensity of heat islands in urban areas and improve the global environmental quality of open areas (Santamouris et al., 2012).

19.4 CONCLUSIONS AND FUTURE PROSPECTS

The phenomenon of heat islands is one of the more documented aspects of climate change. Recent research has highlighted the main characteristics of the phenomenon in many cities of the world. Heat island increases the consumption of energy for cooling purposes, decreases comfort conditions in indoor and outdoor spaces, increases the concentration of specific pollutants and puts the population, especially the more vulnerable, at risk.

To counterbalance the impact of heat island, several mitigation techniques and technologies have been proposed, employed and optimized, and are in general available to the scientific community. The use of reflective materials for roofs, pavements and urban infrastructure, together with the use of green spaces for open areas and roofs, and the use of heat sinks such as ground and water, are the more efficient means among the available mitigation techniques. Important research has been carried out in recent times and the quality of the systems and materials has been improved significantly. Most of the proposed systems are now available on the market as industrial products.

Important applications of advanced mitigation techniques have been designed, implemented and monitored during the recent past. Their overall performance has been found to be quite high and significant decreases in heat island intensity have been measured.

Although the actual progress of mitigation technologies is important and has been spectacular, further research is necessary in order to improve the efficiency of the systems and products, as well as to develop new materials and technologies that may further enhance the climatic quality of urban spaces.

Finally, synergies among urban climatologists, energy experts and urban planners should be developed in order to effectively promote future low-carbon cities and regions.

Nomenclatures/Abbreviations

CFD Computational Fluid Dynamics
DI Discomfort index
ITS Index of thermal stress

PET   Physiologically equivalent temperature
PMV   Predicted mean vote
SET   Standard effective temperature
SVF   Sky view factor
TSV   Thermal sensation
UTCI Universal Thermal Climate Index

REFERENCES

Ahmed, K.S. (2003) Comfort in urban spaces: Defining the boundaries of outdoor thermal comfort for the tropical urban environments. *Energy and Buildings*, 35, 103–110.

Akbari, H. (2002) Shade trees reduce building energy use and CO2 emissions from power plants. *Environmental Pollution*, 116, S119–S126.

Akbari, H., Davis, S., Dorsano, S., Huang, J. and Winett, S. (1992) Cooling our Communities – A Guidebook on Tree Planting and Light Colored Surfacing'. *U.S. Environmental Protection Agency. Office of Policy Analysis, Climate Change Division*.

Ali-Toudert, F. and Mayer, H. (2006) Numerical study on the effects of aspect ratio and orientation of an urban street canyon on outdoor thermal comfort in hot and dry climate. *Building and Environment*, 41, 94–108.

Al-Sallal, K.A. and Al-Rais, L. (2012) Outdoor airflow analysis and potential for passive cooling in the modern urban context of Dubai. *Renewable Energy*, 38, 40–49.

Bedford, T. (1961) Researches on thermal Comfort. *Ergonomics*, 4: 280–310.

Bourbia, F. and Boucheriba, F. (2010) Impact of street design on urban microclimate for semi-arid climate (Constantine). *Renewable Energy*, 35, 343–347.

Bretz, S., Akbari, H. and Rosenfeld, A. (1989) Implementation of solar reflective surfaces: Materials and utility programs, *LBL Report 32467*, University of California.

Bröde, P., Krüger, E.L., Rossi, F.A. and Fiala, D. (2011) Predicting urban outdoor thermal comfort by the universal thermal climate index UTCI-a case study in southern Brazil. *International Journal of Biometeorology*, 1–10.

Campra, P., Garcia, M., Canton, Y. and Palacios-Orueta, A. (2008) Surface temperature cooling trends and negative radiative forcing due to land use change toward greenhouse farming in southeastern Spain. *Journal of Geophysical Research*, 113 D18109.

Chen, Y. and Wong, N.H. (2006) Thermal benefit of city parks, *Energy and buildings*, 38, 105–120.

CRES (2011) *Scientific Guide of the Program "Bioclimatic Renewal of Urban Open Spaces"*. www.cres.gr

Doulos, L., Santamouris, M. and Livada, I. (2004) Passive Cooling of outdoor urban spaces. The role of materials, *Solar Energy*, 77, 231–249.

EPA1 (Content Source); Cutler Cleveland (2007) Environmental effects of urban trees and vegetation. *In: Encyclopedia of Earth*. http://www.eoearth.org/article/Environmental_effects_of_urban_trees_and_vegetation.

EPA2. *Reducing Urban Heat Islands: Compendium of Strategies, Trees and vegetation*. http://www.epa.gov/heatisld/mitigation/coolroofs.htm

EPA3. *Reducing urban heat islands: Compendium of strategies, Green Roofs*. http://www.epa.gov/heatisld/resources/pdf/GreenRoofsCompendium.pdf

EPA4. *Reducing Urban Heat Islands: Compendium of Strategies, Cool Roofs*. http://www.epa.gov/heatisld/mitigation/coolroofs.htm

EPA5. *Reducing Urban Heat Islands: Compendium of Strategies, Cool Pavements*. http://www.epa.gov/heatisld/mitigation/pavements.htm

Fintikakis, N., Gaitani, N., Santamouris, M., Assimakopoulos, M., Assimakopoulos, D.N., Fintikaki, M., Albanis, G., Papadimitriou, K., Chryssochoides, E., Katopodi, K. and Doumas, P. (2011) Bioclimatic design of open public spaces in the historic centre of Tirana, Albania, *Sustainable Cities and Society*, 1, 54–62.

Gaitani, N., Spanou, A., Saliari, M., Synnefa, A., Vassilakopoulou, K., Papadopoulou, K., Pavlou, K., Santamouris, M., Papaioanou, M. and Lagoudaki, A. (2011) Improving the microclimate in urban areas: A case study in the centre of Athens. Building Services Engineering Research and Technology, 32, 53–71.

Georgi, J. N. and Dimitriou, D. (2010) The contribution of urban green spaces to the improvement of environment in cities: Case study of Chania, Greece. *Building and Environment*, 45, 1401–1414.

Getter, K.L. and Rowe, D.B. (2006) The role of extensive green roofs in sustainable development. *Hortscience*, 41, 1276–1285.

Hassid, S., Santamouris, M., Papanikolaou, N., Linardi, A., Klitsikas, N., Georgakis, C. and Assimakopoulos, D.N. (2000) The Effect of the Athens Heat Island on Air Conditioning Load. *Journal Energy and Buildings*, 32, 131–141.

Kántor, N. and Unger, J. (2010) Benefits and opportunities of adopting GIS in thermal comfort studies in resting places: An urban park as an example. *Landscape and Urban Planning*, 98, 36–46.

Kolokotsa, D. (2008) *Cool Roofs Council.* http://www.coolroofs-eu.eu/.

Lenzholzer, S. and van der Wulp, N.Y. (2010) Thermal experience and perception of the built environment in Dutch urban squares. *Journal of Urban Design*, 15, 375–401.

Mahmoud, A.H.A. (2011) Analysis of the microclimatic and human comfort conditions in an urban park in hot and arid regions. *Building and Environment*, 46, 2641–2656.

Makaremi, N., Salleh, E., Jaafar, M.Z. and Ghaffarian Hoseini, A. (2012) Thermal comfort conditions of shaded outdoor spaces in hot and humid climate of Malaysia. *Building and Environment*, 48, 7–14.

McPherson, E.G. (1994) Cooling urban heat islands with sustainable landscapes, *Urbanization and Terrestrial Ecosystems*, 151–171.

Millstein, D. and Menon, S. (2011) Regional climate consequences of large-scale cool roof and photovoltaic array deployment. *Environmental Research Letters*, 6: 9.

Ng, E. and Cheng, V. (2011) Urban human thermal comfort in hot and humid Hong Kong. Energy and Buildings, Available online (http://www.sciencedirect.com/science/article/pii/S0378778811004130).

Ng, E., Chen, L., Wang, Y. and Yuan, C. (2012) A study on the cooling effects of greening in a high-density city: An experience from Hong Kong. *Building and Environment*, 47, 256–271.

Niachou, A., Papakonstantinou, K., Santamouris, M., Tsangrassoulis, A. and Mihalakakou, G. (2001) Analysis of the green roof thermal properties and investigation of its energy performance, *Energy and Buildings*, 33, 719–729.

Nikolopoulou, M. and Lykoudis, S. (2006) Thermal comfort in outdoor urban spaces: Analysis across different European countries. *Building and Environment*, 41, 1455–1470.

Oke, T.R, Johnson, G.T., Steyn, D.G. and Watson, I.D. (1991), Simulation of Surface Urban Heat Islands under 'Ideal' Conditions at Night – Part 2: Diagnosis and Causation. *Boundary Layer Meteorology*, 56, 339–358.

Papadakis, G., Tsamis, P. and Kiritsis, S. (2001) An experimental investigation of the effect of shading with plants for solar control of buildings. *Energy and Buildings*, 33, 831–836.

Picot, X. (2004) Thermal comfort in urban spaces: Impact of vegetation growth. Case study: Piazza della Scienza, Milan, Italy. *Energy and Buildings*, 36, 329–334.

Rosenfeld, A.H., Akbari, H., Bretz, S., Fishman, B.L., Kurn, D.M., Sailor, D. and Taha, H. (1995) Mitigation of urban heat islands: Materials, utility programs, updates. *Energy and Buildings*, 22, 255–265.

Rosenzweig, C., Solecki, W., Parshall, L., Gaffin, S., Lynn, B., Goldberg, R., Cox, J. and Hodges, S. (2006) Mitigating New York City's heat island with urban forestry, living roofs, and light surfaces, *Sixth Symposium on the Urban Environment*, AMS Forum: Managing our Physical and Natural Resources: Successes and Challenges.

Santamouris, M., Asimakopoulos, D.N., Asimakopoulos, V.D., Chrisomallidou, N., Klitsikas, N., Mangold, D., Michel, P. and Tsangrassoulis, A. (2001), Energy and Climate in the Urban Built Environment, Earthscan Ltd.

Santamouris, M., Gaitani, N., Spanou, A., Saliari, M., Giannopoulou, K., Vasilakopoulou, K. and Kardomateas, T. (2012) Using cool paving materials to improve microclimate of urban areas – Design realization and results of the Flisvos project. *Building and Environment*, 53, 128–136.

Santamouris, M., Papanikolaou, N., Livada, I., Koronakis, I., Georgakis, C. and Assimakopoulos, D. N. (2001) On the Impact of Urban Climate to the Energy Consumption of Buildings. *Solar Energy*, 70, 3, 201–216.

Santamouris, M., Paraponiaris, K. and Mihalakakou, G. (2007a) Estimating the Ecological Footprint of the Heat Island effect over Athens, Greece, *Climate Change*; 80: 265–276.

Santamouris, M., Pavlou, K., Doukas, P., Mihalakakou, G., Synnefa, A., Hatzibiros, A. and Patargias, P. (2007c) Investigating and Analysing the Energy and Environmental Performance of an Experimental Green Roof System Installed in a Nursery School Building in Athens, Greece. *Energy*, 32, 1781–1788.

Santamouris, M, Pavlou, K., Synnefa A., Niachou, K. and Kolokotsa, D. (2007b) Recent Progress on Passive Cooling Techniques. Advanced Technological Developments to Improve Survivability levels in Low – Income Households. *Energy and Buildings*, 39, 859–866.

Sfakianaki, E., Pagalou, E., Pavlou, K., Santamouris, M. and Assimakopoulos, M.N. (2009) Theoretical and experimental analysis of the thermal behaviour of a green roof system installed in two residential buildings in Athens, Greece, *International Journal Energy Research*, 33, 1059–1069.

Shashua-Bar, L., Pearlmutter, D. and Erell, E. (2011) The influence of trees and grass on outdoor thermal comfort in a hot-arid environment. *International Journal of Climatology*, 31, 1498–1506.

Soutullo, S., Olmedo, R., Sánchez, M.N. and Heras, M.R. (2011) Thermal conditioning for urban outdoor spaces through the use of evaporative wind towers. *Building and Environment*, 46, 2520–2528.

Stathopoulou, E, Mihalakakou, G, Santamouris, M. and Bagiorgas, H. S. (2008) Impact of Temperature on Tropospheric Ozone Concentration Levels in Urban Environments. *Journal of Earth System Science*, 117, 227–236.

Sustainability Victoria. *Landscape Design, Info Fact Sheet*. http://www.sustainability.vic.gov.au/www/html/1517-home-page.asp

Walton, D., Dravitzki, V. and Donn, M. (2007) The relative influence of wind, sunlight and temperature on user comfort in urban outdoor spaces. *Building and Environment*, 42, 3166–3175.

Wong, N.H., Tan, P.Y. and Chen, Y. (2007) Study of thermal performance of extensive rooftop greenery systems in the tropical climates. *Building and Environment*, 42, 25–54.

Zinzi, M. (2010) Cool materials and cool roofs: Potentialities in Mediterranean buildings. *Advances in Building Energy Research*, 4, 201–266.

Chapter 20

Legislation to foment the use of renewable energies and solar thermal energy in building construction: The case of Spain

Javier Ordoñez
Department of Civil Engineering, Higher Technical College of Civil Engineering, University of Granada, Granada, Spain

20.1 INTRODUCTION

In the European Union (EU), buildings consume 40% of the total energy produced. Of this amount, 69% is used for heating and cooling systems as well as for domestic hot water. The EU has issued a series of directives the objectives of which are to enhance energy efficiency in buildings and to foment the use of renewable energies.

In this regard, *Directive 2002/91/EC of the European Parliament and of the Council of 16 December 2002 on the energy performance of buildings* states that buyers or tenants of buildings must be provided with an energy performance certificate. The energy performance in buildings can be enhanced by improving building materials to make them more energy efficient and/or by eliminating existing obstacles to the use of thermal installations with renewable energies.

Directive 2009/28/EC of the European Parliament and of the Council of 23 April 2009 on the promotion of the use of energy from renewable sources contemplates a series of targets for renewable energies that must be achieved by 2020 in each of the member states of the EU. For example, in Spain it is estimated that the contribution of renewable energies to the total energy consumption will increase from 8.7% in 2005 to 20% in 2020.

This chapter analyses the impact of these EU directives, which have been implemented in Spain, and which have made the use of solar thermal energy a reality in the building construction sector. This legislation is reflected in the development of a regulatory framework as well as in the creation of government economic incentives or subsidies for the production and use of this type of energy.

20.2 EUROPEAN REGULATORY FRAMEWORK FOR RENEWABLE ENERGY SOURCES IN THE CONTEXT OF THE ENERGY PERFORMANCE OF BUILDINGS

Renewable energy sources are defined as renewable non-fossil energy sources (wind, solar, geothermal, wave, tidal, hydropower, biomass, landfill gas, sewage treatment plant gas, and biogases) (*Directive 2003/54/EC*)[1]. *Council Directive 93/76/EEC of 13*

---

[1] *Directive 2003/54/EC. Article 2.*

September 1993 to limit carbon dioxide emissions by improving energy efficiency includes a series of considerations on the energy performance of buildings which have been further specified and developed in subsequent directives. The main environmental objective of this directive is the rational use of natural energy resources. Given the fact that buildings are responsible for 40% of the total energy consumption in the EU, this directive established a set of programmes in the EU to achieve the following:

- Efficient thermal insulation in buildings;
- Energy certification of buildings in all member states[2];
- Regular inspection of heating installations of an effective rated output of more than 15 kW[3].

In addition, particular reference is made to public buildings that should be examples of good energy performance for others to follow.

In this sense, *Directive 2002/91/EC of the European Parliament and of the Council of 16 December 2002 on the energy performance of buildings* aims to improve the energy performance in buildings. Accordingly, it describes the measures that should to be taken in the building construction sector to attain this goal. More specifically, this directive states:

> The energy performance of buildings should be calculated on the basis of a methodology, which may be differentiated at regional level, that includes, in addition to thermal insulation other factors that play an increasingly important role such as heating and air-conditioning installations, application of renewable energy sources and design of the building.

The EU has elaborated measures to support the requirements in *Directive 2002/91/EC*. The purpose is to harmonize at the European level methods to enhance the energy performance of buildings. *European Standard EN 15603:2008* is one of these measures. According to *Directive 2002/91/EC*, for the project design of new buildings with a total useful floor area exceeding 1000 m^2, member states should ensure the technical, environmental and economic feasibility of alternative systems, such as the following: (i) decentralized energy supply systems based on renewable energy: (ii) CHP[4]; (iii) district or block heating or cooling; (iv) heat pumps, under certain conditions. Furthermore, when buildings are constructed, sold or rented out, an energy performance certificate should be made available to the owner or by the owner to the prospective buyer or tenant. The validity of the certificate should not exceed 10 years.

Another example of EU legislation is *Directive 2006/32/EC on energy end-use efficiency and energy services*, which repeals *Council Directive 93/76/EEC*. It emphasizes

---

[2]*Directive 93/76/EC.* Article 2. "Energy certification of buildings, which shall consist of a description of their energy characteristics, must provide information for prospective uses concerning a building's energy efficiency".

[3]*Directive 93/76/EC.* Article 6.

[4]*Directive 2002/91/EC.* Article 2. Definitions: "CHP (combined heat and power): the simultaneous conversion of primary fuels into mechanical or electrical and thermal energy, meeting certain quality criteria of energy efficiency".

the potentially exemplary role of the public sector by stating that this sector should endeavour to use energy-efficiency criteria in tendering procedures for public procurement. Among other measures that enhance energy performance in the residential and tertiary sector, this directive explicitly supports policies geared to the domestic generation of renewable energy sources, whereby the amount of purchased energy is reduced (e.g. solar thermal applications, domestic hot water and solar-assisted space heating and cooling[5]). Each member state should thus draw up programmes and implement measures to improve energy efficiency[6].

Directive 2009/28/EC on the promotion of the use of energy from renewable sources sets up a common framework for all EU countries with the objective of fomenting this type of energy. Mandatory national targets specify quotas for the overall share of energy that must be generated from renewable sources in relation to the total energy consumption in each country. *Directive 2009/28/EC* defines a series of requirements related to the following:

- Statistical transfers between member states;
- Joint projects between EU member states and third countries;
- Administrative procedures;
- Information and training;
- Access to and operation of the grids for energy from renewable sources.

By 2020, the EU aims to achieve a 20% share of energy from renewable sources. The year 2020 was selected as a target in order to give this sector the long-term stability needed to make reasonable and sustainable investments in renewable energies. The objective is to reduce dependence on imported fossil fuels and foment the use of new technologies in the energy sector (Commission of the European Communities, 2007). In 2009, energy dependence reached 53.9% (Eurostat, 2012). This situation is even more serious in countries such as Spain where energy dependence is considerably higher (79.4%).

In that same year, the percentage of renewable energies in relation to the total energy consumption in the EU was 11.7% (see Graph 20.2.1). In order to achieve the target of 20%, *Directive 2009/28/EC* establishes the following:

> Each Member State shall adopt a national renewable energy action plan. The national renewable energy action plans shall set out Member States' national targets for the share of energy from renewable sources consumed in transport, electricity and heating and cooling in 2020, taking into account the effects of other policy measures relating to energy efficiency on final consumption of energy, and adequate measures to be taken to achieve those national overall targets, including cooperation between local, regional and national authorities, planned statistical transfers or joint projects, national policies to develop existing biomass resources and mobilise new biomass resources for different uses[7].

---

[5] *Directive 2006/32/EC.* Annex III.
[6] *Directive 2006/32/EC.* Article 4.3.
[7] *Directive 2009/28/EC.* Article 4.

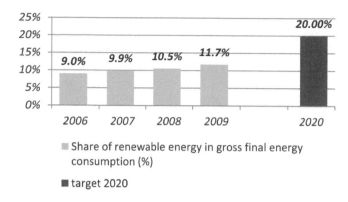

■ Share of renewable energy in gross final energy consumption (%)

■ target 2020

Graph 20.2.1 Share of renewable energy in gross final consumption in CE. *Source:* Eurostat.

Figure 20.2.1 Thermal solar collectors (courtesy of Solaris S.L.).

In relation to the use of solar energy for heating and cooling systems in buildings, this directive states:

- Member States shall ensure that new public buildings and existing public buildings that are subject to major renovation, at national, regional and local level, fulfil an exemplary role in the context of this Directive from 1 January 2012 onwards. Member States may, *inter alia*, allow that obligation to be fulfilled by complying with standards for zero energy housing, or by providing that the roofs of public or mixed private-public buildings are used by third parties for installations that produce energy from renewable sources[8] [see an example of thermal solar collectors in Figure 20.2.1].
- With respect to their building regulations and codes, Member States shall promote the use of renewable energy heating and cooling systems and equipment that achieve a significant reduction of energy consumption. Member States shall use

---

[8]*Directive 2009/28/EC.* Article 13.5.

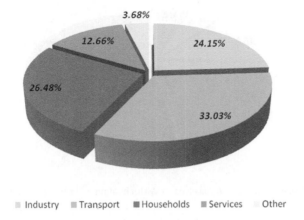

Graph 20.2.2 Final energy consumption in 2009, according to sector (Eurostat).

energy or eco-labels or other appropriate certificates or standards developed at national or Community level, where these exist, as the basis for encouraging such systems and equipment[9].

In 2009, the final energy consumption of buildings in the EU continued to be approximately 40% (see Graph 20.2.2). This total of 39.14% was mostly due to energy use by households and services, whose consumption was largely dependent on buildings (Eurostat, 2011a,b).

On 1 February 2012, *Directive 2002/91/C* was definitively repealed and replaced by *Directive 2010/31/EU on the energy performance of buildings*. This directive reaffirms the target of reducing energy consumption in the EU by 20% and justifies the new regulations by highlighting the fact that buildings are responsible for 40% of the total energy consumption[10].

This obliges European countries to take measures to increase the number of buildings which not only fulfil current energy performance requirements, but which are also more energy efficient. This will reduce energy consumption as well as carbon dioxide emissions. EU member states are thus required to draw up national plans for increasing the number of nearly zero-energy buildings[11] and to regularly report such plans to the Commission.

---

[9]*Directive 2009/28/EC.* Article 13.6.

[10]The Council of the European Union (2007). In the Council of March 2007, the European Community decided to reduce greenhouse gas emissions by 20% compared to those in 1990. The energy and environmental objectives to be achieved by 2020 are known as the 20-20-20 targets. EU member states must thus do the following: (i) reduce by 20% the emissions of greenhouse gases; (ii) increase by 20% the energy efficiency in the EU; (iii) reach 20% of renewables in total energy consumption in the EU.

[11]According to *Directive 2010/31/EU*, a nearly zero-energy building is a building that has a very high energy performance, as defined in Annex I. The nearly zero or very low amount of energy required should be covered to a very significant extent by energy from renewable sources, including energy from renewable sources produced on-site or nearby.

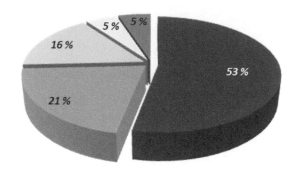

■ Space heating ■ Appliances ■ Water heating ░ Lighting ■ Cooking

Graph 20.2.3 Household energy use. Worldwide trends in energy use and efficiency (International Energy Agency, 2005).

EU member states should thus take measures such that, by 31 December 2020, all new constructions will be nearly zero-energy buildings.

In households, energy is consumed for different purposes. Graph 20.2.3 shows the general distribution of energy use in households.

Energy consumption due to thermal processes in buildings accounts for 69% of the total consumption and is primarily for heating and cooling as well as domestic hot water. These statistics reflect the importance of initiatives to improve the efficiency of energy consumption specifically for these residential uses and also to encourage the use of renewable energy sources for household consumption.

20.3 APPLICATION OF EU REGULATIONS IN MEMBER STATES: THE CASE IN SPAIN

Spain is an especially interesting context since the successful application of EU regulations has led to the design of policies that promote renewable energies. More specifically, in 2009, 9.5% of the total energy consumption (see Graph 20.3.1) and 28.2% of the electricity demand were covered with renewable energies (REE, 2011). Since the target specified in the European directive is for renewables to account for 20% of the total energy consumption in 2020, Spain has to increase its use of renewable energy from 9.5% in 2009 to 20% in 2020.

As previously mentioned, energy consumption in building construction accounts for 40% of the total consumption in the EU (see Graph 20.2.2). For this reason, *Directives 2009/28/EC and 2010/28/EC* regard building construction as a strategic sector and establish measures to foment energy efficiency and the use of renewable energies in this area[12]. To achieve the targets in these directives, EU countries, including Spain, implemented corresponding action plans. In Spain, this framework took the

---

[12]*Directive 2009/28/EC*, Article 13.4: "By 31 December 2014, Member States shall, in their building regulations and codes or by other means with equivalent effect, where appropriate,

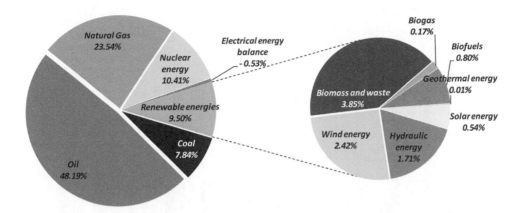

Graph 20.3.1 Primary energy consumption in 2009, depending on energy generation source (National Statistics Institute).

form of the *Plan de Acción Nacional de Energías Renovables* [National Action Plan of Renewable Energies]. The measures in this plan were also transposed into the following set of national laws and codes:

- *Royal Decree 314/2006 of 17 March*, which enacts the Technical Building Code;
- *Royal Decree 47/2007 of 19 January*, which specifies the basic procedure for the energy performance certification of new buildings;
- *Royal Decree 1027/2007 of 20 July*, which regulates heating installations in buildings.

20.3.1   National action plan for renewable energies

Article 4 of the *Directive 2009/28/EC of the European Parliament and of the Council of 23 April 2009 on the promotion of the use of energy from renewable sources* requires each EU member state to adopt a national action plan that defines targets concerning the renewable energy shares consumed in transportation, electricity, as well as heating and cooling. Following these guidelines, the Spanish government elaborated a document titled *Plan de Acción Nacional de Energías Renovables 2011–2020* (PANER) [National Action Plan of Renewable Energies 2011–2020],which was supplemented by the *Plan de Energías Renovables 2011–2020* (PER) [Plan of Renewable Energies 2011–2020] (2010).

The general objective of these plans is to exceed the 20% target in 2020 for renewable energies in the EU directive. This would mean an energy share of 18.9% for heating and cooling, 40% for electricity and 13.6% for transportation. The estimated evolution of renewable energy consumption is reflected in Graph 20.3.2.

---

require the use of minimum levels of energy from renewable sources in new buildings and in existing buildings that are subject to major renovation. Member States shall permit those minimum levels to be fulfilled, *inter alia*, through district heating and cooling produced using a significant proportion of renewable energy sources."

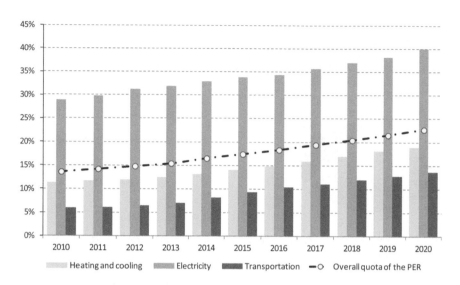

Graph 20.3.2 Objectives of the generation of renewable energies, depending on use (PANER 2011–2020).

In order to reach these targets, Spain has articulated a series of measures geared to regulate and subsidize the production of renewables, as well as to raise awareness of the benefits of their use among the general public. These measures included the following[13]:

- Development of a simplified procedure to obtain authorizations for renewable energy projects for thermal applications (PANER);
- Funding for research and development projects on energy storage systems (PANER);
- Continued support of active public participation in research and development in the renewable energy sector. This includes annual subsidies for high-priority industrial initiatives involving technological development, which will reduce energy generation costs mainly in the wind and solar energy sectors (PANER);
- Financial incentives for the implementation of highly-specialized experimental platforms with international recognition (PANER);
- Modification and improvement of aspects related to thermal renewable energies in technical codes and regulations regarding thermal installations in building construction as well as the inclusion of thermal renewable energies and heating networks in the energy certification systems of buildings (PANER);
- The promotion of energy service companies within the domain of renewable energies[14] (PANER);

---

[13](PANER) pp. 49–53.

[14]*Directive 2006/32/EC*, Article 3.i. An energy service company (ESCO) is defined as a natural or legal person that delivers energy services and/or other energy efficiency improvement measures in a user's facility or premises, and accepts some degree of financial risk in so doing. The

- Strengthening of the national consumption of renewable energy through mechanisms to achieve a zero balance (PER).

This final item is defined as a compensatory energy system that allows consumers who produce part of their own electricity to use the same system to store the surplus. Such a system is especially interesting for electricity generation systems with non-manageable energy sources, such as wind or solar power, since it avoids accumulation of energy in the installation itself (M&I, 2011).

In the case of the use of renewable energy for heating and cooling as well as for domestic hot water, these measures include financial support for the following:

- The diffusion, promotion, and adaptation of solar (photovoltaic, thermal and thermoelectric) installations that foment the horizontal penetration of this renewable energy in all sectors (building construction, agriculture and livestock, industry and services);
- The development of the necessary mechanisms to foment desalination plants based on solar technologies (i.e. low-temperature thermal, photovoltaic and thermoelectric installations);
- The promotion of projects to optimize solar thermal installations, which include comprehensive global solutions (domestic hot water, heating and cooling).

Other measures include awareness campaigns to change social perceptions of solar energy. These include programmes to disseminate the benefits of solar energy as well as the rights and obligations of its users. In addition, these measures encourage the professionalization of the sector.

20.3.2 Basic procedure for the certification of energy efficiency

As previously mentioned, *Directive 2002/91/EC of the European Parliament on the energy performance of buildings* establishes the obligation to provide buyers or tenants of buildings with an energy performance certificate (EPC) in all EU countries.

In Spain, energy performance certificates were transposed into national legislation in *Royal Decree 47/2007*. As stated in this law, such certificates in Spain should include objective information regarding the energy characteristics of buildings to enable the evaluation and comparison of their energy performance. The purpose is to favour energy-efficient buildings and encourage investments in energy economy. In this respect, energy labels (see Figure 20.3.1) provide users with clear information about the energy performance of buildings, which goes beyond a strict compliance with building codes.

Moreover, a building's energy efficiency can be assessed by means of computer programs that are capable of simulating the thermal performance of a construction[15].

---

payment for the services delivered is based (either wholly or in part) on the achievement of energy efficiency improvements and on the meeting of the other agreed performance criteria.

[15]In Spain, the software program used for this purpose is CALENDER, which is available at: http://www.minetur.gob.es/energia/desarrollo/eficienciaenergetica/certificacionenergetica/programacalener/paginas/documentosreconocidos.aspx

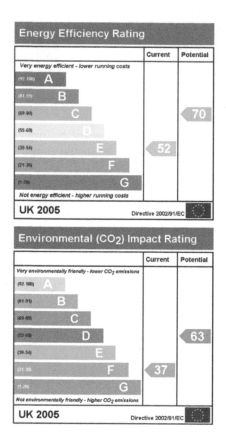

Figure 20.3.1 Energy labels in the European Union (example from the United Kingdom).

The software application is capable of calculating the annual final energy consumption of the building as well as the energy source (fossil fuel, renewable energies, etc.). It then evaluates the building's energy performance by comparing it with that of a reference building. Once the level of energy efficiency is obtained, the building is assigned to an Energy Performance Category, which can vary from Category A (highest energy efficiency) to Class G (lowest energy efficiency). When a building obtains an EPC, the energy performance label is valid during the period of validity of the certificate (see Figure 20.3.1)[16].

20.3.3 The spanish technical building code

From a regulatory perspective, the PANER (National Action Plan of Renewable Energies) proposes the implementation of thermal renewable energy and central heating systems through city ordinances, as well as the modification and improvement of the

---

[16]For further information about the implementation of energy labels in Europe, see Brounen and Kok (2011) and Mlecnik et al. (2010)

Figure 20.3.2 Solar collectors on a building rooftop (courtesy of Solaris S.L.).

articles pertaining to thermal renewable energy in the technical codes and regulations for heating installations in building construction.

Building regulations in Spain are published in the *Código Técnico de la Edificación* [Technical Building Code] (RD 314/2006). The Spanish Technical Building Code is the legal framework that establishes the safety and habitability requirements of buildings. It consists of seven documents. The basic document is the *DB HE* on energy saving[17]. Its purpose is to foster the rational use of the energy in a building, which also signifies that part of a building's energy consumption should come from renewable energy sources. In what follows, Section 20.4 describes low-temperature solar thermal installations for domestic hot water, and Section 20.5 analyses the Spanish Technical Building Code regarding thermal solar energy in buildings.

20.3.4 Spanish regulations for thermal installations in buildings

As a transposition of *Directive 2002/91/EC*, in 2007 the Spanish government enacted the *Reglamento de Instalaciones Térmicas en los Edificios* [Regulation of Thermal Installations in Buildings]. According to this regulation, thermal installations should be designed, implemented, maintained and used so as to achieve a significant reduction of conventional energy consumption and, consequently, a parallel reduction of greenhouse gas emissions and other atmospheric pollutants. This makes it imperative to use energy-efficient systems, systems capable of recovering energy and systems that generate energy based on renewables. The thermal installations should take advantage of the renewable energies available, which must cover part of the building's energy consumption.

The Spanish Regulation of Thermal Installations in Buildings explicitly states the energy performance and safety requirements for thermal installations in buildings. When these requirements are met, a building then complies with welfare and hygiene

---

[17] Available at: http://www.codigotecnico.org/web/

needs during its design and dimensioning, installation, maintenance and use. This regulation also establishes the certification procedure to verify this compliance[18].

20.4 THE SOLAR THERMAL SYSTEM

Low-temperature thermal solar systems[19] use solar energy as a direct heat source for thermal uses such as space heating, domestic hot water, swimming pool heating, etc. These systems capture the sun's energy by means of solar collectors and store it so that it can be used when needed (see Figure 20.3.2). For those periods when there is not sufficient solar radiation and/or energy consumption is greater than expected, the system can also include a supplementary back-up system that automatically switches on and which uses more conventional energy sources such as gas, electricity or diesel fuel.

Figure 20.4.3 shows the diagram of a solar installation for a multi-dwelling building with a central energy storage system and a back-up system with a boiler that runs on a more conventional energy source. The installation consists of an array of solar collectors that transform solar radiation directly into thermal energy, releasing it into a working fluid. The collectors are arranged in rows, each with the same number of elements (see Figure 20.4.2). The rows of collectors can be interconnected in parallel, in series or a combination of the two (see Figure 20.4.1). Cut-off valves should be installed at the entry and exit points of the rows of collectors and between the pumps so that they can be used to isolate these components during maintenance and repair work.

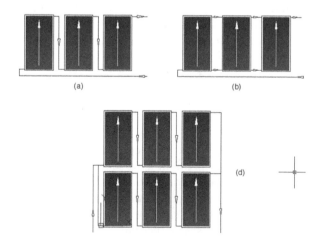

Figure 20.4.1 Connection between solar collectors: a) in series; b) in parallel and c) In series-parallel.

---

[18] *Reglamento de Instalaciones Térmicas en los Edificios.* Article 1.
[19] The energy from solar radiation is transformed into thermal energy by means of a heat collection process. It is regarded as low temperature when the solar energy obtained is used for temperatures lower than 80°C.

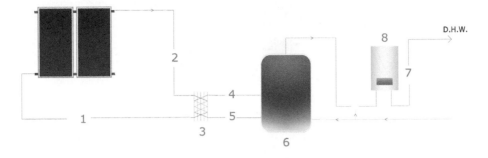

SOLAR CIRCUIT WITH A STORAGE TANK.
1.- PRIMARY CIRCUIT. FLUID OUTLET TOWARDS COLLECTOR
2.- PRIMARY CIRCUIT. FLUID OUTLET TOWARDS COLLECTORS
3.- PLATE HEAT EXCHANGER
4.- SECONDARY CIRCUIT. FLUID OUTLET TO THE STORAGE TANK. FLUID VOLUME IS THE SAME AS THAT OF THE PRIMARY CIRCUIT
5.- SECONDARY CIRCUIT. FLUID RETURN FROM THE STORAGE TANK. FLUID VOLUME IS THE SAME AS THAT OF THE PRIMARY CIRCUIT
6.- SOLAR STORAGE TANK
7.- TERTIARY CIRCUIT. DHW VOLUME
8.- BACKUP SYSTEM

Figure 20.4.2 Hot water system with external heat exchangers (courtesy of Solaris S.L.).

The collectors can be installed on the rooftop of the building or in an obstacle-free area where the building is located. The collectors are mounted on a supporting structure that should be sufficiently sturdy to withstand the action of external forces (e.g. snow, wind). The system should allow thermal expansion without transmitting loads that could affect the integrity of the collectors of the hydraulic circuit.

The system consists of one or various tanks that store hot water until it is needed. The connections of the heat exchangers with the main installation should permit their individual disconnection by means of cut-off valves without interrupting the system's operation. In Europe, solar storage tanks should be certified according to *Directive 97/23/EC*[20]. The installation has a heat exchange system that transfers thermal energy captured by the thermal collectors or the primary circuit to the hot water for user consumption. In certain cases, the heat exchanger is a coil of tubing directly inside the storage tank (see Figure 20.4.2). In this case, the heat exchanger is installed on the inner wall of the lower section of the tank.

When the heat exchanger is located outside the storage tank (see Figure 20.4.2), it is made of copper or stainless steel plates and should be able to withstand maximum operation pressures and temperatures. The minimum design power of an external heat exchanger, *P*, in watts, depending on the solar collector area *A* in square metres, should fulfil the following condition: $P \geq 500A$.

This installation generally has a supplementary back-up system running on an alternate more conventional energy source to ensure that there is no interruption in the heating supply. For a better energy performance, it is not advisable to use an electric back-up system that generates heat by the Joule effect. The back-up system generally consists of a boiler that runs on gas, biomass or diesel fuel.

---

[20] *Directive 97/23/EC of the European Parliament and of the Council of 29 May 1997* on the approximation of the laws of the Member States concerning pressure equipment.

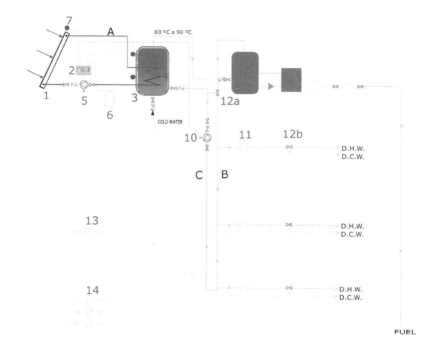

CENTRALIZED SOLAR STORAGE INSTALLATION WITH AN AUXILIARY BACK-UP BOILER

1.- ARRAY OF SOLAR COLLECTORS
2.- REGULATION AND CONTROL UNIT
3.- SOLAR DHW STORAGE TANK
4.- INSIDE COIL
5.- CIRCULATION PUMP
6.- EXPANSION VESSEL
7.- TEMPERATURE SENSOR
8.- AUXILIARY RESERVOIR
9.- BACK-UP BOILER
10.- RECIRCULATION PUMP
11.- DHW METER
12.- REGULATION OF WATER CONSUMPTION TEMPERATURE
    Option a. Centralized in the water consumption circuit
    Option b. Individual regulation in the dwelling

13.- CENTRAL COLD WATER METER (Which only measures the water used to fill the primary circuit for thermal solar collection and storage)
14.- INDIVIDUAL COLD WATER METER FOR EACH DWELLING

A. PRIMARY CIRCUIT (solar, closed)
B. SECONDARY CIRCUIT (outgoing, distribution to dwellings, open)
C. SECONDARY CIRCUIT (return)

    Regulator entry point
    Regulatory exit point. Variable control of the process

Figure 20.4.3 Diagram of a centralized solar storage system with an auxiliary back-up boiler for a multi-dwelling building (courtesy of Solaris S.L.).

The design of a supplementary back-up system depends on the projected use (or uses) of the installation. It is imperative that this system should switch on only when strictly necessary so as to maximally exploit the energy obtained from the array of solar collectors. All system components are interconnected by a series of hydraulic circuits made up of pipes, pumps and other elements such as valves, expansion tanks, deaerators, etc. The primary circuit connects the solar collectors to the heat exchange system and transfers the solar energy to a fluid (e.g. water with anti-freeze) that transports the heat. The secondary circuit connects the heat exchanger to the storage tank. Finally, the consumption circuit carries the water to each consumer (see Figure 20.4.3)[21].

---

[21]The heat exchanger in Figure 20.4.3 is a coil located inside the storage tank. This diagram does not include the circuit that connects the heat exchanger to the storage tank.

The installation should have a regulation and control unit that guarantees that it will operate correctly. Optimal system performance means obtaining maximum benefit from the solar energy collected and only using the back-up system when absolutely necessary. The control unit is composed of the following subsystems:

• Subsystem to assure the smooth operation of the primary and secondary circuits;
• Temperature control subsystems for the safe operation of the installation which prevent it from overheating, freezing, etc.

The regulation and control unit assures that the system will not operate at temperatures higher than can be supported by the materials, components and treatment of the circuits. In installations with a surface area greater than $20 \, m^2$, a monitoring system is installed that provides the values of a wide range of variables, such as the incoming cold water temperature, outgoing water temperature of the solar tank, cold water volume in the system, thermal solar energy stored, etc. The control unit should guarantee that the working fluid temperature never falls below a temperature that is three degrees higher than the freezing temperature of the fluid.

The installation also includes a series of elements designed to protect users and the system itself from possible accidents. For this purpose, it has pressure-relief valves that prevent the system from exceeding maximum operating pressures. Additionally, there are also deaerators that eliminate the air in the circuits. The system also has sensors that prevent it from freezing as well as overheating. To absorb the expansion and contraction of the fluid in a closed circuit (e.g. primary circuit), there is also an expansion tank. Measures are also taken in both the use and design of the system to eliminate the risk of Legionnaires' disease.

20.5 THE SPANISH TECHNICAL BUILDING CODE AS A LEGAL MEANS TO FOMENT THE USE OF RENEWABLE ENERGIES IN BUILDING CONSTRUCTION

The *Código Técnico de la Edificación* [Technical Building Code] (see Section 20.3.3) specifies the requirements in Spain for the design, construction, use, maintenance and conservation of buildings. Compliance with this code foments building quality, guarantees the health and safety of occupants and protects the surrounding environment.

In regards to the functionality and habitability of buildings, the Spanish Technical Building Code specifies basic requirements for each of the following: (i) structural safety; (ii) fire safety; (iii) building use safety; (iv) hygiene, health and environmental protection; (v) noise protection; and (vi) energy economy and heat retention. This building code contains the following types of provision: objectives, requirements, verification methods and consensual solutions. The code is divided into two parts. The first part concerns the general premises, objectives and regulations of the code, and the second part focuses on its implementation documents.

Section one of the Technical Building Code establishes the general conditions in which the code is applied. It also defines the project design characteristics required for buildings. Compliance with these regulations guarantees that a building is suitable for its designated use. Three of the six articles in the code focus on building safety, and the other three on its habitability. The safety requirements are the following: (i) structural safety; (ii) fire safety; and (iii) building use safety. The habilitability requirements include: (i) hygiene; (ii) noise protection; and (iii) energy economy. The second part of the code contains six implementation documents (one for each requirement), which guarantee the building's compliance with the code.

As a specific example, the implementation document on energy economy promotes the rational use of energy in the building and a reduction in energy consumption to sustainable limits. More specifically, its goal is to achieve the following:

- To limit energy demand. For this purpose, measures are taken that directly affect the building envelope (i.e. façades, roofs and zones in contact with the ground). Thermal bridges[22] should be treated in order to limit heat losses and avoid hygrothermal problems;
- To improve the performance of thermal systems;
- To improve the energy performance of lighting installations.

The Spanish Technical Building Code requires that a certain percentage of energy consumption should come from renewable energy sources, particularly thermal and photovoltaic solar energy. For this reason, it states that it is necessary to install systems for solar energy collection and transformation into electrical energy by means of photovoltaic processes for personal use or for the electricity grid[23]. This is obligatory for certain buildings, such as hypermarkets, with a surface area greater than $5000\,m^2$, hotels with more than 100 rooms, etc.

Furthermore, both newly constructed and renovated buildings must cover energy needs for domestic hot water supply with low-temperature systems for the collection, storage and use of solar energy, depending on the average level of solar radiation at the building location. The annual minimum solar contribution[24] depends on the back-up energy source of the system, the hot water demand and the annual mean solar radiation at the building location. This can range from 30%[25] to 70%[26]. The minimum solar contribution can be reduced when the required energy demand for domestic hot water is covered by other renewable energies or cogeneration processes.

Special attention should be paid so that the orientation and tilt of the generation system, as well as the potential shading, are such that the energy losses are lower than the limits established in the Spanish Technical Building Code and which, generally

---

[22] *Código Técnico de la Edificación* [Technical Building Code]. Document on energy economy. Thermal bridge: zone of the building envelope that reflects variation in the uniformity of the construction because of a change in the thickness of enclosure walls, in the materials used, in the penetration of construction elements with different conductivity, etc. This necessarily involves the reduction of the thermal resistance in regards to the rest of the enclosure walls. The thermal bridges are the sensitive parts of the buildings where there is an increased possibility of surface condensations in winter and cold periods.

[23] *Código Técnico de la Edificación* [Tehcnical Building Code], Article 15.1.

[24] Annual minimum solar contribution: ratio of the required annual values of the solar energy contribution and the annual energy demand based on monthly values.

[25] The 30% value occurs in the case when the demand <15000 l domestic hot water/day; back-up energy source: diesel fuel and an annual daily mean solar radiation $<3.8\,kWh/m^2$.

[26] This value is required for zones in which the annual daily mean solar radiation $>5.0\,kWh/m^2$.

speaking, should not exceed 15%. In building projects, it is obligatory to include a report that states the calculation method. It should specify, at least on a monthly basis, the daily mean values of the energy demand and the solar contribution. The calculation method should include the annual overall capabilities defined by the thermal energy demand, the thermal solar energy contribution, the annual and monthly solar fractions, and the annual mean yield.

It is necessary to ascertain whether there is any month of the year in which the energy theoretically produced by the solar installation exceeds the demand corresponding to the actual building occupancy or to any other period of time in which overheating conditions might occur. In such cases, measures should be taken to protect the installation. The performance of the solar collector, independently of the application and technology used, should always be equal to or greater than 40%. In addition, the mean performance of the installation during its period of operation should be greater than 20%.

In order to assure system operation and prolong the useful life of the installation, a monitoring plan and a programme of preventive maintenance are implemented. The monitoring plan involves actions to guarantee that the operational values of the installation remain within normal range and thus guarantee its smooth operation. This simple plan includes, among other things, the visual inspection of installation components and maintaining the collectors in a clean condition.

The maintenance programme involves a series of actions to ensure optimal operating conditions, capabilities, protection and the general durability of the installation. Such maintenance should be performed by trained personnel, and includes all repairs and replacement of consumable units as well as parts that have suffered deterioration during the useful life of the system.

20.6 MEASURES TO FOMENT THE USE OF RENEWABLE ENERGIES: GOVERNMENT INCENTIVES

The EU (e.g. Regulation (EC) No 397/2009), (EP&C, 2009b) as well as the governments of EU member states have provided incentives for initiatives that improve energy efficiency and/or the use of renewable energies in building construction. These incentives are often tax-related (i.e. tax exemptions, deductions or refunds) or are in the form of investment aids and project loans at low interest rates. The beneficiaries of this financial aid can be either the users of the installations, who can use it to purchase solar equipment, or the companies that develop and manufacture renewable energy systems[27].

For example, for the production of electrical power with photovoltaic solar installations, the Spanish government has included these installations in a special regime. This regime, as specified in the *Ley del Sector Eléctrico* (1997) [Electrical Sector Regime Law], was conceived to foment power generation by energy sources whose special characteristics deserve higher financial returns than they would ordinarily obtain. Special regime energy sources receive higher remuneration than those belonging to the ordinary regime through a system of feed-in tariffs and premium payments.

---

[27]Businesses that market, install, service and sell this equipment or the energy produced by it.

In the first case, the owners of renewable energy installations receive a feed-in tariff for the energy produced. The amount of this payment is set by the government and depends on the technology. In the case of premium payments, the electricity produced is freely sold on the electrical energy market. It is bought at the market price (or at a freely negotiated price), supplemented by a premium payment from the government. In the same way as the tariff, the amount of the premium depends on the type of renewable energy.

Regarding the promotion of renewable energies for heating and cooling in EU countries, there are two types of financial incentives: (i) direct investment aids; and (ii) specific funding programmes for solar thermal installations (M&I, 2010). For example, in 2003–2006, subsidies were given in the following conditions:

- *Direct aid.* 40% of the investment cost of the thermal installation and an added 10% if the applicant was a small- or medium-sized company. The investment was calculated by fixing a maximum of 500 euros/m^2 in the case of thermal solar installations for the production of hot water, whatever its use (CDET, 2003).
- *Financing of installations.* Other programmes offered financial aid in the form of low-interest loans that covered up to 100% of the investment costs (thermal solar collectors, storage tanks, heat exchangers, circulation pumps, pipes, valves and connections, expansion tanks, insulation, construction work, etc.), the engineering costs associated with the project, as well as the funds necessary to transact the permits and aids (up to 675 €/m^2), (IDAE, 2004).

20.7 ECONOMIC IMPACT OF SOLAR THERMAL ENERGY

The residential building sector in Spain experienced spectacular growth in 1996-2007 (see Graph 20.7.1). This was accompanied by a significant increase in the investment in housing. In fact, from a little more than 5% of the gross domestic product (GDP) in the 1990s, it rose to 7.4% in 2007. Not surprisingly, construction was also an

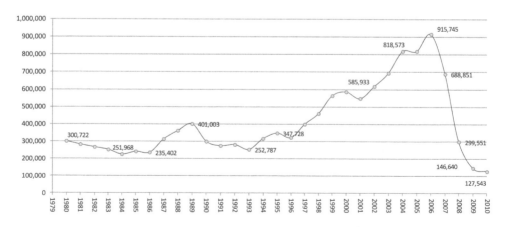

Graph 20.7.1 Evolution of the number of residential buildings (Ministry of Development. Spain).

Figure 20.7.1 Solar collectors on a building rooftop (courtesy of Solaris S.L.).

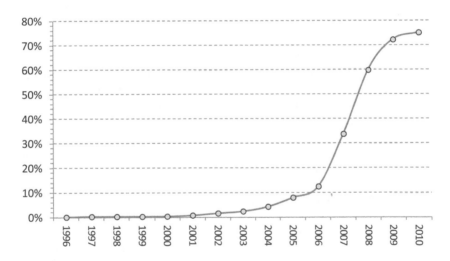

Graph 20.7.2 Percentage of buildings with thermal solar energy installed (Ministry of Development. Spain).

important source of employment during this period. Of the more than six million jobs created from 1996 to 2007, 23% were in this sector (Domenech, 2011).

After 2007, there was a sharp drop in the number of houses built, and even today the residential sector is still in the process of adjusting to this drastic change. In 2011, the investment in housing construction was less than 4% of GDP. Various studies coincide in affirming that the housing demand will eventually begin to increase again and finally stabilize at around 300,000 houses in future years.

The weight of the renewable energy sector in the economy has steadily increased. From 0.47% of GDP in 2005, the percentage rose to 0.70% in 2009. Wind and hydroelectric energy are the most important renewable technologies reflected in the GDP, and together they make up 44.84% of the total contribution of renewables.

In contrast, photovoltaic solar energy accounts for 38.02% and thermal solar energy for 0.78% (Deloitte, 2011).

In 2010, the area in Spain covered by thermal solar panels was approximately 2,400,000 m^2(see Figure 20.7.1). If Spain meets the targets in the PANER, the energy generated in the solar thermal sector for heating and cooling will predictably rise from 61 ktoe in 2005 to 644 ktoe in 2020 (M&I, 2010).

The development of legislation and especially the requirement in the Spanish Technical Building Code[28], enacted in 2006, has made thermal solar technology an integral part of the majority of buildings currently under construction. Although in 2005, less than 10% of the buildings had this technology, in 2010, the percentage of buildings with thermal solar systems soared to almost 80% (see Graph 20.7.2). In 2007 and 2008, the expectations of future growth in this sector were extremely positive despite the drop in the construction of new residential buildings (see Graph 20.7.1). Thermal solar energy has thus become a viable energy alternative in the residential building sector.

20.8 CONCLUSIONS

In recent years, the European Union has made a significant effort to develop a regulatory framework to enhance the energy performance of buildings and to foment the use of renewable energies to increase energy efficiency. The goal is to reduce both energy consumption and energy dependence in the EU by regulating a sector responsible for 40% of greenhouse gas emissions.

Since energy consumption related to thermal processes in buildings accounts for 69% of the total energy consumption, this means that the use of thermal solar energy in building construction has become a priority for the achievement of energy performance objectives. To implement this technology in building construction and/or renovation in the EU, the following measures have been approved.

a) Development of a legislative framework in the EU:
    – *Directive 2002/91/EC of the European Parliament and of the Council of 16 December 2002 on the energy performance of buildings.* This directive states that buyers or tenants of buildings should be provided with an energy performance certificate. The rationale behind these certificates is to thus obtain buildings with a high energy performance by improving their thermal properties and fomenting the investment in energy-saving systems.
    – *Directive 2009/28/EC of the European Parliament and of the Council of 23 April 2009 on the promotion of the use of energy from renewable sources.* This directive specifies a common framework for all EU countries with a view to fostering the use of energy from renewable sources.
    – *Directive 2010/31/EU on the energy performance of buildings.* The objectives of this directive include increasing "the number of buildings which not only

---

[28] According to Article 1.5 of this building code, part of the energy should come from thermal solar energy in all new and renovated buildings in which there is a supply of domestic hot water.

fulfil current minimum energy performance requirements, but are also more energy efficient, thereby reducing both energy consumption and carbon dioxide emissions". For this purpose, the directive requires member states to draw up national plans for increasing the number of nearly zero-energy buildings.

b)   Development of legislation in the member states of the European Union. As an example, the Spanish Technical Building Code requires all new and renovated buildings in which domestic hot water must be installed to use thermal solar energy for part of the energy demand.

This legislation is supplemented by a series of programmes for the funding of installations that run on renewable energy. The financial incentives offered can be tax-related (i.e. tax exemptions, deductions or refunds) or in the form of investment aids and low-interest project loans. Thanks to these legislative measures,, as well as to government subsidies, in 2010, 80% of the new and renovated buildings in Spain had thermal solar energy systems installed.

REFERENCES

Brounen, D. and Kok, N. (2011) On the economics of energy labels in the housing market. *Journal of Environmental Economics and Management*, 62, 166–179.

Commission of the European Communities (2007) *Communication from the Commission to the Council and the European Parliament: Renewable Energy Road Map*. Renewable energies in the 21st century: building a more sustainable future, COM (2006) 848 final, Brussels; 2007. 3–18.

CDET (2003) Council Directive 93/76/EEC of 13 September 1993 to limit carbon dioxide emissions by improving energy efficiency (SAVE) *European Official Journal*, 22 September 1993.

Council of the European Union (2007) *Brussels European Council*, 8/9 march 2007. Presidency Conclusion.

Deloitte (2011) *Impacto económico de las energías renovables en el sistema productivo Español. Estudio Técnico PER 2011–2020*. (In Spanish). IDAE. Madrid.

Doménech, R., (2011) El impacto económico de la construcción y de la actividad inmobiliaria. *XXX Coloquio Nacional APCE*. Madrid, 17 October 2011.

European Parliament and the Council (EP&C) (2002) Directive 20-4/EC of the European Parliament and of the Council of 16 December 2002 on the energy performance of buildings. *Official Journal of the European Union*, 4.1.2003.

European Parliament and the Council (EP&C) (2003) Directive 2003/54/EC of the European Parliament and of the Council of 26 June 2003 concerning common rules for the internal market in electricity and repealing Directive 96/92/EC. *Official Journal of the European Union*, 15.7.2003.

European Parliament and the Council (EP&C) (2006) Directive 2006/32/EC of the European Parliament and of the Council on energy end-use efficiency and energy services and repealing Council Directive 93/76/EEC. *Official Journal of the European Union*, 27.4.2006.

European Parliament and the Council (EP&C) (2009a) Directive 2009/28/EC of the European Parliament and of the Council of 23 April 2009 on the promotion of the use of energy from renewable sources and amending and subsequently repealing Directives 2001/77/EC and 2003/30/EC. *Official Journal of the European Union*, 5.6.2009.

European Parliament and the Council (EP&C) (2010) Directive 2010/31/EU of the European Parliament and of the Council of 19 May 2010 on the energy performance of buildings. *Official Journal of the European Union*, 18.6.2010.

European Parliament and the Council (EP&C), (2009b) Regulation (EC) No 397/2009 of the European Parliament and of the Council of 6 May 2009 amending Regulation (EC) No 1080/2006 on the European Regional Development Fund as regards the eligibility of energy efficiency and renewable energy investments in housing. *Official Journal of the European Union*, 21.5.2009.

European Standard EN 15603 (2008) *Energy performance of buildings – Overall energy use and definition of energy ratings.*

Eurostat (2011a) *Energy balance sheets – 2008–2009.* Luxembourg: Publications. Office of the European Union.

Eurostat (2011b) *Energy, transport and enviroment indicators.* 2010 edition. Luxembourg: Publications. Office of the European Union.

Eurostat (2012) Statistics database. http://epp.eurostat.ec.europa.eu/portal/page/portal/statistics/search-database.

IDAE (2004) *Línea de financiación ICO-IDAE para proyectos de energías renovables y eficiencia energética.*

International Energy Agency (2008) *Worldwide Trends in Energy Use and Efficiency. Key Insights from IEA Indicator Analysis.* France.

Ley 54/1997 (1997) *de 27 de noviembre, del Sector Eléctrico.* (In Spanish). BOE.; 28.11.1997. Num. 285.

Ministry of Economy & IDEA (M&I) (2010) *Plan de Acción Nacional de Energías Renovables de España (PANER) 2011–2020.* (In Spanish). http://www.mityc.es/energia/desarrollo/Energia Renovable/Paginas/paner.aspx

Ministry of Economy & IDEA (M&I) (2011) *Plan de Energías Renovables (PER) 2011–2020.* (In Spanish). http://www.idae.es/index.php/id.670/relmenu.303/mod.pags/mem.detalle

Mlecnik, E., Visscher, H. and van Hal, A. (2010) Barriers and opportunities for labels for highly energy-efficient houses. *Energy Policy*, 38, 4592–4603.

National Statistics Institute. http://www.ine.es/jaxi/menu.do?type=pcaxis&path=/t04/a082/a1998/&file=pcaxis.

Consejeria de Empleo y Desarrollo Tecnologico (CEDT) (2003) Orden de 24 de enero de 2003, por la que se establecen las normas reguladoras y se realiza la convocatoria para el periodo 2003–2006 para el ámbito de la Comunidad Autónoma de Andalucía, del régimen de ayudas del Programa Andaluz de Promoción de Instalaciones de Energías Renovables (PROSOL). (In Spanish.) *Boletín núm.* 33 de 18/02/2003.

REAL DECRETO [Royal Decree] 314/2006 *de 17 de marzo, por el que se aprueba el Código Técnico de la Edificación.* (In Spanish). 28.03.2006 BOE.; num. 74

Real Decreto [Royal Decree] 47/2007 (2007) *de 19 de enero, por el que se aprueba el procedimiento básico para la certificación de eficiencia energética de edificios de nueva construcción.* (In Spanish). 31.01.2007 BOE.; num. 27

REAL DECRETO [Royal Decree] 1027/2007 *de 20 de julio, por el que se aprueba el Reglamento de Instalaciones Térmicas en los Edificios.* (In Spanish). 28.08.2007 BOE.; num. 207

REE (2011) *Informe del sistema eléctrico en 2010.* (In Spanish.) http://www.ree.es/sistema_electrico/informeSEE.asp

Subject index